TOM FARRIS

DYNAMIC SYSTEMS CONTROL

DYNAMIC SYSTEMS CONTROL

LINEAR SYSTEMS ANALYSIS AND SYNTHESIS

Robert E. Skelton
Purdue University

WILEY

JOHN WILEY & SONS

New York • Chichester • Brisbane • Toronto • Singapore

Library of Congress Cataloging in Publication Data:

Skelton, Robert E.
Dynamic systems control.
Includes indexes.
1. Control theory. 2. System analysis. I. Title.
QA402.3.S578 1988 629.8′312 87-27944
ISBN 0-471-83779-2

Printed in the United States of America

10 9 8 7 6 5 4 3 2 1

To
Judy, David, Buzz,
Leigh, and Jeff,
who paid the full price
of the text.

Preface

This text evolved from lecture notes for a two-year course sequence at Purdue and is intended for both self-study and for a dual-level graduate–undergraduate classroom use. The features of this book are briefly mentioned here to help the student and instructor plan their reading. The homework problems appear as exercises at appropriate places in the text as opposed to a collection of problems at the end of each chapter. This, I believe, will better involve the reader in the real time development of the ideas so that the reader may determine his or her level of understanding of each concept before introducing another. In fact, the whole book is arranged in order of increasing sophistication of the engineering concepts. As a result, the topics do not appear in order of sophistication of mathematics. For example, the linear algebra in Chapter 2 involves the simplest engineering concepts but not the simplest mathematics. The building of engineering concepts is our focus, and the mathematical tools are required to fit where needed and are summarized in theorem form for easy reference.

The reader is presumed to have some background in linear algebra and an undergraduate course in the Laplace transform and the single-input/output methods of Evans (root locus), Nyquist, and Bode.

The chapters are arranged as follows. Chapter 2 reviews the needed background in linear algebra and related mathematics. This is a study of linear systems of algebraic equations and is the precursor to our study of linear dynamic systems in Chapters 3 and 4. Chapter 2 should be used as a reference of handy facts and may be skipped or read lightly on first reading. The instructor may choose selected lectures from this chapter to set the stage for notation, but it may be boring for the engineering student to digest *all* the topics of linear algebraic systems before receiving some of the motivations from linear dynamic systems (Chapters 3 and 4). Nonetheless, these are the most fundamental concepts of the book and are presented first to be consistent with the goal of presentation in order of increasing sophistication of the engineering concepts. Education is an *iterative* process, and the student will truly learn Chapter 2 and the power of its applications only by repeated referrals back to Chapter 2 during the study of each of the subsequent chapters of the book. In this way the mathematics is learned in the context of its engineering use. The student may therefore expect that the difficult Chapter 2 will come alive at the end of the book rather than at the beginning.

Chapter 3 develops the equations of motion for some simple dynamic systems by using the language of Chapter 2. This chapter is simply a collection of systems to be used as examples throughout the text as we develop each new concept. There is no methodology taught in this chapter.

The first discussion of the properties of linear dynamic systems begins in Chapter 4, and it is possible to begin instruction here if time is short and if the reader has a good background in linear algebra. After the treatment of the time-varying systems, this chapter discusses the relationships between state variable and transfer function descriptions of dynamic systems. This chapter introduces the deterministic concept of "time correlation" between two vectors, which ties together all remaining chapters.

Chapter 5 focuses on the fundamental controllability and observability properties of linear systems.

Chapter 6 provides the concepts of equivalent systems and develops three different types of equivalence: equivalence with respect to transfer functions, with respect to output correlations, and with respect to a quadratic performance metric called the "cost function." The relationships between these descriptions of a dynamic system are described. Even though this book is entirely deterministic, the introduction of a deterministic version of "time correlation" (a simple integral of two variables) allows the development of mathematical machinery and results that are strikingly similar to the covariance analysis of stochastic systems. This is one advantage of this type of treatment of deterministic systems, even though the student will not realize this advantage until a later course on stochastic processes.

Chapter 7 focuses primarily on two types of stability definitions: Liapunov stability and bounded-input, bounded-output stability. Connections with Chapter 6 will be obvious by the equivalence of stability properties between the original system and its simpler cost-equivalent realization.

Chapter 8 uses the least squares theory of Chapter 2 to solve the simplest of all optimal control problems: quadratic performance measures and linear dynamics. The method chosen for the derivation of these results is based upon the matrix calculus and trace identities of Chapter 2, which allow many different control problems to be viewed within a *common* framework: state feedback control, output feedback control, and dynamic controllers. The instructor will find some new results in Chapter 8, including a completely deterministic theory of optimal dynamic controllers, which eliminates the prior ad hoc practice of putting observers together with optimal state feedback control laws.

Chapter 9 introduces the concept of state estimation from measurement data.

Chapter 10 is a most important chapter because it cautions the reader against misuse of the other nine chapters. Until the tenth chapter the mathematical models of the underlying dynamic system are presumed accurate. Control in the presence of inaccurate models is the necessity of every practical application of control theory. In fact, the limitation in performance of every control design is eventually due to the effect of modeling errors. Hence, great care must be used in the application of

Chapters 1 through 9. Chapter 10 charts the care to be taken, and without Chapter 10 the book would fall short of its goal to provide practical and yet theoretically sound control design techniques. Chapter 10 is an attempt to formally introduce the very fundamental notion that *the modeling problem and the control problem are not independent*. This seemingly innocent concept leads one into a virtual minefield of potential pitfalls in the use of linear systems theory (Chapters 2 through 9). Hence, an introduction to linear systems is incomplete (and perhaps even deceitful) without some elementary introduction to model error concepts and consequences.

The general linear dynamic system is first treated in vector first-order (state) form. However, there are many engineering applications in circuits, acoustics, and mechanics in which the dynamics naturally take on a vector second-order form, $\mathscr{M}\ddot{q} + \mathscr{D}\dot{q} + \mathscr{K}q = f$. Utilizing the special structure of these systems allows many results to be greatly simplified, and these results deserve textbook presentation, in addition to the standard vector first-order theory. For courses that focus only on the basic theory of state space (vector first-order form), the following sections on vector second-order systems can be omitted without loss of continuity: Chapter 3, Sections 5.2.1.4, 5.4.4, 8.7, and 10.2.2.

If this introductory book is used in a transitional course where the students expect to continue their graduate education afterward, the topics will require two semesters and a slower more thoughtful pace in which all theorems are proved. If this book is used in a "terminal" course where students do not expect to continue their controls education, then the instructor may choose to focus on a presentation of the facts without developing the proofs so that the topics can be finished in one semester.

Many Purdue students in A & AE 564 solved numerous example problems. Their feedback influenced the final text in many helpful ways, and without the interaction with them and our joint desire to make that interaction successful, this book would not have been written. It is my pleasure to acknowledge helpful reviews and suggestions from P. Likins, A. Frazho, M. Corless, T. Dwyer, P. Kabamba, and S. Meerkov.

ROBERT E. SKELTON

About the Author

Robert E. Skelton is a Professor of Aeronautical and Astronautical Engineering at Purdue University. He received his B.S. degree in electrical engineering at Clemson University in 1963, his M.S. degree in electrical engineering at the University of Alabama, Huntsville in 1970. He worked in Huntsville Alabama from 1963 to 1973 on such spacecraft modeling and control problems as SKYLAB, project THERMO, Large Space Telescope, and shuttle engine control. He obtained his Ph.D. in mechanics and structures from UCLA in 1976. He serves on the National Research Council's Aeronautics and Space Engineering Board and is a Vice Chairman of the Applications Committee for the International Federation of Automatic Control. He is the recipient of a 1986 Fellowship of the Japan Society for the Promotion of Science and several international research and travel awards from the National Science Foundation. His short courses on the dynamics and control of flexible structures have been given in Germany, Italy, Japan, Australia, and many times in the United States. Most of his 120 journal and conference papers focus on modeling errors and their effect in control problems, with a special emphasis on errors of model order and the attendant model reduction problems. His contributions to control engineering include placement strategies for sensors and actuators, disturbance accommodation, controller reduction, and covariance control.

Contents

1. *Introduction* 1

2. *Mathematical Preliminaries* 8

2.1 Matrix Representations and Linear Spaces 8

2.2 Linear Independence of Vectors 13

2.3 Geometrical Interpretations of Linear Independence 16

2.4 Tests for Determining Linear Independence 18

2.4.1 VECTOR INNER PRODUCTS 19

2.4.2 VECTOR OUTER PRODUCTS 19

2.4.3 VECTOR NORMS 20

2.4.4 THE CAUCHY–SCHWARZ INEQUALITY 21

2.4.5 GEOMETRICAL INTERPRETATION OF ρ: ANGLES BETWEEN VECTORS 22

2.4.6 SIGN DEFINITENESS OF MATRICES 23

2.4.7 A TEST FOR LINEAR INDEPENDENCE OF TIME-VARYING VECTORS 25

2.4.8 A TEST FOR LINEAR INDEPENDENCE OF CONSTANT VECTORS 27

2.5 Orthogonal Bases 28

2.5.1 ORTHOGONAL FUNCTIONS 29

2.5.2 THE GRAM–SCHMIDT PROCEDURE 31

2.6 Solutions of the Liner Algebra Problem **AXB** = **Y** *and Pseudo Inverses* 35

2.7 Linear Matrix Equations and Least Squares Theory 38

2.7.1 CALCULUS OF MATRICES 39

2.7.1.1 Differentiation of a Scalar with Respect to a Vector 39

2.7.1.2 Differentiation of a Scalar with Respect to a Matrix 40

2.7.1.3 Trace Identities 44

2.7.2 PROBLEM: $\min_{x} \|Ax - y\|^2$ 45

2.7.3 PROBLEM: $\min \|x\|^2$ SUBJECT TO $Ax = y$ 47

2.7.4 A GEOMETRICAL APPROACH TO LEAST SQUARES PROBLEMS 50

2.7.5 PROBLEM: $\min_{A} \|Ax - y\|^2$ 52

2.7.5.1 Bessel's Inequality 55

2.7.5.2 Parseval's Equality 56

2.8 *Decomposition of Matrices* 58
2.8.1 SPECTRAL DECOMPOSITION 58
2.8.1.1 Hermitian Matrices 68
2.8.1.2 Skew-Hermitian Matrices 71
2.8.1.3 Unitary Matrices 72
2.8.2 SINGULAR VALUE DECOMPOSITION 72
2.8.3 SQUARE ROOT DECOMPOSITION 80

3. *Models of Dynamic Systems* 84
3.1 *State Space Realizations* 85
3.2 *Linearization* 86
3.3 *A Rocket in Flight* 91
3.4 *A Space Backpack* 94
3.5 *An Orbiting Spacecraft* 101
3.6 *A Rigid Body in Space* 103
3.7 *Elastic Structures* 104
3.7.1 THE CONTINUUM MODEL 105
3.7.2 THE PINNED ELASTIC BEAM 106
3.8 *Electrical Circuits* 116

4. *Properties of State Space Realizations* 121
4.1 *Developing State Space Realizations for Time-Invariant Systems* 121
4.1.1 STATE SPACE REALIZATIONS FOR MIXED-ORDER DIFFERENTIAL EQUATIONS 122
4.1.2 STATE SPACE REALIZATIONS FROM BLOCK DIAGRAMS 123
4.1.3 PROPERTIES OF TRANSFER FUNCTIONS 128
4.1.4 STATE SPACE REALIZATIONS FROM TRANSFER FUNCTIONS OF SINGLE-INPUT / SINGLE-OUTPUT SYSTEMS 135
4.1.4.1 Rectangular Form 137
4.1.4.2 Phase-Variable Form 139
4.1.4.3 Modal Form 141
4.2 *Coordinate Transformations* 143
4.2.1 INVARIANCE OF THE CHARACTERISTIC POLYNOMIAL 144
4.2.2 INVARIANCE OF THE MARKOV PARAMETERS 145
4.2.3 INVARIANCE OF THE TRANSFER FUNCTION 145
4.2.4 INVARIANCE OF THE RESIDUES 146
4.2.5 INVARIANCE OF FUNCTIONS OF THE OUTPUT 148
4.3 *Disturbance Models* 148
4.4 *Solutions of State Equations* 155
4.4.1 COMPUTATION OF THE STATE TRANSITION MATRIX FOR CONSTANT A 159
4.4.2 STATE SPACE TRAJECTORIES FOR SECOND-ORDER SYSTEMS 169
4.4.3 PERIODIC LINEAR SYSTEMS 171

4.5 *Evaluating System Responses for Time-Invariant Systems* 173
4.5.1 TIME CORRELATION OF VECTORS 174
4.5.2 THE COST FUNCTION AND ITS DECOMPOSITION 182
4.5.3 EVALUATION OF THE COST FUNCTION IN THE FREQUENCY DOMAIN 189

5. *Controllability and Observability* 200

5.1 *Output Controllability* 201
5.2 *State Controllability* 210
5.2.1 CONTROLLABILITY OF TRANSFORMED COORDINATES 212
5.2.1.1 Controllability of Modal Coordinates 213
5.2.1.2 Controllability of Controllable Canonical Coordinates 219
5.2.1.3 Other Controllable Canonical Forms 220
5.2.1.4 Controllability of Vector Second-Order Systems 222
5.3 *Observability* 226
5.4 *Observability of Transformed Coordinates* 230
5.4.1 OBSERVABILITY OF MODAL COORDINATES 230
5.4.2 OBSERVABILITY OF OBSERVABLE CANONICAL COORDINATES 232
5.4.3 OTHER OBSERVABLE CANONICAL COORDINATES 234
5.4.4 OBSERVABILITY OF VECTOR SECOND-ORDER SYSTEMS 235
5.5 *Adjoint Systems and Duality* 238
5.6 *On Relative Controllability, Observability* 239
5.6.1 MODAL COST ANALYSIS 240
5.6.2 BALANCED COORDINATES 250
5.6.3 HESSENBERG COORDINATES 255
5.6.3.1 Controllable Hessenberg Coordinates 255
5.6.3.2 Observable Hessenberg Coordinates 259

6. *Equivalent Realizations and Model Reduction* 266

6.1 *Transfer Equivalent Realizations (TERs)* 266
6.2 *Output Correlation Equivalent Realizations (COVER)* 271
6.3 *Cost-Equivalent Realizations* 279
6.4 *Balanced Model Reduction* 286
6.5 *Model Reduction by Singular Perturbation* 289
6.6 *Realizing Models from Output Data* 293
6.6.1 IDENTIFICATION THEORY — A NAIVE APPROACH 293
6.6.2 1-COVER IDENTIFICATION 296
6.6.3 FREQUENCY DOMAIN IDENTIFICATION 299

7. *Stability* 301

7.1 *Liapunov Stability* 301
7.2 *Bounded-Input / Bounded-Output Stability* 311

7.3 *Exponential and Uniform Stability* 314
7.4 *Stabilizability and Detectability* 315
7.5 *Pole Assignment* 318
7.6 *Covariance Assignment* 321

8. *Optimal Control of Time-Invariant Systems* 328

8.1 *Optimal Covariance Control* 329
8.2 *Linear Quadratic Impulse (LQI) Optimal Control* 331
8.3 *The Linear Quadratic (LQ) Optimal Measurement Feedback Control* 344
8.4 *Modal Methods for Solving Riccati Equations* 349
8.5 *Root Locus of the LQ Optimal State Feedback Controller* 351
8.6 *Nyquist Plot and Stability Margin of the LQ Controller* 356
8.7 *Optimal Control of Vector Second-Order Systems* 366
8.8 *Disturbance Accommodation* 377
8.8.1 DISTURBANCE UTILIZATION CONTROL 378
8.8.2 DISTURBANCE CANCELLATION CONTROL 381
8.9 *Optimal Tracking and Servo Mechanisms* 383
8.10 *Weight Selection in the LQ and LQI Problems* 385

9. *State Estimation* 398

9.1 *Optimal State Estimation* 401
9.2 *Minimal-Order State Estimators* 406
9.3 *Closed-Loop Behavior with Estimator-Based Controllers* 410

10. *Model Error Concepts and Compensation* 414

10.1 *The Structure of Modeling Errors* 421
10.1.1 THE MODEL ERROR SYSTEM 424
10.1.2 THE STRUCTURE OF ERRORS IN THE CLOSED-LOOP SYSTEM 430
10.1.3 STABILITY MARGINS IN CONTROL 436
10.2 *First-Order Perturbations of Modal Data* 442
10.2.1 ROOT PERTURBATIONS IN STATE MODELS 443
10.2.2 ROOT PERTURBATIONS IN VECTOR SECOND-ORDER SYSTEMS 448
10.3 *Sensitivity Analysis and Control* 453
10.3.1 MINIMAL ROOT SENSITIVITY IN LINEAR SYSTEMS 453
10.3.2 TRAJECTORY SENSITIVITY ANALYSIS AND CONTROL 457
10.4 *Model Error Estimation* 461
10.5 *Compensation for a Class of Model Errors* 466

Appendix A: Axiomatic Definition of a Linear Space 475
Appendix B: The Four Fundamental Subspaces of Matrix Theory 476
Appendix C: Calculus of Complex Vectors and Matrices 483
Appendix D: Solution of the Linear Matrix Equations $\mathbf{0} = \mathbf{AX} + \mathbf{XB} + \mathbf{Q}$ 490
Appendix E: Laplace Transforms 493

Index 497

CHAPTER 1

Introduction

There is only an ill-defined boundary between that body of knowledge called "control theory" and that called "systems analysis." Systems analysis explains *why* a system response behaves the way it does. Control theory deals with modifications to the system which will alter the response in a desirable manner. It is no surprise then that one's degree of success in the latter task (control) depends critically upon the relative completeness of an understanding of the first task (systems analysis). The system dynamics may be modified (to improve the response) in two fundamentally different ways: (i) by modifying parameters of the system or by (ii) modifying the forcing functions in the system differential equations. The second approach (ii) is commonly understood to be the purpose of *control* theory, but in the discussion of systems with equivalent behavior (from *systems analysis*) it may be possible to obtain the same response by technique (i) or by a simpler combination of (i) and (ii). A suitable introduction to systems analysis is therefore desirable prior to the introduction of control methods. Chapters 2–7 focus on system analysis and Chapters 7–10 introduce control design.

To illustrate these points consider the rocket depicted in Fig. 1.1 whose linear dynamics are described by these differential equations derived in Chapter 3:

$$
\begin{aligned}
J\ddot{\alpha} - \rho L V^2 \alpha + \rho L V \dot{r}_x &= F D \theta, \\
m\ddot{r}_x + \rho V \dot{r}_x - F\alpha &= -F\theta, \\
m\ddot{r}_y &= F - \rho V^2 - mG,
\end{aligned}
\tag{1.1}
$$

$\dot{r}_x$	Horizontal speed
$\dot{r}_y$	Vertical speed (total speed $V = \sqrt{\dot{r}_x^2 + \dot{r}_y^2}$)
α	Attitude of vehicle with respect to inertial space
θ	Gimbal angle of rocket engine
F	Magnitude of thrust (assumed constant)
D	Distance from mass center to engine gimbal

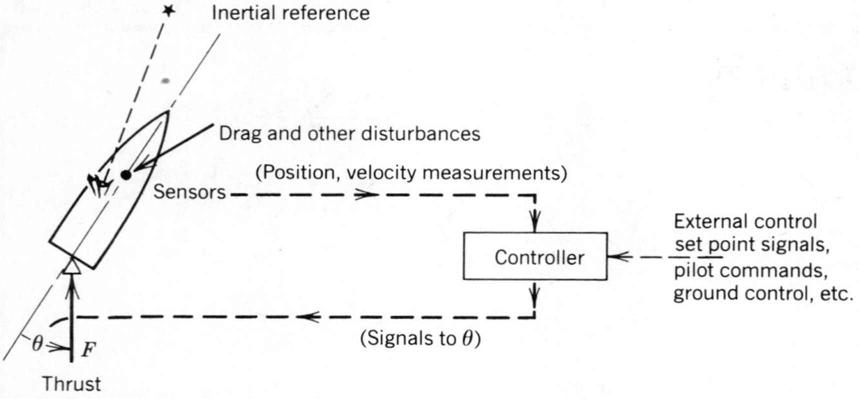

Figure 1.1 *Feedback Control Concepts*

L	Distance from mass center to aerodynamic pressure center
G	Gravity constant
ρV^2	Magnitude of drag force
m	Vehicle mass
J	Moment of inertia

The typical control problem is find a forcing function $\theta(t)$ so that the responses $\alpha(t)$, $r_x(t)$, $r_y(t)$ are acceptable. If $\theta(t)$ is specified as a function of time, then $\theta(t)$ is called an *open loop* control policy. If θ is specified as a function of the system responses, then $\theta(\alpha, \dot{\alpha}, r_x, \dot{r}_x, r_y, \dot{r}_y)$ is called a feedback, or a *closed loop* control policy. The physical and mathematical device that computes the desired θ, given the responses $(\alpha, \dot{\alpha}, r_x, \dot{r}_x, r_y, \dot{r}_y)$, is called the *controller* in Fig. 1.1.

A study of the physical sciences (electricity, mechanics, thermodynamics, chemistry, etc.) prepares the student to apply known physical laws in the derivation of mathematical models of the physical phenomenon such as shown in Fig. 1.1. This is the substance of the undergraduate engineering experience: learning of *how to model* physical systems; whereas the study of control is concerned with *what to do with the model* after it is available. This popular view is much too narrow, of course. It presumes that the modeling problem and the control problem are separable. They are not. One cannot know what level of detail is required in the model prior to knowledge of the accuracy required of the controlled performance and knowledge of the nature of the forces (or control inputs) required to achieve this performance. For example, one cannot know whether the coupling between the translational dynamics (involving $\dot{r}_x$) and the rotational dynamics (involving α) in equations (1.1) is important prior to knowledge of the precision required of the controlled performance and prior to knowledge of the forcing function $\theta(t)$ in (1.1). In fact, for the same reasons one cannot even know *a priori* whether the rigid body in (1.1) and Fig. 1.1 should be modeled as an elastic structure. This depends upon the relative magnitude (and frequency content) of the applied forces. Yet in feedback control the forces applied from the control policy for regulating θ *depend* upon the model chosen. The control problem is to find an appropriate θ, *given* the model (1.1). Hence, the model that is most appropriate for the control design and the control design itself must evolve in an *iterative* fashion. (Do not fall in love with your model!) Control theory and practice still have not, and cannot, produce a fool-proof procedure for accomplishing this [1.1], but the value of seeking a sound theoretical base for the engineer is to help accelerate the convergence of these iterations, with a blend of theory (in the text) and the engineer's judgment and insight (developed on the job). In the first nine chapters of this book we too shall make the traditional assumptions of absolute correctness of our model and the separability of model development from the control design problem. The flaw in this unwritten but commonly evoked *separation* principle will not be corrected until Chapter 10. Yet it is important to learn the traditional wisdom of control theory (Chapters 2 through 9) in full light of its premises, which we here state.

Suppose the horizontal speed of the rocket in (1.1) were truly negligible, then the rotation is governed by

$$J\ddot{\alpha} - \rho L V^2 \alpha = FD\theta \tag{1.2}$$

and the Laplace transform of this equation leads to the transfer function $H(s)$:

$$\alpha(s) = \left[\frac{FD}{Js^2 - \rho L V^2}\right]\theta(s) = H(s)\theta(s), \tag{1.3}$$

where J, ρ, F, and D are positive quantities. The two methods of altering system response mentioned above were (i) modifying parameters of the system and (ii) choosing the forcing function, $\theta(s)$ in this case. Method (i) may be illustrated by changing L, which is positive if the center of pressure is *forward* of the center of mass and negative if the center of pressure is *aft* of the center of mass. The parameter L may be reduced or made negative by making the nose pointed and the tail section larger in diameter, or by adding fins on the tail of the rocket. The center of mass can be made even more forward of the center of pressure (making L more negative) by moving heavy objects (payload) as far forward as possible. Thus, the native Indian's design of an arrow [with rock (heaviest item) on the nose and feathers (large area) as fins on the tail] makes L as negative as possible. This native wisdom is verified by noting that this changes the poles of the transfer function $H(s)$ from the right half-plane ($\lambda = \pm\sqrt{\rho L V^2/J}$ when $L > 0$) to the imaginary axis ($\lambda = \pm j\sqrt{\rho(-L)V^2/J}$ when $L < 0$). This is an improvement since it changes the response from unstable to stable. In fact, aerodynamic damping effects ignored here in (1.1) would actually place the poles slightly inside the left half-plane yielding asymptotic stability of the open-loop system [$\theta(t) \to 0$]. That is, α returns to zero from some nonzero initial condition. However, the price paid in approach (i) is possibly to increase the weight of the rocket, which in turn reduces payload capability. Hence, one possible advantage of feedback control [method (ii) above] is the modification of system behavior without adding the weight or other undesirable features that a structural parameter change might require. Clearly, a trade-off is evident between increasing structural design modifications (parameter changes), which beef up the structure weight, and increasing control sophistication, which would be required as the basic structure degenerates to "cheapest" to build, lightest in weight. The arguments taken to the limit of this latter extreme would yield an absurd design. The control system would fly all the payload, engine, and instrument components in formation with hardly any structure holding them together at all, all of the required interaction forces to hold things together being provided by the addition of multiple control forces (besides the control of θ). Hence, even though this example points to an absurd extreme, it is nonetheless true that as the control requirements become more stringent in modern systems, it is usually necessary to add more "actuators" or control variables since those detailed things we wish to control can be "uncontrollable" with a single actuator. The time domain methods readily accommodate multiple inputs and outputs, and this is the focus of this text, whereas transform methods, equation (1.3), more readily treat the single-input/output class of systems.

Having anticipated that multiple actuators and sensors will be needed in a research frontier pressing for better performance capabilities of a controlled system, we shall become aware of fundamental limitations and dangers in pressing too far. As the controller becomes more complex, it may become less reliable, both from the possibility of failures and from the reliance upon increasing detail of mathematical models (upon which the control policy is based). Chapter 10 sorts out some of these difficulties and guards against a misuse of the first theory we learn in Chapters 2 through 9 (where correctness of the model is presumed).

The classical control tools of Bode, Nyquist, and Evans [1.2–1.4], all developed prior to 1950, are basically graphical in nature, with the following underlying design strategy: "Design for stability, then check for performance." These tools function in the frequency domain. The time domain tools of state space optimal control developed rapidly two decades later, with an eye on the mean squared performance: "Design for performance, then check for stability." These techniques are numerical rather than graphical. Thus, the early time domain design philosophy and the frequency domain design philosophy were *opposite*, and complementary insights are available by studying both. Of course, this is an oversimplification of the methods and their power, but nonetheless it is important for a student in control to enthusiastically embrace both points of view. This text focuses primarily on time domain methods.

While time domain techniques began their rapid development in the 1960s with the vector first-order (state) form of differential equations, it was Sir William Rowan Hamilton in 1835 who introduced state form equations with the presentation of his theory of "generalized momenta" [1.5]. This replaced the Euler–Lagrange equations

$$\frac{d}{dt}\frac{\partial \mathscr{T}}{\partial \dot{q}_i} - \frac{\partial \mathscr{T}}{\partial q_i} = Q_i, \qquad i = 1, \dots, N,\ T = \text{kinetic energy},\ Q_i = \text{generalized forces}, \tag{1.4}$$

which always resulted in systems of equations of a vector second-order form,

$$\mathscr{M}\ddot{\mathbf{q}} + \mathscr{D}\dot{\mathbf{q}} + \mathscr{K}\mathbf{q} = \mathbf{f}, \tag{1.5}$$

by a set of generalized momenta equations,

$$p_i \triangleq \frac{\partial \mathscr{T}}{\partial \dot{q}_i}, \tag{1.6}$$

which always resulted in systems of equations of a vector first-order form,

$$\dot{\mathbf{p}} = \mathbf{A}\mathbf{p} + \bar{\mathbf{f}}. \tag{1.7}$$

For a century, dynamicists largely shunned Hamilton's generalized momenta equations (1.7) until modern computers were available to provide practical computations of solutions. His first-order "Hamilton's canonical" equations,

$$\dot{p}_i \triangleq -\frac{\partial \mathscr{H}}{\partial q_i}, \qquad \dot{q}_i = \frac{\partial \mathscr{H}}{\partial p_i},$$

where $\mathscr{H} \triangleq \mathbf{p}^T\mathbf{q} - \mathscr{L}$, $\mathscr{L} \triangleq \mathscr{T} - \mathscr{U} + \mathscr{F}$, $\mathscr{U}$ = potential energy, and $\mathscr{F}$ = work done by nonconservative forces, became popular [1.6] many years after their introduction. Thus, it is no coincidence that the modern and rapid development of state space techniques closely followed the development of computers and efficient numerical methods.

If equations (1.1) were placed in the vector second-order from (1.5), we would have

$$\begin{bmatrix} J & 0 & 0 \\ 0 & m & 0 \\ 0 & 0 & m \end{bmatrix}\begin{pmatrix} \ddot{\alpha} \\ \ddot{r}_x \\ \ddot{r}_y \end{pmatrix} + \begin{bmatrix} 0 & \rho LV & 0 \\ 0 & \rho V & 0 \\ 0 & 0 & 0 \end{bmatrix}\begin{pmatrix} \dot{\alpha} \\ \dot{r}_x \\ \dot{r}_y \end{pmatrix} + \begin{bmatrix} -\rho LV^2 & 0 & 0 \\ -F & 0 & 0 \\ 0 & 0 & 0 \end{bmatrix}\begin{pmatrix} \alpha \\ r_x \\ r_y \end{pmatrix}$$

$$= \begin{bmatrix} FD \\ -F \\ 0 \end{bmatrix}\theta + \begin{bmatrix} 0 \\ 0 \\ F - \rho V^2 - mG \end{bmatrix}. \tag{1.8}$$

And if placed in the vector first-order form (state form), we would have, from the generalized momenta equations,

$$\begin{bmatrix} \dot{q}_1 \\ \dot{q}_2 \\ \dot{q}_3 \end{bmatrix} = \begin{bmatrix} 1/J & 0 & 0 \\ 0 & 1/m & 0 \\ 0 & 0 & 1/m \end{bmatrix}\begin{bmatrix} p_1 \\ p_2 \\ p_3 \end{bmatrix}, \qquad \begin{bmatrix} q_1 \\ q_2 \\ q_3 \end{bmatrix} \triangleq \begin{bmatrix} \alpha \\ r_x \\ r_y \end{bmatrix}, \tag{1.9a}$$

$$\begin{bmatrix} \dot{p}_1 \\ \dot{p}_2 \\ \dot{p}_3 \end{bmatrix} = \begin{bmatrix} 0 & -\rho LV/m & 0 \\ 0 & -\rho V/m & 0 \\ 0 & 0 & 0 \end{bmatrix}\begin{bmatrix} p_1 \\ p_2 \\ p_3 \end{bmatrix} + \begin{bmatrix} \rho LV^2 q_1 + FD\theta \\ Fq_1 - F\theta \\ F - \rho V^2 - mG \end{bmatrix}, \tag{1.9b}$$

leading to the state form

$$\begin{bmatrix} \dot{q}_1 \\ \dot{q}_2 \\ \dot{q}_3 \\ \dot{p}_1 \\ \dot{p}_2 \\ \dot{p}_3 \end{bmatrix} = \begin{bmatrix} 0 & 0 & 0 & 1/J & 0 & 0 \\ 0 & 0 & 0 & 0 & 1/m & 0 \\ 0 & 0 & 0 & 0 & 0 & 0 \\ \rho L V^2 & 0 & 0 & 0 & -\rho L V/m & 0 \\ F & 0 & 0 & 0 & -\rho V/m & 0 \\ 0 & 0 & 0 & 0 & 0 & 0 \end{bmatrix} \begin{bmatrix} q_1 \\ q_2 \\ q_3 \\ p_1 \\ p_2 \\ p_3 \end{bmatrix}$$

$$+ \begin{bmatrix} 0 \\ 0 \\ 0 \\ FD \\ -F \\ 0 \end{bmatrix} \theta + \begin{bmatrix} 0 \\ 0 \\ 0 \\ 0 \\ 0 \\ F - \rho V^2 - mG \end{bmatrix}. \tag{1.10}$$

If placed in a transfer function form, regarding $\alpha(s)$, $r_x(s)$, $r_y(s)$ as the outputs, then taking the Laplace transform of (1.1), the model takes the form defining $\Delta(s) \triangleq s[\rho LVF + (Js^2 - LV^2)(ms + \rho V)]$,

$$\begin{bmatrix} q_1(s) \\ q_2(s) \\ q_3(s) \end{bmatrix} = \begin{bmatrix} F(Dms + \rho V(L + D))s/\Delta(s) \\ F(FD - Js^2 + \rho LV^2)/\Delta(s) \\ 0 \end{bmatrix} \theta(s) + \begin{bmatrix} 0 \\ 0 \\ (F - \rho V^2 - mG)/ms^3 \end{bmatrix}$$

$$+ \frac{1}{\Delta(s)} \begin{bmatrix} J(\rho V + ms)s^2 & 0 & 0 & J(\rho V + ms)s & -\rho VLms & 0 \\ JFs & \rho LVF + (\rho V + ms)(Js^2 - \rho LV^2) & 0 & JF & m(Js^2 - \rho LV^2) & 0 \\ 0 & 0 & \Delta(s)/s & 0 & 0 & \Delta(s)/s^2 \end{bmatrix} \begin{bmatrix} q_1(0) \\ q_2(0) \\ q_3(0) \\ p_1(0) \\ p_2(0) \\ p_3(0) \end{bmatrix}. \tag{1.11}$$

Now, (1.11) has the form

$$\mathbf{q}(s) = \mathbf{G}(s)\theta(s) + \mathbf{G}_0(s)\mathbf{x}(0) + \mathbf{d}(s),$$

where $\mathbf{G}$ and $\mathbf{G}_0$ are matrices in (1.11), and

$$\mathbf{q}(s) = \begin{bmatrix} q_1(s) \\ q_2(s) \\ q_3(s) \end{bmatrix}, \quad \mathbf{x}(0) = \begin{bmatrix} q_1(0) \\ q_2(0) \\ q_3(0) \\ p_1(0) \\ p_2(0) \\ p_3(0) \end{bmatrix}, \quad \mathbf{d}(s) = \begin{bmatrix} 0 \\ 0 \\ (F - \rho V^2 - mG)/ms^2 \end{bmatrix}.$$

Now, consider the net upward force $(F - \rho V^2 - mG)$ to be slightly uncertain but constant. The system response $\mathbf{q}(t)$ must be made appropriate by choice of θ, *in the presence* of uncertain initial conditions $\mathbf{x}(0)$ and disturbances $\mathbf{d}(t)$. Transform methods typically ignore $\mathbf{x}(0)$ for convenience. This is unfortunate since the response $\mathbf{q}(t)$ can be influenced by $\mathbf{x}(0)$ in ways that cannot be predicted by examining the transfer functions $\mathbf{G}(s)$ alone. The concepts of controllability and observability in Chapter 5 will expose such circumstances. Hence, if $\mathbf{x}(0)$ should not be ignored, then the most tractable way to include it is to work directly with the state equations (1.10). Hence, this text focuses on time domain methods with a presumption that the reader already has some familiarity with elementary transform methods to appreciate our frequent reference to equivalent transform interpretations of our results.

The goal of Chapters 2 through 6 is to provide the insights and the mathematical machinery to be later used in the design-oriented Chapters 7 through 10. In other words, this author thinks of applications and design as the *highest* level of control engineering activity, not to be short-circuited by poor preparation for this high calling. (Hence, the theoretical work of Chapters 2 through 6 is a necessary beginning).

References

1.1 R. E. Skelton and D. H. Owens, *Model Error Concepts and Compensation*, Pergamon Press, Elmsford, N.Y., 1986.

1.2 W. R. Evans, "Graphical Analysis of Control Systems," *AIEE Trans.*, Part II, 67, 1948, pp. 547–551.

1.3 H. Nyquist, "Regeneration Theory," *Bell Systems Tech. J.*, 11, 1932, pp. 126–147.

1.4 H. W. Bode, "Network Analysis and Feedback Amplifier Design," D. Van Nostrand Co., Inc., Princeton, N.J., 1945.

1.5 W. R. Hamilton, "Second Essay on a General Method in Dynamics," *Philosophical Transactions of the Royal Society of London*, 1835, *p*. 98.

1.6 H. Rund, "The Hamilton–Jacobi Theory in the Calculus of Variations," D. Van Nostrand Co., Inc., Princeton, N.J., 1966.

CHAPTER 2

Mathematical Preliminaries

Studies of differential equations and linear algebra are fundamental to the analysis and design of control systems. There are two opposing points of view of the educational process, however. Those seeking to make advancements in the mathematical theory of control must take a slower, more detailed trajectory through linear algebra. The typical engineering student, however, might wish to initially spend less time on fundamental building blocks to get a quicker perspective of the applications.

Some authors have taken the ultimate step in this direction by altogether avoiding a separate chapter on the linear algebra building blocks, integrating the development of the mathematical tools with the development of the control concepts and applications [2.1]. Other texts go to great lengths to treat linear algebra in a completely formal manner before advancing to control concepts and applications [2.2], [2.3]. Both approaches serve important but different needs. The approach of this chapter is a compromise between the two. The decision to provide a separate chapter to summarize linear algebra results is based upon a desire to provide convenient places for repetitive reference to standard facts. But on a first reading, the chapter should be treated with a minimum of detail in order not to discourage those who are impatient to get to the control applications. More importantly, readers may wait to sharpen their specialized knowledge of linear algebra within the context of the controls applications.

2.1 Matrix Representations and Linear Spaces

Much of linear algebra is centered around the study of such simultaneous linear equations as

$$\begin{aligned} A_{11}x_1 + A_{12}x_2 + A_{13}x_3 &= y_1, \\ A_{21}x_1 + A_{22}x_2 + A_{23}x_3 &= y_2, \end{aligned} \tag{2.1}$$

which is conveniently studied in the compact notation of matrices. Hence, (2.1) may be written

$$\begin{bmatrix} A_{11} & A_{12} & A_{13} \\ A_{21} & A_{22} & A_{23} \end{bmatrix} \begin{pmatrix} x_1 \\ x_2 \\ x_3 \end{pmatrix} = \begin{pmatrix} y_1 \\ y_2 \end{pmatrix} \tag{2.2}$$

or simply

$$[\mathbf{A}](\mathbf{x}) = (\mathbf{y}). \tag{2.3}$$

The still simpler notation

$$\mathbf{A}\mathbf{x} = \mathbf{y} \tag{2.4}$$

will be used throughout to study properties of the system of equations (2.1).

The dimension of a $k \times n$ matrix $\mathbf{A}$ will be denoted by the notation $\mathbf{A} \in \mathscr{R}^{k \times n}$ or $\mathbf{A} \in \mathscr{C}^{k \times n}$, where these and other notations that will be needed later are catalogued in Table 2.1. Note that the most elementary element of $\mathbf{A}$ is each element A_{ik}. It will prove convenient, however, to discuss slightly more sophisticated elements of $\mathbf{A}$: its rows and columns. If $\mathbf{A} \in \mathscr{R}^{k \times n}$, then the n columns of $\mathbf{A}$ are denoted $\mathbf{a}_i \in \mathscr{R}^{k \times 1}$, $i = 1, \ldots, n$. For the example (2.2), $k = 2$, $n = 3$. Hence, $\mathbf{A}$ is expressed in terms of its columns by

$$\begin{aligned} \mathbf{A} &= [\mathbf{a}_1, \mathbf{a}_2, \mathbf{a}_3] = \left[\begin{pmatrix} A_{11} \\ A_{21} \end{pmatrix}, \quad \begin{pmatrix} A_{12} \\ A_{22} \end{pmatrix}, \quad \begin{pmatrix} A_{13} \\ A_{23} \end{pmatrix} \right], \\ \mathbf{a}_i &= \begin{pmatrix} A_{1i} \\ A_{2i} \end{pmatrix}, \end{aligned} \tag{2.5}$$

and $\mathbf{a}_i \in \mathscr{R}^{2 \times 1}$.

If $\mathbf{A} \in \mathscr{R}^{k \times n}$, then the k rows of $\mathbf{A}$ are denoted by $\mathbf{b}_\alpha^T \in \mathscr{R}^{1 \times n}$, $\alpha = 1, \ldots, k$. For the example (2.2), $k = 2$, $n = 3$. Hence, $\mathbf{A}$ is expressed in terms of its rows by

$$\mathbf{A} = \begin{bmatrix} \mathbf{b}_1^T \\ \mathbf{b}_2^T \end{bmatrix} = \begin{bmatrix} (A_{11} A_{12} A_{13}) \\ (A_{21} A_{22} A_{23}) \end{bmatrix}, \qquad \mathbf{b}_i^T \triangleq (A_{i1} A_{i2} A_{i3}). \tag{2.6}$$

When one of the dimensions of the matrix is 1, it is simpler to omit the 1 and write $\mathbf{b} \in \mathscr{R}^k$ to describe the $k \times 1$ matrix $\mathbf{b}$. Hence, we shall henceforth describe the $\mathbf{a}_i$ in (2.5) and the $\mathbf{b}_i$ in (2.6) by $\mathbf{a}_i \in \mathscr{R}^2$ and $\mathbf{b}_i \in \mathscr{R}^3$ in lieu of the earlier notation $\mathbf{a}_i \in \mathscr{R}^{2 \times 1}$ and $\mathbf{b}_i \in \mathscr{R}^{3 \times 1}$. The ordered array of $k \times n$ numbers $\mathbf{A} \in \mathscr{R}^{k \times n}$ is called "matrix," but the ordered array of k numbers $\mathbf{b} \in \mathscr{R}^k$ is called a "vector." This is a slight abuse of language since the word "vector" was originally used by Gibbs [2.4] to denote a more restricted entity with magnitude and direction in a three-dimensional space. To illustrate the differences, consider an example.

TABLE 2.1 Notation

NOTATION	MEANING
iff	If and only if
$\{A\} \Rightarrow \{B\}$	Event $\{A\}$ *implies* event $\{B\}$
$\{A\} \Leftarrow \{B\}$	Event $\{A\}$ *is implied by* event $\{B\}$
$\{A\} \Leftrightarrow \{B\}$	Event $\{A\}$ and event $\{B\}$ *are equivalent* (event $\{A\}$; happens iff event $\{B\}$ happens)
$\forall i$	For every i
$t \in [0, \tau]$	$0 \le t \le \tau$ (t lies in the *closed interval* $[0, \tau]$)
$t \in [0, \tau)$	$0 \le t < \tau$ (t lies in the *semiclosed* interval $[0, \tau)$)
$\overline{\mathbf{A}}$	The complex conjugate of $\mathbf{A} = \mathscr{Re}[\mathbf{A}] + j\mathscr{Im}[\mathbf{A}]$ is $\overline{\mathbf{A}} \triangleq \mathscr{Re}[\mathbf{A}] - j\mathscr{Im}[\mathbf{A}]$
$\mathbf{A}^T$	$[\mathbf{A}^T]_{ik} = A_{ki}$ = the jith element of matrix $\mathbf{A}$
$\mathbf{A}^*$	The complex conjugate transpose of $\mathbf{A}$; $\mathbf{A}^* \triangleq \overline{\mathbf{A}}^T \triangleq \mathscr{Re}[\mathbf{A}]^T - j\mathscr{Im}[\mathbf{A}]^T$
j	$j \triangleq \sqrt{-1}$
$\mathbf{A} \in \mathscr{R}^{k \times n}$	$\mathbf{A}$ is composed of real elements A_{ik}; ($\mathbf{A}$ has k rows, n columns)
$\mathbf{A} \in \mathscr{C}^{k \times n}$	$\mathbf{A}$ is composed of complex elements A_{ij}; ($\mathbf{A}$ has k rows, n columns)
$\mathbf{AB} = \mathbf{C}$	The product of matrices $\mathbf{A}$ and $\mathbf{B}$ is equal to the matrix $\mathbf{C}$ ($\Sigma_j A_{ij} B_{jk} = C_{ik}$)
$\mathbf{A} = \mathbf{B}$	$\mathbf{A}$ equal $\mathbf{B}$ ($A_{ij} = B_{ij}$)
$\mathbf{A}(t) \equiv \mathbf{B}(t)$	$\mathbf{A}$ equal $\mathbf{B}$ for all time t
$\mathbf{A} \triangleq \mathbf{B}$	$\mathbf{A}$ is *defined* to be $\mathbf{B}$
tr $\mathbf{A}$	*tr*ace of $A \triangleq \Sigma_{i=1} A_{ii}$
$\mathbf{I}_n$	$\mathbf{I}_n = \begin{bmatrix} 1 & 0 & 0 & \\ 0 & 1 & 0 & \\ 0 & 0 & 1 & \\ & & & \ddots \end{bmatrix}_{n \times n}$ identity matrix
Symmetric matrix $\mathbf{A}$	$\mathbf{A} = \mathbf{A}^T$
Skew-symmetric matrix $\mathbf{A}$	$\mathbf{A} = -\mathbf{A}^T$
Hermitian matrix $\mathbf{A}$	$\mathbf{A} = \mathbf{A}^*$
Skew-Hermitian matrix $\mathbf{A}$	$\mathbf{A} = -\mathbf{A}^*$
$\alpha\beta$ Minor of $\mathbf{A}$ ($= M_{\alpha\beta}$)	$M_{\alpha\beta}$ = determinant of matrix $\mathbf{A}$ after deleting α row, β column
$\alpha\beta$ Cofactor of A ($= C_{\alpha\beta}$)	$C_{\alpha\beta} = (-1)^{\alpha+\beta} M_{\alpha\beta}$
Cofactor matrix of $\mathbf{A}$ (= Co A)	$[\text{Co}\,\mathbf{A}]_{\alpha\beta} = C_{\alpha\beta}$
Adjoint matrix of $\mathbf{A}$ (= adj A)	adj $\mathbf{A} \triangleq [\text{Co}\,\mathbf{A}]^T$
Determinant of A ($\triangleq \lvert A \rvert$)	$\lvert\mathbf{A}\rvert = \Sigma_{j=1}^{n} A_{\alpha j} C_{\alpha j}$ (for any selected row α); also $\lvert\mathbf{A}\rvert = \Sigma_{k=1}^{n} A_{k\beta} C_{k\beta}$ (for any selected column of β)

continued

TABLE 2.1 Notation

NOTATION	MEANING
Inverse of **A**	$\mathbf{A}^{-1} = \frac{1}{\lvert A\rvert}[\text{adj}\,\mathbf{A}]$
Singular matrix **A**	**A** has property $\lvert\mathbf{A}\rvert = 0$
Nonsingular matrix **A**	**A** has property $\lvert\mathbf{A}\rvert \neq 0$
Principal minors of A	$[A_{11}], \begin{bmatrix} A_{11} & A_{12} \\ A_{21} & A_{22} \end{bmatrix}, \quad \ldots \quad \begin{bmatrix} A_{11} & \cdots & A_{1n} \\ \vdots & & \\ A_{n1} & \cdots & A_{nn} \end{bmatrix}$
Normal matrix A	$AA^* = A^*A$
Orthonormal matrix A	$A^{-1} = A^T$
Unitary matrix A	$A^{-1} = A^*$
Eigenvalues of A	Scalars λ_i, $i = 1, \ldots, n$, that satisfy $Ae_i = e_i\lambda_i$ for nontrivial vectors e_i
Eigenvectors of A	Nontrivial vectors e_i that satisfy $Ae_i = e_i\lambda_i$
Left eigenvector of A	Nontrivial vectors l_i that satisfy $l_i^*A = \lambda_i l_i^*$

EXAMPLE 2.1

The spring–mass system shown in Fig. 2.1 has motion in the $\mathbf{e}_1$ direction described by the sum of force vectors

$$m\frac{d^2}{dt^2}[\mathbf{r}] = \mathbf{f}_1 + \mathbf{f}_2, \tag{2.7}$$

where the position vector **r** is

$$\mathbf{r} = z\mathbf{e}_1 + y\mathbf{e}_2 + 0\mathbf{e}_3 \tag{2.8}$$

and $\mathbf{e}_1$, $\mathbf{e}_2$, and $\mathbf{e}_3$ are the orthogonal unit vectors fixed in inertial space. The force vectors $\mathbf{f}_1$ and $\mathbf{f}_2$ having magnitude (f_1) and (kz) and direction $(\mathbf{e}_1)$ and $(-\mathbf{e}_1)$ are described by

$$\mathbf{f}_1 = f_1\mathbf{e}_1, \qquad \mathbf{f}_2 = kz(-\mathbf{e}_1). \tag{2.9}$$

Equation (2.7) leads to

$$m\frac{d^2}{dt^2}(z\mathbf{e}_1 + y\mathbf{e}_2) = m\ddot{z}\mathbf{e}_1 + m\ddot{y}\mathbf{e}_2 = f_1\mathbf{e}_1 - kz\mathbf{e}_1. \tag{2.10}$$

Since $\mathbf{e}_1$ and $\mathbf{e}_2$ are not zero, and since

$$(m\ddot{z} + kz - f_1)\mathbf{e}_1 = 0, \tag{2.11a}$$

$$(m\ddot{y})\mathbf{e}_2 = 0 \tag{2.11b}$$

must hold for all $f_1(t)$, $z(t)$, it follows from the physical constraint $\dot{y} = 0$ that

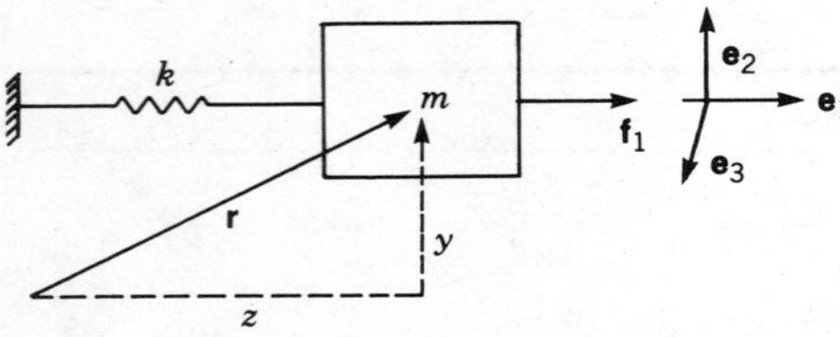

Figure 2.1 A Spring–Mass System

$y(t) \equiv$ constant and

$$m\ddot{z} + kz - f_1 = 0 \tag{2.12}$$

is the proper equation of motion. However, the *direction* of the motion z and applied forces must be understood for complete understanding of (2.12). That is, having a given *basis* $\{\mathbf{e}_1, \mathbf{e}_2, \mathbf{e}_3\}$ in the three-*dimensional space*, (2.12) describes the motion in that basis. The vectors $\mathbf{e}_1$, $\mathbf{e}_2$, $\mathbf{e}_3$, $\mathbf{f}_1$, $\mathbf{f}_2$, and $\mathbf{r}$ are vectors in the sense of Gibbs [2.4], and are entities with magnitude and direction in a three-dimensional space.

Now, consider the equivalent of (2.12),

$$\frac{d}{dt}\begin{pmatrix} z \\ \dot{z} \end{pmatrix} \begin{bmatrix} 0 & 1 \\ -k/m & 0 \end{bmatrix} \begin{pmatrix} z \\ \dot{z} \end{pmatrix} + \begin{pmatrix} 0 \\ 1 \end{pmatrix} \frac{f_1}{m}, \tag{2.13}$$

which can be written in the compact form

$$\dot{\mathbf{x}} = \mathbf{A}\mathbf{x} + \mathbf{B}u \tag{2.14}$$

where

$$\mathbf{x} \triangleq \begin{pmatrix} z \\ \dot{z} \end{pmatrix}, \quad \mathbf{A} \triangleq \begin{bmatrix} 0 & 1 \\ -k/m & 0 \end{bmatrix}, \quad \mathbf{B} \triangleq \begin{pmatrix} 0 \\ 1 \end{pmatrix}, \quad u \triangleq \frac{f_1}{m}. \quad \blacksquare$$

In the language of Section 2.1, $\mathbf{A} \in \mathscr{R}^{2\times 2}$ in (2.14) is called a matrix and $\mathbf{x} \in \mathscr{R}^2$ in (2.14) is called a vector. Note that this kind of "vector" is a 2×1 matrix where n is *not* limited to 3 as is the case for the Gibbs vectors $\mathbf{e}_i$ in Fig. 2.1 and equations (2.8)–(2.11). If Fig. 2.1 had included two masses instead of one, then the final equations of motion arranged in the form (2.14) would have a "vector" x of dimension 4. Many types of systems studied in this book have the linear form (2.14).

In dynamics, the variables needed to *construct* the differential equations of motion (2.12) are called the "configuration variables" [2.5], and for Example 2.1 there is only *one* configuration variable z. In linear systems theory, the focus is on the *solution* of differential equations rather than on their construction, and the variables required to *solve* the differential equations (2.12) or (2.14) are called "state variables." Hence, the system of Fig. 2.1 requires *one* configuration variable and *two* state variables. Systems described by equations of the form (2.14) are said to be in *state-variable* form, and in such a form, x is called the *state vector*. Formal study of the state variables follow in a later section. See Appendix A for a more formal definition of a linear vector space.

The vectors $x(t)$ in (2.14), a_i in (2.5), and b_i in (2.6) can have real or complex values. It is convenient to think of the vector $\mathbf{x} \in \mathscr{C}^n$ as a point in an n-dimensional space. (Of course, one cannot *visualize* such a space if $n > 3$.) The notation $\mathscr{C}^n$ is used to denote a complex n-dimensional linear vector space, and $\mathscr{R}^n$ is used to denote a real n-dimensional linear vector space. We consider the space of all rational functions and the n-dimensional linear vector spaces $\mathscr{C}^n$ and $\mathscr{R}^n$ are special cases of the space of all rational functions.

2.2 Linear Indepedence of Vectors

A most fundamental concept in linear systems theory is the concept of linear independence.

Definition 2.1

The vectors $\mathbf{a}_i(t)$, $i = 1, \ldots, n$, are "linearly independent" on the interval $[t_1, t_2]$ iff

$$\sum_{i=1}^{n} \mathbf{a}_i(t) x_i = 0, \qquad t \in [t_1, t_2] \tag{2.15}$$

holds only for the trivial values $x_i = 0$, $\forall i = 1, \ldots, n$. Otherwise, the vectors are "linearly dependent on the interval $t \in [t_1, t_2]$."

Note that the vectors $\mathbf{a}_1(t) \in \mathscr{C}^k$, $\mathbf{a}_2(t) \in \mathscr{C}^k, \ldots, \mathbf{a}_n(t) \in \mathscr{C}^k$ are "linearly independent on the interval $t \in [t_1, t_2]$" if none of the $\mathbf{a}_i$ can be written as a linear combination of the remaining vectors.

EXAMPLE 2.2

Consider the two scalar functions $a_1(t)$, $a_2(t)$ shown in Fig. 2.2. Show that $\mathbf{a}_1(t)$, $a_2(t)$ are (i) linearly *dependent* on the interval $t \in [0, 1)$, (ii) linearly *dependent* on the interval $t \in (1, 2]$, and (iii) linearly *independent* on the interval $t \in [0, 2]$.

Solution:

(i) Choose the nontrivial values $x_1 = 1$, $x_2 = 1$ in (2.15) to get

$$a_1(t)x_1 + a_2(t)x_2 = (t)(1) + (-t)(+1) = 0, \qquad t \in [0, 1).$$

(ii) Now show that $a_1(t)$ and $a_2(t)$ are linearly dependent on the interval $t \in (1, 2]$. To show this, choose $x_1 = 1$, $x_2 = -1$ in (2.15) to get

$$a_1(t)x_1 + a_2(t)x_2 = (t)(1) + (t)(-1) = 0, \qquad t \in (1, 2].$$

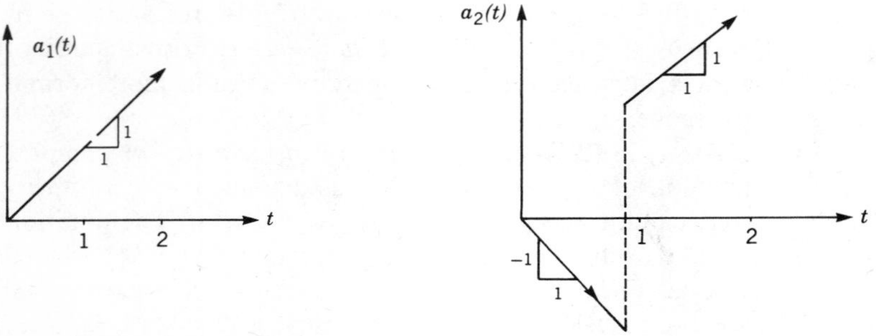

Figure 2.2 *Linear Independence of Functions*

(iii) No nontrivial values of x_1, x_2 exist, however, that satisfy (2.15) on the interval $t \in [0, 2]$. That is, $a_2(t)$ *cannot* be written as a linear function of $a_1(t)$ on the interval $t \in [0, 2]$, but $a_2(t)$ *can* be written as a linear function of $a_1(t)$ on both the intervals $t \in [0, 1)$ and $t \in (1, 2]$. To illustrate linear independence, the above choices of x_1, x_2 were *ad hoc*. A systematic test for determining existence of such x's will appear in Section 2.4.

If the a_i, $i = 1, \ldots, n$, in Definition 2.1 are *constant*, then the linear independence question is not a function of t. Hence, the phrase "on the interval $t \in [t_1, t_2]$" may be deleted in Definition 2.1. Thus, we have the following.

Definition 2.2
The constant vectors $\mathbf{a}_i \in \mathscr{C}^k$, $i = 1, \ldots, n$, are said to be linearly independent iff

$$\sum_{i=1}^{n} \mathbf{a}_i x_i = 0 \tag{2.16}$$

can be satisfied only for the trivial scalars $x_i = 0$, $i = 1, \ldots, n$. Otherwise, the vectors are "linearly dependent."

The conclusions from Definitions 2.1 and 2.2 are now written in a much more compact and easy-to-remember matrix form. Define $\mathbf{A} \in \mathscr{C}^{k \times n}$ such that

$$\mathbf{A} = [\mathbf{a}_1, \mathbf{a}_2, \ldots, \mathbf{a}_n]$$

and $\mathbf{x} \in \mathscr{C}^n$. Then, for time-varying $\mathbf{A}$, Definition 2.1 allows the conclusion that the columns of $\mathbf{A}(t)$ are linearly independent on the interval $t \in [t_1, t_2]$ iff

$$\mathbf{A}(t)\mathbf{x} = \mathbf{0} \Rightarrow \mathbf{x} = 0, \qquad t \in [t_1, t_2]; \tag{2.17}$$

and for constant matrix $\mathbf{A}$, Definition 2.2 allows the conclusion that the columns of

A are linearly independent iff

$$\mathbf{Ax} = \mathbf{0} \Rightarrow \mathbf{x} = \mathbf{0}. \tag{2.18}$$

EXAMPLE 2.3

Show that

$$\mathbf{a}_1 = \begin{pmatrix} 1 \\ 0 \end{pmatrix}, \qquad \mathbf{a}_2 = \begin{pmatrix} 1 \\ 1 \end{pmatrix}$$

are linearly independent.

Solution: From (2.18), this requires showing that

$$\begin{bmatrix} 1 & 1 \\ 0 & 1 \end{bmatrix} \begin{pmatrix} x_1 \\ x_2 \end{pmatrix} = \mathbf{0} \Rightarrow \begin{pmatrix} x_1 \\ x_2 \end{pmatrix} = \mathbf{0},$$

which result follows immediately from the scalar equations

$$\left\{ \begin{aligned} x_1 + x_2 &= 0, \\ x_2 &= 0. \end{aligned} \right\}$$

■

EXAMPLE 2.4

Show that the *rows* of

$$B = \begin{bmatrix} 1 & 1 & -1 \\ 1 & -1 & -2 \end{bmatrix}$$

are linearly independent.

Solution: To use the result (2.18), define $\mathbf{A} \triangleq \mathbf{B}^*$ and ask if

$$\begin{bmatrix} 1 & 1 \\ 1 & -1 \\ -1 & -2 \end{bmatrix} \begin{pmatrix} x_1 \\ x_2 \end{pmatrix} = \mathbf{0} \Rightarrow \mathbf{x} = \mathbf{0}?$$

The scalar equations

$$x_1 + x_2 = 0$$

$$x_1 - x_2 = 0$$

$$-x_1 - 2x_2 = 0$$

have no solution except $x_1 = x_2 = 0$, as follows from the first two equations. The second equation yields $x_1 = x_2$. Substitution into the first equation yields $2x_2 = 0$, hence $x_2 = 0$. Hence, the rows of **B** are linearly independent. ■

2.3 Geometrical Interpretations of Linear Independence

Consider whether the columns of **B** in Example 2.4 are linearly independent. Since the columns of **B** lie in a two-dimensional real space $\mathscr{R}^2$, they can be plotted on the plane of the paper as in Fig 2.3. For the vectors to be linearly independent, it must be impossible to express one of the vectors as a linear combination of the others. Clearly, any two of the vectors in Fig. 2.3 form a basis with which to express any other vector in the same plane. Therefore, three vectors in the plane cannot be linearly independent.

This geometrical idea is further enhanced by the fact that the determinant of $[\mathbf{A}^T\mathbf{A}]$ is the volume squared of the space formed by the parallelopiped of the columns of **A**. Before pursuing this point, a proper definition is needed for "spanning a space."

Definition 2.3

The set of vectors $\mathbf{a}_i \in \mathscr{C}^k$, $i = 1, \ldots, n$, is said to span the space $\mathscr{C}^k$ if every vector in that space, $\mathbf{b} \in \mathscr{C}^k$, can be expressed as a linear combination of the set of $\mathbf{a}_i$, $i = 1, \ldots, n$.

Exercise 2.1

Consider the two vectors $\mathbf{a}_1$ and $\mathbf{a}_2$:

$$\mathbf{A} = [\mathbf{a}_1 \quad \mathbf{a}_2] = \begin{bmatrix} 1 & 1 \\ 0 & 2 \end{bmatrix}.$$

Prove that the determinant of $\mathbf{A}^*\mathbf{A}$ is the square of the area of the parallelogram spanned by $\mathbf{a}_1, \mathbf{a}_2$ in Fig. 2.4a.

Exercise 2.2

Prove that the determinant $\mathbf{A}^*\mathbf{A}$ is the square of the volume spanned by the vectors $\mathbf{a}_1$, $\mathbf{a}_2$, and $\mathbf{a}_3$ in Fig. 2.4b.

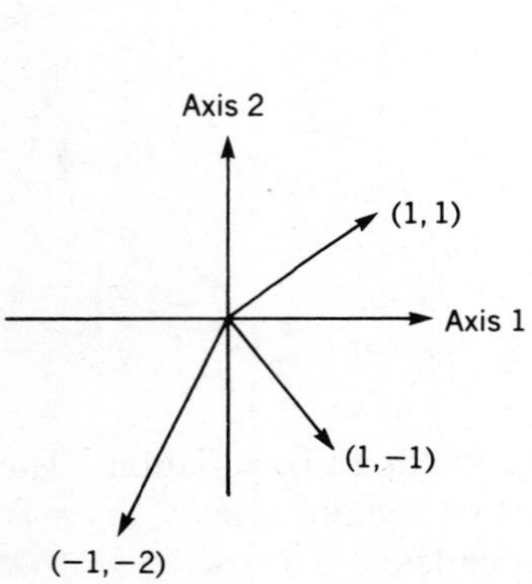

Figure 2.3 *Linear Independence of Vectors*

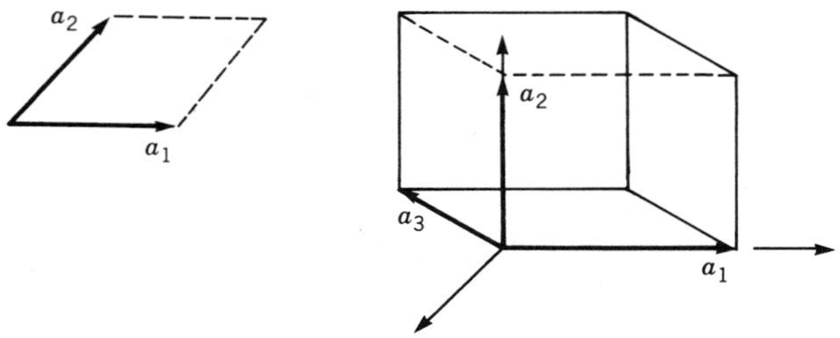

Figure 2.4

These exercises give the student the idea that two vectors cannot be linearly independent if the area they span is zero and that three vectors cannot be linearly independent if the volume they span is zero.

Exercise 2.3
Is the set of vectors that *span* a space necessarily linearly independent?

Definition 2.4
A set of vectors $\mathbf{a}_i \in \mathscr{C}^k$, $i = 1, \ldots, n$, is said to form a "basis" for the space $\mathscr{C}^n$ if n is the smallest number for which the vectors a_i span the space.

Since the number of basis vectors needed to span a space is no greater than the dimension of that space, n must be equal or less than k. Furthermore, since *every* vector in that space must be included within the span of the basis, clearly $n = k$, and Definition 2.4 could have written k for n throughout. However, Definition 2.4 is written in terms of different symbols, k and n, to emphasize the distinction between Definitions 2.3 and 2.4.

Exercise 2.4
Is the set of vectors that form a *basis* for a space necessarily linearly independent?

Exercise 2.5
Plot the vectors $\mathbf{a}_i \in \mathscr{R}^2$, $i = 1, \ldots 4$,

$$\mathbf{A} = [\mathbf{a}_1 \quad \mathbf{a}_2 \quad \mathbf{a}_3 \quad \mathbf{a}_4] = \begin{bmatrix} 1 & 1 & -2 & 1 \\ 0 & 1 & 1 & -0.5 \end{bmatrix}.$$

From this set of $\mathbf{a}_i$, determine all the possible combinations of vectors that form a *basis* for the space $\mathscr{R}^2$.

Definition 2.5
The rank of a matrix is the number of linearly independent columns or rows.

Define the row (column) rank of a matrix **A** as the number of linearly independent rows (columns). Then,

Theorem 2.1
Row rank **A** = *column rank* **A**.

Consider Definition 2.5 in terms of *columns*. The discussion below Definition 2.4 explains why the number of linearly independent rows must *equal* the number of linearly independent columns, hence Theorem 2.1. It is therefore proper as written in Definition 2.5 to assume they are the same.

Exercise 2.6
Using the above discussions and definitions, prove the following fact for $A \in R^{k\times n}$, $\mathbf{B} \in \mathscr{R}^{n\times p}$.

$$\text{rank}\,[\mathbf{AB}] \le \text{minimum}\,\{\text{rank}\,\mathbf{A}, \text{rank}\,\mathbf{B}\}.$$

Under what conditions does the equality hold? Give an example.

Exercise 2.7
Employing the tests of linear independence and Definition 2.5, determine the rank of **A** of these four matrices.

(i) $\mathbf{A} = \mathbf{bb}^*, \qquad \mathbf{b} \in \mathscr{C}^n,$

(ii) $\mathbf{A} = \mathbf{B}^T\mathbf{B}, \qquad \mathbf{B} = \begin{bmatrix} 1 & 1 & 1 \\ 1 & 2 & 3 \end{bmatrix},$

(iii) $\mathbf{A} = \mathbf{BB}^T, \qquad \mathbf{B}$ as in (ii),

(iv) $$\mathbf{A} = \begin{bmatrix} 1 & 1 & -2 & 0 & 1 \\ 0 & -1 & 2 & 1 & 0 \\ 0 & \alpha & -2 & 1 & 2 \end{bmatrix}.$$

For what range of α is rank $\mathbf{A} = 3$?

2.4 Tests for Determining Linear Independence

The methods used to determine solutions of (2.17) and (2.18) in the above examples are *ad hoc* and too cumbersome to use for a large number of vectors. To develop more systematic approaches, some definitions of vector products are required.

2.4.1 VECTOR INNER PRODUCTS

Definition 2.6

Two vectors $\mathbf{x}(t) \in \mathscr{C}^n$ *and* $\mathbf{y}(t) \in \mathscr{C}^n$ *have an "inner product on the linear space of continuous functions in the interval* $t \in [t_1, t_2]$*" defined by*

$$\langle \mathbf{x}(t), \mathbf{y}(t) \rangle \triangleq \frac{1}{t_2 - t_1} \int_{t_1}^{t_2} \mathbf{x}^*(t)\mathbf{y}(t)\, dt = \frac{1}{t_2 - t_1} \int_{t_1}^{t_2} \sum_{i=1}^{n} \bar{\mathbf{x}}_i(t)\mathbf{y}_i(t)\, dt. \tag{2.19}$$

Suppose $x(t)$, $y(t)$ were *constant* in (2.19). This leads to the following result: *The "inner product" of* ***constant*** *vectors* $x \in \mathscr{C}^n$ *and* $y \in \mathscr{C}^n$ *is given by*

$$\langle \mathbf{x}, \mathbf{y} \rangle = \mathbf{x}^*\mathbf{y} = \sum_{i=1}^{n} \bar{x}_i y_i. \tag{2.20}$$

Also see that

$$\langle \mathbf{x}, \mathbf{y} \rangle = \overline{\langle \mathbf{y}, \mathbf{x} \rangle}. \tag{2.21}$$

The notation $\langle \mathbf{x}, \mathbf{y} \rangle$ will often be used to denote the inner product of $\mathbf{x}$ and $\mathbf{y}$. The above inner product definitions suffice for this book. Other definitions of inner products are available in ref. [2.7].

Definition 2.7

The vectors $\mathbf{x} \in \mathscr{C}^n$ *and* $\mathbf{y} \in \mathscr{C}^n$ *are said to be "orthogonal" if*

$$\langle \mathbf{x}, \mathbf{y} \rangle = 0. \tag{2.22}$$

2.4.2 VECTOR OUTER PRODUCTS

Definition 2.8

The notation $\rangle \mathbf{x}, \mathbf{y} \langle$ *will be used to denote the product*

$$\rangle \mathbf{x}(t), \mathbf{y}(t) \langle \triangleq \frac{1}{t_2 - t_1} \int_{t_1}^{t_2} \mathbf{x}(t)\mathbf{y}^*(t)\, dt$$

$$= \frac{1}{t_2 - t_1} \begin{bmatrix} \int_{t_1}^{t_2} x_1(t)\,\bar{y}_1(t)\, dt & \int_{t_1}^{t_2} x_1(t)\,\bar{y}_2(t)\, dt & \cdots \\ \int_{t_1}^{t_2} x_2(t)\,\bar{y}_1(t)\, dt & \int_{t_1}^{t_2} x_2(t)\,\bar{y}_2(t)\, dt & \cdots \\ \vdots & \vdots & \end{bmatrix}, \tag{2.23a}$$

which will be called the "outer product of $\mathbf{x}(t)$ *and* $\mathbf{y}(t)$ *on the interval* $t \in [t_1, t_2]$*."*

Note that the outer product of a k-dimensional vector with a p-dimensional vector yields a $k \times p$ matrix. Now, suppose **x** and **y** are *constant* k and p-dimensional vectors, respectively. Then, (2.23a) yields

$$\rangle \mathbf{x}, \mathbf{y} \langle = \mathbf{x}\mathbf{y}^* = \begin{bmatrix} x_1\bar{y}_1 & x_1\bar{y}_2 \\ x_2\bar{y}_1 & x_2\bar{y}_2 \\ \vdots & \vdots \end{bmatrix}_{k \times p}. \tag{2.23b}$$

Equation (2.23a) involves the integral of a matrix which notation we now formally define.

Definition 2.9
The notation $\int_{t_1}^{t_2} \mathbf{A}(t)\,dt$ will be used to denote the integration of the matrix $\mathbf{A}(t) \in \mathscr{C}^{k \times n}$ as follows:

$$\int_{t_1}^{t_2} \mathbf{A}(t)\,dt \triangleq \begin{bmatrix} \int_{t_1}^{t_2} A_{11}(t)\,dt & \int_{t_1}^{t_2} A_{12}(t)\,dt \\ \int_{t_1}^{t_2} A_{21}(t)\,dt & \int_{t_1}^{t_2} A_{22}(t)\,dt & \ddots \end{bmatrix}_{k \times n}. \tag{2.24}$$

Hence, the outer product of $x(t)$ and $y(t)$ involves the integral of the matrix $\mathbf{x}(t)\mathbf{y}^*(t)$.

2.4.3 VECTOR NORMS

Definition 2.10
The norm of the vector $\mathbf{x}(t) \in \mathscr{C}^n$ on the interval $t \in [t_1, t_2]$ is defined by the positive square root of the inner product of $\mathbf{x}(t)$ with itself. Thus, the norm-squared of $\mathbf{x}(t)$ is

$$\|\mathbf{x}(t)\|^2 \triangleq \langle \mathbf{x}(t), \mathbf{x}(t) \rangle \triangleq \frac{1}{t_2 - t_1} \int_{t_1}^{t_2} \mathbf{x}^*(t)\mathbf{x}(t)\,dt. \tag{2.25}$$

Thus, the norm-squared of a constant is

$$\|\mathbf{x}\|^2 \triangleq \mathbf{x}^*\mathbf{x} = \langle \mathbf{x}, \mathbf{x} \rangle. \tag{2.26a}$$

The weighted norm of **x** *is defined by**

$$\|\mathbf{x}\|_{\mathbf{Q}}^2 \triangleq \langle \mathbf{x}, \mathbf{Q}\mathbf{x} \rangle, \tag{2.26b}$$

where **Q** *is a symmetric matrix.*

*Other vector norms may be defined [2.8]. However, the norm (2.26a) is the only one of interest in this book.

EXAMPLE 2.15

For the vector $\mathbf{x}^* \triangleq (1, 1 + j1)$ and $\mathbf{Q} = \begin{bmatrix} 1 & 2 \\ 2 & 10 \end{bmatrix}$, find (i) the weighted norm-squared of $\mathbf{x}$, and (ii) the outer product of $\mathbf{x}$ with itself.

Solution:

(i) $$\|\mathbf{x}\|_{\mathbf{Q}}^2 = \mathbf{x}^*\mathbf{Q}\mathbf{x} = (1, 1 + j1)\begin{bmatrix} 1 & 2 \\ 2 & 10 \end{bmatrix}\begin{pmatrix} 1 \\ 1 - j1 \end{pmatrix} = 25,$$

(ii) $$\rangle \mathbf{x}, \mathbf{x} \langle = \mathbf{x}\mathbf{x}^* = \begin{pmatrix} 1 \\ 1 - j1 \end{pmatrix}(1, 1 + j1) = \begin{bmatrix} 1 & 1 + j1 \\ 1 - j1 & 2 \end{bmatrix}.$$

■

EXAMPLE 2.16

For the vector $\mathbf{x}^* \triangleq (1, 2, 3)$, find (i) the norm, and (ii) the outer product of $\mathbf{x}$ with itself.

Solution:

(i) $$\|\mathbf{x}\| = \sqrt{\mathbf{x}^*\mathbf{x}} = \left[(1, 2, 3)\begin{pmatrix} 1 \\ 2 \\ 3 \end{pmatrix}\right]^{1/2} = (1 + 4 + 9)^{1/2} = \sqrt{14},$$

(ii) $$\mathbf{x}, \mathbf{x} \langle = \mathbf{x}\mathbf{x}^* = \begin{pmatrix} 1 \\ 2 \\ 3 \end{pmatrix}(1 \quad 2 \quad 3) = \begin{bmatrix} 1 & 2 & 3 \\ 2 & 4 & 6 \\ 3 & 6 & 9 \end{bmatrix}.$$

■

2.4.4 THE CAUCHY – SCHWARZ INEQUALITY

A most useful result called the Cauchy–Schwarz inequality is the following.

Theorem 2.2

The magnitude of the inner product of two vectors $\mathbf{x} \in \mathscr{C}^n$ *and* $\mathbf{y} \in \mathscr{C}^n$ *obeys the inequality*

$$|\langle \mathbf{x}, \mathbf{y} \rangle| \leq \|\mathbf{x}\| \, \|\mathbf{y}\|, \tag{2.27}$$

and the equality holds iff $\mathbf{x}$ *is colinear with* $\mathbf{y}$ ($\mathbf{x} = \beta\mathbf{y}$ *for some scalar* β).

The proof is postponed momentarily.

Exercise 2.8

Verify that the vectors $\mathbf{x}^T = (1, 1)$, $\mathbf{y}^T = (2, 7)$ obey the Cauchy–Schwarz inequality. Verify that $\mathbf{x}^T = (4, 14)$ and $\mathbf{y}^T = (2, 7)$ satisfy the Cauchy–Schwarz inequality. In this case, why does the *equality* hold?

Suppose we introduce a multiplier ρ on the right-hand side of (2.27) to convert it to an equality. Thus, define ρ such that

$$\langle \mathbf{x}, \mathbf{y} \rangle = \|\mathbf{x}\|\,\|\mathbf{y}\|\rho, \qquad 0 \le |\rho| \le 1. \tag{2.28}$$

Hence, ρ may be calculated by

$$\rho = \frac{\langle \mathbf{x}, \mathbf{y} \rangle}{\|\mathbf{x}\|\,\|\mathbf{y}\|}. \tag{2.29a}$$

2.4.5 GEOMETRICAL INTERPRETATION OF ρ: ANGLES BETWEEN VECTORS

Consider for $\mathbf{x} \in \mathscr{R}^n$, $\mathbf{y} \in \mathscr{R}^n$ that ρ becomes

$$\rho = \frac{\mathbf{x}^T\mathbf{y}}{\|\mathbf{x}\|\,\|\mathbf{y}\|} = \left(\frac{\mathbf{x}}{\|\mathbf{x}\|}\right)^T\left(\frac{\mathbf{y}}{\|\mathbf{y}\|}\right), \tag{2.29b}$$

which is simply the inner product of two unit vectors in the directions of $\mathbf{x}$ and $\mathbf{y}$. Consider that the unit vectors $\mathbf{x}/\|\mathbf{x}\|$ and $\mathbf{y}/\|\mathbf{y}\|$ define a plane that contains both vectors, and let θ be the angle between these vectors. Within this plane, recall the law of cosines, which relates the three sides of the triangle in the plane. The sides are $\mathbf{x}$, $\mathbf{y}$, and $(\mathbf{x} - \mathbf{y})$ as shown in Fig. 2.5. The law of cosines states that

$$\|\mathbf{x} - \mathbf{y}\|^2 = \|\mathbf{x}\|^2 + \|\mathbf{y}\|^2 - 2\|\mathbf{x}\,\|\mathbf{y}\|\cos\theta, \tag{2.30}$$

whereupon, using the norm (2.26a) in this real space $\mathscr{R}^n$, (2.30) becomes

$$(\mathbf{x} - \mathbf{y})^T(\mathbf{x} - \mathbf{y}) = \mathbf{x}^T\mathbf{x} + \mathbf{y}^T\mathbf{y} - 2\|\mathbf{x}\|\,\|\mathbf{y}\|\cos\theta,$$

or

$$\mathbf{x}^T\mathbf{x} + \mathbf{y}^T\mathbf{y} - \mathbf{x}^T\mathbf{y} - \mathbf{y}^T\mathbf{x} = \mathbf{x}^T\mathbf{x} + \mathbf{y}^T\mathbf{y} - 2\|\mathbf{x}\|\,\|\mathbf{y}\|\cos\theta. \tag{2.31}$$

But since $\mathbf{x}^T\mathbf{y} = \mathbf{y}^T\mathbf{x}$, (2.31) leads to

$$\langle \mathbf{x}, \mathbf{y} \rangle = \mathbf{x}^T\mathbf{y} = \|\mathbf{x}\|\,\|\mathbf{y}\|\cos\theta, \tag{2.32a}$$

or

$$\cos\theta = \frac{\mathbf{x}^T\mathbf{y}}{\|\mathbf{x}\|\,\|\mathbf{y}\|}. \tag{2.32b}$$

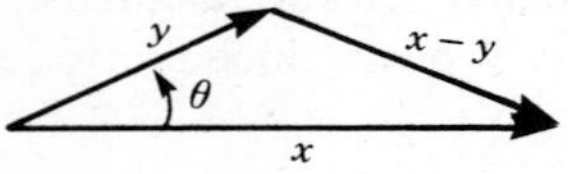

Figure 2.5 Inner Products of Vectors

A comparison of (2.29b) and (2.32b) leads to the conclusion that ρ has the physical meaning $\rho \triangleq \cos\theta$. Result (2.32b) is now useful in the proof of Theorem 2.2.

Proof of Theorem 2.2: Using the fact $|\cos\theta| \le 1$, the absolute magnitude of (2.32a) immediately yields the result (2.27). Furthermore, the equality in (2.27) holds if and only if $\cos\theta = 1$, in (2.32a), in which case $\mathbf{x}$ and $\mathbf{y}$ are parallel. (Hence, $\mathbf{x} = \beta\mathbf{y}$ for some scalar β).

This proof is rigorous only for real vectors $\mathbf{x} \in \mathscr{R}^n$, $\mathbf{y} \in \mathscr{R}^n$. A more general proof is left to the reader in the following exercise. ■

Exercise 2.9

Using the well-known triangle inequality from plane geometry and Fig. 2.5,

$$\|\mathbf{y}\| + \|\mathbf{x} - \mathbf{y}\| \ge \|\mathbf{x}\|, \tag{2.33}$$

square both sides of (2.33) to prove (2.27) *without* any notion of angles.

2.4.6 SIGN DEFINITENESS OF MATRICES

The weighted norm of vectors plays an important role in linear systems theory. Some important definitions related to the weighted norm follow.

Definition 2.12

The matrix $\mathbf{Q}(t) \in \mathscr{C}^{n\times n}$ is called "positive definite on the interval $t \in [t_1, t_2]$" if

$$\langle \mathbf{x}, \mathbf{Q}(t)\mathbf{x}\rangle = \|\mathbf{x}\|^2_{\mathbf{Q}(t)} > 0 \qquad \text{for all constant } \mathbf{x} \ne 0,\ t \in [t_1, t_2] \tag{2.34}$$

and "positive semidefinite on the interval $t \in [t_1, t_2]$" if

$$\|\mathbf{x}\|^2_{\mathbf{Q}(t)} \ge 0 \qquad \text{for all constant } \mathbf{x},\ t \in [t_1, t_2] \tag{2.35}$$

and "negative definite on the interval $t \in [t_1, t_2]$" if

$$\|\mathbf{x}\|^2_{\mathbf{Q}(t)} < 0 \qquad \text{for all constant } \mathbf{x} \ne 0,\ t \in [t_1, t_2] \tag{2.36}$$

and "negative semidefinite on the interval $t \in [t_1, t_2]$" if

$$\|\mathbf{x}\|^2_{\mathbf{Q}(t)} \le 0 \qquad \text{for all constant } \mathbf{x},\ t \in [t_1, t_2]. \tag{2.37}$$

From (2.25) and (2.26b),

$$\|\mathbf{x}\|^2_{\mathbf{Q}(t)} = \langle \mathbf{x}, \mathbf{Q}(t)\mathbf{x}\rangle = \frac{1}{t_2 - t_1}\int_{t_1}^{t_2} \mathbf{x}^*\mathbf{Q}(t)\mathbf{x}\, dt = \frac{1}{t_2 - t_1} x^*\left[\int_{t_1}^{t_2} \mathbf{Q}(t)\, dt\right]\mathbf{x}. \tag{2.38}$$

Hence, (2.34) becomes

$$\|\mathbf{x}\|^2_{\mathbf{Q}(t)} = \|\mathbf{x}\|^2_{\mathbf{K}(t_1, t_2)} > 0 \qquad \text{for all } \mathbf{x} \neq 0, \tag{2.39}$$

$$\mathbf{K}(t_1, t_2) \triangleq \frac{1}{t_2 - t_1} \int_{t_1}^{t_2} \mathbf{Q}(t)\, dt.$$

Note from (2.34) and (2.39) the equivalent statements

$$\left\{ \|\mathbf{x}\|^2_{Q(t)} > 0 \text{ for all } \mathbf{x} \neq 0,\ t \in [t_1, t_2] \right\} \Leftrightarrow \left\{ \|\mathbf{x}\|^2_{\mathbf{K}(t_1, t_2)} > 0 \text{ for all } \mathbf{x} \neq 0 \right\}. \tag{2.40}$$

The reader is reminded that the weight $\mathbf{Q}(t)$ is a *time varying* matrix and $\mathbf{K}(t_1, t_2)$ is a *constant* matrix.

Definition 2.13
The constant matrix $\mathbf{K} \in \mathscr{C}^{n \times n}$ *is "positive definite" if*

$$\|\mathbf{x}\|^2_{\mathbf{K}} > 0 \qquad \textit{for all } \mathbf{x} \neq 0, \tag{2.41}$$

and the shorthand notation "$\mathbf{K} > \mathbf{0}$*" will be used to mean that* $\mathbf{K}$ *has the property* (2.41).

Since the definiteness of a *time-varying* matrix $\mathbf{Q}(t)$ reduces to the determination of the definiteness of a *constant* matrix $\mathbf{K}(t_1, t_2)$, we only need a *test* for determining definiteness of a constant matrix $\mathbf{K}$.

Theorem 2.3
The constant Hermitian matrix $\mathbf{K} \in \mathscr{C}^{n \times n}$ *is*

(a) *Positive definite* ($\mathbf{K} > \mathbf{0}$) *if either of these equivalent conditions holds*:
1. *All eigenvalues of* $\mathbf{K}$ *are positive*,
2. *All principal minors of* $\mathbf{K}$ *have positive determinants*;

(b) *Positive semidefinite* ($\mathbf{K} \geq \mathbf{0}$) *if either of these equivalent conditions holds*:
1. *All eigenvalues of* $\mathbf{K}$ *are zero or positive*,
2. *All principal minors of* $\mathbf{K}$ *have zero or positive determinants*;

(c) *Negative definite* ($\mathbf{K} < \mathbf{0}$) *if either of these equivalent conditions holds*:
1. *All eigenvalues of* $(-\mathbf{K})$ *are positive*
2. *All principal minors of* $(-\mathbf{K})$ *have positive determinants*;

(d) *Negative semidefinite* ($\mathbf{K} \leq \mathbf{0}$) *if either of these equivalent conditions holds*:
1. *All eigenvalues of* $(-\mathbf{K})$ *are zero or positive*,
2. *All principal minors of* $(-\mathbf{K})$ *have zero or positive determinants.*

Parts (a1), (b1), (c1), and (d1) will be proved in Section 2.8.1 after a more formal introduction to eigenvalues. The remaining parts are proved in ref. [2.9].

EXAMPLE 2.7

Determine whether

$$\mathbf{Q}(t)\begin{bmatrix} 1 & \sin t \\ \sin t & 1 \end{bmatrix}$$

is positive definite on the interval $t \in [0, \pi/2]$.

Solution: From (2.39) construct $\mathbf{K}(t_1, t_2)$.

$$\mathbf{K}(0, \pi/2) = \frac{1}{(\pi/2) - 0}\int_0^{\pi/2}\begin{bmatrix} 1 & \sin t \\ \sin t & 1 \end{bmatrix} dt = \begin{bmatrix} \pi/2 & 1 \\ 1 & \pi/2 \end{bmatrix}\frac{1}{\pi/2}. \tag{2.42}$$

The principal minors of $\mathbf{K}$ have determinants 1 and $(\pi^2/4 - 1)(2/\pi)$, which are both positive. Thus, $\mathbf{K}(0, \pi/2)$ is a positive definite matrix. Hence, from (2.40) and Definition 2.13, $\mathbf{Q}(t)$ is positive definite on the interval $t \in [0, \pi/2]$. ■

EXAMPLE 2.8

Determine whether the function

$$x_1^2 + 2x_1x_2 + x_2^2 \tag{2.43}$$

is positive for all nontrivial values of x_1, x_2.

Solution: The quadratic function (2.43) can be written in the form

$$(x_1 \quad x_2)\begin{bmatrix} 1 & 1 \\ 1 & 1 \end{bmatrix}\begin{pmatrix} x_1 \\ x_2 \end{pmatrix} = \mathbf{x}^T\mathbf{K}\mathbf{x}. \tag{2.44}$$

From Definition 2.13 and Theorem 2.3, $\mathbf{K}$ is *not* positive definite since both principal minor determinants (having values 1 and 0) are *not* positive. Hence, (2.43) is *not* positive for all $\mathbf{x} \neq 0$. Verify this by the nontrivial choice $x_1 = 1$, $x_2 = -1$. (Note that $\mathbf{K}$ is positive *semidefinite*.) ■

With these concepts of vector norms in hand, we must return now to the development of a test for linear independence of vectors.

2.4.7 A TEST FOR LINEAR INDEPENDENCE OF TIME-VARYING VECTORS

Clearly, the norm $\|\mathbf{Ax}\|$ is zero iff $\mathbf{Ax} = \mathbf{0}$. Hence, from (2.17) it follows that

$$\left\{\|\mathbf{A}(t)\mathbf{x}\|^2 = 0\right\} \Rightarrow \left\{\mathbf{x} = \mathbf{0},\ t \in [t_1, t_2]\right\} \tag{2.45}$$

if the columns of $\mathbf{A}$ are linearly independent on the interval $t \in [t_1, t_2]$. From (2.25),

$$\|\mathbf{A}(t)\mathbf{x}\|^2 = \frac{1}{t_2 - t_1}\int_{t_1}^{t_2}\mathbf{x}^*\mathbf{A}^*(t)\mathbf{A}(t)\mathbf{x}\, dt = \mathbf{x}^*\left[\frac{1}{t_2 - t_1}\int_{t_1}^{t_2}\mathbf{A}^*(t)\mathbf{A}(t)\, dt\right]\mathbf{x}. \tag{2.46}$$

Thus, (2.45) reduces to the statement

$$\|\mathbf{x}\|^2\mathbf{K}(t_1, t_2) = \mathbf{0} \Rightarrow \mathbf{x} = \mathbf{0}, \qquad \mathbf{K}(t_1, t_2) \triangleq \frac{1}{t_2 - t_1}\int_{t_1}^{t_2} \mathbf{A}^*(t)\mathbf{A}(t)\, dt \quad (2.47)$$

if the columns of $\mathbf{A}(t)$ are linearly independent on the interval $t \in [t_1, t_2]$. Equation (2.47) holds if $\mathbf{K}(t_1, t_2)$ is nonsingular. (This type of result will be proved in Section 2.8.)

Theorem 2.4

The columns of the matrix $\mathbf{A}(t) \in \mathscr{C}^{k\times n}$ are linearly independent on the interval $t \in [t_1, t_2]$ iff

$$\frac{1}{t_2 - t_1}\int_{t_1}^{t_2} \mathbf{A}^*(t)\mathbf{A}(t)\, dt \tag{2.48}$$

is a nonsingular matrix. The rows of $\mathbf{A}(t) \in \mathscr{C}^{k\times n}$ are linearly independent on the interval $t \in [t_1, t_2]$ iff

$$\frac{1}{t_1 - t_2}\int_{t_1}^{t_2} \mathbf{A}(t)\mathbf{A}^*(t)\, dt \tag{2.49}$$

is a nonsingular matrix.

EXAMPLE 2.9

Determine whether the functions $(t, t - t^2, t^2 - 2t)$ are linearly independent on the interval $t \in [0, 1]$.

Solution: Since Theorem 2.4 is set up to determine independence of *columns* of $\mathbf{A}$ (rather than rows), the appropriate definition of $A \in R^{k\times n}$ is

$$\mathbf{A}(t) = [t, t - t^2, t^2 - 2t], \tag{2.50}$$

with $k = 1$, $n = 3$.

According to Theorem 2.4,

$$\mathbf{K}(0,1) = \int_0^1 \begin{pmatrix} t \\ t - t^2 \\ t^2 - 2t \end{pmatrix} (t, t - t^2, t^2 - 2t)\, dt$$

must be a nonsingular matrix. The integration gives

$$\mathbf{K}(0,1) = \begin{bmatrix} \frac{1}{3} & \frac{1}{12} & -\frac{5}{12} \\ \frac{1}{12} & \frac{1}{30} & \frac{-7}{60} \\ -\frac{5}{12} & \frac{-7}{60} & \frac{8}{15} \end{bmatrix},$$

which has a nonzero determinant. Hence, the functions $(t, t - t^2, t^2 - 2t)$ are linearly independent on the interval $t \in [0, 1]$. ■

Exercise 2.10

Use Theorem 2.4 to answer the questions in Example 2.2.

2.4.8 A TEST FOR LINEAR INDEPENDENCE OF CONSTANT VECTORS

Clearly, the norm $\|\mathbf{Ax}\|$ is zero iff $\mathbf{Ax} = \mathbf{0}$. Hence, from (2.18) it follows that

$$\|\mathbf{Ax}\|^2 = 0 \Rightarrow \mathbf{x} = \mathbf{0} \tag{2.51}$$

if the columns of $\mathbf{A}$ are linearly independent. The appropriate norm for constant $\mathbf{A}$ according to (2.20) is

$$\|\mathbf{Ax}\|^2 = \mathbf{x}^*\mathbf{A}^*\mathbf{Ax}. \tag{2.52}$$

For constant A, both (2.52) and (2.48) lead to the conclusion that $\mathbf{A}^*\mathbf{A}$ must be nonsingular. Since $\mathbf{A}^*\mathbf{A}$ is either positive semidefinite or positive definite, it can be nonsingular only if it is positive definite. Hence, the following conclusion follows immediately from (2.48) and (2.49).

Corollary To Theorem 2.4

The columns of the constant matrix $\mathbf{A} \in \mathscr{C}^{k \times n}$ *are linearly independent iff* $[\mathbf{A}^*\mathbf{A}]$ *is a nonsingular matrix. The rows of* $\mathbf{A} \in \mathscr{C}^{k \times n}$ *are linearly independent iff* $[\mathbf{AA}^*]$ *is a nonsingular matrix*

EXAMPLE 2.10

Determine if the *rows* of the matrix

$$\mathbf{B} = \begin{bmatrix} 1 & 1 \\ 1 & -1 \\ -1 & -2 \end{bmatrix}$$

are linearly independent.

Solution: From the corollary to Theorem 2.4,

$$\mathbf{BB}^* = \begin{bmatrix} 1 & 1 \\ 1 & -1 \\ -1 & -2 \end{bmatrix} \begin{bmatrix} 1 & 1 & -1 \\ 1 & -1 & -2 \end{bmatrix} = \begin{bmatrix} 2 & 0 & -3 \\ 0 & 2 & 1 \\ -3 & 1 & 5 \end{bmatrix}.$$

The determinant of $\mathbf{BB}^*$ is zero. (The principal minors have determinants 2, 4, 0. Hence, $\mathbf{BB}^*$ is positive semidefinite). Since $\mathbf{BB}^*$ is singular, the rows of $\mathbf{B}$ are not linearly independent. ■

Theorem 2.4a

$$|\mathbf{A}^*\mathbf{A}| \neq 0 \qquad \textit{iff}\ \mathbf{A}^*\mathbf{A} > 0. \tag{2.52a}$$

Proof: First, we show that

$$\mathbf{A}^*\mathbf{A} > 0 \Rightarrow |\mathbf{A}^*\mathbf{A}| \neq 0.$$

Then, we will show that

$$|\mathbf{A}^*\mathbf{A}| \neq 0 \Rightarrow \mathbf{A}^*\mathbf{A} > 0.$$

First, prove that $\mathbf{A}^*\mathbf{A} > 0 \Rightarrow |\mathbf{A}^*\mathbf{A}|$:

$$\mathbf{x}^*\mathbf{A}^*\mathbf{A}\mathbf{x} > 0 \qquad \forall x \neq 0 \text{ (given)}.$$

Hence,

$$\mathbf{A}\mathbf{x} \neq 0, \qquad \forall \mathbf{x} \neq \mathbf{0}.$$

Hence,

$$\mathbf{A}\mathbf{x} = \mathbf{0} \Rightarrow \mathbf{x} = \mathbf{0},$$

which *means* that columns of $\mathbf{A}$ are linearly independent, (2.18). The test for linear independence of columns of $\mathbf{A}$ is $|\mathbf{A}^*\mathbf{A}| \neq 0$. This proves that $\mathbf{A}^*\mathbf{A} > \mathbf{0} \Rightarrow |\mathbf{A}^*\mathbf{A}| \neq 0$. Now, suppose $|\mathbf{A}^*\mathbf{A}| \neq 0$ and try to show that this implies $\mathbf{A}^*\mathbf{A} > \mathbf{0}$.

$$|\mathbf{A}^*\mathbf{A}| = \prod_{i=1}^{n} \lambda_i[\mathbf{A}^*\mathbf{A}] \neq 0 \Rightarrow \lambda_i \neq 0 \qquad \forall i,$$

$$\lambda_i = \mathbf{e}_i^*\mathbf{A}^*\mathbf{A}\mathbf{e}_i \neq 0, \qquad \text{where } [\mathbf{A}^*\mathbf{A}]\mathbf{e}_i = \lambda_i\mathbf{e}_i,$$

$$\mathbf{e}_i^*\mathbf{e}_i = 1,$$

$$\lambda_i = \mathbf{y}^*\mathbf{y} = \|\mathbf{y}\|^2, \qquad \mathbf{y} \triangleq \mathbf{A}\mathbf{e}_i,$$

$$\|\mathbf{y}\|^2 = \mathbf{e}_i^*\mathbf{A}^*\mathbf{A}\mathbf{e}_i > 0 \qquad \forall \text{eigenvectors } \mathbf{e}_i.$$

But since the set of eigenvectors span the entire space of $[\mathbf{A}^*\mathbf{A}]$, then *any* linear combination of the $\mathbf{e}_i$ will also have this property. Hence,

$$\mathbf{x}^*\mathbf{A}^*\mathbf{A}\mathbf{x} > 0 \qquad \forall \mathbf{x} \neq \mathbf{0}.$$ ■

2.5 Orthogonal Bases

Example 2.9 shows that the functions $\{t, t - t^2, t^2 - 2t\}$ are linearly independent on the interval $t \in [0, 1]$. The following corollary is a generalization of Theorem 2.2 in one respect (by adding a weight to the integral) and a special case of Theorem 2.4 in another (by making $\mathbf{A}(t)$ a row matrix, $\mathbf{A}(t) = \boldsymbol{\gamma}^*(t)$).

Second Corollary to Theorem 2.4
The real or complex scalar functions $\{\gamma_1(t), \ldots, \gamma_n(t)\}$ *are linearly independent on the interval* $t \in [t_1, t_2]$ *subject to the weight* $g(t)$ *iff the matrix*

$$\int_{t_1}^{t_2} \gamma(t) g(t) \gamma^*(t)\, dt, \qquad \gamma^* = (\gamma_1, \ldots, \gamma_n) \tag{2.53}$$

is nonsingular.

Proof: Note in (2.48) that a weight $\mathbf{Q}(t)$ could be added to read

$$\mathbf{K}(t_1, t_2) \triangleq \frac{1}{t_2 - t_1} \int_{t_1}^{t_2} \mathbf{A}^*(t)\mathbf{Q}(t)\mathbf{A}(t)\, dt, \tag{2.54}$$

and if $\mathbf{K}(t_1, t_2)$ is nonsingular, the columns of $\mathbf{A}(t)$ would be called "linearly independent on the interval $t \in [t_1, t_2]$ subject to the weight $\mathbf{Q}(t)$." See that in this case (2.45) and (2.46) would be modified only by the addition of the weight $\mathbf{Q}(t)$ in the manner $\|\mathbf{A}(t)\mathbf{x}\|^2_{\mathbf{Q}(t)}$, similar to the notation (2.38). In the present circumstance, let $\mathbf{A}(t) \in \mathscr{C}^{k \times n}$ be a row matrix ($k = 1$), which is renamed as the vector γ^*,

$$\mathbf{A}(t) \triangleq \gamma(t)^*, \qquad \gamma \in \mathscr{C}^n.$$

The proof of the corollary then follows immediately from the modification (2.54) required of Theorem 2.4, with the 1×1 matrix $\mathbf{Q}(t)$ written as $q(t)$. ■

The functions in Example 2.9 are said to *span* a three-dimensional space $\mathscr{R}^3$ on the interval $t \in [0, 1]$. In fact, since the linear independence "test" matrix $\mathbf{K}(0, 1)$ is nonsingular, the columns or rows of the symmetric matrix $\mathbf{K}(0, 1)$ in Example 2.16 may be taken as basis vectors for the space spanned by these functions. Exercise 2.11 shows these three basis vectors taken from the rows of $\mathbf{K}(0, 1)$.

Exercise 2.11
Show that the basis vectors

$$\begin{pmatrix} 1/3 \\ 1/12 \\ -5/12 \end{pmatrix}, \begin{pmatrix} 1/12 \\ 1/30 \\ 23/60 \end{pmatrix}, \begin{pmatrix} -5/12 \\ 23/60 \\ 8/15 \end{pmatrix}$$

are not orthogonal (see Definition 2.8).

2.5.1 ORTHOGONAL FUNCTIONS

When these vectors are orthogonal, it is common to give the corresponding functions of Example 2.9 a special name ("orthogonal functions"). To obtain the following meaning, let $\mathbf{x}$ and $\mathbf{y}$ be *scalar functions of time* in Definition 2.8.

Definition 2.14

The n functions $\{\gamma_1(t), \ldots, \gamma_n(t)\}$ *are said to be "orthogonal functions on the interval* $t \in [t_1, t_2]$ *subject to the weight* $q(t)$*" if* $\langle \gamma_i(t), q(t)\gamma_j(t)\rangle = 0,\ i \neq j$.

One consequence of this definition is the following. Suppose the functions $\gamma_i(t)$, $i = 1, \ldots, n$, are orthogonal such that

$$\mathbf{K}(t_1, t_2) = \operatorname{diag}\{\sigma_1^2, \ldots, \sigma_n^2\}. \tag{2.55}$$

Then, the $\alpha\beta$ element of $\mathbf{K}(t_1, t_2)$ in (2.41) is equal to

$$K_{\alpha\beta}(t_1, t_2) = \frac{1}{t_2 - t_1}\int_{t_1}^{t_2} \gamma_\alpha(t) q(t) \bar{\gamma}_\beta(t)\, dt$$

$$= \langle \gamma_\beta(t), q(t)\gamma_\alpha(t)\rangle = \sigma_\alpha^2 \delta_{\alpha\beta}, \qquad \delta_{\alpha\beta} = \begin{cases} 1 & \text{if } \alpha = \beta \\ 0 & \text{if } \alpha \neq \beta \end{cases} \tag{2.56}$$

where, by using the inner product notation of (2.19), it is again clear that the function $\gamma_i(t)$ is indeed orthogonal to the function $\gamma_j(t)$ if $i \neq j$, subject to the weight $q(t)$. The functions are called *orthonormal* if $\mathbf{K}(t_1, t_2) = \mathbf{I}_n$.

Some common orthogonal polynomials are listed in Table 2.2 with the interval $[t_1, t_2]$ and the weight $q(t)$ for which they are orthogonal.

EXAMPLE 2.11

Determine whether the Fourier series

$$\gamma_i(t) = e^{j(i-1)\omega_0 t}, \qquad i = 1, \ldots, n, \tag{2.57}$$

is an orthogonal set of functions on the interval $t \in [0, \pi/\omega_0]$ subject to the weight $q(t) = 1$.

TABLE 2.2 Some Orthogonal Polynomials

ORTHOGONAL FUNCTION	$[t_1, t_2]$	$q(t)$
Legendre	$[-1, 1]$	1
Jacobi	$[-1, 1]$	$(1 - t)^\alpha (1 + t)^\beta$
Laguerre	$[0, \infty]$	$t^\alpha e^{-t}$
Hermite	$[-\infty, \infty]$	e^{-t^2}
Chebyshev	$[-1, 1]$	$(1 - t^2)^{-1/2}$

Solution: From (2.56),

$$K_{\alpha\beta}\left(0, \frac{\pi}{\omega_0}\right) = \left(\frac{\pi}{\omega_0}\right)^{-1} \int_0^{\pi/\omega_0} e^{j(\alpha-1)\omega_0 t} e^{-j(\beta-1)\omega_0 t}\, dt$$

$$= \begin{cases} \dfrac{e^{j(\alpha-\beta)\pi} - 1}{j(\alpha-\beta)\pi}, & \text{if } \alpha \neq \beta; \\ 1, & \text{if } \alpha = \beta. \end{cases}$$

The reader should verify that the matrix $\mathbf{K}(0, \pi/\omega_0)$ is not diagonal. Hence, the functions (2.57) are *not* orthogonal on the interval $t \in [0, \pi/\omega_0]$ subject to the weight $q(t) = 1$. The reader may wish to verify that (2.57) is a set of orthogonal functions on the interval $t \in [0, 2\pi/\omega_0]$ subject to the weight 1.

Exercise 2.12

Show that the polynomials $\{1, t, t^2, t^3, \ldots, t^n\}$ are not orthogonal on the interval $t \in [0, 1]$ subject to the weight $q(t) = 1$.

Exercise 2.13

Walsh functions are orthogonal functions convenient for use in digital systems. The first three Walsh functions are plotted in Fig. 2.6. (i) Determine whether these functions are orthogonal on the interval $t \in [0, 1]$ subject to the weight $q(t) = 1$. (ii) Determine the values for (A, B, C) which will make them orthonormal functions.

2.5.2 THE GRAM – SCHMIDT PROCEDURE

Up to now, we have asked whether a given set of vectors or functions (a) span a space, (b) are linearly independent, (c) are a basis for the space, or (d) are

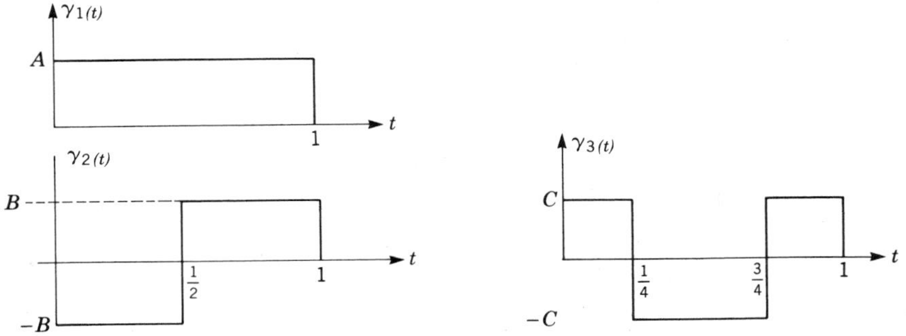

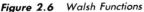

Figure 2.6 *Walsh Functions*

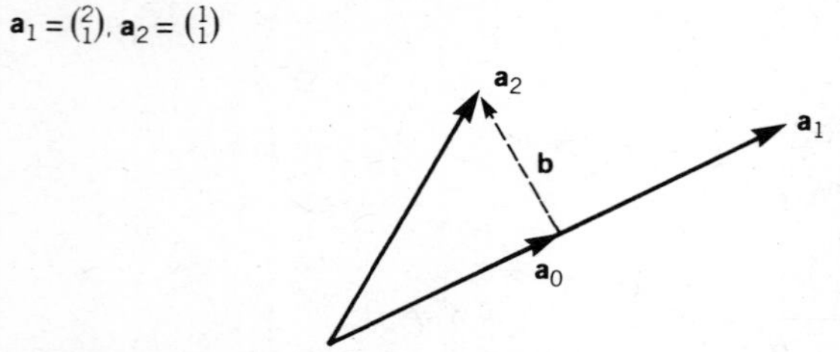

Figure 2.7 Illustration of Gram–Schmidt Procedure

orthogonal. In addition, one should know how to construct an orthogonal basis if one is not given. In this section, we give a standard "recipe" for constructing an orthogonal set of basis vectors or functions from a given nonorthogonal set.

To illustrate the idea first with the simplest example, consider the two nonorthogonal vectors $\mathbf{a}_1$ and $\mathbf{a}_2$, both in $\mathscr{R}^2$. Now, $\mathbf{a}_1$ and $\mathbf{a}_2$ do form a basis for $\mathscr{R}^2$ since they are linearly independent and span the space. To construct an *orthogonal* basis for $\mathscr{R}^2$, we need only to begin with either $\mathbf{a}_1$ or $\mathbf{a}_2$ (say, $\mathbf{a}_1$) and then construct a vector orthogonal to $\mathbf{a}_1$. To take this last step, note that $\mathbf{a}_2$ can be written as the sum of a vector along $\mathbf{a}_1$ and a vector perpendicular to $\mathbf{a}_1$, as $\mathbf{b}$ in Fig. 2.7. Thus, an orthogonal basis is $\{\mathbf{a}_1, \mathbf{b}\}$. The details of constructing $\mathbf{b}$ remain. Write from Fig. 2.7

$$\mathbf{a}_2 = \mathbf{a}_0 + \mathbf{b}. \tag{2.58}$$

Note that $\mathbf{a}_0$ is the projection of $\mathbf{a}_2$ onto the direction of $\mathbf{a}_1$ and that it can be described by

$$\mathbf{a}_0 = [\|\mathbf{a}_2\| \cos\theta] \frac{\mathbf{a}_1}{\|\mathbf{a}_1\|}.$$

Using (2.32),

$$\mathbf{a}_0 = \|\mathbf{a}_2\| \frac{\mathbf{a}_1^T\mathbf{a}_2}{\|\mathbf{a}_2\|\,\|\mathbf{a}_1\|} \frac{\mathbf{a}_1}{\|\mathbf{a}_1\|} = \frac{\mathbf{a}_1^T\mathbf{a}_2}{\|\mathbf{a}_1\|^2}\mathbf{a}_1.$$

Summarizing, we state the general result.

Theorem 2.6

The projection $\mathbf{a}_0$ of the vector $\mathbf{a}_2 \in \mathscr{C}^n$ onto the line containing $a_1 \in \mathscr{C}^n$ is given by

$$\mathbf{a}_0 = \frac{\langle \mathbf{a}_1, \mathbf{a}_2 \rangle}{\langle \mathbf{a}_1, \mathbf{a}_1 \rangle}\mathbf{a}_1. \tag{2.59}$$

Now, the vector perpendicular to $\mathbf{a}_1$ can be calculated from (2.58) and (2.59):

$$\mathbf{b} = \mathbf{a}_2 - \mathbf{a}_0 = \mathbf{a}_2 - \frac{\langle \mathbf{a}_1, \mathbf{a}_2 \rangle}{\langle \mathbf{a}_1, \mathbf{a}_1 \rangle} \mathbf{a}_1. \tag{2.60}$$

Note that the magnitude of the vector $\mathbf{b}$ is the distance between a point $\mathbf{a}_2$ and a line along $\mathbf{a}_1$. Take the norm of (2.59) to get this distance

$$\|\mathbf{b}\| = \frac{(\langle \mathbf{a}_1, \mathbf{a}_1 \rangle)(\langle \mathbf{a}_2, \mathbf{a}_2 \rangle) - (\langle \mathbf{a}_1, \mathbf{a}_2 \rangle)^2}{\langle \mathbf{a}_1, \mathbf{a}_1 \rangle}.$$

To prove that $\mathbf{b}$ is orthogonal to $\mathbf{a}_1$, construct $\langle \mathbf{a}_1, \mathbf{b} \rangle = \mathbf{a}_1^* \mathbf{b}$, using (2.60):

$$\mathbf{a}_1^* \mathbf{b} = \mathbf{a}_1^* \mathbf{a}_2 - \mathbf{a}_1^* \mathbf{a}_1 \frac{\mathbf{a}_1^* \mathbf{a}_2}{\mathbf{a}_1^* \mathbf{a}_1} = 0.$$

In a three-dimensional space, the third basis vector would be constructed in this manner to be orthogonal to $\mathbf{a}_1$ and $\mathbf{b}$. In an n-dimensional space, this orthogonalization procedure continues as summarized in the following steps, where one step is added to normalize the orthogonal base vectors to unit length. This step eliminates the need to divide by the norm of $\mathbf{a}_1$ as in (2.60), since this norm will be unity.

The Gram–Schmidt Orthonormalization

PROBLEM

From the given set of independent vectors $\{\mathbf{y}_1, \ldots, \mathbf{y}_n\}$, construct an orthonormal set of vectors $\{\mathbf{v}_1, \ldots, \mathbf{v}_n\}$.

Solution

STEP 1: Set $\mathbf{v}_1 \triangleq \dfrac{\mathbf{y}_1}{\|\mathbf{y}_1\|}$,

STEP 2: $\mathbf{v}_2 = \dfrac{\mathbf{y}_2 - \langle \mathbf{v}_1, \mathbf{y}_2 \rangle \mathbf{v}_1}{\|\mathbf{y}_2 - \langle \mathbf{v}_1, \mathbf{y}_2 \rangle \mathbf{v}_1\|}$,

⋮

STEP i: $\mathbf{v}_i = \dfrac{\mathbf{y}_i - \sum_{\alpha=1}^{i-1} \langle \mathbf{v}_\alpha, \mathbf{y}_i \rangle \mathbf{v}_\alpha}{\left\| \mathbf{y}_i - \sum_{\alpha=1}^{i-1} \langle \mathbf{v}_\alpha, \mathbf{y}_i \rangle \mathbf{v}_\alpha \right\|}$,

for $i = 1, \ldots, n$. ■

EXAMPLE 2.12

Generate an orthonormal basis for the space spanned by

$$\mathbf{y}_1 = \begin{pmatrix} 1 \\ 1 \end{pmatrix}, \qquad \mathbf{y}_2 = \begin{pmatrix} 1 \\ 10 \end{pmatrix}.$$

Solution: From the Gram–Schmidt process:

STEP 1: $$\mathbf{v}_1 = \begin{pmatrix} 1 \\ 1 \end{pmatrix} \frac{1}{\sqrt{2}},$$

STEP 2: $$\mathbf{v}_2 = \left\{ \begin{pmatrix} 1 \\ 10 \end{pmatrix} - \frac{(1,1)}{\sqrt{2}} \begin{pmatrix} 1 \\ 10 \end{pmatrix} \begin{pmatrix} 1 \\ 1 \end{pmatrix} \frac{1}{\sqrt{2}} \right\} \Big/ \|\{\cdot\}\|$$

$$= \left\{ \begin{pmatrix} 1 \\ 10 \end{pmatrix} - \frac{11}{2} \begin{pmatrix} 1 \\ 1 \end{pmatrix} \right\} \Big/ \|\{\cdot\}\|$$

$$= \begin{pmatrix} 1 - 5.5 \\ 10 - 5.5 \end{pmatrix} \frac{1}{\sqrt{(1-5.5)^2 + (10-5.5)^2}}. \qquad \blacksquare$$

EXAMPLE 2.13

Now, interpret the inner products of the Gram–Schmidt process as appropriate for time-varying functions; see (2.19). Generate an orthogonal set of functions $v_1(t)$, $v_2(t)$, and $v_3(t)$ on the interval $t \in [-1, 1]$ subject to the weight $q(t) = 1$, from the set of given functions $y_1(t) \triangleq 1$, $y_2(t) \triangleq t$, $y_3(t) = t^2$.

Solution: From the Gram–Schmidt process and (2.19):

STEP 1: $$v_1(t) = \frac{y_1(t)}{\left(\langle y_1(t), y_1(t)\rangle\right)^{1/2}} = \frac{1}{\left[(1/2)\int_{-1}^{1} dt\right]^{1/2}} = \frac{1}{\sqrt{2}/\sqrt{2}} = 1,$$

STEP 2: $$\langle v_1, y_2 \rangle = \tfrac{1}{2}\int_{-1}^{1} (t)\, dt = 0.$$

Hence,

$$v_2 = \frac{\{y_2 - (0)v_2\}}{\|y_2 - (0)v_1\|} = \frac{y_2}{\|y_2\|} = \frac{t}{\left[(1/2)\int_{-1}^{1} t^2\, dt\right]^{1/2}} = \sqrt{3}\,t.$$

STEP 3: $$\langle v_1, y_3 \rangle = \tfrac{1}{2}\int_{-1}^{1} t^2\, dt = \tfrac{1}{3},$$

$$\langle v_2, y_3 \rangle = \tfrac{1}{2}\int_{-1}^{1} (\sqrt{3}\,t)t^2\, dt = 0.$$

Hence,

$$v_3 = \left\{ y_3 - \sum_{\alpha=1}^{2} \langle v_\alpha, y_3 \rangle v_\alpha \right\} \| \{ \cdot \} \|$$

$$= \left\{ t^2 = \frac{1}{3} \right\} \Big/ \left\| t^2 - \frac{1}{3} \right\| = \frac{t^2 - 1/3}{\left[(1/2) \int_{-1}^{1} (t^2 - 1/3)^2 \, dt \right]^{1/2}}$$

$$= \sqrt{\frac{45}{4}} \left(t^2 - \frac{1}{3} \right).$$

■

2.6 *Solution of the Linear Algebra Problem AXB = Y and Pseudo Inverses*

The result of this section is the solution of a more general linear algebra problem that includes $\mathbf{Ax} = \mathbf{y}$ as a special case. Hence, we shall characterize many special cases of the following theorem.

Theorem 2.7
Let $\mathbf{A}$ *be* $n_1 \times n_2$, $\mathbf{X}$ *be* $n_2 \times n_3$, $\mathbf{B}$ *be* $n_3 \times n_4$, *and* $\mathbf{Y}$ *be* $n_1 \times n_4$. *Then, for a given matrix triple* $\mathbf{A}, \mathbf{B}, \mathbf{Y}$, *the equation*

$$\mathbf{AXB} = \mathbf{Y} \tag{2.61}$$

can be solved for $\mathbf{X}$ *iff* $\mathbf{A}$, $\mathbf{B}$, *and* $\mathbf{Y}$ *have this property*:

$$\mathbf{AA}^{(1)}\mathbf{YB}^{(1)}\mathbf{B} = \mathbf{Y}, \tag{2.62}$$

where $\mathbf{A}^{(1)}$ *and* $\mathbf{B}^{(1)}$ *are any matrices satisfying*

$$\mathbf{AA}^{(1)}\mathbf{A} = \mathbf{A}, \tag{2.63}$$

$$\mathbf{BB}^{(1)}\mathbf{B} = \mathbf{B}. \tag{2.64}$$

Furthermore, if (2.61) *has any solutions, then all of them are given by*

$$\mathbf{X} = \mathbf{A}^{(1)}\mathbf{YB}^{(1)} - \mathbf{A}^{(1)}\mathbf{AZBB}^{(1)} + \mathbf{Z}, \tag{2.65}$$

where $\mathbf{Z}$ *is an arbitrary matrix of dimension* $n_2 \times n_3$.

Proof: Substitute (2.65) into (2.61) to get

$$\mathbf{A}[\mathbf{A}^{(1)}\mathbf{YB}^{(1)} - \mathbf{A}^{(1)}\mathbf{AZBB}^{(1)} + \mathbf{Z}]\mathbf{B} = \mathbf{Y}.$$

Hence,

$$\mathbf{AA}^{(1)}\mathbf{YB}^{(1)}\mathbf{B} - \mathbf{AA}^{(1)}\mathbf{AZBB}^{(1)}\mathbf{B} + \mathbf{AZB} = \mathbf{Y}$$

holds by virtue of equations (2.62) through (2.64). This proves the "if" part. Now, to prove "only if" we must show that there is no other possibility. Suppose there was another solution not given by (2.65). Label this solution $\mathbf{X}_1$ and write $\mathbf{X}_1$ as the above solution (2.65) (which we now label $\mathbf{X}_0$ instead of $\mathbf{X}$), plus a new term $\tilde{\mathbf{X}}$ (whatever is needed to make $\mathbf{X}_1$ a solution but different from $\mathbf{X}_0$):

$$\mathbf{X}_1 = \mathbf{X}_0 + \tilde{\mathbf{X}}, \qquad \tilde{\mathbf{X}} \neq \mathbf{0}.$$

To see if this is possible, substitute $\mathbf{X}_1$ into (2.61) to get

$$\mathbf{AA}^{(1)}\mathbf{YB}^{(1)}\mathbf{B} - \mathbf{AA}^{(1)}\mathbf{AZBB}^{(1)}\mathbf{B} + \mathbf{AZB} + \mathbf{A}\tilde{\mathbf{X}}\mathbf{B} = \mathbf{Y}.$$

But given the conditions (2.66)–(2.68), this equation reduces to

$$\mathbf{A}\tilde{\mathbf{X}}\mathbf{B} = \mathbf{0}.$$

Hence, $\mathbf{AX}_1\mathbf{B} = \mathbf{Y}$ leads to

$$\mathbf{AX}_0\mathbf{B} + \mathbf{0} = \mathbf{Y},$$

where the zero matrix is the contribution of *any* other solution besides (2.65). Thus, (2.65) captures the entire class of nontrivial solutions to (2.61). ■

To consider special cases, define

$$r_A \triangleq \operatorname{rank} \mathbf{A}, \quad r_B \triangleq \operatorname{rank} \mathbf{B}$$

Definition 2.15
A matrix $\mathbf{A}_L$ satisfying $\mathbf{A}_L\mathbf{A} = \mathbf{I}$ is said to be a "left inverse" of $\mathbf{A}$. A matrix $\mathbf{A}_R$ satisfying $\mathbf{AA}_R = \mathbf{I}$ is said to be "right inverse" of $\mathbf{A}$.

Now, if the rank of the $n_1 \times n_2$ matrix $\mathbf{A}$ is $r_A = n_1$, then a right inverse of $\mathbf{A}$ always exists. If $r_A = n_2$, then a left inverse of $\mathbf{A}$ always exists. To show this, note that $r_A = n_1$ means that $\mathbf{A}$ has linearly independent rows. The *test* for linearly independent rows is given by the corollary to Theorem 2.4, namely, $|\mathbf{AA}^*| \neq 0$. Use this fact to assure that this calculation is possible,

$$\mathbf{A}_R = \mathbf{A}^*[\mathbf{AA}^*]^{-1}, \tag{2.66}$$

and see that this satisfies $\mathbf{AA}_R = \mathbf{I}$. To show that $r_A = n_2$ implies the existence of a left inverse, note that the test for linear independence of the columns of $\mathbf{A}$ is given (corollary to Theorem 2.4) by $|\mathbf{A}^*\mathbf{A}| \neq 0$. Hence, this calculation is possible,

$$\mathbf{A}_L = [\mathbf{A}^*\mathbf{A}]^{-1}\mathbf{A}^*, \tag{2.67}$$

and it is readily verified that $\mathbf{A}_L\mathbf{A} = \mathbf{I}$.

Pseudo Inverses

For every finite matrix A there exists a *unique* matrix A^+ satisfying these four conditions:

1. $\mathbf{AA}^+\mathbf{A} = \mathbf{A}$
2. $\mathbf{A}^+\mathbf{AA}^+ = \mathbf{A}^+$
3. $(\mathbf{AA}^+)^* = \mathbf{AA}^+$
4. $(\mathbf{A}^+\mathbf{A})^* = \mathbf{A}^+\mathbf{A}$

This matrix $\mathbf{A}^+$ is called the "Moore–Penrose inverse" of $\mathbf{A}$. Note that condition (2.63) on $\mathbf{A}^{(1)}$ requires *only* the first property 1 of the Moore–Penrose inverse. Hence, an $\mathbf{A}^{(1)}, \mathbf{B}^{(1)}$ satisfying (2.63) and (2.64) always exists (but $\mathbf{A}^{(1)}, \mathbf{B}^{(1)}$ need not be chosen as the Moore–Penrose inverses). When $\mathbf{A}$ and $\mathbf{B}$ do have maximal rank (that is, whenever $\mathbf{A}$ and $\mathbf{B}$ have either left or right inverses), their Moore–Penrose inverses are given by (2.66) or (2.67). (But note that Moore–Penrose inverses always exist even when matrices (2.66) or (2.67) do not.)

With this insight we return to (2.65) and tabulate the special cases in Table 2.3. This table is useful for handy reference on linear algebra problems.

If one prefers a geometrical interpretation of the problem $\mathbf{Ax} = \mathbf{y}$, then the following corollary to Theorem 2.7 is useful:

Corollary to Theorem 2.7

$\mathbf{Ax} = \mathbf{y}$ *has a solution* $\mathbf{x}$ *iff*

$$\text{rank}[\mathbf{A}] = \text{rank}[\mathbf{A}, \mathbf{y}]. \tag{2.68}$$

Proof: Since $\mathbf{y}$ is a linear combination of the columns of $\mathbf{A}$, that is, $\Sigma \mathbf{a}_i x = \mathbf{y}$, it is clearly necessary that these columns $\mathbf{a}_i$, $i = 1, 2, \ldots, n_x$, must span the space in which $\mathbf{y}$ resides. To see the sufficiency of (2.68), note that the space spanned by the columns of $\mathbf{A}$ is of dimension = rank[$\mathbf{A}$]. The space spanned by the columns of $\mathbf{A}$ and the vector $\mathbf{y}$ is of dimension = rank[$\mathbf{A}, \mathbf{y}$]. Hence, if they are the *same* spaces, then there is a linear combination of the $\mathbf{a}_i$ that will yield $\mathbf{y}$.

Now, Table 2.3, case $n_3 = 1$, $r_A = ?$, $r_B = n_4 = 1$, gives results equivalent to the corollary (since both are necessary and sufficient). Hence, the statement "$[\mathbf{I} - \mathbf{AA}^{(1)}]\mathbf{y} = \mathbf{0}$" and the statement "rank[$\mathbf{A}$] = rank[$\mathbf{A}, \mathbf{y}$]" are equivalent statements.

TABLE 2.3 Linear Algebra, Solutions of $\mathbf{AXB} = \mathbf{Y}$*

GIVEN CONDITIONS: CASES			PROPERTIES OF $\mathbf{A}(n_1 \times n_2)$, $\mathbf{B}(n_3 \times n_4)$	SOLUTION $\mathbf{X} =$	EXISTENCE CONDITION (iff)
	r_A	r_B			
	?	?	?	$\mathbf{X}_a$	$\mathbf{AA}^{(1)}\mathbf{YB}^{(1)}\mathbf{B} = \mathbf{Y}$
	0	0	$\mathbf{A} = \mathbf{0}, \mathbf{B} = \mathbf{0}$,	$\mathbf{Z}$ (arbitrary)	(always exists)
	n_1	?	$\mathbf{AA}^{(1)} = \mathbf{I}$	$\mathbf{X}_a$	$\mathbf{Y}[\mathbf{I} - \mathbf{B}^{(1)}\mathbf{B}] = \mathbf{0}$
	n_2	?	$\mathbf{A}^{(1)}\mathbf{A} = \mathbf{I}$	$\mathbf{X}_b$	$\mathbf{AA}^{(1)}\mathbf{YB}^{(1)}\mathbf{B} = \mathbf{Y}$
	?	n_3	$\mathbf{BB}^{(1)} = \mathbf{I}$	$\mathbf{X}_c$	$\mathbf{AA}^{(1)}\mathbf{YB}^{(1)}\mathbf{B} = \mathbf{Y}$
	?	n_4	$\mathbf{B}^{(1)}\mathbf{B} = \mathbf{I}$	$\mathbf{X}_a$	$[\mathbf{I} - \mathbf{AA}^{(1)}]\mathbf{Y} = \mathbf{0}$
	n_2	n_3	$\mathbf{A}^{(1)}\mathbf{A} = \mathbf{I}$, $\mathbf{B}^{(1)} = \mathbf{I}$	$\mathbf{A}^{(1)}\mathbf{YB}^{(1)}$ (unique)	$\mathbf{AA}^{(1)}\mathbf{YB}^{(1)}\mathbf{B} = \mathbf{Y}$
	n_1	n_4	$\mathbf{AA}^{(1)} = \mathbf{I}$, $\mathbf{B}^{(1)}\mathbf{B} = \mathbf{I}$	$\mathbf{X}_a$	(always exists)
	n_2	n_4	$\mathbf{A}^{(1)}\mathbf{A} = \mathbf{I}$, $\mathbf{B}^{(1)}\mathbf{B} = \mathbf{I}$	$\mathbf{X}_b$	$[\mathbf{I} - \mathbf{AA}^{(1)}]\mathbf{Y} = \mathbf{0}$
	n_1	n_3	$\mathbf{AA}^{(1)} = \mathbf{I}$, $\mathbf{BB}^{(1)} = \mathbf{I}$	$\mathbf{X}_c$	$\mathbf{Y}[\mathbf{I} - \mathbf{B}^{(1)}\mathbf{B}] = \mathbf{0}$
$n_3 = 1$	?	$n_4 = 1$	$B = 1$	$\mathbf{A}^{(1)}\mathbf{Y} + [\mathbf{I} - \mathbf{A}^{(1)}\mathbf{A}]\mathbf{Z}$	$[\mathbf{I} - \mathbf{AA}^{(1)}]\mathbf{Y} = \mathbf{0}$
$n_3 = 1$	n_1	$n_4 = 1$	$B = 1$, $\mathbf{AA}^{(1)} = \mathbf{I}$	$\mathbf{A}^{(1)}\mathbf{Y} + [\mathbf{I} - \mathbf{A}^{(1)}\mathbf{A}]\mathbf{Z}$	(always exists)
$n_3 = 1$	n_2	$n_4 = 1$	$B = 1$, $\mathbf{A}^{(1)}\mathbf{A} = \mathbf{I}$	$\mathbf{A}^{(1)}\mathbf{Y}$ (unique)	$[\mathbf{I} - \mathbf{AA}^{(1)}]\mathbf{Y} = \mathbf{0}$
$n_1 = n_2$	n_2	$n_4 = 1$, $n_3 = 1$	$B = 1$, $\mathbf{A}^{(1)} = \mathbf{A}^{-1}$	$\mathbf{A}^{-1}\mathbf{Y}$ (unique)	(always exists)

*$\mathbf{X}_a \triangleq \mathbf{A}^{(1)}\mathbf{YB}^{(1)} - \mathbf{A}^{(1)}\mathbf{AZBB}^{(1)} + \mathbf{Z}$, $\mathbf{X}_b \triangleq \mathbf{A}^{(1)}\mathbf{YB}^{(1)} + \mathbf{Z}[\mathbf{I} - \mathbf{BB}^{(1)}]$, $\mathbf{X}_c \triangleq \mathbf{A}^{(1)}\mathbf{YB}^{(1)} + [\mathbf{I} - \mathbf{A}^{(1)}\mathbf{A}]\mathbf{Z}$.

2.7 *Linear Matrix Equations and Least Squares Theory*

The objective of the previous section was to study properties of two types of linear algebraic problems:

I. For a given $(\mathbf{A}, \mathbf{y})$, solve $\mathbf{Ax} = \mathbf{y}$ for $\mathbf{x}$.
II. For a given $(\mathbf{x}, \mathbf{y})$, solve $\mathbf{Ax} = \mathbf{y}$ for $\mathbf{A}$.

It was shown that problems I and II may not have solutions, in which case it is often of interest to modify the problem as follows:

I′. For a given $(\mathbf{A}, \mathbf{y})$, solve: $\min_{\mathbf{x}} \|\mathbf{Ax} - \mathbf{y}\|^2$.
II′. For a given $(\mathbf{x}, \mathbf{y})$, solve: $\min_{\mathbf{A}} \|\mathbf{Ax} - \mathbf{y}\|^2$.

Clearly, these are more general problems to study than I and II, since the solutions of I and II will be obtained when they exist by solving I′ and II′. Such notions as I′ and II′ suggest a prior understanding of vector norms and calculus of matrices. Hence, the next section is needed as background prior to our return to problems I′ and II′.

2.7.1 CALCULUS OF MATRICES

The purpose of this section is to develop rules for the differentiation of matrices. The integral calculus is straightforward and has already been given by (2.24). Therefore, this section focuses on differential calculus. These results will be used in Chapter 8 and later in Chapter 2 in the discussion of least squares approximation.

2.7.1.1 Differentiation of a Scalar with Respect to a Vector

Suppose $f(\mathbf{x})$ is a real scalar function of a real vector $\mathbf{x} \in \mathscr{R}^n$. The first three terms of the Taylor's series expansion of $f(\mathbf{x})$ about $\mathbf{x}_0$ (in terms of $\delta\mathbf{x} \triangleq \mathbf{x} - \mathbf{x}_0$) is

$$\delta f(\mathbf{x}) \triangleq f(\mathbf{x}) - f(\mathbf{x}_0) = \sum_{\alpha=1}^{n} \frac{\partial f(\mathbf{x})}{\partial x_\alpha} \delta x_\alpha + \frac{1}{2} \sum_{\beta,\gamma=1}^{n} \frac{\partial^2 f(\mathbf{x})}{\partial x_\beta\, \partial x_\gamma} \delta x_\beta\, \delta x_\gamma. \tag{2.69}$$

By defining the gradient and variational vectors by

$$\left(\frac{\partial f(\mathbf{x})}{\partial \mathbf{x}}\right)^T \triangleq \left(\frac{\partial f(\mathbf{x})}{\partial x_1}, \ldots, \frac{\partial f(\mathbf{x})}{\partial x_n}\right), \tag{2.70}$$

$$(\delta x)^T \triangleq (\delta x_1, \ldots, \delta x_n) \tag{2.71}$$

and the second derivative, called the *Hessian* matrix, by

$$\frac{\partial^2 f(\mathbf{x})}{\partial \mathbf{x}^2} \triangleq \frac{\partial}{\partial \mathbf{x}}\left(\frac{\partial f}{\partial \mathbf{x}}\right)^T = \begin{bmatrix} \dfrac{\partial}{\partial x_1}\left(\dfrac{\partial f}{\partial \mathbf{x}}\right)^T \\ \vdots \\ \dfrac{\partial}{\partial x_n}\left(\dfrac{\partial f}{\partial \mathbf{x}}\right)^T \end{bmatrix} = \begin{bmatrix} \dfrac{\partial^2 f}{\partial x_1^2} & \cdots & \dfrac{\partial^2 f}{\partial x_1\, \partial x_n} \\ \vdots & & \vdots \\ \dfrac{\partial^2 f}{\partial x_n\, \partial x_1} & \cdots & \dfrac{\partial^2 f}{\partial x_n^2} \end{bmatrix}, \tag{2.72}$$

(2.69) may be written in the compact vector matrix form

$$\delta f(\mathbf{x}) = \left(\frac{\partial f}{\partial \mathbf{x}}\right)^T \delta\mathbf{x} + \frac{1}{2}\, \delta\mathbf{x}^T \left[\frac{\partial^2 f}{\partial \mathbf{x}^2}\right] \delta\mathbf{x}. \tag{2.73}$$

Suppose one wishes to choose $\mathbf{x}$ so as to minimize a scalar function $f(\mathbf{x})$. A necessary condition is that small perturbations in $f(\mathbf{x})$ are not negative; $df(\mathbf{x}) \geq 0$

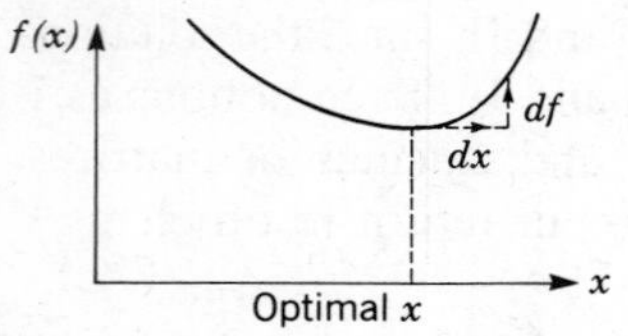

Figure 2.8 Function Minimization

in the vicinity of the optimal value of **x**. See Fig. 2.8. Now, if **x** is complex, then the necessary condition is (Appendix C equation (C.13a))

$$df(\mathbf{x}) \triangleq \left(\frac{\partial f}{\partial \mathbf{x}}\right)^T d\mathbf{x} + \left(\frac{\partial f}{\partial \bar{\mathbf{x}}}\right)^T d\bar{\mathbf{x}} \geq 0. \tag{2.74}$$

Now, in this course the only functions which we shall minimize are norms of vectors or the inner products of vectors. Appendix C proves that condition (C.19) for the function $f(\mathbf{x}) = \langle \mathbf{x}, \mathbf{Qx} \rangle$ becomes $d\langle \mathbf{x}, \mathbf{Qx} \rangle = \langle \mathbf{x}, \mathbf{Q}\, d\mathbf{x} \rangle + \langle d\mathbf{x}, \mathbf{Qx} \rangle \geq 0$, and also proves the useful results

$$\frac{\partial}{\partial \mathbf{x}}(\mathbf{y}^T\mathbf{Qx}) = \mathbf{Q}^T\mathbf{y} \qquad \text{if } \mathbf{x}, \mathbf{y}, \mathbf{Q} \text{ are real,} \tag{2.75a}$$

$$\frac{\partial}{\partial \mathbf{x}_R}(\mathbf{y}^*\mathbf{Qx}) = \mathbf{Q}^T\bar{\mathbf{y}} \qquad \text{if } \mathbf{Q} \text{ is real, } \mathbf{x}, \mathbf{y} \text{ are complex (and } \mathbf{x} = \mathbf{x}_R + j\mathbf{x}_I\text{),} \tag{2.75b}$$

$$\frac{\partial}{\partial \mathbf{x}}(\mathbf{x}^*\mathbf{Qx}) = \mathbf{Q}^T\bar{\mathbf{x}} \qquad \text{if } \mathbf{Q} \text{ is real, } \mathbf{x} \text{ is complex,} \tag{2.75c}$$

$$\frac{\partial}{\partial \bar{\mathbf{x}}}(\mathbf{x}^*\mathbf{Qx}) = \mathbf{Qx} \qquad \text{if } \mathbf{Q} \text{ is real, } \mathbf{x} \text{ is complex.} \tag{2.75d}$$

2.7.1.2 Differentiation of a Scalar with Respect to a Matrix

From the convention (2.70) it follows that the derivative of a scalar $f(\mathbf{A})$ with respect to a matrix $\mathbf{A} = [\mathbf{a}_1, \ldots, \mathbf{a}_n]$ has the structure

$$\frac{\partial f}{\partial \mathbf{A}} = \left[\frac{\partial f}{\partial \mathbf{a}_1}, \ldots, \frac{\partial f}{\partial \mathbf{a}_n}\right] = \begin{bmatrix} \dfrac{\partial f}{\partial A_{11}} & \cdots & \dfrac{\partial f}{\partial A_{1n}} \\ \vdots & \cdots & \vdots \\ \dfrac{\partial f}{\partial A_{n1}} & \cdots & \dfrac{\partial f}{\partial A_{nn}} \end{bmatrix}. \tag{2.76}$$

The differential of a real or complex scalar $f(\mathbf{A})$ that is a function of a complex matrix $\mathbf{A} \in \mathscr{C}^{k\times n}$ is

$$df(\mathbf{A}) = \sum_{\alpha=1}^{n} \left[\frac{\partial f}{\partial \mathbf{a}_\alpha}\right]^T d\mathbf{a}_\alpha + \sum_{\alpha=1}^{n} \left[\frac{\partial f}{\partial \bar{\mathbf{a}}_\alpha}\right]^T d\bar{\mathbf{a}}_\alpha, \tag{2.77}$$

which follows directly from Appendix C by considering the n columns of $\mathbf{A}$ as vectors.

Exercise 2.14

Verify that (2.77) is equivalent to

$$df(\mathbf{A}) = \sum_{\alpha=1}^{n} \sum_{\beta=1}^{n} \left[\left(\frac{\partial f}{\partial A_{\alpha\beta}}\right) dA_{\alpha\beta} + \left(\frac{\partial f}{\partial \bar{A}_{\alpha\beta}}\right) d\bar{A}_{\alpha\beta}\right]. \tag{2.78}$$

By employing results (C2) through (C4) applied to the column vectors of $\mathbf{A}$,

$$\mathbf{a}_\alpha \triangleq \mathbf{a}_{\alpha R} + j\mathbf{a}_{\alpha I} \qquad \alpha = 1, \ldots, n,$$

we get, from (C5),

$$\frac{\partial f(\mathbf{A})}{\partial \mathbf{a}_{\alpha R}} = \frac{\alpha f(\mathbf{A})}{\partial \mathbf{a}_\alpha} + \frac{\partial f(\mathbf{A})}{\partial \bar{\mathbf{a}}_\alpha}, \tag{2.79}$$

$$\frac{\partial f(\mathbf{A})}{\partial \mathbf{a}_{\alpha I}} = j\left(\frac{\partial f(\mathbf{A})}{\partial \mathbf{a}_\alpha} - \frac{\partial f(\mathbf{A})}{\partial \bar{\mathbf{a}}_\alpha}\right), \tag{2.80}$$

and, from (C7),

$$\frac{\partial f(\mathbf{A})}{\partial \mathbf{a}} = \frac{1}{2}\left(\frac{\partial f(\mathbf{A})}{\partial \mathbf{a}_{\alpha R}} - j\frac{\partial f(\mathbf{A})}{\partial \mathbf{a}_{\alpha I}}\right),$$

$$\frac{\partial f(\mathbf{A})}{\partial \bar{\mathbf{a}}_\alpha} = \frac{1}{2}\left(\frac{\partial f(\mathbf{A})}{\partial \mathbf{a}_{\alpha R}} + j\frac{\partial f(\mathbf{A})}{\partial \mathbf{a}_{\alpha I}}\right).$$

EXAMPLE 2.14

Write the differential of the scalar function

$$f = \operatorname{tr} \mathbf{AB}, \quad \mathbf{A} \in \mathscr{C}^{k\times n}, \quad \mathbf{B} \in \mathscr{C}^{n\times k}.$$

Solution: From (2.77),

$$d(\operatorname{tr} \mathbf{AB}) = \sum_{\alpha=1}^{n} \left[\frac{\partial}{\partial \mathbf{a}_\alpha}(\operatorname{tr} \mathbf{AB})\right]^T d\mathbf{a}_\alpha + \sum_{\alpha=1}^{n} \left[\frac{\partial}{\partial \bar{\mathbf{a}}_\alpha}(\operatorname{tr} \mathbf{AB})\right]^T d\bar{\mathbf{a}}_\alpha. \tag{2.81}$$

■

Note that

$$\operatorname{tr}\mathbf{AB} = \sum_{\alpha=1}^{k}\sum_{\beta=1}^{n} A_{\alpha\beta}B_{\beta\alpha} = \sum_{\beta=1}^{n}\sum_{\alpha=1}^{k} B_{\beta\alpha}A_{\alpha\beta} = \operatorname{tr}\mathbf{BA} \tag{2.82}$$

proves the identity

$$\operatorname{tr}\mathbf{AB} = \operatorname{tr}\mathbf{BA}. \tag{2.83}$$

Exercise 2.15

Prove another useful identity,

$$\operatorname{tr}\mathbf{AB} = \operatorname{tr}\mathbf{A}^T\mathbf{B}^T. \tag{2.84}$$

Now, the right-hand side of (2.83) can be expanded in terms of the columns of **A**,

$$\operatorname{tr}\mathbf{BA} = \operatorname{tr}\begin{bmatrix}\mathbf{b}_1^*\\ \vdots \\ \mathbf{b}_n^*\end{bmatrix}[\mathbf{a}_1,\ldots,\mathbf{a}_n] = \sum_{\beta=1}^{n} \mathbf{b}_\beta^*\mathbf{a}_\beta, \tag{2.85}$$

where $\mathbf{b}_i^*$ is defined as the ith row of **B**. Equation (2.85) readily leads to the conclusions

$$\left[\frac{\partial(\operatorname{tr}\mathbf{AB})}{\partial\mathbf{a}_\alpha}\right]^T = \mathbf{b}_\alpha^* \qquad \text{or} \qquad \frac{\partial(\operatorname{tr}\mathbf{AB})}{\partial\mathbf{a}_\alpha} = \bar{\mathbf{b}}_\alpha \tag{2.86}$$

and

$$\left[\frac{\partial(\operatorname{tr}\mathbf{AB})}{\partial\bar{\mathbf{a}}_\alpha}\right]^T = \mathbf{0}. \tag{2.87}$$

The facts (C12) and (2.87) prove that ($\operatorname{tr}\mathbf{AB}$) is an analytic function. See from (2.86) and (2.76) that the structure of $\partial(\operatorname{tr}\mathbf{AB})/\partial\mathbf{A}$ is

$$\frac{\partial(\operatorname{tr}\mathbf{AB})}{\partial\mathbf{A}} = \left[\frac{\partial(\operatorname{tr}\mathbf{AB})}{\partial\mathbf{a}_1},\ldots,\frac{\partial(\operatorname{tr}\mathbf{AB})}{\partial\mathbf{a}_n}\right]$$

$$= \left[\bar{\mathbf{b}}_1,\ldots,\bar{\mathbf{b}}_n\right] = \mathbf{B}^T.$$

This proves an identity worth remembering:

$$\frac{\partial(\operatorname{tr}\mathbf{AB})}{\partial\mathbf{A}} = \mathbf{B}^T. \tag{2.88}$$

From (2.86) and (2.87) the differential (2.81) is

$$d(\operatorname{tr}\mathbf{AB}) = \sum_{\alpha=1}^{n} \mathbf{b}_\alpha^* \, d\mathbf{a}_\alpha = \sum_{\alpha=1}^{n} \langle \mathbf{b}_\alpha, d\mathbf{a}_\alpha \rangle. \tag{2.89}$$

Exercise 2.16

Prove that for any $\mathbf{A} \in \mathscr{C}^{k\times n}$, $\mathbf{B} \in \mathscr{C}^{n\times k}$

$$\frac{\partial(\operatorname{tr}\mathbf{AB})}{\partial \mathbf{B}} = \mathbf{A}^T. \tag{2.90}$$

Exercise 2.17

Derive result (2.88) by beginning with

$$d(\operatorname{tr}\mathbf{AB}) = \sum_{\alpha=1}^{n} \left[\frac{\partial(\operatorname{tr}\mathbf{AB})}{\partial \mathbf{a}_{\alpha R}}\right]^T d\mathbf{a}_{\alpha R} + \sum_{\alpha=1}^{n} \left[\frac{\partial(\operatorname{tr}\mathbf{AB})}{\partial \mathbf{a}_{\alpha I}}\right]^T d\mathbf{a}_{\alpha I} \tag{2.91}$$

in lieu of the starting point (2.81).

EXAMPLE 2.15

Prove (C.25) again by considering the function

$$f(\mathbf{x}) \triangleq \|\mathbf{x}\|_{\mathbf{Q}}^2 = \mathbf{x}^*\mathbf{Q}\mathbf{x} = \operatorname{tr}\mathbf{x}^*\mathbf{Q}\mathbf{x} \tag{2.92}$$

by the use of identities (2.83) and (2.88).

Solution: Equation (2.83) allows $\operatorname{tr}(\mathbf{x}^*\mathbf{Q})(\mathbf{x}) = \operatorname{tr}(\mathbf{x})(\mathbf{x}^*\mathbf{Q})$; hence, (2.92) becomes

$$f(\mathbf{x}) = \|\mathbf{x}\|_{\mathbf{Q}}^2 = \operatorname{tr}\mathbf{x}\mathbf{x}^*\mathbf{Q}. \tag{2.93}$$

Now, use result (2.88) to get

$$\frac{\partial\left[\operatorname{tr}(\mathbf{x})(\bar{\mathbf{x}}^T\mathbf{Q})\right]}{\partial \mathbf{x}} = \mathbf{Q}^T\bar{\mathbf{x}}. \tag{2.94}$$

And since (2.84) allows $\operatorname{tr}(\mathbf{x})(\bar{\mathbf{x}}^T\mathbf{Q}) = \operatorname{tr}(\mathbf{x}^T)(\mathbf{Q}^T\bar{\mathbf{x}})$, the application again of (2.88) yields

$$\frac{\partial\left[\operatorname{tr}(\mathbf{x}^T\mathbf{Q}^T)(\bar{\mathbf{x}})\right]}{\partial \bar{\mathbf{x}}} = \mathbf{Q}\mathbf{x}, \tag{2.95}$$

which verifies (2.75c) and (2.75d). The differential of (2.93) is now written

$$df = \left[\frac{\partial f}{\partial \mathbf{x}}\right]^T d\mathbf{x} + \left[\frac{\partial f}{\partial \bar{\mathbf{x}}}\right]^T d\bar{\mathbf{x}}$$

$$= \bar{\mathbf{x}}^T\mathbf{Q}\,d\mathbf{x} + \mathbf{x}^T\mathbf{Q}^T\,d\bar{\mathbf{x}} = \bar{\mathbf{x}}^T\mathbf{Q}\,d\mathbf{x} + (d\bar{\mathbf{x}})^T\mathbf{Q}\mathbf{x}$$

$$= \mathbf{x}^*\mathbf{Q}\,d\mathbf{x} + d\mathbf{x}^*\mathbf{Q}\mathbf{x} = \langle \mathbf{x}, \mathbf{Q}\,d\mathbf{x}\rangle + \langle d\mathbf{x}, \mathbf{Q}\mathbf{x}\rangle,$$

which verifies (C25) in Appendix C. ■

EXAMPLE 2.16

Derive (2.88) by appealing directly to the right-hand side of (2.82).

Solution

$$\left[\frac{\partial(\operatorname{tr}\mathbf{AB})}{\partial\mathbf{A}}\right]_{\alpha\beta} = \frac{\partial(\operatorname{tr}\mathbf{AB})}{\partial A_{\alpha\beta}} = \frac{\partial}{\partial A_{\alpha\beta}}\left(\sum_{\Omega=1}^{k}\sum_{\gamma=1}^{n} A_{\Omega\gamma}B_{\gamma\Omega}\right)$$

$$= \sum_{\Omega=1}^{k}\sum_{\gamma=1}^{n}\left(\frac{\partial A_{\Omega\gamma}}{\partial A_{\alpha\beta}}\right)B_{\gamma\Omega} = \sum_{\Omega=1}^{k}\sum_{\gamma=1}^{n}\delta_{\alpha\Omega}\delta_{\gamma\beta}B_{\gamma\Omega}.$$

Hence,

$$\left[\frac{\partial(\operatorname{tr}\mathbf{AB})}{\partial\mathbf{A}}\right]_{\alpha\beta} = B_{\beta\alpha} \Rightarrow \frac{\partial(\operatorname{tr}\mathbf{AB})}{\partial\mathbf{A}} = \mathbf{B}^T. \tag{2.96}$$

■

2.7.1.3 Trace Identities

For convenient reference the following identities are recorded. Each identity can be derived by repeated application of (2.83), (2.84), and (2.88). The elements of **A** are assumed to be independent:

$$\frac{\partial(\operatorname{tr}\mathbf{AB})}{\partial\mathbf{A}} = \frac{\partial(\operatorname{tr}\mathbf{A}^T\mathbf{B}^T)}{\partial\mathbf{A}} = \frac{\partial(\operatorname{tr}\mathbf{B}^T\mathbf{A}^T)}{\partial\mathbf{A}} = \frac{\partial(\operatorname{tr}\mathbf{BA})}{\partial\mathbf{A}} = \mathbf{B}^T, \tag{2.97}$$

$$\frac{\partial(\operatorname{tr}\mathbf{BAC})}{\partial\mathbf{A}} = \frac{\partial(\operatorname{tr}\mathbf{B}^T\mathbf{C}^T\mathbf{A}^T)}{\partial\mathbf{A}} = \frac{\partial(\operatorname{tr}\mathbf{C}^T\mathbf{A}^T\mathbf{B}^T)}{\partial\mathbf{A}} = \frac{\partial(\operatorname{tr}\mathbf{ACB})}{\partial\mathbf{A}}$$

$$= \frac{\partial(\operatorname{tr}\mathbf{CBA})}{\partial\mathbf{A}} = \frac{\partial(\operatorname{tr}\mathbf{A}^T\mathbf{B}^T\mathbf{C}^T)}{\partial\mathbf{A}} = \mathbf{B}^T\mathbf{C}^T, \tag{2.98}$$

$$\frac{\partial(\operatorname{tr}\mathbf{A}^T\mathbf{BA})}{\partial\mathbf{A}} = \frac{\partial(\operatorname{tr}\mathbf{BAA}^T)}{\partial\mathbf{A}} = \frac{\partial(\operatorname{tr}\mathbf{AA}^T\mathbf{B})}{\partial\mathbf{A}} = (\mathbf{B} + \mathbf{B}^T)\mathbf{A}. \tag{2.99}$$

Exercise 2.18

Verify (2.97) through (2.99).

Exercise 2.19

The elements of **A** are not always independent. Show for symmetric **A** that

$$\frac{\partial (\mathrm{tr}\,\mathbf{AB})}{\partial \mathbf{A}} = \mathbf{B}^T + \mathbf{B} - \mathrm{diag}\,[\mathbf{B}]. \tag{2.100}$$

where $\mathrm{diag}\,[\mathbf{B}] \triangleq \mathrm{diag}\,[\cdots \; B_{ii} \; \cdots]$.

2.7.2 PROBLEM: $\min_{\mathbf{x}} \|\mathbf{Ax} - \mathbf{y}\|^2$

Consider now the linear equation

$$\mathbf{Ax} = \mathbf{y} \tag{2.101}$$

for a given $\mathbf{y} \in \mathscr{C}^k$ and a given $\mathbf{A} \in \mathscr{C}^{k \times n}$. It is desired to find a solution $\mathbf{x} \in \mathscr{C}^n$ if one exists, and if one does not exist to find an approximate solution that comes as close as possible to a solution. To make this objective more precise, suppose a solution **x** to (2.101) does not exist. Then, (2.101) cannot be written with an equality. For any chosen **x**, define an "equation error" vector $\mathbf{e} \in \mathscr{C}^k$ by

$$\mathbf{e} \triangleq \mathbf{Ax} - \mathbf{y}. \tag{2.102}$$

By choosing **x** so as to minimize the norm of **e**, a "least squares" solution to the problem I′ is obtained. This minimization problem is written as follows:

$$\min_{\mathbf{x}} \|\mathbf{e}\|^2 \qquad \text{or} \qquad \min_{\mathbf{x}} \|\mathbf{Ax} - \mathbf{y}\|^2. \tag{2.103}$$

Now, suppose some equations in (2.101) are more important to satisfy than others. That is, if *all* of the equations cannot be satisfied, the analyst may desire *some* of the equations to be satisfied, or nearly satisfied, when larger errors in the remaining equations are acceptable. To reflect this flexibility in the problem, equation (2.103) is modified to include a *weighted* norm. This yields the final problem statement

$$\min_{\mathbf{x}} \|\mathbf{Ax} - \mathbf{y}\|^2_{\mathbf{Q}}. \tag{2.104}$$

Now, the necessary and sufficient conditions are to be developed for problem (2.104), rewritten here

$$\min_{\mathbf{x}} \mathscr{V}, \qquad \mathscr{V} \triangleq \langle \mathbf{Ax} - \mathbf{y}, \mathbf{Q}(\mathbf{Ax} - \mathbf{y}) \rangle = (\mathbf{x}^*\mathbf{A}^* - \mathbf{y}^*)\mathbf{Q}(\mathbf{Ax} - \mathbf{y}), \tag{2.105a}$$

$$\mathscr{V} = (\mathbf{x}^*\mathbf{A}^*\mathbf{QAx} + \mathbf{y}^*\mathbf{Qy} - \mathbf{x}^*\mathbf{A}^*\mathbf{Qy} - \mathbf{y}^*\mathbf{QAx}). \tag{2.105b}$$

Exercise 2.20

From (2.74) the differential of (2.105b) is

$$d\mathscr{V} = \left(\frac{\partial V}{\partial \mathbf{x}}\right)^T d\mathbf{x} + \left(\frac{\partial V}{\partial \bar{\mathbf{x}}}\right)^T d\bar{\mathbf{x}}. \tag{2.106}$$

Show from (2.75c) and (2.75a) that the partial $\partial \mathscr{V}/\partial \mathbf{x}$ of (2.75b) is

$$\left(\frac{\partial \mathscr{V}}{\partial \mathbf{x}}\right)^T = \mathbf{x}^*(\mathbf{A}^*\mathbf{Q}\mathbf{A}) - \mathbf{y}^*\mathbf{Q}\mathbf{A} \tag{2.107a}$$

and from (2.75d) and (2.75a) that

$$\left(\frac{\partial \mathscr{V}}{\partial \bar{\mathbf{x}}}\right)^T = \mathbf{x}^T(\mathbf{A}^*\mathbf{Q}\mathbf{A})^T - \mathbf{y}^T(\mathbf{A}^*\mathbf{Q})^T. \tag{2.107b}$$

From Theorem C.1 and (2.107), note that the necessary conditions for a minimum are

$$\mathbf{x}^*\mathbf{A}^*\mathbf{Q}\mathbf{A} - \mathbf{y}^*\mathbf{Q}\mathbf{A} = 0,$$

$$\mathbf{x}^T(\mathbf{A}^*\mathbf{Q}\mathbf{A})^T - \mathbf{y}^T(\mathbf{A}^*\mathbf{Q})^T = 0,$$

which are both satisfied by the single condition

$$[\mathbf{A}^*\mathbf{Q}\mathbf{A}]\mathbf{x} = [\mathbf{A}^*\mathbf{Q}]\mathbf{y} \qquad \mathbf{Q} = \mathbf{Q}^* > \mathbf{0}. \tag{2.108}$$

The interesting conclusion thus far is that our attempt to solve the linear equation (2.101) with minimal weighted error has only led us to *another* linear equation, (2.108), to solve. There are some notable differences between (2.101) and (2.108), however. The matrix coefficient of $\mathbf{x}$ in (2.108) has two properties that make it easier to solve than (2.101); the matrix coefficient of $\mathbf{x}$ in (2.108) is square and *Hermitian* $((\mathbf{A}^*\mathbf{Q}\mathbf{A})^* = (\mathbf{A}^*\mathbf{Q}\mathbf{A}))$.

We are now prepared to summarize the solution to problem (2.104).

Theorem 2.8

If $\mathbf{A} \in \mathscr{C}^{k \times n}$ *and* $\operatorname{rank} \mathbf{A} = n$, $\mathbf{Q} = \mathbf{Q}^* > \mathbf{0}$, *then*

$$\min_{\mathbf{x}} \|\mathbf{A}\mathbf{x} - \mathbf{y}\|_{\mathbf{Q}}^2 \Rightarrow \mathbf{x} = (\mathbf{A}^*\mathbf{Q}\mathbf{A})^{-1}\mathbf{A}^*\mathbf{Q}\mathbf{y}, \tag{2.109}$$

and the minimum value of the error function is

$$\min_{\mathbf{x}} \|\mathbf{A}\mathbf{x} - \mathbf{y}\|_{\mathbf{Q}}^2 = \mathbf{y}^*\left[\mathbf{Q} - \mathbf{Q}\mathbf{A}(\mathbf{A}^*\mathbf{Q}\mathbf{A})^{-1}\mathbf{A}^*\mathbf{Q}\right]\mathbf{y}. \tag{2.110}$$

To prove (2.110), merely substitute $\mathbf{x}$ from (2.109) into (2.104) and simplify.

Exercise 2.21

Show that if $k = n$, then, under the conditions of Theorem 2.8,

$$\min_{\mathbf{x}} \|\mathbf{Ax} - \mathbf{y}\|_{\mathbf{Q}}^2 = 0. \tag{2.111}$$

2.7.3 PROBLEM: $\min_{\mathbf{x}} \|\mathbf{x}\|^2$ SUBJECT TO $\mathbf{Ax} = \mathbf{y}$

Theorem 2.8 summarizes the case when $\min_{\mathbf{x}} \|\mathbf{Ax} - \mathbf{y}\|_{\mathbf{Q}}^2 \geq 0$ and $k \geq n$. We now consider the *remaining* special case when $\min_{\mathbf{x}} \|\mathbf{Ax} - \mathbf{y}\|_{\mathbf{Q}}^2 = 0$ and $k < n$. The peculiar thing about this case is that there are more unknowns (x_i, $i = 1, \ldots, n$) than equations ($e_i = 1, \ldots, k$), and the condition rank $A = n$ of Theorem 2.8 is not satisfied (since rank $\mathbf{A} \leq \min\{k, n\}$ and $k < n$). This suggests that the $\mathbf{x}$ which yields $\min_{\mathbf{x}} \|\mathbf{Ax} - \mathbf{y}\|_{\mathbf{Q}}^2 = 0$ is not unique. The additional freedom in the choice of $\mathbf{x}$ will be used to minimize the norm of $\mathbf{x}$. We present this problem as

$$\min_{\mathbf{x}} \|\mathbf{x}\|_{\mathbf{Q}}^2, \qquad \text{subject the constraint } \mathbf{Ax} - \mathbf{y} = 0.$$

This problem is a constrained optimization problem. A standard practice is to convert a constrained optimization problem to an unconstrained one by use of Lagrange multipliers (an elementary explanation is in ref. [2.11], more advanced ones in [2.7] and [2.12]). That is, if a differential function $f(\mathbf{x})$ is to be minimized subject to the differentiable constraint $g(\mathbf{x}) = 0$, the necessary conditions are $(\partial/\partial\lambda)(f(\mathbf{x}) + \lambda g(\mathbf{x})) = 0$, $(\partial/\partial\mathbf{x})(f(\mathbf{x}) + \lambda g(\mathbf{x})) = 0$, where λ is called a "Lagrange multiplier." To see how this is done, consider the minimization of x_2 subject to the constraint $x_1^2 + 4x_2^2 = 1$. See from Fig. 2.9 that the minimum value is $x_2 = -\frac{1}{2}$. To show this using Lagrange multipliers, compute [using $f(\mathbf{x}) = x_2$, $g(\mathbf{x}) = x_1^2 + 4x_2^2 - 1$]

$$\frac{\partial}{\partial x_1}[f(\mathbf{x}) + \lambda g(\mathbf{x})] = 2\lambda x_1 = 0,$$

$$\frac{\partial}{\partial x_2}[f(\mathbf{x}) + \lambda g(\mathbf{x})] = 1 + \lambda 8 x_2 = 0,$$

$$\frac{\partial}{\partial \lambda}[f(\mathbf{x}) + \lambda g(\mathbf{x})] = g(\mathbf{x}) = x_1^2 + 4x_2^2 - 1 = 0,$$

which solution yields

$$x_1 = 0, \quad x_2 = \pm\tfrac{1}{2}, \quad \lambda = \mp\tfrac{1}{4}.$$

Thus, there are two candidates: $x_2 = +\frac{1}{2}$ yields a maximum, and $x_2 = -\frac{1}{2}$ yields a minimum.

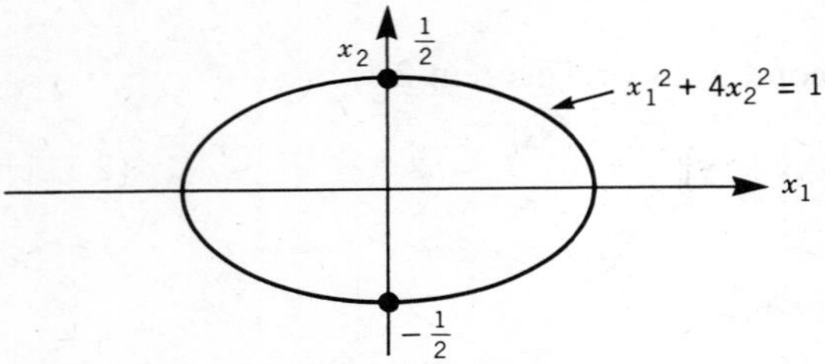

Figure 2.9 Constrained Optimization Example

Return now to the problem

$$\min_{\mathbf{x}} \|\mathbf{x}\|^2_{\mathbf{Q}} \qquad \text{subject to } \mathbf{Ax} - \mathbf{y} = \mathbf{0}.$$

By use of the k Lagrange-type multipliers $\mathbf{p}^T \triangleq (p_1, \ldots, p_k)$, the constrained optimization problem is converted to the unconstrained optimization problem. Suppose now that $\mathbf{x}$, $\mathbf{A}$, and $\mathbf{y}$ are all real. The unconstrained optimization problem is

$$\min_{\mathbf{x},\mathbf{p}} \hat{\mathscr{V}}, \quad \hat{\mathscr{V}} \triangleq \mathbf{x}^T\mathbf{Qx} + \mathbf{p}^T(\mathbf{Ax} - \mathbf{y}), \quad \mathbf{Q} = \mathbf{Q}^T > 0. \tag{2.112}$$

The necessary conditions are

$$\frac{\partial \hat{\mathscr{V}}}{\partial \mathbf{x}} = \mathbf{0} = \mathbf{Q}^T\mathbf{x} + \mathbf{A}^T\mathbf{p} + \mathbf{Qx} = 2\mathbf{Qx} + \mathbf{Q}^T\mathbf{p}, \tag{2.113a}$$

$$\frac{\partial \hat{\mathscr{V}}}{\partial \mathbf{p}} = \mathbf{0} = \mathbf{Ax} - \mathbf{y}. \tag{2.113b}$$

Solve (2.113a) for $\mathbf{x}$ to get

$$\mathbf{x} = -\tfrac{1}{2}\mathbf{Q}^{-1}\mathbf{A}^T\mathbf{p}. \tag{2.114}$$

Premultiply (2.114) by $\mathbf{A}$ and use (2.113b) to get

$$\mathbf{Ax} = -\tfrac{1}{2}\mathbf{AQ}^{-1}\mathbf{A}^T\mathbf{p} = \mathbf{y}. \tag{2.115}$$

Hence, from (2.115)

$$\mathbf{p} = -2(\mathbf{AQ}^{-1}\mathbf{A}^T)^{-1}\mathbf{y} \qquad (\text{if rank } \mathbf{A} = k). \tag{2.116}$$

Substituting (2.116) into (2.114),

$$\mathbf{x} = \mathbf{Q}^{-1}\mathbf{A}^T(\mathbf{AQ}^{-1}\mathbf{A}^T)^{-1}\mathbf{y}. \tag{2.117}$$

The results are now summarized for the general case with complex $\mathbf{x}$, $\mathbf{A}$, and $\mathbf{y}$.

Theorem 2.9
If $\mathbf{A} \in \mathscr{C}^{k\times n}$ *and* rank $\mathbf{A} = k \le n$, *the solution to*

$$\mathbf{Ax} = \mathbf{y} \tag{2.118}$$

that has minimum norm $\|\mathbf{x}\|_{\mathbf{Q}}^2$ *is given by*

$$\mathbf{x} = \mathbf{Q}^{-1}\mathbf{A}^*(\mathbf{AQ}^{-1}\mathbf{A}^*)^{-1}\mathbf{y}. \tag{2.119}$$

The resulting value of the norm $\|\mathbf{x}\|_{\mathbf{Q}}^2$ *is*

$$\|\mathbf{x}\|_{\mathbf{Q}}^2 = \mathbf{y}^*(\mathbf{AQ}^{-1}\mathbf{A}^*)^{-1}\mathbf{y}. \tag{2.120}$$

Exercise 2.22
Prove Theorem 2.18, assuming **x**, **A**, and **y** are complex. (*Hint*: The constraint $\mathbf{Ax} - \mathbf{y} = \mathbf{0}$ is now a complex vector. Two Lagrange multipliers are needed: one for the constraint $\mathscr{Re}\,(\mathbf{Ax} - \mathbf{y}) = \mathbf{0}$ and another for the constraint $\mathscr{Im}\,(\mathbf{Ax} - \mathbf{y}) = \mathbf{0}$.)

Exercise 2.23
Under the conditions of Exercise 2.22, $\mathbf{Ax} = \mathbf{y}$ and rank $\mathbf{A} = k = n$. Show in this case that the norm of **x** is also given by (2.120).

EXAMPLE 2.17
Solve for x if

$$\begin{bmatrix} 1.0 \\ 1.5 \\ 0 \end{bmatrix} x = \begin{bmatrix} -0.5 \\ 1.5 \\ -0.5 \end{bmatrix}.$$

Solution: In this problem,

$$\mathbf{A} = \begin{pmatrix} 1.0 \\ 1.5 \\ 0 \end{pmatrix}, \quad \mathbf{y} = \begin{pmatrix} -0.5 \\ 1.5 \\ -0.5 \end{pmatrix}, \quad \mathbf{Q} = \begin{bmatrix} 100 & & \\ & 200 & \\ & & 400 \end{bmatrix},$$

and rank $\mathbf{A} = n = 1 < k = 3$. Hence, Theorem 2.8 applies. Equation (2.109) provides the solution:

$$x = \left\{ (1.0, 1.5, 0) \begin{bmatrix} 100 & & \\ & 200 & \\ & & 400 \end{bmatrix} \begin{pmatrix} 1.0 \\ 1.5 \\ 0 \end{pmatrix} \right\}^{-1}$$

$$\times (1.0, 1.5, 0) \begin{bmatrix} 100 & & \\ & 200 & \\ & & 400 \end{bmatrix} \begin{pmatrix} -0.5 \\ 1.5 \\ -0.5 \end{pmatrix}$$

$$= \frac{400}{550} = \frac{8}{11}.$$

■

EXAMPLE 2.18
Determine x that solves

$$\min_x \|1.0x - (-0.5)\|_{100}^2$$

and hence

$$A = 1.0, \quad y = -0.5, \quad Q = 100.$$

Solution: Rank $A = k = n = 1$, and Theorem 2.8 applies. Since A is square, from (2.109)

$$x = (A^*QA)^{-1}A^*Qy = A^{-1}Q^{-1}A^{*-1}A^*Qy = A^{-1}y = -0.5.$$

Theorem 2.9 also applies, giving

$$x = Q^{-1}A^*(AQ^{-1}A^*)^{-1}y = Q^{-1}A^*A^{*-1}QA^{-1}y = A^{-1}y = -0.5. \quad \blacksquare$$

EXAMPLE 2.19
Verify that there are an infinite number of values of $\mathbf{x}$ that will satisfy $\mathbf{A}\mathbf{x} = y$ if

$$\mathbf{A} = [1 - \Gamma, \Gamma],$$

and find the solution that minimizes $\|\mathbf{A}\mathbf{x} - y\|^2$, with $Q = \begin{bmatrix} 1 & 0 \\ 0 & 10 \end{bmatrix}$.

Solution: In this case, rank $\mathbf{A} = k < n$, and Theorem 2.9 applies. From (2.119)

$$\mathbf{x} = \begin{pmatrix} 1 & 0 \\ 0 & 10 \end{pmatrix}^{-1} \begin{pmatrix} 1 - \Gamma \\ \Gamma \end{pmatrix} \left\{ (1 - \Gamma, \Gamma) \begin{pmatrix} 1 & 0 \\ 0 & 10 \end{pmatrix}^{-1} \begin{pmatrix} 1 - \Gamma \\ \Gamma \end{pmatrix} \right\}^{-1} y$$

$$= \begin{pmatrix} 1 - \Gamma \\ 0.1\Gamma \end{pmatrix} \frac{y}{(1 - 2\Gamma + 1.1\Gamma^2)}.$$

2.7.4 A GEOMETRICAL APPROACH TO LEAST SQUARES PROBLEMS

This section may be omitted without loss of continuity. In this section we shall provide alternate proofs for Theorems 2.8 and 2.9, using the tools of Appendix B:

Problem I′: $\min_x \|\mathbf{A}\mathbf{x} - \mathbf{y}\|^2$, rank $\mathbf{A} = \mathbf{n}$.

For *any* $\mathbf{x}$ we might choose, $\mathbf{A}\mathbf{x}$ lies in the column space of $\mathbf{A}$, denoted by $\mathscr{R}[\mathbf{A}]$ in the simplified Fig. 2.10. Theorem B.2 states that the space orthogonal to $\mathscr{R}[\mathbf{A}]$ is $\mathscr{N}[\mathbf{A}^*]$, as shown in Fig. 2.10. The problem is to find the shortest length vector $\mathbf{y} - \mathbf{A}\mathbf{x}$ given that $\mathbf{A}\mathbf{x}$ must lie in $\mathscr{R}[\mathbf{A}]$. The shortest distance from a line ($\mathscr{R}[\mathbf{A}]$ in Fig. 2.10) to a point ($\mathbf{y}$ in Fig. 2.10) is perpendicular to the line $\mathscr{R}[\mathbf{A}]$. Since $\mathscr{N}[\mathbf{A}^*]$

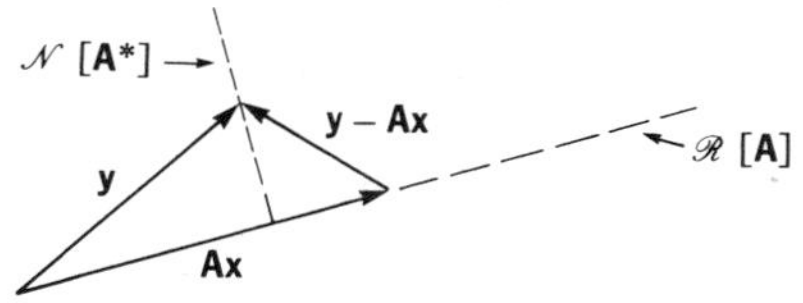

Figure 2.10 Orthogonal Projections for Problem I′

is perpendicular to the line $\mathscr{R}[\mathbf{A}]$, note that the optimal $\mathbf{y} - \mathbf{Ax}$ must lie in $\mathscr{N}[\mathbf{A}^*]$. Confirm this visually by inspection of Fig. 2.10. This means that $(y - Ax)$ must lie in the null space of A^*

$$\mathbf{A}^*(\mathbf{y} - \mathbf{Ax}) = \mathbf{0}. \tag{2.121}$$

This in turn yields

$$\mathbf{x} = (\mathbf{A}^*\mathbf{A})^{-1}\mathbf{A}^*\mathbf{y} \qquad \text{if rank } \mathbf{A} = n. \tag{2.122}$$

The rank $\mathbf{A} = n$ condition is necessary and sufficient for $(\mathbf{A}^*\mathbf{A})^{-1}$ to exist. This is the result of Theorem 2.8 with $\mathbf{Q} = \mathbf{I}$. This latter proof is somewhat shorter than the original proof of Theorem 2.8, although the proof of our geometrical result, equation (2.122), is not as rigorous as that developed in books more devoted to geometrical approaches [2.6], [2.7].

In problem I′, the answer was unique since rank $\mathbf{A} = n$. To address problem I_2', now suppose that the rank of $\mathbf{A} \in \mathscr{C}^{k \times n}$ is equal to its row dimension, rank $\mathbf{A} = k \leq n$. If $k = n$, the solution is unique. If $k < n$, there are an infinite number of solutions for $\mathbf{x}$. The total $\mathbf{x}$ is decomposed into the orthogonal subspaces $\mathbf{x} = \mathbf{x}_1 + \mathbf{x}_2$, where $\mathbf{x}_1 \in \mathscr{N}[\mathbf{A}]$, $\mathbf{x}_2 \in \mathscr{R}[\mathbf{A}^*]$, and we wish to satisfy $\mathbf{Ax} = \mathbf{y}$ with the shortest length for $\mathbf{x}$. See Fig. 2.11.

Clearly, our chosen $\mathbf{x}$ cannot lie totally in $\mathscr{N}[\mathbf{A}]$ since $\mathbf{Ax} = \mathbf{0}$ does *not* satisfy $\mathbf{Ax} = \mathbf{y}$. Therefore, our chosen $\mathbf{x}$ must lie in $\mathscr{R}[\mathbf{A}^*]$ $(\mathbf{x} = \mathbf{x}_1)$, and to have minimum norm $\|\mathbf{x}\|^2$,

$$\|\mathbf{x}\|^2 = \|\mathbf{x}_1\|^2 + \|\mathbf{x}_2\|^2, \qquad \text{recall } \mathbf{x}_1^*\mathbf{x}_2 = 0,$$

we must choose $\mathbf{x}_2 = \mathbf{0}$. Now, compute $\mathbf{x}_1$. By definition of $\mathscr{R}[\mathbf{A}^*]$,

$$\mathbf{A}^*\boldsymbol{\beta} = \mathbf{x}_1. \tag{2.123}$$

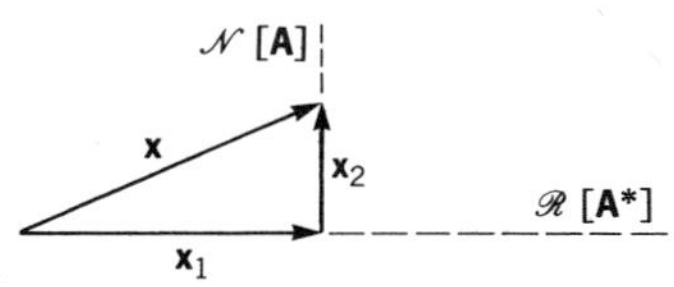

Figure 2.11 The Minimum Norm Problem

And to satisfy $\mathbf{Ax} = \mathbf{y}$,

$$\mathbf{Ax} = \mathbf{A}(\mathbf{x}_1 + \mathbf{x}_2) = \mathbf{Ax}_1 = \mathbf{y} \tag{2.124}$$

since $\mathbf{Ax}_2 = \mathbf{0}$ (even if we choose $\mathbf{x}_2 \neq \mathbf{0}$). Substitute (2.123) into (2.124) to obtain

$$\mathbf{AA}^*\beta = \mathbf{y} \Rightarrow \boldsymbol{\beta} = (\mathbf{AA}^*)^{-1}\mathbf{y}.$$

Substitute this into (2.123). Then,

$$\mathbf{x}_1 = \mathbf{A}^*(\mathbf{AA}^*)^{-1}\mathbf{y}.$$

We choose $\mathbf{x}_2 = \mathbf{0}$ to have minimum norm $\|\mathbf{x}\|$. Hence, $\mathbf{x} = \mathbf{x}_1 + \mathbf{x}_2 = \mathbf{x}_1$, and

$$\mathbf{x} = \mathbf{A}^*(\mathbf{AA}^*)^{-1}\mathbf{y} \tag{2.125}$$

verifies Theorem 2.9, with $\mathbf{Q} = \mathbf{I}$.

2.7.5 PROBLEM: $\min_{\mathbf{A}} \|\mathbf{Ax} - \mathbf{y}\|^2$

Now, suppose $\mathbf{x}(t)$ and $\mathbf{y}(t)$ are *given* vector functions of time, and it is desired to find a constant matrix $\mathbf{A}$ such that the equation error $\mathbf{e}(t) = \mathbf{Ax}(t) - \mathbf{y}(t)$ is smallest over a specified interval of time $t \in [t_1, t_2]$. The optimization problem is stated as follows:

$$\min_{\mathbf{A}} \|\mathbf{Ax}(t) - \mathbf{y}(t)\|_{\mathbf{Q}}^2, \qquad t \in [t_1, t_2], \quad \mathbf{Q} = \mathbf{Q}^* > \mathbf{0}.$$

To develop the solution, write the weighted norm of the error $\mathbf{e}$ in the notation of Section 2.4:

$$\begin{aligned}\|\mathbf{Ax} - \mathbf{y}\|_{\mathbf{Q}}^2 &= \langle \mathbf{Ax} - \mathbf{y}, \mathbf{Q}(\mathbf{Ax} - \mathbf{y})\rangle \\ &\triangleq \frac{1}{t_2 - t_1}\int_{t_1}^{t_2}(\mathbf{Ax}(t) - \mathbf{y}(t))^*\mathbf{Q}(\mathbf{Ax}(t) - \mathbf{y}(t))\,dt\end{aligned}$$

But since

$$\langle \mathbf{a}, \mathbf{b}\rangle = \operatorname{tr}[\rangle \mathbf{b}, \mathbf{a}\langle],$$

write the vector norm as

$$\begin{aligned}\langle \mathbf{Ax} - \mathbf{y}, \mathbf{Q}(\mathbf{Ax} - \mathbf{y})\rangle &= \operatorname{tr}\rangle \mathbf{Q}(\mathbf{AX} - \mathbf{y}), (\mathbf{Ax} - \mathbf{y})\langle \\ &= \operatorname{tr}\rangle \mathbf{QAx}, \mathbf{Ax}\langle + \operatorname{tr}\rangle \mathbf{Qy}, \mathbf{y}\langle - \operatorname{tr}\rangle \mathbf{Qy}, \mathbf{Ax}\langle - \operatorname{tr}\rangle \mathbf{QAx}, \mathbf{y}\langle.\end{aligned} \tag{2.126}$$

Note from the outer product definition (2.23), that

$$\rangle \mathbf{QAx}, \mathbf{Ax}\langle = \mathbf{QA}(\rangle \mathbf{x}, \mathbf{x}\langle)(\mathbf{A})^*$$

$$\rangle \mathbf{Qy}, \mathbf{Ax}\langle = \mathbf{Q}(\rangle \mathbf{y}, \mathbf{x}\langle)(\mathbf{A})^*$$

$$\rangle \mathbf{QAx}, \mathbf{y}\langle = \mathbf{QA}(\rangle \mathbf{x}, \mathbf{y}\langle).$$

Hence, (2.126) may be expressed in the form

$$\|\mathbf{Ax}(t) - \mathbf{y}(t)\|_{\mathbf{Q}}^2 = \operatorname{tr}\mathbf{QA}(\rangle \mathbf{x}, \mathbf{x}\langle)\mathbf{A}^* + \operatorname{tr}\mathbf{Q}(\rangle \mathbf{y}, \mathbf{y}\langle) - \operatorname{tr}\mathbf{Q}(\rangle \mathbf{y}, \mathbf{x}\langle)\mathbf{A}^* - \operatorname{tr}\mathbf{QA}(\rangle \mathbf{x}, \mathbf{y}\langle). \tag{2.127}$$

The optimization problem II′ then becomes

$$\min_{\mathbf{A}} \hat{\mathscr{V}}(\mathbf{A}), \hat{\mathscr{V}}(\mathbf{A}) \triangleq \operatorname{tr}\mathbf{A}(\rangle \mathbf{x}, \mathbf{x}\langle)\mathbf{A}^*\mathbf{Q} - \operatorname{tr}(\rangle \mathbf{y}, \mathbf{x}\langle)\mathbf{A}^*\mathbf{Q} - \operatorname{tr}\mathbf{A}(\rangle \mathbf{x}, \mathbf{y}\langle)\mathbf{Q}. \tag{2.128}$$

For the differential of $\hat{\mathscr{V}}(\mathbf{A})$, see (2.77),

$$d\hat{\mathscr{V}}(\mathbf{A}) = \sum_{\alpha=1}^{n} \left[\left(\frac{\partial \hat{\mathscr{V}}(\mathbf{A})}{\partial \mathbf{a}_{\alpha_R}} \right)^T d\mathbf{a}_{\alpha_R} + \left(\frac{\partial \hat{\mathscr{V}}(\mathbf{A})}{\partial \mathbf{a}_{\alpha_I}} \right)^T d\mathbf{a}_{\alpha_I} \right] \tag{2.129}$$

to be non-negative $d\hat{\mathscr{V}}(\mathbf{A}) \geq 0$, it is required that

$$\left[\frac{\partial \hat{\mathscr{V}}(\mathbf{A})}{\partial \mathbf{a}_{1_R}} \cdots \frac{\partial \hat{\mathscr{V}}(\mathbf{A})}{\partial \mathbf{a}_{n_R}} \right] = \frac{\partial \hat{\mathscr{V}}(\mathbf{A})}{\partial \mathbf{A}_R} = \mathbf{0}, \tag{2.130}$$

$$\left[\frac{\partial \hat{\mathscr{V}}(\mathbf{A})}{\partial \mathbf{a}_{1_I}} \cdots \frac{\partial \hat{\mathscr{V}}(\mathbf{A})}{\partial \mathbf{a}_{n_I}} \right] = \frac{\partial \hat{\mathscr{V}}(\mathbf{A})}{\partial \mathbf{A}_I} = \mathbf{0}. \tag{2.131}$$

From (2.79) and (2.80) see that

$$\frac{\partial \hat{\mathscr{V}}(\mathbf{A})}{\partial \mathbf{A}_R} = \left(\frac{\partial \hat{\mathscr{V}}(\mathbf{A})}{\partial \mathbf{A}} + \frac{\partial \hat{\mathscr{V}}(\mathbf{A})}{\partial \bar{\mathbf{A}}} \right), \tag{2.132}$$

$$\frac{\partial \hat{\mathscr{V}}(\mathbf{A})}{\partial \mathbf{A}_I} = j\left(\frac{\partial \hat{\mathscr{V}}(\mathbf{A})}{\partial \mathbf{A}} - \frac{\partial \hat{\mathscr{V}}(\mathbf{A})}{\partial \bar{\mathbf{A}}} \right), \tag{2.133}$$

where, from the definition of $\hat{\mathscr{V}}(\mathbf{A})$ in (2.128) and the trace identities (2.97) through

(2.99), we obtain

$$\frac{\partial \hat{\mathscr{V}}(\mathbf{A})}{\partial \mathbf{A}} = [()\mathbf{x},\mathbf{x}\langle)\mathbf{A}^*\mathbf{Q}^*]^T - \overline{\mathbf{Q}}()\mathbf{x},\mathbf{y}\langle)^T, \tag{2.134}$$

$$\frac{\partial \hat{\mathscr{V}}(\mathbf{A})}{\partial \overline{\mathbf{A}}} = \mathbf{Q}^*\mathbf{A}()\mathbf{x},\mathbf{x}\langle) - \mathbf{Q}^*()\mathbf{y},\mathbf{x}\langle). \tag{2.135}$$

Thus, from (2.130) through (2.135)

$$\frac{\partial \hat{\mathscr{V}}(\mathbf{A})}{\partial \mathbf{A}_R} = \overline{\mathbf{Q}}\overline{\mathbf{A}}()\mathbf{x},\mathbf{x}\langle)^T - \overline{\mathbf{Q}}()\mathbf{x},\mathbf{y}\langle)^T + \overline{\mathbf{Q}}^T\mathbf{A}()\mathbf{x},\mathbf{x}\langle) - \overline{\mathbf{Q}}^T()\mathbf{y},\mathbf{x}\langle) = \mathbf{0}, \tag{2.136}$$

$$\frac{1}{j}\frac{\partial \hat{\mathscr{V}}(\mathbf{A})}{\partial \mathbf{A}_I} = \overline{\mathbf{Q}}\overline{\mathbf{A}}()\mathbf{x},\mathbf{x}\langle)^T - \overline{\mathbf{Q}}()\mathbf{x},\mathbf{y}\langle)^T - \overline{\mathbf{Q}}^T\mathbf{A}()\mathbf{x},\mathbf{x}\langle) + \overline{\mathbf{Q}}^T()\mathbf{y},\mathbf{x}\langle) = \mathbf{0}. \tag{2.137}$$

Subtract (2.137) from (2.136) to obtain

$$2\mathbf{Q}^*\mathbf{A}()\mathbf{x},\mathbf{x}\langle) - 2\mathbf{Q}^*()\mathbf{y},\mathbf{x}\langle) = \mathbf{0}.$$

Solving for **A** gives

$$\mathbf{A} = ()\mathbf{y},\mathbf{x}\langle)()\mathbf{x},\mathbf{x}\langle)^{-1}. \tag{2.138}$$

Now, recalling the definitions of the outer products (2.23a), the result (2.138) is written now in literal form:

$$\mathbf{A} = \left[\frac{1}{t_2 - t_1}\int_{t_1}^{t_2}\mathbf{y}(t)\mathbf{x}^*(t)\,dt\right]\left[\frac{1}{t_2 - t_1}\int_{t_1}^{t_2}\mathbf{x}(t)\mathbf{x}^*(t)\,dt\right]^{-1}. \tag{2.139}$$

These results are now summarized.

Theorem 2.10

If the given functions of time $x_i(t)$, $i = 1,\ldots,n$, are linearly independent on the interval $t \in [t_1,t_2]$, subject to the weight $g(t) = 1$, then, for a given set of functions $y_i(t)$, $i = 1,\ldots,k$, the matrix **A** *that solves*

$$\min_{\mathbf{A}} \|\mathbf{A}\mathbf{x}(t) - \mathbf{y}(t)\|_{\mathbf{Q}}^2, \qquad t \in [t_1, t_2] \tag{2.140}$$

is

$$\mathbf{A} = \left[\int_{t_1}^{t_2}\mathbf{y}(t)\mathbf{x}^*(t)\,dt\right]\left[\int_{t_1}^{t_2}\mathbf{x}(t)\mathbf{x}^*(t)\,dt\right]^{-1}. \tag{2.141}$$

Now, interpret (2.138) in the following sense. Suppose the given data $\mathbf{x}(t)$ have these properties: $x_i(t)$ are orthogonal functions on the interval $t \in [t_1, t_2]$ subject to the weight $g(t)$; that is,

$$\rangle \mathbf{x}, \mathbf{x} \langle \triangleq \frac{1}{t_2 - t_1} \int_{t_1}^{t_2} \mathbf{x}(t) g(t) \mathbf{x}^*(t) \, dt = \Lambda = \operatorname{diag}\left(\lambda_1^2, \ldots, \lambda_n^2\right), \tag{2.142}$$

$\lambda_i^2 > 0$, $i = 1, \ldots, n$. The weighted outer product is defined by

$$\rangle \mathbf{y}, \mathbf{x} \langle \triangleq \frac{1}{t_2 - t_1} \int_{t_1}^{t_2} \mathbf{y}(t) g(t) \mathbf{x}^*(t) \, dt. \tag{2.143}$$

Under these conditions the following is evident.

Corollary to Theorem 2.10

If the $x_i(t)$, $i = 1, \ldots, n$, are a set of orthogonal functions on the interval $t \in [t_1, t_2]$ subject to the weight $g(t)$, then the matrix $\mathbf{A}$ that solves

$$\min_{\mathbf{A}} \|\mathbf{A}\mathbf{x}(t) - \mathbf{y}(t)\|_{\mathbf{Q}}^2 \tag{2.144}$$

is

$$\mathbf{A} = (\rangle \mathbf{y}, \mathbf{x} \langle) \Lambda^{-1} = \int_{t_1}^{t_2} \mathbf{y}(t) g(t) \mathbf{x}^*(t) \, dt \, \Lambda^{-1}, \tag{2.145}$$

where Λ is defined by (2.142).

2.7.5.1 Bessel's Inequality

To gain insight concerning the error (2.144), note from (2.127) that if $x_i(t)$ are orthogonal in the sense of (2.142), then (2.127) becomes

$$\|\mathbf{A}\mathbf{x} - \mathbf{y}\|_{\mathbf{Q}}^2 = \operatorname{tr}\left[\mathbf{A}\Lambda\mathbf{A}^*\mathbf{Q}^* + (\rangle \mathbf{y}, \mathbf{y} \langle)\mathbf{Q}^* - (\rangle \mathbf{y}, \mathbf{x} \langle)\mathbf{A}^*\mathbf{Q}^* - \mathbf{A}(\rangle \mathbf{y}, \mathbf{x} \langle)^*\mathbf{Q}^*\right]. \tag{2.146}$$

Note from (2.145) that $\rangle \mathbf{y}, \mathbf{x} \langle = \mathbf{A}\Lambda$. Substitute this expression into (2.146) to obtain

$$\|\mathbf{A}\mathbf{x} - \mathbf{y}\|_{\mathbf{Q}}^2 = \operatorname{tr}\left[\mathbf{A}\Lambda\mathbf{A}^* + (\rangle \mathbf{y}, \mathbf{y} \langle) - \mathbf{A}\Lambda\mathbf{A}^* - \mathbf{A}\Lambda^*\mathbf{A}^*\right]\mathbf{Q}^*. \tag{2.147}$$

But since $\Lambda = \Lambda^*$, $\mathbf{Q} = \mathbf{Q}^*$,

$$\|\mathbf{A}\mathbf{x} - \mathbf{y}\|_{\mathbf{Q}}^2 = \operatorname{tr}\left[(\rangle \mathbf{y}, \mathbf{y} \langle) - \mathbf{A}\Lambda\mathbf{A}^*\right]\mathbf{Q}. \tag{2.148}$$

Since $\|\mathbf{Ax} - \mathbf{y}\|_{\mathbf{Q}}^2 \geq 0$, it follows from (2.148) that

$$\min_{\mathbf{A}} \|\mathbf{Ax} - \mathbf{y}\|_{\mathbf{Q}}^2 = \operatorname{tr}[\rangle \mathbf{y}, \mathbf{y} \langle - \mathbf{A \Lambda A^*}]\mathbf{Q} \geq 0. \tag{2.149}$$

This is known as "Bessel's inequality" [2.12]. Written in literal form, (2.148) becomes

$$\min_{\mathbf{A}} \|\mathbf{Ax} - \mathbf{y}\|_{\mathbf{Q}}^2 = \operatorname{tr}\left[\int_{t_1}^{t_2} \mathbf{y}(t)\mathbf{y}^*(t)\, dt - \mathbf{A \Lambda A^*}\right]\mathbf{Q} \geq 0. \tag{2.150}$$

Now, using the facts that Λ is diagonal and that $\operatorname{tr} \rangle \mathbf{y}, \mathbf{y} \langle = \langle \mathbf{y}, \mathbf{y} \rangle$, (2.149) may be written

$$\min_{\mathbf{A}} \|\mathbf{Ax} - \mathbf{y}\|_{\mathbf{Q}}^2 = \langle \mathbf{y}, \mathbf{Qy} \rangle - \sum_{\alpha=1}^{n} \langle \mathbf{a}_\alpha, \mathbf{Qa}_\alpha \rangle \lambda_\alpha^2, \qquad \mathbf{A} = [\mathbf{a}_1, \ldots, \mathbf{a}_n]. \tag{2.151}$$

But since $\langle \mathbf{a}_\alpha, \mathbf{Qa}_\alpha \rangle = \|\mathbf{a}_\alpha\|_{\mathbf{Q}}^2$, Bessel's inequality (2.149) may be written

$$\min_{\mathbf{A}} \|\mathbf{Ax}(t) - \mathbf{y}(t)\|_{\mathbf{Q}}^2 = \|\mathbf{y}(t)\|_{\mathbf{Q}}^2 - \sum_{\alpha=1}^{n} \|\mathbf{a}_\alpha\|_{\mathbf{Q}}^2 \lambda_\alpha^2 \geq 0. \tag{2.152}$$

2.7.5.2 Parseval's Equality

To continue our investigation of (2.144), consider the following definition.

Definition 2.16
The functions $x_i(t)$, $i = 1, \ldots, n$, are called "complete" with respect to the given $y(t)$ on the interval $t \in [t_1, t_2]$ if the equality holds in (2.149) *and* (2.152). *That is, the $x_i(t)$ are complete if* $\min_{\mathbf{A}} \|\mathbf{Ax}(t) - \mathbf{y}(t)\|_{\mathbf{Q}}^2 = 0$.

Clearly, from (2.152) if the functions $x_i(t)$ are "complete," then $\min_{\mathbf{A}} \|\mathbf{Ax}(t) - \mathbf{y}(t)\|_{\mathbf{Q}}^2 = 0$, and

$$\|\mathbf{y}\|_{\mathbf{Q}}^2 = \sum_{\alpha=1}^{n} \|\mathbf{a}_\alpha\|_{\mathbf{Q}}^2 \lambda_\alpha^2. \tag{2.153}$$

Equation (2.153) is called "Parseval's equality" [2.12]. Thus, Bessel's name is associated with the *inequality* (2.149), and Parseval's name is associated with the same expression when the *equality* (2.153) holds. The left-hand side of (2.153) is always equal to or greater than the right-hand side, but the equality holds *only if* the functions $x_i(t)$ are *complete*.

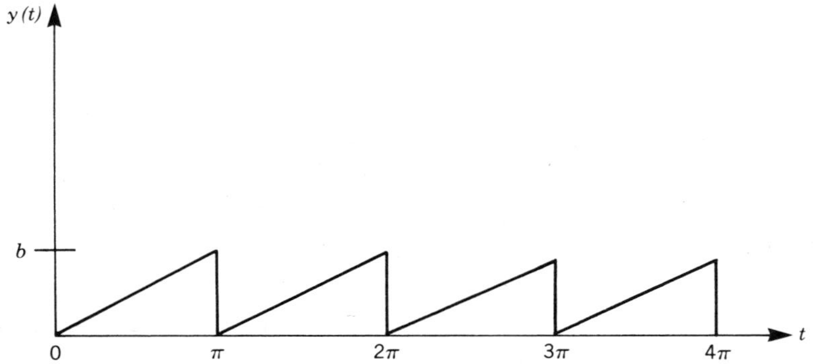

Figure 2.12 Fourier Series Expansion of a Sawtooth

EXAMPLE 2.20

Find the Fourier series expansion of the sawtooth function of Fig. 2.12.

Solution: The Fourier basis functions

$$x_i(t) = e^{j(i\omega t)}, \qquad \omega = \frac{2\pi}{T} = \frac{2\pi}{\pi} = 2, \tag{2.154}$$

are orthogonal on the interval $t \in [k\pi, (k+1)\pi]$ subject to the weight $g(t) = 1$. That is, from (2.142),

$$\Lambda_{\alpha\beta} = \int_{k\pi}^{(k+1)\pi} e^{j(\alpha\omega t)} e^{-j(\beta\omega t)}\, dt = \pi\delta_{\alpha\beta}.$$

From Fig. 2.12,

$$y(t) = \frac{b}{\pi}t - bk, \qquad k\pi \le t \le (k+1)\pi.$$

Hence, from (2.145), the first $n+1$ Fourier coefficients are

$$\mathbf{A} = \int_{k\pi}^{(k+1)\pi} y\mathbf{x}^*\, dt\, \mathbf{\Lambda}^{-1} = \left[\frac{b}{2}, \frac{b}{2}\frac{j}{\pi}, \ldots, \frac{b}{2}\frac{j}{\pi n}\right]. \tag{2.155}$$

To determine if this set of basis functions $x_i(t)$, $i = 1, \ldots, n-1$, is complete with respect to $y(t)$, one must determine if (2.153) is satisfied. ■

Exercise 2.24

Verify (2.155) and determine whether the basis functions are complete for the function in Fig. 2.12.

2.8 Decomposition of Matrices

The analysis of a system of equations is often facilitated if the matrices can be put into a diagonal form, a triangular form, or some other special form that can be exploited to simplify the analysis and to gain insight into the underlying physical phenomena. The purpose of this section is to examine several canonical structures of matrices. We shall show that a square matrix **A** can sometimes be decomposed into the product of three matrices of the form

$$\mathbf{A} = \mathbf{E}\boldsymbol{\Lambda}\mathbf{E}^{-1}, \tag{2.156}$$

where Λ is diagonal. We also show that when Λ cannot be diagonal, it takes on a form that has entries only on the diagonal and the superdiagonal. Any matrix **A** can be decomposed into the product of three matrices of the form

$$\mathbf{A} = \mathbf{U}\boldsymbol{\Sigma}\mathbf{V}^* \tag{2.157}$$

and sometimes into the product of two matrices of the form

$$\mathbf{A} = \mathbf{C}\mathbf{C}^*. \tag{2.158}$$

These standard decompositions have special names: (2.156) is called the "spectral decomposition" of **A**, and the diagonal entries of Λ are eigenvalues and the columns of **E** are eigenvectors of **A**. Equation (2.157) is called "singular value decomposition." We shall call (2.158) the "square root decomposition" of **A** since **C** plays a role similar to the square root in the analysis of scalars. Each of these special cases will be addressed in turn.

2.8.1 SPECTRAL DECOMPOSITION

In this section the matrix $\mathbf{A} \in \mathscr{C}^{n\times n}$ is always square, and this section always deals with the modal data of matrix **A**. The *modal data* of **A** is composed of its eigenvalues and its eigenvectors. We repeat the earlier definitions of Section 2.1. The *eigenvalues* λ_i and the *eigenvectors* $\mathbf{e}_i$ of a square matrix **A** are defined as nontrivial solutions to

$$\mathbf{A}\mathbf{e}_i = \mathbf{e}_i\lambda_i, \tag{2.159}$$

where $\mathbf{A} \in \mathscr{C}^{n\times n}$, $\mathbf{e}_i \in \mathscr{C}^n$, $\lambda_i \in \mathscr{C}^{1\times 1}$. The *left eigenvectors* $\mathbf{l}_i$ of **A** are defined by

$$\mathbf{l}_i^*\mathbf{A} = \lambda_i\mathbf{l}_i^*. \tag{2.160}$$

The *modal matrix* of **A** is defined by

$$\mathbf{E} \triangleq [\mathbf{e}_1, \dots, \mathbf{e}_n], \qquad \mathbf{E} \in \mathscr{C}^{n\times n}, \tag{2.161}$$

assuming that n linearly independent eigenvectors $\mathbf{e}_i$, $i = 1, \ldots, n$, can be determined from (2.159). We shall return later to this detail.

The eigenvalues are computed as follows. From (2.159), note that a solution of

$$[\mathbf{A} - \lambda_i \mathbf{I}]\mathbf{e}_i = 0 \tag{2.162}$$

is sought for $\mathbf{e}_i \neq 0$. In terms of the language of Section 2.7.4, we say that eigenvectors $\mathbf{e}_i$ lie in the null space of the matrix $[\mathbf{A} - \lambda_i \mathbf{I}]$, denoted by $\mathbf{e}_i \in \mathcal{N}[\mathbf{A} - \lambda_i \mathbf{I}]$. Since from Appendix B the dimension of this null space is $n - r_i$ (where $r_i \triangleq \operatorname{rank}[\mathbf{A} - \lambda_i \mathbf{I}]$), there are only $n - r_i$ linearly independent eigenvectors associated with λ_i. Now, if $\operatorname{rank}[\mathbf{A} - \lambda_i \mathbf{I}] = n$, then the columns of $[\mathbf{A} - \lambda_i \mathbf{I}]$ are linearly independent and (2.162) has a *unique* solution, $\mathbf{e}_i = \mathbf{0}$; the trivial solution we do not seek. Hence, *by definition of the eigenvalue problem* (2.162), $[\mathbf{A} - \lambda_i \mathbf{I}]$ does *not* have linearly independent columns, $r_i \triangleq \operatorname{rank}[\mathbf{A} - \lambda_i \mathbf{I}] < n$, and (2.162) has an infinite number of solutions (but only $n - r_i$ linearly independent ones). Stated another way, the null space of $[\mathbf{A} - \lambda_i \mathbf{I}]$ is not empty. Since $[\mathbf{A} - \lambda_i \mathbf{I}]$ is square, Theorem 2.3 guarantees that the test for $r_i < n$ is

$$|\mathbf{A} - \lambda_i \mathbf{I}| = 0. \tag{2.163}$$

Therefore, (2.163) is used to calculate λ_i. The determinant (2.163) always leads to a polynomial in λ_i of degree n, called the *characteristic polynomial*. Hence, there are n values λ_i, $i = 1, \ldots, n$, which satisfy (2.163). In general,

$$|\mathbf{A} - \lambda \mathbf{I}| = (-1)^n (\lambda - \lambda_1)^{m_1} (\lambda - \lambda_2)^{m_2} \ldots (\lambda - \lambda_p)^{m_p} = 0, \tag{2.164}$$

where m_i indicates the multiplicity of root λ_i, and $m_1 + m_2 + \cdots + m_p = n$.

Exercise 2.25

Compute the two eigenvalues of each $\mathbf{A}$:

$$\mathbf{A} = \begin{bmatrix} 1 & 2 \\ 0 & 4 \end{bmatrix}, \begin{bmatrix} 1 & 2 \\ 2 & 4 \end{bmatrix}, \begin{bmatrix} -2 & 3 \\ -1 & 1 \end{bmatrix}, \begin{bmatrix} 0 & 1 \\ -a_0 & -a_1 \end{bmatrix}, \begin{bmatrix} a_{11} & a_{12} \\ a_{21} & a_{22} \end{bmatrix}.$$

Exercise 2.26

Verify that the characteristic equation of any real 2×2 matrix $\mathbf{A}$ is $\lambda^2 - (\operatorname{tr} \mathbf{A})\lambda + |\mathbf{A}| = 0$.

Now, suppose a system with characteristic polynomial (2.164) has $n - r_i = m_i$, $i = 1, \ldots, p$, where $r_i \triangleq \operatorname{rank}[\mathbf{A} - \lambda_i \mathbf{I}]$. Then, the number of linearly independent eigenvectors is equal to n, the total number of eigenvalues. Matrices $\mathbf{A}$ with this property will be called "nondefective."

Definition 2.17
The matrix **A** *is called "nondefective" iff it has a set of n linearly independent eigenvectors.*

Thus, the test for nondefective **A** is $n - r_i = m_i$, $i = 1, \ldots, p$.
If A is nondefective, the matrix of eigenvectors

$$\mathbf{E} = [\mathbf{e}_1, \ldots, \mathbf{e}_n] \tag{2.165}$$

is nonsingular, since $\mathbf{e}_1, \ldots, \mathbf{e}_n$ form a linearly independent set.

EXAMPLE 2.21
Find the modal data for

$$\mathbf{A} = \begin{bmatrix} 1 & 0 \\ 1 & -1 \end{bmatrix}.$$

Solution: Eigenvalues:

$$|\mathbf{A} - \lambda \mathbf{I}| = \begin{bmatrix} 1-\lambda & 0 \\ 1 & -1-\lambda \end{bmatrix} = (1-\lambda)(-1-\lambda) = 0 \Rightarrow \lambda_1 = 1,\ \lambda_2 = -1.$$

Eigenvectors:

$$[\mathbf{A} - \lambda_1 \mathbf{I}]e_1 = \begin{bmatrix} 0 & 0 \\ 1 & -2 \end{bmatrix}\begin{pmatrix} e_{11} \\ e_{12} \end{pmatrix} = \begin{pmatrix} 0 \\ e_{11} - 2e_{12} \end{pmatrix} = \begin{pmatrix} 0 \\ 0 \end{pmatrix},$$

$$e_{11} = 2e_{12} \Rightarrow \mathbf{e}_1 = \begin{pmatrix} 2e_{12} \\ e_{12} \end{pmatrix} \quad \text{for arbitrary } e_{12},$$

$$[\mathbf{A} - \lambda_2 I]\mathbf{e}_1 = \begin{bmatrix} 2 & 0 \\ 1 & 0 \end{bmatrix}\begin{pmatrix} e_{21} \\ e_{22} \end{pmatrix} = \begin{pmatrix} 2e_{21} \\ e_{21} \end{pmatrix} = 0 \Rightarrow \mathbf{e}_2 = \begin{pmatrix} 0 \\ e_{22} \end{pmatrix}$$

for arbitrary $e_{22} \neq 0$.

Thus, the modal matrix is

$$\mathbf{E} = [\mathbf{e}_1 \mathbf{e}_2] = \begin{bmatrix} 2e_{12} & 0 \\ e_{12} & e_{22} \end{bmatrix}.$$

If we choose the arbitrary parameters to have value 1, then

$$\mathbf{E} = \begin{bmatrix} 2 & 0 \\ 1 & 1 \end{bmatrix}.$$

Note that $n - \text{rank}[\mathbf{A} - \lambda_1 \mathbf{I}] = 2 - 1 = 1$ and $n - \text{rank}[\mathbf{A} - \lambda_2 \mathbf{I}] = 2 - 1 = 1$. Hence, there is one eigenvector for each λ_i, $i = 1, 2$, and **A** is nondefective. ■

EXAMPLE 2.22

Determine whether this **A** is nondefective,

$$\mathbf{A} = \begin{bmatrix} 0 & 1 \\ 0 & 0 \end{bmatrix}.$$

Solution: $|\mathbf{A} - \lambda\mathbf{I}| = \lambda^2 = 0 \Rightarrow \lambda_1 = 0,\ \lambda_2 = 0 \Rightarrow m_1 = 2$. Thus, **A** is nondefective if $n - r_1 = m_1 = 2$. But $n - \operatorname{rank}[\mathbf{A} - \lambda_1\mathbf{I}] = 2 - 1 = 1$. Since $n - r_1 = 1 \neq 2 = m_1$, **A** is defective. ■

Many matrix-algebraic concepts are numerically ill-conditioned and should be used only for conceptual analysis rather than practical computations. To illustrate the point, consider the four matrices

$$A_1 = \begin{bmatrix} 1 & 1 \\ 0 & 1 \end{bmatrix},\quad \mathbf{A}_2 = \begin{bmatrix} 1+\varepsilon & 1 \\ 0 & 1 \end{bmatrix},\quad \mathbf{A}_3 = \begin{bmatrix} 1 & 1 \\ 0 & 1+\varepsilon \end{bmatrix},\quad \mathbf{A}_4 = \begin{bmatrix} 1 & 1+\varepsilon \\ 0 & 1 \end{bmatrix}.$$

For ε small enough, $\mathbf{A}_1$ and $\mathbf{A}_4$ are defective, whereas $\mathbf{A}_2$ and $\mathbf{A}_3$ are diagonalizable. However, in simple precision, $10^{-7} \cong 0$, and in double precision, $10^{-14} \cong 0$. Therefore, if $\varepsilon = 10^{-10}$, the matrices $\mathbf{A}_2$ and $\mathbf{A}_3$ are defective in simple precision and nondefective in double precision.

The existence of nondefective **A** is determined by the test $n - r_i \overset{?}{=} m_i$ as shown in Definition 16. But such tests need not be conducted when $m_i = 1$ (the case of distinct eigenvalues) due to the following theorem.

Theorem 2.11

Matrices **A** *with distinct eigenvalues* $(m_i = 1,\ i = 1, \ldots, n)$ *are nondefective. Hence, in this case* $r_i \triangleq \operatorname{rank}(\mathbf{A} - \lambda_i\mathbf{I}) = n - 1,\ i = 1, \ldots, n$.

Proof: By definition of linear independence, (2.16), we must show that

$$\sum_{i=1}^{n} c_i\mathbf{e}_i = 0, \qquad \Rightarrow c_i = 0, \quad \forall i = 1, \ldots, n. \tag{2.166}$$

Multiply (2.166) by $\{[\mathbf{A} - \lambda_1\mathbf{I}][\mathbf{A} - \lambda_2\mathbf{I}]\ldots[\mathbf{A} - \lambda_n\mathbf{I}]\}$, but omit $[\mathbf{A} - \lambda_\alpha\mathbf{I}]$ in the product. Use $[\mathbf{A} - \lambda_i\mathbf{I}]\mathbf{e}_i = 0$ and the identity

$$\left[\mathbf{A} - \lambda_k\mathbf{I} + \lambda_j\mathbf{I} - \lambda_j\mathbf{I}\right]\mathbf{e}_k = 0,$$
$$\Rightarrow \left[\mathbf{A} - \lambda_j\mathbf{I}\right]\mathbf{e}_k = \left(\lambda_k - \lambda_j\right)\mathbf{e}_k$$

to reduce (2.166) to

$$c_\alpha(\lambda_\alpha - \lambda_1)(\lambda_\alpha - \lambda_2)\ldots(\lambda_\alpha - \lambda_n)\mathbf{e}_\alpha = \mathbf{0}. \tag{2.167}$$

Since $\mathbf{e}_\alpha \neq \mathbf{0}$, $(\lambda_\alpha - \lambda_i) \neq 0$ for $\alpha \neq i$, this proves that $c_\alpha = 0$. Repeat for any α. ■

Now, multiply (2.167) one more time by the missing $[A - \lambda_\alpha I]$ to obtain

$$c_\alpha(\lambda_\alpha - \lambda_1)(\lambda_\alpha - \lambda_2)\ldots(\lambda_\alpha - \lambda_n)[\mathbf{A} - \lambda_\alpha \mathbf{I}]\mathbf{e}_\alpha = \mathbf{0},$$

since $[A - \lambda_\alpha I]c_\alpha = 0$. In other words, we have multiplied (2.166) by $\prod_{k=1}^{n}[\mathbf{A} - \lambda_k \mathbf{I}] \triangleq [\mathbf{A} - \lambda_1 \mathbf{I}]\ldots[\mathbf{A} - \lambda_n \mathbf{I}]$, where the linear combination of eigenvectors in (2.166) may now represent *any* vector $\mathbf{x} = \sum_{i=1}^{n} c_i \mathbf{e}_i$. Thus, we have shown that

$$\prod_{k=1}^{n} [\mathbf{A} - \lambda_k \mathbf{I}]\mathbf{x} = \mathbf{0}$$

for *any* $\mathbf{x}$. Hence,

$$\prod_{k=1}^{n} [\mathbf{A} - \lambda_k \mathbf{I}] = \mathbf{0}. \tag{2.168}$$

Note from (2.164) that the characteristic polynomial for the distinct root case is

$$|\mathbf{A} - \lambda \mathbf{I}| = (-1)^n \prod_{k=1}^{n} (\lambda - \lambda_k) = 0. \tag{2.169}$$

A comparison of (2.168) and (2.169) proves the following theorem for matrices $\mathbf{A}$ with distinct eigenvalues.

Theorem 2.12 (The Cayley–Hamilton Theorem)
The square matrix $\mathbf{A}$ *satisfies its own characteristic equation.*

This theorem is true even if the eigenvalues are not distinct. For the more general result, Theorem 2.12 states that if

$$|\mathbf{A} - \lambda \mathbf{I}| = \prod_{k=1}^{p} (\lambda - \lambda_k)^{m_k} = 0,$$

then

$$\prod_{k=1}^{p} (A - \lambda_k I)^{m_k} = 0. \tag{2.170}$$

Equivalently, if the characteristic polynomial is written in the form

$$|\mathbf{A} - \lambda \mathbf{I}| = \lambda^n + a_{n-1}\lambda^{n-1} + \cdots a_1\lambda + a_0,$$

then

$$\mathbf{A}^n + a_{n-1}\mathbf{A}^{n-1} + \cdots + a_1\mathbf{A} + a_0\mathbf{I} = 0.$$

For a proof, see refs. [2.6] and [2.7]. The important use of the Cayley–Hamilton theorem is that $\mathbf{A}^m$ for any $m \geq n$ is expressible as a linear combination of $\mathbf{I}, \mathbf{A}, \ldots, \mathbf{A}^{n-1}$.

Now, suppose $n - r_i < m_i$ for some i. Then, the eigenvectors are less than n in number and cannot span the complete n-dimensional space. Yet it is the sole purpose of modal analysis to use eigenvectors as a *basis* for this space. To patch up this deficiency of the eigenvectors to span the space, certain *additional vectors* are added to make the modal matrix $\mathbf{E}$ nonsingular, in which event the columns of E *can* be used as a basis for the n-dimensional space. From the above discussion it should be clear that the *number* of additional vectors needed to make $\mathbf{E}$ nonsingular is $\sum_{i=1}^{p}(m_i - (n - r_i))$. We shall call these additional vectors *generalized eigenvectors*, and the number of generalized eigenvectors needed for each λ_i is $m_i - (n - r_i)$. These vectors may be generated as follows. Find $n - r_i$ eigenvectors $\mathbf{e}_i$ satisfying

$$\mathbf{A}\mathbf{e}_i = \lambda_i \mathbf{e}_i \tag{2.171}$$

and find $m_i - (n - r_i)$ generalized eigenvectors $\mathbf{e}_{i+1}$ satisfying the recursive equation

$$\mathbf{A}\mathbf{e}_{i+1} = \lambda_i \mathbf{e}_{i+1} + \mathbf{e}_i. \tag{2.172}$$

In this way a complete set of n linearly independent vectors can be generated and placed in the modal matrix $\mathbf{E}$ as in (2.165). Now, the construction of the matrix form of equations (2.171) and (2.172) yields

$$\mathbf{A}\mathbf{E} = \mathbf{E}\Lambda, \tag{2.173}$$

where Λ is the so-called "Jordan form" of $\mathbf{A}$,

$$\Lambda = \begin{bmatrix} \Lambda_1 & & & \\ & \Lambda_2 & & \\ & & \ddots & \\ & & & \Lambda_p \end{bmatrix}, \tag{2.174}$$

and Λ_i is the ith "Jordan block,"

$$\Lambda_i = \begin{bmatrix} \lambda_i & 1 & & & & \\ & \lambda_i & \ddots & & & \\ & & \ddots & 1 & & \\ & & & \lambda_i & \ddots & \\ & & & & \ddots & 1 \\ & & & & & \lambda_i \end{bmatrix}, \qquad \Lambda_i \in \mathscr{C}^{(m_i-(n-r_i)+1)\times(m_i-(n-r_i)+1)}. \tag{2.175}$$

The 1's above the diagonal appear at each location of generalized eigenvectors within **E** as follows:

$$\Lambda = \left[\begin{array}{ccc|cc|c} 2 & 1 & 0 & & & \\ 0 & 2 & 1 & & 0 & 0 \\ 0 & 0 & 2 & & & \\ \hline & 0 & & 2 & 1 & 0 \\ & & & 0 & 2 & \\ \hline & 0 & & & 0 & 3 \end{array}\right],$$

where the **E** corresponding to this Λ is

$$\mathbf{E} = [\,\mathbf{e}_1\mathbf{g}_2\mathbf{g}_3 \mid \mathbf{e}_4\mathbf{g}_5 \mid \mathbf{e}_6\,],$$

where the generlized eigenvectors are written as **g**'s. Note the location of the **g**'s, and note that each Jordan block has one eigenvector (the **e**'s).

EXAMPLE 2.23
Find the Jordan form for the matrix

$$\mathbf{A} = \left[\begin{array}{ccc|cc|c} 2 & 3 & 0 & & & \\ 0 & 2 & 4 & & 0 & 0 \\ 0 & 0 & 2 & & & \\ \hline & 0 & & 2 & -1 & 0 \\ & & & 0 & 2 & \\ \hline & 0 & & & 0 & 3 \end{array}\right].$$

Solution: The eigenvalues are

$$|\mathbf{A} - \lambda\mathbf{I}| = (\lambda - 2)^5(\lambda - 3) \Rightarrow \lambda_1 = 2, \qquad m_1 = 5$$

$$\lambda_6 = 3, \qquad m_6 = 1,$$

rank $[\mathbf{A} - \lambda_1\mathbf{I}] = 4 \Rightarrow n - r_1 = 6 - 4 = 2$ eigenvectors for λ_1, $m_1 - (n - \lambda_1) = 5 - 2 = 3$ generalized eigenvectors for λ_1. Since $m_6 = 1$, we have $n - r_6 = 6 -$ rank $[\mathbf{A} - \lambda_6\mathbf{I}] = 1$ eigenvector for λ_6. Since there is only one eigenvector for each Jordan block, there must be two Jordan blocks associated with λ_1 (since $n - r_1 = 2$). From (2.171) and (2.172), compute

$$\mathbf{Ae}_1 = \lambda_1\mathbf{e}^1,$$

$$\mathbf{Ag}_2 = \lambda_1\mathbf{g}_2 + \mathbf{e}_1,$$

$$\mathbf{Ag}_3 = \lambda_1\mathbf{g}_3 + \mathbf{g}_2$$

to get

$$\mathbf{e}_1 = \begin{pmatrix} 1 \\ 0 \\ 0 \\ 0 \\ 0 \\ 0 \end{pmatrix}, \quad \mathbf{g}_2 = \begin{pmatrix} 1 \\ \frac{1}{3} \\ 0 \\ 0 \\ 0 \\ 0 \end{pmatrix}, \quad \mathbf{g}_3 = \begin{pmatrix} 1 \\ \frac{1}{3} \\ \frac{1}{12} \\ 0 \\ 0 \\ 0 \end{pmatrix}.$$

An attempt to solve $\mathbf{A}\mathbf{g}_4 = \lambda_1\mathbf{g}_4 + \mathbf{g}_3$ yields no solution because there are only two generalized eigenvectors associated with the first Jordan block. The second Jordan block is generated from

$$\begin{aligned} \mathbf{A}\mathbf{e}_4 &= \lambda_1\mathbf{e}_4, \\ \mathbf{A}\mathbf{g}_5 &= \lambda_1\mathbf{g}_5 + \mathbf{e}_4, \end{aligned} \tag{2.176}$$

yielding

$$\mathbf{e}_4 = \begin{pmatrix} 0 \\ 0 \\ 0 \\ 1 \\ 0 \\ 0 \end{pmatrix}, \quad \mathbf{g}_5 = \begin{pmatrix} 0 \\ 0 \\ 0 \\ 1 \\ -1 \\ 0 \end{pmatrix}.$$

Note that there are two linearly independent vectors satisfying (2.176) because $n - r_1 = 2$. These vectors have been labeled e_1 and e_4. Finally,

$$\mathbf{A}\mathbf{e}_6 = \lambda_6\mathbf{e}_6$$

yields

$$\mathbf{e}_6 = \begin{pmatrix} 0 \\ 0 \\ 0 \\ 0 \\ 0 \\ 1 \end{pmatrix}.$$

By arranging the modal matrix as

$$\mathbf{E} = [\mathbf{e}_1 \quad \mathbf{g}_2 \quad \mathbf{g}_3 \quad \mathbf{e}_4 \quad \mathbf{g}_5 \quad \mathbf{e}_6], \tag{2.177}$$

it is clear where the 1's belong on the superdiagonal in the Jordan form

$$\Lambda = \left[\begin{array}{ccc|cc|c} 2 & 1 & 0 & & & \\ 0 & 2 & 1 & & & \\ 0 & 0 & 2 & & & \\ \hline & & & 2 & 1 & \\ & & & 0 & 2 & \\ \hline & & & & & 3 \end{array}\right]. \tag{2.178}$$

■

Exercise 2.27

Verify that the above modal matrix calculations are correct by computing Λ from (2.173). That is, compute

$$\Lambda = \mathbf{E}^{-1}\mathbf{A}\mathbf{E} \tag{2.179}$$

to check the result (2.178).

It has now been established that a modal matrix $\mathbf{E}$ can always be found satisfying (2.173), where Λ is the Jordan form given by (2.174) and (2.175). Recall also that Λ is *diagonal* if $\mathbf{A}$ is nondefective. Now, the purpose of all this formality is to derive certain results from (2.173). First, note that since $\mathbf{E}$ is always nonsingular, (2.179) holds in general. Now, multiply (2.173) from the right by $\mathbf{E}^{-1}$ to get

$$\mathbf{A} = \mathbf{E}\Lambda\mathbf{E}^{-1}, \tag{2.180}$$

which is called the "spectral decomposition" of the matrix $\mathbf{A}$. This is an expansion of $\mathbf{A}$ in terms of its modal data, the left and right eigenvectors and eigenvalues.

Since $\mathbf{E}$ is nonsingular, its columns span the n-dimensional space. Thus, the eigenvectors can be used as a basis for this space. Note also that $\mathbf{E}^{-1}$ has linearly independent rows and columns. We shall often find occassion to use the rows of $\mathbf{E}^{-1}$ as a basis for the space. Some properties of these rows are explained below. Denote l_i^* as the ith row of $\mathbf{E}^{-1}$,

$$\mathbf{E}^{-1} = \begin{bmatrix} \mathbf{l}_i^* \\ \vdots \\ \mathbf{l}_n^* \end{bmatrix}. \tag{2.181}$$

If we call the ith column of $\mathbf{E}$ a *base vector*, then $\mathbf{l}_i$ is called a *reciprocal base vector*. Multiply (2.173) from the right and left by $\mathbf{E}^{-1}$ to obtain

$$\mathbf{E}^{-1}\mathbf{A} = \Lambda\mathbf{E}^{-1}. \tag{2.182}$$

Suppose Λ is diagonal. Then, the ith row of equation (2.182) is simply

$$\mathbf{l}_i^* \mathbf{A} = \lambda_i \mathbf{l}_i^*, \tag{2.183}$$

which is identical to (2.160), the defining equation for a left eigenvector of **A**.

Thus, the columns of **E** are (right) eigenvectors, and the rows of $\mathbf{E}^{-1}$ are left eigenvectors. Now, using left and right eigenvectors, expand the spectral decomposition (2.180) for a nondefective **A** matrix:

$$\mathbf{A} = \mathbf{E}\Lambda\mathbf{E}^{-1} = [\mathbf{e}_1, \ldots, \mathbf{e}_n]\begin{bmatrix} \lambda_1 & & \\ & \ddots & \\ & & \lambda_n \end{bmatrix}\begin{bmatrix} \mathbf{l}_i^* \\ \vdots \\ \mathbf{l}_n^* \end{bmatrix} = \sum_{i=1}^{n} \lambda_i \mathbf{e}_i \mathbf{l}_i^*, \tag{2.184}$$

which is the detailed *spectral decomposition* of a nondefective **A**.

Exercise 2.28

Use identity $\operatorname{tr}\mathbf{AB} = \operatorname{tr}\mathbf{BA}$ and the spectral decomposition of **A** in (2.180) to prove that for any square **A**,

$$\operatorname{tr}\mathbf{A} = \sum_{i=1}^{n} \lambda_i. \tag{2.185}$$

Exercise 2.29

Use identity $\operatorname{tr}\mathbf{AB} = \operatorname{tr}\mathbf{BA}$ and (2.180) to prove that

$$|\mathbf{A}| = \prod_{i=1}^{n} \lambda_i. \tag{2.186}$$

Exercise 2.30

Let $\mathbf{E} = [\mathbf{e}_1, \ldots, \mathbf{e}_{2n}]$ be the modal matrix for nondefective $2n \times 2n$ real matrix **A** with no real eigenvalues. The complex eigenvalues and eigenvectors may be written

$$\mathbf{e}_i = \mathbf{a}_i + j\mathbf{b}_i, \qquad i = 1, \ldots, n, \tag{2.187a}$$

$$\lambda_i = \sigma_i + j\omega_i. \tag{2.187b}$$

Define a real $2n \times 2n$ matrix

$$\mathscr{E} = [\mathbf{a}_1, \mathbf{b}_1, \mathbf{a}_2, \mathbf{b}_2, \mathbf{b}_3, \ldots, \mathbf{a}_n, \mathbf{b}_n]. \tag{2.188a}$$

Verify that **A** and $\mathscr{E}$ satisfy

$$\mathbf{A}\mathscr{E} = \mathscr{E}\mathscr{J}, \tag{2.188b}$$

where

$$\mathcal{J} \triangleq \left[\begin{array}{cc|cc|c} \sigma_1 & \omega_1 & & & \\ -\omega_1 & \sigma_1 & & & \\ \hline & & \sigma_2 & \omega_2 & \\ & & -\omega_2 & \sigma_2 & \\ \hline & & & & \ddots \end{array}\right]. \tag{2.189}$$

Note that $\mathcal{J}$ is a *block diagonal* matrix of 2×2 blocks instead of the *diagonal* Jordan form. $\mathcal{J}$ has the advantage of involving only real numbers, however.

2.8.1.1 Hermitian Matrices

The previous discussion treats the general case when **A** has no specific structure. If **A** has special structure, then this structure can often be used to simplify the above general calculations. We treat here the special structure with the property $\mathbf{A} = \mathbf{A}^*$. Such matrices are called *Hermitian*. If **A** is real, then $\mathbf{A}^* = \mathbf{A}^T$, and $(\mathbf{A} = \mathbf{A}^*) \Rightarrow (\mathbf{A} = \mathbf{A}^T)$. Such real matrices are called *symmetric*. Thus, if a matrix is Hermitian and real, it is simply called symmetric. A matrix with the property $\mathbf{A} = -\mathbf{A}^*$ is called "skew-Hermitian" or "skew-symmetric" if **A** is real. The following four properties of Hermitian matrices are presented. Keep in mind that the results also apply to the important special case when **A** is symmetric.

Theorem 2.13

If $\mathbf{A} = \mathbf{A}^*$, then for all complex vectors $\mathbf{x} \in \mathscr{C}^n$, $\mathbf{x}^*\mathbf{A}\mathbf{x}$ is real.

Proof: The scalar $\mathbf{x}^*\mathbf{A}\mathbf{x}$ is a complex number in general. Therefore, in general the complex number $\mathbf{x}^*\mathbf{A}\mathbf{x}$ will *not* equal its conjugate $\overline{\mathbf{x}^*\mathbf{A}\mathbf{x}}$, unless the number is real. Thus, the theorem is proved if we can show that $\mathbf{x}^*\mathbf{A}\mathbf{x} = \overline{\mathbf{x}^*\mathbf{A}\mathbf{x}}$.

This follows immediately from the given facts $\mathbf{A} = \mathbf{A}^*$, and $\overline{\mathbf{x}^*\mathbf{A}\mathbf{x}} = \mathbf{x}^*\mathbf{A}^*\mathbf{x}$. ■

Theorem 2.14

All eigenvalues of a Hermitian matrix A are real.

Proof: Multiply the eigenvalue equation $\mathbf{A}\mathbf{e}_i = \lambda_i\mathbf{e}_i$ from the left by $\mathbf{e}_i^*$ to get

$$\mathbf{e}_i^*\mathbf{A}\mathbf{e}_i = \lambda_i\mathbf{e}_i^*\mathbf{e}_i \quad \text{or} \quad \lambda_i = \frac{\mathbf{e}_i^*\mathbf{A}\mathbf{e}_i}{\mathbf{e}_i^*\mathbf{e}_i}. \tag{2.190}$$

Theorem 2.13 guarantees that $\mathbf{e}_i^*\mathbf{B}\mathbf{e}_i$ is real for any Hermitian **B**, including the two choices $\mathbf{B} = \mathbf{A}$ and $\mathbf{B} = \mathbf{I}$. Thus, since both $\mathbf{e}_i^*\mathbf{A}\mathbf{e}_i$ and $\mathbf{e}_i^*\mathbf{e}_i$ are real by Theorem 2.13, it follows from (2.190) that λ_i must be real. ■

Theorem 2.15
The eigenvectors associated with the distinct eigenvalues of a Hermitian matrix **A** *are orthogonal.*

Proof: Consider two eigenvectors $\mathbf{e}_i$ and $\mathbf{e}_j$ with $\lambda_i \neq \lambda_j$:

$$\mathbf{A}\mathbf{e}_i = \mathbf{e}_i\lambda_i, \tag{2.191a}$$

$$\mathbf{A}\mathbf{e}_j = \mathbf{e}_j\lambda_j. \tag{2.191b}$$

Multiply (2.191a) from the left by $\mathbf{e}_j^*$,

$$\mathbf{e}_j^*\mathbf{A}\mathbf{e}_i = \lambda_i\mathbf{e}_j^*\mathbf{e}_i, \tag{2.192}$$

and multiply (2.191b) from the left by $\mathbf{e}_i^*$,

$$\mathbf{e}_i^*\mathbf{A}\mathbf{e}_j = \lambda_j\mathbf{e}_i^*\mathbf{e}_j. \tag{2.193}$$

The complex conjugate of (2.193) is

$$\mathbf{e}_j^*\mathbf{A}^*\mathbf{e}_i = \mathbf{e}_j^*\mathbf{A}\mathbf{e}_i = \lambda_j\mathbf{e}_j^*\mathbf{e}_i, \tag{2.194}$$

where the facts $\mathbf{A} = \mathbf{A}^*$ and $\overline{\lambda_j} = \lambda_j$ have been used. From (2.192) and (2.194), it follows that

$$\lambda_i\mathbf{e}_j^*\mathbf{e}_i = \lambda_j\mathbf{e}_j^*\mathbf{e}_i \tag{2.195}$$

for $\lambda_i \neq \lambda_j$. This is impossible unless $\mathbf{e}_j^*\mathbf{e}_i = 0$, which is the orthogonality condition. ■

Since the eigenvectors $\mathbf{e}_i$ and $\mathbf{e}_j$ are orthogonal, and since one can always choose $\mathbf{e}_i$ and $\mathbf{e}_j$ to be unit length, the modal matrix for a Hermitian matrix **A** is *unitary* ($\mathbf{E}^*\mathbf{E} = \mathbf{I}$), if **A** is nondefective. Finally, we state without proof the last property of Hermitian matrices.

Theorem 2.16
Hermitian matrices are nondefective.

On the strength of this theorem and Definition 2.17, it follows that for Hermitian matrices n-rank $[\mathbf{A} - \lambda_i\mathbf{I}] = m_i$. This fact guarantees that for Hermitian matrices the Jordan form will always be diagonal, regardless of the number of repeated eigenvalues. Furthermore, since **E** is unitary ($\mathbf{E}^* = \mathbf{E}^{-1}$) for Hermitian matrices, the spectral decomposition (2.184) becomes

$$\mathbf{A} = \mathbf{E}\mathbf{\Lambda}\mathbf{E}^* = \sum_{i=1}^{n} \lambda_i\mathbf{e}_i\mathbf{e}_i^* \tag{2.196a}$$

and (2.179) becomes

$$\Lambda = \mathbf{E}^*\mathbf{A}\mathbf{E}. \tag{2.196b}$$

It should be noted that *the left and right eigenvectors of a Hermitian matrix are the same*, $\mathbf{l}_i = \mathbf{e}_i$.

Exercise 2.31

Prove that $\mathbf{x}^*\mathbf{A}\mathbf{x}$ is an imaginary number if $\mathbf{A}$ is skew-Hermitian ($\mathbf{A} = -\mathbf{A}^*$). If $\mathbf{A}$ is skew-symmetric, show that $\mathbf{x}^*\mathbf{A}\mathbf{x} = 0$.

Exercise 2.32

(i) Prove that every square matrix $\mathbf{A}$ can be decomposed into Hermitian and skew-Hermitian parts $\mathbf{A} = \mathbf{A}_{\mathrm{H}} + \mathbf{A}_{\mathrm{SH}}$, where

$$\mathbf{A}_{\mathrm{H}} \triangleq \tfrac{1}{2}(\mathbf{A} + \mathbf{A}^*), \tag{2.197}$$

$$\mathbf{A}_{\mathrm{SH}} \triangleq \tfrac{1}{2}(\mathbf{A} - \mathbf{A}^*). \tag{2.198}$$

(ii) Show that for any square matrix $\mathbf{A}$, $\mathbf{x}^*\mathbf{A}\mathbf{x} = \mathbf{x}^*\mathbf{A}_{\mathrm{H}}\mathbf{x} + j\mathbf{x}^*\mathbf{A}_{\mathrm{SH}}\mathbf{x}$.

Exercise 2.33

Show that, for Hermitian $\mathbf{A}$ with eigenvalues λ_i,

$$\mathbf{x}^*\mathbf{A}\mathbf{x} = \sum_{i=1}^{n} \lambda_i z_i^2, \qquad \mathbf{z} \triangleq \mathbf{E}^*\mathbf{x}. \tag{2.199}$$

Use this result to prove parts (a1), (b1), (c1), and (d1) of Theorem 2.3.

EXAMPLE 2.24

Prove the statement: "$\{\|\mathbf{x}\|_{\mathbf{K}}^2 = 0 \Rightarrow x = 0\}$ requires $\mathbf{K}$ to be nonsingular." This will prove (a1) of Theorem 2.3.

Solution: Since $\mathbf{K}$ may be assumed Hermitian, use the spectral decomposition (2.196a) to write

$$\|\mathbf{x}\|_{\mathbf{K}}^2 = \mathbf{x}^*\mathbf{K}\mathbf{x} = \mathbf{x}^*\mathbf{E}\Lambda\mathbf{E}^*\mathbf{x}.$$

Now, let $\mathbf{z} \triangleq \mathbf{E}^*\mathbf{x}$, $\Lambda = \operatorname{diag}\{\ldots\lambda_i\ldots\}$. Then,

$$\|\mathbf{x}\|_{\mathbf{K}}^2 = \|\mathbf{z}\|_{\Lambda}^2 = \mathbf{z}^*\Lambda\mathbf{z} = \sum_{i=1}^{n} \lambda_i z_i^2.$$

The nonsingular transformation matrix $\mathbf{E}^*$ allows an equivalent statement to be considered: "$\{\|\mathbf{z}\|_{\Lambda}^2 = 0 \Rightarrow \mathbf{z} = \mathbf{0}\}$ requires Λ to be nonsingular." For Λ to be nonsingular, it is required that $\lambda_i \neq 0$ for any $i = 1, \ldots, n$. On the contrary, it is easy to see that if there exists $\lambda_i = 0$ for some i, then for this i the value z_i does *not* influence the norm $\|\mathbf{z}\|_\Lambda^2$. Hence, $\|\mathbf{z}\|_\Lambda^2 = 0 \Rightarrow \mathbf{z} = \mathbf{0}$ unless Λ is nonsingular. The relations $\mathbf{z} = \mathbf{E}^*\mathbf{x}$ and $\mathbf{z}\Lambda\mathbf{z} = \mathbf{x}^*\mathbf{K}\mathbf{x}$ establish that $\mathbf{x}^*\mathbf{E}\Lambda\mathbf{E}^*\mathbf{x} = \mathbf{x}^*\mathbf{K}\mathbf{x}$ for all $\mathbf{x}$, and hence $\mathbf{K} = \mathbf{E}\Lambda\mathbf{E}^*$. Thus, $\mathbf{K}$ is nonsingular, since $\mathbf{E}$ and Λ are both nonsingular. ■

Exercise 2.34

On a plane of x_1 and x_2, plot the function $\mathbf{x}^*\mathbf{A}\mathbf{x} = 1$ if

(i) $$\mathbf{A} = \begin{bmatrix} 1 & 0 \\ 0 & 1 \end{bmatrix},$$

(ii) $$\mathbf{A} = \begin{bmatrix} 1 & 0 \\ 0 & 10 \end{bmatrix},$$

(iii) $$\mathbf{A} = \begin{bmatrix} 1 & 2 \\ 2 & 5 \end{bmatrix}.$$

Show in each case that $\mathbf{x}^*\mathbf{A}\mathbf{x} = 1$ is an ellipsoid whose major and minor axes are the eigenvectors of $\mathbf{A}$ [i.e., plot the eigenvectors on the plots in (i), (ii), and (iii)].

Exercise 2.35

Repeat the above plots of $\mathbf{x}^*\mathbf{A}\mathbf{x} = 1$ if (i) $\mathbf{A} = \begin{bmatrix} 1 & 5 \\ 5 & 10 \end{bmatrix}$ and (ii) $\mathbf{A} = \begin{bmatrix} 0 & 5 \\ 5 & 10 \end{bmatrix}$. Sketch the eigenvectors on this plot.

2.8.1.2 Skew-Hermitian Matrices

Skew-Hermitian matrices are those which satisfy $\mathbf{A} = -\mathbf{A}^*$. Note from Exercise 2.32 that any square matrix $\mathbf{A}$ can be decomposed into the sum of a Hermitian and a skew-Hermitian matrix. here, we mention some properties of skew-Hermitian matrices $\mathbf{A}_{SH}$.

Theorem 2.17

The skew-Hermitian matrix $\mathbf{A}$ *has these properties*:

(i) $j\mathbf{A}$ *is Hermitian*;

(ii) $\mathbf{A}$ *is nondefective*;

(iii) *All eigenvalues of* $\mathbf{A}$ *have zero real parts*;

(iv) $\mathbf{x}^*\mathbf{A}\mathbf{x}$ = *imaginary number for any* $\mathbf{x} \in \mathscr{C}^n$;

(v) *Eigenvectors associated with distinct eigenvalues are orthogonal.*

Exercise 2.36

A spacecraft freely spinning about an axis of symmetry can be described by

$$\dot{\omega}_1 + \lambda\omega_2 = 0,$$

$$\dot{\omega}_2 - \lambda\omega_1 = 0,$$

where ω_1 and ω_2 are the angular velocities about two orthogonal axes both orthogonal to the spin axis; λ is the nutation (wobble) frequency. Show that these equations can be put into the form

$$\begin{pmatrix} \dot{\omega}_1 \\ \dot{\omega}_2 \end{pmatrix} = \mathbf{A}\begin{pmatrix} \omega_1 \\ \omega_2 \end{pmatrix},$$

where **A** is skew-symmetric. Compute the eigenvalues and eigenvectors of A. Form the modal matrix **E**. Write the differential equations in terms of the vector

$$\mathbf{x} \triangleq \mathbf{E}\begin{pmatrix} \omega_1 \\ \omega_2 \end{pmatrix}.$$

2.8.1.3 Unitary Matrices

Unitary matrices are square matrices with the property $\mathbf{A}^*\mathbf{A} = \mathbf{I}$ or, equivalently, $\mathbf{A}^* = \mathbf{A}^{-1}$. If **A** is real, the unitary matrix is described by $\mathbf{A}^T\mathbf{A} = \mathbf{I}$ or $\mathbf{A}^T = \mathbf{A}^{-1}$. Matrices with the property $\mathbf{A}^T = \mathbf{A}^{-1}$ are often called *orthogonal* matrices. Thus, a real unitary matrix is an orthogonal matrix. Unitary matrices have the following properties.

Theorem 2.18

Unitary matrices **A** *are defined by the property* $\mathbf{A}^* = \mathbf{A}^{-1}$ *and, in addition, unitary matrices have these properties*:

(i) $\|\mathbf{A}\mathbf{x}\|^2 = \|\mathbf{x}\|^2$.

(ii) *All eigenvalues of* **A** *have modulus* $|\lambda| = 1$.

(iii) *Eigenvectors associated with distinct eigenvalues are orthogonal.*

(iv) **A** *is nondefective.*

(v) *The modal matrix for a unitary matrix is also unitary.*

Exercise 2.37

Prove Theorems 2.17 and 2.18.

2.8.2 SINGULAR VALUE DECOMPOSITION

The computational burdens are great for the computation of modal data, except for Hermitian matrices. The matrix decompositions in this section have only one purpose—to choose decompositions that require modal data computations for matrices which are *always* Hermitian. In this regard this section is a special case of the previous section. In another way it is more general, since we shall *not* restrict A to be square. The main result is as follows.

Theorem 2.19
Let $\mathbf{A} \in \mathscr{C}^{k \times n}$ *be any* $k \times n$ *matrix of rank* r, $\mathbf{U} \in \mathscr{C}^{k \times k}$ *be unitary* ($\mathbf{U}^*\mathbf{U} = \mathbf{I}$), *and let* $\mathbf{V} \in \mathscr{C}^{n \times n}$ *be unitary* ($\mathbf{V}^*\mathbf{V} = \mathbf{I}$). *For any* $\mathbf{A}$ *there exists a unitary* $\mathbf{U}$ *and* $\mathbf{V}$ *such that*

$$\mathbf{A} = \mathbf{U}\mathbf{\Sigma}\mathbf{V}^*, \tag{2.200}$$

where $\mathbf{U}$ *satisfies*

$$\mathbf{A}\mathbf{A}^*\mathbf{U} = \mathbf{U}\mathbf{\Sigma}\mathbf{\Sigma}^* \tag{2.201}$$

and $\mathbf{V}$ *satisfies*

$$\mathbf{A}^*\mathbf{A}\mathbf{V} = \mathbf{V}\mathbf{\Sigma}^*\mathbf{\Sigma}, \tag{2.202}$$

where $\mathbf{\Sigma}$ *has the canonical structure*

$$\mathbf{\Sigma} = \begin{bmatrix} \mathbf{\Sigma}_0 & \mathbf{0} \\ \mathbf{0} & \mathbf{0} \end{bmatrix}, \quad \mathbf{\Sigma}_0 = \operatorname{diag}\{\sigma_1, \ldots, \sigma_r\}, \qquad \sigma_i > 0,\ i = 1, \ldots, r. \tag{2.203}$$

The numbers σ_i, $i = 1, \ldots, r$, *are called the nonzero singular values of* $\mathbf{A}$.

Proof: Assume that the decomposition (2.200) exists. Substituting (2.200) into (2.201) yields

$$\mathbf{U}\mathbf{\Sigma}\mathbf{V}^*\mathbf{V}\mathbf{\Sigma}^*\mathbf{U}^*\mathbf{U} = \mathbf{U}\mathbf{\Sigma}\mathbf{\Sigma}^*, \tag{2.204}$$

which is obviously satisfied if $\mathbf{V}$ and $\mathbf{U}$ are unitary. Substituting (2.200) into (2.202) yields

$$\mathbf{\Sigma}^*\mathbf{U}^*\mathbf{U}\mathbf{\Sigma}\mathbf{V}^*\mathbf{V} = \mathbf{V}\mathbf{\Sigma}^*\mathbf{\Sigma}, \tag{2.205}$$

which is obviously satisfied if $\mathbf{V}$ and $\mathbf{U}$ are unitary. Therefore, we must now prove the existence of unitary solutions $\mathbf{U}$ and $\mathbf{V}$ of (2.201) and (2.202). Since both $\mathbf{\Sigma}\mathbf{\Sigma}^*$ and $\mathbf{\Sigma}^*\mathbf{\Sigma}$ are diagonal matrices, it is clear from (2.201) and (2.202) that $\mathbf{U}$ is the modal matrix for the Hermitian matrix ($\mathbf{A}\mathbf{A}^*$) and $\mathbf{V}$ is the modal matrix for the Hermitian matrix ($\mathbf{A}^*\mathbf{A}$). The Jordan forms of ($\mathbf{A}\mathbf{A}^*$) and ($\mathbf{A}^*\mathbf{A}$) are ($\mathbf{\Sigma}\mathbf{\Sigma}^*$) and ($\mathbf{\Sigma}^*\mathbf{\Sigma}$), respectively. From the previous section recall that the modal matrix for a Hermitian matrix is unitary (more precisely, we should say that it *can be made* unitary by normalizing the eigenvectors to unit length). Hence, $\mathbf{U}$ and $\mathbf{V}$ are the unitary modal matrices of the Hermitian matrices ($\mathbf{A}\mathbf{A}^*$) and ($\mathbf{A}^*\mathbf{A}$), respectively.

Now, to show that $\sigma_i > 0$, let us solve (2.201) for $\mathbf{U}$ and (2.202) for $\mathbf{V}$. The columns of these unitary matrices are unique within a multiplication by the unit complex number $e^{j\theta}$. That is, if $\mathbf{U}$ satisfies (2.201), then so does

$$\mathbf{U}\begin{bmatrix} e^{j\theta_1} & & & \\ & e^{j\theta_2} & & \\ & & \ddots & \\ & & & e^{j\theta_n} \end{bmatrix} = \begin{bmatrix} e^{j\theta_1}\mathbf{u}_1 & e^{j\theta_2}\mathbf{u}_2 & e^{j\theta_3}\mathbf{u}_3 & \cdots \end{bmatrix},$$

$$\mathbf{U} = \begin{bmatrix} \mathbf{u}_1 & \mathbf{u}_2 & \mathbf{u}_3 \end{bmatrix}$$

for any choice of θ. (If $\mathbf{u}_i$ is real, we can choose $\theta = 0$ or 180° to get $\pm\mathbf{u}_i$). Now, for any initial choice of θ's for $\mathbf{U}$ (and similarly α's for $\mathbf{V}$), solve (2.200) for $\boldsymbol{\Sigma} = \mathbf{U}^*\mathbf{A}\mathbf{V}$:

$$\sigma_i = \mathbf{u}_i^*\mathbf{A}\mathbf{v}_i,$$

where our freedom in the choice of signs is indicated by

$$\sigma_i = \left(e^{j\theta_i}\mathbf{u}_i\right)^*\mathbf{A}\left(e^{j\alpha_i}\mathbf{v}_i\right).$$

Now, if our initial choice of $\mathbf{U}$ and $\mathbf{V}$ gives $\sigma_i < 0$, then either change the θ_i of $\mathbf{u}_i$ or the α_i of $\mathbf{v}_i$. This shows that σ_i can always be made positive or zero.

Now, to prove that the decomposition (2.200) exists, see that $\|\mathbf{A}\mathbf{x}\|^2 = \|\mathbf{U}'\mathbf{A}\mathbf{x}\|^2$ for a unitary matrix $\mathbf{U}'$ [see Theorem 2.17, part (i)]. Now, define the notation $\sqrt{\mathbf{C}}$ to mean $\sqrt{\mathbf{C}}\sqrt{\mathbf{C}} = \mathbf{C}$ for a positive semidefinite C. Let $\sqrt{\mathbf{A}^*\mathbf{A}}\sqrt{\mathbf{A}^*\mathbf{A}} = \mathbf{A}^*\mathbf{A}$; then,

$$\mathbf{x}^*\mathbf{A}^*\mathbf{A}\mathbf{x} = \|\mathbf{A}\mathbf{x}\|^2 = \|\mathbf{U}'\mathbf{A}\mathbf{x}\|^2 = \mathbf{x}^*\sqrt{\mathbf{A}^*\mathbf{A}}\,\mathbf{U}'^*\mathbf{U}'\sqrt{\mathbf{A}^*\mathbf{A}}\,\mathbf{x}.$$

This means that there exists a unitary $\mathbf{U}'$ such that

$$\mathbf{A} = \mathbf{U}'\sqrt{\mathbf{A}^*\mathbf{A}}\,.$$

Now, denote the spectral decomposition of $\sqrt{\mathbf{A}^*\mathbf{A}}$ by

$$\sqrt{\mathbf{A}^*\mathbf{A}} = \mathbf{P}\boldsymbol{\Sigma}\mathbf{P}.$$

Then,

$$\mathbf{A} = \mathbf{U}'\mathbf{P}\boldsymbol{\Sigma}\mathbf{P}^* = \mathbf{U}\boldsymbol{\Sigma}\mathbf{V}',$$

where $\mathbf{U} \triangleq \mathbf{U}'\mathbf{P}$, $\mathbf{V} = \mathbf{P}$.

We note that the eigenvalues of $\mathbf{A}\mathbf{A}^*$ and $\mathbf{A}^*\mathbf{A}$ are computed first to obtain the σ_i^2, $i = 1, \ldots, r$,

$$\begin{bmatrix} \sigma_1^2 & & & \mathbf{0} \\ & \ddots & & \\ & & \sigma_r^2 & \\ \mathbf{0} & & & \mathbf{0} \end{bmatrix} = \begin{bmatrix} \boldsymbol{\Sigma}_0\boldsymbol{\Sigma}_0^* & \mathbf{0} \\ \mathbf{0} & \mathbf{0} \end{bmatrix}. \tag{2.206}$$

Then, the σ_i are computed from $\sigma_i = +\sqrt{\sigma_i^2}$. ■

Exercise 2.38

Prove that there are no negative eigenvalues of $(\mathbf{A}^*\mathbf{A})$ or $(\mathbf{A}\mathbf{A}^*)$.

EXAMPLE 2.25

Compute the singular value decomposition of matrices $\mathbf{A}_1$, $\mathbf{A}_2$, and $\mathbf{A}_3$:

$$\mathbf{A}_1 = [1 \quad 1], \quad \mathbf{A}_2 = \begin{bmatrix} 1 \\ 1 + j1 \end{bmatrix}, \quad \mathbf{A}_3 = \begin{bmatrix} 1 & 1 \\ 2 & 0 \end{bmatrix}.$$

Solution: For $\mathbf{A}_1$,

$$\mathbf{A}_1\mathbf{A}_1^* = [1 \quad 1]\begin{bmatrix} 1 \\ 1 \end{bmatrix} = 2.$$

Hence, $U = 1$ and $\sigma_1^2 = 2 = \mathbf{\Sigma\Sigma}^*$,

$$\mathbf{A}_1^*\mathbf{A}_1 = \begin{bmatrix} 1 \\ 1 \end{bmatrix}[1 \quad 1] = \begin{bmatrix} 1 & 1 \\ 1 & 1 \end{bmatrix} \qquad \text{(has eigenvalues 2, 0)},$$

$$\begin{bmatrix} 1 & 1 \\ 1 & 1 \end{bmatrix}[\mathbf{v}_1 \quad \mathbf{v}_2] = [\mathbf{v}_1 \quad \mathbf{v}_2]\begin{bmatrix} 2 & 0 \\ 0 & 0 \end{bmatrix} \Rightarrow [\mathbf{v}_1 \quad \mathbf{v}_2] = \begin{bmatrix} 1 & 1 \\ 1 & -1 \end{bmatrix}.$$

After normalizing the columns,

$$\mathbf{V} = \frac{1}{\sqrt{2}}\begin{bmatrix} 1 & 1 \\ 1 & -1 \end{bmatrix}, \qquad \mathbf{\Sigma}^*\mathbf{\Sigma} = \begin{bmatrix} 2 & 0 \\ 0 & 0 \end{bmatrix}.$$

From the expressions of both $\mathbf{\Sigma\Sigma}^*$ and $\mathbf{\Sigma}^*\mathbf{\Sigma}$, we conclude that $\mathbf{\Sigma} = [\sqrt{2}, 0]$ and $\mathbf{\Sigma}_0 = \sqrt{2}$. The singular value decomposition of $\mathbf{A}_1$ is therefore

$$\mathbf{A}_1 \triangleq [1 \quad 1] = (1)[\sqrt{2}, 0]\begin{bmatrix} 1 & 1 \\ 1 & -1 \end{bmatrix}\frac{1}{\sqrt{2}}.$$

For $\mathbf{A}_2$,

$$\mathbf{A}_2\mathbf{A}_2^* = \begin{bmatrix} 1 \\ 1 + j1 \end{bmatrix}[1, 1 - j1]$$

$$= \begin{bmatrix} 1 & 1 - j1 \\ 1 + j1 & 2 \end{bmatrix} \Rightarrow \left(\text{has eigenvalues } \sigma_1^2 = 3,\ \sigma_2^2 = 0\right).$$

The computation for $\mathbf{U}$,

$$\begin{bmatrix} 1 & 1 - j1 \\ 1 + j1 & 2 \end{bmatrix}[u_1 \quad u_2] = [u_1 \quad u_2]\begin{bmatrix} 3 & 0 \\ 0 & 0 \end{bmatrix},$$

leads to

$$\mathbf{\Sigma\Sigma}^* = \begin{bmatrix} 3 & 0 \\ 0 & 0 \end{bmatrix}$$

and to the column-normalized unitary modal matrix

$$\mathbf{U} = \begin{bmatrix} 1/\sqrt{3} & (-1 + j1)/\sqrt{3} \\ (1 + j1)/\sqrt{3} & 1/\sqrt{3} \end{bmatrix},$$

$$\mathbf{A}_2^*\mathbf{A}_2 = [1, 1 - j1]\begin{bmatrix} 1 \\ 1 + j1 \end{bmatrix} = 3 = \boldsymbol{\Sigma}^*\boldsymbol{\Sigma}.$$

Hence, $V = 1$, and from the expressions for $\boldsymbol{\Sigma}^*\boldsymbol{\Sigma}$ and $\boldsymbol{\Sigma}\boldsymbol{\Sigma}^*$ we conclude that $\boldsymbol{\Sigma} = \begin{bmatrix} \sqrt{3} \\ 0 \end{bmatrix}$. Thus, the singular value decomposition of $\mathbf{A}_2$ is

$$\mathbf{A}_2 = \begin{bmatrix} 1 \\ 1 + j1 \end{bmatrix}\begin{bmatrix} 1/\sqrt{3} & (-1 + j1)/\sqrt{3} \\ (1 + j1)\sqrt{3} & 1/\sqrt{3} \end{bmatrix}\begin{bmatrix} \sqrt{3} \\ 0 \end{bmatrix}[1].$$

For A_3,

$$\mathbf{A}_3\mathbf{A}_3^* = \begin{bmatrix} 1 & 1 \\ 2 & 0 \end{bmatrix}\begin{bmatrix} 1 & 2 \\ 1 & 0 \end{bmatrix} = \begin{bmatrix} 2 & 2 \\ 2 & 4 \end{bmatrix} \Rightarrow (\text{eigenvalues } 3 + \sqrt{5}, 3 - \sqrt{5})$$

has modal matrix

$$\mathbf{U} = \begin{bmatrix} \dfrac{\sqrt{2}}{\sqrt{5 + \sqrt{5}}} & \dfrac{\sqrt{2}}{\sqrt{5 - \sqrt{5}}} \\ 1/2(1 + \sqrt{5})\dfrac{\sqrt{2}}{\sqrt{5 + \sqrt{5}}} & 1/2(1 - \sqrt{5})\dfrac{\sqrt{2}}{\sqrt{5 - \sqrt{5}}} \end{bmatrix},$$

$$\boldsymbol{\Sigma}\boldsymbol{\Sigma}^* = \begin{bmatrix} 3 + \sqrt{5} & 0 \\ 0 & 3 - \sqrt{5} \end{bmatrix}.$$

$$\mathbf{A}_3^* = \mathbf{A}_3 = \begin{bmatrix} 1 & 2 \\ 1 & 0 \end{bmatrix}\begin{bmatrix} 1 & 1 \\ 2 & 0 \end{bmatrix} = \begin{bmatrix} 5 & 1 \\ 1 & 1 \end{bmatrix}, \qquad \boldsymbol{\Sigma}^*\boldsymbol{\Sigma} = \begin{bmatrix} 3 + \sqrt{5} & 0 \\ 0 & 3 - \sqrt{5} \end{bmatrix},$$

has modal matrix

$$\mathbf{V} = \begin{bmatrix} [2(5 - 2\sqrt{5})]^{-1/2} & [2(5 + 2\sqrt{5})]^{-1/2} \\ (-2 + \sqrt{5})[2(5 - 2\sqrt{5})]^{-1/2} & (-2 - \sqrt{5})[2(5 + 2\sqrt{5})]^{-1/2} \end{bmatrix}.$$

Hence, the singular value decomposition

$$\mathbf{A}_3 = \begin{bmatrix} 1 & 1 \\ 2 & 0 \end{bmatrix} = \mathbf{U\Sigma V}^*,$$

where **U** and **V** are given above and

$$\boldsymbol{\Sigma} = \begin{bmatrix} \sqrt{3+\sqrt{5}} & 0 \\ 0 & \sqrt{3-\sqrt{5}} \end{bmatrix}. \qquad \blacksquare$$

Exercise 2.39

Show that the unitary matrix

$$\mathbf{U}' = \begin{bmatrix} (1-j1)/\sqrt{6} & (-1+j1)/\sqrt{3} \\ \sqrt{2}/\sqrt{3} & 1/\sqrt{3} \end{bmatrix} = \mathbf{U}\begin{bmatrix} e^{j\theta_1} & 0 \\ 0 & e^{j\theta_2} \end{bmatrix}$$

will also diagonalize $\mathbf{A}_2\mathbf{A}_2^*$ and that this choice of $\mathbf{U}^*$ corresponds to $\theta_1 = 7\pi/4$ rad, $\theta_2 = 0$.

One important use of singular value decomposition is in the solution of the linear set of equations

$$\mathbf{Ax} = \mathbf{y}, \quad r = \operatorname{rank}\mathbf{A}, \qquad \mathbf{x} \in \mathscr{C}^n, \quad \mathbf{y} \in \mathscr{C}^k. \tag{2.207}$$

Substituting the singular value decomposition of A yields

$$\mathbf{U\Sigma V}^*\mathbf{x} = \mathbf{y}, \quad \mathbf{U} = [\mathbf{U}_1, \mathbf{U}_2], \quad \boldsymbol{\Sigma} = \begin{bmatrix} \boldsymbol{\Sigma}_0 & \mathbf{0} \\ \mathbf{0} & \mathbf{0} \end{bmatrix}, \quad \mathbf{V} = [\mathbf{V}_1, \mathbf{V}_2]. \tag{2.208}$$

Premultiply by $\mathbf{U}^*$ and define new variables $\mathbf{z} \triangleq \mathbf{V}^*\mathbf{x}$, $\mathbf{w} \triangleq \mathbf{U}^*\mathbf{y}$ to write

$$\boldsymbol{\Sigma}\mathbf{V}^*\mathbf{x} = \mathbf{U}^*\mathbf{y} \quad \text{or} \quad \begin{bmatrix} \boldsymbol{\Sigma}_0 & 0 \\ 0 & 0 \end{bmatrix}\begin{pmatrix} \mathbf{z}_1 \\ \mathbf{z}_2 \end{pmatrix} = \begin{pmatrix} \mathbf{w}_1 \\ \mathbf{w}_2 \end{pmatrix}, \tag{2.209}$$

where $\boldsymbol{\Sigma}_0$ is a diagonal $r \times r$ matrix, $r = \operatorname{rank}\mathbf{A}$.

Solutions are easily written for

$$\begin{aligned} (\boldsymbol{\Sigma}_0)\mathbf{z}_1 &= \mathbf{w}_1 \\ (\mathbf{0})\mathbf{z}_2 &= \mathbf{w}_2 \end{aligned} \;\Rightarrow\; \begin{aligned} \mathbf{z}_1 &= \boldsymbol{\Sigma}_0^{-1}\mathbf{w}_1, \\ \mathbf{z}_2 &= \text{arbitrary}, \end{aligned} \tag{2.210}$$

where $\mathbf{w}_2 = \mathbf{U}_2^*\mathbf{y} = \mathbf{0}$. To recover the solution in terms of the original variables, write $\mathbf{x} = \mathbf{Vz}$. Writing this solution out in matrix form gives

$$\mathbf{x} = [\mathbf{V}_1 \quad \mathbf{V}_2]\begin{bmatrix} \boldsymbol{\Sigma}_0^{-1} & \mathbf{0} \\ \mathbf{0} & \mathbf{0} \end{bmatrix}\begin{bmatrix} \mathbf{V}_1^* \\ \mathbf{V}_2^* \end{bmatrix}\mathbf{y} = \mathbf{V}_1\boldsymbol{\Sigma}_0^{-1}\mathbf{U}_1^*\mathbf{y}. \tag{2.211}$$

The solution (2.211) is now shown to be a solution of problem I′ in Section 2.7.2. The solution to this problem cited previously is (2.109) and (2.122), but the assumption rank **A** = n was employed in this earlier result. No assumption is made about the size of $r \triangleq \operatorname{rank} \mathbf{A}$ below.

Theorem 2.20
The result (2.211) *solves problem I′*:

$$\min_{\mathbf{x}} \|\mathbf{A}\mathbf{x} - \mathbf{y}\|^2.$$

Proof: Recall in Section 2.7.4 that the solution to problem I′ requires the error $\mathbf{y} - \mathbf{A}\mathbf{x}$ to lie in the null space of $\mathbf{A}^*$; see (2.122). Our proof will be completed by showing this property of (2.211):

$$\mathbf{A}^*(\mathbf{y} - \mathbf{A}\mathbf{x}) = \mathbf{0}. \tag{2.212}$$

But since from the singular value decomposition (SVD) of **A**, (2.207) and (2.208),

$$\mathbf{A} = \mathbf{U}_1\mathbf{\Sigma}_0\mathbf{V}_1^*, \tag{2.213}$$

(2.212) reduces to

$$\mathbf{V}_1\mathbf{\Sigma}_0^*\mathbf{U}_1^*\left(\mathbf{y} - \mathbf{U}_1\mathbf{\Sigma}_0\mathbf{V}_1^*\mathbf{V}_1\mathbf{\Sigma}_0^{-1}\mathbf{U}_1^*\mathbf{y}\right) = \mathbf{0},$$

or

$$[\mathbf{V}_1\mathbf{\Sigma}_0^*\mathbf{U}_1^* - \mathbf{V}_1\mathbf{\Sigma}_0^*\mathbf{U}_1^*]\mathbf{y} = \mathbf{0}.$$

This completes the proof. ■

Now, consider the problem of Section 2.7.3.

Theorem 2.21
If rank [**A**] = rank [**A**, **y**], *the result* (2.211) *solves*: $\min_{\mathbf{x}} \|\mathbf{x}\|^2$ *subject to* $\mathbf{A}\mathbf{x} = \mathbf{y}$.

Proof: The necessary and sufficient condition for the existence of a solution to $\mathbf{A}\mathbf{x} = \mathbf{y}$ is that **y** lies in the column space of **A**. This is tested by

$$r \triangleq \operatorname{rank}[\mathbf{A}] = \operatorname{rank}[\mathbf{A}, \mathbf{y}]. \tag{2.214}$$

We shall assume that this is satisfied. The theorem will be proved by showing that **x** lies in the row space of **A**, as required in Section 2.7.4; see Fig. 2.11. We must show from the Definition B.1 that

$$\mathbf{A}^*\boldsymbol{\beta} = \mathbf{x} \qquad \text{for some } \boldsymbol{\beta}. \tag{2.215}$$

Using (2.213), (2.215) becomes

$$\mathbf{V}_1\boldsymbol{\Sigma}_0\mathbf{U}_1^*\boldsymbol{\beta} = \mathbf{A}^*\boldsymbol{\beta} = \mathbf{x} = \mathbf{V}_1\boldsymbol{\Sigma}_0^{-1}\mathbf{U}_1^*\mathbf{y},$$

where the right-hand side comes from (2.211). Premultiply by $\boldsymbol{\Sigma}_0^{-1}\mathbf{V}_1^*$ to get

$$\mathbf{U}_1^*\boldsymbol{\beta} = \boldsymbol{\Sigma}_0^{-2}\mathbf{U}_1^*\mathbf{y},$$

which has a solution

$$\boldsymbol{\beta} = U_1\boldsymbol{\Sigma}_0^{-2}\mathbf{U}_1^*\mathbf{y}.$$

Hence, $\boldsymbol{\beta}$ in (2.215) exists as required to complete the proof. ■

See refs. [2.13] and [2.14] for examples of further uses of SVD in the study of linear dynamic systems.

Exercise 2.40

Repeat Exercise 2.25 using singular value decomposition.

Exercise 12.41

Show that the norm of the $k \times n$ matrix $\mathbf{A}$ defined by $\|\mathbf{A}\| = \operatorname{tr} \mathbf{A}^*\mathbf{A}]^{1/2}$ is the root mean square of the singular values of $\mathbf{A}$.

Later chapters will find frequent use of the following.

Theorem 2.22

$$|\operatorname{tr} \mathbf{AB}| \leq \|\mathbf{A}\|\,\|\mathbf{B}\|, \qquad \|\mathbf{A}\| \triangleq (\operatorname{tr} \mathbf{A}^*\mathbf{A})^{1/2} \tag{2.216}$$

Proof: Define $\mathbf{a}_i$ as the ith column of $\mathbf{A}$ and b_i as the ith column of $\mathbf{B}^T$. Then,

$$\operatorname{tr} \mathbf{AB} = \operatorname{tr}[\mathbf{a}_1 \ldots \mathbf{a}_n]\begin{bmatrix} \mathbf{b}_1^T \\ \vdots \\ \mathbf{b}_n^T \end{bmatrix} = \operatorname{tr} \sum_{i=1}^{n} \mathbf{a}_i\mathbf{b}_i^T = \sum_{i=1}^{n} \mathbf{b}_i^T\mathbf{a}_i.$$

But from the Cauchy–Schwarz inequality (2.27),

$$|\mathbf{b}_i^T\mathbf{a}_i| \leq \|\mathbf{a}_i\|\,\|\mathbf{b}_i\|.$$

Therefore,

$$|\operatorname{tr} \mathbf{AB}| = \left|\sum_{i=1}^{n} \mathbf{b}_i^T\mathbf{a}_i\right| \leq \sum_{i=1}^{n} |\mathbf{b}_i^T\mathbf{a}_i| \leq \sum_{i=1}^{n} \|\mathbf{a}_i\|\,\|\mathbf{b}_i\|. \tag{2.217}$$

Now, recall that

$$
\begin{aligned}
\|\mathbf{A}\|^2\|\mathbf{B}\|^2 = (\operatorname{tr}\mathbf{A}\mathbf{A}^T)(\operatorname{tr}\mathbf{B}^T\mathbf{B}) &= \left(\operatorname{tr}\sum_{i=1}^{n}\mathbf{a}_i\mathbf{a}_i^T\right)\left(\operatorname{tr}\sum_{i=1}^{n}\mathbf{b}_i\mathbf{b}_i^T\right)\\
&= \left(\sum_{i=1}^{n}\|\mathbf{a}_i\|^2\right)\left(\sum_{i=1}^{n}\|\mathbf{b}_i\|^2\right)\\
&= \sum_{i=1}^{n}\|\mathbf{a}_i\|^2\|\mathbf{b}_i\|^2 + \sum_{i\neq j}\|\mathbf{a}_i\|^2\|\mathbf{b}_j\|^2 \geq \sum_{i=1}^{n}\|\mathbf{a}_i\|^2\|\mathbf{b}_i\|^2. \qquad (2.218)
\end{aligned}
$$

From (2.217) and (2.218), it follows that (2.216) holds. ■

Exercise 2.42

Show that the equality in (2.216) holds when $\mathbf{B} = \mathbf{A}^*$.

2.8.3 SQUARE ROOT DECOMPOSITION

In this section we consider the decomposition of a positive semidefinite Hermitian matrix $\mathbf{A}$. The matrix $\sqrt{\mathbf{A}}$ is called the "square root" of $\mathbf{A}$ if

$$\mathbf{A} = \sqrt{\mathbf{A}}\sqrt{\mathbf{A}}. \qquad (2.219)$$

The matrix $\mathbf{C}$ is called a "factor" of $\mathbf{A}$ if

$$A = CC^*. \qquad (2.220)$$

Note that $\sqrt{\mathbf{A}}$ must be square, whereas $\mathbf{C}$ might not be square. That is, if $r = \operatorname{rank}\mathbf{A}$ and $\mathbf{A} \in \mathscr{C}^{k\times n}$, then $\mathbf{C}$ must have rank r but its dimension can be $n \times \alpha$ for *any* $\alpha \geq r$. By a slight abuse of language, we shall refer to $\mathbf{C}$ as the "square root" of A. There are several differences between this matrix intepretation of the square root idea and the standard square root of a scalar. For example, with a scalar the square root of 16 has only two possible solutions, $+4$ or -4, whereas the square root of the matrix

$$\begin{bmatrix} 16 & 0 \\ 0 & 0 \end{bmatrix}$$

according to the definition (2.220) has an infinite number of solutions. We mention five: $\mathbf{C}_1$, $\mathbf{C}_2$, $\mathbf{C}_3$, $\mathbf{C}_4$, and $\mathbf{C}_5$:

$$\mathbf{C}_1 = \begin{bmatrix} 4 \\ 0 \end{bmatrix}, \quad \mathbf{C}_2 = \begin{bmatrix} -4 \\ 0 \end{bmatrix}, \quad \mathbf{C}_3 = \begin{bmatrix} 4 & 0 \\ 0 & 0 \end{bmatrix},$$

$$\mathbf{C}_4 = \begin{bmatrix} -4 & 0 \\ 0 & 0 \end{bmatrix}, \quad \mathbf{C}_5 = \begin{bmatrix} 4 & 0 & 0 & 0 \\ 0 & 0 & 0 & 0 \end{bmatrix}.$$

That is, the column dimension of $\mathbf{C}$ is not unique, and the entries within $\mathbf{C}$ are not unique. Obviously, some conventions are needed to make the most effective use of square root ideas for matrices. The first convention we shall use is to restrict **C** to be maximally ranked. That is, the rank of **C** should be equal to its column dimension. This convention makes the dimension of the square root matrix unique, and would eliminate $\mathbf{C}_3$, $\mathbf{C}_4$, and $\mathbf{C}_5$ in the above example, leaving $\mathbf{C}_1$ and $\mathbf{C}_2$ as the only admissible solutions.

Note from (2.157) and (2.158) that **C** could be constructed quite simply from **U** and **Σ** if **V** = **U**. Furthermore, this condition (**V** = **U**) is guaranteed for Hermitian matrices. Take the square root of the positive diagonal entries of the matrix

$$\mathbf{\Sigma} = \begin{bmatrix} \mathbf{\Sigma}_0 & \mathbf{0} \\ \mathbf{0} & \mathbf{0} \end{bmatrix}$$

and form this new square root matrix

$$\sqrt{\mathbf{\Sigma}} \triangleq \begin{bmatrix} \sqrt{\mathbf{\Sigma}_0} \\ \mathbf{0} \end{bmatrix}, \qquad \sqrt{\mathbf{\Sigma}_0} \triangleq \operatorname{diag}\left\{\sqrt{\sigma_1}, \ldots, \sqrt{\sigma_r}\right\},$$

and define

$$\mathbf{C} \triangleq \mathbf{U}\sqrt{\mathbf{\Sigma}}.$$

(Note that $\mathbf{\Sigma}_0 = \overline{\mathbf{\Sigma}}_0$ since $\mathbf{\Sigma}_0$ contains the eigenvalues of a Hermitian matrix.) Hence,

$$\mathbf{A} = \mathbf{C}\mathbf{C}^* = \mathbf{U}\sqrt{\mathbf{\Sigma}}\sqrt{\mathbf{\Sigma}^*}\,\mathbf{U}^* = \mathbf{U}\mathbf{\Sigma}\mathbf{U}^*$$

as the equivalent to (2.157) when **V** = **U**.

Now, consider the possibility of computing **C** in (2.158) from **E** and Λ in (2.156). This task requires that Λ be diagonal and $\mathbf{E}^{-1} = \mathbf{E}^*$. These properties are immediate if **A** is Hermitian. Hence, for positive semidefinite Hermitian matrices **A**,

$$\begin{aligned} \mathbf{A} &= \mathbf{C}\mathbf{C}^*, \\ \mathbf{C} &= \mathbf{E}\sqrt{\Lambda}, \end{aligned} \tag{2.221}$$

where $\sqrt{\Lambda}$ is the positive square root of Λ. Since Λ is square (and diagonal), the square root matrix is square with diagonal entries that are the square root of the diagonal entries λ_i in Λ. These square roots do not exist such that

$$\left(\sqrt{\lambda_i}\right)\left(\sqrt{\lambda_i}\right)^* = \lambda_i,$$

unless $\lambda_i \geq 0$. This explains the positive semidefinite requirement of **A** in this section. However, the square root matrix $\sqrt{\Lambda}$ will *not* have maximal rank if there are any zero eigenvalues. This detail is easily resolved by arranging all nonzero eigenvalues into Λ_0 and defining $\sqrt{\Lambda}$ as follows:

$$\sqrt{\Lambda} = \begin{bmatrix} \sqrt{\Lambda_0} \\ \mathbf{0} \end{bmatrix}, \quad \Lambda = \begin{bmatrix} \Lambda_0 & \mathbf{0} \\ \mathbf{0} & \mathbf{0} \end{bmatrix}, \quad \Lambda_0 = \left(\sqrt{\Lambda_0}\right)\left(\sqrt{\Lambda_0}\right), \tag{2.222}$$

where Λ_0 is a diagonal matrix containing all the positive eigenvalues of $\mathbf{A}$. Then, (2.221) satisfies our requirements for square root matrices.

Theorem 2.23
A square root of the positive semidefinite Hermitian matrix $\mathbf{A}$ may be constructed by

$$\mathbf{A} = \mathbf{C}\mathbf{C}^*, \qquad \mathbf{C} = \mathbf{E}\sqrt{\Lambda}, \tag{2.223}$$

where $\mathbf{E}$ and Λ are the modal data for A satisfying $\mathbf{AE} = \mathbf{E}\Lambda$ and

$$\mathbf{E}^*\mathbf{E} = \mathbf{I}, \quad \sqrt{\Lambda} = \begin{bmatrix} \sqrt{\Lambda_0} \\ \mathbf{0} \end{bmatrix}, \quad \Lambda = \begin{bmatrix} \Lambda_0 & \mathbf{0} \\ \mathbf{0} & \mathbf{0} \end{bmatrix}, \tag{2.224}$$

where Λ_0 contains all positive eigenvalues of $\mathbf{A}$.

Matrix square roots can be obtained with much less work than required by Theorem 2.23. The "overkill" in this theorem is the computation of the diagonal matrix Λ. We now show that matrix square roots can be computed from transformations that only *triangularize* a matrix instead of *diagonalizing* it.

Linear algebra [2.8], [2.9] provides us with these facts: a square matrix $\mathbf{A}$ can be decomposed uniquely into a form $\mathbf{A} = \mathbf{LDU}$, where $\mathbf{L}$ is lower triangular with 1's on the diagonal, $\mathbf{U}$ is upper triangular with 1's on the diagonal, and $\mathbf{D}$ is diagonal. These matrices may be found by Gaussian elimination [2.8]. If $\mathbf{A}$ is symmetric, then $\mathbf{U} = \mathbf{L}^T$. By taking the square root of the diagonal entries of $\mathbf{D}$ (here, we require $\mathbf{A} > \mathbf{0}$ so that these entries are positive) and placing them in $\sqrt{\mathbf{D}}$ so that $\mathbf{D} = \sqrt{\mathbf{D}}\sqrt{\mathbf{D}}$, we have

$$\mathbf{A} = (\mathbf{L}\sqrt{\mathbf{D}})(\sqrt{\mathbf{D}}\mathbf{L}^T) = \mathbf{C}\mathbf{C}^T, \qquad \mathbf{C} = \mathbf{L}\sqrt{\mathbf{D}}, \tag{2.225}$$

for a symmetric positive-definite $\mathbf{A}$. This is called the Cholesky decomposition, (2.225) [2.8].

Exercise 2.43
For the matrices

(i)
(ii)

$$\mathbf{A} = \begin{bmatrix} 1 & 0 \\ 0 & 10 \end{bmatrix}, \quad \begin{bmatrix} 1 & 2 \\ 2 & 5 \end{bmatrix},$$

find both the Cholesky decomposition and the square root from (2.223).

References

2.1 T. Kailath, *Linear Systems*, Prentice Hall, Englewood Cliffs, N.J., 1980.

2.2 C. T. Chen, *Introduction to Linear Systems Theory*, Holt, Rinehart and Winston, New York, 1970.

2.3 L. A. Zadeh and C. A. Desoer, *Linear Systems Theory*, McGraw-Hill, New York, 1963.

2.4 P. W. Likins, *Elements of Engineering Mechanics*, McGraw-Hill, New York, 1973.

2.5 L. Meirovitch, *Methods of Analytical Dynamics*, McGraw-Hill, New York, 1970.

2.6 C. Nelson Dorney, *A Vector Space Approach to Models & Optimization*, John Wiley, New York, 1975.

2.7 L. Padulo and M. A. Arbib, *System Theory A Unified State Space Approach and Discrete Systems*, Hemisphere Publishing, Washington, D.C., 1974.

2.8 G. Strang, *Linear Algebra and its Applications*, Academic Press, New York, 1976.

2.9 B. Noble and J. Daniel, *Applied Linear Algebra*, Prentice Hall, Englewood Cliffs, N.J., 1977.

2.10 L. Nachbin, *Holomorphic Functions, Domains of Holomorphy and Local Properties*, North-Holland, Amsterdam, 1972.

2.11 I. S. Sokolnikoff and R. M. Redheffer, *Mathematics of Physics and Modern Engineering*, McGraw-Hill, New York, 1958.

2.12 E. K. Blum, *Numerical Analysis and Computation: Theory and Practice*, Addison-Wesley, Reading, Mass., 1972.

2.13 B. C. Moore, "Singular Value Analysis of Linear Systems," *IEEE Proceedings of the Control and Decision Conference*, Dec. 1978, pp. 66–73.

2.14 B. C. Moore, "Principal Component Analysis in Linear Systems, Controllability, Observability, and Model Reduction," *IEEE Trans. Auto. Control*, AC-26, 1981, pp. 17–32.

CHAPTER 3

Models of Dynamic Systems

The purpose of this chapter is to develop mathematical models for several dynamic systems. These models will be used throughout the book to illustrate control concepts. It is hoped that repeated use of the same examples will further enhance the learning process. By using examples that have *physical* significance and by using the *same* examples to illustrate each new control concept (introduced later in the text), a bridge between mathematics and control engineering is to be built.

For decades, control theory has been treated as a discipline to be applied *after* the model is developed, and the model was developed independently of the control policy. This point of view must eventually be changed since *the mathematical models never describe the physical phenomena exactly*. There are many different models that could be developed for a given physical phenomenon, and appropriate modeling decisions can be made only if the particular *purpose* of the model is kept in mind. If the purpose of the model is to develop a control policy, then the inevitable conclusion is that *the modeling problem and the control problem are not independent*. This leads to a discomforting reality: *An appropriate control policy cannot be developed without prior knowledge of the model, and the best model for the situation cannot be developed without prior knowledge of the control policy*. This "chicken and egg" dilemma occupies much of the current research on control theory and will be motivated and justified in more detail in Chapter 10. For the introductory purposes of Chapters 2 to 9, however, we shall present simple models of physical dynamic systems, and we shall pretend that these models are correct and hence that their accuracy is *not* dependent upon the control policy. In other words, we shall momentarily invoke the traditional "separation" between the modeling and control problems, and we will not correct this attitude until Chapter 10.

3.1 State Space Realizations

A friend of mine down at Purdue
Said, "Here is the ultimate clue:
My analysis shows
That the universe grows
Like $\dot{x} = Ax + Bu$."

P. C. Hughes

The most rapid growth in linear control theory occurred *after* it was generally recognized that all linear systems of differential equations could be represented in a standard form. Powerful theoretical tools could then be developed for the special form. This special form is a vector of first-order equations called the "state" form. The definition of the "state," however, is not restricted to linear systems.

Definition 3.1
The "state" $\mathbf{x} \in \mathscr{C}^n$ of a dynamic system is any set of variables which, if specified at t_0, provide all the necessary information to solve the differential equations for any $t > t_0$, for any specified set of forcing functions.

The linear equation

$$\mathscr{M}\ddot{\mathbf{q}} + \mathscr{D}\dot{\mathbf{q}} + \mathscr{K}\mathbf{q} = \mathscr{B}\mathbf{u}, \qquad \mathbf{q} \in \mathscr{R}^N \tag{3.1}$$

is also a special form. It has a second-order appearance, but we cannot call them "second-order" systems, since $\mathbf{q}$ is a vector. These types of models will be called *vector second-order* (VSO) systems.

Note from (3.1) that the number of variables $(q_1, q_2, \ldots, q_N)$ required to *write* the differential equations is N, whereas the number of pieces of information required to *solve* the differential equations is $2N$ [both $\mathbf{q}(0)$ and $\dot{\mathbf{q}}(0)$ must be specified]. That is, the *order* of the system of equation (3.1) is $2N$ and the order of the system dictates the number of necessary initial conditions required to solve the differential equations (3.1). The variables q_i in (3.1) are called *configuration variables* in dynamics [3.6]. If the system (3.1) is put into the vector first-order form $\dot{\mathbf{x}} = \mathbf{Ax} + \mathbf{Bu}$, then the $2N$ variables $\mathbf{x}^T = (q_1, \dot{q}_1, q_2, \dot{q}_2, \ldots, q_N, \dot{q}_N)$ are called the *state variables*.

It should be clear that the choice of state variables is not unique. If the vector $\mathbf{x}(0)$ provides all the initial conditions needed to solve the differential equations (together with knowledge of the inputs $\mathbf{u}(t)$, $t \geq 0$), then so does the vector $\boldsymbol{x}(0) = \mathbf{Tx}(0)$, for *any* nonsingular matrix $\mathbf{T}$. Therefore, if one state vector $(\mathbf{x})$ is known, then an entire class of vectors which qualify as state vectors is described by $\mathbf{Tx}$, for any nonsingular $\mathbf{T}$.

3.2 Linearization

Consider the nonlinear system

$$\mathcal{M}(\mathbf{q})\ddot{\mathbf{q}} + \mathcal{D}(\mathbf{q}, \dot{\mathbf{q}})\dot{\mathbf{q}} + \mathcal{K}\mathbf{q} = \mathcal{B}\mathbf{u}, \qquad \mathbf{y} = \mathcal{P}\mathbf{q} + \mathcal{R}\dot{\mathbf{q}}. \tag{3.2}$$

The purpose of studying the *solution* of the model (3.2) is to understand the *motion* of the physical system; and when we ignore modeling errors, we assign a one-to-one correspondence to these two concepts. Generally, infinitely many solutions exist, one for each set of initial conditions. Hence, we must focus on the particular solution $\bar{\mathbf{q}}(t)$ of interest. Recall from undergraduate mathematics that a Taylor's series expansion about $\bar{\mathbf{z}}$ of any differentiable scalar function $f(\mathbf{z}, t)$ of the vector $\mathbf{z}$ is

$$f(\mathbf{z}, t) = \mathbf{f}(\bar{\mathbf{z}}, t) + \left(\frac{\partial f}{\partial \mathbf{z}}\right)_{\bar{\mathbf{z}}}^{T}(\mathbf{z} - \bar{\mathbf{z}}) + \frac{1}{2}(\mathbf{z} - \bar{\mathbf{z}})^T\left(\frac{\partial^2 f}{\partial \mathbf{z}^2}\right)_{\bar{\mathbf{z}}}(\mathbf{z} - \bar{\mathbf{z}}) + h, \tag{3.3}$$

where the $\bar{\mathbf{z}}$ subscript denotes evaluation at the values $\mathbf{z} = \bar{\mathbf{z}}$ and where h contains all the higher-order terms (which do not admit to convenient matrix notation). If the function we desire to expand is given by (3.2), then we may use (3.3) by substituting $\mathbf{z}^T \triangleq (\ddot{\mathbf{q}}^T, \dot{\mathbf{q}}^T, \mathbf{q}^T, \mathbf{u}^T)$ and $f(\mathbf{z}, t) \triangleq \mathcal{M}\ddot{\mathbf{q}} + \mathcal{D}\dot{\mathbf{q}} + \mathcal{K}\mathbf{q} - \mathcal{B}\mathbf{u}$. The first two terms in (3.3) constitute a *linearization* of the function $f(\mathbf{z}, t)$ about the trajectory $\bar{\mathbf{z}}(t)$. The third term is included briefly in our discussion due to its later role in least squares theory.

It is much easier to discuss more general situations by using state space realizations. Suppose a set of state variables x has been found for a system. Then, the nonlinear differential equations describing the system can be written

$$\begin{aligned} \dot{\mathscr{x}} &= \mathbf{f}(\mathscr{x}, \mathscr{u}, t), \\ \mathscr{y} &= \mathbf{g}(\mathscr{x}, \mathscr{u}, t). \end{aligned} \tag{3.4a}$$

Let $\bar{\mathscr{x}}(t), \bar{\mathscr{u}}(t)$ be the solution of interest. Hence, by definition of a solution, $\bar{\mathbf{x}}(t), \bar{\mathbf{u}}(t)$ must satisfy

$$\begin{aligned} \dot{\bar{\mathscr{x}}} &= \mathbf{f}(\bar{\mathscr{x}}, \bar{\mathscr{u}}, t), \\ \bar{\mathscr{y}} &= \mathbf{g}(\bar{\mathscr{x}}, \bar{\mathscr{u}}, t). \end{aligned} \tag{3.4b}$$

A Taylor's series expansion of $\mathbf{f}(\mathscr{x}, \mathscr{u}, t)$ about the solution $\bar{\mathscr{x}}(t), \bar{\mathscr{u}}(t)$ is developed as follows. The Taylor's series of the scalar $f(\mathbf{z})$ in (3.5) can be applied to each

$f_i(\boldsymbol{x}, \boldsymbol{u}, t)$ in (3.4) if we let $(\boldsymbol{x}^T, \boldsymbol{u}^T)$ play the role of $\mathbf{z}^T$ in (3.3). Hence, the first three terms in the Taylor's series expansion of (3.4) are

$$\dot{\boldsymbol{x}} = f(\bar{\boldsymbol{x}}, \bar{\boldsymbol{u}}, t) + \left[\left(\frac{\partial f}{\partial \boldsymbol{x}}\right)^T, \left(\frac{\partial f}{\partial \boldsymbol{u}}\right)^T\right]_{\bar{\boldsymbol{x}}, \bar{\boldsymbol{u}}} \begin{pmatrix} \boldsymbol{x} - \bar{\boldsymbol{x}} \\ \boldsymbol{u} - \bar{\boldsymbol{u}} \end{pmatrix}$$

$$+ \frac{1}{2} \begin{bmatrix} \left[(\boldsymbol{x} - \bar{\boldsymbol{x}})^T, (\boldsymbol{u} - \bar{\boldsymbol{u}})^T\right]\mathbf{F}_1 \\ \vdots \\ \left[(\boldsymbol{x} - \bar{\boldsymbol{x}})^T, (\boldsymbol{u} - \bar{\boldsymbol{u}})^T\right]\mathbf{F}_n \end{bmatrix}_{\bar{\boldsymbol{x}}, \bar{\boldsymbol{u}}} \begin{pmatrix} \boldsymbol{x} - \bar{\boldsymbol{x}} \\ \boldsymbol{u} - \bar{\boldsymbol{u}} \end{pmatrix} + \mathbf{h}, \tag{3.5a}$$

where $\mathbf{h}$ represents higher-order terms and the "Hessian" matrices $\mathbf{F}_i$ are defined by

$$\mathbf{F}_i \triangleq \begin{bmatrix} \dfrac{\partial^2 f_i}{\partial \boldsymbol{x}^2} & \dfrac{\partial^2 f_i}{\partial \boldsymbol{x}\, \partial \boldsymbol{u}} \\ \dfrac{\partial^2 f_i}{\partial \boldsymbol{u}\, \partial \boldsymbol{x}} & \dfrac{\partial^2 f_i}{\partial \boldsymbol{u}^2} \end{bmatrix}_{\bar{\boldsymbol{x}}, \bar{\boldsymbol{u}}} \tag{3.5b}$$

Some compactness in notation is achieved by defining

$$\mathbf{A}(t) \triangleq \left(\frac{\partial \mathbf{f}}{\partial \boldsymbol{x}}\right)^T_{\bar{\boldsymbol{x}}, \bar{\boldsymbol{u}}}, \quad \mathbf{B}(t) \triangleq \left(\frac{\partial \mathbf{f}}{\partial \boldsymbol{u}}\right)^T_{\bar{\boldsymbol{x}}, \bar{\boldsymbol{u}}}, \quad \mathbf{x} \triangleq \boldsymbol{x} - \bar{\boldsymbol{x}}, \quad \mathbf{u} \triangleq \boldsymbol{u} - \bar{\boldsymbol{u}},$$

$$\mathbf{C}(t) \triangleq \left(\frac{\partial \mathbf{g}}{\partial \boldsymbol{x}}\right)^T_{\bar{\boldsymbol{x}}, \bar{\boldsymbol{u}}}, \quad \mathbf{H}(t) \triangleq \left(\frac{\partial \mathbf{g}}{\partial \boldsymbol{u}}\right)^T_{\bar{\boldsymbol{x}}, \bar{\boldsymbol{u}}}, \quad \mathbf{y} = \boldsymbol{y} - \bar{\boldsymbol{y}}.$$

Rewriting (3.5) and adding the output equations,

$$\begin{aligned} \dot{\mathbf{x}} &= \mathbf{A}(t)\mathbf{x} + \mathbf{B}(t)\mathbf{u} + \frac{1}{2}\begin{pmatrix} (\mathbf{x}^T, \mathbf{u}^T)\mathbf{F}_1 \\ \vdots \\ (\mathbf{x}^T, \mathbf{u}^T)\mathbf{F}_n \end{pmatrix} \begin{pmatrix} \mathbf{x} \\ \mathbf{u} \end{pmatrix} + \mathbf{h}, \\ \mathbf{y} &= \mathbf{C}(t)\mathbf{x} + \mathbf{H}(t)\mathbf{u} + \frac{1}{2}\begin{pmatrix} (\mathbf{x}^T, \mathbf{u}^T)\mathbf{G}_1 \\ \vdots \\ (\mathbf{x}^T, \mathbf{u}^T)\mathbf{G}_k \end{pmatrix} \begin{pmatrix} \mathbf{x} \\ \mathbf{u} \end{pmatrix} + \mathbf{h}', \end{aligned} \tag{3.6}$$

where $\mathbf{G}_i$ is similar in form to (3.5b) with f_i replaced by g_i. Note that $\mathbf{A}$ and $\mathbf{B}$ in (3.6) are usually time varying if $\bar{\boldsymbol{x}}, \bar{\boldsymbol{u}}$ are time varying, and are constant if $\bar{\boldsymbol{x}}, \bar{\boldsymbol{u}}$ are constant. The easiest way to construct the linearized model is to expand each nonlinear scalar function in a Taylor's series [including all the required terms in (3.5) and (3.6)] and leave alone the terms that are already linear. The brute force application of recipes (3.5) and (3.6) will yield correct answers with a lot of unnecessary derivative calculations.

Now, suppose that the matrices $\left[\dfrac{\partial^2 f_i}{\partial \boldsymbol{x}^2}\right]_{\bar{\boldsymbol{x}}, \bar{\boldsymbol{u}}}$ and $\left[\dfrac{\partial^2 f_i}{\partial \boldsymbol{u}^2}\right]_{\bar{\boldsymbol{x}}, \bar{\boldsymbol{u}}}$ in (3.5b) are zero for $i = 1, \ldots, n$. For this special class of nonlinear systems, the first three terms in the Taylor's series expansion (3.6) give

$$\dot{\mathbf{x}} = \mathbf{A}(t)\mathbf{x} + \mathbf{B}(t)\mathbf{u} + \begin{bmatrix} \mathbf{u}^T\mathbf{N}_1(t) \\ \vdots \\ \mathbf{u}^T\mathbf{N}_n(t) \end{bmatrix}\mathbf{x}, \qquad \mathbf{N}_i \triangleq \left[\frac{\partial^2 f_i}{\partial \boldsymbol{u}\, \partial \boldsymbol{x}}\right]_{\bar{\boldsymbol{x}}, \bar{\boldsymbol{u}}},$$

$$\mathbf{y} = \mathbf{C}(t)\mathbf{x} + \mathbf{H}(t)\mathbf{u} + \begin{bmatrix} \mathbf{u}^T\mathbf{N}_1' \\ \vdots \\ \mathbf{u}^T\mathbf{N}_k' \end{bmatrix}\mathbf{x}, \qquad \mathbf{N}_i' = \left(\frac{\partial^2 g_i}{\partial \boldsymbol{u}\, \partial \boldsymbol{x}}\right)_{\bar{\boldsymbol{x}}, \bar{\boldsymbol{u}}}. \tag{3.7}$$

Such systems are called *bilinear systems* and have been studied extensively in the literature [3.7]–[3.9].

Exercise 3.1

Derive equation (3.7) from (3.6) if $\left(\dfrac{\partial^2 f_i}{\partial \boldsymbol{x}^2}\right)_{\bar{\boldsymbol{x}}, \bar{\boldsymbol{u}}}$ and $\left(\dfrac{\partial^2 f_i}{\partial \boldsymbol{u}^2}\right)_{\bar{\boldsymbol{x}}, \bar{\boldsymbol{u}}}$ are both zero for $i = 1, \ldots, n$.

For that special class of nonlinear systems with the property

$$\left[\frac{\partial^2 f_i}{\partial \boldsymbol{u}^2}\right]_{\bar{\boldsymbol{x}}, \bar{\boldsymbol{u}}} = 0, \qquad i = 1, \ldots, n, \tag{3.8}$$

it is always possible to put the system approximation (3.6) in a bilinear form! Using the higher dimensional vector $x^{(2)}$, which contains all products of the type $x_i x_j$, such that

$$\boldsymbol{h}_i^T\mathbf{x}^{(2)} \triangleq \mathbf{x}^T\mathbf{H}_i\mathbf{x}, \qquad \mathbf{H}_i \triangleq \left[\frac{\partial^2 f_i}{\partial \boldsymbol{x}^2}\right]_{\bar{\boldsymbol{x}}, \bar{\boldsymbol{u}}},$$

where $\boldsymbol{h}_i$ is a vector of the coefficients in $\mathbf{H}_i$, the system (3.6) can be put in a bilinear form in the vector $(\mathbf{x}^T, \mathbf{x}^{(2)T})$. This new state vector has dimension $\geq 3n$, however. By similar increases in the dimension of the state vector, it has been shown [3.8] that a Taylor's series expansion up to *any* specified number of higher-order terms can be put in a bilinear form, provided equation (3.8) holds; the accommodation of higher-order terms in the Taylor series expansion will require correspondingly higher-dimensional state vectors in the resulting bilinear form.

Bilinear systems have the form (3.7), and *linear* dynamic systems of this text have the form

$$\begin{aligned}\dot{\mathbf{x}} &= \mathbf{A}(t)\mathbf{x} + \mathbf{B}(t)\mathbf{u} + \mathbf{D}(t)\mathbf{w},\\ \mathbf{y} &= \mathbf{C}(t)\mathbf{x} + \mathbf{H}(t)u + \mathbf{J}(t)\mathbf{w},\end{aligned} \tag{3.9}$$

where we have added the input **w**, which will be called the *disturbance vector*, while **u** will be called the *control vector*. The distinction is that $\mathbf{u}(t)$ can be manipulated by the designer and $\mathbf{w}(t)$ cannot be directly manipulated but is a function dictated by the natural environmental disturbances of the system.

Some texts begin with a definition of linearity based upon a *superposition* property by stating that

If $y_1(t)$ is the response from $u_1(t)$ and $y_2(t)$ is the response from $u_2(t)$, then the system is linear if for every $u_1(t)$, $u_2(t)$ the response from $u_1(t) + u_2(t)$ is $y_1(t) + y_2(t)$.

However, there are some subtleties in such formal definitions of linear systems that we shall ignore. For example, a system may have nonlinear elements interacting in such a way to have responses which satisfy this definition of a linear system. This superposition definition also has some difficulties in its extension to an infinite number of inputs [3.10]. These subtleties surface in pathological contrivances which do not relate to such physical systems as described in Chapter 3. We shall soon deal with solutions of equations (3.9), and then we shall show that (3.9) does indeed satisfy the superposition property. Without further ado, we take as our definition of a linear *causal* system a system of equations that can be put in the form (3.9). We shall later elaborate on the word causal.

The existence of the linearized model (3.9) of the nonlinear system (3.4) is not assured. Clearly, if the function $\mathbf{f}(\mathscr{x}, \mathscr{u}, t)$ is not infinitely differentiable, the Taylor's series (3.5a) does not exist. The function $\mathbf{f}(\mathscr{x}, \mathscr{u}, t)$ must be twice differentiable for the bilinear form of the model and once differentiable for the linear form of the model. The nonlinear system (3.4) has a unique solution, $\bar{\mathscr{x}}(t)$, $\bar{\mathscr{u}}(t)$, passing through $\mathscr{x}(0)$, $\mathscr{u}(0)$ if $\mathbf{f}(\mathscr{x}, \mathscr{u}, t)$ is continuous and satisfies a Lipschitz condition

$$\left\|\mathbf{f}(\mathscr{x}_1, \mathscr{u}_1, t) - \mathbf{f}(\mathscr{x}_2, \mathscr{u}_2, t)\right\| \le \alpha \left\|\begin{matrix}\mathscr{x}_1 - \mathscr{x}_2\\ \mathscr{u}_1 - \mathscr{u}_2\end{matrix}\right\|, \qquad \text{for some constant } \alpha. \tag{3.10}$$

Linearized equations (3.9) have meaning *only* if the higher-order terms in $(\mathscr{x} - \bar{\mathscr{x}}), (\mathscr{u} - \bar{\mathscr{u}})$ are smaller than the linear terms. Otherwise, the higher-order terms **h** in the Taylor's series expansion would dominate the system behavior. This means that the linearization is only valid when $\mathscr{x}(t)$ remains *close* to $\bar{\mathscr{x}}(t)$, in the sense that

$$\lim_{\left\|\begin{smallmatrix}\mathscr{x}\\ \mathscr{u}\end{smallmatrix}\right\| \to 0} \left\{ \frac{h(\mathscr{x}, \mathscr{u}, t)}{\left\|\begin{matrix}\mathscr{x}\\ \mathscr{u}\end{matrix}\right\|} \right\} = 0. \tag{3.11}$$

Hence, for the linearization to make sense, the trajectory $\bar{\mathscr{x}}(t)$ about which the

Taylor's series is expanded *must* be a *solution* of the nonlinear system (3.4). Time-varying solutions $\bar{x}(t), \bar{u}(t)$ usually lead to matrices $\mathbf{A}(t)$, $\mathbf{B}(t)$, $\mathbf{C}(t)$, $\mathbf{D}(t)$, $\mathbf{H}(t)$, and $J(t)$ in (3.9) that are functions of time. Such systems will be called *time-varying* linear systems. If these matrices are *constant*, the system is called *time-invariant*.

We hedged on a formal definition of linear systems because of the following example. Consider the nonlinear operations indicated in the boxes of the block diagram

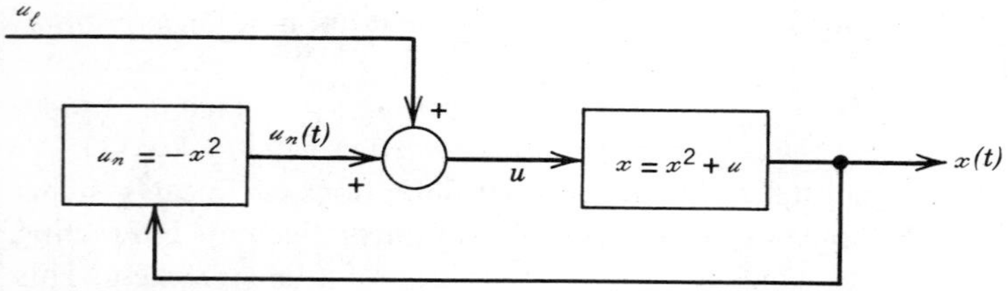

A Linear System ?

Should this system be called linear or nonlinear? The following calculations suggest that it is linear between signals $u_\ell(t)$ and $x(t)$.

$$\left.\begin{aligned} \dot{x} &= x^2 + u \\ u &= u_n + u_\ell \\ u_n &= -x^2 \end{aligned}\right\} \Rightarrow \dot{x} = u_\ell.$$

Clearly, the relationship between $u(t)$ and $x(t)$ is nonlinear. If $u(t)$ is the control for the system $\dot{x} = x^2 + u$, then such control policies as described in the block diagram have been called "feedback linearization." This suggests that linear systems may be composed of nonlinear subsystems that interact in a way that cancels the nonlinear effects, leaving a linear relationship between certain inputs and outputs. However, if $u_n(t)$ were calculated (in a computer) slightly incorrectly ($u = -\bar{x}^2$, $\bar{x} \neq x$), say, from errors in the initial value of $x(0)$, then the cancellation of the nonlinearity is not perfect, leaving a resulting *nonlinear* term

$$\dot{x} = x^2 + u_n + u_\ell = x^2 - \bar{x}^2 + u_\ell,$$

which yields a totally different class of properties than linear systems. On the other hand, if similar computational errors were made on *linear* calculations of linear subsystems, the resulting system equations would still be *linear* (albeit different trajectories $x(t)$ than we may desire). Hence, this author would not choose to call systems as in the above block diagram *linear systems* because linear analysis does not apply to the perturbed system.

3.3 A Rocket in Flight

Consider a rocket of constant mass m with the following assumptions:

(i) The motion is constrained to the plane of the page.
(ii) The lift forces are zero.
(iii) The drag force is proportional to speed squared: $V^2 \triangleq \dot{r}_x^2 + \dot{r}_y^2$.
(iv) The *linearized* model will assume small α, β, θ and constant V.

Figure 3.1 describes the kinematic variables, from which Newton's laws for a rigid body,

$$\text{(mass)(acceleration vector)} = \text{sum of applied forces}$$

and

$$\text{time rate of change of angular momentum vector} = \text{sum of applied torques}$$

yield, respectively,

$$m\ddot{\mathbf{r}} = \mathbf{f} + \mathbf{w} + m\mathbf{g} + \mathbf{f}_w, \tag{3.12a}$$

$$\dot{\mathbf{h}} = \mathbf{d} \times \mathbf{f} + \mathbf{l} \times (\mathbf{w} + \mathbf{f}_w), \tag{3.12b}$$

where $\mathbf{g}$ is the gravity vector $\mathbf{g} = -gi_1$; $(\mathbf{i}_1, \mathbf{i}_2, \mathbf{i}_3)$ is an orthogonal set of unit vectors fixed in inertial space; and the momentum vector is

$$\mathbf{h} = -J\dot{\alpha}\mathbf{i}_3.$$

The velocity vector is

$$\dot{\mathbf{r}} = \dot{\mathbf{r}}_1\mathbf{i}_1 + \dot{\mathbf{r}}_2\mathbf{i}_2.$$

All other necessary vectors are given in Figure 3.1. The "cross" product of two vectors, expressed in common reference frames, is defined by

$$\mathbf{c} \triangleq \mathbf{a} \times \mathbf{b} = (a_1\mathbf{i}_1 + a_2\mathbf{i}_2 + a_3\mathbf{i}_3) \times (b_1\mathbf{i}_1 + b_2\mathbf{i}_2 + b_3\mathbf{i}_3)$$

$$\triangleq (-a_3b_2 + a_2b_3)\mathbf{i}_1 + (a_3b_1 - a_1b_3)\mathbf{i}_2 + (-a_2b_1 + a_1b_2)\mathbf{i}_3,$$

$$\mathbf{c} = \begin{pmatrix} c_1 \\ c_2 \\ c_3 \end{pmatrix} = \begin{bmatrix} 0 & -a_3 & a_2 \\ a_3 & 0 & -a_1 \\ -a_2 & a_1 & 0 \end{bmatrix} \begin{pmatrix} b_1 \\ b_2 \\ b_3 \end{pmatrix} = \tilde{\mathbf{a}}\mathbf{b},$$

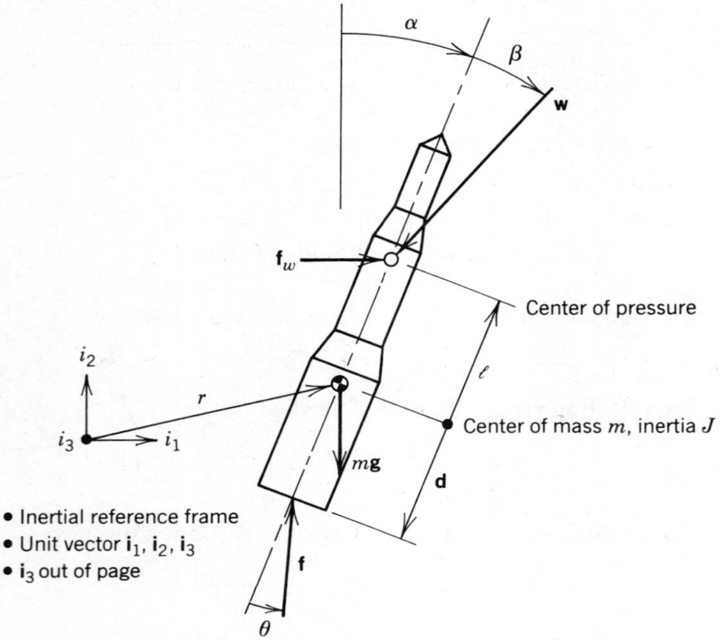

$\mathbf{f} = F\sin(\alpha - \theta)\mathbf{i}_1 + F\cos(\alpha - \theta)\mathbf{i}_2$	thrust force
$\mathbf{w} = -W\sin(\alpha + \beta)\mathbf{i}_1 - W\cos(\alpha + \beta)\mathbf{i}_2$	drag force
$\mathbf{l} = L\sin\alpha\mathbf{i}_1 + L\cos\alpha\mathbf{i}_2$	location of center of pressure
$\mathbf{d} = -D\sin\alpha\mathbf{i}_1 - D\cos\alpha\mathbf{i}_2$	location of rocket engine
$\mathbf{h} = -J\dot{\alpha}\mathbf{i}_3$	angular momentum
$m\mathbf{g} = -mG\mathbf{i}_l$	gravity force
$\mathbf{f}_w = F_w\mathbf{i}_1$	wind disturbance force

Figure 3.1 *Rocket Control Dynamics*

where $\tilde{\mathbf{a}}$ is a skew-symmetric matrix defined in Section 2.8.1.2. In the inertial coordinates $\mathbf{i}_1, \mathbf{i}_2, \mathbf{i}_3$, equations (3.12) yield

Motion about the $\mathbf{i}_3$ axis: $J\ddot{\alpha} = -LW\sin\beta + FD\sin\theta - LF_w\cos\alpha$, (3.13a)

Motion in the $\mathbf{i}_1$ direction: $m\ddot{r}_1 = F\sin(\alpha - \theta) - W\sin(\alpha + \beta) + F_w$, (3.13b)

Motion in the $\mathbf{i}_2$ direction: $m\ddot{r}_2 = F\cos(\alpha - \theta) - W\cos(\alpha + \beta) - mG$, (3.13c)

where $\tan(\alpha + \beta) = \dot{r}_1/\dot{r}_2$ follows from the fact that the drag force $\mathbf{w}$ is assumed to be parallel (and opposite in sense) to the velocity vector $\dot{\mathbf{r}}$. Now, to analyze the linear behavior *near* the values $\{\alpha = \beta = \theta = \dot{r}_1 = 0,\ \dot{r}_2 = V = \text{constant},\ F = \rho V^2 + mG\}$ with the assumption that the drag is proportional to velocity squared, $W = \rho V^2$, the above nonlinear equations reduce to the linear ones (define $\dot{\tilde{r}}_2 \triangleq \dot{r}_2 - V$, $\tilde{F} \triangleq F - \rho V^2 - mG$),

$$J\ddot{\alpha} = \rho L V^2\left(\alpha - \frac{\dot{r}_1}{V}\right) + FD\theta = \rho VL(V\alpha - \dot{r}_1) + FD\theta - F_w L$$

$$= \rho LV(V\alpha - \dot{r}_1) + (\rho V^2 + mG)D - F_w L, \tag{3.14a}$$

$$m\ddot{r}_1 = F(\alpha - \theta) - \rho V\dot{r}_1 + F_w, \tag{3.14b}$$

$$m\ddot{\tilde{r}}_2 = F - \rho V^2 - mG = \tilde{F}. \tag{3.14c}$$

If one is interested in the transfer functions between the outputs $(\alpha, r_1, \tilde{r}_2)$ and the control variable θ, then the Laplace transform (see Appendix E) of these three equations leads immediately to the relationships (for $F_w = 0$):

$$\begin{bmatrix} \alpha(s) \\ r_1(s) \\ r_2(s) \end{bmatrix} = \begin{bmatrix} sF[Dms + \rho V(L+D)]\dfrac{1}{\Delta(s)} \\ F[FD - (Js^2 - \rho LV^2)]\dfrac{1}{\Delta(s)} \\ \tilde{F}\dfrac{1}{ms^2} \end{bmatrix} \theta(s) \triangleq G(s)\theta(s),$$

$$\Delta(s) \triangleq s[(ms + \rho V)(Js^2 - \rho LV^2) + \rho LVF].$$

If one chooses to put the equations of motion (3.14) in a matrix form, then a straightforward arrangement of (3.14) yields (for $F_w = 0$)

$$\begin{bmatrix} J & 0 & 0 \\ 0 & m & 0 \\ 0 & 0 & m \end{bmatrix}\begin{bmatrix} \ddot{\alpha} \\ \ddot{r}_1 \\ \ddot{\tilde{r}}_2 \end{bmatrix} + \begin{bmatrix} 0 & \rho LV & 0 \\ 0 & \rho V & 0 \\ 0 & 0 & 0 \end{bmatrix}\begin{bmatrix} \dot{\alpha} \\ \dot{r}_1 \\ \dot{\tilde{r}}_2 \end{bmatrix} + \begin{bmatrix} -\rho LV^2 & 0 & 0 \\ -F & 0 & 0 \\ 0 & 0 & 0 \end{bmatrix}\begin{bmatrix} \alpha \\ r_1 \\ \tilde{r}_2 \end{bmatrix}$$

$$= \begin{bmatrix} FD \\ -F \\ 0 \end{bmatrix}\theta + \begin{bmatrix} 0 \\ 0 \\ \tilde{F} \end{bmatrix}, \tag{3.15}$$

which is a *vector second-order* set of differential equations. It is possible also to arrange these equations in a *vector first-order* form by writing

$$\begin{bmatrix} \dot{\alpha} \\ \dot{r}_1 \\ \dot{\tilde{r}}_2 \\ \ddot{\alpha} \\ \ddot{r}_1 \\ \ddot{\tilde{r}}_2 \end{bmatrix} = \begin{bmatrix} 0 & 0 & 0 & 1 & 0 & 0 \\ 0 & 0 & 0 & 0 & 1 & 0 \\ 0 & 0 & 0 & 0 & 0 & 1 \\ \rho L V^2/J & 0 & 0 & 0 & -\rho L V/J & 0 \\ F/m & 0 & 0 & 0 & -\rho V/m & 0 \\ 0 & 0 & 0 & 0 & 0 & 0 \end{bmatrix} \begin{bmatrix} \alpha \\ r_1 \\ \tilde{r}_2 \\ \dot{\alpha} \\ \dot{r}_1 \\ \dot{\tilde{r}}_2 \end{bmatrix} + \begin{bmatrix} 0 \\ 0 \\ 0 \\ FD/J \\ -F/m \\ 0 \end{bmatrix} \theta + \begin{bmatrix} 0 \\ 0 \\ 0 \\ 0 \\ 0 \\ \tilde{F}/m \end{bmatrix}, \tag{3.16}$$

which has the form $\dot{\mathbf{x}} = \mathbf{A}\mathbf{x} + \mathbf{B}\theta + \boldsymbol{\gamma}$, where $\mathbf{x}$ is a 6×1 matrix (i.e., a 6-vector), $\mathbf{A}$ is a 6×6 matrix, and $\mathbf{B}$ and $\boldsymbol{\gamma}$ are both 6×1 matrices. The vector first-order form is called the *state* form. In the vector second-order form $\mathcal{M}\ddot{\mathbf{q}} + \mathcal{D}\dot{\mathbf{q}} + \mathcal{K}\mathbf{q} = \mathcal{B}\theta + \Gamma$ in (3.15), $\mathbf{q}$ is a 3×1 matrix (i.e., a 3-vector) and $\mathcal{M}$, $\mathcal{D}$, and $\mathcal{K}$ are all 3×3 matrices. In this form the variables $q_1 \triangleq \alpha$, $q_2 \triangleq r_1$, $q_3 \triangleq \tilde{r}_2$ are called in dynamics the *configuration* variables.

Exercise 3.2

Linearize the model (3.13) about the solution $\alpha = \beta = \theta = \dot{r}_1 = 0$, $\dot{r}_2 = V =$ constant, $F =$ constant to obtain (3.15) and (3.16).

Exercise 3.3

Find the three transfer functions between $\theta(s)$ and $\alpha(s)$, $r_1(s)$ and $\tilde{r}_2(s)$, respectively. Derive your results in two ways: directly from (3.14) and directly from (3.16).

3.4 *A Space Backpack*

For our second dynamic system example, consider the space backpack of Fig. 3.2. The system is idealized as three rigid bodies with center of mass *CM*. Rigid body B is the astronaut torso and "backpack" combination. Body B has pitch moment of inertia J_B about CM_B, mass m_B, and center of mass at point CM_B. Rigid body W is a wheel mounted in the backpack, with a spin axis perpendicular to the page. It has mass m_W, and the moment of inertia about the spin axis is J_W. By applying a torque $-T_c$ to the wheel, there results an equal and opposite reaction torque applied to the torso. This torque will be used for attitude control of the astronaut torso, body B. Rigid body L is the astronaut's pair of legs. The center of mass of body L is at CM_L, and the hip joint is at point h. The system center of mass is *CM*. The mass of body L is m_L, and the pitch moment of inertia about CM_L is J_L. The hip joint is assumed to exhibit a damped spring restraint with spring constant k and a coefficient of viscous friction c. The backpack also contains a thruster providing force f, which acts at an angle β at point p on the backpack. The attitude of the

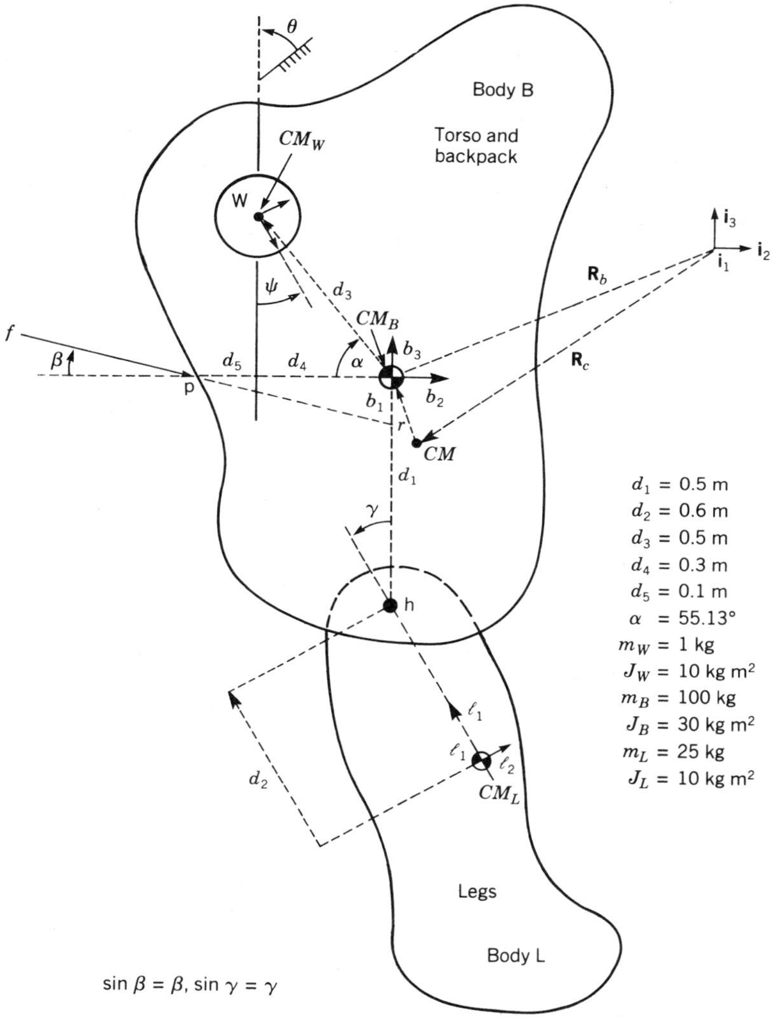

Figure 3.2 *Astronaut Backpack*

torso with respect to an inertially referenced set of axes is θ. The center of mass CM is located at (y_0, z_0) relative to the inertial frame $(\mathbf{i}_1, \mathbf{i}_2, \mathbf{i}_3)$.

Using an independent set of generalized coordinates q_i, one can use Lagrange's equation,

$$\frac{d}{dt}\frac{\partial \mathscr{L}}{\partial \dot{\mathbf{q}}} - \frac{\partial \mathscr{L}}{\partial \mathbf{q}} = \mathscr{Q}, \qquad (3.17)$$

to arrive at similar results for the rigid bodies of Fig. 3.2, where the independent generalized coordinates for this example are

$$\mathbf{q}^T = (q_1, q_2, q_3, q_4, q_5) = (\theta, \gamma, \psi, y_0, z_0)$$

and

$$\mathscr{L} = \mathscr{T} - \mathscr{U},$$

where the potential energy $\mathscr{U}$ is the virtual work done by conservative forces, and $\mathscr{Q}$ is from the generalized nonconservative forces,

$$\mathscr{Q}^T = (\mathscr{Q}_1, \ldots, \mathscr{Q}_5)$$

obtained from ref. [3.5], for each generalized coordinate $i = 1 \to 5$,

$$Q_i = \sum_{k=1}^{3} \mathbf{F}^k \frac{\partial \dot{\mathbf{R}}^k}{\partial \dot{q}_i} + \mathbf{M}^k \frac{\partial \boldsymbol{\omega}^k}{\partial \dot{q}_i}.$$

$\mathbf{F}^k$ is the force applied to rigid body k, $\mathbf{M}^k$ is the torque applied to rigid body k, and the

$$\dot{\mathbf{R}}^L = \dot{\mathbf{R}}_c + \dot{\mathbf{r}} + \dot{\mathbf{d}}_1 + \dot{\mathbf{d}}_2$$

$$\dot{\mathbf{R}}^B = \dot{\mathbf{R}}_c + \dot{\mathbf{r}}$$

$$\dot{\mathbf{R}}^W = \dot{\mathbf{R}}_c + \dot{\mathbf{r}} + \dot{\mathbf{d}}_3$$

From center of mass definition $\mathbf{r}m_B + (\mathbf{r} + \mathbf{d}_3)m_W + (\mathbf{r} + \mathbf{d}_1 + \mathbf{d}_2)m_L = 0$, see that

$$\dot{\mathbf{r}} = -(\dot{\mathbf{d}}_1 + \dot{\mathbf{d}}_2)\frac{m_L}{m} - \dot{\mathbf{d}}_3\frac{m_W}{m},$$

where $m \triangleq m_W + m_B + m_L$. The angular velocities are

$$\boldsymbol{\omega}^L = (\dot{\theta} + \dot{\gamma})\mathbf{i}_1,$$

$$\boldsymbol{\omega}^B = \dot{\theta}\mathbf{i}_1,$$

$$\boldsymbol{\omega}^W = (\dot{\theta} + \dot{\psi})\mathbf{i}_1,$$

and the $\mathbf{F}^i$ and $\mathbf{M}^i$ are

$$\mathbf{F}^L = 0,$$

$$\mathbf{F}^W = 0,$$

$$\mathbf{F}^B = f\cos\beta\mathbf{b}_2 - f\sin\beta\mathbf{b}_3,$$

$$\mathbf{M}^L = (-c\dot{\gamma})\mathbf{i}_1,$$

$$\mathbf{M}^B = T_c\mathbf{i}_1 + c\dot{\gamma}\mathbf{i}_1 + f\sin\beta(d_4 + d_5)\mathbf{i}_1,$$

$$\mathbf{M}^W = -T_c\mathbf{i}_1.$$

The kinetic energy is

$$\mathcal{T} = \tfrac{1}{2}m_W\dot{\mathbf{R}}^W \cdot \dot{\mathbf{R}}^W + \tfrac{1}{2}m_B\dot{\mathbf{R}}^B \cdot \dot{\mathbf{R}}^B + \tfrac{1}{2}m_L\dot{\mathbf{R}}^L \cdot \dot{\mathbf{R}}^L$$
$$+ \tfrac{1}{2}(\dot{\theta} + \dot{\psi})^2 J_W + \tfrac{1}{2}\dot{\theta}^2 J_B + \tfrac{1}{2}(\dot{\theta} + \dot{\gamma})^2 J_L,$$

where $\dot{\mathbf{R}}^W \cdot \dot{\mathbf{R}}^W$ denotes dot product of the vector $\dot{\mathbf{R}}^W$ with itself. The "dot" product of two vectors expressed in a common orthogonal references frame is defined by

$$d \triangleq \mathbf{a} \cdot \mathbf{b} = (a_1\mathbf{i}_1 + a_2\mathbf{i}_2 + a_3\mathbf{i}_3) \cdot (b_1\mathbf{i}_1 + b_2\mathbf{i}_2 + b_3\mathbf{i}_3)$$
$$= a_1 b_1 + a_2 b_2 + a_3 b_3.$$

Note that this can be expressed as the inner product defined in Section 2.4.1:

$$d = (a_1 \quad a_2 \quad a_3)\begin{bmatrix} b_1 \\ b_2 \\ b_3 \end{bmatrix} = \mathbf{a}^T\mathbf{b}.$$

The potential energy is

$$\mathcal{U} = \tfrac{1}{2}k\gamma^2. \tag{3.18}$$

Finally, the coordinate transformation needed is

$$\mathbf{b}_1 = \mathbf{i}_1,$$
$$\mathbf{b}_2 = \cos\theta\mathbf{i}_2 + \sin\theta\mathbf{i}_3, \tag{3.19}$$
$$\mathbf{b}_3 = -\sin\theta\mathbf{i}_2 + \cos\theta\mathbf{i}_3.$$

Using these data, the equations of motion (3.17) become lengthy and nonlinear:

$$\begin{aligned}
\mathscr{M}_{11}\ddot{\theta} + \mathscr{M}_{12}\ddot{\gamma} + \mathscr{M}_{13}\ddot{\psi} + \mathscr{C}_{12}^{(1)}\dot{\theta}\dot{\gamma} + \mathscr{C}_{22}^{(1)}\dot{\gamma}^2 &= \mathscr{B}_{22}f, \\
\mathscr{M}_{12}\ddot{\theta} + \mathscr{M}_{22}\ddot{\gamma} + \mathscr{C}_{21}^{(2)}\dot{\theta}^2 + \mathscr{K}_{22}\gamma + \mathscr{C}_{22}\dot{\gamma} &= \mathscr{B}_{22}f, \\
\mathscr{M}_{13}\ddot{\theta} + \mathscr{M}_{33}\ddot{\psi} &= \mathscr{B}_{31}T_c, \\
\mathscr{M}_{14}\ddot{y}_0 &= \mathscr{B}_{42}f, \\
\mathscr{M}_{15}\ddot{z}_0 &= \mathscr{B}_{52}f,
\end{aligned} \tag{3.20}$$

where

$$\begin{aligned}
&\mathscr{M}_{11} \triangleq J_B + J_W + J_L + D_3 + D_4\cos\gamma + D_8\sin\gamma, \\
&\mathscr{M}_{12} \triangleq J_L + D_6 + D_5\cos\gamma + D_9\sin\gamma, \\
&\mathscr{M}_{13} \triangleq J_W, \\
&\mathscr{M}_{22} \triangleq J_L + D_6, \\
&\mathscr{M}_{33} \triangleq J_W, \\
&\mathscr{M}_{14} \triangleq m \triangleq \mathscr{M}_{15}, \qquad m \triangleq m_B + m_L + m_W, \\
&\mathscr{C}_{12}^{(1)} \triangleq D_8\cos\gamma - D_4\sin\gamma, \\
&\mathscr{C}_{22}^{(1)} \triangleq D_9\cos\gamma - D_5\sin\gamma, \\
&\mathscr{C}_{21}^{(2)} = D_5\sin\gamma - D_9\cos\gamma, \\
&\mathscr{C}_{22} = c, \\
&\mathscr{K}_{22} = k, \\
&\mathscr{B}_{12} = \left(\frac{m_W}{m}d_3\sin\alpha - \frac{m_L}{m}d_1\right)\cos\beta - \frac{m_W}{m}d_4\sin\beta \\
&\qquad - \frac{m_L}{m}d_2\cos(\beta+\gamma), \\
&\mathscr{B}_{22} = -\frac{m_L}{m}d_2\cos(\beta+\gamma), \\
&\mathscr{B}_{31} = 1, \\
&\mathscr{B}_{42} = \cos(\theta - \beta),
\end{aligned} \tag{3.21}$$

$$\mathscr{B}_{52} = \sin(\theta - \beta),$$

$$D_3 = m_W d_4^2\left(1 - \frac{m_W}{m}\right) + m_W d_3^2 \sin^2\alpha\left(1 - \frac{m_W}{m}\right)$$

$$+ 2d_1 d_3 \sin\alpha \frac{m_L m_W}{m} + m_L d_1^2\left(1 - \frac{m_L}{m}\right) + m_L d_2^2\left(1 - \frac{m_L}{m}\right),$$

$$D_4 = 2m_L d_2\left[\frac{m_W}{m} d_3 \sin\alpha + d_1\left(1 - \frac{m_L}{m}\right)\right],$$

$$D_5 = \tfrac{1}{2}D_4,$$

$$D_6 = m_L d_2^2\left(1 - \frac{m_L}{m}\right),$$

$$D_8 = 2d_2 d_4 \frac{m_W m_L}{m},$$

$$D_9 = \tfrac{1}{2}D_8.$$

Our interest centers on the relatively small motion about the solution

$$\bar{\theta}(t) = \bar{\gamma}(t) = \bar{\dot{\psi}}(t) = \bar{f}(t) = \bar{T}_c(t) = \bar{\beta}(t) = \bar{y}_0(t) = \bar{z}_0(t) = 0. \quad (3.22)$$

Such nonlinear differential equations can be "linearized" about the solution (3.22). Such results are

$$\begin{bmatrix} \bar{\mathscr{M}}_{11} & \bar{\mathscr{M}}_{12} & \bar{\mathscr{M}}_{13} \\ \bar{\mathscr{M}}_{12} & \bar{\mathscr{M}}_{22} & 0 \\ \bar{\mathscr{M}}_{13} & 0 & \bar{\mathscr{M}}_{33} \end{bmatrix}\begin{pmatrix} \ddot{\theta} \\ \ddot{\gamma} \\ \ddot{\psi} \end{pmatrix} + \begin{bmatrix} 0 & 0 & 0 \\ 0 & \bar{\mathscr{C}}_{22} & 0 \\ 0 & 0 & 0 \end{bmatrix}\begin{pmatrix} \dot{\theta} \\ \dot{\gamma} \\ \dot{\psi} \end{pmatrix} + \begin{bmatrix} 0 & 0 & 0 \\ 0 & \bar{\mathscr{K}}_{22} & 0 \\ 0 & 0 & 0 \end{bmatrix}\begin{pmatrix} \theta \\ \gamma \\ \psi \end{pmatrix}$$

$$= \begin{bmatrix} 0 & \bar{\mathscr{B}}_{12} \\ 0 & \bar{\mathscr{B}}_{22} \\ \bar{\mathscr{B}}_{31} & 0 \end{bmatrix}\begin{pmatrix} T_c \\ f \end{pmatrix}, \quad (3.23a)$$

where the translational equations in y_0 and z_0 in (3.20) have been discarded since our interest is in the attitude motion of the system and since y_0 and z_0 are not coupled into the equations for θ, γ, and ψ. It should be noted that such decoupling did not naturally occur from (3.9), but some rearrangement of the equations was necessary to get the form (3.23a). The parameters in (3.23a) are

$$\overline{\mathscr{M}}_{11} = J_B + J_W + J_L + D_3 + D_4$$

$$\overline{\mathscr{M}}_{12} = J_L + D_6 + D_5$$

$$\overline{\mathscr{M}}_{13} = \mathscr{M}_{13} = J_W$$

$$\overline{\mathscr{M}}_{22} = J_L + D_6 = \mathscr{M}_{22}$$

$$\overline{\mathscr{M}}_{33} = J_W = \mathscr{M}_{33}$$

$$\overline{\mathscr{C}}_{22} = c = \mathscr{C}_{22}$$

$$\overline{\mathscr{K}}_{22} = k = \mathscr{K}_{22}$$

$$\overline{\mathscr{B}}_{12} = \left(\frac{m_W}{m} d_3 \sin\alpha - \frac{m_L}{m}(d_1 + d_2)\right)\cos\beta - \frac{m_W}{m} d_4 \sin\beta$$

$$\overline{B}_{22} = -\frac{m_L}{m} d_2 \cos\beta$$

$$\overline{\mathscr{B}}_{31} = 1 = \mathscr{B}_{31}.$$

Even though the *dynamics* of $y_0(t)$ and $z_0(t)$ are uncoupled from the rotational motion, one may not discard these equations prior to a statement of the *outputs* of interest. We presume a primary interest in torso rate, so we define the output to be

$$y \triangleq \dot{\theta} = [0 \quad 0 \quad 0]\begin{pmatrix}\theta\\ \gamma\\ \psi\end{pmatrix} + [1 \quad 0 \quad 0]\begin{pmatrix}\dot{\theta}\\ \dot{\gamma}\\ \dot{\psi}\end{pmatrix}, \tag{3.23b}$$

which, together with (3.23a), forms the complete model describing the dynamic relationship between the inputs (T_c, f) and the output $\dot{\theta}$.

Exercise 3.4

Determine whether $\boldsymbol{x}^T \triangleq (\mathbf{p}^T, \dot{\mathbf{p}}^T)$,

$$\mathbf{p} \triangleq \begin{pmatrix}\theta + \gamma + \Psi\\ \gamma - \Psi\\ -5\theta + \gamma - 11\Psi\end{pmatrix} = \begin{bmatrix}1 & 1 & 1\\ 0 & 1 & -1\\ -5 & 1 & -11\end{bmatrix}\begin{pmatrix}\theta\\ \gamma\\ \Psi\end{pmatrix},$$

qualifies as a state vector for the nonlinear system (3.20).

Exercise 3.5

Determine whether $\boldsymbol{x}^T = (\mathbf{p}^T, \dot{\mathbf{p}}^T)$

$$\mathbf{p} \triangleq \begin{pmatrix} \theta - \Psi \\ \Psi + \gamma \\ 2\theta + \gamma \end{pmatrix} = \begin{bmatrix} 1 & 0 & -1 \\ 0 & 1 & 1 \\ 2 & 1 & 0 \end{bmatrix} \begin{pmatrix} \theta \\ \gamma \\ \Psi \end{pmatrix}$$

qualifies as a state vector for the linear system (3.23).

Exercise 3.6

Verify that $\bar{x}$ is indeed a solution of (3.20), if

$$\bar{\boldsymbol{x}}^T = \left(\overline{\Theta}, \overline{\dot{\Theta}}, \bar{\gamma}, \bar{\dot{\gamma}}, \overline{\dot{\Psi}}\right) = 0, \qquad \bar{\boldsymbol{u}}^T = \left(\overline{T}_c, \bar{f}\right) = 0.$$

Place the backpack example (3.20) in the form (3.6b). That is, compute **A**, **B**, **C**, **H**, and $\mathbf{F}_1, \ldots, \mathbf{F}_5, \mathbf{G}_1$.

3.5 *An Orbiting Spacecraft*

Consider in Fig. 3.3 a satellite as a point mass in an inverse square gravitational field. Let the satellite of mass m have the capability of thrusting in the radial direction with a thrust u_1 and in the tangential direction with a thrust u_2. Newton's law yields

$$m\ddot{\mathbf{r}} = \mathbf{u}_1 + \mathbf{u}_2 + \mathbf{f}_G, \tag{3.24}$$

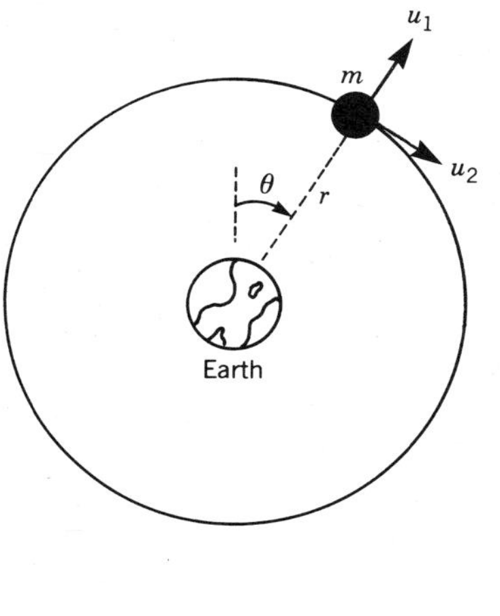

Figure 3.3 *An Orbiting Spacecraft*

where the force of gravity is [3.3]

$$\mathbf{f}_G = -\frac{km}{r^2}\left(\frac{\mathbf{r}}{|r|}\right) = -\frac{km}{r^3}\mathbf{r} \tag{3.25}$$

for some constant k. Define coordinates centered on the spacecraft such that $\mathbf{b}_1$ is a unit vector parallel to $\mathbf{r}$, $\mathbf{b}_2$ is a perpendicular unit vector, and $\mathbf{b}_3$ completes the triad perpendicular into the page. We develop the equations (3.24) as follows:

$$\mathbf{r} = r\mathbf{b}_1, \qquad \boldsymbol{\omega} = \dot{\theta}\mathbf{b}_3,$$

$$\dot{\mathbf{r}} = \dot{r}\mathbf{b}_1 + \boldsymbol{\omega}\times\mathbf{r} = \dot{r}\mathbf{b}_1 + \dot{\theta}\mathbf{b}_3 \times r\mathbf{b}_1$$

$$= \dot{r}\mathbf{b}_1 + r\dot{\theta}\mathbf{b}_2,$$

$$\ddot{\mathbf{r}} = \ddot{r}\mathbf{b}_1 + \dot{r}\dot{\theta}\mathbf{b}_2 + r\ddot{\theta}\mathbf{b}_2 + \boldsymbol{\omega}\times\dot{\mathbf{r}}$$

$$= \ddot{r}\mathbf{b}_1 + \dot{r}\dot{\theta}\mathbf{b}_2 + r\ddot{\theta}\mathbf{b}_2 + \dot{\theta}\mathbf{b}_3 \times \left(\dot{r}\mathbf{b}_1 + r\dot{\theta}\mathbf{b}_2\right).$$

But since $\mathbf{b}_3 \times \mathbf{b}_1 = \mathbf{b}_2$ and $\mathbf{b}_3 \times \mathbf{b}_2 = -\mathbf{b}_1$, this leads to

$$\ddot{\mathbf{r}} = (\ddot{r} - r\dot{\theta}^2)\mathbf{b}_1 + (2\dot{r}\dot{\theta} + r\ddot{\theta})\mathbf{b}_2.$$

Hence, (3.24) yields, in the $\mathbf{b}_1$ and $\mathbf{b}_2$ directions, respectively,

$$m(\ddot{r} - r\dot{\theta}^2) = u_1 - \frac{km}{r^2}, \tag{3.26a}$$

$$m(2\dot{r}\dot{\theta} + r\ddot{\theta}) = u_2. \tag{3.26b}$$

If $u_1 = u_2 = 0$, then one equilibrium solution is

$$\begin{aligned} r(t) &= R_0 \qquad \text{a constant,} \\ \theta(t) &= \omega_0 t \qquad (\omega_0 = \text{a constant}), \end{aligned} \tag{3.27}$$

where the gravity constant is approximately $k = R_0^3\omega_0^2$. That is, circular orbits are possible. From Section 3.1, it may be shown that the linearized equations may be written

$$\begin{pmatrix}\ddot{r}\\ \ddot{\theta}\end{pmatrix} + \begin{bmatrix} 0 & -2R_0\omega_0 \\ +2\omega_0/R_0 m & 0 \end{bmatrix}\begin{pmatrix}\dot{r}\\ \dot{\theta}-\omega_0\end{pmatrix} + \begin{bmatrix} -3\omega_0^2 & 0 \\ 0 & 0\end{bmatrix}\begin{pmatrix} r - R_0 \\ \theta - \omega_0 t\end{pmatrix}$$

$$= \begin{bmatrix} 1/m & 0 \\ 0 & 1/R_0 m\end{bmatrix}\begin{pmatrix}u_1\\ u_2\end{pmatrix}. \tag{3.28}$$

Exercise 3.7

Verify (3.28) from (3.26) and (3.27). From (3.28), find the transfer function between $u(s)$ and $\theta(s)$. Determine whether there are any pole-zero cancellations in these transfer functions.

3.6 *A Rigid Body in Space*

The rotational motion of a three-dimensional rigid body in Fig. 3.4 may be described by

$$\frac{d}{dt}\mathbf{h} = \mathbf{T}, \tag{3.29}$$

where $\mathbf{h}$ is the angular momentum vector and $\mathbf{T}$ is the sum of applied torques; $\mathbf{h}$ may be described in terms of the dot product of the inertia dyadic $\underline{\mathbf{J}} = \Sigma_{\alpha,\beta} J_{\alpha\beta}\mathbf{b}_\alpha\mathbf{b}_\beta$ and the angular velocity vector $\boldsymbol{\omega} = \omega_1\mathbf{b}_1 + \omega_2\mathbf{b}_2 + \omega_3\mathbf{b}_3$, where $(\mathbf{b}_1, \mathbf{b}_2, \mathbf{b}_3)$ is a set of unit vectors coincident with the body-fixed axes of the spacecraft. The inertia matrix and angular velocity components associated with these axes are

$$\mathbf{J} = \begin{bmatrix} J_{11} & J_{12} & J_{13} \\ J_{12} & J_{22} & J_{23} \\ J_{13} & J_{23} & J_{33} \end{bmatrix}, \qquad \omega = \begin{pmatrix} \omega_1 \\ \omega_2 \\ \omega_3 \end{pmatrix}; \tag{3.30}$$

and hence,

$$\frac{d}{dt}(\mathbf{J}\boldsymbol{\omega}) = (\mathbf{J}\dot{\boldsymbol{\omega}})_1\mathbf{b}_1 + (\mathbf{J}\dot{\boldsymbol{\omega}})_2\mathbf{b}_2 + (\mathbf{J}\dot{\boldsymbol{\omega}})_3\mathbf{b}_3 + \boldsymbol{\omega} \times J\boldsymbol{\omega}. \tag{3.31}$$

Let the applied torques be (u_1, u_2, u_3) about the body axes $(\mathbf{b}_1, \mathbf{b}_2, \mathbf{b}_3)$. In the $\mathbf{b}_i$ coordinates, equation (3.31) becomes

$$\mathbf{J}\dot{\boldsymbol{\omega}} + \tilde{\boldsymbol{\omega}}\mathbf{J}\boldsymbol{\omega} = \mathscr{B}\mathbf{u}, \tag{3.32}$$

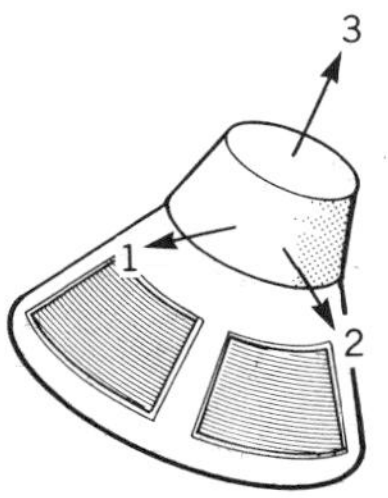

Figure 3.4 *Rigid Spacecraft*

where

$$\tilde{\omega} = \begin{bmatrix} 0 & -\omega_3 & \omega_2 \\ \omega_3 & 0 & -\omega_1 \\ -\omega_2 & \omega_1 & 0 \end{bmatrix}, \quad \mathscr{B} = \begin{bmatrix} 1 & 0 & 0 \\ 0 & 1 & 0 \\ 0 & 0 & 1 \end{bmatrix}, \quad \mathbf{u} = \begin{pmatrix} u_1 \\ u_2 \\ u_3 \end{pmatrix}. \tag{3.33}$$

Exercise 3.8

Let the spacecraft be axisymmetric and let $\mathbf{b}_i$ represent principal axes (i.e., $J_{11} = J_{22}$, $J_{ij} = 0$, $i \neq j$) and let $u_3(t) \equiv 0$. Show that the nonlinear equations (3.32) reduce to the linear ones

$$\mathbf{J}\dot{\boldsymbol{\omega}} + \mathbf{S}\boldsymbol{\omega} = \mathscr{B}_T\mathbf{u}_T, \tag{3.34a}$$

where

$$\mathscr{B}_T \triangleq \begin{bmatrix} 1 & 0 \\ 0 & 1 \\ 0 & 0 \end{bmatrix}, \qquad \mathbf{u}_T \triangleq \begin{pmatrix} u_1 \\ u_2 \end{pmatrix} \tag{3.34b}$$

and $\mathbf{S}$ is the skew-symmetric matrix

$$\mathbf{S} = \begin{bmatrix} 0 & -(J_{11} - J_{33})\bar{\omega}_3 & 0 \\ (J_{11} - J_{33})\bar{\omega}_3 & 0 & 0 \\ 0 & 0 & 0 \end{bmatrix}. \tag{3.34c}$$

3.7 *Elastic Structures*

In many applications of control and estimation theories, the physical process is described by a set of partial differential equations for the elastic structure of Fig. 3.5. The reference frame in which the generic position vector $\mathbf{r}$ is described is inertially fixed; and the constant vector $\mathbf{r}$ is chosen so that $\boldsymbol{\mu}(\mathbf{r}, t)$ is a vector of *short*

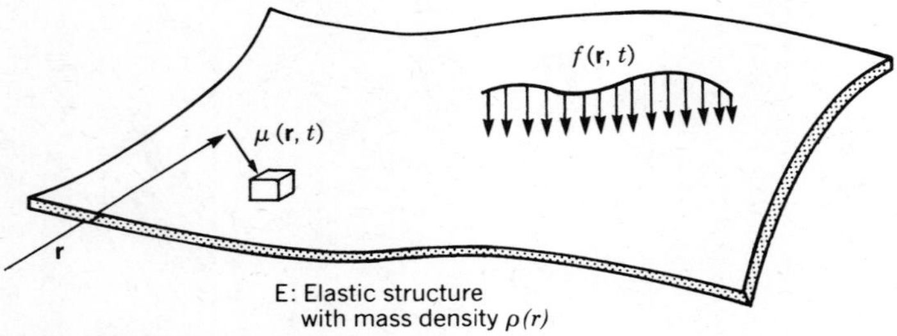

Figure 3.5 *An Elastic Structure*

length to the elemental volume. In this way the equations of motion can later be linearized in the small variable μ.

The derivation of such equations of motion uses Hamilton's principle [3.1], which is beyond the scope of this course. However, we shall briefly discuss the concepts and present the results for a simple beam. This is useful to illustrate applications of the notation and concepts of Chapter 2.

3.7.1 THE CONTINUUM MODEL

Hamilton's principle [3.1] states that the first variation of the "action" $\mathscr{L}$ is zero along the actual time-varying path of $\boldsymbol{\mu}(\mathbf{r}, t)$, where the "action" $\mathscr{L}$ is the difference between the kinetic energy $\mathscr{T}$ and the potential energy $\mathscr{U}$, plus the work done by nonconservative forces $\mathscr{W}$. Thus, from Hamilton's principle without elaboration we indicate that the first variation is zero by

$$\delta \int_{t_1}^{t_2} \mathscr{L}(\underline{\boldsymbol{\mu}}, \dot{\underline{\boldsymbol{\mu}}}, t)\, dt = 0, \qquad \mathscr{L} \triangleq \mathscr{T} - \mathscr{U} + \mathscr{W}. \tag{3.35}$$

We shall now assume that all vectors $\boldsymbol{\mu}$ and $\mathbf{r}$ are described in the same given reference frame, and we shall drop the basis vectors. Hence, we denote $\boldsymbol{\mu}$ and $\mathbf{r}$ as

$$\begin{aligned} \underline{\mathbf{r}} &= \mathbf{e}^T\mathbf{r} = \mathbf{e}_1 r_1 + \mathbf{e}_2 r_2 + \mathbf{e}_3 r_3, \\ \underline{\boldsymbol{\mu}} &= \mathbf{e}^T\boldsymbol{\mu} = \mathbf{e}_1\mu_1 + \mathbf{e}_2\mu_2 + \mathbf{e}_3\mu_3 \end{aligned} \tag{3.36}$$

simply by $\boldsymbol{\mu} \in \mathscr{R}^3$ and $\mathbf{r} \in \mathscr{R}^3$, with the basis $\mathbf{e}$ understood. The kinetic energy is the norm of the velocity vector $\dot{\boldsymbol{\mu}}(\mathbf{r}, t)$ over the entire structure, weighted with $\frac{1}{2}\rho(\mathbf{r})$. (By contrast, refer to (2.29) for norms on a *time* interval.)

$$\mathscr{T} = \|\dot{\boldsymbol{\mu}}(\mathbf{r}, t)\|^2_{1/2\rho(r)} = \tfrac{1}{2}\int_{\mathscr{E}} \dot{\boldsymbol{\mu}}^T(\mathbf{r}, t)\rho(\mathbf{r})\dot{\boldsymbol{\mu}}(\mathbf{r}, t)\, d\mathbf{r}, \tag{3.37}$$

where dr denotes a volume increment. The potential energy is the norm of the displacement vector $\boldsymbol{\mu}(\mathbf{r}, t)$ over the entire structure, weighted with $\frac{1}{2}\tilde{\mathscr{K}}$, where $\tilde{\mathscr{K}}$ is a symmetric matrix operator defined to include the boundary conditions. We shall later construct $\tilde{\mathscr{K}}$ for an example. Now, write the potential energy

$$\mathscr{U} \|\boldsymbol{\mu}(\mathbf{r}, t)\|^2_{1/2\tilde{\mathscr{K}}} = \tfrac{1}{2}\int_{\mathscr{E}} \boldsymbol{\mu}^T(\mathbf{r}, t)\tilde{\mathscr{K}}\boldsymbol{\mu}(\mathbf{r}, t)\, dr. \tag{3.38}$$

Without proof, we shall borrow the first variation of (3.35) from the calculus of variations [3.2]:

$$\begin{aligned} \delta \int_{t_1}^{t_2} \mathscr{L}(\boldsymbol{\mu}, \dot{\boldsymbol{\mu}}, t)\, dt = {} & \left(\frac{\partial \mathscr{L}}{\partial \dot{\boldsymbol{\mu}}}\right)^T \delta\boldsymbol{\mu}\Big|_{t_1}^{t_2} + \left(\mathscr{L} - \left(\frac{\partial \mathscr{L}}{\partial \dot{\boldsymbol{\mu}}}\right)^T \dot{\boldsymbol{\mu}}\right) \Delta t\Big|_{t_1}^{t_2} \\ & + \int_{t_1}^{t_2} \left(\frac{\partial \mathscr{L}}{\partial \mu} - \frac{d}{dt}\frac{\partial \mathscr{L}}{\partial \dot{\mu}}\right)^T \Delta\mu\, dt, \end{aligned} \tag{3.39}$$

where

$$\mathscr{L} = \frac{1}{2}\int_{\mathscr{E}} \dot{\mu}^T \dot{\mu} \rho \, d\mathbf{r} - \frac{1}{2}\int_{\mathscr{E}} \mu^T \tilde{\mathscr{K}} \mu \, d\mathbf{r} + \int_{\mathscr{E}} \mathbf{f}^T \mu \, d\mathbf{r}. \tag{3.40}$$

Hence, from the rules of differentiation (2.67),

$$\frac{\partial \mathscr{L}}{\partial \dot{\mu}} = \int_{\mathscr{E}} \rho \dot{\mu} \, d\mathbf{r}, \tag{3.41}$$

$$\frac{\partial \mathscr{L}}{\partial \mu} = -\int_{\mathscr{E}} \tilde{\mathscr{K}} \mu \, dr + \int_{\mathscr{E}} \mathbf{f} \, d\mathbf{r}. \tag{3.42}$$

By defining variations that satisfy

$$\left(\frac{\partial \mathscr{L}}{\partial \dot{\mu}}\right)^T \delta\mu\Big|_{t_1}^{t_2} = 0, \qquad \left(\mathscr{L} - \left(\frac{\partial \mathscr{L}}{\partial \dot{\mu}}\right)^T \dot{\mu}\right) \Delta t\Big|_{t_1}^{t_2} = 0,$$

then, for arbitrary variations $\Delta\mu$ between t_1 and t_2, the requirement (3.35), using (3.39), dictates that

$$\frac{\partial \mathscr{L}}{\partial \mu} - \frac{d}{dt}\frac{\partial \mathscr{L}}{\partial \dot{\mu}} = \mathbf{0}, \tag{3.43}$$

which, using (3.40) through (3.42), leads to

$$-\int_{\mathscr{E}} \tilde{\mathscr{K}} \mu \, d\mathbf{r} + \int_{\mathscr{E}} \mathbf{f} \, d\mathbf{r} - \int_{\mathscr{E}} \rho \ddot{\mu} \, d\mathbf{r} = \mathbf{0} \tag{3.44}$$

or, equivalently,

$$\rho(r)\ddot{\mu}(\mathbf{r}, t) + \tilde{\mathscr{K}} \mu(\mathbf{r}, t) = \mathbf{f}(\mathbf{r}, t), \tag{3.45}$$

with initial conditions $\mu(\mathbf{r}, 0) = \mu_0(\mathbf{r})$, $\dot{\mu}(\mathbf{r}, 0) = \dot{\mu}_0(\mathbf{r})$. This is the partial differential equation describing the dynamics of the elastic structure of Fig. 3.5.

3.7.2 THE PINNED ELASTIC BEAM

The above equation of motion, (3.44) or (3.45), will be illustrated for the simply supported beam of Fig. 3.6. The beam has deflection $\mu(r, t)$ only in the plane of the paper, where r is the position from the left end of the beam, r_0 is the location of the displacement of interest y, and r_m is the location of the displacement measured by a sensor z. (Due to physical constraints in real systems, we often cannot directly

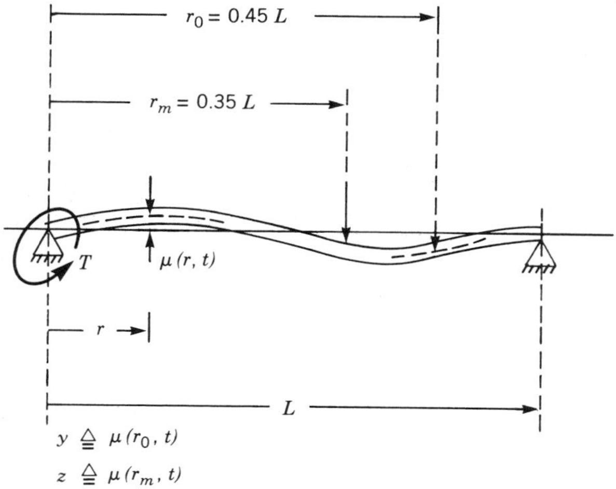

Figure 3.6 *Simply Supported Beam*

measure the outputs of interest.) The kinetic energy of such beams is given by (3.37):

$$\mathcal{T} = \tfrac{1}{2}\int_0^L \rho(r)\dot{\mu}^2(r,t)\rho\,dr. \tag{3.46}$$

A uniform mass density will be assumed so that $\rho(r) = \rho =$ constant. The potential energy for the beam is*

$$\mathcal{U} = \tfrac{1}{2}\int_0^L EI\left[\frac{\partial^2\mu(r,t)}{\partial r^2}\right]^2 dr, \tag{3.47}$$

where the modulus of elasticity EI will also be assumed constant. Integrating (3.47) by parts twice leads to

$$\mathcal{U} = \left\{\left[\frac{\partial\mu(r,t)}{\partial r}\frac{\partial^2\mu(r,t)}{\partial r^2}\right]\Bigg|_0^L - \left[\mu(r,t)\frac{\partial^3\mu(r,t)}{\partial r^3}\right]\Bigg|_0^L + \int_0^L \mu(r,t)\frac{\partial^4\mu(r,t)}{\partial r^4}\,dr\right\}\tfrac{1}{2}EI. \tag{3.48}$$

*The assumptions made here are that both the shear deformation and rotary inertia effects are negligible. This is the so-called Euler–Bernoulli beam [3.4].

But the beam in Fig. 3.6 has the boundary conditions (in fact, these constitute a *definition* of the "simply supported" beam)

$$\frac{\partial^2\mu(L,t)}{\partial r^2} = \frac{\partial^2\mu(0,t)}{\partial r^2} = 0, \tag{3.49a}$$

$$\mu(L,t) = \mu(0,t) = 0. \tag{3.49b}$$

Then, (3.48) reduces to

$$\mathscr{U} = \tfrac{1}{2}EI\int_0^L \mu(r,t)\frac{\partial^4\mu(r,t)}{\partial r^4}\,dr. \tag{3.50}$$

Thus, comparison of (3.38) with (3.50) leads to the conclusion that the operator $\tilde{\mathscr{K}}$ for the beam of Fig. 3.6 is

$$\tilde{\mathscr{K}} = EI\frac{\partial^4}{\partial r^4}. \tag{3.51}$$

By comparing (3.38) with (3.48), the earlier statement is now clear that the definition of the operator $\tilde{\mathscr{K}}$ must include boundary conditions. If the forces $\mathbf{f}(\mathbf{r}, t)$ are applied only at discrete points $(\mathbf{r}_1, \mathbf{r}_2, \dots, \mathbf{r}_m)$ on the structure, then

$$f(r,t) = \sum_{i=1}^{m} f_i\delta(r - r_i), \qquad \mathbf{f}_i = f_i\mathbf{n}_i, \tag{3.52}$$

where f_i is the force applied in the $\mathbf{n}_i$ direction at the spatial location $\mathbf{r}_i$ and δ is the Dirac delta.

For the beam example, $\mathbf{n}_i$ is vertical in the plane of the paper. Now, the equation of motion (3.45) can be written using (3.51) and the notation $[\mu]'''' \triangleq (\partial^4[\mu]/\partial r^4)$:

$$\begin{gathered} \rho\ddot{\mu}(r,t) + EI\mu''''(r,t) = f(r,t) = f(r_c,t)\delta(r - r_c), \\ y = \mu(r_0, t), \end{gathered} \tag{3.53}$$

where $f(r_c, t)$ is the applied force at location r_c in Fig. 3.6 and $y \triangleq \mu(r_0, t)$ is the displacement at point r_0. The boundary conditions on the partial differential equation (3.53) are given by (3.49).

A separation of the spatial variable r and the temporal variable t will now be illustrated to convert the partial differential equations (3.45) and (3.53) to ordinary differential equations. Let the set of basis functions $\boldsymbol{\Psi}_i(\mathbf{r})$, $i = 1, 2, \dots, N$, be *complete* (in the sense of Section 2.7.5.2, namely, for arbitrary "square-integrable" $\mu(r)$: $\min_{\mathbf{q}} \|\boldsymbol{\mu}(r) - \boldsymbol{\Psi}(r)\mathbf{q}\|^2 = 0$). Then,

$$\boldsymbol{\Psi}(r) \triangleq \left[\boldsymbol{\Psi}_1(r), \boldsymbol{\Psi}_2(r), \dots, \boldsymbol{\Psi}_N(r)\right]. \tag{3.54}$$

Hence, in the mean squared sense we have

$$\boldsymbol{\mu}(\mathbf{r}, t) = \boldsymbol{\Psi}(r)\mathbf{q}(t). \tag{3.55}$$

More details about this procedure due to Ritz are found in ref. [3.11]. Substitution of (3.54) and (3.55) into (3.45) yields

$$\rho(\mathbf{r})\boldsymbol{\Psi}(\mathbf{r})\ddot{\mathbf{q}} + \tilde{\mathscr{K}}\boldsymbol{\Psi}(\mathbf{r})\mathbf{q} = \mathbf{f}(\mathbf{r}, t). \tag{3.56}$$

Multiply (3.56) from the left by $\boldsymbol{\Psi}^T(\mathbf{r})$ and integrate (3.56) over the volume of the structure $\mathscr{E}$ to obtain

$$\left[\int_{\mathscr{E}} \boldsymbol{\Psi}^T(\mathbf{r})\rho(\mathbf{r})\boldsymbol{\Psi}(\mathbf{r})\, d\mathbf{r}\right]\ddot{\mathbf{q}} + \left[\int_{\mathscr{E}} \boldsymbol{\Psi}^T(\mathbf{r})\tilde{\mathscr{K}}\boldsymbol{\Psi}(\mathbf{r})\, d\mathbf{r}\right]\mathbf{q} = \int_{\mathscr{E}} \boldsymbol{\Psi}^T(\mathbf{r})\mathbf{f}(\mathbf{r}, t)\, d\mathbf{r}. \tag{3.57}$$

But by defining

$$\mathscr{M} \triangleq \int_{\mathscr{E}} \boldsymbol{\Psi}^T(\mathbf{r})\rho(\mathbf{r})\boldsymbol{\Psi}(\mathbf{r})\, d\mathbf{r}, \tag{3.58a}$$

$$\mathscr{K} \triangleq \int_{\mathscr{E}} \boldsymbol{\Psi}^T(\mathbf{r})\tilde{\mathscr{K}}\boldsymbol{\Psi}(\mathbf{r})\, d\mathbf{r}, \tag{3.58b}$$

$$\mathscr{f}(t) \triangleq \int_{\mathscr{E}} \boldsymbol{\Psi}^T(\mathbf{r})\mathbf{f}(\mathbf{r}, \mathbf{t})\, d\mathbf{r}, \tag{3.58c}$$

equation (3.57) can be written simply

$$\mathscr{M}\ddot{\mathbf{q}} + \mathscr{K}\mathbf{q} = \mathscr{f}(t), \tag{3.58d}$$

where $\mathscr{M}$ is commonly referred to as the "mass matrix" of the structure and $\mathscr{K}$ is referred to as the "stiffness matrix" of the structure. Now, for the beam, the volume integral of (3.53) also reduces to (3.57) by defining

$$\mathbf{u}^T \triangleq \left[f_1^T, f_2^T, \ldots, f_m^T\right], \tag{3.59a}$$

$$\mathscr{B} \triangleq \left[\boldsymbol{\Psi}^T(r_1), \boldsymbol{\Psi}^T(r_2), \ldots, \boldsymbol{\Psi}^T(r_m)\right]. \tag{3.59b}$$

Employing forces $f_i(t)$ at a finite number m of locations on the beam, then (3.52) and (3.58a, b) reduce (3.58d) to

$$\mathscr{M}\ddot{\mathbf{q}} + \mathscr{K}\mathbf{q} = \mathscr{B}\mathbf{u}. \tag{3.59c}$$

Note that if torques rather than forces are applied to the structure, the right-hand side of (3.59c) is changed as follows. Let the torque T applied at r_1 be described (in the limit as $\Delta r \to 0$) as a couple applied at $r_1 \pm \frac{1}{2}\Delta r$ (that is, equal and opposite forces of magnitude f_1 are separated by the small distance Δr, hence $T = f_1\,\Delta r$). Thus, for this couple the right-hand side of (3.59c) becomes

$$\begin{aligned}\mathscr{B}\mathbf{u} &= \lim_{\Delta r\to 0}\left[\boldsymbol{\Psi}^T(r_1)f_1 + \boldsymbol{\Psi}^T(r_2)f_2\right], \qquad f_1 = -f_2,\ r_2 = r_1 + \Delta r,\ T = f_1\,\Delta r \\ &= \lim_{\Delta r\to 0}\left\{\left[\boldsymbol{\Psi}^T(r_1+\Delta r) - \boldsymbol{\Psi}^T(r_1)\right]\frac{T}{\Delta r}\right. \\ &= \left[\frac{\partial \boldsymbol{\Psi}^T}{\partial r}(r_1)\right]T(r_1,t) = \boldsymbol{\Phi}^T(r_1)T(r_1,t) = \boldsymbol{\Phi}^T(r_1)T_1,\end{aligned} \tag{3.59d}$$

where $\boldsymbol{\Phi} \triangleq \partial\boldsymbol{\Psi}/\partial r$. Therefore, we shall interpret the right-hand side of (3.59c) as applying to either forces or torques simply by substituting T_i for f_i and $\boldsymbol{\Phi}_i$ for Ψ_i where a torque is applied.

We shall study the solution of ordinary differential equations in Chapter 4. Now, we wish to show how to calculate an acceptable choice of $\boldsymbol{\Psi}(r)$ in (3.58) for the Euler-Bernoulli beam. Consider first the homogeneous solution of (3.53):

$$\rho\boldsymbol{\Psi}(r)\ddot{\mathbf{q}}(t) + EI\boldsymbol{\Psi}(r)''''\mathbf{q}(t) = \mathbf{0}. \tag{3.60}$$

For separation of variables, the ratio $q_i(t)/\ddot{q}_i(t) = -\rho\Psi_i(r)/EI\Psi_i''''(r)$ must be constant. Hence, this constant is labeled $-\omega_i^2$, yielding

$$\rho\omega_i^2\Psi_i(r) = EI\Psi_i''''(r), \qquad \Psi_i(L) = \Psi_i(0) = \Psi_i''(L) = \Psi_i''(0) = 0, \tag{3.61}$$

where the boundary conditions on the Ψ_i satisfy the physical constraints (3.49). The matrix $\boldsymbol{\Omega}^2$ will be defined by

$$\boldsymbol{\Omega}^2 \triangleq \begin{bmatrix}\omega_1^2 & & \\ & \omega_2^2 & \\ & & \ddots\end{bmatrix}. \tag{3.62}$$

Now, (3.60) may be written

$$\rho\boldsymbol{\Psi}(r)\left[\ddot{\mathbf{q}}(t) + \boldsymbol{\Omega}^2\mathbf{q}(t)\right] = \mathbf{0} = \rho\sum_{i=1}^{\infty}\Psi_i(r)\left[\ddot{q}_i + \omega_i^2 q_i\right] = 0. \tag{3.63}$$

But since the $\Psi_i(r)$ are linearly independent on the interval $r \in [0, L]$, (3.63) requires that for all $i = 1, 2, \ldots,$

$$\ddot{q}_i(t) + \omega_i^2 q_i(t) = 0 \qquad (q_i(0),\ \dot{q}_i(0)\ \text{specified}). \tag{3.64}$$

Note that the Ritz method (3.55) has allowed a separation of variables, with (3.61) to be solved in the r domain and (3.64) to be solved in the time domain. We leave the time domain solution of linear differential equations to Chapter 4, but we must solve (3.61) to find $\boldsymbol{\Psi}(r)$.

Our first trial solution for (3.61) is

$$\Psi_i(r) = A_i \cosh \beta_i r + B_i \sinh \beta_i r + C_i \cos \beta_i r + D_i \sin \beta_i r,$$

where $\beta_i^4 \triangleq \rho\omega_i^2/EI$. Now, to check to see if this is an admissible solution, we must satisfy (3.61). The four boundary conditions require

$$\Psi_i(0) = 0 = A_i \cosh 0 + B_i \sinh 0 + C_i \cos 0 + D_i \sin 0$$

$$= A_i + C_i.$$

Likewise,

$$\Psi_i''(0) = 0 = A_i - C_i.$$

Hence, $A_i = C_i = 0$. Now,

$$\Psi_i(L) = 0 = B_i \sinh \beta_i L + D_i \sin \beta_i L,$$

$$\Psi_i''(L) = 0 = B_i \sinh \beta_i L - D_i \sin \beta_i L.$$

Add these two equations to get $2B_i \sinh \beta_i L = 0$, $\Rightarrow B_i = 0$, and subtract the two equations to get $D_i \sin \beta_i L = 0$, the nontrivial solution of which is $\sin \beta_i L = 0$ or

$$\beta_i = \frac{i\pi}{L}, \qquad i = 1, 2, 3, \ldots, \infty. \tag{3.65}$$

Hence,

$$\begin{aligned} \Psi_i(r) &= D_i \sin \frac{i\pi}{L} r, \\ \boldsymbol{\Psi}(r) &= [\Psi_1(r), \ldots, \Psi_\infty(r)]. \end{aligned} \tag{3.66}$$

Using (3.66), (3.61) becomes

$$\rho\omega_i^2 \Psi_i(r) = EI\left(\frac{i\pi}{L}\right)^4 \Psi_i(r),$$

which must be satisfied for nonzero $\Psi_i(r)$. Hence, ω_i must be

$$\omega_i = \sqrt{\frac{EI}{\rho}}\,\beta_i^2 = \sqrt{\frac{EI}{\rho}}\left(\frac{i\pi}{L}\right)^2, \qquad i = 1, 2, \ldots, \infty, \tag{3.67}$$

which establishes the frequencies of unforced vibration of the simply supported beam of Fig. 3.6. The functions $\Psi_i(r)$ are called the "mode shapes," and ω_i^2 are called the "mode frequencies."

We shall make the columns of $\boldsymbol{\Psi}(r)$ orthogonal by proper choice of D_i in (3.66). That is, columns of $\boldsymbol{\Psi}(r)$ are orthogonal on $r \in \mathscr{E}$ (r ranges throughout the elastic structure $\mathscr{E}$) subject to the weight $g(r)$ if

$$\Lambda = \int_{\mathscr{E}} \boldsymbol{\Psi}^*(\mathbf{r}) g(\mathbf{r}) \boldsymbol{\Psi}(r)\, d\mathbf{r} \tag{3.68}$$

is a diagonal matrix (see Definition 2.14). Note from (3.58a) and (3.58b) that (3.68) is simply the *mass* matrix of the structure if we choose the weight $g(\mathbf{r}) = \rho(\mathbf{r})$, and (3.68) is the *stiffness* matrix if we choose the weight $g(r) = \tilde{\mathscr{K}}$. We shall choose $g(r) = \rho(r)$ so that Λ is the mass matrix. For the present example, $\Psi_i(r)$ is given by (3.66) for some unspecified D_i. Substituting (3.66) into (3.68) yields

$$\Lambda_{ij} = \int_0^L \rho D_i^2 \sin\frac{i\pi}{L} r \sin\frac{j\pi}{L} r\, dr = \rho \frac{D_i^2 L}{2} \delta_{ij}. \tag{3.69}$$

If we choose $\Lambda_{ii} = 1$ (i.e., normalize the mass matrix to identity $\mathscr{M} = \boldsymbol{I}$), then (3.69) yields

$$D_i = \sqrt{\frac{2}{\rho L}}, \qquad i = 1, 2, \ldots. \tag{3.70}$$

Hence, from (3.66) and (3.70), the ith "mode shape" is

$$\Psi_i(r) = \sqrt{\frac{2}{\rho L}} \sin\frac{i\pi}{L} r, \qquad i = 1, 2, \ldots. \tag{3.71}$$

The first five mode shapes (3.71) are plotted in Fig. 3.7. The homogeneous solution of (3.61) is now completed, and the solution is described by the mode shapes (3.71) and Fig. 3.7 and frequencies (3.67).

The homogeneous solution of (3.64) is

$$q_i(t) = Q_i \sin(\omega_i t + \phi_i), \tag{3.72}$$

as can easily be checked by substituting into (3.64) if Q_i and ϕ_i are chosen to satisfy the initial conditions $Q_i \sin\phi_i = q_{i0}$, $\omega Q_i \cos\phi_i = \dot{q}_{i0}$. That is,

$$Q_i = \sqrt{q_{i0}^2 + \frac{\dot{q}_{i0}^2}{\omega_i^2}}, \qquad \phi_i = \tan^{-1} q_{i0} \frac{\omega_i}{\dot{q}_{i0}}. \tag{3.73}$$

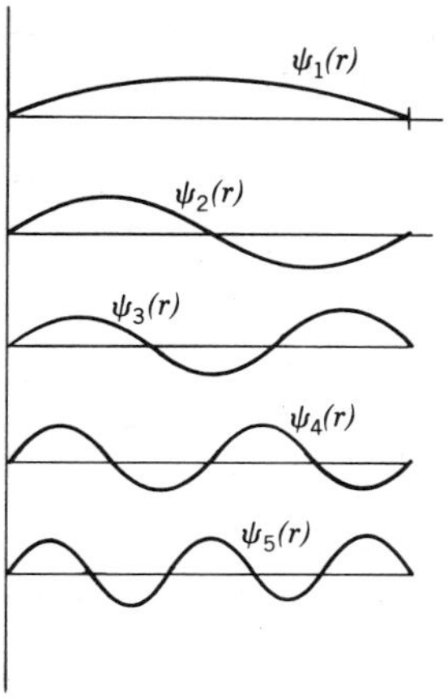

Figure 3.7 *Mode Shapes for Pinned Elastic Beam*

Thus, the complete solution of *unforced* (3.53) is

$$\mu(r,t) = \boldsymbol{\Psi}(r)\mathbf{q}(t) = \sum_{i=1}^{\infty} \Psi_i(r) q_i(t)$$

$$= \sum_{i=1}^{\infty} \sqrt{\frac{2}{\rho L}} \sin \frac{i\pi r}{L} Q_i \sin(\omega_i t + \phi_i) \tag{3.74}$$

written in terms of the mode shape $\Psi_i(r)$ and the modal amplitude $q_i(t)$.

Exercise 3.9

Show whether the $q_i(t)$ modal amplitudes given by (3.72) and (3.67) are a set of functions orthogonal on the interval $t \in [0, k\pi/\omega_1]$. Let $q_i(0) = 0$.

Exercise 3.10

If the $q_i(t)$ are orthogonal and $\lambda_i^2 \triangleq \int_{t_1}^{t_2} q_i^2(t)\, dt$, then employ Parseval's equality to show that for the complete set of functions $\Psi_i(r)$, $i = 1, 2, \ldots, \infty$,

$$\| \mu(r,t) \|^2 = \sum_{i=1}^{\infty} \| \Psi_i(r) \|^2 \lambda_i^2. \tag{3.75}$$

To investigate the forced solution of (3.30), first assume a solution of the form

$$\mu(r,t) = \boldsymbol{\Psi}(r)\mathbf{q}(t), \tag{3.76}$$

where $\boldsymbol{\Psi}(r)$ is given as before, (3.71), and $\mathbf{q}(t)$ is unknown. Since the $\Psi_i(r)$,

$i = 1, 2, \ldots, \infty$, form a *complete* set, they may be used as basis functions for the forced response as well. This task is reserved for Chapter 4, but we finalize the form of the differential equations for a special case. Multiplying (3.53) by $\Psi^T(r)$ and integrating with respect to r yields the earlier result (3.57) repeated here for the beam example with a torque applied at $r = r_c = 0$ [see Fig. 3.6 and equations (3.59)]:

$$\mathscr{M}\ddot{\mathbf{q}} + \mathscr{K}\mathbf{q} = \mathbf{\Phi}^T(r_c)T(r_c, t). \tag{3.77}$$

The unit normalization of the mode shapes (3.68) yields $\mathscr{M} = \mathbf{I}$. Hence,

$$\begin{aligned} \ddot{q}_i(t) + \omega_i^2 q_i(t) &= \phi_i(r_c)T(r_c, t), \\ y = \mu(r_0, t) &= \mathbf{\Psi}(r_0)\mathbf{q}(t), \end{aligned} \tag{3.78}$$

where the definition of ω_i^2 is given by (3.61) and (3.67). Note from (3.71) that the ith "mode slope" at the point of application r_c of the torque,

$$\phi_i(r_c) \triangleq \left.\frac{\partial \psi_i}{\partial r}\right|_{r=r_c} = \frac{i\pi}{L}\sqrt{\frac{2}{\rho L}}\cos\frac{i\pi}{L}r_c, \tag{3.79}$$

is zero if $r_c = [(2k+1)L]/2i$ for any integer k. These points are called "nodes" of the ith mode slope ψ_i. Thus, if a torque is applied at a node of the ith mode slope, then the torque cannot excite mode i. A systematic way to detect these "uncontrollable" events will be developed in Chapter 5.

Let the parameters for the beam example be given by

$$L = \pi, \quad \rho = \frac{2}{L}, \quad EI = \rho, \quad r_0 = 0.45L, \quad r_c = 0. \tag{3.80a}$$

Hence, from (3.67) and (3.71),

$$\omega_i = i^2, \quad \psi_i(r_0) \triangleq \sin(0.45\pi i), \quad b_i \triangleq i, \tag{3.80b}$$

where the definitions (p_i, m_i, b_i) are needed in the following exercise.

Exercise 3.11

For the simply supported beam (3.80), show that the equations (3.77) may be put into this form:

$$\begin{aligned} \dot{\mathbf{x}} &= \mathbf{A}\mathbf{x} + \mathbf{B}u, \quad \mathbf{x} \in \mathscr{R}^{2N}, \qquad u = T \text{ applied at } r = 0, \\ y &= \mathbf{C}\mathbf{x} \triangleq \mu(r_0, t), \end{aligned} \tag{3.81}$$

where

$$\mathbf{A} = \begin{bmatrix} \begin{matrix} 0 & 1 \\ -\omega_1^2 & 0 \end{matrix} & & & \\ & \begin{matrix} 0 & 1 \\ -\omega_2^2 & 0 \end{matrix} & & \\ & & \ddots & \\ & & & \ddots \end{bmatrix}, \quad \mathbf{B} = \begin{bmatrix} 0 \\ \mathbf{b}_1^T \\ 0 \\ \mathbf{b}_2^T \\ \vdots \end{bmatrix}, \quad \mathbf{x} = \begin{pmatrix} q_1 \\ \dot{q}_1 \\ q_2 \\ \dot{q}_2 \\ \vdots \end{pmatrix},$$

$$\mathbf{C} = [\Psi_1(r_0)\ 0\ \Psi_2(r_0)\ 0\ \Psi_3(r_0) \ldots].$$

Eigenvectors and eigenvalues of vector second-order systems have a special structure. We shall calculate them for the flexible beam example (3.78) through (3.80), with "modal damping" $\zeta_i < 1$ added,

$$\ddot{\mathbf{q}} + 2\boldsymbol{\zeta}\boldsymbol{\Omega}\dot{\mathbf{q}} + \boldsymbol{\Omega}^2\mathbf{q} = \mathscr{B}\mathbf{u}, \qquad q \in \mathscr{R}^N, \tag{3.82}$$

where $\boldsymbol{\zeta} = \text{diag}[\ldots \zeta_i \ldots]$ and $\Omega_{ij} = 0$, $\Omega_{ii} = \omega_i$. Define $\mathbf{x}^T = (\mathbf{q}^T, \dot{\mathbf{q}}^T)$. Then, $\dot{\mathbf{x}} = \mathbf{Ax} + \mathbf{Bu}$ yields

$$\mathbf{A} = \begin{bmatrix} \mathbf{0} & \mathbf{I} \\ -\boldsymbol{\Omega}^2 & -2\boldsymbol{\zeta}\boldsymbol{\Omega} \end{bmatrix}, \qquad \mathbf{B} = \begin{bmatrix} \mathbf{0} \\ \mathscr{B} \end{bmatrix}, \tag{3.83}$$

and $\mathbf{AE} = \mathbf{E}\Lambda$ yields

$$\mathbf{E} = \begin{bmatrix} \mathbf{I} & \mathbf{I} \\ \Lambda_{\mathbf{c}} & \overline{\Lambda}_{\mathbf{c}} \end{bmatrix}, \qquad \begin{aligned} \Lambda_{\mathbf{c}} &\triangleq \boldsymbol{\Omega}\left[-\boldsymbol{\zeta} + j\sqrt{\mathbf{I} - \boldsymbol{\zeta}^2}\right] \\ \overline{\Lambda}_{\mathbf{c}} &\triangleq \boldsymbol{\Omega}\left[-\boldsymbol{\zeta} - j\sqrt{\mathbf{I} - \boldsymbol{\zeta}^2}\right] \end{aligned}, \tag{3.84}$$

with the Jordan form for $\mathbf{A}$,

$$\Lambda = \begin{bmatrix} \Lambda_{\mathbf{c}} & \mathbf{0} \\ 0 & \overline{\Lambda}_{\mathbf{c}} \end{bmatrix} = \text{diagonal}.$$

The left eigenvectors are computed as follows:

$$\begin{aligned} \mathbf{E}^{-1} &= \begin{bmatrix} \mathbf{I} + [\overline{\Lambda}_{\mathbf{c}} - \Lambda_{\mathbf{c}}]^{-1}\Lambda_{\mathbf{c}} & -[\overline{\Lambda}_{\mathbf{c}} - \Lambda_{\mathbf{c}}]^{-1} \\ -[\overline{\Lambda}_{\mathbf{c}} - \Lambda_{\mathbf{c}}]^{-1}\Lambda_{\mathbf{c}} & [\overline{\Lambda}_{\mathbf{c}} - \Lambda_{\mathbf{c}}]^{-1} \end{bmatrix} \\ &= \begin{bmatrix} [\overline{\Lambda}_{\mathbf{c}} - \Lambda_{\mathbf{c}}]^{-1}\overline{\Lambda}_{\mathbf{c}} & -[\overline{\Lambda}_{\mathbf{c}} - \Lambda_{\mathbf{c}}]^{-1} \\ -[\overline{\Lambda}_{\mathbf{c}} - \Lambda_{\mathbf{c}}]^{-1}\Lambda_{\mathbf{c}} & [\overline{\Lambda}_{\mathbf{c}} - \Lambda_{\mathbf{c}}]^{-1} \end{bmatrix}. \end{aligned} \tag{3.85}$$

It may be shown that the spectral decomposition of **A** is

$$\mathbf{A} = \mathbf{E}\boldsymbol{\Lambda}\mathbf{E}^{-1} = \begin{bmatrix} \mathbf{0} & \mathbf{I} \\ -\bar{\boldsymbol{\Lambda}}_{\mathbf{c}}\boldsymbol{\Lambda}_{\mathbf{c}} & \boldsymbol{\Lambda}_{\mathbf{c}} + \bar{\boldsymbol{\Lambda}}_{\mathbf{c}} \end{bmatrix}.$$

3.8 Electrical Circuits

Previous examples describe mechanical dynamic systems. Other dynamic systems such as industrial processes, chemical reactions, economics, and so on, are not described in this text due to limitations in the author's background, although the theories to be developed for dynamic systems apply to such problems. This section will describe simple electrical dynamic systems.

Consider an electrical circuit described in Fig. 3.8. Kirchoff's laws may be applied by summing currents into each mode of the circuit or by summing the voltage drops around each current loop. Applying the latter yields for loop 1 with current i_1

$$-V_1 + V_2 + V_3 + V_4 = 0 \tag{3.86}$$

and for loop 2 with current i_2

$$-V_4 + V_5 = 0, \tag{3.87}$$

where V_1 is the voltage from the voltage source, V_2 is the voltage drop across the resistor R_1,

$$V_2 = R_1 i_1, \qquad R_1 = \text{resistance due to } R_1, \tag{3.88}$$

V_3 is the voltage drop across the inductor L,

$$V_3 = L\frac{di_1}{dt}, \qquad L = \text{inductance}, \tag{3.89}$$

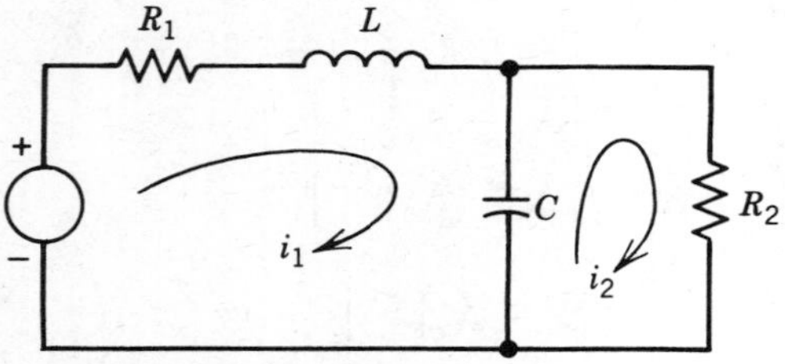

Figure 3.8 *RLC Circuit*

V_4 is the voltage drop across the capacitor from i_1,

$$V_4 = \frac{1}{C}\int_0^t (i_1 - i_2)\,dt + V_4(0), \qquad C = \text{capacitance}, \tag{3.90}$$

and V_5 is the voltage drop across the resistor R_2 due to i_2,

$$V_5 = R_2 i_2. \tag{3.91}$$

Hence, the equations

$$-V_1 + R_1 i_1 + L\frac{di_1}{dt} + \frac{1}{C}\int_0^t (i_1 - i_2)\,dt + V_4(0) = 0, \tag{3.92a}$$

$$-\frac{1}{C}\int_0^t (i_1 - i_2)\,dt - V_4(0) + R_2 i_2 = 0 \tag{3.92b}$$

are of differential, integral form. To convert them to purely differential equations, differentiate each with respect to time:

$$-\frac{dV_1}{dt} + R_1\frac{di_1}{dt} + L\frac{d^2 i_1}{dt^2} + \frac{1}{C}(i_1 - i_2) = 0, \tag{3.93a}$$

$$-\frac{1}{C}(i_1 - i_2) + R_2\frac{di_2}{dt} = 0. \tag{3.93b}$$

Define state variables $x_1 \triangleq i_1$, $x_2 \triangleq \dfrac{di_1}{dt}$, $x_3 \triangleq i_2$ and output $y \triangleq V_5 = R_2 i_2$ and input $u \triangleq \dot{V}_1$ to obtain

$$-u + R_1 x_2 + L\dot{x}_2 + \frac{1}{C}(x_1 - x_3) = 0, \tag{3.94a}$$

$$-\frac{1}{C}(x_1 - x_3) + R_2\dot{x}_3 = 0, \tag{3.94b}$$

$$y = R_2 x_3, \tag{3.94c}$$

or, in matrix form,

$$\begin{pmatrix}\dot{x}_1\\ \dot{x}_2\\ \dot{x}_3\end{pmatrix} = \begin{bmatrix} 0 & 1 & 0 \\ -\dfrac{1}{LC} & -\dfrac{R_1}{L} & \dfrac{1}{LC} \\ \dfrac{1}{R_2 C} & 0 & -\dfrac{1}{R_2 C}\end{bmatrix}\begin{pmatrix}x_1\\ x_2\\ x_3\end{pmatrix} + \begin{bmatrix}0\\1\\0\end{bmatrix}u, \tag{3.95}$$

$$y = [0 \quad 0 \quad R_2]\begin{pmatrix}x_1\\ x_2\\ x_3\end{pmatrix}.$$

The transfer function from $u(s)$ to $y(s)$ is

$$\frac{y(s)}{u(s)} = \frac{C(R_2Cs+1)}{s\left[LR_2C^2s^2 + (LC + R_1R_2C^2)s + (R_1+R_2)C\right]}. \tag{3.96}$$

Exercise 3.12

Show that if $u(t) \triangleq \dot{V}_1(t)$, $y(t) = V_4(t)$, and $R_2 = \infty$ in Fig. 3.8, then the transfer function from $u(s)$ to $y(s)$ is

$$y(s) = \frac{C}{LCs^2 + R_1Cs + 1}u(s). \tag{3.97}$$

Consider now the parallel circuit of Fig. 3.9. Apply Kirchoff's nodal equations (the sum of currents into a node is zero) yields

$$i_1 - i_L - i_C = 0, \tag{3.98}$$

where u is the current source $u = i_1$ and the output is the voltage across the circuit as shown in the Fig. 3.9. Hence,

$$y = L\frac{di_L}{dt} + R_1 i_L \tag{3.99}$$

and thus,

$$i_L(s) = \frac{1}{sL + R_1}y(s). \tag{3.100}$$

Also,

$$y = \frac{1}{C}\int i_C\,dt + R_2 i_C \tag{3.101}$$

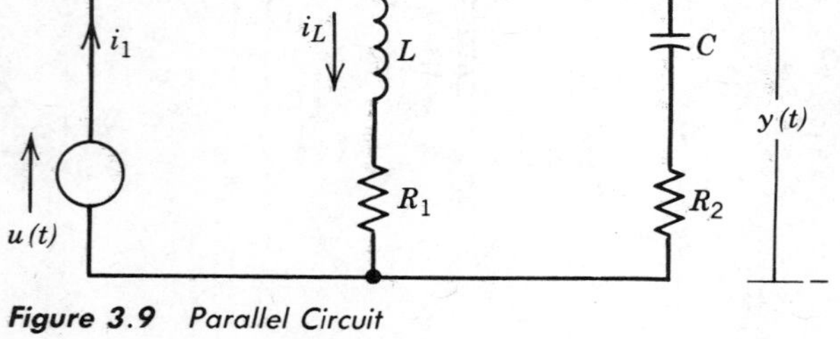

Figure 3.9 *Parallel Circuit*

and thus,

$$i_C(s) = \frac{1}{1/sC + R_2} y(s) = \frac{sC}{R_2Cs + 1} y(s). \tag{3.102}$$

The equation of motion becomes

$$i_1(s) = \left[\frac{1}{sL + R_1} + \frac{sC}{R_2Cs + 1}\right] y(s) = \frac{LCs^2 + [R_1 + R_2]Cs + 1}{R_2CLs^2 + (L + R_1R_2C)s + R_1} y(s) \tag{3.103}$$

or

$$\frac{y(s)}{u(s)} = \frac{R_2\{s^2 + [(L + R_1R_2C)/R_2LC]s + (R_1/R_2LC)\}}{s^2 + [(R_1 + R_2)/L]s + 1/LC}. \tag{3.104}$$

Exercise 3.13

Show that if $R_1 = R_2 = \sqrt{L/C}$, then for the circuit in Fig. 3.9,

$$\frac{y(s)}{u(s)} = R_2 = \sqrt{\frac{L}{C}}.$$

Hence, all frequencies pass through this circuit without attenuation. This is an example of an "all-pass" network to be discussed in Chapter 6.

Exercise 3.14

For the circuit in Fig. 3.10, find values of R, L, and C that will make this an all-pass network between current input $u(t)$ and voltage output $y(t)$.

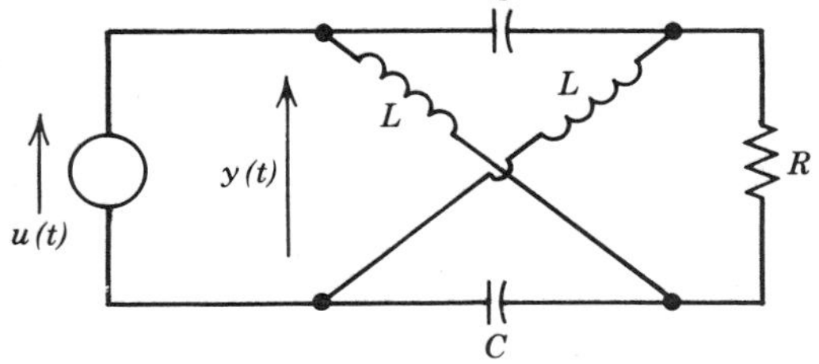

Figure 3.10 *RLC Circuit, Exercise 3.14*

References

3.1 W. R. Hamilton, *Philosophical Transactions*, 1834, pp. 247–308.

3.2 G. A. Bliss, *Lectures on the Calculus of Variations*. University of Chicago Press, 1946.

3.3 P. C. Hughes, *Spacecraft Attitude Dynamics*. Wiley, New York, 1986.

3.4 L. Meirovitch, *Elements of Vibration Analysis*. McGraw-Hill, New York, 1975.

3.5 P. W. Likins, *Elements of Engineering Mechanics*. McGraw-Hill, New York, 1973.

3.6 L. Meirovitch, *Methods of Analytical Dynamics*. McGraw-Hill, New York, 1970.

3.7 A. E. Frazho, "Abstract Bilinear Systems: The Forward Shift Approach," *Mathematical Systems Theory*, 14, 83–94, 1981.

3.8 C. Brani, G. Di Pillo, and G. Koch, "Bilinear Systems an Appealing Class of 'Nearly Linear' Systems in Theory and Applications," *IEEE Trans. Autom. Control*, AC-19(4), 1974.

3.9 S. G. Tzafestas, K. E. Anagnostou, and T. G. Pimenides, "Stabilizing Optimal Control of Bilinear Systems with a Generalized Cost," *Optimal Control Applications and Methods*, 5, 111–117, 1984.

3.10 T. Kailath, *Linear Systems*. Prentice-Hall, Englewood Cliffs, 1983.

3.11 G. Strang and G. J. Fix, *An Analysis of the Finite Element Method*. Prentice-Hall, Englewood Cliffs, N.J., 1973.

CHAPTER 4

Properties of State Space Realizations

The definition of state variables for a system was given in Chapter 3. Chapter 4 seeks to illustrate useful properties of the state space, first for single-input/single-output time-invariant systems (Section 4.1) and then for the general multi-input/multioutput system in Sections 4.2 through 4.6.

4.1 Developing State Space Realizations for Time-Invariant Systems

In this section we assume that the system of interest is linear and time-invariant, and we wish to practice the art of constructing state equations starting from three basic descriptions of a linear system:

(a) A set of mixed-order differential equations;

(b) A block diagram of the system;

(c) A Laplace transfer function between the system inputs and outputs.

Since there are many equivalent descriptions of a linear system, we refer to a particular state space model as just one "realization" of the system, hence the phrase "state space realization."

4.1.1 STATE SPACE REALIZATIONS FROM MIXED-ORDER DIFFERENTIAL EQUATIONS

Consider the system (3.14), which is a collection of equations of different order. A state space realization of (3.14) requires some work, and this task can be generally outlined as follows:

STEP 1: *Solve the equations explicitly for the highest derivatives of all variables appearing in the equations.* This is always possible for linear equations, not necessarily otherwise. For example, solve (3.14) for

$$\begin{pmatrix} \ddot{\alpha} \\ \ddot{r}_1 \\ \ddot{r}_2 \end{pmatrix} = \begin{pmatrix} f_1(\alpha, \dot{\alpha}, r_1, \dot{r}_1, r_2, \dot{r}_2, \theta) \\ f_2(\alpha, \dot{\alpha}, r_1, \dot{r}_1, r_2, \dot{r}_2, \theta) \\ f_3(\alpha, \dot{\alpha}, r_1, \dot{r}_1, r_2, \dot{r}_2, \theta) \end{pmatrix},$$

where f_1, f_2, and f_3 are each linear functions of $(\theta, \dot{\theta}, \gamma, \dot{\gamma}, \dot{\psi}, T_c, f)$.

STEP 2: *Define states to be the integrals of these left-hand side variables (we need at least as many integrals as appear in the right-hand side).* That is, let $(\alpha, \dot{\alpha}, r_1, \dot{r}_1, r_2, \dot{r}_2)$ be state variables for the example in step 1, and let θ and $\tilde{F} \triangleq (F - \rho V^2 - mG)$ be inputs.

STEP 3: *Write the differential equations for these state variables in the form*

$$\begin{aligned} \dot{\mathbf{x}} &= \mathbf{A}\mathbf{x} + \mathbf{B}\mathbf{u} + \mathbf{D}\mathbf{w}, \\ \mathbf{y} &= \mathbf{C}\mathbf{x} + \mathbf{H}\mathbf{u} + \mathbf{J}\mathbf{w}. \end{aligned} \tag{4.1}$$

For the example, verify that (3.16) *is a state space realization of* (3.14).

EXAMPLE 4.1

Write a state space realization of the system described by

$$\ddot{y} + \dot{z} + y + z = 0,$$

$$\dot{z} + \dot{y} = u.$$

Solution: From Step 1:

$$\begin{bmatrix} 1 & 1 \\ 0 & 1 \end{bmatrix} \begin{pmatrix} \ddot{y} \\ \dot{z} \end{pmatrix} = \begin{pmatrix} -y & -z \\ -\dot{y} & +u \end{pmatrix} \Rightarrow \begin{pmatrix} \ddot{y} \\ \dot{z} \end{pmatrix} = \begin{pmatrix} -y - z + \dot{y} - u \\ -\dot{y} + u \end{pmatrix}.$$

From Step 2:

$$\mathbf{x} = \begin{pmatrix} \dot{y} \\ y \\ z \end{pmatrix}.$$

From Step 3:

$$\dot{\mathbf{x}} = \begin{pmatrix} \ddot{y} \\ \dot{y} \\ \dot{z} \end{pmatrix} = \begin{pmatrix} -y - z + \dot{y} - u \\ \dot{y} \\ -\dot{y} + u \end{pmatrix}$$

$$= \begin{bmatrix} 1 & -1 & -1 \\ 1 & 0 & 0 \\ -1 & 0 & 0 \end{bmatrix} \begin{pmatrix} \dot{y} \\ y \\ z \end{pmatrix} + \begin{bmatrix} -1 \\ 0 \\ 1 \end{bmatrix} u = \mathbf{A}\mathbf{x} + \mathbf{B}u. \qquad \blacksquare$$

4.1.2 STATE SPACE REALIZATIONS FROM BLOCK DIAGRAMS

Block diagrams for linear dynamic systems need only three basic building blocks:

(a) A summer, denoted by the symbol

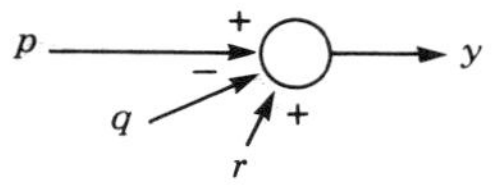

means $y = p - q + r$.

(b) A constant multiplier denoted by the symbol

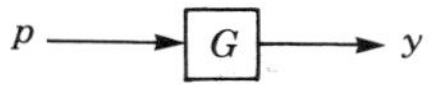

means $y = Gp$, $G =$ constant.

(c) An integrator, denoted by the symbol

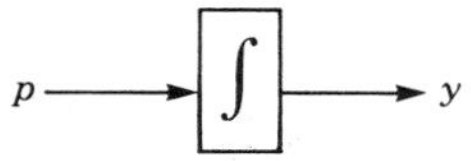

means $\dot{y} = p$, or, equivalently, $y(t) = \int_0^t p(\sigma)\, d\sigma + y(0)$.

The state space realization from block diagrams requires the following three steps:

STEP 1: *Expand the block diagram to its elementary form (employing only the basic elements* (a) *through* (c) *described above).* Examples will follow.

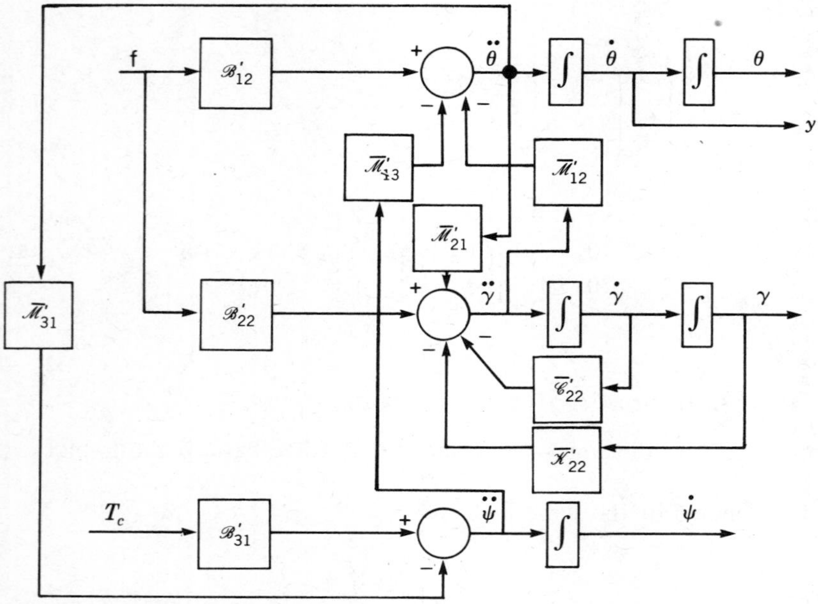

Figure 4.1 *Block Diagram of Backpack System (3.23)*

STEP 2: *Define the state variables as the outputs of each of the integrators.*

STEP 3: *Write the differential equations from the block diagram, employing definitions* (a), (b), *and* (c).

Exercise 4.1

Apply steps 2 and 3 to the block diagram of Fig 4.1 and verify (3.23). Develop a state space realization from the block diagram.

The Laplace transform is useful because it allows us to study differential equations using the tools of *algebra* instead of *calculus*. See Appendix E for a review of Laplace transforms.

From the earlier block diagram algebra, we wish to show that the same basic building blocks (summer, constant multiplier, and integrator) carry over to the Laplace domain simply by changing the integrator symbol from $\int$ to $1/s$. To see this, note from Table E.1 in Appendix E that $\dot{x}(s) = \mathscr{L}[\dot{x}(t)] = s\mathscr{L}[x(t)[= sx(s)$. Hence, the block diagram

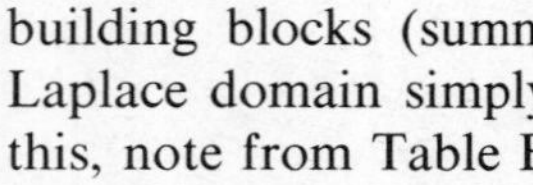

can be replaced by

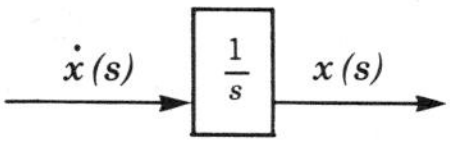

if we ignore initial conditions.

Exercise 4.2

Draw a detailed block diagram (using only summers, constant multipliers, and integrators) of (3.23) in the Laplace domain.

Now, after working through Exercise 4.2, the reader may wonder "why draw such a detailed block diagram" when the Laplace transform of (3.23) suggests a much simpler diagram. That is, the Laplace transform of (3.23) leads to (ignoring initial conditions)

$$\left\{\begin{bmatrix} \mathscr{M}_{11} & \mathscr{M}_{12} & \mathscr{M}_{13} \\ \mathscr{M}_{12} & \mathscr{M}_{22} & 0 \\ \mathscr{M}_{13} & 0 & \mathscr{M}_{33} \end{bmatrix} s^2 + \begin{bmatrix} 0 & 0 & 0 \\ 0 & \mathscr{C}_{22} & 0 \\ 0 & 0 & 0 \end{bmatrix} s + \begin{bmatrix} 0 & 0 & 0 \\ 0 & \mathscr{K}_{22} & 0 \\ 0 & 0 & 0 \end{bmatrix}\right\} \begin{pmatrix} \theta(s) \\ \gamma(s) \\ \psi(s) \end{pmatrix}$$

$$= \begin{bmatrix} 0 & \mathscr{B}_{12} \\ 0 & \mathscr{B}_{22} \\ \mathscr{B}_{31} & 0 \end{bmatrix} \begin{pmatrix} T_c(s) \\ f(s) \end{pmatrix} \tag{4.2}$$

or, if $u_1(s) \triangleq T_c(s)$, $u_2(s) \triangleq f(s)$, then

$$\begin{pmatrix} \theta(s) \\ \gamma(s) \\ \psi(s) \end{pmatrix} = [\mathscr{M}s^2 + \mathscr{D}s + \mathscr{K}]^{-1}\mathscr{B}\mathbf{u}(s) = \mathbf{G}(s)\mathbf{u}(s)$$

$$= \begin{bmatrix} G_{11}(s) & G_{12}(s) \\ G_{21}(s) & G_{22}(s) \\ G_{31}(s) & G_{32}(s) \end{bmatrix} \begin{pmatrix} u_1(s) \\ u_2(s) \end{pmatrix},$$

$$y(s) = [s \quad 0 \quad 0] \begin{pmatrix} \theta(s) \\ \gamma(s) \\ \psi(s) \end{pmatrix} = \mathbf{C}(s)\mathbf{q}(s),$$

and the relationship between the inputs $\mathbf{u}(s)$ and the output $\mathbf{y}(s)$ is simply

$$y(s) = s\theta(s) = s[G_{11}(s)u_1(s) + G_{12}(s)u_2(s)] \tag{4.3}$$

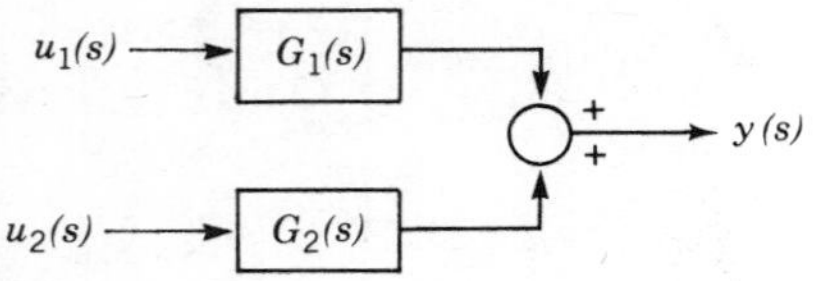

Figure 4.2 *Simple Block Diagrams*

or, in block diagram form in Fig. 4.2, where $\mathcal{G}_1(s) = sG_{11}(s)$, $\mathcal{G}_2(s) = sG_{12}(s)$. This is a simpler diagram than that in Fig. 4.1 or Exercise 4.2.

The more detailed block diagram is needed to specify the *states* of the system, whereas the simple diagram of Fig. 4.2 is all that is needed to specify the *input/output* relationships. These simple input/output relationships are called *transfer functions*.

For the *time-invariant* version of (3.9) (**A**, **B**, **C**, **D**, **H**, **J** all constant), the Laplace transform gives

$$\begin{aligned} s\mathbf{x}(s) &= \mathbf{Ax}(s) + \mathbf{Bu}(s) + \mathbf{Dw}(s) + \mathbf{x}(0), \\ y(s) &= \mathbf{Cx}(s) + \mathbf{Hu}(s) + \mathbf{Jw}(s), \end{aligned} \tag{4.4}$$

hence, $\mathbf{x}(s) = (s\mathbf{I} - \mathbf{A})^{-1}[\mathbf{Bu}(s) + \mathbf{Dw}(s) + \mathbf{x}(0)]$. Therefore,

$$\mathbf{y}(s) = \mathbf{C}(s\mathbf{I} - \mathbf{A})^{-1}[\mathbf{Bu}(s) + \mathbf{Dw}(s) + \mathbf{x}(0)] + \mathbf{Hu}(s) + \mathbf{Jw}(s), \tag{4.5}$$

which has the block diagram shown in Fig. 4.3.

Now, the *transfer function* between any particular input and any particular output is determined by *ignoring all other inputs and outputs*. The idea is best conveyed by illustration. The *transfer function* between $\mathbf{x}(0)$ and $\mathbf{y}(s)$ in Fig. 4.3 and

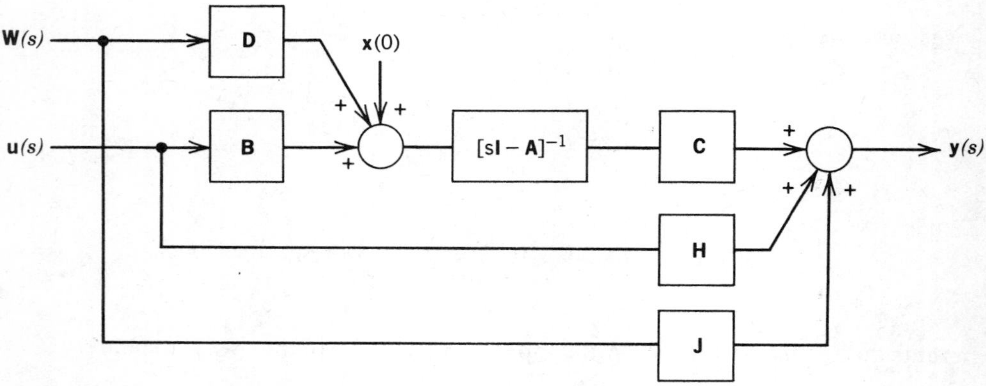

Figure 4.3 *Block Diagram of State Equations*

equation (4.5) is $\mathbf{C}(s\mathbf{I} - \mathbf{A})^{-1}$. The transfer function between $\mathbf{u}(s)$ and $\mathbf{y}(s)$ in Fig. 4.3 and equation (4.5) is $[\mathbf{C}(s\mathbf{I} - \mathbf{A})^{-1}\mathbf{B} + \mathbf{H}]$. The *transfer function* between $\mathbf{w}(s)$ and $\mathbf{y}(s)$ is $[\mathbf{C}(s\mathbf{I} - \mathbf{A})^{-1}\mathbf{D} + \mathbf{J}]$. These are *matrix* functions of s, but by a slight abuse of language we call them transfer functions instead of the more correct but cumbersome label *transfer matrices*. Clearly, (4.5) shows that the *total* response is the sum of the responses from $\mathbf{x}(0)$, $\mathbf{w}(t)$, and $\mathbf{u}(t)$,

$$\mathbf{y}(s) = \mathbf{G}(s)\mathbf{u}(s) + \mathbf{G}_w(s)\mathbf{w}(s) + \mathbf{G}_0(s)\mathbf{x}(0),$$

$$\mathbf{G}(s) = \mathbf{C}(s\mathbf{I} - \mathbf{A})^{-1}\mathbf{B} + \mathbf{H}, \quad \mathbf{G}_0(s) = \mathbf{C}(s\mathbf{I} - \mathbf{A})^{-1} \tag{4.6}$$

$$\mathbf{G}_w(s) = \mathbf{C}(s\mathbf{I} - \mathbf{A})^{-1}\mathbf{D} + \mathbf{J}.$$

EXAMPLE 4.2

Suppose y and u are vectors of different dimensions in (3.9) and (4.5), $\mathbf{y} \in \mathscr{R}^{n_y}$, $\mathbf{u} \in \mathscr{R}^{n_u}$. Find the transfer function between y_i and u_j.

Solution: By definition, the *transfer function* may be found by considering all other inputs to be zero. Note from (4.6) that if all excitation but u is zero, then,

$$y_i(s) = \sum_{j=1}^{n_u} G_{ij}(s)u_j(s), \qquad i = 1, \ldots, k. \tag{4.7}$$

Hence, the transfer function between $y_i(s)$ and $u_j(s)$ is simply $G_{ij}(s) = [\mathbf{C}(s\mathbf{I} - \mathbf{A})^{-1}\mathbf{B} + \mathbf{H}]_{ij}$. ■

EXAMPLE 4.3

Show that the transfer function between $u_j(s)$ and $y_i(s)$ is the *impulse response*; that is, $y_i(s)$ when $u_j(s) = \mathscr{L}[\delta(t)]$.

Solution: The transfer function between $u_j(s)$ and $y_i(s)$ is shown above to be $G_{ij}(s)$. Now, when $u_k(s) = 0\ \forall k \neq j$ and $u_j(s) = \mathscr{L}[\delta(t)] = 1$ from property 2 of Table E.1 (Appendix E), the response $y_i(s)$ is $y_i(s) = G_{ij}(s)u_j(s) = G_{ij}(s)$. ■

Thus, these three statements are equivalent definitions of a transfer function between $u_j(s)$ *and* $y_i(s)$:

1. $y_i(s)$ when $u_j(t) = \delta(t)$, and $u_\alpha(s) = 0$, $\alpha \neq j$, $\mathbf{x}(0) = 0$, $\mathbf{w}(s) = 0$;
2. $\dfrac{y_i(s)}{u_j(s)}$ when $u_\alpha(s) = 0$, $\alpha \neq j$, $\mathbf{x}(0) = 0$, $\mathbf{w}(s) = 0$;
3. $\dfrac{\partial y_i(s)}{\partial u_j(s)}$.

4.1.3 PROPERTIES OF TRANSFER FUNCTIONS

In this section it is only necessary to describe properties of the transfer functions $\mathbf{C}(s\mathbf{I} - \mathbf{A})^{-1}\mathbf{B}$ in (4.6). All other transfer functions in (4.6) can be interpreted from the following results by setting $\mathbf{B} = \mathbf{I}$ or $\mathbf{B} = \mathbf{D}$. Note that $\mathbf{C}(s\mathbf{I} - \mathbf{A})^{-1}\mathbf{B} = \mathbf{N}(s)/\Delta(s)$, where $\mathbf{N}(s)$ is a matrix polynomial of degree $Z \leq n_x - 1$ and $\Delta(s)$ is a scalar polynomial of degree n_x. Hence, $\Delta(s) = |sI - A|$, since

$$\mathbf{C}(s\mathbf{I} - \mathbf{A})^{-1}\mathbf{B} = \frac{\mathbf{N}(s)}{|sI - A|}, \qquad \mathbf{N} = \mathbf{C}[\mathrm{Co}(s\mathbf{I} - \mathbf{A})]^T\mathbf{B} = \mathbf{C}[\mathrm{adj}(s\mathbf{I} - \mathbf{A})]\mathbf{B}, \tag{4.8}$$

where $\mathrm{Co}(s\mathbf{I} - \mathbf{A})$ means the "cofactor matrix of $(s\mathbf{I} - \mathbf{A})$," and $\mathrm{Co}(s\mathbf{I} - \mathbf{A})^T$ is commonly called the "adjoint matrix," $\mathrm{adj}(s\mathbf{I} - \mathbf{A})$. Since (4.8) clearly shows that the denominator, or "characteristic polynomial" $\Delta(s)$, is the determinant of $[s\mathbf{I} - \mathbf{A}]$, it should also be clear that this is a polynomial of degree n_x where $\mathbf{A} \in \mathscr{R}^{n_x \times n_x}$. However, if there have been common factors in $\mathbf{N}(s)$ and $\Delta(s)$ that have been cancelled, then these pole-zero cancellations will reduce the degree of Δ below n_x. The degree of $\mathbf{N}(s)$ is Z and $Z \triangleq \deg\{\mathbf{N}(s)\} = \deg\{\mathbf{C}[\mathrm{adj}(s\mathbf{I} - \mathbf{A})]\mathbf{B}\} \leq n_x - 1$. This verifies that a system with the property $Z > n_x$ cannot be written in state space form. For example,

$$G(s) = \frac{s^2 + 1}{s + 2} = s - 2 + \frac{5}{s + 2} \tag{4.9}$$

cannot result from a system of the form (4.4). By taking the inverse Laplace, note that

$$G(s) = \frac{y(s)}{u(s)} = \frac{s^2 + 1}{s + 2} \Rightarrow sy(s) + 2y(s) = s^2u(s) + u(s)$$

$$\Rightarrow \dot{y}(t) + 2y(t) = \ddot{u}(t) + u(t), \tag{4.10}$$

where the number of derivatives of the input exceeds the number of derivatives of the output. Now, (4.9) describes a linear *noncausal* system. Two definitions help clarify the situation.

Definition 4.1

A system is said to be "causal" if the output $\mathbf{y}(t)$ *does not depend upon future inputs,* $\mathbf{u}(\tau)$ *for* $\tau > t$. *A system is said to be "noncausal" if it is not causal.*

Definition 4.2

A "dynamic" system is a causal system for which there exists a real output $\mathbf{y}(t)$ *for every real input* $\mathbf{u}(t)$.

EXAMPLE 4.4

Prove that (4.9) is a noncausal system.

Solution: Define

$$G_u(s) = s - 2, \qquad G_c(s) = \frac{5}{s+2}.$$

Then, from (4.9),

$$\frac{y(s)}{u(s)} = G_u(s) + G_c(s).$$

First, the $G_c(s)$ part is causal, since for this term only,

$$y(t) = \mathscr{L}^{-1}\left[\frac{5}{s+2}u(s)\right] = \int_0^t u(\sigma)5e^{-2(t-\sigma)}\,d\sigma.$$

Hence, $y(t)$ depends only on the integral of $u(\sigma)$, $\sigma \leq t$. For the $G_u(s)$ part, see that

$$\begin{aligned} y(t) &= \mathscr{L}^{-1}[(s-2)u(s)] = \mathscr{L}^{-1}[\dot{u}(s) - 2u(s)] \\ &= \dot{u}(t) - 2u(t). \end{aligned}$$

But the *excess* derivatives of $u(t)$ over derivatives of $y(t)$ lead to noncausality, since the time derivative of the input requires knowledge of $u(\tau)$ in the *neighborhood* of $\tau = t$ and not just for $\tau < t$. For example, the function $u(t)$ might not be differentiable at $\tau = t$, and one cannot know this except to look at the function on both sides of $\tau = t$. A less rigorous view is to look at the definition of the derivative,

$$\dot{u}(t^+) \triangleq \lim_{\Delta t \to 0}\left[\frac{u(t+\Delta t) - u(t)}{\Delta t}\right],$$

which involves $u(\tau)$, $\tau = t + \Delta t > t$. Therefore, $G_u(s)$ represents the noncausal part of the system, by Definition 4.1. Hence, a *causal* system requires that the degree of the numerator polynomial is equal to or less than the denominator polynomial in every transfer function of the system. Such transfer functions are said to be *proper*. If the degree of the numerator polynomial is *strictly less* than the degree of the denominator polynomial, the transfer function is said to be *strictly proper*. Whenever $H \neq 0$ in (4.4) and (4.6), then the system (i.e., the system transfer function) cannot be strictly proper. One may realistically argue that all real-world dynamic systems are strictly proper, due to (perhaps neglected) dynamics in input devices which prevent a direct feedforward path as shown by H in Fig. 4.3. ∎

This book is concerned with only linear dynamic systems, and the transfer functions will be proper. The complete dynamic system cannot have such forms as (4.10), although mathematical *subsystems* may have such forms. A subsystem is any *subset* of the differential equations which make up the complete system.

EXAMPLE 4.5

For the backpack problem of Fig. 3.2, let the measurement devices be described by:

Sensor 1:

$z_1 \triangleq \dot{\theta}$ (a rate gyroscope might realistically provide this measurement z_1),

Sensor 2:

$z_2 = \ddot{\theta}$ (an angular accelerometer might realistically provide this measurement).

Draw two block diagrams of the complete system with outputs z_1 and z_2, such that one has noncausal subsystems and the other has only causal subsystems. Show that the complete system satisfies Definition 4.2 for a dynamic system.

Solution: The block diagram of Fig. 4.4a shows only the causal subsystems $G(s)$, $H_1(s)$, and $H_2(s)$. Alternately, the block diagram of Fig. 4.4b represents exactly the same system but with different subsystems $G(s)$, $H_1'(s)$, and $H_2'(s)$, and $H_2'(s)$ is a noncausal subsystem. Of course, the difference here is that $H_1(s)$ and $H_2(s)$ represent real dynamic subsystems (an accelerometer and a rate gyro), whereas $H_2'(s)$ in Fig. 4.4b does *not* represent a real dynamic subsystem, since z_2 in the statement of the problem is obtained by a *direct* measurement of the physical quality $\ddot{\theta}$ *rather* than a differentiation of $\dot{\theta}$ as Fig. 4.4b might suggest. To show that the complete system is causal, we have only to show that the transfer functions from $u(s)$ to $z_1(s)$, $z_2(s)$ have numerator polynomials in s of equal to

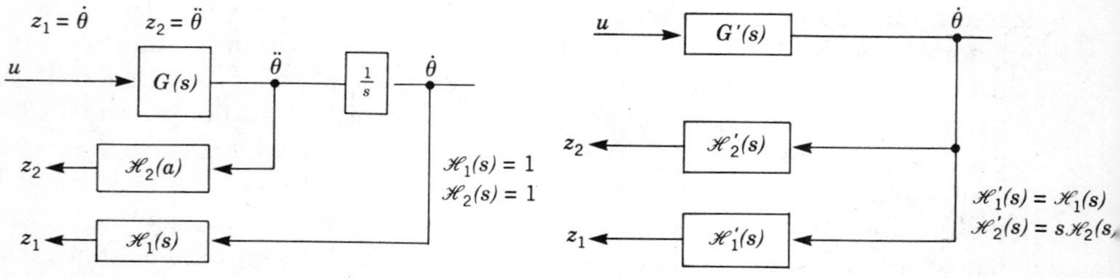

Figure 4.4 *Block Diagram with Causal, Noncausal Subsystems*

or less than the degree of the denominator polynomials. From Fig. 4.4a,

$$z_1(s) = \left[\frac{1}{s}G(s)H_1(s)\right]u(s) = \left[\frac{1}{s}G(s)\right]u(s),$$

$$z_2(s) = [G(s)H_2(s)]\,u(s) = [G(s)]\,u(s),$$

$$G(s) = \frac{\overline{\mathscr{B}}_{12}\left(\overline{\mathscr{M}}_{11}s^2 + \overline{\mathscr{C}}_{22}s + \overline{\mathscr{K}}_{22}\right) + \overline{\mathscr{M}}_{12}\overline{\mathscr{B}}_{22}}{\overline{\mathscr{M}}_{11}\left(\overline{\mathscr{M}}_{22}s^2 + \overline{\mathscr{C}}_{22}s + \overline{\mathscr{K}}_{22}\right) - \overline{\mathscr{M}}_{12}^2}.$$

From Fig. 4.4b,

$$z_1(s) = \left[\frac{1}{s}G(s)H_1'(s)\right]u(s) = \left[\frac{1}{s}G(s)\right]u(s),$$

$$z_2(s) = \left[\frac{1}{s}G(s)H_2'(s)\right]u(s) = [G(s)]\,u(s).$$

Thus, either way the overall result is the same, and $G(s)$ and $(1/s)G(s)$ are causal by Definition 4.1 and Example 4.5.

The dynamic systems we shall study, then, have state space descriptions $\dot{\mathbf{x}} = \mathbf{Ax} + \mathbf{Bu} + \mathbf{Dw}$, $\mathbf{y} = \mathbf{Cx} + \mathbf{Hu} + \mathbf{Jw}$; but it is possible to define mathematical *subsystems* that are not causal, and therefore have no state space description of this form. Note that linear systems (disturbances not shown) of the form

$$\dot{\mathbf{x}} = \mathbf{Ax} + \mathbf{Bu} + \sum_{i=1}^{n \geq n_x} \hat{\mathbf{B}}_i\left[\frac{d^i}{dt^i}\mathbf{u}\right],$$

$$\mathbf{y} = \mathbf{Cx} + \mathbf{Hu} + \sum_{i=1}^{n \geq n_x} \hat{\mathbf{H}}_i\left[\frac{d^i}{dt^i}\mathbf{u}\right]$$

are not causal, and their transfer functions are not proper.

The transfer function $\mathbf{G}(s)$ can be expanded into a power series in $(1/s)$ and takes the form

$$\mathbf{G}(s) = \mathbf{H} + \mathbf{C}(s\mathbf{I} - \mathbf{A})^{-1}\mathbf{B} = \sum_{i=0}^{\infty} \frac{1}{s^{i+1}}\mathbf{M}_i + \mathbf{H}. \tag{4.11}$$

The matrices $\mathbf{M}_i$, $i = 0, 1, \ldots, \infty$, are called the *Markov parameters* and have the value

$$\mathbf{M}_i = \mathbf{CA}^i\mathbf{B}. \tag{4.12}$$

The partial fraction expansion of $\mathbf{G}(s)$ yields

$$\mathbf{G}(s) \triangleq \mathbf{H} + \mathbf{C}(s\mathbf{I} - \mathbf{A})^{-1}\mathbf{B} = \sum_{i=1}^{p}\left[\frac{\mathbf{J}_i}{(s-\lambda_i)^{m_i}} + \sum_{\alpha=1}^{m_i-1}\frac{\mathbf{P}_{m_i-\alpha}}{(s-\lambda_i)^{m_i-\alpha}}\right] + \mathbf{H}, \tag{4.13}$$

where $\mathbf{J}_i$ is the "residue" of the root λ_i, m_i is the multiplicity of the root λ_i of the *characteristic polynomial* $|s\mathbf{I} - \mathbf{A}| = 0$ [see (2.163)], and $\mathbf{P}_{m_i-\alpha}$ are coefficients required in the case of repeated roots $m_i > 1$. From the partial fraction expansion (4.13),

$$\mathbf{J}_i = \mathbf{G}(s)(s-\lambda_i)^{m_i}\big|_{s=\lambda_i}, \tag{4.14a}$$

$$\mathbf{P}_{m_i-1} = \left\{\frac{d}{ds}\left[\mathbf{G}(s)(s-\lambda_i)^{m_i}\right]\right\}_{s=\lambda_i}$$

$$\vdots$$

$$P_{m_i-\alpha} = \frac{1}{\alpha!}\left\{\frac{d^\alpha}{ds^\alpha}\left[\mathbf{G}(s)(s-\lambda_i)^{m_i}\right]\right\}_{s=\lambda_i} \tag{4.14b}$$

$$\vdots$$

$$\mathbf{P}_1 = \frac{1}{(m_i-1)!}\left\{\frac{d^{m_i-1}}{ds^{m_i-1}}\left[\mathbf{G}(s)(s-\lambda_i)^{m_i}\right]\right\}_{s=\lambda_i}$$

Since $\mathbf{A}$ is $n_x \times n_x$, there are n_x roots of the characteristic polynomial. Hence, $\sum_{i=1}^{p} m_i = n_x$.

Finally, $\mathbf{G}(s)$ will frequently be decomposed into the form

$$[\mathbf{G}(s)]_{\alpha\beta} = \left[\mathbf{H} + \mathbf{C}(s\mathbf{I} - \mathbf{A})^{-1}\mathbf{B}\right]_{\alpha\beta} = \frac{\prod_{j=1}^{Z_{\alpha\beta}}(s-\psi_{\alpha\beta j})}{\prod_{i=1}^{p}(s-\lambda_i)^{m_i}}, \tag{4.15}$$

where $\psi_{\alpha\beta j}$ is the jth root of the total of $Z_{\alpha\beta}$ roots of the numerator polynomial of $[\mathbf{G}(s)]_{\alpha\beta}$. The $\psi_{\alpha\beta j}$, $j = 1, \ldots, Z_{\alpha\beta}$, are called the "zeros" of the transfer function $[\mathbf{G}(s)]_{\alpha\beta}$. Note that the "poles" of $\mathbf{G}(s)$ are the zeros of the characteristic polynomial $|s\mathbf{I} - \mathbf{A}| = 0$, and this is influenced only by $\mathbf{A}$.

Transmission Zeros

The zeros of the transfer function $[\mathbf{G}(s)]_{\alpha\beta}$ are not necessarily zeros of the transfer function matrix $\mathbf{G}(s)$. To define zeros for the matrix case and explain this curious situation, we need the following:

Definition 4.3

The complex number ψ is a transmission zero of the transfer function $\mathbf{G}(s) \triangleq \mathbf{C}(s\mathbf{I} - \mathbf{A})^{-1}\mathbf{B} + \mathbf{H}$ iff ψ satisfies

$$\mathbf{Q} \triangleq \begin{bmatrix} \psi\mathbf{I} - \mathbf{A} & -\mathbf{B} \\ \mathbf{C} & \mathbf{H} \end{bmatrix}, \qquad |\mathbf{Q}^*\mathbf{Q}| = 0. \tag{4.16}$$

The physical significance of this definition is as follows. For some $\mathbf{u}$ there exists an input $\mathbf{u}(t) = \boldsymbol{\mu} e^{\psi t}$ such that the output $\mathbf{y}(t)$ is identically zero, $\mathbf{y}(t) \equiv 0$, if ψ is a zero of $\mathbf{C}(s\mathbf{I} - \mathbf{A})^{-1}\mathbf{B} + \mathbf{H}$. To verify this, we require $\mathbf{y}(t) \equiv 0$ and

$$\begin{aligned} \dot{\mathbf{x}} &= \mathbf{A}\mathbf{x} + \mathbf{B}\mathbf{u}, \qquad \mathbf{u} = \boldsymbol{\mu} e^{\psi t}, \\ \mathbf{y} &= \mathbf{C}\mathbf{x} + \mathbf{H}\mathbf{u}. \end{aligned} \tag{4.17}$$

Now, if $\mathbf{y}(t) \equiv 0$, then

$$\mathbf{y}^{(k)}(t) \triangleq \frac{d^k}{dt^k}\mathbf{y}(t) = 0 \qquad \text{for all } k > 0.$$

Hence,

$$\mathbf{y}(0) = \mathbf{C}\mathbf{x}(0) + \mathbf{H}\mathbf{u}(0) = \mathbf{0},$$

$$\dot{\mathbf{y}}(0) = \mathbf{C}[\mathbf{A}\mathbf{x}(0) + \mathbf{B}\mathbf{u}(0)] + \mathbf{H}\dot{\mathbf{u}}(0) = \mathbf{0},$$

$$\vdots$$

$$\mathbf{y}^{(k)}(0) = \mathbf{C}\mathbf{A}^k\mathbf{x}(0) + \mathbf{C}\sum_{i=1}^{k} \mathbf{A}^{k-i}\mathbf{B}\mathbf{u}^{(i-1)}(0) + \mathbf{H}\mathbf{u}^{(k)}(0) = \mathbf{0}.$$

But since $\mathbf{u}^{(k)}(t) = \boldsymbol{\mu}\psi^k e^{\psi t}$, $\mathbf{u}^{(k)}(0) = \boldsymbol{\mu}\psi^k$, and this substitution into the $\mathbf{y}^{(k)}(0) = \mathbf{0}$ equation yields

$$\mathbf{y}^{(k)}(0) = \mathbf{C}\left\{\mathbf{A}^k\mathbf{x}(0) + \sum_{i=1}^{k} \mathbf{A}^{k-i}\psi^{i-1}\mathbf{B}\boldsymbol{\mu}\right\} + \psi^k\mathbf{H}\boldsymbol{\mu} = \mathbf{0} \tag{4.18}$$

for $k \geq 1$; and for $k = 0$,

$$\mathbf{y}(0) = \mathbf{C}\mathbf{x}(0) + \mathbf{H}\boldsymbol{\mu} = 0. \tag{4.19}$$

From (4.18), note that

$$\psi^{k-1}\mathbf{H}\boldsymbol{\mu} = -\mathbf{C}\left\{\mathbf{A}^{k-1}\mathbf{x}(0) + \sum_{i=1}^{k-1} \mathbf{A}^{k-1-i}\psi^{i-1}\mathbf{B}\boldsymbol{\mu}\right\}. \tag{4.20}$$

Now write the last term in (4.18) as $\psi(\psi^{k-1}\mathbf{H}\boldsymbol{\mu})$, using (4.20) to reduce (4.18) to

$$\mathbf{C}\left\{\mathbf{A}^{k}\mathbf{x}(0) + \sum_{i=1}^{k} \mathbf{A}^{k-i}\psi^{i-1}\mathbf{B}\boldsymbol{\mu}\right\} - \psi\mathbf{C}\left\{\mathbf{A}^{k-1}\mathbf{x}(0) + \sum_{i=1}^{k-1} \mathbf{A}^{k-1-i}\psi^{i-1}\mathbf{B}\boldsymbol{\mu}\right\} = \mathbf{0}$$

or simply

$$-\mathbf{C}\mathbf{A}^{k-1}\{(\psi\mathbf{I} - \mathbf{A})\mathbf{x}(0) - \mathbf{B}\boldsymbol{\mu}\} = 0. \tag{4.21}$$

Now, assume that the columns of the matrix

$$\mathbf{W}_0 = \begin{bmatrix} \mathbf{C} \\ \mathbf{CA} \\ \vdots \\ \mathbf{CA}^{k-1} \end{bmatrix}$$

are linearly independent (for physical interpretations of this assumption we shall wait for Chapter 5). Then, (4.21) and (4.19) together are equivalent to the statement

$$\begin{bmatrix} \psi\mathbf{I} - \mathbf{A} & -\mathbf{B} \\ \mathbf{C} & \mathbf{H} \end{bmatrix}\begin{pmatrix} \mathbf{x}(0) \\ \boldsymbol{\mu} \end{pmatrix} = \mathbf{0} \tag{4.22}$$

for nontrivial $\mathbf{x}(0)$, $\boldsymbol{\mu}$. This requires the columns of $\mathbf{Q}$ to be linearly dependent. From Chapter 2 linear algebra, this implies equation (4.16). ∎

Exercise 4.3

If $\mathbf{H}^{-1}$ exists, show that the transmission zeros of $(\mathbf{A}, \mathbf{B}, \mathbf{C}, \mathbf{H})$ are the eigenvalues of the matrix $[\mathbf{A} - \mathbf{B}\mathbf{H}^{-1}\mathbf{C}]$ and that ψ is a transmission zero if $|\mathbf{Q}| = 0$.

Exercise 4.4

For the system described by

$$\mathbf{A} = \begin{bmatrix} 0 & -2 & 0 \\ 0 & 0 & -1 \\ 0 & 0 & 0 \end{bmatrix}, \quad \mathbf{B} = \begin{bmatrix} 1 & 0 \\ 0 & 0 \\ 0 & 1 \end{bmatrix}, \quad \mathbf{H} = 0,$$

$$\mathbf{C} = \begin{bmatrix} 1 & 0 & 0 \\ 0 & 1 & 0 \end{bmatrix},$$

show that $\mathbf{G}(s) = \mathbf{C}(s\mathbf{I} - \mathbf{A})^{-1}\mathbf{B}$ has no zeros.

Exercise 4.5

Show that the system described by

$$\mathbf{B} = \begin{bmatrix} 0 & 0 \\ 1 & 0 \\ 0 & 1 \end{bmatrix}$$

and (**A**, **C**) as above has an infinite number of zeros, even though *each* transfer function within the matrix $\mathbf{G}(s) = \mathbf{C}(s\mathbf{I} - \mathbf{A})^{-1}\mathbf{B}$ has no zero.

Exercise 4.6

Prove that the steady-state response, if it exists, to unit step inputs $u_i(t) = 1$, $i = 1, \ldots, n_u$, is

$$\lim_{t \to \infty} \mathbf{y}(t) = \sum_{i=1}^{n_u} \left(\mathbf{C}\mathbf{A}^{-1}\mathbf{b}_i + \mathbf{h}_i\right) = \sum_{i=1}^{p} \left[\frac{J_i}{(-\lambda_i)^{m_i}} + \sum_{\alpha=1}^{m_i - 1} \frac{\mathbf{P}_{m_i - \alpha}}{-\lambda_i^{m_i - \alpha}} \right] \mathbf{1} + \sum_{i=1}^{n_u} \mathbf{h}_i, \tag{4.23}$$

where $\mathbf{h}_i$ is the ith column of **H**, and $\mathbf{1}^T = (1, \ldots, 1)$. (*Hint*: Use final value theorem item 10 in Table E.1 Appendix E.)

4.1.4 STATE SPACE REALIZATIONS FROM TRANSFER FUNCTIONS OF SINGLE-INPUT / SINGLE-OUTPUT SYSTEMS

State space realizations are not needed to provide the input/output descriptions represented by transfer functions.

EXAMPLE 4.6

Find the transfer function between the output $y(s)$ and $u(s)$ in Example 4.1.

Solution: Taking Laplace transforms (ignoring initial conditions, since only the transfer function between $y(s)$ and $z(s)$ is required),

$$\begin{aligned} s^2 y(s) + sz(s) + y(s) + z(s) &= 0, \\ sy(s) + sz(s) &= u(s). \end{aligned} \tag{4.24}$$

These are *algebraic* equations (thanks to the Laplace transforms) to be solved for the output $y(s)$ in terms of $u(s)$. Hence, putting the equations in the form $y(s) = G(s)u(s)$ will immediately give the required transfer function $G(s)$. From (4.24),

$$\begin{aligned} \begin{bmatrix} s^2 + 1 & s + 1 \\ s & s \end{bmatrix} \begin{pmatrix} y(s) \\ z(s) \end{pmatrix} &= \begin{pmatrix} 0 \\ 1 \end{pmatrix} u(s), \\ y(s) &= [1 \quad 0] \begin{pmatrix} y(s) \\ z(s) \end{pmatrix}. \end{aligned} \tag{4.25}$$

Hence,

$$y(s) = [1 \quad 0]\begin{bmatrix} s^2+1 & s+1 \\ s & s \end{bmatrix}^{-1}\begin{pmatrix} 0 \\ 1 \end{pmatrix} u(s) = G(s)u(s), \tag{4.26}$$

where

$$G(s) = [1 \quad 0]\begin{bmatrix} s & -(s+1) \\ -s & s^2+1 \end{bmatrix}\begin{pmatrix} 0 \\ 1 \end{pmatrix}\frac{1}{s^3-s^2}$$

$$= \frac{(s+1)}{-s^3+s^2}.$$

We have intentionally used matrix methods in (4.25), and (4.26) to illustrate the systematic technique needed for working with larger, more complex problems, even though this simple example could have been solved by simply eliminating $z(s)$ in (4.24). ■

So the point is this. Only a transfer function might be available, and one might need to *develop* a state space realization from the given transfer function. Such a development is the purpose of this section, and the process is depicted by Fig. 4.5.

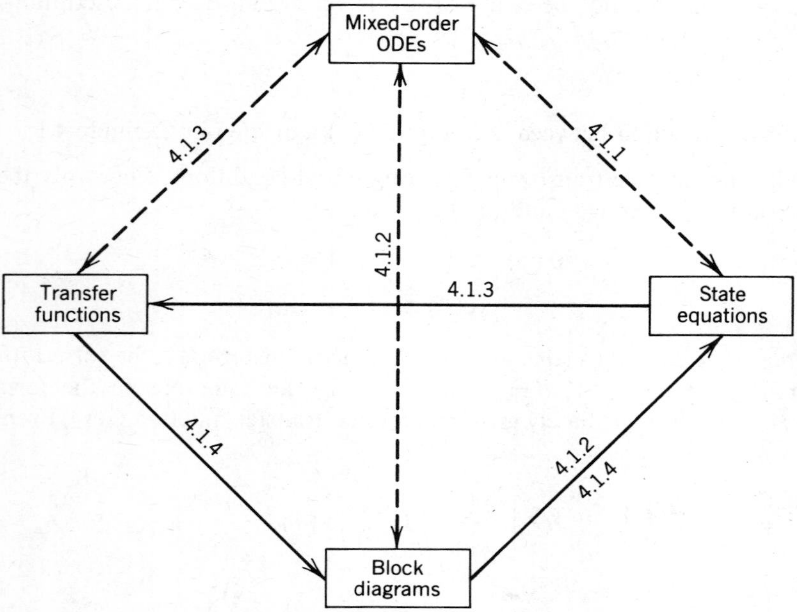

Figure 4.5 *Constructions of State Space Realizations*

The text which treats the transition from one form of the system description to another is identified on the lines connecting the blocks in Fig. 4.5. Note from Section 4.1.3 the ease with which one can go from a state space description to a transfer function. Figure 4.5 indicates a more complicated procedure for the reverse operation. To go from transfer functions to a state space realization, we shall always proceed in a two-step process: from the *transfer function* to a *block diagram* to the *state equations*, as illustrated by the solid lines and arrows in the lower half of Fig. 4.5.

4.1.4.1 Rectangular Form

This method takes its name from the form of the block diagram that emerges. An example illustrates the method.

EXAMPLE 4.7

Find a state space realization of the linear dynamic system described by

$$y(s) = G(s)u(s), \qquad G(s) = \frac{c_4 s^4 + c_3 s^3 + c_2 s^2 + c_1 s + c_0}{s^4 + d_3 s^3 + d_2 s^2 + d_1 s + d_0}. \tag{4.27}$$

Solution: The rectangular method proceeds as follows:

1. Solve for the highest time derivative of y in the time domain. This is equivalent in the Laplace domain to solving for $[y(s)s^4]$ in (4.27). This leads to

$$\begin{aligned} y(s)\left(s^4 + d_3 s^3 + d_2 s^2 + d_1 s + d_0\right) \\ = u(s)\left(c_4 s^4 + c_3 s^3 + c_2 s^2 + c_1 s + c_0\right), \end{aligned} \tag{4.28a}$$

$$\begin{aligned} y(s)s^4 = u(s)\left(c_4 s^4 + c_3 s^3 + c_2 s^2 + c_1 s + c_0\right) \\ -y(s)\left(d_3 s^3 + d_2 s^2 + d_1 s + d_0\right). \end{aligned} \tag{4.28b}$$

2. Group terms having the same derivative in (4.28b);

$$\begin{aligned} y(s) = c_4 u(s) + [c_3 u(s) - d_3 y(s)]\frac{1}{s} + [c_2 u(s) - d_2 y(s)]\frac{1}{s^2} \\ + [c_1 u(s) - d_1 y(s)]\frac{1}{s^3} + [c_0 u(s) - d_0 y(s)]\frac{1}{s^4}. \end{aligned} \tag{4.28c}$$

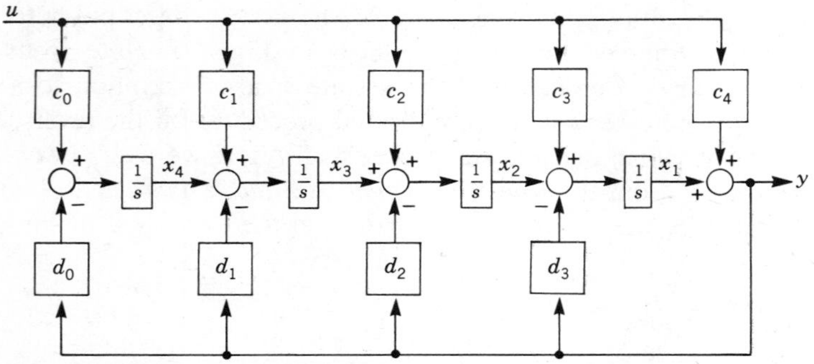

Figure 4.6 Example 4.8, The Rectangular Method

3. Draw a block diagram of (4.28c). Figure 4.6 is a block diagram of (4.28c). The states are the outputs of the integrators. Hence, from the diagram,

$$
\begin{aligned}
\dot{x}_1 &= -d_3 y + c_3 u + x_2, \\
\dot{x}_2 &= -d_2 y + c_2 u + x_3, \\
\dot{x}_3 &= -d_1 y + c_1 u + x_4, \\
\dot{x}_4 &= -d_0 y + c_0 u,
\end{aligned} \tag{4.29}
$$

and $y = x_1 + c_4 u$. Now, in matrix form,

$$
\dot{\mathbf{x}} = \begin{bmatrix} -d_3 & 1 & 0 & 0 \\ -d_2 & 0 & 1 & 0 \\ -d_1 & 0 & 0 & 1 \\ -d_0 & 0 & 0 & 0 \end{bmatrix} \mathbf{x} + \begin{bmatrix} c_3 - d_3 c_4 \\ c_2 - d_2 c_4 \\ c_1 - d_1 c_4 \\ c_0 - d_0 c_4 \end{bmatrix} u, \tag{4.30}
$$

$$
y = [1 \quad 0 \quad 0 \quad 0] x + [c_4] u.
$$

■

Note from Fig. 4.6 that $c_4 \neq 0$ serves to introduce a "feedforward" term since this term is not integrated before finding its way to the output. Hence, for systems that do not have such feedforward terms, the matrix **H** in (4.17) is zero. The reader should now be sufficiently comfortable with these ideas that (4.30) can now be obtained immediately by inspection of Fig. 4.6, without laboring over the preliminary steps above (4.30).

EXAMPLE 4.8

A state vector is defined in (4.30) for the system (4.27). But for the system (4.27), suppose the only initial conditions given are $y(0)$, $\dot{y}(0)$, $\ddot{y}(0)$, $\dddot{y}(0)$, $u(0)$, $\dot{u}(0)$, $\ddot{u}(0)$, and $\dddot{u}(0)$. Determine the initial state in (4.30).

Solution: From (4.30), differentiate $y(t)$ three times and set $t = 0$:

$$\mathbf{y}(0) = \mathbf{Cx}(0) + \mathbf{Hu}(0),$$

$$\dot{\mathbf{y}}(0) = \mathbf{C}[\mathbf{Ax}(0) + \mathbf{Bu}(0)] = \mathbf{H}\dot{\mathbf{u}}(0),$$

$$\ddot{\mathbf{y}}(0) = \mathbf{C}\{\mathbf{A}[\mathbf{Ax}(0) + \mathbf{Bu}(0)] + \mathbf{B}\dot{\mathbf{u}}(0)\} + \mathbf{H}\ddot{\mathbf{u}}(0),$$

$$\dddot{\mathbf{y}}(0) = \mathbf{C}\{\mathbf{A}^2[\mathbf{Ax}(0) + \mathbf{Bu}(0)] + \mathbf{AB}\dot{\mathbf{u}}(0) + \mathbf{B}\ddot{\mathbf{u}}(0)\} + \mathbf{H}\dddot{\mathbf{u}}(0).$$

Or, in matrix form, putting only unknowns on the right-hand side,

$$\mathbf{Y}(0) \triangleq \begin{pmatrix} \mathbf{y}(0) \\ \dot{\mathbf{y}}(0) \\ \ddot{\mathbf{y}}(0) \\ \dddot{\mathbf{y}}(0) \end{pmatrix} - \begin{bmatrix} \mathbf{H} & \mathbf{0} & \mathbf{0} & \mathbf{0} \\ \mathbf{CB} & \mathbf{H} & \mathbf{0} & \mathbf{0} \\ \mathbf{CAB} & \mathbf{CB} & \mathbf{H} & \mathbf{0} \\ \mathbf{CA}^2\mathbf{B} & \mathbf{CAB} & \mathbf{CB} & \mathbf{H} \end{bmatrix} \begin{pmatrix} \mathbf{u}(0) \\ \dot{\mathbf{u}}(0) \\ \ddot{\mathbf{u}}(0) \\ \dddot{\mathbf{u}}(0) \end{pmatrix} = \begin{bmatrix} \mathbf{C} \\ \mathbf{CA} \\ \mathbf{CA}^2 \\ \mathbf{CA}^3 \end{bmatrix} \mathbf{x}(0) = \mathbf{W}_0\mathbf{x}(0),$$

where we must solve for $\mathbf{x}(0)$. This is a linear algebra problem from Chapter 2. Denote the left-hand side by the vector $\mathbf{Y}(0)$. Then, from Table 2.3 (case $n_3 = 1$, $n_2 = r_A$, $n_4 = 1$, $B = 1$) a unique solution exists iff rank $\mathbf{W}_0 = 3$, in which case

$$\mathbf{x}(0) = \mathbf{W}_0^{-1}\mathbf{Y}(0). \qquad \blacksquare$$

4.1.4.2 Phase-Variable Form

Consider the system (4.27) again, but this time define an *intermediate output* y' by

$$y(s) = (c_4s^4 + c_3s^3 + c_2s^2 + c_1s + c_0)y'(s), \tag{4.31}$$

where

$$y'(s) \triangleq \left[\frac{1}{s^4 + d_3s^3 + d_2s^2 + d_1s + d_0}\right]u(s); \tag{4.32}$$

and, for the system described by (4.32) repeat steps (1) through (3) in Example 4.7. Solve (4.32) for the highest-output "derivative," that is,

$$s^4y'(s) = u(s) - d_3s^3y'(s) - d_2s^2y'(s) - d_1sy'(s) - d_0y'(s) \tag{4.33}$$

and draw the diagram of Fig. 4.7a. Again, the outputs of the integrators are selected as the state variables. Thus, from Fig. 4.7a,

$$\begin{aligned} \dot{x}_1 &= x_2, \\ \dot{x}_2 &= x_3, \\ \dot{x}_3 &= x_4, \\ \dot{x}_4 &= -d_0x_1 - d_1x_2 - d_2x_3 - d_3x_4 + u, \\ y' &= x_1. \end{aligned} \tag{4.34}$$

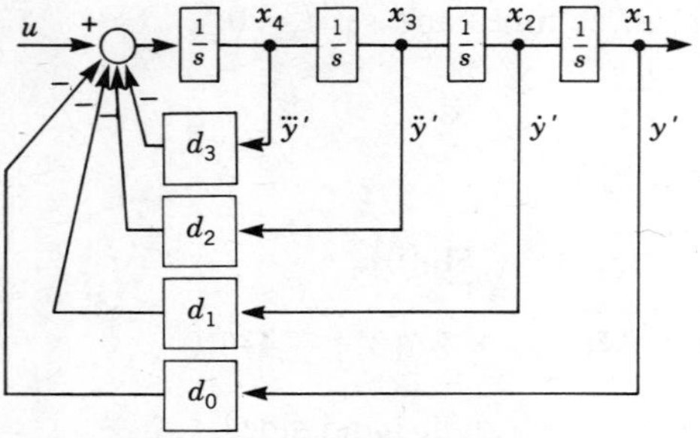

Figure 4.7a *Phase Variable Diagram, Step 1*

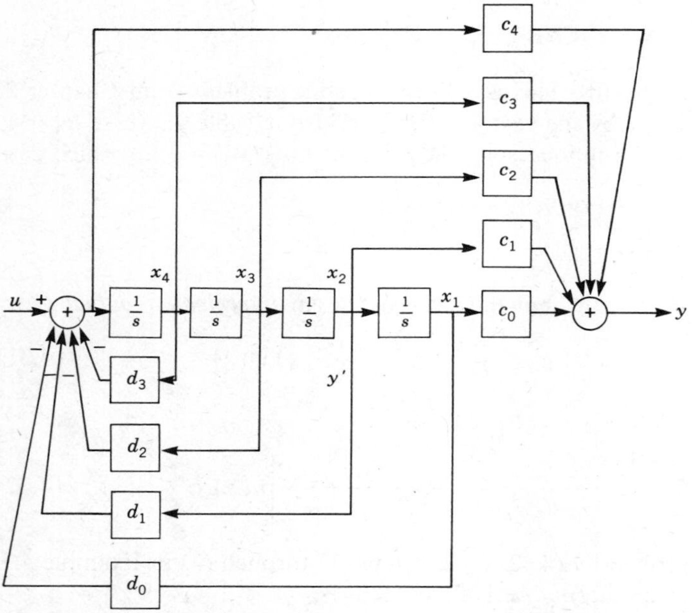

Figure 4.7b *Phase Variable Diagram*

Now, in matrix form,

$$\dot{x} = \begin{bmatrix} 0 & 1 & 0 & 0 \\ 0 & 0 & 1 & 0 \\ 0 & 0 & 0 & 1 \\ -d_0 & -d_1 & -d_2 & -d_3 \end{bmatrix} x + \begin{bmatrix} 0 \\ 0 \\ 0 \\ 1 \end{bmatrix}, \tag{4.35}$$

$$y' = [1 \quad 0 \quad 0 \quad 0]x.$$

The reader can now write the state form (4.35) immediately by inspection of Fig. 4.7a. To complete our task, we must find y from y'. Equation (4.31) and Fig. 4.7a leads to

$$\begin{aligned} y &= c_4 \ddddot{y}' + c_3 \dddot{y}' + c_2 \ddot{y}' + c_1 \dot{y}' + c_0 y' \\ &= c_4 \dot{x}_4 + c_3 x_4 + c_2 x_3 + c_1 x_2 + c_0 x_1, \end{aligned} \tag{4.36}$$

as diagrammed in Fig. 4.7b. The final state equations are obtained from (4.35) and (4.36):

$$\begin{aligned} \dot{\mathbf{x}} &= \begin{bmatrix} 0 & 1 & 0 & 0 \\ 0 & 0 & 1 & 0 \\ 0 & 0 & 0 & 1 \\ -d_0 & -d_1 & -d_2 & -d_3 \end{bmatrix} \mathbf{x} + \begin{bmatrix} 0 \\ 0 \\ 0 \\ 1 \end{bmatrix} u, \\ y &= [c_0 - c_4 d_0, c_1 - c_4 d_1, c_2 - c_4 d_2, c_3 - c_4 d_3]\mathbf{x} + [c_4]u. \end{aligned} \tag{4.37}$$

The reader should now observe the correspondence between Fig. 4.7b and the final form of the state equations (4.37). Also, note the *different* state space realizations (4.30) and (4.37), obtained for the *same* system.

Exercise 4.7

For the backpack problem with two controls, $u_1 = T_c$ and $u_2 = f$ in (3.23), find a state space realization in which the **A** matrix is in the phase-variable form as in (4.37). Compare your **B** matrix with (4.37).

4.1.4.3 Modal Form

Now, consider expanding the transfer function $G(s)$ in (4.27) into partial fraction form

$$\begin{aligned} \frac{y(s)}{u(s)} = G(s) &= \frac{c_4 s^4 + c_3 s^3 + c_2 s^2 + c_1 s + c_0}{s^4 + d_3 s^3 + d_2 s^2 + d_1 s + d_0} \\ &= \sum_{i=1}^{4} \frac{J_i}{s - \lambda_i} + c_4, \end{aligned} \tag{4.38}$$

where distinct roots λ_i are assumed for simplicity. Hence, the residue J_i computation (4.14a) becomes

$$J_i = [G(s) - c_4](s - \lambda_i)_{s=\lambda_i}. \tag{4.39}$$

The block diagram of (4.38) appears in Fig. 4.8, where c_i and $\ell_i{}^*$ are *any* values

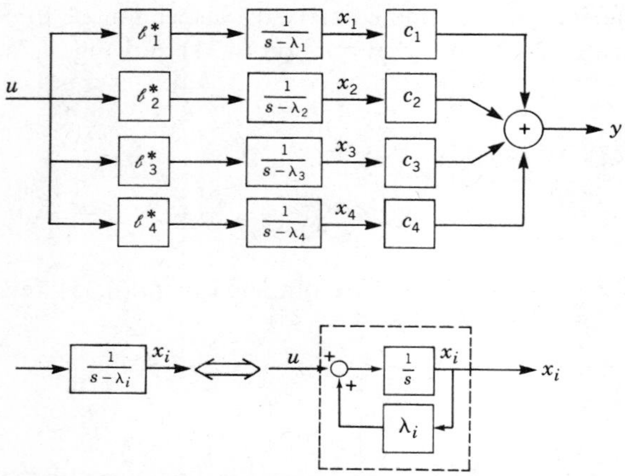

Figure 4.8 Modal Form Block Diagram

such that $c_i b_i^* \triangleq J_i$. We'll say more about c_i and b_i^* later. The states are again selected as the integrator outputs. The corresponding state equations are

$$
\begin{aligned}
\dot{x}_1 &= \lambda_1 x_1 + b_1^* u, \\
\dot{x}_2 &= \lambda_2 x_2 + b_2^* u, \\
\dot{x}_3 &= \lambda_3 x_3 + b_3^* u, \\
\dot{x}_4 &= \lambda_4 x_4 + b_4^* u, \\
y &= c_1 x_1 + c_2 x_2 + c_3 x_3 + c_4 x_4,
\end{aligned}
\tag{4.40}
$$

and, in matrix form,

$$
\begin{aligned}
\dot{\mathbf{x}} &= \begin{bmatrix} \lambda_1 & 0 & 0 & 0 \\ 0 & \lambda_2 & 0 & 0 \\ 0 & 0 & \lambda_3 & 0 \\ 0 & 0 & 0 & \lambda_4 \end{bmatrix} \mathbf{x} + \begin{bmatrix} b_1^* \\ b_2^* \\ b_3^* \\ b_4^* \end{bmatrix} u, \\
y &= [c_1 \quad c_2 \quad c_3 \quad c_4] \mathbf{x}
\end{aligned}
\tag{4.41}
$$

Comparing the three state space realizations (4.30), (4.37), and (4.41), it is clear that

(4.30) has the simplest output matrix **C**, (4.37) has the simplest input matrix **B**, and (4.41) has the simplest matrix **A**. From Section 2.8.1 of Chapter 2 we know that **A** in (4.41) is in its Jordan form, which is the simplest possible form for **A**.

4.2 Coordinate Transformations

In the discussion of state variable definitions, it has been established that for a particular system, a set of vectors (**x**) that qualify as state vectors can be obtained by a nonsingular matrix multiplication (**T**) of a given state vector ($\boldsymbol{\mathscr{x}}$). Hence, $\mathbf{x} = \mathbf{T}\boldsymbol{\mathscr{x}}$ will be called a *coordinate transformation* from coordinates (state) **x** to $\boldsymbol{\mathscr{x}}$.

Applying the coordinate transformation

$$\mathbf{x} = \mathbf{T}\boldsymbol{\mathscr{x}} \tag{4.42}$$

to the system

$$\begin{aligned} \dot{\mathbf{x}} &= \mathbf{Ax} + \mathbf{Bu} + \mathbf{Dw}, \\ \mathbf{y} &= \mathbf{Cx} + \mathbf{Hu} + \mathbf{Jw} \end{aligned} \tag{4.43}$$

leads to

$$\begin{aligned} \mathbf{T}\dot{\boldsymbol{\mathscr{x}}} &= \mathbf{AT}\boldsymbol{\mathscr{x}} + \mathbf{Bu} + \mathbf{Dw}, \\ \mathbf{y} &= \mathbf{CT}\boldsymbol{\mathscr{x}} + \mathbf{Hu} + \mathbf{Jw} \end{aligned} \tag{4.44}$$

or

$$\begin{aligned} \dot{\boldsymbol{\mathscr{x}}} &= \mathbf{T}^{-1}\mathbf{AT}\boldsymbol{\mathscr{x}} + \mathbf{T}^{-1}\mathbf{Bu} + \mathbf{T}^{-1}\mathbf{Dw} = \mathscr{A}\boldsymbol{\mathscr{x}} + \mathscr{B}\mathbf{u} + \mathscr{D}\mathbf{w}, \\ \mathbf{y} &= \mathbf{CT}\boldsymbol{\mathscr{x}} + \mathbf{Hu} + \mathbf{Jw} = \mathscr{C}\boldsymbol{\mathscr{x}} + \mathbf{Hu} + \mathbf{Jw}. \end{aligned} \tag{4.45}$$

Thus, we see that the system (4.43) transforms to the system (4.45). In terms of the new matrices

$$\mathscr{A} = \mathbf{T}^{-1}\mathbf{AT}, \quad \mathscr{B} = \mathbf{T}^{-1}\mathbf{B}, \quad \mathscr{C} = \mathbf{CT}, \quad \mathscr{D} = \mathbf{T}^{-1}\mathbf{D}, \quad \mathbf{H} = \mathbf{H}, \quad \mathbf{J} = \mathbf{J}, \tag{4.46}$$

we say that the matrix **A** has undergone a *similarity* transformation, ($\mathscr{A}$ is *similar* to **A**). We say that system (4.45) is a transformation of (4.43).

The purpose of this section is to determine the properties of linear dynamic systems that remain *invariant* under coordinate transformation (i.e., properties that do not depend upon the choice of T). Specifically, we will show that these properties are invariant under coordinate transformation:

1. The characteristic polynomial (and hence its roots)
2. The Markov parameters
3. The transfer function from $u(s)$ to $y(s)$
4. The residues
5. Functions of the output such as time correlation

$$\mathscr{C}_{yy}(t) \triangleq \int_0^\infty [\mathbf{y}(t+\tau)\mathbf{y}^*(\tau)]\, d\tau \qquad \text{or the cost function } \left(\mathscr{V} \triangleq \int_0^\infty \mathbf{y}^*\mathbf{Q}\mathbf{y}\, dt\right)$$

The power of these conclusions is that these certain properties of linear systems may be studied in any convenient set of coordinates (states) without altering the results.

Exercise 4.8

Find a coordinate transformation matrix **T** that will transform (4.37) to (4.41).

4.2.1 INVARIANCE OF THE CHARACTERISTIC POLYNOMIAL

The characteristic polynomial for a particular state space realization is given by $|s\mathbf{I} - \mathbf{A}|$. The fact to be resolved is the following.

Theorem 4.1

For any $\mathbf{A} \in \mathscr{R}^{n \times n}$, *the polynomial* $|s\mathbf{I} - \mathbf{A}|$ *is invariant under similarity transformation of* **A**.

Proof: The characteristic polynomial of the similarly transformed **A** may be written

$$|s\mathbf{I} - \mathbf{T}^{-1}\mathbf{A}\mathbf{T}| = |s\mathbf{T}^{-1}\mathbf{T} - \mathbf{T}^{-1}\mathbf{A}\mathbf{T}| = \left|\mathbf{T}^{-1}[s\mathbf{I} - \mathbf{A}]\mathbf{T}\right|. \tag{4.47}$$

Now, the determinant of the product of square matrices is the product of the determinants, so

$$\begin{aligned} \left|\mathbf{T}^{-1}[s\mathbf{I} - \mathbf{A}]\mathbf{T}\right| &= |\mathbf{T}^{-1}|\,|s\mathbf{I} - \mathbf{A}|\,|\mathbf{T}| \\ &= |\mathbf{T}^{-1}|\,|\mathbf{T}|\,|s\mathbf{I} - \mathbf{A}| \\ &= |s\mathbf{I} - \mathbf{A}|. \end{aligned} \tag{4.48}$$

Thus, $|s\mathbf{I} - \mathbf{T}^{-1}\mathbf{A}\mathbf{T}| = |s\mathbf{I} - \mathbf{A}|$. ■

Exercise 4.9

Verify that (4.30) and (4.37) have the same characteristic polynomial.

4.2.2 INVARIANCE OF THE MARKOV PARAMETERS

The Markov parameters $\mathbf{M}_i \triangleq \mathbf{CA}^i\mathbf{B}$ are defined from the series (4.11),

$$\mathbf{C}(s\mathbf{I} - \mathbf{A})^{-1}\mathbf{B} = \sum_{i=0}^{\infty} \frac{\mathbf{M}_i}{s^{i+1}}. \tag{4.49}$$

The coordinate transformation of the parameters is given by (4.46). Hence, the required result is as follows:

Theorem 4.2

For any $\mathbf{A} \in \mathscr{C}^{n\times n}$, $\mathbf{B} \in \mathscr{C}^{k\times n}$, and $\mathbf{C} \in \mathscr{C}^{k\times n}$, the Markov parameters $\mathbf{M}_i \triangleq \mathbf{CA}^i\mathbf{B}$ remain invariant under coordinate transformation of the state.

Proof: The transformed Markov parameters, from (4.46), are

$$\begin{aligned}\mathscr{C}\mathscr{A}^i\mathscr{B} &= \mathbf{CT}(\mathbf{T}^{-1}\mathbf{AT})^i\mathbf{T}^{-1}\mathbf{B}\\ &= \mathbf{CTT}^{-1}\mathbf{A}^i\mathbf{TT}^{-1}\mathbf{B}\\ &= \mathbf{CA}^i\mathbf{B},\end{aligned} \tag{4.50}$$

and the theorem is proved. ■

Exercise 4.10

For the scalar input system (4.30), show that the impulse response $u(t) = \delta(t)$ may be written

$$\mathbf{y}(t) = \sum_{i=0}^{\infty} \frac{\mathbf{M}_i t^i}{i!} \tag{4.51}$$

and that all the "time moments" $\mathbf{M}_i$ are the same for systems (4.30), (4.37), and (4.41). Show also that the Markov parameters are

$$\left[\frac{d^{k-1}}{dt^{k-1}} y(t)\right]_{t=0} = \mathbf{M}_k$$

when $u(t) = \delta(t)$ and $\mathbf{H} = \mathbf{0}$.

4.2.3 INVARIANCE OF THE TRANSFER FUNCTION

Consider the transfer function $G(s)$:

$$\mathbf{y}(s) = \mathbf{G}(s)\mathbf{u}(s) = \left[\mathbf{C}(s\mathbf{I} - \mathbf{A})^{-1}\mathbf{B} + \mathbf{H}\right]\mathbf{u}(s). \tag{4.52}$$

The similar parameters (4.46) yield for the transformed system

$$\begin{aligned}
\mathbf{y}(s) &= \left[\mathscr{C}(s\mathbf{I} - \mathscr{A})^{-1}\mathscr{B} + \mathbf{H}\right]\mathbf{u}(s) \\
&= \left[\mathbf{CT}(s\mathbf{I} - \mathbf{T}^{-1}\mathbf{AT})^{-1}\mathbf{T}^{-1}\mathbf{B} + \mathbf{H}\right]\mathbf{u}(s) \\
&= \left[\mathbf{CT}(\mathbf{T}^{-1}(s\mathbf{I} - \mathbf{A})\mathbf{T})^{-1}\mathbf{T}^{-1}\mathbf{B} + \mathbf{H}\right]\mathbf{u}(s) \\
&= \left[\mathbf{C}(s\mathbf{I} - \mathbf{A})^{-1}\mathbf{B} + \mathbf{H}\right]u(s) \\
&= \mathbf{G}(s)\mathbf{u}(s).
\end{aligned} \tag{4.53}$$

This situation is summarized as follows.

Theorem 4.3

The transfer function between the input $\mathbf{u}(s)$ *and the output* $\mathbf{y}(s)$ *is invariant under coordinate transformation of the state.*

4.2.4 INVARIANCE OF THE RESIDUES

Now, consider the residue formula (4.14). We treat here only the distinct root case $m_i = 1$ for simplicity:

$$\mathbf{J}_i = \left[\mathbf{C}(s\mathbf{I} - \mathbf{A})^{-1}\mathbf{B} + \mathbf{H}\right](s - \lambda_i)|_{s=\lambda_i}. \tag{4.54}$$

By Theorem 4.1 the characteristic roots λ_i are invariant under similarity transformation, and by Theorem 4.3 the transfer function remains invariant. Since the residue $\mathbf{J}_i$ is the product involving these two quantities, the residue is also invariant.

Theorem 4.4

The residues associated with a linear dynamic system remain invariant under state transformation.

An expression for the residues is expecially simple when computed in modal coordinates, and we shall now do this, since by Theorem 4.4 this result will generally be useful. From Chapter 2, the modal matrix $\mathbf{E}$ for $\mathbf{A}$ satisfies

$$\mathbf{AE} = \mathbf{E}\Lambda, \qquad \Lambda = \text{block diag}[\ldots \Lambda_i \ldots], \tag{4.55}$$

where the ith column of $\mathbf{E}$ is the ith eigenvector $\mathbf{e}_i$, Λ is the Jordan form of $\mathbf{A}$, and λ_i is the ith eigenvalue. Recall also that the reciprocal base vectors $\mathbf{l}_i^*$ are defined by

$$\mathbf{E}^{-1} = \begin{bmatrix} \mathbf{l}_1^* \\ \vdots \\ \mathbf{l}_n^* \end{bmatrix} \tag{4.56}$$

and satisfy

$$\mathbf{E}^{-1}\mathbf{A} = \Lambda\mathbf{E}^{-1}. \tag{4.57}$$

Hence, $\mathbf{l}_i^*$ are also left eigenvectors, but they are subject to the normalization of $\mathbf{e}_i$ and cannot be separately normalized. Now, use $\mathbf{T} = \mathbf{E}$ for a state space coordinate transformation $\mathbf{x} = \mathbf{T}\mathscr{x}$. The result is

$$\begin{aligned} \dot{\mathscr{x}} &= \mathbf{E}^{-1}\mathbf{A}\mathbf{E}\mathscr{x} + \mathbf{E}^{-1}\mathscr{B}\mathbf{u}, \\ \mathbf{y} &= \mathbf{C}\mathbf{E}\mathscr{x} + \mathbf{H}\mathbf{u}. \end{aligned} \tag{4.58}$$

But since $\mathbf{E}^{-1}\mathbf{A}\mathbf{E} = \Lambda$, the system (4.58) is as dynamically uncoupled as possible. Hence, the ith component of $\mathscr{x}_i$ satisfies

$$\begin{aligned} \dot{\mathscr{x}}_1 &= \Lambda_i \mathscr{x}_i + \mathscr{B}_i^*\mathbf{u}, \qquad \mathscr{B}^* = \left[\mathscr{B}_1, \ldots, \mathscr{B}_p\right], \\ \mathbf{y} &= \sum_{i=1}^{p} \mathscr{C}_i \mathscr{x}_i + Hu, \qquad \mathscr{C} = \left[\mathscr{C}_1, \ldots, \mathscr{C}_p\right]. \end{aligned} \tag{4.59}$$

The transfer function has the form

$$\begin{aligned} \mathscr{C}(s\mathbf{I} - \mathscr{A})^{-1}\mathscr{B} + \mathbf{H}\mathbf{u} &= \mathbf{C}\mathbf{E}(s\mathbf{I} - \mathbf{E}^{-1}\mathbf{A}\mathbf{E})^{-1}\mathbf{E}^{-1}\mathbf{B} + \mathbf{H} \\ &= \left[\mathscr{C}_1 \ldots \mathscr{C}_p\right] \begin{bmatrix} s\mathbf{I} - \Lambda_1 & & \\ & \ddots & \\ & & s\mathbf{I} - \Lambda_p \end{bmatrix}^{-1} \begin{bmatrix} \mathscr{B}_i^* \\ \vdots \\ \mathscr{B}_n^* \end{bmatrix} + \mathbf{H} \\ &= \sum_{i=1}^{p} \mathscr{C}_i [s\mathbf{I} - \Lambda_i]^{-1} \mathscr{B}_i^* + \mathbf{H}. \end{aligned} \tag{4.60}$$

Now, if λ_i, $i = 1, \ldots, n_4$, are distinct, then $\Lambda_i = \lambda_i$ and the partial fraction expansion

$$\mathscr{C}(s\mathbf{I} - \mathscr{A})^{-1}\mathscr{B} = \sum_{i=1}^{n} \frac{\mathbf{c}_i\mathbf{b}_i^*}{s - \lambda_i} = \sum_{i=1}^{n} \frac{\mathbf{J}_i}{s - \lambda_i} \tag{4.61}$$

makes it clear that whenever $\mathbf{H} = \mathbf{0}$, the residue $\mathbf{J}_i$ of mode i is related to $\mathbf{c}_i, \mathbf{b}_i$ by

$$\mathbf{J}_i = \mathbf{c}_i\mathbf{b}_i^*. \tag{4.62}$$

We shall find further use of (4.62) in the next section.

4.2.5 INVARIANCE OF FUNCTIONS OF THE OUTPUT

In the transformed coordinates, the output vector

$$\mathbf{y} = \mathscr{C}\mathscr{x} + \mathbf{Hu}, \qquad \mathbf{x} = \mathbf{T}\mathscr{x} \tag{4.63}$$

has this value in the original coordinates:

$$\mathbf{y} = \mathbf{CTT}^{-1}\mathbf{x} + \mathbf{Hu} = \mathbf{Cx} + \mathbf{Hu}. \tag{4.64}$$

Hence, the *output y is unaffected by coordinate transformations on the state*, and the analyst is free to choose **T** for the convenience of other calculations that may be of interest.

4.3 Disturbance Models

The dynamic systems of interest have the form

$$\begin{aligned} \dot{\mathbf{x}} &= \mathbf{Ax} + \mathbf{Bu} + \mathbf{Dw}, \\ \mathbf{y} &= \mathbf{Cx} + \mathbf{Hu} + \mathbf{Jw}, \end{aligned} \tag{4.65}$$

where u is the "control" vector and $\mathbf{w}(t)$ is a vector of "disturbances." The purpose of this section is to describe modeling aspects of $\mathbf{w}(t)$. The word "disturbance" defies precise mathematical definition, although the physical source of $\mathbf{w}(t)$ may be clear. To illustrate the complexity of the disturbance modeling problem, consider the sinusoidal disturbance

$$w(t) = \alpha \sin \omega t + \beta \cos \omega t. \tag{4.66a}$$

Modeled this way, $w(t)$ is a function of t only, but $w(t)$ could also be described by a set of differential equations:

$$\begin{aligned} w(t) &= \gamma, \\ \ddot{\gamma} + \omega^2\gamma = 0,\quad \gamma(0) &= \beta,\quad \dot{\gamma}(0) = \omega\alpha. \end{aligned} \tag{4.66b}$$

(To verify this, differentiate (4.66a) twice.) Equivalently, this "disturbance model" has the state form

$$\begin{pmatrix} \dot{\gamma} \\ \ddot{\gamma} \end{pmatrix} = \begin{bmatrix} 0 & 1 \\ -\omega^2 & 0 \end{bmatrix} \begin{pmatrix} \gamma \\ \dot{\gamma} \end{pmatrix}, \qquad \begin{pmatrix} \gamma(0) \\ \dot{\gamma}(0) \end{pmatrix} = \begin{pmatrix} \beta \\ \omega\alpha \end{pmatrix} \tag{4.66c}$$

or simply

$$\dot{\mathbf{x}}_w = \mathbf{A}_w\mathbf{x}_w, \quad \mathbf{x}_w(0) = \begin{pmatrix} \beta \\ \omega\alpha \end{pmatrix}, \quad w = \mathbf{C}_w\mathbf{x}_w. \tag{4.66d}$$

The reader will be asked in the next section to verify that $\mathbf{x}_w(t)$ is the same vector function of time whether $\mathbf{x}_w(0)$ excites the homogeneous system ($\dot{\mathbf{x}}_w = \mathbf{A}_w\mathbf{x}_w$, $\mathbf{x}_w(0) = \mathbf{x}_{w0}$) or whether the Dirac delta with strength $\mathbf{u}_w(t) = \mathbf{x}_{w0}\delta(t)$ excites the system $\dot{\mathbf{x}}_w = \mathbf{A}_w x_w + \mathbf{u}_w$ with zero initial condition $\mathbf{x}_w(0) = \mathbf{0}$. We use this fact now to write the total system as described by an augmentation of (4.66d) and (4.65) to give

$$\begin{pmatrix} \dot{\mathbf{x}} \\ \dot{\mathbf{x}}_w \end{pmatrix} = \begin{bmatrix} \mathbf{A} & \mathbf{DC}_w \\ \mathbf{0} & \mathbf{A}_w \end{bmatrix} \begin{pmatrix} \mathbf{x} \\ \mathbf{x}_w \end{pmatrix} + \begin{bmatrix} \mathbf{B} \\ \mathbf{0} \end{bmatrix} \mathbf{u} + \begin{bmatrix} \mathbf{0} \\ \mathbf{I} \end{bmatrix} \mathbf{u}_w,$$

$$\mathbf{y} = [\mathbf{C} \quad \mathbf{JC}_w] \begin{pmatrix} \mathbf{x} \\ \mathbf{x}_w \end{pmatrix} + \mathbf{Hu}, \tag{4.67}$$

where we may choose either $\left[\mathbf{u}_w = 0, \mathbf{x}_w(0) = \mathbf{x}_{w0} \triangleq \begin{pmatrix} \beta \\ \omega\alpha \end{pmatrix}\right]$ or $[\mathbf{u}_w(t) = \mathbf{x}_{w0}\delta(t)$, $\mathbf{x}_w(0) = \mathbf{0}]$. But the form of this system is simply

$$\dot{\mathbf{x}} = \mathscr{A}x + \mathscr{B}\mathbf{u} + \mathscr{D}\mathbf{u}_w,$$

$$\mathbf{y} = \mathscr{C}x + \mathbf{Hu}, \tag{4.68}$$

with obvious definitions for $\mathscr{A}, \mathscr{B}, \mathscr{C}, \mathscr{D}$.

The disturbance model we allow in this text is a generalization of $\dot{\mathbf{x}}_w = \mathbf{A}_w\mathbf{x}_w + \mathbf{u}_w$, namely,

$$\dot{\mathbf{x}}_w = \mathbf{A}_w\mathbf{x}_w + \mathbf{A}_x\mathbf{x} + \mathbf{B}_w\mathbf{u} + \mathbf{D}_w\mathbf{u}_w, \qquad \mathbf{w} = \mathbf{C}_w\mathbf{x}_w + \mathbf{C}_x\mathbf{x} + \mathbf{H}_w\mathbf{u}, \tag{4.69}$$

where for $i = 1, 2, \ldots, n_{xw}$, $(\mathbf{u}_w)_i = \delta(t)$ is the Dirac delta function of Appendix E. In this case, (4.67) and (4.68) generalize to

$$\begin{pmatrix} \dot{\mathbf{x}} \\ \dot{\mathbf{x}}_w \end{pmatrix} = \begin{bmatrix} \mathbf{A} + \mathbf{DC}_x & \mathbf{DC}_w \\ \mathbf{A}_x & \mathbf{A}_w \end{bmatrix} \begin{pmatrix} \mathbf{x} \\ \mathbf{x}_w \end{pmatrix} + \begin{bmatrix} \mathbf{B} + \mathbf{DH}_w \\ \mathbf{B}_w \end{bmatrix} \mathbf{u} + \begin{bmatrix} \mathbf{0} \\ \mathbf{D}_w \end{bmatrix} \mathbf{u}_w,$$

$$\mathbf{y} = [\mathbf{C} + \mathbf{JC}_x \quad \mathbf{JC}_w] \begin{pmatrix} \mathbf{x} \\ \mathbf{x}_w \end{pmatrix} + (\mathbf{H} + \mathbf{JH}_w)\mathbf{u},$$

or

$$\dot{x} = \mathscr{A}x + \mathscr{B}\mathbf{u} + \mathscr{D}\mathbf{u}_w,$$

$$\mathbf{y} = \mathscr{C}x + \mathscr{H}\mathbf{u}, \tag{4.70}$$

where $\mathbf{u}_w$ is a vector of Dirac delta functions. Thus, for disturbances that admit a state space description (4.69) (this is a large class of disturbances which includes *every* disturbance of interest in this text), the form (4.70) is without loss of

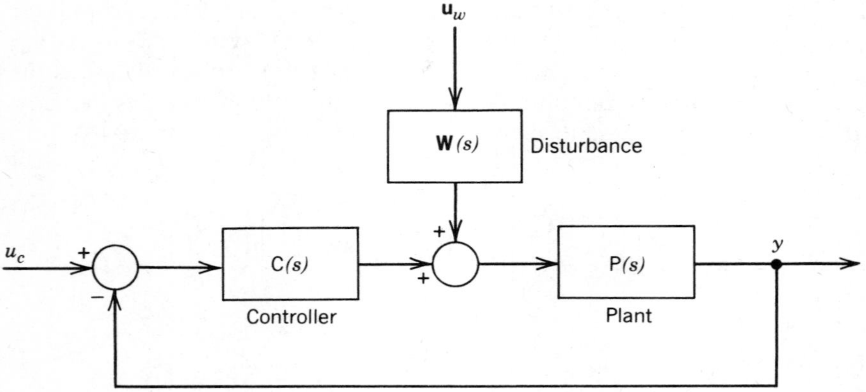

$$C(s) = \frac{C_n(s)}{\tilde{C}_d(s)\, W_d(s)}, \quad \mathbf{W}(s) = \frac{\mathbf{W}_n(s)}{W_d(s)}$$

Figure 4.9 The Disturbance Modeling Principle

generality. We refer to (4.69) as the *disturbance model*, or the *internal model of the disturbances*.

It is premature to discuss detailed control design since this is the subject of later chapters. However, a good motivation for using disturbance models (4.69) is the following observation about control design in the presence of disturbances. Consider the single-input/output system of Fig. 4.9 which is subject to a disturbance modeled by (4.69), where the system model is (4.70). The indicated transfer functions in Fig. 4.9 are

$$P(s) = \left[\mathscr{C}(s\mathbf{I} - \mathscr{A})^{-1}\mathscr{B} + \mathscr{H}\right], \tag{4.71a}$$

$$\mathbf{W}(s) = P(s)^{-1}\mathscr{C}(s\mathbf{I} - \mathscr{A})^{-1}\mathscr{D}, \tag{4.71b}$$

as the reader may verify from (4.70), and the controller is $u(s) = C(s)[u_c(s) - y(s)]$ shown in Fig. 4.9. Note that $\mathbf{u}_w$ is a vector impulse function of some strength $\mathbf{u}_w(t) = \mathbf{x}_{w0}\delta(t)$. Hence, $\mathbf{u}_w(s) = \mathbf{x}_{w0}$ and the closed-loop system is described by

$$y(s) = [I + P(s)C(s)]^{-1}[P(s)C(s)u_c(s) + P(s)\mathbf{W}(s)\mathbf{x}_{w0}].$$

In practice, it is often required that the tracking error $e(t) \triangleq y(t) - u_c(t)$ go to zero as $t \to \infty$. In our example, let $u_c(t) = 1$. Using the final value theorem, we must therefore require

$$se(s)\big|_{s=0} = 0.$$

Since

$$e(s) = y(s) - u_c(s),$$

we have

$$\begin{aligned} e(s) &= \{[I + P(s)C(s)]^{-1}P(s)C(s) - I\}u_c(s) \\ &\quad + [I + P(s)C(s)]^{-1}P(s)\mathbf{W}(s)\mathbf{x}_{w0} \\ &= [I + P(s)C(s)]^{-1}[-u_c(s) + P(s)\mathbf{W}(s)\mathbf{x}_{w0}]. \end{aligned}$$

The poles of $e(s)$ must lie in the left half-plane for the final value theorem to apply. To see the nature of these poles more clearly, let the scalar functions $\mathbf{W}(s)\mathbf{x}_{w0}$, $P(s)$, and $C(s)$ be the ratio of polynomials:

$$\mathbf{W}(s)\mathbf{x}_{w0} = \frac{W_n(s)}{W_d(s)}, \quad P(s) = \frac{P_n(s)}{P_d(s)}, \quad C(s) = \frac{C_n(s)}{C_d(s)}.$$

Then, multiplying numerator and denominator of $e(s)$ by $W_d P_d C_d(s)$ yields

$$e(s) = \frac{C_d(s)[W_n(s)P_n(s) - u_c(s)W_d(s)P_d(s)]}{W_d(s)[C_n(s)P_n(s) + C_d(s)P_d(s)]}.$$

Hence, the poles of $e(s)$ include the poles of $W(s)$ (i.e., any value of s that makes $W_d(s) = 0$ will make $e(s) = \infty$). This is almost always unacceptable since this implies that the tracking properties are dictated by the disturbance properties. Disturbances do not generally behave "nicely," so it is necessary to modify this situation. See that $e(s)$ contains the poles of $W(s)$ *unless* we choose the compensator to *include the disturbance model* as a factor, $C_d(s) = W_d(s)\tilde{C}_d(s)$, for some $\tilde{C}_d(s)$. In this case,

$$e(s) = \frac{\tilde{C}_d(s)[W_n(s)P_n(s) - u_c(s)W_d(s)P_d(s)]}{C_n(s)P_n(s) + \tilde{C}_d(s)P_d(s)W_d(s)},$$

where the poles of $e(s)$ can now be placed as desired by proper choice of $C_n(s)$ and $\tilde{C}_d(s)$, subject to the tracking requirement

$$se(s)|_{s=0} = \left.\frac{\tilde{C}_d(s)[sW_n(s)P_n(s) - su_c(s)W_d(s)P_d(s)]}{C_n(s)P_n(s) + \tilde{C}_d(s)P_d(s)W_d(s)}\right|_{s=0} = 0.$$

In summary, we conclude as follows:

The disturbance modeling principle: *Linear dynamic systems can be designed to track desired inputs in the presence of disturbances only if the control designs employ a model of the disturbance dynamics.*

Hence, rather than treat $w(t)$ in (4.65) as a function of *time*, as in (4.66a), we shall treat $w(t)$ as a function of *state* as in (4.69) and (4.70), since the state embodies the dynamics of the disturbance model which we shall need in any controller design. Note that the "disturbance dynamics" $\mathbf{W}(s)$ in Fig. 4.9 and equation (4.71b) are coupled with the plant dynamics $P(s)$. Hence, the impulsive input $\mathbf{u}_w(t)$ in (4.70) does not constitute the "disturbance" in the way the input $\mathbf{w}(t)$ constitutes the "disturbance." When we talk of "disturbances" in this text, we shall usually only speak of impulsive inputs as in (4.70), assuming that the state equations have already incorporated the actual disturbance model (4.69).

When $\mathbf{w}(t)$ is not a function of the plant state $\mathbf{x}$, the task of converting $\mathbf{w}(t)$ into a state model deserves a brief discussion. Quite often, a time series for $\mathbf{w}(t)$ is given (or hypothesized) from previous experiments. In this case, $\mathbf{w}(t)$ may be modeled by any set of basis functions $x_{wi}(t)$ which are "complete" in the sense of Chapter 2. That is, in a mean squared sense,

$$\mathbf{w}(t) = \sum_{i=1}^{d} \mathbf{C}_{wi} x_{wi}(t) = \mathbf{C}_w \mathbf{x}_w(t), \tag{4.72a}$$

where $\mathbf{C}_w = [C_{w1}, C_{w2}, \ldots, C_{wd}]$, $\mathbf{x}_w^T = [x_{w1}, x_{w2}, \ldots, x_{wd}]$, and $x_{wi}(t)$, $i = 1, 2, \ldots, d$, form an independent set of (basis) functions. Once the basis functions $x_{wi}(t)$ are selected, then the differential equations that generate $x_{wi}(t)$ may be constructed in the state form:

$$\dot{\mathbf{x}}_w = \mathbf{A}_w \mathbf{x}_w + \mathbf{u}_w, \quad \mathbf{u}_w = \mathbf{x}_{w0}\delta(t), \quad \mathbf{x}_w(0) = \mathbf{0}. \tag{4.72b}$$

For example, suppose we choose polynomials in time, $x_{wi}(t) = t^{i-1}$, $i = 1, 2, \ldots, d$. Then

$$\dot{x}_{wi} = (i-1)t^{i-2} = (i-1)x_{w,i-1},$$

in which case $\mathbf{A}_w$ in (4.72b) becomes

$$\mathbf{A}_w = \begin{bmatrix} 0 & 0 & 0 & 0 & \cdots & 0 \\ 1 & 0 & 0 & 0 & \cdots & 0 \\ 0 & 2 & 0 & 0 & \cdots & 0 \\ 0 & 0 & 3 & 0 & \cdots & 0 \\ \vdots & \vdots & \vdots & & & \\ 0 & 0 & 0 & 0 & d-1 & 0 \end{bmatrix}.$$

As another example, suppose we choose the sinusoidal basis functions

$$x_{wi}(t) = \sin \omega_i t, \quad x_{w,i+1}(t) = \cos \omega_i t, \quad i = 1, 3, 5, \ldots.$$

Then,

$$\dot{\mathbf{x}}_w = \mathbf{A}_w \mathbf{x}_w + \mathbf{u}_w, \tag{4.73}$$

where

$$\mathbf{A}_w = \begin{bmatrix} 0 & \omega_1 & & & & & \\ -\omega_1 & 0 & & & & & \\ & & 0 & \omega_2 & & & \\ & & -\omega_2 & 0 & & & \\ & & & & 0 & \omega_3 & \\ & & & & -\omega_3 & 0 & \\ & & & & & & \ddots \end{bmatrix}.$$

Finally, suppose the $x_{wi}(t)$ were chosen to be functions that are *orthogonal* on the interval $t \in [t_1, t_2]$ subject to weight $g(t)$. From the discussions of orthogonal functions in Section 2.5.1, this means that

$$\int_{t_1}^{t_2} \mathbf{x}_w(t) g(t) \mathbf{x}_w^*(t)\, dt = \Lambda, \tag{4.74}$$

where Λ is a diagonal matrix.

Now, *if* a sample $\mathbf{w}(t)$ function were given for study, then the least squares fitting on the interval $[t_1, t_2]$ yields a choice of $\mathbf{C}_w$ in (4.72a). Again, the results of Chapter 2, Section 2.7.5, give the solution, where it is proved that the solution of

$$\min_{\mathbf{C}_w} \| \mathbf{C}_w \mathbf{x}_w(t) - \mathbf{w}(t) \|_{g(t)}^2$$

for a given $\mathbf{x}_w(t)$, $\mathbf{w}(t)$ yields

$$\begin{aligned} \mathbf{C}_w &= (\rangle \mathbf{w}(t), g(t)\mathbf{x}_w(t)\langle)(\rangle \mathbf{x}_w(t), g(t)\mathbf{x}_w(t)\langle)^{-1} \\ &= \int_{t_1}^{t_2} g(t)\mathbf{w}(t)\mathbf{x}_w^*(t)\, dt \left[\int_{t_1}^{t_2} \mathbf{x}_w(t) g(t) \mathbf{x}_w^*(t)\, dt \right]^{-1}, \end{aligned}$$

where (4.74) leads to

$$\mathbf{C}_w = \int_{t_1}^{t_2} g(t)\mathbf{w}(t)\mathbf{x}_w^*(t)\, dt\, \Lambda^{-1}. \tag{4.75}$$

However, a reliable sample or estimate of $\mathbf{w}(t)$ is not always available, so $\mathbf{C}_w$ in (4.72) is often chosen more simply, so that all $x_{wi}(t)$ are "visible" in the vector $\mathbf{C}_w \mathbf{x}_w(t)$. The precise condition for this choice of $\mathbf{C}_w$ is discussed in Chapter 5 under "observability" concepts, so we leave further discussion on the choice of $\mathbf{C}_w$ to Chapter 5.

Some common orthogonal functions are given in Table 2.2. As an illustration of disturbance modeling with Chebyshev polynomials $x_{wi}(t) \triangleq \cos[(i-1)\cos^{-1} t]$, $t \in [-1, 1]$, $i = 1, 2, \ldots, n_{xw}$, the corresponding $\mathbf{A}_w$ matrix is given as follows:

$$\mathbf{A}_w(\text{Chebyshev}) = \begin{bmatrix} 0 & 0 & 0 & 0 & 0 & \\ 1 & 0 & 0 & 0 & 0 & \\ 0 & 4 & 0 & 0 & 0 & \\ 3 & 0 & 6 & 0 & 0 & \\ 0 & 8 & 0 & 8 & 0 & \\ & & & & & \ddots \end{bmatrix}_{n_{xw} \times n_{xw}}, \tag{4.76}$$

and Λ defined by (4.74) is

$$\Lambda(\text{Chebyshev}) = \frac{\pi}{2}\begin{bmatrix} 2 & \mathbf{0} \\ \mathbf{0} & \mathbf{I}_{n_{xw}-1} \end{bmatrix}.$$

The time interval $t \in [-1, 1]$ may be converted to any other interval of interest by a change of variable. For example, the interval $t' \in [0, \tau]$ corresponds to the change of variable

$$t = 2\left(\frac{t'}{\tau}\right) - 1, \qquad \text{where } t' \in [0, \tau] \text{ and } t \in [-1, 1].$$

To write (4.72) in the t' domain, the reader may show that $\mathbf{A}_w$(Chebyshev) in (4.76) must be multiplied by $2/\tau$. For further discussion of the choice of orthogonal basis functions in the modeling of disturbances, see [4.1] and [4.2]. For additional discussion of disturbance modeling in control problems, see refs [4.3] and [4.4].

Exercise 4.11

Find a state space model of the form (4.72) for the disturbance function

$$w(t) = a + bt + ct^2, \qquad t \in [-1, 1],$$

using (i) the first three Chebyshev polynomials, (ii) the first two Chebyshev polynomials. Discuss whether the basis functions in (i) and (ii) are "complete." Find the least squares fit [i.e., find $\mathbf{C}_w$ in (4.75)]. In both cases (i) and (ii), determine whether Parseval's equality holds,

$$\|w(t)\|^2 = \sum_{i=1}^{3} \|C_{wi}\|^2 \Lambda_{ii}.$$

Which side of this questioned equality is guaranteed to be equal to or larger than the other side? (*Hint*: Review Section 2.7.5.)

4.4 Solutions of State Equations

Thus far, we have focused on the *construction* of the state equations. Chapter 5 will focus on important *properties* of the solutions of these equations, but first we need to establish these solutions. Thus, the objective of this section is to solve the state equation

$$\begin{aligned} \dot{\mathbf{x}} &= \mathbf{A}\mathbf{x} + \mathbf{B}\mathbf{u} + \mathbf{D}\mathbf{w}, \\ \mathbf{y} &= \mathbf{C}\mathbf{x} + \mathbf{H}\mathbf{u} + \mathbf{J}\mathbf{w}, \end{aligned} \tag{4.77}$$

where **A**, **B**, **C**, **H**, and **J** might be functions of time.

Apply a time-varying $\mathbf{T}(t)$ to (4.77) as follows:

$$\mathbf{x}(t) = \mathbf{T}(t)\mathscr{x}(t), \tag{4.78}$$

where $\mathbf{T}(t)$ is a nonsingular matrix that satisfies

$$\dot{\mathbf{T}}(t) = \mathbf{A}(t)\mathbf{T}(t), \qquad |\mathbf{T}(0)| \neq 0. \tag{4.79}$$

From (4.77) and (4.78),

$$\dot{\mathbf{x}}(t) = \dot{\mathbf{T}}(t)\mathscr{x}(t) + \mathbf{T}(t)\dot{\mathscr{x}}(t) = \mathbf{A}(t)\mathbf{T}(t)\mathscr{x}(t) + \mathbf{B}(t)\mathbf{u}(t) + \mathbf{D}(t)\mathbf{w}(t)$$

or

$$\dot{\mathscr{x}}(t) = \mathbf{T}^{-1}(t)\left[\mathbf{A}(t)\mathbf{T}(t) - \dot{\mathbf{T}}(t)\right]\mathscr{x}(t) + \mathbf{T}^{-1}(t)\mathbf{B}(t)\mathbf{u}(t) + \mathbf{T}^{-1}(t)\mathbf{D}(t)\mathbf{w}(t).$$

But by (4.79), this special choice of $\mathbf{T}(t)$ yields

$$\dot{\mathscr{x}}(t) = \mathbf{T}^{-1}(t)[\mathbf{B}(t)\mathbf{u}(t) + \mathbf{D}(t)\mathbf{w}(t)]. \tag{4.80}$$

Integrating (4.80) from t_0 to t yields

$$\mathscr{x}(t) = \mathscr{x}(t_0) + \int_{t_0}^{t} \mathbf{T}^{-1}(\tau)[\mathbf{B}(\tau)\mathbf{u}(\tau) + \mathbf{D}(\tau)\mathbf{w}(\tau)]\, d\tau.$$

Now, apply (4.78) to get

$$\mathbf{x}(t) = \mathbf{T}(t)\mathbf{T}^{-1}(t_0)\mathbf{x}(t_0) + \int_{t_0}^{t} \mathbf{T}(t)\mathbf{T}^{-1}(\tau)[\mathbf{B}(\tau)\mathbf{u}(\tau) + \mathbf{D}(\tau)\mathbf{w}(\tau)]\, d\tau. \tag{4.81}$$

Define

$$\boldsymbol{\Phi}(t, \tau) \triangleq \mathbf{T}(t)\mathbf{T}^{-1}(\tau), \tag{4.82}$$

where the invertibility of $\boldsymbol{\Phi}(t, \tau)$ is now established by the invertibility guaranteed for $\mathbf{T}(t)$ and $\mathbf{T}(\tau)$. Hence, $\mathbf{\Phi}^{-1}(t, \tau) = \mathbf{T}(\tau)\mathbf{T}^{-1}(t) = \mathbf{\Phi}(\tau, t)$ exists. Note from (4.82) that $(d/dt)\mathbf{\Phi}(t, \sigma) = \mathbf{A}(t)\mathbf{\Phi}(t, \sigma)$.

The matrix $\mathbf{T}(t)$ satisfying (4.79) is called a *fundamental matrix*, and one way to find a fundamental matrix is to take

$$\mathbf{T}(t) = \left[\mathbf{x}^1(t), \dots, \mathbf{x}^n(t)\right],$$

where $\mathbf{x}^i(t)$, $i = 1, \dots, n$, are any linearly independent solutions of $\dot{\mathbf{x}} = \mathbf{A}(t)\mathbf{x}$ [i.e. resulting from a linearly independent set of $\mathbf{x}(0)$]. Clearly, this choice of $\mathbf{T}(t)$ satisfies $\dot{\mathbf{T}}(t) = \mathbf{A}(t)\mathbf{T}(t)$ and hence also (4.79). Such transformations $\mathbf{T}(t)$ are called *Liapunov transformations*. Note that they transform the system to a set of equations (4.80), which has no homgoeneous part!

Definition 4.4

A matrix $\mathbf{\Phi}(t, \tau)$ that satisfies the differential equation

$$\frac{d}{dt}\mathbf{\Phi}(t, \tau) = \mathbf{A}(t)\mathbf{\Phi}(t, \tau), \qquad \mathbf{\Phi}(\tau, \tau) = \mathbf{I} \tag{4.83}$$

will be called a state transition matrix for the matrix $\mathbf{A}(t)$.

Note that the fundamental matrix $\mathbf{T}(t)$ and the state transition matrix $\mathbf{\Phi}(t, \tau)$ satisfy the same differential equation, but do *not* have the same initial condition.

The general solution of (4.77) can be stated in terms of $\mathbf{\Phi}(t, \tau)$ as follows.

Theorem 4.5

The unique solution of (4.77) is given by

$$\mathbf{y}(t) = \mathbf{C}\mathbf{\Phi}(t, \tau)\mathbf{x}(\tau) + \mathbf{C}\int_\tau^t \mathbf{\Phi}(t, \sigma)[\mathbf{B}\mathbf{u}(\sigma) + \mathbf{D}\mathbf{w}(\sigma)]\, d\sigma + \mathbf{H}\mathbf{u}(t) + \mathbf{J}\mathbf{w}(t), \tag{4.84}$$

where $\mathbf{\Phi}(t, \tau)$ is the state transition matrix for $\mathbf{A}$, and t and τ are any specified times.

Proof: First, see from the output equation of (4.77) that $\mathbf{y}(t) = \mathbf{C}\mathbf{x}(t) + \mathbf{H}\mathbf{u}(t) + \mathbf{J}\mathbf{w}(t)$. By comparison with (4.84), we see that the only thing left to prove is that

$$\mathbf{x}(t) = \mathbf{\Phi}(t, \tau)\mathbf{x}(\tau) + \int_\tau^t \mathbf{\Phi}(t, \sigma)[\mathbf{B}\mathbf{u}(\sigma) + \mathbf{D}\mathbf{w}(\sigma)]\, d\sigma. \tag{4.85}$$

This is evident from (4.82) and (4.83). However, we may prove (4.85) directly. To prove that any proposed solution of a differential equation is indeed a solution, differentiate the proposed solution $\mathbf{x}(t)$ and see if it satisfies (4.77). Using the Leibniz rule for differentiation under the integral, the derivative of (4.85) is

$$\dot{\mathbf{x}}(t) = \frac{d}{dt}[\mathbf{\Phi}(t, \tau)]\mathbf{x}(\tau) + \mathbf{\Phi}(t, t)[\mathbf{B}\mathbf{u}(t) + \mathbf{D}\mathbf{w}(t)]$$
$$+ \int_\tau^t \frac{d}{dt}[\mathbf{\Phi}(t, \sigma)][\mathbf{B}\mathbf{u}(\sigma) + \mathbf{D}\mathbf{w}(\sigma)]\, d\sigma.$$

Using the property of the state transition matrix (4.83),

$$\dot{\mathbf{x}}(t) = \mathbf{A}\boldsymbol{\Phi}(t,\tau)\mathbf{x}(\tau) + \mathbf{Bu}(t) + \mathbf{Dw}(t) + \int_{\tau}^{t} \mathbf{A}\boldsymbol{\Phi}(t,\sigma)[\mathbf{Bu}(\sigma) + \mathbf{Dw}(\sigma)]\,d\sigma.$$

$$= \mathbf{A}\left[\boldsymbol{\Phi}(t,\tau)\mathbf{x}(\tau) + \int_{\tau}^{t} \boldsymbol{\Phi}(t,\sigma)[\mathbf{Bu}(\sigma) + \mathbf{Dw}(\sigma)]\,d\sigma\right] + \mathbf{Bu}(t) + \mathbf{Dw}(t)$$

$$= \mathbf{Ax}(t) + \mathbf{Bu}(t) + \mathbf{Dw}(t)$$ ■

Now, to prove uniqueness, we use the Lipschitz condition (3.10) (which is sufficient but not necessary for uniqueness). See that

$$\|\mathbf{A}(t)\mathbf{x}_1(t) - \mathbf{A}(t)\mathbf{x}_2(t)\| \le \max_{\tau}\|\mathbf{A}(\tau)\|\,\|\mathbf{x}_1(t) - \mathbf{x}_2(t)\|$$

does indeed satisfy (3.10). Hence, the Lipschitz condition is satisfied and the solution (4.85) is unique.

Note also that (4.84) may be written

$$\mathbf{y}(t) = \mathbf{C}(t)\boldsymbol{\Phi}(t,\tau)\mathbf{x}(\tau) + \int_{\tau}^{t}[\mathbf{C}(t)\boldsymbol{\Phi}(t,\sigma)\mathbf{B}(\sigma) + \mathbf{H}(\sigma)\delta(t-\sigma)]\mathbf{u}(\sigma)\,d\sigma$$

$$+ \int_{\tau}^{t}[\mathbf{C}(t)\boldsymbol{\Phi}(t,\sigma)\mathbf{D}(\sigma) + \mathbf{J}(\sigma)\delta(t-\sigma)]\mathbf{w}(\sigma)\,d\sigma$$

using the screening property of the Dirac delta function.

The state transition matrix $\boldsymbol{\Phi}(t,\tau)$ describes the "transition of the state" from $\mathbf{x}(\tau)$ to $\mathbf{x}(t)$. There is no restriction on τ, and hence Theorem 4.5 describes the propagation of the state forward *or* backward in time (i.e., $\tau < t$ or $\tau > t$).

Theorem 4.6

The state transition matrix $\boldsymbol{\Phi}(t,\tau)$ has these properties:

1. *Transition property*: $\boldsymbol{\Phi}(t,\tau) = \boldsymbol{\Phi}(t,\sigma)\boldsymbol{\Phi}(\sigma,\tau)$;
2. *Inversion property*: $\boldsymbol{\Phi}^{-1}(t,\tau) = \boldsymbol{\Phi}(\tau,t)$;
3. *Nonsingular property*: $|\boldsymbol{\Phi}(t,\tau)| \neq 0\ \forall t,\tau$.

Proof: Properties 2 and 3 follows immediately from (4.82). To verify 1, substitute $\boldsymbol{\Phi}(t,\tau) = \boldsymbol{\Phi}(t,\sigma)\boldsymbol{\Phi}(\sigma,\tau)$ into the defining equation (4.83) and ask whether the differential equation (4.83) is satisfied:

$$\frac{d}{dt}[\boldsymbol{\Phi}(t,\sigma)\boldsymbol{\Phi}(\sigma,\tau)] \stackrel{?}{=} \mathbf{A}(t)[\boldsymbol{\Phi}(t,\sigma)\boldsymbol{\Phi}(\sigma,\tau)],$$

$$\boldsymbol{\Phi}(\tau,\sigma)\boldsymbol{\Phi}(\sigma,\tau) \stackrel{?}{=} \mathbf{I}.$$

The first of these relations holds by virtue of (4.83) and the second, by virtue of property 2 of Theorem 4.6. ■

Exercise 4.12

Prove the identity

$$\frac{d}{dt}[\mathbf{M}^{-1}(t)] = -\mathbf{M}^{-1}(t)\left[\frac{d}{dt}\mathbf{M}(t)\right]\mathbf{M}^{-1}(t).$$

Hint: Differentiate $\mathbf{M}(t)\mathbf{M}^{-1}(t) = \mathbf{I}$.

Exercise 4.13

Prove that

$$\frac{d}{dt}\mathbf{\Phi}(\sigma, t) = -\mathbf{\Phi}(\sigma, t)\mathbf{A}(t).$$

Hint: Use Exercise 4.12 and property 2 of Theorem 4.6.

Exercise 4.14

Show that any given $\mathbf{x}(0)$ gives a response that is identical to the response obtained from impulses applied simultaneously in all input channels at $t = 0$ iff

$$\mathbf{x}(0) = \mathbf{B}\boldsymbol{u}, \qquad \boldsymbol{u}^T \triangleq (u_1, \ldots, u_{n_u})$$

has a solution $\boldsymbol{u}$, where u_i is the "strength" of the ith impulse, $\mathbf{u}(t) = \boldsymbol{u}\delta(t)$.

EXAMPLE 4.9

If $u_i(t) = \delta(t)$, $\forall i = 1, \ldots, n_u$, and if $\mathbf{b} \triangleq \Sigma_{i=1}^{n_u}\mathbf{b}_i$, then prove that the response $\mathbf{y}(t)$ of the *multiinput* system

$$\dot{\mathbf{x}} = \mathbf{A}\mathbf{x} + \mathbf{B}\mathbf{u}, \qquad \mathbf{u} \in \mathscr{R}^{n_u}, \quad \mathbf{B} = \left[\mathbf{b}_1, \ldots, \mathbf{b}_{n_u}\right],$$

$$\mathbf{y} = \mathbf{C}\mathbf{x}, \qquad \mathbf{x}(0) = \mathbf{x}_0,$$

is equivalent to the response $\mathbf{y}(t)$ of the *single-input* system

$$\dot{\boldsymbol{x}} = \mathbf{A}\boldsymbol{x} + \mathbf{b}\delta(t),$$

$$\mathbf{y} = \mathbf{C}\boldsymbol{x}, \qquad \mathbf{x}(0) = \mathbf{x}_0.$$

Proof: From the given data,

$$\mathbf{Bu} = [b_1, \ldots, b_{n_u}] \begin{pmatrix} 1 \\ 1 \\ \vdots \\ 1 \end{pmatrix} \delta(t)$$

$$= \left[\sum_{i=1}^{n_u} \mathbf{b}_i \right] \delta(t)$$

$$= \mathbf{b}\delta(t).$$

Hence, the two systems satisfy the same differential equation and the same initial condition. The uniqueness of linear system solutions establishes the result. ∎

For *time-varying* matrices $\mathbf{A}(t)$, there is usually no choice but to solve (4.83) by numerical integration. For *constant* $\mathbf{A}$, there is a brighter promise.

4.4.1 COMPUTATION OF THE STATE TRANSITION MATRIX FOR CONSTANT A

Theorem 4.7

The solution of (4.83) *for constant* $\mathbf{A}$ *is*

$$\mathbf{\Phi}(t, \tau) = \sum_{i=0}^{\infty} \mathbf{A}^i \left[\frac{(t-\tau)^i}{i!} \right]. \tag{4.86}$$

Proof: We must show that this solution satisfies (4.83):

$$\frac{d}{dt}\left[\sum_{i=0}^{\infty} \mathbf{A}^i \frac{(t-\tau)^i}{i!} \right] \stackrel{?}{=} \mathbf{A}\left[\sum_{i=0}^{\infty} \mathbf{A}^i \frac{(t-\tau)^i}{i!} \right],$$

where

$$\sum_{i=1}^{\infty} \mathbf{A}^i \frac{(t-\tau)^{i-1}}{(i-1)!} = \sum_{i=0}^{\infty} \mathbf{A}^{i+1} \frac{(t-\tau)^i}{i!}.$$

By replacing the right-hand side index from $i+1$ to i with appropriate change of limits, the question mark can be removed, and the theorem is proved. ∎

Suppose $\mathbf{A}$ is a scalar $\mathbf{A} = a$. Then, (4.86) becomes

$$\mathbf{\Phi}(t, \tau) = \sum_{i=0}^{\infty} a^i \frac{(t-\tau)^i}{i!} = e^{a(t-\tau)},$$

where the infinite series is recognized as the series for the exponential $e^{a(t-\tau)}$. Now, in the matrix case we shall *define* the matrix $e^{\mathbf{A}(t-\tau)}$ by the series

$$e^{\mathbf{A}(t-\tau)} \triangleq \sum_{i=0}^{\infty} \mathbf{A}^i \frac{(t-\tau)^i}{i!}. \tag{4.87}$$

To show that the power series (4.87) converges, use the Cauchy–Schwarz inequality to see that

$$\left\| \sum_{i=k}^{\infty} \frac{\mathbf{A}^i}{i!} \right\|^2 \leq \sum_{i=k}^{\infty} \|\mathbf{A}^i\| \frac{1}{i!} \leq \sum_{i=k}^{\infty} \frac{\|\mathbf{A}\|^i}{i!},$$

where the right-hand side is arbitrarily small for large enough k.

The Laplace transform of the state transition matrix e^{At} is as follows:

$$\mathscr{L}[e^{\mathbf{A}t}] = [s\mathbf{I} - \mathbf{A}]^{-1} = \frac{1}{s}\left[\mathbf{I} - \frac{\mathbf{A}}{s}\right]^{-1} = \frac{1}{s}\left[\sum_{i=0}^{\infty} \frac{\mathbf{A}^i}{s^i}\right].$$

Thus

$$\mathscr{L}[\mathbf{C}e^{\mathbf{A}t}\mathbf{B}] = \mathbf{C}[s\mathbf{I} - \mathbf{A}]^{-1}\mathbf{B}.$$

The matrix $[s\mathbf{I} - \mathbf{A}]^{-1}$ is called the *resolvent* matrix.

The Cayley–Hamilton Theorem 2.12 allows $e^{\mathbf{A}t}$ to be computed with a *finite* series instead of the infinite series (4.87). This result is developed as follows. Write

$$\mathscr{L}[e^{\mathbf{A}t}] = [s\mathbf{I} - \mathbf{A}]^{-1} = \frac{\mathbf{S}_{n-1}s^{n-1} + \mathbf{S}_{n-2}s^{n-2} + \cdots \mathbf{S}_1 s + \mathbf{S}_0}{s^n + a_{n-1}s^{n-1} + \cdots a_1 s + a_0}, \tag{4.88}$$

where $\mathbf{S}_i$ are $n \times n$ matrices. Note that the cofactor matrix $\operatorname{Co}(s\mathbf{I} - \mathbf{A})$ is composed of $(n-1) \times (n-1)$ determinants. Thus, the matrix $(s\mathbf{I} - \mathbf{A})^{-1}$ is a strictly proper ratio of polynomials, since the highest degree polynomial possible in the numerator is $n - 1$. Hence,

$$[\operatorname{Co}(s\mathbf{I} - \mathbf{A})]^T = \operatorname{adj}(s\mathbf{I} - \mathbf{A}) = \mathbf{S}_{n-1}s^{n-1} + \mathbf{S}_{n-2}s^{n-2} + \cdots \mathbf{S}_0. \tag{4.89}$$

A recursive algorithm has been developed to compute the $\mathbf{S}_i$ and a_i simultaneously.

The resolvent algorithm [4.7] is as follows:

$$
\begin{aligned}
a_{n-1} &= -\operatorname{tr}[\mathbf{S}_{n-1}\mathbf{A}], & \mathbf{S}_{n-1} &= \mathbf{I}, \\
a_{n-2} &= -\frac{1}{2}\operatorname{tr}[\mathbf{S}_{n-2}\mathbf{A}], & \mathbf{S}_{n-2} &= \mathbf{S}_{n-1}\mathbf{A} + a_{n-1}\mathbf{I}, \\
a_{n-3} &= -\frac{1}{3}\operatorname{tr}[\mathbf{S}_{n-3}\mathbf{A}], & \mathbf{S}_{n-3} &= \mathbf{S}_{n-2}\mathbf{A} + a_{n-2}\mathbf{I}, \\
a_{n-4} &= -\frac{1}{4}\operatorname{tr}[\mathbf{S}_{n-4}\mathbf{A}], & \mathbf{S}_{n-4} &= \mathbf{S}_{n-3}\mathbf{A} + a_{n-3}\mathbf{I}, \\
&\vdots & &\vdots \\
a_{n-i+1} &= \frac{-1}{i-1}\operatorname{tr}[\mathbf{S}_{n-i+1}\mathbf{A}], & \mathbf{S}_{n-i} &= \mathbf{S}_{n-i+1}\mathbf{A} + a_{n-i+1}\mathbf{I}, \\
&\vdots & &\vdots \\
a_1 &= -\frac{1}{n-1}\operatorname{tr}[\mathbf{S}_1\mathbf{A}], & \mathbf{S}_0 &= \mathbf{S}_1\mathbf{A} + a_1\mathbf{I}, \\
a_0 &= -\frac{1}{n}\operatorname{tr}[\mathbf{S}_0\mathbf{A}], & \mathbf{0} &= \mathbf{S}_0\mathbf{A} + a_0\mathbf{I}.
\end{aligned}
\tag{4.90}
$$

Exercise 4.15

Verify the above formulas for $\mathbf{S}_i$. *Hint*: Verify (4.89) by multiplying both sides by $[s\mathbf{I} - \mathbf{A}]$ and use the fact that

$$[s\mathbf{I} - \mathbf{A}]\operatorname{adj}[s\mathbf{I} - \mathbf{A}] = \mathbf{I}|s\mathbf{I} - \mathbf{A}|.$$

EXAMPLE 4.10

Using the resolvent algorithm, compute the resolvent matrix for

$$\mathbf{A} = \begin{bmatrix} 1 & 0 & 0 \\ 0 & 0 & 1 \\ 1 & -1 & 0 \end{bmatrix}.$$

Solution:

$$\mathbf{S}_2 = \mathbf{I}, \qquad a_2 = -\operatorname{tr}\mathbf{A} = -1,$$

$$\mathbf{S}_1 = \mathbf{A} - \mathbf{I} = \begin{bmatrix} 0 & 0 & 0 \\ 0 & -1 & 1 \\ 1 & -1 & -1 \end{bmatrix}, \qquad a_1 = -\tfrac{1}{2}\operatorname{tr}(\mathbf{A} - \mathbf{I})\mathbf{A} = 1,$$

$$\mathbf{S}_0 = (\mathbf{A} - \mathbf{I})\mathbf{A} + \mathbf{I} = \mathbf{I}, \qquad a_0 = -\tfrac{1}{3}\operatorname{tr}[(\mathbf{A} - \mathbf{I})\mathbf{A} + \mathbf{I}]\mathbf{A} = -1.$$

Hence,

$$(s\mathbf{I} - \mathbf{A})^{-1} = \frac{\mathbf{I}s^2 + \mathbf{S}_1 s + \mathbf{I}}{s^3 - s^2 + s - 1}. \qquad \blacksquare$$

Exercise 4.16

Compute $(s\mathbf{I} - \mathbf{A})^{-1}$ for

$$\mathbf{A} = \begin{bmatrix} 0 & 1 \\ -1 & 0 \end{bmatrix}$$

using the resolvent algorithm and check your answer by a direct calculation.

The above results lead to the following useful fact.

Theorem 4.8

The resolvent matrix and the matrix exponential may be expressed as

$$[s\mathbf{I} - \mathbf{A}]^{-1} = \sum_{i=0}^{n-1} \mathbf{A}^i \alpha_i(s), \tag{4.91a}$$

$$e^{\mathbf{A}t} = \sum_{i=0}^{n-1} \mathbf{A}^i \alpha_i(t), \qquad \alpha_i(t) \triangleq \mathscr{L}^{-1}[\alpha_i(s)], \tag{4.91b}$$

where

$$\begin{aligned} \alpha_{n-1}(s) &= |s\mathbf{I} - \mathbf{A}|^{-1}, \\ \alpha_{n-2}(s) &= |s\mathbf{I} - \mathbf{A}|^{-1}(s + a_{n-1}), \\ \alpha_{n-3}(s) &= |s\mathbf{I} - \mathbf{A}|^{-1}(s^2 + a_{n-1}s + a_{n-2}), \\ &\vdots \\ \alpha_0(s) &= |s\mathbf{I} - \mathbf{A}|^{-1}(s^{n-1} + a_{n-1}s^{n-2} + \cdots a_2 s + a_1). \end{aligned} \tag{4.92}$$

Exercise 4.17

Prove Theorem 4.8. *Hint*: Substitute (4.90) into (4.88).

Exercise 4.18

Using the resolvent algorithm, prove that

$$|\mathbf{A}| = (-1)^n a_0, \tag{4.93}$$

$$\mathbf{A}^{-1} = -\mathbf{S}_0 a_0^{-1} \qquad \text{if } a_0 \neq 0. \tag{4.94}$$

Then, find the inverse of

$$\mathbf{A} = \begin{bmatrix} 0 & 1 \\ 1 & 2 \end{bmatrix}.$$

Exercise 4.19

Use Theorem 4.8 to compute $e^{\mathbf{A}t}$ for the $\mathbf{A}$ in Example 4.10.

Exercise 4.20

Use Theorem 4.8 to compute $e^{\mathbf{A}t}$ for the $\mathbf{A}$ in Exercise 4.16. Note that (4.91) is a considerably reduced computational burden than (4.87).

Exercise 4.21

Use the final value theorem $\lim_{s\to 0}\{s[\mathbf{C}(sI - \mathbf{A})^{-1}\mathbf{B} + \mathbf{H}]u(s)\}$ and (4.90) and (4.92) to compute the steady-state step response

$$u_i(t) = 1, \qquad i = 1, \ldots, n_u,$$

$$\lim_{t\to\infty} y(t) = ?$$

The next method for computing $e^{\mathbf{A}t}$ is described as follows.

Theorem 4.9

If Λ is the Jordan form of $\mathbf{A}$, then

$$e^{\mathbf{A}t} = \mathbf{E}e^{\Lambda t}\mathbf{E}^{-1}, \tag{4.95}$$

where $\mathbf{E}$ is the modal matrix of $\mathbf{A}$,

$$\mathbf{A}\mathbf{E} = \mathbf{E}\Lambda. \tag{4.96}$$

Proof: Use the spectral decomposition of $\mathbf{A}$, equation (2.180),

$$\mathbf{A} = \mathbf{E}\Lambda\mathbf{E}^{-1},$$

to write

$$e^{\mathbf{A}t} \triangleq \sum_{i=0}^{\infty} \mathbf{A}^i \frac{t^i}{i!} = \sum_{i=0}^{\infty} (\mathbf{E}\Lambda\mathbf{E}^{-1})^i \frac{t^i}{i!}$$

$$= \mathbf{E}\left[\sum_{i=0}^{\infty} \Lambda^i \frac{t^i}{i!}\right]\mathbf{E}^{-1}$$

$$= \mathbf{E}e^{\Lambda t}\mathbf{E}^{-1}. \qquad \blacksquare$$

Exercise 4.22

Show that if $\Lambda_i \in \mathscr{C}^{n_i \times n_i}$ is given by equation (2.175), then

$$e^{\Lambda_i t} = e^{\lambda_i t} \begin{bmatrix} 1 & t & \cdots & \dfrac{t^{n_i - 1}}{(n_i - 1)!} \\ & \ddots & \ddots & \vdots \\ & & \ddots & t \\ 0 & & & 1 \end{bmatrix}. \tag{4.97}$$

Hint: Write

$$\Lambda_i = \lambda_i \mathbf{I} + \mathscr{E}, \qquad \mathscr{E} \triangleq \begin{bmatrix} 0 & 1 & 0 & \cdots & 0 \\ & \ddots & \ddots & & \vdots \\ & & \ddots & \ddots & 0 \\ & & & \ddots & 1 \\ & & & & 0 \end{bmatrix}.$$

Exercise 4.23

Use Theorem 4.9 to compute $e^{\mathbf{A}t}$ if

$$\mathbf{A} = \begin{bmatrix} 0 & 1 \\ 0 & 0 \end{bmatrix}, \qquad \mathbf{A} = \begin{bmatrix} 0 & 1 \\ -1 & 0 \end{bmatrix}.$$

Compare computational burdens by using the definition (4.87) to compute $e^{\mathbf{A}t}$.

Exercise 4.24

Show that if $\mathbf{A}$ has distinct eigenvalues, then (4.95) is equivalent to

$$e^{\mathbf{A}t} = \sum_{i=1}^{n} \mathscr{I}_i e^{\lambda_i t}, \qquad \mathscr{I}_i \triangleq \mathbf{e}_i \mathbf{l}_i^*, \tag{4.98}$$

where $\mathbf{e}_i$ is the ith column of $\mathbf{E}$ and $\mathbf{l}_i^*$ is the ith row of $\mathbf{E}^{-1}$, and

$$\sum_{i=1}^{n} \mathscr{I}_i = \mathbf{I} = \mathbf{E}\mathbf{E}^{-1}. \tag{4.99}$$

Show also that $\mathscr{I}_i$ is a residue of the resolvent matrix

$$\mathscr{L}[e^{\mathbf{A}t}] = (s\mathbf{I} - \mathbf{A})^{-1} = \sum_{i=1}^{n} \frac{\mathscr{I}_i}{s - \lambda_i},$$

and hence $\mathscr{I}_i$ may also be computed by

$$\mathscr{I}_i = (s - \lambda_i)(s\mathbf{I} - \mathbf{A})^{-1}\Big|_{s=\lambda_i} = \frac{\text{adj}\,[\lambda_i \mathbf{I} - \mathbf{A}]}{\prod\limits_{j \neq i}^{n} (\lambda_i - \lambda_j)}. \tag{4.100}$$

Show also that

$$\mathscr{I}_i = \prod_{\substack{j=1 \\ j \neq 1}}^{n} \frac{[\mathbf{A} - \lambda_j \mathbf{I}]}{\lambda_i - \lambda_j} \tag{4.101}$$

and that

$$\mathscr{I}_i = \mathbf{E}\mathbf{I}^{(i)}\mathbf{E}^{-1}, \quad \mathbf{I}^{(i)} = \begin{bmatrix} 0 & & 0 \\ & 1 & \\ 0 & & 0 \end{bmatrix}, \quad 1 \text{ in } ii \text{ position.} \tag{4.102}$$

We now summarize these several methods for computing $e^{\mathbf{A}t}$.

1. $e^{\mathbf{A}t} \triangleq \sum\limits_{i=0}^{\infty} \mathbf{A}^i \dfrac{t^i}{i!}$,
2. $e^{\mathbf{A}t} = \mathscr{L}^{-1}[s\mathbf{I} - \mathbf{A}]^{-1}$,
3. $e^{\mathbf{A}t} = \sum\limits_{i=0}^{n-1} \mathbf{A}^i \alpha_i(t)$, $\alpha_i(t)$ from (4.92),
4. $e^{\mathbf{A}t} = \mathbf{E}e^{\Lambda t}\mathbf{E}^{-1}$, $\mathbf{AE} = \mathbf{E}\Lambda$,
5. $e^{\mathbf{A}t} = \sum\limits_{i=1}^{n} \mathscr{I}_i e^{\lambda_i t}$ [if Λ diagonal, $\mathscr{I}_i$ from (4.98) or (4.100)].

It should be clear by now that even though the inputs u_i, $i = 1, \ldots, n_u$, and the outputs y_i, $i = 1, \ldots, n_y$ variables are real, the state variables x_j, $i = 1, \ldots, n_x$, may be complex.

Exercise 4.25

For the system described by

$$\frac{y(s)}{u(s)} = G(s) = \frac{1}{s^2 + 0.1s + 10},$$

write state space realizations from the (a) phase variable, (b) rectangular, and (c) modal form of the block diagram. In which cases are the state variables complex?

This exercise illustrates the need for the complex algebra and calculus introduced in Chapter 2. Hereafter, in our choice of notation we shall consider the state to be complex. This means that **A**, **B**, and **C** are all considered to be possibly complex matrices, but **u**, **y**, **H**, and **J** all remain real.

State Transition Matrices for Vector Second-Order Systems

Now, we use the result $e^{\mathbf{A}t} = \mathbf{E}e^{\Lambda t}\mathbf{E}^{-1}$ to compute the state transition matrix for vector second-order systems of the form

$$\mathcal{M}\ddot{\mathbf{q}} + \mathcal{D}\dot{\mathbf{q}} + \mathcal{K}\mathbf{q} = \mathcal{B}\mathbf{u}, \qquad \mathcal{M} = \mathcal{M}^T > \mathbf{0},$$
$$\mathcal{D} = \alpha\mathcal{M} + \beta\mathcal{K}, \qquad \mathcal{K} = \mathcal{K}^T \geq \mathbf{0} \tag{4.103}$$

The following theorem is useful here.

Theorem 4.10

Given any two real symmetric matrices $\mathcal{M}$ and $\mathcal{K}$ with $\mathcal{M}$ positive definite, there always exists a nonsingular matrix $\mathcal{E}$ such that

$$\mathcal{E}^T\mathcal{M}\mathcal{E} = \mathbf{I}, \qquad \mathcal{E}^T\mathcal{K}\mathcal{E} = \Omega^2 = \text{diag.} \tag{4.104}$$

Proof: Since $\mathcal{M}$ is symmetric, there always exists a unitary modal matrix for $\mathcal{M}$ such that

$$\mathcal{M} = \mathcal{E}_m\Lambda_m\mathcal{E}_m^T, \qquad \mathcal{E}_m^T\mathcal{E}_m = \mathbf{I}, \quad \Lambda_m = \text{diag.} > \mathbf{0} \tag{4.105}$$

(see below Theorem 2.17). Define $\sqrt{\Lambda_m}$ by

$$\Lambda_m = \sqrt{\Lambda_m}\sqrt{\Lambda_m} \tag{4.106}$$

and $(\sqrt{\Lambda_m})^{-1}$ exists since $\Lambda_m > 0$. Finally, define $\mathcal{E}_k$ as the unitary modal matrix of the matrix

$$\left(\sqrt{\Lambda_m}\right)^{-1}\mathcal{E}_m^T\mathcal{K}\mathcal{E}_m\left(\sqrt{\Lambda_m}\right)^{-1} = \mathcal{E}_k\Omega^2\mathcal{E}_k^T, \qquad \Omega = \text{diag.} \tag{4.107}$$

Now, to conclude the proof, we have only to show that the choice

$$\mathcal{E} = \mathcal{E}_m\left(\sqrt{\Lambda_m}\right)^{-1}\mathcal{E}_k \tag{4.108}$$

does the trick, since

$$\mathcal{E}^T\mathcal{M}\mathcal{E} = \mathbf{I} = \mathcal{E}_k^T\left(\sqrt{\Lambda_m}\right)^{-1}\mathcal{E}_m^T\left(\mathcal{E}_m\Lambda_m\mathcal{E}_m^T\right)\mathcal{E}_m\left(\sqrt{\Lambda_m}\right)^{-1}\mathcal{E}_k, \tag{4.109}$$

$$\mathcal{E}^T\mathcal{K}\mathcal{E} = \Omega^2 = \mathcal{E}_k^T\left(\sqrt{\Lambda_m}\right)^{-1}\mathcal{E}_m^T\mathcal{K}\mathcal{E}_m\left(\sqrt{\Lambda_m}\right)^{-1}\mathcal{E}_k = \mathcal{E}_k^T\left(\mathcal{E}_k\Omega^2\mathcal{E}_k^T\right)\mathcal{E}_k. \quad \blacksquare$$

Now, write (4.103) in η coordinates where $\mathbf{q} = \mathscr{E}\eta$. Hence,

$$\mathscr{E}^T[\mathscr{M}\mathscr{E}\ddot{\eta} + \mathscr{D}\mathscr{E}\dot{\eta} + \mathscr{K}\mathscr{E}\eta = \mathscr{B}\mathbf{u}]. \tag{4.110}$$

Using the choice of $\mathscr{E}$ in (4.108) yields

$$\ddot{\eta} + \mathscr{E}^T\mathscr{D}\mathscr{E}\dot{\eta} + \Omega^2\eta = \mathscr{E}^T\mathscr{B}\mathbf{u}. \tag{4.111}$$

Now, $\mathscr{E}^T\mathscr{D}\mathscr{E}$ will not be diagonal for every $\mathscr{D}$. Assume $\mathscr{D} = \alpha\mathscr{M} + \beta\mathscr{K}$ for some scalars α, β. Within this restricted class of problems (Rayleigh damping [4.1]), we then shall write the diagonal matrix $\mathscr{E}^T\mathscr{D}\mathscr{E} = 2\zeta\Omega$, where

$$\zeta \triangleq \begin{bmatrix} \zeta_1 & & & \\ & \zeta_2 & & \\ & & \ddots & \\ & & & \zeta_N \end{bmatrix} \tag{4.112}$$

with assumption $\mathscr{D} = \alpha\mathscr{M} + \beta\mathscr{K}$ the "damping" matrix ζ is determined from

$$2\zeta\Omega = \alpha\mathbf{I} + \beta\Omega^2 = \mathscr{E}^T\mathscr{D}\mathscr{E}. \tag{4.113}$$

Hence,

$$\zeta = \tfrac{1}{2}\left[\alpha\Omega^{-1} + \beta\Omega\right] \Rightarrow \zeta_i = \tfrac{1}{2}\left(\alpha\omega_i^{-1} + \beta\omega_i\right). \tag{4.114}$$

We restrict our attention to systems with $\zeta_i < 1$, for otherwise we may examine the state space (vector first-order) form directly.

The eigenvalues of $\mathbf{A}$ in

$$\dot{\mathbf{x}} = \mathbf{A}\mathbf{x} + \mathbf{B}\mathbf{u}, \qquad \mathbf{x} = \begin{pmatrix} \eta \\ \dot{\eta} \end{pmatrix} \tag{4.115}$$

are those of (4.111), which are shown to be

$$\Lambda_c = -\zeta\Omega + j\Omega\sqrt{\mathbf{I} - \zeta^2}, \qquad \overline{\Lambda}_c = -\zeta\Omega - j\Omega\sqrt{\mathbf{I} - \zeta^2}, \tag{4.116}$$

and the Jordan form of $\mathbf{A}$ is

$$\mathbf{E}^{-1}\mathbf{A}\mathbf{E} = \begin{bmatrix} \Lambda_c & \mathbf{0} \\ \mathbf{0} & \overline{\Lambda}_c \end{bmatrix}, \tag{4.117}$$

where $\mathbf{E}$ satisfies

$$\mathbf{A}\mathbf{E} = \mathbf{E}\Lambda, \qquad \mathbf{A} = \begin{bmatrix} \mathbf{0} & \mathbf{I} \\ -\Omega^2 & -2\zeta\Omega \end{bmatrix}. \tag{4.118}$$

Now, it is easy to show that

$$\mathbf{E} = \begin{bmatrix} \mathbf{I} & \mathbf{I} \\ \Lambda_c & \bar{\Lambda}_c \end{bmatrix}. \tag{4.119}$$

Now, the state transition matrix of $\mathbf{A}$ is computed as follows:

$$e^{\mathbf{A}t} = \mathbf{E}e^{\Lambda t}\mathbf{E}^{-1}, \tag{4.120}$$

where (3.85) is needed to show that

$$e^{\mathbf{A}t} = \left[\begin{array}{c|c} e^{\Lambda_c t} + [e^{\Lambda_c t} - e^{\bar{\Lambda}_c t}][\bar{\Lambda}_c - \Lambda_c]^{-1}\Lambda_c & [e^{\bar{\Lambda}_c t} - e^{\Lambda_c t}][\bar{\Lambda}_c - \Lambda_c]^{-1} \\ \Lambda_c e^{\Lambda_c t} + [\Lambda_c e^{\Lambda_c t} - \bar{\Lambda}_c e^{\bar{\Lambda}_c t}][\bar{\Lambda}_c - \Lambda_c]^{-1}\Lambda_c & [\bar{\Lambda}_c e^{\bar{\Lambda}_c t} - \Lambda_c e^{\Lambda_c t}][\bar{\Lambda}_c - \Lambda_c]^{-1} \end{array}\right]. \tag{4.121}$$

Now, (4.121) can also be written as

$$e^{\mathbf{A}t} = \left[\begin{array}{c|c} [\bar{\Lambda}_c e^{\Lambda_c t} - \Lambda_c e^{\bar{\Lambda}_c t}][\bar{\Lambda}_c - \Lambda_c]^{-1} & [e^{\bar{\Lambda}_c t} - e^{\Lambda_c t}][\bar{\Lambda}_c - \Lambda_c]^{-1} \\ [e^{\Lambda_c t} - e^{\bar{\Lambda}_c t}][\bar{\Lambda}_c - \Lambda_c]^{-1}\Lambda_c\bar{\Lambda}_c & [\bar{\Lambda}_c e^{\bar{\Lambda}_c t} - \Lambda_c e^{\Lambda_c t}][\bar{\Lambda}_c - \Lambda_c]^{-1} \end{array}\right],$$

where

$$e^{\Lambda_c t} + e^{\bar{\Lambda}_c t} = 2e^{-\zeta\Omega t}\cos\Omega\sqrt{\mathbf{I} - \zeta^2}\,t$$

$$e^{\Lambda_c t} - e^{\bar{\Lambda}_c t} = 2je^{-\zeta\Omega t}\sin\Omega\sqrt{\mathbf{I} - \zeta^2}\,t$$

where

$$\left[\sin\Omega\sqrt{\mathbf{I} - \zeta^2}\,t\right]_{ii} \triangleq \sin\omega_i\sqrt{1 - \zeta_i^2}\,t.$$

Substitute these into (4.121) to get

$$e^{\mathbf{A}t} = \left[\begin{array}{c|c} e^{-\zeta\Omega t}\left[\cos\Omega\sqrt{\mathbf{I} - \zeta^2}\,t + \zeta[\mathbf{I} - \zeta^2]^{-1/2}\sin\Omega[\mathbf{I} - \zeta^2]^{1/2}t\right] & e^{-\zeta\Omega t}\Omega^{-1}[\mathbf{I} - \zeta^2]^{-1/2}\sin\Omega\sqrt{\mathbf{I} - \zeta^2}\,t \\ -e^{-\zeta\Omega t}\Omega[\mathbf{I} - \zeta^2]^{-1/2}\sin\Omega\sqrt{\mathbf{I} - \zeta^2}\,t & e^{-\zeta\Omega t}\left[\cos\Omega\sqrt{\mathbf{I} - \zeta^2}\,t - \zeta[\mathbf{I} - \zeta^2]^{-1/2}\sin\Omega\sqrt{\mathbf{I} - \zeta^2}\,t\right] \end{array}\right]. \tag{4.122}$$

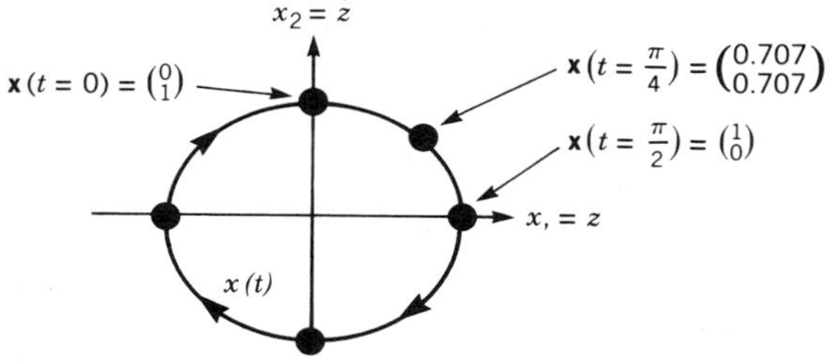

***Figure 4.10** Two dimensional State Space*

For lightly damped structures, $\zeta \ll \mathbf{I}$. Then,

$$e^{\mathbf{A}t} = \begin{bmatrix} e^{-\zeta\Omega t}[\cos\Omega t + \zeta\sin\Omega t] & e^{-\zeta\Omega t}\Omega^{-1}\sin\Omega t \\ e^{-\zeta\Omega t}\Omega\sin\Omega t & e^{-\zeta\Omega t}[\cos\Omega t - \zeta\sin\Omega t] \end{bmatrix}. \tag{4.123}$$

Now, for this case, with $\mathbf{y} = \mathbf{Cx}$, $\mathbf{C} = \begin{bmatrix} \mathscr{P} & \mathbf{0} \\ \mathbf{0} & \mathscr{R} \end{bmatrix}$,

$$\mathbf{C}e^{\mathbf{A}t}\mathbf{B} = \begin{bmatrix} \mathscr{P} & 0 \\ 0 & \mathscr{R} \end{bmatrix} e^{\mathbf{A}t} \begin{bmatrix} 0 \\ \mathscr{B} \end{bmatrix} = \begin{bmatrix} \mathscr{P}\mathscr{B}e^{-\zeta\Omega t}\Omega^{-1}\sin\Omega t \\ \mathscr{R}\mathscr{B}e^{-\zeta\Omega t}[\cos\Omega t - \zeta\sin\Omega t] \end{bmatrix}. \tag{4.124}$$

4.4.2 STATE SPACE TRAJECTORIES FOR SECOND-ORDER SYSTEMS

The formal solution (4.84) of the state equations lacks insight into the geometrical properties in the state space. Several examples of low-order systems give the general idea of the shape of state space trajectories.

EXAMPLE 4.11

For the system (2.14) with $k = 1$, $m = 1$, $f_1 = 0$, $z(0) = 0$, and $\dot{z}(0) = 1$, plot the trajectory of $\mathbf{x}(t)$ in the linear vector space $x \in \mathscr{R}^2$.

Solution: Note that

$$\mathbf{x}(t) = \begin{pmatrix} z(t) \\ \dot{z}(t) \end{pmatrix} = \begin{pmatrix} \dot{z}(0)\sin t \\ \dot{z}(0)\cos t \end{pmatrix} = \begin{pmatrix} \sin t \\ \cos t \end{pmatrix}.$$

Hence, the state trajectory is as sketched in Fig. 4.10. See that the state vector $\mathbf{x}(t)$ lies in the two-dimensional space $\mathscr{R}^2$ for all $t \in [0, \infty]$. ■

EXAMPLE 4.12

Plot in the (x_1, x_2) plane the motion of the linear system

$$\dot{\mathbf{x}} = \begin{bmatrix} -\frac{3}{2} & \frac{3}{2} \\ \frac{1}{6} & -\frac{3}{2} \end{bmatrix}\mathbf{x}, \qquad \mathbf{x}(0) = \begin{pmatrix} 3 \\ 1 \end{pmatrix}, \begin{pmatrix} 3 \\ -1 \end{pmatrix}, \begin{pmatrix} 2 \\ 4 \end{pmatrix}.$$

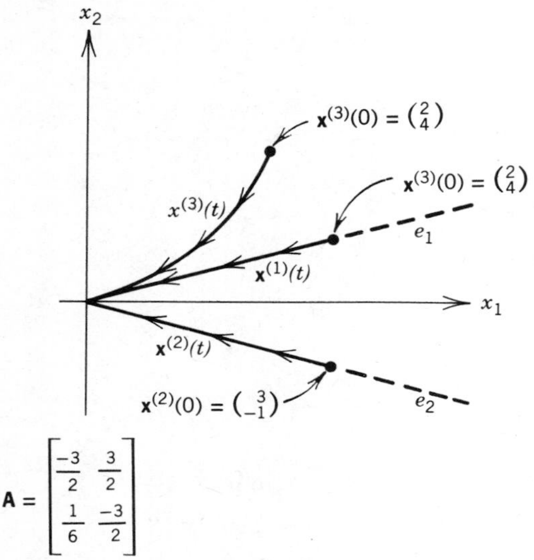

Figure 4.11 *State Trajectories for a Second-Order System*

Solution: First, we should plot the eigenvectors to see the relationship of the state trajectories to the eigenvectors. From $\mathbf{AE} = \mathbf{E}\Lambda$ we find

$$[\mathbf{e}_1 \ \ \mathbf{e}_2] = \mathbf{E} = \begin{bmatrix} 1 & 1 \\ \frac{1}{3} & -\frac{1}{3} \end{bmatrix}, \qquad \Lambda = \begin{bmatrix} -1 & 0 \\ 0 & -2 \end{bmatrix},$$

and the eigenvectors $\mathbf{e}_1$ and $\mathbf{e}_2$ are sketched in Fig. 4.11. In modal coordinates, $\mathbf{x}(t) = \mathbf{E}\mathscr{x}(t) = \mathbf{e}_1\mathscr{x}_1(t) + \mathbf{e}_2\mathscr{x}_2(t)$. Since $\dot{\mathscr{x}} = \Lambda\mathscr{x}$, we have $\mathscr{x}_1(t) = e^{-t}\mathscr{x}_1(0)$ and $\mathscr{x}_2(t) = e^{-2t}\mathscr{x}_2(0)$. Hence, the total $\mathbf{x}(t)$ is a linear combination of motions in the e_1 and e_2 directions. Now,

$$\mathbf{x}(t) = \mathbf{e}_1\left(e^{\lambda_1 t}\mathscr{x}_1(0)\right) + \mathbf{e}_2\left(e^{\lambda_2 t}\mathscr{x}_2(0)\right) = \begin{pmatrix} 1 \\ \frac{1}{3} \end{pmatrix} e^{-t}\mathscr{x}_1(0) + \begin{pmatrix} 1 \\ -\frac{1}{3} \end{pmatrix} e^{-2t}\mathscr{x}_2(0). \tag{4.125}$$

Hence, if $\mathscr{x}_2(0) = 0$, then the entire state trajectory $\mathbf{x}(t)$ *remains* on the eigenvector $\mathbf{e}_1$! Conversely, if $\mathscr{x}_1(0) = 0$, then the state trajectory remains on the eigenvector $\mathbf{e}_2$. For $\mathbf{x}(0) = \begin{pmatrix} 3 \\ 1 \end{pmatrix}$, see that $\mathscr{x}_1(0) = 3$, and $\mathscr{x}_2(0) = 0$, from $\mathscr{x}(0) = \mathbf{E}^{-1}\mathbf{x}(0)$. Recall that rows of $\mathbf{E}^{-1}$ are reciprocal base vectors which are orthogonal to eigenvectors (columns of $\mathbf{E}$). Hence, if $\mathbf{x}(0)$ lies along an eigenvector, say, $\mathbf{e}_2$ $[\mathbf{x}(0) = \mathbf{e}_2\alpha]$, then $\mathscr{x}_1(0) = 0$ since

$$\mathscr{x}(0) = \mathbf{E}^{-1}\mathbf{x}(0) = \begin{pmatrix} \mathbf{l}_1^* & \mathbf{x}(0) \\ \mathbf{l}_2^* & \mathbf{x}(0) \end{pmatrix} = \begin{pmatrix} \mathbf{l}_1^*\mathbf{e}_2\alpha \\ \mathbf{l}_2^*\mathbf{x}(0) \end{pmatrix} = \begin{pmatrix} 0 \\ \mathbf{l}_2^*\mathbf{x}(0) \end{pmatrix}.$$

See that $\mathbf{x}^{(1)^T}(0) \triangleq (3, 1)$ lies along $\mathbf{e}_1$ and that the resulting state trajectory $\mathbf{x}^{(1)}(t)$ lies along $\mathbf{e}_1$ in Fig. 4.11. See also that $\mathbf{x}^{(2)^T}(0) = (3, -1)$ lies along $\mathbf{e}_2$ in Fig. 4.11. Note also from Fig. 4.11 that $\mathbf{x}^{(1)}(t)$ goes *toward* the origin if $\lambda_1 < 0$ and that $\mathbf{x}^{(1)}(t)$ goes *away* from the origin if $\lambda_1 > 0$. Now, see that $\mathbf{x}^{(3)^T}(0) = (2, 4)$ is a point that does not lie on either eigenvector. Hence, equation (4.125) holds with neither $\mathscr{x}_1(0) = 0$ nor $\mathscr{x}_2(0) = 0$. This trajectory is shown in Fig. 4.11. ■

Exercise 4.26

For

$$\dot{\mathbf{x}} = \begin{bmatrix} -1 & 2 \\ -5 & 1 \end{bmatrix} \mathbf{x},$$

show that the "real Jordan form" $\mathscr{J}$ in (2.189) is

$$\mathscr{J} = \begin{bmatrix} \sigma & \omega \\ -\omega & \sigma \end{bmatrix} = \begin{bmatrix} 0 & 3 \\ -3 & 0 \end{bmatrix}, \qquad \mathscr{E} = \begin{bmatrix} 2 & 0 \\ 1 & 3 \end{bmatrix}.$$

4.4.3 PERIODIC LINEAR SYSTEMS

The solution of the general time-varying state equation $\dot{\mathbf{x}} = \mathbf{A}(t)\mathbf{x}$ was given earlier by $\mathbf{x}(t) = \mathbf{\Phi}(t, t_0)\mathbf{x}(t_0)$, with $\mathbf{\Phi}$ satisfying $\dot{\mathbf{\Phi}}(t, t_0) = \mathbf{A}(t)\mathbf{\Phi}(t, t_0)$, $\mathbf{\Phi}(t_0, t_0) = \mathbf{I}$. Now, we study a very special time-varying state equation, where $\mathbf{A}(t)$ is *periodic*. That is, for some constant $\mathbf{T}$,

$$\dot{\mathbf{x}}(t) = \mathbf{A}(t)x(t), \qquad \mathbf{A}(t + T) = \mathbf{A}(t),$$

where $\mathbf{\Phi}(t, t_0)$ is the state transition matrix for $\mathbf{A}(t)$. Hence,

$$\dot{\mathbf{\Phi}}(t, t_0) = \mathbf{A}(t)\mathbf{\Phi}(t, t_0), \qquad \mathbf{\Phi}(t_0, t_0) = \mathbf{I},$$

and

$$\mathbf{\Phi}(t + T, t_0 + T) = \mathbf{\Phi}(t, t_0).$$

Define a matrix $\mathbf{R}$ by

$$e^{\mathbf{R}T} \triangleq \mathbf{\Phi}(T, 0) \tag{4.126}$$

and a matrix $\mathbf{P}(t)$ by

$$\mathbf{P}^{-1}(t) \triangleq \mathbf{\Phi}(t, 0)e^{-\mathbf{R}t}. \tag{4.127}$$

Then,

$$\mathbf{P}^{-1}(t + T) \triangleq \mathbf{\Phi}(t + T, 0)e^{-\mathbf{R}(t+T)} \tag{4.128}$$

can be written, using the transition property of Theorem 4.6, as

$$\begin{aligned}\mathbf{P}^{-1}(t+T) &= \boldsymbol{\Phi}(t+T,T)\boldsymbol{\Phi}(T,0)e^{-\mathbf{R}T}e^{-\mathbf{R}t}\\ &= \boldsymbol{\Phi}(t+T,T)e^{\mathbf{R}T}e^{-\mathbf{R}T}e^{-\mathbf{R}t}\\ &= \boldsymbol{\Phi}(t+T,T)e^{-\mathbf{R}t}\\ &= \boldsymbol{\Phi}(t,0)e^{-\mathbf{R}t}.\end{aligned} \tag{4.129}$$

But from (4.126), $\boldsymbol{\Phi}(t,0) = \mathbf{P}^{-1}(t)e^{\mathbf{R}t}$. Hence,

$$\mathbf{P}^{-1}(t+T) = \boldsymbol{\Phi}(t,0)e^{-\mathbf{R}t} = \mathbf{P}^{-1}e^{\mathbf{R}t}e^{-\mathbf{R}t} = \mathbf{P}^{-1}(t) \tag{4.130}$$

Therefore, the state transition matrix for a periodic matrix $\mathbf{A}(t)$ can, from (4.127) and (4.130), be written

$$\boldsymbol{\Phi}(t,t_0) = \mathbf{P}^{-1}(t)e^{\mathbf{R}(t-t_0)}\mathbf{P}(t_0), \qquad \mathbf{P}(t) = \mathbf{P}(t+T). \tag{4.131}$$

Now, consider the coordinate transformation

$$\mathbf{x}(t) = \mathbf{P}^{-1}(t)\boldsymbol{x}(t). \tag{4.132}$$

Then,

$$\begin{aligned}\dot{\boldsymbol{x}}(t) &= \left[\mathbf{P}(t)\mathbf{A}(t)\mathbf{P}^{-1}(t) - \mathbf{P}(t)\dot{\mathbf{P}}^{-1}(t)\right]\boldsymbol{x}(t) + \mathbf{P}(t)\mathbf{B}(t)\mathbf{u},\\ \mathbf{y}(t) &= \mathbf{C}(t)\mathbf{P}^{-1}(t)\boldsymbol{x}(t).\end{aligned} \tag{4.133}$$

But

$$\begin{aligned}&\mathbf{P}(t)\mathbf{A}(t)\mathbf{P}^{-1}(t) - \mathbf{P}(t)\dot{\mathbf{P}}^{-1}(t)\\ &\quad= e^{\mathbf{R}(t-t_0)}\mathbf{P}(t_0)\boldsymbol{\Phi}(t_0,t)\mathbf{A}(t)\boldsymbol{\Phi}(t,t_0)\mathbf{P}^{-1}(t_0)e^{-\mathbf{R}(t-t_0)}\\ &\quad\quad - e^{\mathbf{R}(t-t_0)}\mathbf{P}(t_0)\boldsymbol{\Phi}(t_0,t)\\ &\quad\quad \times\left[\mathbf{A}(t)\boldsymbol{\Phi}(t,t_0)\mathbf{P}^{-1}(t_0)e^{-\mathbf{R}(t-t_0)} - \boldsymbol{\Phi}(t,t_0)\mathbf{P}^{-1}(t_0)e^{-\mathbf{R}(t-t_0)}\mathbf{R}\right] = \mathbf{R}.\end{aligned} \tag{4.134}$$

This proves the following:

Theorem 4.11
The periodic linear system

$$\begin{aligned}\dot{\mathbf{x}} &= \mathbf{A}(t)\mathbf{x} + \mathbf{B}(t)\mathbf{u} + \mathbf{D}(t)\mathbf{w}, \qquad \mathbf{A}(t) = \mathbf{A}(t+T),\\ \mathbf{y} &= \mathbf{C}(t)\mathbf{x} + \mathbf{H}(t)\mathbf{u} + \mathbf{J}(t)\mathbf{w},\end{aligned} \tag{4.135}$$

is equivalent to the system

$$\dot{x} = \mathbf{R}x + \mathscr{B}(t)\mathbf{u} + \mathscr{D}(t)\mathbf{w},$$
$$\mathbf{y} = \mathscr{C}(t)x + \mathbf{H}(t)\mathbf{u} + \mathbf{J}(t)\mathbf{w}, \tag{4.136}$$

where **R** *is a constant defined by*

$$e^{\mathbf{R}T} \triangleq \Phi(T,0), \tag{4.137}$$

where

$$\dot{\Phi}(t, t_0) = \mathbf{A}(t)\Phi(t, t_0), \qquad \Phi(t_0, t_0) = \mathbf{I} \tag{4.138}$$

and

$$\mathscr{B}(t) = \mathbf{P}(t)\mathbf{B}(t), \qquad \mathscr{D}(t) = \mathbf{P}(t)\mathbf{D}(t),$$
$$\mathscr{C}(t) = \mathbf{C}(t)\mathbf{P}^{-1}(t), \tag{4.139}$$
$$\mathbf{P}^{-1}(t) \triangleq \Phi(t,0)\, e^{-\mathbf{R}t}.$$

Exercise 4.27
Examine the solution of

$$\dot{\mathbf{x}} = \mathbf{A}(t)\mathbf{x}, \qquad \mathbf{A}(t) = \begin{bmatrix} \sin t & 0 \\ 0 & \cos 2t \end{bmatrix}.$$

The period T over which $A(t)$ is periodic is $T = ?$ Find the matrix **R**.

4.5 *Evaluating System Responses for Time-Invariant Systems*

The previous section has been concerned about the *detailed* behavior (the internal dynamics) of the system. The state response, therefore, describes the most detailed information about system performance. Sometimes, this much detail is not required. To make value judgements about the quality of the response, we give three levels of decreasing detail:

1. State response $\mathbf{x}(t)$ (most detailed information)
2. Output response $\mathbf{y}(t)$ (less detailed information than 1)
3. Output time correlation $\mathscr{C}_{yy}$ (less detailed information than 2)
4. Norm of output response $\|\mathbf{y}\|^2$ (less detailed information than 3)

The responses $\mathbf{y}(t)$ and $\mathbf{x}(t)$ have already been developed. Now, we concentrate on levels 3 and 4 in this list.

4.5.1 TIME CORRELATION OF VECTORS

The definitions of *time correlations** will now be advanced. The correlation between two vectors **x** and **y** is the matrix of their outer product, defined as follows.

Definition 4.5

The time correlation $\mathscr{C}_{xy}(t)$ between two vector functions of time, $\mathbf{x}(t+\tau)$ and $\mathbf{y}(\tau)$, is defined by

$$\mathscr{C}_{xy}(t) \triangleq \int_0^{\infty} \mathbf{x}(t+\tau)\mathbf{y}^*(\tau)\, d\tau.$$

From this definition, it follows that the *output correlation* matrix is given by

$$\mathscr{C}_{yy}(t) \triangleq \int_0^{\infty} \mathbf{y}(t+\tau)\mathbf{y}^*(\tau)\, d\tau, \tag{4.140}$$

the *state correlation* matrix is given by

$$\mathscr{C}_{xx}(t) \triangleq \int_0^{\infty} \mathbf{x}(t+\tau)\mathbf{x}^*(\tau)\, d\tau, \tag{4.141}$$

the *input correlation* matrix is given by

$$\mathscr{C}_{uu}(t) \triangleq \int_0^{\infty} \mathbf{u}(t+\tau)\mathbf{u}^*(\tau)\, d\tau, \tag{4.142}$$

and the correlation between $\mathbf{y}(t+\tau)$ and $\mathbf{u}(\tau)$ is given by

$$\mathscr{C}_{yu}(t) \triangleq \int_0^{\infty} \mathbf{y}(t+\tau)\mathbf{u}^*(\tau)\, d\tau. \tag{4.143}$$

Time correlations will be most important in our study of the properties of linear systems, due to the fact that certain information about the linear system $\dot{\mathbf{x}} = \mathbf{A}\mathbf{x} + \mathbf{B}\mathbf{u}$ is embodied in $\mathscr{C}_{xx}(t)$ that cannot be recovered from $\mathbf{x}(t)$. For example, examination of a particular trajectory $\mathbf{x}(t)$ will not reveal to the analyst whether there exists a $\mathbf{u}(t)$ that will take $\mathbf{x}(t)$ to any other desired point. We shall discover in Chapter 5 that indeed $\mathscr{C}_{xx}(t)$ does embody such information. Hence, $\mathscr{C}_{xx}(t)$ is a more fundamental entity for our studies than $\mathbf{x}(t)$.

We begin with the derivation of the state correlation $\mathscr{C}_{xx}(t)$ for the case of impulsive inputs of the system:

$$\dot{\mathbf{x}} = \mathbf{A}\mathbf{x} + \mathbf{B}\mathbf{u} + \mathbf{D}\mathbf{w}, \tag{4.144}$$

where $u_i(t) = \mu_i\delta(t)$ and $w_j(t) = \omega_j\delta(t)$ indicate impulses of strength μ_i and ω_j, respectively. [We have already discussed the generality of impulsive "disturbances"

*Not to be confused with correlation of *random* variables, which is beyond the scope of this book.

in the augmented form (4.70).] Let the impulse be applied at $t = 0$. Hence, the solution $\mathbf{x}(t)$ satisfies

$$\begin{aligned}\mathbf{x}(t) &= e^{\mathbf{A}t}\mathbf{x}(0) + \int_0^t e^{\mathbf{A}(t-\sigma)}(\mathbf{B}\boldsymbol{u} + \mathbf{D}\boldsymbol{\omega})\delta(\sigma)\, d\sigma \\ &= e^{\mathbf{A}t}[\mathbf{x}(0) + \mathbf{B}\boldsymbol{u} + \mathbf{D}\boldsymbol{\omega}],\end{aligned} \tag{4.145}$$

where $\boldsymbol{u}^T = (u_1, u_2, \ldots, u_{n_u})$, $\boldsymbol{\omega}^T = (\omega_1, \omega_2, \ldots, \omega_{n_w})$. Also, note that if $0 < \tau < t + \tau$, then

$$\begin{aligned}\mathbf{x}(t+\tau) &= e^{\mathbf{A}(t+\tau-\tau)}\mathbf{x}(\tau) + \int_\tau^t e^{\mathbf{A}(t-\sigma)}(\mathbf{B}\boldsymbol{u} + \mathbf{D}\boldsymbol{\omega})\delta(\sigma)\, d\sigma \\ &= e^{\mathbf{A}t}\mathbf{x}(\tau).\end{aligned}$$

Hence, the state correlation is given by

$$\begin{aligned}\mathscr{C}_{xx}(t) &= \int_0^\infty \mathbf{x}(t+\tau)\mathbf{x}^*(\tau)\, d\tau \\ &= \int_0^\infty e^{\mathbf{A}t}\mathbf{x}(\tau)\mathbf{x}^*(\tau)\, d\tau \\ &= e^{\mathbf{A}t}\int_0^\infty \mathbf{x}(\tau)\mathbf{x}^*(\tau)\, d\tau \\ &= e^{\mathbf{A}t}\mathscr{C}_{xx}(0).\end{aligned}$$

It follows immediately that $\mathscr{C}_{xx}(t)$ obeys the differential equation

$$\dot{\mathscr{C}}_{xx}(t) = \mathbf{A}\mathscr{C}_{xx}(t), \tag{4.146}$$

with the initial condition $\mathscr{C}_{xx}(0)$, which we have yet to compute. (By definition of the fundamental matrix (4.79), we observe that $\mathscr{C}_{xx}(t)$ qualifies as a fundamental matrix.) Now, to compute $\mathscr{C}_{xx}(0)$, we use (4.145) to obtain

$$\begin{aligned}\mathscr{C}_{xx}(0) &\triangleq \int_0^\infty \mathbf{x}(t)\mathbf{x}^*(t)\, dt \\ &= \int_0^\infty e^{\mathbf{A}t}(\mathbf{x}(0) + \mathbf{B}\boldsymbol{u} + \mathbf{D}\boldsymbol{\omega})(\mathbf{x}(0) + \mathbf{B}\boldsymbol{u} + \mathbf{D}\boldsymbol{\omega})^* e^{\mathbf{A}^*t}\, dt.\end{aligned} \tag{4.147}$$

It is instructive now to show that this is the solution of a certain matrix differential equation of the form

$$\dot{\mathbf{Z}}(t) = \mathbf{Z}(t)\mathbf{A}_2^*(t) + \mathbf{A}_1(t)\mathbf{Z}(t) + \mathbf{F}(t), \qquad \mathbf{Z}(t_0) = \mathbf{Z}(0). \tag{4.148}$$

Theorem 4.12
The linear matrix differential equation (4.148) *has the solution*

$$\mathbf{Z}(t) = \mathbf{\Phi}_1(t,0)\mathbf{Z}(0)\mathbf{\Phi}_2^*(t,0) + \int_0^t \mathbf{\Phi}_1(t,\sigma)\mathbf{F}(\sigma)\mathbf{\Phi}_2^*(t,\sigma)\,d\sigma, \tag{4.149}$$

where $\mathbf{\Phi}_1(t,0)$ *and* $\mathbf{\Phi}_2(t,0)$ *are the state transition matrices for* $\mathbf{A}_1(t)$ *and* $\mathbf{A}_2(t)$, *respectively. Moreover, if* $\lim_{t\to\infty}\mathbf{\Phi}_1(t,0)\mathbf{Z}(0)\mathbf{\Phi}_2^*(t,0) = 0$ *and* $\mathbf{A}_1$, $\mathbf{A}_2$, *and* $\mathbf{F}$ *are constant, then, if* $\lim_{t\to\infty}\mathbf{Z}(t)$ *exists, it is given by*

$$\mathbf{Z}(\infty) = \int_0^\infty \mathbf{\Phi}_1(t,0)\mathbf{F}\mathbf{\Phi}_2^*(t,0)\,dt, \tag{4.150}$$

which satisfies

$$\mathbf{0} = \mathbf{Z}(\infty)\mathbf{A}_2^* + \mathbf{A}_1\mathbf{Z}(\infty) + \mathbf{F}. \tag{4.151}$$

Proof: The proof of (4.150) proceeds as follows. The first term in (4.149) goes to zero by assumption of stability of $\mathbf{A}_1(t), \mathbf{A}_2(t)$, leaving

$$\mathbf{Z}(\infty) = \lim_{t\to\infty}\int_0^t \mathbf{\Phi}_1(t,\sigma)\mathbf{F}(\sigma)\mathbf{\Phi}_2^*(t,\sigma)\,d\sigma = \lim_{t\to\infty}\int_0^t e^{\mathbf{A}_1(t-\sigma)}\mathbf{F}(\sigma)e^{\mathbf{A}_2^*(t-\sigma)}\,d\sigma.$$

Define a new dummy variable of integration

$$t - \sigma = \tau,$$

$$-\,d\sigma = d\tau,$$

to get

$$\mathbf{Z}(\infty) = \lim_{t\to\infty}\left[-\int_t^0 e^{\mathbf{A}_1\tau}\mathbf{F}(\tau)e^{\mathbf{A}_2^*\tau}\,d\tau\right] = \int_0^\infty e^{\mathbf{A}_1\tau}\mathbf{F}(\tau)e^{\mathbf{A}_2^*\tau}\,d\tau.$$

See also that the steady-state solution of (4.148), if it exists, must satisfy (4.151). ■

Proof that (4.149) solves (4.148) is left as an exercise for the reader.

Thus, we know that if the matrix $\mathbf{X} \triangleq \int_0^\infty e^{\mathbf{A}t}\mathbf{W}e^{\mathbf{A}^*t}\,dt$ exists, it satisfies $\mathbf{0} = \mathbf{XA}^* + \mathbf{AX} + \mathbf{W}$. Hence, if $\mathscr{C}_{xx}(0)$ in (4.147) exists, it satisfies

$$\mathbf{0} = \mathscr{C}_{xx}(0)\mathbf{A}^* + \mathbf{A}\mathscr{C}_{xx}(0) + \mathbf{W},$$

where

$$\mathbf{W} \triangleq [\mathbf{x}(0) + \mathbf{B}\boldsymbol{u} + \mathbf{D}\boldsymbol{w}][\mathbf{x}(0) + \mathbf{B}\boldsymbol{u} + \mathbf{D}\boldsymbol{w}]^*.$$

Now, we consider special cases which will be of lasting value. Let the system be driven only by $\mathbf{u}$ so that $\mathbf{x}(0) = \mathbf{D}\boldsymbol{\omega} = \mathbf{0}$, and define $\mathscr{C}^i_{xx}(0)$ to be the state correlation of $\mathbf{x}^i(t)$, when $\boldsymbol{u}^* = (0 \ldots 0, u_i, 0, \ldots, 0)$. This implies that an impulsive input is inserted in the ith input channel while all other inputs are zero, $u_j(t) = 0$, $i \neq j$, $u_i(t) = u_i\delta(t)$. Also, define the state *covariance matrix*† due to impulsive inputs in $\mathbf{u}$, by

$$\mathbf{X} \triangleq \sum_{i=1}^{n_u} \int_0^{\infty} \mathbf{x}^i(t)\mathbf{x}^{i*}(t)\, dt \triangleq \sum_{i=1}^{n_u} \mathscr{C}^i_{xx}(0)$$

[$\mathbf{x}^i(t)$ is the state impulse response due to $u_i = u_i\delta(t)$], and observe that $\mathbf{X}$ satisfies

$$\mathbf{0} = \mathbf{X}\mathbf{A}^* + \mathbf{A}\mathbf{X} + \mathbf{B}\mathscr{U}\mathbf{B}^*, \qquad \mathscr{U} \triangleq \begin{bmatrix} u_1^2 & & \mathbf{0} \\ & \ddots & \\ \mathbf{0} & & u_{n_u}^2 \end{bmatrix}. \tag{4.152}$$

To see this, merely note that $\mathscr{C}^i_{xx}(0)$ obeys

$$\mathbf{0} = \mathscr{C}^i_{xx}(0)\mathbf{A}^* + \mathbf{A}\mathscr{C}^i_{xx}(0) + \mathbf{b}_i u_i^2 \mathbf{b}_i^*, \qquad \mathbf{B} = \left[\mathbf{b}_1, \ldots, \mathbf{b}_{n_u}\right], \tag{4.153}$$

and that

$$\sum_{i=1}^{n_u} \mathbf{b}_i u_i^2 \mathbf{b}_i^* = \mathbf{B}\mathscr{U}\mathbf{B}^*.$$

Sum (4.153) over i to get (4.152). In a similar manner, it is clear that if inputs are applied one at a time in the *disturbance* channels [$w_i(t) = \omega_i\delta(t)$, $w_j(t) = 0$, $i \neq j$, $\mathbf{u}(t) \equiv 0$, $\mathbf{x}(0) = \mathbf{0}$] and the resulting *new* $\mathscr{C}^i_{xx}(0)$ matrices are summed, then the state *covariance* due to disturbance impulses is

$$\mathbf{X}_w = \sum_{i=1}^{n_w} \mathscr{C}^i_{xx}(0),$$

which satisfies

$$\mathbf{0} = \mathbf{X}_w\mathbf{A}^* + \mathbf{A}\mathbf{X}_w + \mathbf{D}\mathscr{W}\mathbf{D}^*, \qquad \mathscr{W} \triangleq \begin{bmatrix} \omega_1^2 & & 0 \\ & \ddots & \\ 0 & & \omega_{n_w}^2 \end{bmatrix}. \tag{4.154}$$

To complete the possibilities, let initial conditions be applied one at a time

† Not to be confused with covariance of *random* variables, which is beyond the scope of this book.

$[x_i(0) = x_{i0},\ x_j(0) = 0,\ i \neq j]$ with $\mathbf{u}(t) \equiv 0$, $\mathbf{w}(t) \equiv 0$. The resulting *new* $\mathscr{C}^i_{xx}(0)$ matrices were summed, yielding the state *covariance* due to initial conditions,

$$\mathbf{X}_x \triangleq \sum_{i=1}^{n_x} \mathscr{C}^i_{xx},$$

which satisfies

$$\mathbf{0} = \mathbf{X}_x\mathbf{A}^* + \mathbf{A}\mathbf{X}_x + \mathbf{X}_0, \qquad \mathbf{X}_0 \triangleq \begin{bmatrix} x_1^2(0) & & 0 \\ & \ddots & \\ 0 & & x_{n_x}^2(0) \end{bmatrix}. \tag{4.155}$$

Finally, note that $\mathbf{X}_{xwu} \triangleq \mathbf{X}_x + \mathbf{X}_w + \mathbf{X}$ satisfies

$$\mathbf{0} = \mathbf{X}_{xwu}\mathbf{A}^* + \mathbf{A}\mathbf{X}_{xwu} + \mathbf{X}_0 + \mathbf{D}\mathscr{W}\mathbf{D}^* + \mathbf{B}\mathscr{U}\mathbf{B}^*, \tag{4.156}$$

which describes the summed effect of excitations from initial conditions, impulsive inputs in $\mathbf{u}(t)$ and $\mathbf{w}(t)$, all applied one at a time. Matrix $\mathbf{X}_x$ contains information about the excitation of the system due to initial conditions, $\mathbf{X}_w$ contains information about the excitation of the system due to impulsive disturbances, and $\mathbf{X}$ likewise contains information about the excitation of the system due to impulsive inputs in $\mathbf{u}(t)$. Much more physical significance will be attached to $\mathbf{X}$ in Chapter 5. This is a reminder that $\mathbf{X}$ as defined by (4.147) represents the *sum* of responses from different "experiments," where each "experiment" is an impulsive input in only one input channel. We suggest that this sum of responses as measured by $\mathbf{X}$ provides a much more reasonable and reliable prediction of real-world performance than the assumption of *simultaneous* impulses applied in all channels.

We shall also discover valuable information in $\mathbf{X}$ about linear system behavior that *cannot* be obtained from $\mathscr{C}_{xx}(t)$ or $\mathscr{C}_{xx}(0)$ or even $\mathbf{x}(t)$. The matrix $\mathbf{X}$ will be shown to relate to controllability, stability, and optimality, in later chapters where these concepts will be made more precise. Henceforth, we shall consider $\mathbf{X}$ as a fundamental entity in linear systems to which practically every result in this book will be related.

The output correlation is computed as follows. Note that for the impulsive input (applied at $t = 0$, with $0 \leq \tau \leq t + \tau$),

$$\mathbf{y}(t+\tau) = \mathbf{C}e^{\mathbf{A}(t)}\mathbf{x}(\tau) + \mathbf{H}\mathbf{u}(t+\tau) + \mathbf{J}\mathbf{w}(t+\tau),$$

where $\mathbf{w}(\sigma) = \boldsymbol{\omega}\delta(\sigma)$, $\mathbf{u}(\sigma) = \boldsymbol{\mu}\delta(\sigma)$. For strictly proper systems, $\mathbf{H} = \mathbf{0}$, $\mathbf{J} = \mathbf{0}$. Otherwise, impulsive inputs yield impulsive outputs, a situation not found in practice. Hence, in *this* section we consider only $\mathbf{H} = \mathbf{0}$, $\mathbf{J} = \mathbf{0}$. In this case, since $t \geq 0$, $\tau \geq 0$,

$$\mathbf{y}(t+\tau) = \mathbf{C}e^{\mathbf{A}t}\mathbf{x}(\tau)$$

and (4.140) becomes

$$\mathscr{C}_{yy}(t) = \int_0^\infty \mathbf{C}e^{\mathbf{A}t}\mathbf{x}(\tau)\mathbf{x}^*(\tau)\,d\tau\,\mathbf{C}^* = \mathbf{C}e^{\mathbf{A}t}\mathscr{C}_{xx}(0)\mathbf{C}^* = \mathbf{C}\mathscr{C}_{xx}(t)\mathbf{C}^*. \quad (4.157)$$

Note also from (4.146) that $\dot{\mathscr{C}}_{xx}(t) = \mathbf{A}\mathscr{C}_{xx}(t)$. Hence,

$$\dot{\mathscr{C}}_{yy}(t) = \mathbf{C}\dot{\mathscr{C}}_{xx}(t)\mathbf{C}^* = \mathbf{C}\mathbf{A}\mathscr{C}_{xx}(t)\mathbf{C}^*.$$

For the ith impulse response $\mathbf{y}^i(t)$, where only $u_i = u_i\delta(t)$ is acting,

$$\mathscr{C}^i_{yy}(t) \triangleq \int_0^\infty \mathbf{y}^i(t+\tau)\mathbf{y}^{i*}(\tau)\,d\tau.$$

Now write

$$\mathbf{R}(t) \triangleq \sum_{i=1}^{n_u} \mathscr{C}^i_{yy}(t) = \sum_{i=1}^{n_u} \mathbf{C}\mathscr{C}^i_{xx}(t)\mathbf{C}^* = \sum_{i=1}^{n_u} \mathbf{C}e^{\mathbf{A}t}\mathscr{C}^i_{xx}(0)\mathbf{C}^* = \mathbf{C}e^{\mathbf{A}t}\mathbf{X}\mathbf{C}^* \quad (4.158a)$$

and define $\mathbf{R}_0$, the covariance of the output, by

$$\mathbf{R}_0 \triangleq \mathbf{R}(0) = \sum_{i=1}^{n_u} \mathscr{C}^i_{yy}(0) = \mathbf{C}\mathbf{X}\mathbf{C}^*. \quad (4.158b)$$

Additional properties of $\mathbf{R}(t)$ will now be described. The substitution $e^{\mathbf{A}t} \triangleq \sum_{i=0}^\infty \mathbf{A}^i t^i/i!$ yields

$$\mathbf{R}(t) = \mathbf{C}e^{\mathbf{A}t}\mathbf{X}\mathbf{C}^* = \sum_{i=0}^\infty \frac{\mathbf{C}\mathbf{A}^i\mathbf{X}\mathbf{C}^* t^i}{i!} = \sum_{i=0}^\infty \frac{\mathbf{R}_i t^i}{i!}, \qquad \mathbf{R}_i \triangleq \mathbf{C}\mathbf{A}^i\mathbf{X}\mathbf{C}^*, \quad (4.159)$$

where

$$\frac{d^i}{dt^i}[\mathbf{R}(t)] = \mathbf{C}\mathbf{A}^i e^{\mathbf{A}t}\mathbf{X}\mathbf{C}^*, \qquad i = 0, 1, 2, \ldots, \quad (4.160a)$$

and

$$\left[\frac{d^i}{dt^i}\mathbf{R}(t)\right]_{t=0} = \mathbf{R}_i \quad \text{and} \quad [\mathbf{R}(t)]_{t=0} = \mathbf{R}_0. \quad (4.160b)$$

Note that the Laplace transform of $\mathbf{R}(t)$ yields

$$\mathbf{R}(s) \triangleq \mathscr{L}[\mathbf{R}(t)] = \mathbf{C}[s\mathbf{I} - \mathbf{A}]^{-1}\mathbf{X}\mathbf{C}^* = \sum_{i=0}^\infty \frac{\mathbf{R}_i}{s^{i+1}}, \quad (4.160c)$$

as compared with the transfer function expansion

$$\mathbf{G}(s) = \mathbf{C}[s\mathbf{I} - \mathbf{A}]^{-1}\mathbf{B} = \sum_{i=0}^\infty \frac{\mathbf{M}_i}{s^{i+1}}, \qquad \mathbf{M}_i \triangleq \mathbf{C}\mathbf{A}^i\mathbf{B}, \quad (4.160d)$$

where the $\mathbf{M}_i$ are called *Markov parameters* and the $\mathbf{R}_i$ are called *covariance parameters*. Now, define

$$\mathbf{P}_{ij} \triangleq \sum_{\alpha=1}^{n_u} \int_0^{\infty} \left[\frac{d^i}{dt^i}\mathbf{y}^{\alpha}(t)\right]\left[\frac{d^j}{dt^j}\mathbf{y}^{\alpha}(t)\right]^* dt. \tag{4.160e}$$

Using the definition of the state covariance,

$$\mathbf{X} = \sum_{\alpha=1}^{n_u} \int_0^{\infty} \mathbf{x}^{\alpha}(t)\mathbf{x}^{\alpha^*}(t)\, dt,$$

$\mathbf{P}_{ij}$ may be computed by (show this as an exercise)

$$\mathbf{P}_{ij} = \mathbf{C}\mathbf{A}^i\mathbf{X}\mathbf{A}^{j^*}\mathbf{C}^*. \tag{4.160f}$$

The following identity relates the Markov parameters and the $\mathbf{P}_{ij}$.

Theorem 4.13
Given definitions (4.160d, e),

$$\mathbf{P}_{i+1,j} + \mathbf{P}_{i,j+1} + \mathbf{M}_i\mathscr{U}\mathbf{M}_j^* = \mathbf{0}, \qquad i, j = 0,1,2,\ldots. \tag{4.160g}$$

Proof: From (4.160g) and (4.160f),

$$\mathbf{C}\mathbf{A}^{i+1}\mathbf{X}\mathbf{A}^{j^*}\mathbf{C}^* + \mathbf{C}\mathbf{A}^i\mathbf{X}\mathbf{A}^{j+1^*}\mathbf{C}^* + \mathbf{C}\mathbf{A}^i\mathbf{B}\mathscr{U}\mathbf{B}^*\mathbf{A}^{j^*}\mathbf{C}^* = \mathbf{0}$$

or, equivalently,

$$\mathbf{C}\mathbf{A}^i[\mathbf{A}\mathbf{X} + \mathbf{X}\mathbf{A}^* + \mathbf{B}\mathscr{U}\mathbf{B}^*]\mathbf{A}^{j^*}\mathbf{C}^* = \mathbf{0},$$

which follows immediately from (4.152). ∎

Note from the definition of $\mathbf{R}_i$ in (4.159) and (4.160f) that

$$\mathbf{R}_i = \mathbf{P}_{i0}.$$

Hence, setting $i = j = 0$ in (4.160g) yields

$$\mathbf{R}_1 + \mathbf{R}_1^* + \mathbf{M}_0\mathbf{M}_0^* = \mathbf{0}.$$

The matrices $\mathbf{M}_i$, $\mathbf{R}_i$, and $\mathbf{P}_{ij}$ play a fundamental role in the characterization of linear systems, as we shall see in Chapter 6.

Exercise 4.28
For the system $\dot{\mathbf{x}} = \mathbf{A}\mathbf{x} + \mathbf{B}\mathbf{u}$, show that there exists a coordinate transformation $\mathbf{x} = \mathbf{T}\mathscr{x}$ such that the state covariance is identity $\hat{\mathbf{X}} = \mathbf{I}$, where $\hat{\mathbf{X}}$ satisfies

$$\mathbf{0} = \hat{\mathbf{X}}\hat{\mathbf{A}}^* + \hat{\mathbf{A}}\hat{\mathbf{X}} + \hat{\mathbf{B}}\mathscr{U}\hat{\mathbf{B}}, \quad \hat{\mathbf{A}} = \mathbf{T}^{-1}\mathbf{A}\mathbf{T}, \quad \hat{\mathbf{B}} = \mathbf{T}^{-1}\mathbf{B}, \quad \mathscr{U}_{ij} = u_i^2\delta_{ij}, \quad u_i^2 > 0.$$

Exercise 4.29

Suppose the Euler–Bernoulli beam described in (3.78) through (3.80) has a small amount of material damping $\zeta = 0.005$ so that (3.78) is modified to

$$\ddot{q}_i(t) + 2\zeta\omega_i\dot{q}_i(t) + \omega_i^2 q_i(t) = b_i^* u, \qquad i = 1, 2, \ldots,$$

$$y_p = \sum_{i=1}^{2} \psi_i(r_0) q_i(t),$$

$$y_r = \sum_{i=1}^{2} \phi_i(r_m) \dot{q}_i(t),$$

where y_p represents a rectilinear displacement at location r_0 on the beam, and y_r represents a rate gyro measurement (angular velocity measurement) at point r_m on the beam. Find the state space model for this system if $\mathbf{x}^T \triangleq (q_1, \dot{q}_1, q_2, \dot{q}_2)$, $\mathbf{y}^T \triangleq (y_p, y_r)$. Show that the state covariance $\mathbf{X}$ defined by (4.152) for the example is given by

$$\mathbf{X} = \begin{bmatrix} X_{11} & 0 & X_{13} & X_{14} \\ 0 & X_{22} & -X_{14} & X_{24} \\ X_{13} & -X_{14} & X_{33} & 0 \\ X_{14} & X_{24} & 0 & X_{44} \end{bmatrix},$$

where

$$X_{11} = \frac{b_1^* b_1}{4\zeta\omega_1^3},$$

$$X_{13} = \frac{2\zeta b_1^* b_2}{(\omega_1 + \omega_2)\left[\omega_1^2 + \omega_2^2 + 2(2\zeta^2 - 1)\omega_1\omega_2\right]},$$

$$X_{14} = \left(\frac{\omega_1 - \omega_2}{\omega_1 + \omega_2}\right)\left(\frac{b_1^* b_2}{\omega_1^2 + \omega_2^2 + 2(2\zeta^2 - 1)\omega_1\omega_2}\right),$$

$$X_{22} = \frac{b_1^* b_1}{4\zeta\omega_1},$$

$$X_{24} = \frac{2\zeta\omega_1\omega_2 b_1^* b_2}{(\omega_1 + \omega_2)(\omega_1^2 + \omega_2^2 + 2(2\zeta^2 - 1)\omega_1\omega_2)},$$

$$X_{33} = \frac{b_2^* b_2}{4\zeta\omega_2^3},$$

$$X_{44} = \frac{b_2^* b_2}{4\zeta\omega_2}.$$

Exercise 4.30

For the example in Exercise 4.29, compute the Markov and covariance parameters (M_0, M_1, M_2, R_0, R_1, and R_2) and P_{11}, P_{22}, and P_{33} defined by

$$M_i \triangleq \mathbf{CA}^i\mathbf{B}, \quad R_i \triangleq \mathbf{CA}^i\mathbf{XC}^*, \quad P_{ii} = \mathbf{CA}^i\mathbf{XA}^{i*}\mathbf{C}^*.$$

4.5.2 THE COST FUNCTION AND ITS DECOMPOSITION

The scalar function

$$\int_0^\infty \mathbf{y}^*\mathbf{Q}\mathbf{y}\, dt$$

is often used as a measure of system performance where the vector y is composed of those error variables that are of importance to the analyst or designer. The smaller the error variables, the better is the performance. Hence, the norm of the vector of error variables $\mathbf{y}$ is an important measure of performance. Such quadratic forms find frequent use when energy and root mean square (RMS) errors are important. The "cost function" $\mathscr{V}$ is defined by

$$\mathscr{V} \triangleq \sum_{i=1}^{n_u} \int_0^\infty \mathbf{y}^i(t)^*\mathbf{Q}\mathbf{y}^i(t)\, dt, \tag{4.161}$$

where $\mathbf{y}^i(t)$ is the response due to

$$u_i(t) = \mathscr{u}_i\delta(t),$$

with

$$\mathbf{x}(0) = \mathbf{0}, \quad \mathbf{w}(t) \equiv \mathbf{0}, \quad u_j(t) \equiv 0, \quad i \neq j.$$

Note that

$$\begin{aligned}\int_0^\infty \mathbf{y}^*(t)\mathbf{Q}\mathbf{y}(t)\, dt &= \operatorname{tr} \int_0^\infty \mathbf{Q}\mathbf{y}(t)\mathbf{y}^*(t)\, dt \\ &= \operatorname{tr} \mathbf{Q} \int_0^\infty \mathbf{y}(t)\mathbf{y}^*(t)\, dt \\ &= \operatorname{tr} \mathbf{Q}\mathscr{C}_{yy}(0).\end{aligned} \tag{4.162}$$

But from (4.159) and (4.160b), $\mathbf{R}_0 = \sum_{i=1}^{n_u} \mathscr{C}_{yy}^i(0) = \mathbf{CXC}^*$; we have, from (4.152) and (4.162),

$$\mathscr{V} = \operatorname{tr} \mathbf{QR}_0 = \operatorname{tr} \mathbf{QCXC}^*, \qquad \mathbf{0} = \mathbf{XA}^* + \mathbf{AX} + \mathbf{B}\mathscr{U}\mathbf{B}^*. \tag{4.163}$$

An alternate calculation of $\mathscr{V}$ is as follows. Using the facts $\mathbf{X} = \int_0^\infty e^{\mathbf{A}t}\mathbf{B}\mathscr{U}\mathbf{B}^* e^{\mathbf{A}^*t}\,dt$ and $\operatorname{tr}\mathbf{MN} = \operatorname{tr}\mathbf{NM}$, write (4.163) in the form

$$\mathscr{V} = \operatorname{tr}\int_0^\infty \mathbf{QC}e^{\mathbf{A}\tau}\mathbf{B}\mathscr{U}\mathbf{B}^* e^{\mathbf{A}^*\tau}\mathbf{C}^*\,d\tau$$

$$= \operatorname{tr}\int_0^\infty e^{\mathbf{A}^*\tau}\mathbf{C}^*\mathbf{QC}e^{\mathbf{A}\tau}\,d\tau\,\mathbf{B}\mathscr{U}\mathbf{B}^* = \operatorname{tr}\mathbf{KB}\mathscr{U}\mathbf{B}^*, \tag{4.164}$$

where

$$\mathbf{K} \triangleq \int_0^\infty e^{\mathbf{A}^*\tau}\mathbf{C}^*\mathbf{QC}e^{\mathbf{A}\tau}\,d\tau \tag{4.165}$$

satisfies

$$\mathbf{0} = \mathbf{KA} + \mathbf{A}^*\mathbf{K} + \mathbf{C}^*\mathbf{QC}. \tag{4.166}$$

To prove (4.166), substitute $\mathbf{A}^*$, $\mathbf{C}^*\mathbf{QC}$, and $\mathbf{B}\mathscr{U}\mathbf{B}^*$ into (4.152) for $\mathbf{A}$, $\mathbf{B}\mathscr{U}\mathbf{B}^*$, and $\mathbf{C}^*\mathbf{QC}$, respectively. The results are summarized as follows:

Theorem 4.14
Consider the linear system

$$\dot{\mathbf{x}} = \mathbf{Ax} + \mathbf{Bu},$$

$$\mathbf{y} = \mathbf{Cx},$$

where u_i is the strength of the impulse in the ith input channel and $\mathscr{U}_{ij} = u_i^2\delta_{ij}$. The covariance of the output is $\mathbf{R}_0 = \mathbf{CXC}^*$. The value of the cost function

$$\mathscr{V} \triangleq \sum_{i=1}^{n_u}\int_0^\infty \mathbf{y}^i(t)^*\mathbf{Q}\mathbf{y}^i(t)\,dt, \tag{4.167a}$$

where $\mathbf{y}^i(t)$ is the response due to $u_i(t) = u_i\delta(t)$ acting alone, is given by

$$\mathscr{V} = \operatorname{tr}\mathbf{XC}^*\mathbf{QC} = \operatorname{tr}\mathbf{KB}\mathscr{U}\mathbf{B}^*, \tag{4.167b}$$

where $\mathbf{X}$ and $\mathbf{K}$ satisfy

$$\mathbf{0} = \mathbf{KA} + \mathbf{A}^*\mathbf{K} + \mathbf{C}^*\mathbf{QC}, \tag{4.167c}$$

$$\mathbf{0} = \mathbf{XA}^* + \mathbf{AX} + \mathbf{B}\mathscr{U}\mathbf{B}^*. \tag{4.167d}$$

Exercise 4.31
Show that

$$\mathscr{V} \triangleq \sum_{i=1}^{n_x+n_w+n_u}\int_0^\infty \mathbf{y}^{i*}(t)\mathbf{Q}\mathbf{y}^i(t)\,dt$$

satisfies

$$\mathscr{V} = \operatorname{tr} \mathbf{X}_{xwu}\mathbf{C}^*\mathbf{Q}\mathbf{C} = \operatorname{tr} \mathbf{K}(\mathbf{X}_0 + \mathbf{D}\mathscr{W}\mathbf{D}^* + \mathbf{B}\mathscr{U}\mathbf{B}^*), \tag{4.168a}$$

where $\mathbf{y}^i(t)$, $i = 1, 2, \ldots, n_x + n_w + n_u$, are the responses from the sequence of excitations $x_i(0)$, $w_i(t) = \omega_i\delta(t)$, $u_i(t) = \mu_i\delta(t)$ acting alone, and

$$\mathbf{0} = \mathbf{K}\mathbf{A} + \mathbf{A}^*\mathbf{K} + \mathbf{C}^*\mathbf{Q}\mathbf{C}, \tag{4.168b}$$

$$\mathbf{X}_{0ij} = x_i^2(0)\delta_{ij}, \quad \mathscr{W}_{ij} = \omega_i^2\delta_{ij}, \quad U_{ij} = \mu_i^2\delta_{ij},$$

$$\mathbf{0} = \mathbf{X}_{xwu}\mathbf{A}^* + \mathbf{A}\mathbf{X}_{xwu} + \mathbf{X}_0 + \mathbf{D}\mathscr{W}\mathbf{D}^* + \mathbf{B}\mathscr{U}\mathbf{B}^*. \tag{4.168c}$$

Exercise 4.32

Suppose the "actuator" device that generates $u(t)$ has dynamics of its own, described by

$$\dot{u}(t) + \beta u(t) = \beta \hat{u}(t), \qquad \beta > 0.$$

In this event, the transfer function between the input to the actuator device $\hat{u}$ and the actuator output is

$$\frac{u(s)}{\hat{u}(s)} = \frac{\beta}{s + \beta}.$$

For this case of a single input, find the value of the cost function $\mathscr{V} = \int_0^\infty \mathbf{y}^*(t)\mathbf{Q}\mathbf{y}(t)\,dt$ when $\hat{u}(t) = \delta(t)$, $\mathbf{x}(0) = 0$, and $\dot{\mathbf{x}} = \mathbf{A}\mathbf{x} + \mathbf{B}\mathbf{u}$, $\mathbf{y} = \mathbf{C}\mathbf{X}$. Find the amount that the actuator dynamics add to the cost function. That is, compute $\mathscr{V}$ when $\beta =$ finite and $\beta = \infty$. This exercise demonstrates the applicability of the cost function and output correlation theories for any class of excitations.

Cost Decomposition

It is of great value to know how the transfer function decomposes into contributions of each mode,

$$\mathbf{G}(s) = \sum_{i=1}^{n_x} \mathbf{G}_i(s), \qquad \mathbf{G}_i(s) \triangleq \frac{\mathbf{J}_i}{s - \lambda_i}.$$

[For the more general case of repeated roots, see equation (4.13)]. The modal contribution $\mathbf{G}_i(s)$ indicates the relative importance of mode i in the impulse response

$$\mathbf{y}(t) = \sum_{i=1}^{n_x} \mathscr{L}^{-1}\left(\frac{\mathbf{J}_i}{s - \lambda_i}\right) = \sum_{i=1}^{n_x} \mathbf{J}_i e^{\lambda_i t}.$$

It proves equally useful to know how each input, output, or state contributes to the output correlation and the cost function. In what follows, we shall decompose the output correlation matrix $\mathbf{R}(t)$ into contributions from each input and each state.

Input Cost Analysis

Note that $\mathbf{R}(t)$ was defined by

$$\mathbf{R}(t) \triangleq \sum_{i=1}^{n_u} \int_0^\infty \mathbf{y}^i(t+\tau)\mathbf{y}^{i*}(\tau)\,d\tau = \sum_{i=1}^{n_u} \mathscr{C}_{yy}^i(t)$$

$$= \sum_{i=1}^{n_u} \mathbf{C}e^{\mathbf{A}t}\mathscr{C}_{xx}^i(0)\mathbf{C}^*, \tag{4.169}$$

where $\mathscr{C}_{xx}^i(0)$ satisfies $\mathbf{0} = \mathscr{C}_{xx}^i(0)\mathbf{A}^* + \mathbf{A}\mathscr{C}_{xx}^i(0) + \mathbf{b}_i\mathbf{b}_i^*\mathscr{u}_i^2$. Now, the matrix

$$\mathscr{C}_{yy}^i(t) \triangleq \int_0^\infty \mathbf{y}^i(t+\tau)\mathbf{y}^{i*}(\tau)\,d\tau = \mathbf{C}e^{\mathbf{A}t}\mathscr{C}_{xx}^i(0)\mathbf{C}^*$$

describes the contribution from only the ith input $u_i(t) = \mathscr{u}_i\delta(t)$. Define the ith *input cost* by $\mathscr{V}_{ui} \triangleq \operatorname{tr}\mathbf{Q}\mathscr{C}_{yy}^i(0)$. Then, from (4.163) note that the total cost function is a superposition of the "input costs"

$$\mathscr{V} \triangleq \operatorname{tr}\mathbf{Q}\mathbf{R}_0 = \sum_{i=1}^{n_u} \operatorname{tr}\mathbf{Q}\mathscr{C}_{yy}^i(0) = \sum_{i=1}^{n_u} \mathscr{V}_{ui}, \tag{4.170}$$

where the ith "input cost" is defined by

$$\mathscr{V}_{ui} \triangleq \int_0^\infty \mathbf{y}^{i*}(\tau)\mathbf{Q}\mathbf{y}^i(\tau)\,d\tau = \operatorname{tr}\mathbf{Q}\mathscr{C}_{yy}^i(0), \qquad \mathscr{C}_{yy}^i(0) \triangleq \int_0^\infty \mathbf{y}^i(t)\mathbf{y}^{i*}(t)\,dt, \tag{4.171}$$

and, considering (4.167), $\mathscr{V}_{ui}$ can be computed by either

$$\mathscr{V}_{ui} = \operatorname{tr}\mathbf{Q}\mathscr{C}_{yy}^i(0) = \operatorname{tr}\mathbf{Q}\mathbf{C}\mathscr{C}_{xx}^i(0)\mathbf{C}^*, \qquad \mathbf{0} = \mathscr{C}_{xx}^i(0)\mathbf{A}^* + \mathbf{A}\mathscr{C}_{xx}^i(0) + \mathbf{b}_i\mathbf{b}_i^*\mathscr{u}_i^2, \tag{4.172a}$$

or

$$\mathscr{V}_{ui} = [\mathbf{B}^*\mathbf{K}\mathbf{B}\mathscr{U}]_{ii} = \mathbf{b}_i^*\mathbf{K}\mathbf{b}_i\mathscr{u}_i^2, \qquad \mathbf{0} = \mathbf{K}\mathbf{A} + \mathbf{A}^*\mathbf{K} + \mathbf{C}^*\mathbf{Q}\mathbf{C}, \tag{4.172b}$$

where (4.172b) follows from (4.172a) since

$$\mathscr{V}_{ui} = \operatorname{tr}\int_0^\infty \mathbf{Q}\mathbf{C}e^{\mathbf{A}t}\mathbf{b}_i\mathbf{b}_i^*\mathscr{u}_i^2 e^{\mathbf{A}^*t}\mathbf{C}^*\,dt$$

$$= \operatorname{tr}\int_0^\infty e^{\mathbf{A}^*t}\mathbf{C}^*\mathbf{Q}\mathbf{C}e^{\mathbf{A}t}\,dt\,\mathbf{b}_i\mathscr{u}_i^2\mathbf{b}_i = \operatorname{tr}\mathbf{K}\mathbf{b}_i\mathbf{b}_i^*\mathscr{u}_i^2.$$

Note that (4.172b) is the preferred calculation, since **K** has to be calculated only once, as opposed to the calculation in (4.172a) of $\mathscr{C}_{xx}^{i}(0)$ for each i. The input costs $\mathscr{V}_{ui}$ provide a wealth of information about the effect of inputs $i = 1, 2, \ldots, n_u$.

EXAMPLE 4.13

$$\dot{\mathbf{x}} = \begin{bmatrix} -1 & 0 \\ 0 & -10 \end{bmatrix} \mathbf{x} + \begin{bmatrix} 1 & 0 \\ 1 & 70 \end{bmatrix} \begin{Bmatrix} u_1 \\ u_2 \end{Bmatrix},$$

$$y = [1 \quad -0.2]\mathbf{x}.$$

Compute the cost function $\mathscr{V} = \sum_{i=1}^{n_u} \int_0^\infty y^{i^2}(t)\, dt = \operatorname{tr} R_0$ for impulsive inputs and determine which input has the greater effect on $\mathscr{V}$ if each input impulse has unit strength.

Solution: The answer is provided by computing the two input costs $\mathscr{V}_{u1}$ and $\mathscr{V}_{u2}$:

$$\mathscr{V}_{ui} = \mathbf{b}_i^* \mathbf{K} \mathbf{b}_i \mu_i^2, \qquad \mathbf{0} = \mathbf{KA} + \mathbf{A}^*\mathbf{K} + \mathbf{C}^*\mathbf{QC}.$$

For

$$\mathbf{A} = \begin{bmatrix} -1 & 0 \\ 0 & -10 \end{bmatrix}, \quad \mathbf{b}_1^* = (1 \quad 1), \quad \mathbf{b}_2^* = [0 \quad 70],$$

$$Q = 1, \quad \mathbf{C} = [1 \quad -0.2], \quad \mu_i = 1, \quad i = 1, 2,$$

this yields

$$\mathbf{K} = \begin{bmatrix} 0.5 & -0.2/11 \\ -0.2/11 & 0.002 \end{bmatrix},$$

$$\mathscr{V}_{u1} = \mathbf{b}_1^* \mathbf{K} \mathbf{b}_1 = 0.466, \qquad \mathscr{V}_{u2} = \mathbf{b}_2^* \mathbf{K} \mathbf{b}_2 = 9.8.$$

Thus, input u_2 has an order-of-magnitude more impact on the performance measured by the scalar $\mathscr{V}$ than does u_1. Clearly, u_2 is the "critical" input in this problem ■

Exercise 4.33

Prove that the input costs $\mathscr{V}_{ui}$ are invariant under coordinate transformations of the state.

Component Cost Analysis

It proves equally useful to know how each state contributes to the cost function $\mathscr{V}$. Note that the decomposition of any quadratic form

$$\mathscr{H} \triangleq x^T \mathbf{Q} x = x_1 Q_{11} x_1 + x_1 Q_{12} x_2 + x_2 Q_{12} x_1 + x_2 Q_{22} x_2 \qquad (4.173)$$

into contributions from x_1 and x_2 yield

$$\mathscr{H} = \mathscr{H}_1 + \mathscr{H}_2,$$

if we define $\mathscr{H}_i \triangleq \frac{1}{2}\dfrac{\partial \mathscr{H}}{\partial x_i} x_i$, for then

$$\mathscr{H}_1 = \tfrac{1}{2}(2x_1 Q_{11} + 2Q_{12}x_2)x_1 = [xx^T\mathbf{Q}]_{11},$$

$$\mathscr{H}_2 = \tfrac{1}{2}(2x_2 Q_{22} + 2Q_{12}x_1)x_2 = [xx^T\mathbf{Q}]_{22}.$$

Now, extending these ideas to the scalar function $\mathscr{V}$ yields

$$\begin{aligned}\mathscr{V} \triangleq \sum_{i=1}^{n_u} \int_0^\infty \mathbf{y}^{i*}(t)\mathbf{Q}\mathbf{y}^i(t)\,dt &= \sum_{i=1}^{n_u} \operatorname{tr}\mathbf{QC}\mathscr{C}_{xx}^i(0)C^* = \operatorname{tr}\mathbf{QCXC}^* = \operatorname{tr}\mathbf{XC}^*\mathbf{QC}\\ &= \sum_{i=1}^{n_x}[\mathbf{XC}^*\mathbf{QC}]_{ii}\\ &= \sum_{i=1}^{n_x}\mathscr{V}_{xi}\end{aligned} \tag{4.174}$$

where

$$\begin{aligned}\mathscr{V}_{xi} &\triangleq \frac{1}{2}\int_0^\infty \left[\sum_{\alpha=1}^{n_u}\left(\frac{\partial \mathbf{y}^{\alpha*}\mathbf{Q}\mathbf{y}^\alpha}{\partial x_i^\alpha}\right)x_i^\alpha\right] dt = \int_0^\infty \left[\sum_{\alpha=1}^{n_u}\mathbf{x}^\alpha(t)\mathbf{x}^{\alpha*}(t)\mathbf{C}^*\mathbf{QC}\right]_{ii} dt\\ &= \int_0^\infty \sum_{j=1}^{n_x}\left[\sum_{\alpha=1}^{n_u}\mathbf{x}^\alpha(t)\mathbf{x}^{*\alpha}(t)\right]_{ij}[\mathbf{C}^*\mathbf{QC}]_{ji}\,dt = \sum_{j=1}^{n_x}\left[\sum_{\alpha=1}^{n_u}\mathscr{C}_{xx}^\alpha(0)\right]_{ij}[\mathbf{C}^*\mathbf{QC}]_{ji}\\ &= \sum_{\alpha=1}^{n_u}[\mathscr{C}_{xx}^\alpha(0)\mathbf{C}^*\mathbf{QC}]_{ii}\\ &= [\mathbf{XC}^*\mathbf{QC}]_{ii}.\end{aligned} \tag{4.175}$$

V_{xi} is called the ith "component cost" and represents the contribution of state variable x_i in the cost function $\mathscr{V}$. Note from (4.130) that $V_{xi}^{u\alpha} \triangleq [\mathscr{C}_{xx}^\alpha(0)\mathbf{C}^*\mathbf{QC}]_{ii}$ represents the contribution which input α makes on the component cost of the ith state. Clearly, component cost analysis provides a wealth of information about the internal structure of the system. We shall use this to our advantage later in Chapter 6, where a special set of coordinates will be introduced which is especially amenable to component cost analysis. Without this additional care, (4.175) can be misused, since V_{xi} is not invariant under coordinate transformations of the state. Hence, we postpone its use until Chapter 6. However, both input costs and the following output costs are invariant under coordinate transformations, and these will be immediately useful.

Output Cost Analysis

Now, we wish to know how much the jth output $y_j(t)$ contributes to the cost function $\mathscr{V}$. Similarly to (4.175), the output cost is

$$
\begin{aligned}
\mathscr{V}_{yi} &\triangleq \frac{1}{2}\int_0^\infty \sum_{\alpha=1}^{n_u} \left(\frac{\partial \mathbf{y}^{\alpha *}\mathbf{Q}\mathbf{y}^{\alpha}}{\partial y_i^\alpha} \right) y_i^\alpha \, dt \\
&= \frac{1}{2} 2\int_0^\infty \sum_{\alpha=1}^{n_u} \left(\mathbf{y}^{\alpha *}\mathbf{Q} \frac{\partial \mathbf{y}^{\alpha}}{\partial y_i^\alpha} \right) y_i^\alpha \, dt \\
&= \int_0^\infty \sum_{\alpha=1}^{n_u} (\mathbf{y}^{\alpha *}\mathbf{Q}\mathbf{1}_i)\, y_i^\alpha \, dt, \qquad \mathbf{1}_i^T \triangleq (0,0,\ldots,1,\ldots 0,0) \\
&= \int_0^\infty \sum_{\alpha=1}^{n_u} (\mathbf{y}^{\alpha *}\mathbf{Q}_{i\,\mathrm{col}})\, y_i^\alpha \, dt \\
&= \int_0^\infty \sum_{\alpha=1}^{n_u} (\mathbf{x}^{\alpha *}\mathbf{C}^*\mathbf{Q}_{i\,\mathrm{col}})\mathbf{C}_{i\,\mathrm{row}} x^\alpha \, dt \\
&= \int_0^\infty \sum_{\alpha=1}^{n_u} \left[\mathbf{C}_{i\,\mathrm{row}}\mathbf{x}^{\alpha}\mathbf{x}^{\alpha *}\mathbf{C}^*\mathbf{Q}_{i\,\mathrm{col}}\right] dt \\
&= \mathbf{C}_{i\,\mathrm{row}} \left[\sum_{\alpha=1}^{n_u} \int_0^\infty \mathbf{x}^{\alpha}\mathbf{x}^{\alpha *} \, dt \right] \mathbf{C}^*\mathbf{Q}_{i\,\mathrm{col}} = \mathbf{C}_{i\,\mathrm{row}} \sum_{\alpha=1}^{n_u} \mathscr{C}_{xx}^{\alpha}(0)\mathbf{C}^*\mathbf{Q}_{i\,\mathrm{col}} \\
&= \mathbf{C}_{i\,\mathrm{row}}\mathbf{X}\mathbf{C}^*\mathbf{Q}_{i\,\mathrm{col}} \\
&= [\mathbf{C}\mathbf{X}\mathbf{C}^*\mathbf{Q}]_{ii},
\end{aligned} \tag{4.176}
$$

where $\mathscr{V}_{yi}^{u\alpha} \triangleq [\mathbf{C}\mathscr{C}_{xx}^{\alpha}(0)\mathbf{C}^*\mathbf{Q}]_{ii}$ is the contribution of the αth input in the output cost $\mathscr{V}_{yi}$.

Exercise 4.34

The spinning axisymmetric spacecraft described by (3.34) has kinetic energy $\mathscr{T}$,

$$\mathscr{T} = \boldsymbol{\omega}^T\mathbf{J}\boldsymbol{\omega}\tfrac{1}{2}, \tag{4.177}$$

where $\mathbf{J}$ is the inertia matrix (3.30). Let the system output vector be $\mathbf{y} = \begin{pmatrix} \omega_1 \\ \omega_2 \end{pmatrix}$. Find the contributions of ω_k, $k = 1, 2$, in the cost function $\mathscr{V} = \sum_{i=1}^{2}\int_0^\infty \boldsymbol{\omega}^{i^T}(t)\mathbf{J}\boldsymbol{\omega}^i(t)\, dt$. $J_{11} = J_{22}$, $J_{ij} = J_{ij}\delta_{ij}$, and $\boldsymbol{\omega}^i(t)$ is the impulse response due only to u_i.

4.5.3 EVALUATION OF THE COST FUNCTION IN THE FREQUENCY DOMAIN

Some comments can be made about the evaluation of the cost in the frequency domain. In this section we shall consider t on the interval $t \in [-\infty, \infty]$; but for all practical purposes, we assume $\mathbf{y}(t) = 0 \; \forall t < 0$ so that the value of $\mathscr{V}$ is not altered by changing the lower limit of the integral to $-\infty$. As a special case of the bilateral Laplace transform

$$\mathscr{L}[f(t)] = \int_{-\infty}^{\infty} f(t) e^{-st}\, dt, \tag{4.178}$$

let $s = j\omega$. This is called the bilateral Fourier transform

$$\mathscr{F}[f(t)] = \int_{-\infty}^{\infty} f(t) e^{-j\omega t}\, dt. \tag{4.179}$$

In this section we avoid confusion by writing $\mathbf{y}^T$ instead of $\mathbf{y}^*$ since $\mathbf{y}$ is always real. Now, use the property 14 in Table E.1 in Appendix E to show that

$$\mathscr{F}\left[\mathbf{y}^T(t)\mathbf{Q}\mathbf{y}(t)\right] \triangleq \int_{-\infty}^{\infty} \mathbf{y}^T(t)\mathbf{Q}\mathbf{y}(t) e^{-j\omega t}\, dt = \frac{1}{2\pi}\int_{-\infty}^{\infty} \mathbf{Y}^T(j\omega - j\beta)\mathbf{Q}\mathbf{Y}(j\beta)\, d\beta, \tag{4.180}$$

which holds for all $0 \le \omega < \infty$. In particular, let $\omega = 0$. Then, see that

$$\int_{-\infty}^{\infty} \mathbf{y}^T(t)\mathbf{Q}\mathbf{y}(t)\, dt = \frac{1}{2\pi}\int_{-\infty}^{\infty} \mathbf{Y}^T(-j\beta)\mathbf{Q}\mathbf{Y}(j\beta)\, d\beta = \frac{1}{2\pi}\int_{-\infty}^{\infty} \mathbf{Y}^*(j\beta)\mathbf{Q}\mathbf{Y}(j\beta)\, d\beta. \tag{4.181}$$

Thus, the same value of the integral occurs over the time *or* frequency domain! This result is due to Parseval [4.8].

For stable matrices $\mathbf{A}$ the steady-state response of $\mathbf{y} = \mathbf{C}\mathbf{x}$, $\dot{\mathbf{x}} = \mathbf{A}\mathbf{x} + \mathbf{B}\mathbf{u}$, $\mathbf{x}(0) = \mathbf{0}$ is

$$\mathbf{y}_{ss}(t) = \lim_{t\to\infty} \int_{-t}^{t} \mathbf{C} e^{\mathbf{A}(t-\sigma)}\mathbf{B}\mathbf{u}(\sigma)\, d\sigma \tag{4.182}$$

from the convolution integral. Equivalently,

$$\mathbf{y}_{ss}(t) = \lim_{t\to\infty} \int_{-t}^{t} \mathbf{C} e^{\mathbf{A}\sigma}\mathbf{B}\mathbf{u}(t-\sigma)\, d\sigma = \lim_{t\to\infty} \int_{-t}^{t} \mathbf{G}(\sigma)\mathbf{u}(t-\sigma)\, d\sigma, \tag{4.183}$$

where $\mathbf{G}(\sigma) \triangleq \mathbf{C} e^{\mathbf{A}\sigma}\mathbf{B}$.

The time correlation between two vectors $\mathbf{y}(t)$ and $\mathbf{u}(t)$ is defined by

$$\mathscr{C}_{yu}(t) \triangleq \rangle \mathbf{y}(t+\tau), \mathbf{u}(\tau) \langle \triangleq \int_{-\infty}^{\infty} \mathbf{y}(t+\tau)\mathbf{u}^T(\tau)\, d\tau = \int_{-\infty}^{\infty} \mathbf{y}(\sigma)\mathbf{u}^T(\sigma - t)\, d\sigma. \tag{4.184}$$

The Fourier transform of (4.184) is

$$\begin{aligned}\mathscr{F}[\rangle \mathbf{y}(t+\tau), \mathbf{u}(\tau) \langle] &= \mathscr{F}\left[\int_{-\infty}^{\infty} \mathbf{y}(\sigma)\mathbf{u}^T(\sigma - t)\, d\sigma\right] \\ &= \int_{-\infty}^{\infty}\left[\int_{-\infty}^{\infty} \mathbf{y}(\sigma)\mathbf{u}^T(\sigma - t)\, d\sigma\right] e^{-j\omega t}\, dt. \end{aligned} \tag{4.185}$$

Reversing the order of integration,

$$\mathscr{F}[\rangle \mathbf{y}(t+\tau), \mathbf{u}(\tau) \langle] = \int_{-\infty}^{\infty} \mathbf{y}(\sigma)\left[\int_{-\infty}^{\infty} \mathbf{u}^T(\sigma - t) e^{-j\omega t}\, dt\right] d\sigma. \tag{4.186}$$

The right-hand side bracketed term is (see item 6 in Table E.1, Appendix E)

$$\left[\int_{-\infty}^{\infty} \mathbf{u}^T(\sigma - t) e^{-j\omega t}\, dt\right] = \overline{e^{j\omega\sigma}\mathscr{F}\left[\mathbf{u}^T(-t)\right]} = \overline{e^{j\omega\sigma}\mathbf{U}^T(j\omega)} = e^{-j\omega\sigma}\mathbf{U}^*(j\omega). \tag{4.187}$$

Then, from (4.186) and (4.187),

$$\begin{aligned}\mathscr{F}[\rangle \mathbf{y}(t+\tau), \mathbf{u}(\tau) \langle] &= \mathscr{F}\left[\int_{-\infty}^{\infty} \mathbf{y}(\sigma)\mathbf{u}^T(\sigma - t)\, d\sigma\right] = \int_{-\infty}^{\infty} \mathbf{y}(\sigma) e^{-j\omega\sigma}\mathbf{U}^*(j\omega)\, d\sigma \\ &= \left[\int_{-\infty}^{\infty} y(\sigma) e^{-j\omega\sigma}\, d\sigma\right]\mathbf{U}^*(j\omega) \\ &= \mathbf{Y}(j\omega)\mathbf{U}^*(j\omega). \end{aligned} \tag{4.188}$$

In the following theorem, (4.189) and (4.190) hold for any functions $\mathbf{y}(t), \mathbf{u}(t)$, but (4.191) holds only when transients have died out in the $\mathbf{y}(t)$ of (4.189), $\lim_{t\to\infty} \mathbf{y}(t) = \mathbf{y}_{ss}(t)$.

Theorem 4.15

The Fourier transform of the time correlation between $\mathbf{y}_{ss}(t)$ and $\mathbf{u}(t)$ is given by

$$\mathscr{F}[\rangle \mathbf{y}_{ss}(t+\tau), \mathbf{u}(\tau) \langle] = \mathbf{Y}(j\omega)\mathbf{U}^*(j\omega), \tag{4.189}$$

the Fourier transform of the time correlation of $\mathbf{u}(t)$ *with itself is given by*

$$\mathscr{F}[\rangle\mathbf{u}(t+\tau),\mathbf{u}(\tau)\langle] = \mathbf{U}(j\omega)\mathbf{U}^*(j\omega), \tag{4.190}$$

and the relationship between them is

$$\mathbf{Y}(j\omega)\mathbf{U}^*(j\omega) = \mathbf{G}(j\omega)\mathbf{U}(j\omega)\mathbf{U}^*(j\omega), \tag{4.191}$$

where $\mathbf{G}(j\omega) = \mathscr{F}[\mathbf{C}e^{\mathbf{A}t}\mathbf{B}] = \mathbf{C}(j\omega\mathbf{I} - \mathbf{A})^{-1}\mathbf{B}$.

Proof: The proof of (4.189) and (4.190) follows immediately from the development (4.188), where the function $\mathbf{y}(t+\tau)$ is replaced by the function $\mathbf{y}_{ss}(t+\tau)$. The proof of (4.191) follows these lines.

From (4.183) and Definition 4.5,

$$\begin{aligned}\rangle\mathbf{y}_{ss}(t+\tau),\mathbf{u}(\tau)\langle &\triangleq \int_{-\infty}^{\infty} \mathbf{y}_{ss}(t+\tau)\mathbf{u}^T(\tau)\,d\tau \\ &= \int_{-\infty}^{\infty}\left[\int_{-\infty}^{\infty} \mathbf{G}(\sigma)\mathbf{u}(t+\tau-\sigma)\,d\sigma\right]\mathbf{u}^T(\tau)\,d\tau.\end{aligned} \tag{4.192}$$

Reversing the order of integration,

$$\rangle\mathbf{y}_{ss}(t+\tau),\mathbf{u}(\tau)\langle = \int_{-\infty}^{\infty} \mathbf{G}(\sigma)\left[\int_{-\infty}^{\infty} \mathbf{u}(t-\sigma+\tau)\mathbf{u}^T(\tau)\,d\tau\right]d\sigma, \tag{4.193}$$

and by Definition 4.5, the right-hand side becomes

$$\rangle\mathbf{y}_{ss}(t+\tau),\mathbf{u}(\tau)\langle = \int_{-\infty}^{\infty} \mathbf{G}(\sigma)(\rangle\mathbf{u}(t-\sigma+\tau),\mathbf{u}(\tau)\langle)\,d\sigma, \tag{4.194}$$

which is a convolution integral. Hence, by property 13 of Table E.1,

$$\mathscr{F}[\rangle\mathbf{y}_{ss}(t+\tau),\mathbf{u}(\tau)\langle] = \mathscr{F}[\mathbf{G}(t)]\,\mathscr{F}[\rangle\mathbf{u}(t+\tau),\mathbf{u}(\tau)\langle]. \tag{4.195}$$

Using (4.190), (4.195) becomes

$$\mathscr{F}[\rangle\mathbf{y}_{ss}(t+\tau),\mathbf{u}(\tau)\langle] = \mathbf{G}(j\omega)\mathbf{U}(j\omega)\mathbf{U}^*(j\omega), \tag{4.196}$$

which verifies (4.191). ■

The matrix $\mathbf{U}(j\omega)\mathbf{U}^*(j\omega)$ is called the *power spectral density* of the signal $\mathbf{u}(t)$.

Now, return to the computation of the integral in (4.181). The Fourier transform of

$$\mathbf{y}_{ss}(t) = \lim_{t\to\infty}\int_{-t}^{t} \mathbf{G}(\sigma)\mathbf{u}(t-\sigma)\,d\sigma \tag{4.197}$$

is, by property 13 of Table E.1 in Appendix E,

$$\mathbf{Y}_{ss}(j\omega) = \mathbf{G}(j\omega)\mathbf{U}(j\omega)\mathbf{U}(j\omega).$$

Hence, for $\mathbf{y}(t) = 0\ \forall t < 0$ we have

$$\begin{aligned}\int_0^\infty \mathbf{y}^*(t)\mathbf{Q}\mathbf{y}(t)\,dt = \int_{-\infty}^\infty \mathbf{y}^*(t)\mathbf{Q}\mathbf{y}(t)\,dt &= \frac{1}{2\pi}\int_{-\infty}^\infty \mathbf{Y}^*(j\omega)\mathbf{Q}\mathbf{Y}(j\omega)\,d\omega \\ &= \frac{1}{2\pi}\operatorname{tr}\int_{-\infty}^\infty \mathbf{Q}\mathbf{Y}(j\omega)\mathbf{Y}^*(j\omega)\,d\omega \\ &= \frac{1}{2\pi}\operatorname{tr}\int_{-\infty}^\infty \mathbf{Q}\mathbf{G}(j\omega)\mathbf{U}(j\omega)\mathbf{U}^*(j\omega)\mathbf{G}^*(j\omega)\,d\omega. \quad (4.198)\end{aligned}$$

Since $\mathbf{U}(j\omega)\mathbf{U}^*(j\omega)$ is the power spectral density of $\mathbf{u}(t)$, it is useful to think of special inputs that have a *constant* power spectral density

$$\mathbf{U}(j\omega)\mathbf{U}^*(j\omega) = \text{constant} = \mathbf{W}. \quad (4.199)$$

Such signals are called "white noise," since they have equal power at *all* frequencies. A formal introduction to white noise as a random variable is not possible here, since random variables are beyond the scope of this text, so we shall take (4.199) as our "definition" of white noise. A truly constant power spectral density over all frequencies is not possible in practice, but this mathematical convenience is worthwhile simply by assuming that the bandwidth of the input is *much* wider than the bandwidth of the system $\mathbf{G}(j\omega)$. For white noise inputs, (4.198) becomes

$$\int_0^\infty \mathbf{y}^*(t)\mathbf{Q}\mathbf{y}(t)\,dt = \frac{1}{2\pi}\operatorname{tr}\int_{-\infty}^\infty \mathbf{Q}\mathbf{G}(j\omega)\mathbf{W}\mathbf{G}^*(j\omega)\,d\omega. \quad (4.200)$$

Exercise 4.35

Prove that the Fourier transform of the unit impulse is $\mathscr{F}[\delta(t)] = 1$. Use the impulse function defined in Appendix E.

Now, consider the cost function $\mathscr{V}$ defined earlier as the sum of integrals $\int_0^\infty \mathbf{y}^{i*}(t)\mathbf{Q}\mathbf{y}^i(t)\,dt$, where $\mathbf{y}^i(t)$ is the response to impulsive input $u_i(t) = \alpha_i\delta(t)$. Hence, from (4.198),

$$\mathscr{V} = \sum_{i=1}^{n_u}\int_0^\infty \mathbf{y}^{i*}(t)\mathbf{Q}\mathbf{y}^i(t)\,dt = \frac{1}{2\pi}\sum_{i=1}^{n_u}\operatorname{tr}\int_{-\infty}^\infty \mathbf{Q}\mathbf{G}(j\omega)\mathbf{U}^i(j\omega)\mathbf{U}^{i*}(j\omega)\mathbf{G}^*(j\omega)\,d\omega.$$

Now, Exercise 4.34 suggests that for impulsive inputs (applied one at a time as before),

$$\sum_{i=1}^{n_u} \mathbf{U}^i(j\omega)\mathbf{U}^{i*}(j\omega) = \mathscr{U}, \qquad \mathscr{U} \triangleq \begin{bmatrix} u_1^2 & & 0 \\ & \ddots & \\ 0 & & u_{n_u}^2 \end{bmatrix}, \tag{4.201}$$

which qualifies as "white noise" inputs by our limited definition in this text. Now, for impulsive inputs the above expression for $\mathscr{V}$ becomes

$$\mathscr{V} = \frac{1}{2\pi} \operatorname{tr} \int_{-\infty}^{\infty} \mathbf{Q}\mathbf{G}(j\omega)\mathscr{U}\mathbf{G}^*(j\omega)\, d\omega,$$

which is compared to our earlier result

$$\mathscr{V} = \operatorname{tr} \mathbf{Q}\mathbf{R}_0 = \operatorname{tr} \mathbf{Q}\mathbf{C}\mathbf{X}\mathbf{C}^*, \qquad \mathbf{0} = \mathbf{X}\mathbf{A}^* + \mathbf{A}\mathbf{X} + \mathbf{B}\mathscr{U}\mathbf{B}^*,$$

$$= \operatorname{tr} \int_0^{\infty} \mathbf{Q}\mathbf{C}e^{\mathbf{A}t}\mathbf{B}\mathscr{U}\mathbf{B}^*e^{\mathbf{A}^*t}\, dt.$$

Exercise 4.36

For the scalar system

$$\dot{x} = ax + bu, \quad y = cx, \qquad \text{with } u = \delta(t), \quad x(0) = 0,$$

find the output correlation $\mathscr{C}_{yy}(t)$ and compute

$$\mathscr{V} = \int_0^{\infty} y^2(t)\, dt.$$

Repeat if the input is white noise with $U(j\omega)U^*(j\omega) = 1$.

Note from formula (4.190) that the Fourier transform of the time correlation of any vector $y(t)$ with itself is

$$\mathscr{F}[\rangle \mathbf{y}(t+\tau), \mathbf{y}(\tau)\langle] = \mathbf{Y}(j\omega)\mathbf{Y}^*(j\omega). \tag{4.202}$$

Hence, from (4.158) and (4.202), see that

$$\mathscr{F}[\mathbf{R}(t)] = \mathscr{F}[\mathbf{C}e^{\mathbf{A}t}\mathbf{X}\mathbf{C}^*] = \mathbf{C}(j\omega\mathbf{I} - \mathbf{A})^{-1}\mathbf{X}\mathbf{C}^* = \sum_{i=1}^{n_u} \mathbf{Y}^i(j\omega)\mathbf{Y}^{i*}(j\omega), \tag{4.203}$$

where $\mathbf{Y}^i(j\omega)$ is the Fourier transform of $\mathbf{y}^i(t)$. Tables 4.1 and 4.2 record the time and frequency domain formulas for these calculations.

TABLE 4.1 Denitions of Time Correlations

FUNCTION	EQUATION	NAME
$\mathscr{C}^i_{xx}(t) \triangleq \int_0^\infty \mathbf{x}^i(t+\tau)\mathbf{x}^{i*}(\tau)\,d\tau$	(4.141)	State Correlation
$\mathscr{C}^i_{yy}(t) = \int_0^\infty \mathbf{y}^i(t+\tau)\mathbf{y}^{i*}(\tau)\,d\tau$	(4.140)	Output correlation
$\mathbf{x}^i(t), \mathbf{y}^i(t)$	Above (4.152)	State or output response with only input $u_i = u_i\delta(t)$ acting
$\mathbf{X} \triangleq \sum_{i=1}^{n_u} \mathscr{C}^i_{xx}(0)$	Above (4.152)	State covariance
$\mathbf{R}_0 \triangleq \mathbf{CXC}^*$	(4.158b)	Output covariance
$\mathbf{R}(t) \triangleq \sum_{i=1}^{n_u} \mathscr{C}^i_{yy}(t)$	(4.158a)	Output correlation
$\mathbf{R}_i = \mathbf{CA}^i\mathbf{XC}^*$	(4.159)	Covariance parameters
$\mathbf{P}_{ij} = \mathbf{CA}^i\mathbf{XA}^{j*}\mathbf{C}^*$	(4.160e)	

EXAMPLE 4.14

Consider the first two modes of the flexible beam described by (3.78) through (3.80):

$$\ddot{q}_i + 2\zeta\omega_i\dot{q}_i + \omega_i^2 q_i = \phi_i(r_c)T(r_c, t), \qquad i = 1, 2,$$
$$y = \sum_{i=1}^{2} \psi_i(r_0)q_i, \qquad r_0 = 0.45L, \quad \zeta = 0.005, \tag{4.204}$$

with

$$\phi_i = i, \quad \psi_i(r_0) = \sin(0.45\pi i), \qquad \omega_i = i^2.$$

(i) Compute the output correlation

$$\mathscr{C}_{yy}(t) \triangleq \int_0^\infty y(t+\tau)y(\tau)\,d\tau.$$

(ii) Compute the power spectral density of the output $Y(j\omega)Y^*(j\omega) = G(j\omega)U(j\omega)U^*(j\omega)G^*(j\omega)$, when $T(r_c, t) = \delta(t)$.

(iii) Compute the mean squared value of the deflections of the beam at $r = 0.45L$; that is, compute

$$\mathscr{V}^{1/2} = \left[\int_0^\infty y^2(t)\,dt\right]^{1/2}.$$

TABLE 4.2 Impulse Response Calculations (when $H = 0$, $J = 0$)

FUNCTION	TIME DOMAIN (t)	FREQUENCY DOMAIN $(j\omega)$
$\mathbf{G}(s)$ (transfer function)	$\mathbf{C}e^{\mathbf{A}t}\mathbf{B}$	$[\mathbf{Y}(j\omega)\mathbf{U}^*(j\omega)][\mathbf{U}(j\omega)\mathbf{U}^*(j\omega)]^{-1} = \mathbf{C}(j\omega\mathbf{I} - \mathbf{A})^{-1}\mathbf{B}$
$\mathbf{G}(s) = \sum_{i=1}^{n_x} \mathbf{G}_i(s)$ (transfer decomposition)	$\mathbf{G}_i(t) = \mathbf{J}_i e^{\lambda_i t}$	$\mathbf{G}_i(j\omega) = \dfrac{\mathbf{J}_i}{j\omega - \lambda_i}$
$\mathbf{X} \triangleq \sum_{i=1}^{n_u} \mathscr{C}^i_{xx}(0) = \int_0^\infty e^{\mathbf{A}t}\mathbf{B}\mathscr{U}\mathbf{B}^* e^{\mathbf{A}^*t}\, dt$ (state covariance)	$\mathbf{0} = \mathbf{X}\mathbf{A}^* + \mathbf{A}\mathbf{X} + \mathbf{B}\mathscr{U}\mathbf{B}^*$	$\mathbf{X} = \dfrac{1}{2\pi}\int_{-\infty}^{\infty} (j\omega\mathbf{I} - \mathbf{A})^{-1}\mathbf{B}\mathscr{U}\mathbf{B}^*(j\omega\mathbf{I} - \mathbf{A})^{-*}\, d\omega$
$\mathbf{K} = \int_0^\infty e^{\mathbf{A}^*t}\mathbf{C}^*\mathbf{Q}\mathbf{C}e^{\mathbf{A}t}\, dt$	$\mathbf{0} = \mathbf{K}\mathbf{A} + \mathbf{A}^*\mathbf{K} + \mathbf{C}^*\mathbf{Q}\mathbf{C}$	$\mathbf{K} = \dfrac{1}{2\pi}\int_{-\infty}^{\infty} (j\omega\mathbf{I} - \mathbf{A}^*)^{-1}\mathbf{C}^*\mathbf{Q}\mathbf{C}(j\omega\mathbf{I} - \mathbf{A}^*)^{-*}\, d\omega$
$\sum_{i=1}^{n_u} \mathscr{C}^i_{yy}(t) = \mathbf{R}(t)$ (output correlation)	$\mathbf{R}(t) = \mathbf{C}e^{\mathbf{A}t}\mathbf{X}\mathbf{C}^*$	$\mathbf{R}(j\omega) = \sum_{i=1}^{n_u} \mathbf{Y}^i(j\omega)\mathbf{Y}^{i*}(j\omega)$; $\mathbf{y}^{(i)}(t) \triangleq$ response due to $u_i(t) = \mathscr{u}_i\delta(t)$
$\mathscr{V} \triangleq \sum_{i=1}^{n_u} \int_0^\infty \mathbf{y}^{i*}(t)\mathbf{Q}\mathbf{y}^i(t)\, dt$ (cost function)	$\mathscr{V} = \operatorname{tr}\mathbf{X}\mathbf{C}^*\mathbf{Q}\mathbf{C} = \operatorname{tr}\mathbf{Q}\mathbf{R}_0$ $= \operatorname{tr}\mathbf{K}\mathbf{B}\mathscr{U}\mathbf{B}^*$	$\mathscr{V} = \dfrac{1}{2\pi}\operatorname{tr}\int_{-\infty}^{\infty} \mathbf{Q}\mathbf{G}(j\omega)\mathscr{U}\mathbf{G}^*(j\omega)\, d\omega$
$\mathscr{V} = \sum_{i=1}^{n_x} \mathscr{V}_{xi}$ (component cost)	$\mathscr{V}_{xi} = [\mathbf{X}\mathbf{C}^*\mathbf{Q}\mathbf{C}]_{ii}$	$\mathscr{V}_{xi} = \dfrac{1}{2\pi}\left[\int_{-\infty}^{\infty} (j\omega\mathbf{I} - \mathbf{A})^{-1}\mathbf{B}\mathscr{U}\mathbf{B}^*(j\omega\mathbf{I} - \mathbf{A})^{-*}\, d\omega\, \mathbf{C}^*\mathbf{Q}\mathbf{C}\right]_{ii}$

(iv) Compute the mean squared value of the rates of deflection at $r = 0.45L$:

$$\left[\int_0^\infty \dot{y}^2(t)\,dt\right]^{1/2}.$$

Assume that the applied torque is impulsive. Assume $T(r_c, t) = \delta(t)$.

Solution: From (4.158a),

$$\mathbf{R}(t) = \mathbf{C}e^{\mathbf{A}t}\mathbf{X}\mathbf{C}^*$$

where use of $\hat{\mathbf{x}}^{\mathrm{T}} = (q_1, q_2, \dot{q}_1, \dot{q}_2)$ leads to

$$\mathbf{C} = [\psi_1(r_0) \quad \psi_2(r_0) \quad 0 \quad 0],$$

$$\mathbf{B} = \begin{bmatrix} 0 \\ 0 \\ \phi_1 \\ \phi_2 \end{bmatrix}, \quad \mathbf{A} = \begin{bmatrix} 0 & I \\ -\boldsymbol{\Omega}^2 & -2\zeta\boldsymbol{\Omega} \end{bmatrix}, \quad \boldsymbol{\Omega} = \begin{bmatrix} \omega_1 & 0 \\ 0 & \omega_2 \end{bmatrix}, \quad \zeta = 0.005.$$

Now, from (4.123),

$$e^{\mathbf{A}t} = \begin{bmatrix} e^{-\zeta\Omega t}[\cos\Omega t + \zeta\sin\Omega t] & e^{-\zeta\Omega t}\Omega^{-1}\sin\Omega t \\ e^{-\zeta\Omega t}\Omega\sin\Omega t & e^{-\zeta\Omega t}[\cos\Omega t - \zeta\sin\Omega t] \end{bmatrix}, \tag{4.205}$$

where $\zeta = 0.005$ and

$$\boldsymbol{\Omega} = \begin{bmatrix} \omega_1 & 0 \\ 0 & \omega_2 \end{bmatrix} = \begin{bmatrix} 1 & 0 \\ 0 & 4 \end{bmatrix}.$$

Hence,

$$\mathbf{C}e^{\mathbf{A}t} = \big[\psi_i(r_0)e^{-\zeta\omega_1 t}(\cos\omega_1 t + \zeta\sin\omega_1 t), \psi_2(r_0)e^{-\zeta\omega_2 t}(\cos\omega_2 t + \zeta\sin\omega_2 t),$$

$$\psi_1(r_0)e^{-\zeta\omega_1 t}\omega_1^{-1}\sin\omega_1 t, \psi_2(r_0)e^{-\zeta\omega_2 t}\omega_2^{-1}\sin\omega_2 t\big].$$

Note that the coordinate transformation

$$\begin{pmatrix} q_1 \\ q_2 \\ \dot{q}_1 \\ \dot{q}_2 \end{pmatrix} = \left[\begin{array}{cc|cc} 1 & 0 & 0 & 0 \\ 0 & 0 & 1 & 0 \\ \hline 0 & 1 & 0 & 0 \\ 0 & 0 & 0 & 1 \end{array}\right] \begin{pmatrix} q_1 \\ \dot{q}_1 \\ q_2 \\ \dot{q}_2 \end{pmatrix} = \left[\begin{array}{c|c} T_{11} & T_{12} \\ \hline T_{12}^T & T_{22} \end{array}\right] \begin{bmatrix} q_1 \\ \dot{q}_1 \\ \hline q_2 \\ \dot{q}_2 \end{bmatrix} \tag{4.206}$$

allows us to use the previous calculation of $\mathbf{X}$ in Exercise 4.29. Let $\mathbf{X}$ be calculated for coordinates $\mathbf{x}^T = (q_1, \dot{q}_1, q_2, \dot{q}_2)$ as given by Exercise 4.29. Then,

$$\hat{\mathbf{x}} = \mathbf{T}\mathbf{x}$$

transforms

$$\rangle \hat{\mathbf{x}}, \hat{\mathbf{x}} \langle = \int_0^\infty \hat{\mathbf{x}}\hat{\mathbf{x}}^* \, dt \triangleq \hat{\mathbf{X}}$$

into

$$\hat{\mathbf{X}} = \rangle \mathbf{T}\mathbf{x}, \mathbf{T}\mathbf{x} \langle = \int_0^\infty \mathbf{T}\mathbf{x}\mathbf{x}^*\mathbf{T}^* \, dt = \mathbf{T}\mathbf{X}\mathbf{T}^*.$$

Using the $\mathbf{T}$ in (4.206) and the $\hat{\mathbf{X}}$ given by Exercise 4.29, see that

$$\hat{X}_{11} = X_{11} = \frac{b_1^2}{4\zeta\omega_1^3} = \frac{1^2}{4(0.005)} = \frac{1}{0.02} = 50,$$

$$\hat{X}_{12} = X_{13} = \frac{2\zeta b_1 b_2}{(\omega_1 + \omega_2)(\omega_1^2 + \omega_2^2 + 2(2\zeta^2 - 1)\omega_1\omega_2)} = \frac{0.004}{9},$$

$$\hat{X}_{13} = X_{12} = 0,$$

$$\hat{X}_{14} = X_{14} = \frac{(\omega_1 - \omega_2) b_1 b_2}{(\omega_1 + \omega_2)(\omega_1^2 + \omega_2^2 + 2(2\zeta^2 - 1)\omega_1\omega_2)} = -\frac{2}{15},$$

$$\hat{X}_{22} = X_{33} = \frac{b_2^2}{4\zeta\omega_2^3} = \frac{4}{4(0.005)4^3} = \frac{1}{0.32},$$

$$\hat{X}_{23} = X_{23} = \frac{(\omega_2 - \omega_1) b_1 b_2}{(\omega_1 + \omega_2)(\omega_1^2 + \omega_2^2 + 2(2\zeta^2 - 1)\omega_1\omega_2)} = \frac{2}{15},$$

$$\hat{X}_{24} = X_{34} = 0,$$

$$\hat{X}_{33} = X_{22} = \frac{b_1^2}{4\zeta\omega_1} = \frac{1}{4\zeta} = 50,$$

$$\hat{X}_{34} = X_{24} = \frac{2\zeta\omega_1\omega_2 b_1 b_2}{(\omega_1 + \omega_2)(\omega_1^2 + \omega_2^2 + 2(2\zeta^2 - 1)\omega_1\omega_2)} = \frac{8}{4500},$$

$$\hat{X}_{44} = X_{44} = \frac{b_2^2}{4\zeta\omega_2} = \frac{4}{4(0.005)4} = 50.$$

Now, the value of $\mathscr{V}$ in (iii) is

$$\mathscr{V} = \hat{R}_0 = \mathbf{C}\hat{\mathbf{X}}\mathbf{C}^* = \hat{X}_{11}C_1^2 + \hat{X}_{22}C_2^2 + 2\hat{X}_{12}C_1C_2$$

$$= \frac{b_1^2}{4\zeta\omega_1^3}\sin^2(0.45\pi) + \frac{b_2^2}{4\zeta\omega_2^3}\sin^2(0.9\pi)$$

$$+ \frac{4\zeta b_1 b_2 \sin(0.45\pi)\sin(0.9\pi)}{(\omega_1+\omega_2)(\omega_1^2+\omega_2^2+2(2\zeta^2-1)\omega_1\omega_2)} = 50.078.$$

And since $n_u = 1$, (4.205) and (4.206) lead to

$$R(t) = \mathbf{C}e^{\mathbf{A}t}\mathbf{X}\mathbf{C}^* = \psi_1 e^{-\zeta\omega_1 t}(\cos\omega_1 t + \zeta\sin\omega_1 t)\left(50\psi_1 + \frac{0.004}{9}\psi_2\right)$$

$$+ \psi_2 e^{-\zeta\omega_2 t}(\cos\omega_2 t + \zeta\sin\omega_2 t)\left(\frac{0.004}{9}\psi_1 + \frac{1}{0.32}\psi_2\right)$$

$$+ \psi_1 e^{-\zeta\omega_1 t}\omega_1^{-1}\sin\omega_1 + \left(\frac{2}{15}\psi_2\right) + \psi_2 e^{-\zeta\omega_2 t}\omega_2^{-1}\sin\omega_2 t\left(-\frac{2}{15}\psi_1\right).$$

Now, from $\mathbf{B}$ and the $\mathbf{C}e^{\mathbf{A}t}$ calculated above,

$$\mathscr{L}[\mathbf{C}e^{\mathbf{A}t}\mathbf{B}] = G(s) = \frac{\psi_1\phi_1\omega_1^2}{s^2+2\zeta\omega_1 s+\omega_1^2} + \frac{\psi_2\phi_2\omega_2^2}{s^2+2\zeta\omega_2 s+\omega_2^2},$$

and for an impulsive input, $U(j\omega) = 1$. Hence, $Y(j\omega)Y^*(j\omega) = G(j\omega)G^*(j\omega)$. ■

Closure

This chapter shows how to *model* the system using state variables, keeping in mind that the "state of the system" is a misnomer, since the state is not unique. Instead, one must talk about the state of a particular *realization* of the system. We may, therefore, talk about "a" state of the system. The construction of state space realizations is described from these starting points: from a given set of mixed-order equations, from a transfer function, and from another state space realization.

The solution of the state equation is shown to involve the state transition matrix, and five different ways to compute the state transition matrix are developed, using time domain and Laplace techniques.

Both time domain and frequency domain techniques are described for computing a *norm* of the output vector y or for the time *correlation* of y with itself.

Evaluations of system performance are discussed by making three different computations:

1. The transfer function
2. The cost function
3. The output correlation matrix

We return to these different characterizations of the response in later chapters, and we shall find that the output and state correlation matrices contain all the information required to determine controllability and optimality properties for linear systems. This key concept introduced in Chapter 4 is the cornerstone of much of the remainder of the book.

Input cost analysis, output cost analysis, and component cost analysis are introduced to assign a value to each system input, output, and state, respectively. This diagnostic procedure will find important uses in model reduction and control design in subsequent chapters. Some earlier references on this subject are [4.11] and [4.12].

References

4.1 R. E. Skelton and P. W. Likins, "Orthogonal Filters for Model Error Compensation in the Control of Nonrigid Spacecraft," *J. Guidance and Control*, 1(1), 41–49, 1978.

4.2 R. E. Skelton, "Adaptive Orthogonal Filters for Compensation of Model Errors in Matrix-Second-Order Systems," *J. Guidance and Control*, 4(2), 1981.

4.3 C. D. Johnson and R. E. Skelton, "Optimal Desaturation of Momentum Exchange Control Systems," *AIAA J.*, 9(1), 12–22, 1971.

4.4 C. D. Johnson, "Theory of Disturbance Accommodating Controllers," *Advances in Control and Dynamic Systems*, Vol. 12, C. T. Leondes, ed., Academic Press, New York, 1986.

4.5 C. D. Johnson, "Accommodation of External Disturbances in Linear Regulator and Servomechanism Problems," *IEEE Trans. Autom. Control*, AC-16(6), 635–645, 1971.

4.6 A. G. J. MacFarlane, editor, *Frequency-Response Methods in Control Systems*. IEEE Press, 1978.

4.7 T. E. Fortmann and K. L. Hitz, *An Introduction to Linear Control Systems*. Marcel Dekker, New York, 1977.

4.8 E. C. Titchmarsh, *Introduction to the Theory of Fourier Integrals*. Clarendon Press, Oxford, 1948.

4.9 Y. Takahashi, M. J. Rabins, and D. M. Auslander, *Control and Dynamic Systems*, Addison-Wesley, Reading, Mass., 1972.

4.10 S. C. Garg, P. C. Hughes, R. A. Miller, and F. R. Vigneron, "Flight Results on Structural Dynamics from Hermes," *J. Spacecraft*, 16(2), 1979.

4.11 R. E. Skelton and A. Yousuff, "Component Cost Analysis of Large Scale Systems," *Int. J. Control*, 37(2), 285–304, 1983.

4.12 R. E. Skelton and M. Delorenzo, "Space Structure Control by Variance Assignment," *J. Guidance and Control*, 8(4), 454–462, 1985.

CHAPTER 5

Controllability and Observability

The basic issue of control theory is to force a given dynamic system to "do what we want it to do." To make the task tractable, control practitioners first *presume a mathematical model* and then ask *whether one can find a set of control functions* $\mathbf{u}(t)$ such that the response of the model (denoted by $\mathbf{y}$) satisfies a *given set of performance requirements*. The three parts of this scenario (indicated by italics) contain seemingly innocent assumptions which can have serious consequences that may prevent accurate predictions about the response of the *physical* systems to which the chosen controls $u(t)$ are to be applied.

First, we *presume to have a model*, but choosing an appropriate model is not straightforward. It is very important that this model be chosen with due regard to the particular response requirements. A different model of the same system will generally lead to different control functions required to meet the same output response requirements. Efforts to systematically "tailor" the model to the control objectives are for later discussion, and we shall not be more explicit now.

Second, there may *not exist* any control functions $\mathbf{u}(t)$ that will meet the objectives. This chapter deals with this *existence* question, and the accepted name for this study is "controllability." In oversimplified terms, a system is "controllable" if a control $\mathbf{u}(t)$ exists such that any response requirement can be met, and is uncontrollable if response requirements exist that cannot be met. For example, the flexible beam in Fig. 3.6 is not controllable from the torque T applied at r_c if $r_c = (2k + 1)L/2i$ for any integer k, since the right-hand sides of (3.78) and (3.79) are zero in this event. Hence, mode i cannot be excited from T, so there exists a response requirement which could not be met (i.e., "reduce the vibration of mode i to zero").

Third, even if the system is controllable and the response objectives are met, one may not be able to deduce from this response all *internal* properties of the system. Lack of "observability" will characterize this condition, and "observability" will indicate the ability to deduce *all* internal properties, given only the inputs $\mathbf{u}(t)$ and outputs $\mathbf{y}(t)$. The internal properties most often sought are the state variables and stability properties. For example, in the backpack problem of Fig. 3.2, one might ask, "In what sense can one deduce $\dot{\gamma}(t)$, given only the output function $y(t) = \dot{\theta}(t)$? Alternately, one might ask whether the initial state $\mathbf{x}(0)$ can be deduced from output data as in Example 4.8. These questions relate to the concept of "observability."

Without some standard form for the system model, it would be exceedingly difficult to make much progress toward a standard theory for controllability and observability, and no standard efficient numerical test would emerge. For this reason it is sometimes convenient to study these properties *after* models are placed in a standard form, called a "state space realization":

$$\begin{aligned} \dot{\mathbf{x}} &= \mathbf{A}(t)\mathbf{x} + \mathbf{B}(t)\mathbf{u} + \mathbf{D}(t)\mathbf{w}, \\ \mathbf{y} &= \mathbf{C}(t)\mathbf{x} + \mathbf{H}(t)\mathbf{u} + \mathbf{J}(t)\mathbf{w}, \end{aligned} \tag{5.1}$$

where $\mathbf{x} \in \mathscr{C}^{n_x}$, $\mathbf{u} \in \mathscr{R}^{n_u}$, $\mathbf{w} \in \mathscr{R}^{n_w}$, $\mathbf{y} \in \mathscr{R}^{n_y}$. A theory for controllability and observability of vector second-order systems,

$$\begin{aligned} \mathscr{M}\ddot{\mathbf{q}} + \mathscr{D}\dot{\mathbf{q}} + \mathscr{K}\mathbf{q} &= \mathscr{B}\mathbf{u}, \\ \mathbf{y} &= \mathscr{P}\mathbf{q} + \mathscr{R}\dot{\mathbf{q}}, \end{aligned}$$

will also be developed since such systems occur frequently in practice.

5.1 Output Controllability

Given the system (5.1) with possibly time-varying matrices $\mathbf{A}(t)$, $\mathbf{B}(t)$, $\mathbf{C}(t)$, $\mathbf{H}(t)$, and $\mathbf{J}(t)$, it is desired to know whether an input $\mathbf{u}(t)$ exists that will transfer $\mathbf{y}(t)$ from an arbitrary initial value $\mathbf{y}(t_0)$ to a specified final value $\mathbf{y}(t_f)$ in a finite amount of time t_f.

Definition 5.1

For any specified $\mathbf{w}(t)$, *the system* (5.1) *is said to be "output controllable at* t_0*" if there exists a control* $\mathbf{u}(t)$ *and a* t_f, $t_0 \le t \le t_f < \infty$, *such that an arbitrarily specified* $\mathbf{y}(t_f)$ *can be achieved for an arbitrary* $\mathbf{y}(t_0)$.

The value of $\mathbf{y}(t_f)$ is given from (4.84) as

$$\begin{aligned} \mathbf{y}(t_f) = {} & \mathbf{C}(t_f)\boldsymbol{\Phi}(t_f, t_0)\mathbf{x}(t_0) + \int_{t_0}^{t_f} \mathbf{C}(t_f)\boldsymbol{\Phi}(t_f, \sigma)[\mathbf{B}(\sigma)\mathbf{u}(\sigma) + \mathbf{D}(\sigma)\mathbf{w}(\sigma)]\, d\sigma \\ & + \mathbf{H}(t_f)\mathbf{u}(t_f) + \mathbf{J}(t_f)\mathbf{w}(t_f). \end{aligned} \tag{5.2}$$

Suppose we desire a given (finite) value of $\mathbf{y}(t_f)$ and ask if such a $\mathbf{u}(\sigma)$, $t_0 \le \sigma \le t_f$, exists for the specified $\mathbf{y}(t_f)$. To solve (5.2) for $\mathbf{u}(\sigma)$ is not straightforward. First, we must ask what is $\mathbf{x}(t_0)$? Only $\mathbf{y}(t_0)$ is specified in the Definition 5.1. Since the solution of

$$\mathbf{y}(t_0) = \mathbf{C}(t_0)\mathbf{x}(t_0) + \mathbf{H}(t_0)\mathbf{u}(t_0) + \mathbf{J}(t_0)\mathbf{w}(t_0) \tag{5.3}$$

for $\mathbf{x}(t_0)$ is a linear algebra problem with (usually) more unknowns than equations, [$\mathbf{x}(t_0) \in \mathscr{C}^{n_x}$, $\mathbf{y}(t_0) \in \mathscr{R}^{n_y}$, where usually $n_x > n_y$], Theorem 2.10 is useful. Recall that *any* solution of (5.3) will serve our present needs, and the necessary and sufficient condition for the existence of a solution, for $\mathbf{x}(t_0)$, $\mathbf{u}(t_0)$ for arbitrary $\mathbf{y}(t_0)$, and given $\mathbf{w}(t_0)$, is that the matrix $[\mathbf{C}(t_0) \quad \mathbf{H}(t_0)]$,

$$[\mathbf{C}(t_0) \quad \mathbf{H}(t_0)]\begin{pmatrix} \mathbf{x}(t_0) \\ \mathbf{u}(t_0) \end{pmatrix} = \mathbf{y}(t_0) - \mathbf{J}(t_0)\mathbf{w}(t_0), \tag{5.4}$$

has a right inverse (i.e., $\operatorname{rank}[C(t_0), H(t_0)] = n_y$). This will always occur when the elements of the $\mathbf{y}$ vector are linearly independent ($\mathbf{C}(t)$ has full row rank). This we assume. Now, find a $\mathbf{u}(t)$ that satisfies

$$\begin{aligned} \mathscr{y} &\triangleq \mathbf{y}(t_f) - \mathbf{C}(t_f)\mathbf{\Phi}(t_f, t_0)\mathbf{x}(t_0) - \mathbf{J}(t_f)\mathbf{w}(t_f) - \int_{t_0}^{t_f}\mathbf{C}(t_f)\mathbf{\Phi}(t_f, \sigma)\mathbf{D}(\sigma)\mathbf{w}(\sigma)\, d\sigma \\ &= \int_{t_0}^{t_f}\big[\mathbf{C}(t_f)\mathbf{\Phi}(t_f, \sigma)\mathbf{B}(\sigma) + \mathbf{H}(t_f)\delta(t_f - \sigma)\big]\mathbf{u}(\sigma)\, d\sigma \end{aligned}$$

or, simply,

$$\mathscr{y}(t_f) = \int_{t_0}^{t_f}\mathscr{G}(\sigma)\mathbf{u}(\sigma)\, d\sigma, \qquad \mathscr{G}(\sigma) \triangleq \big[\mathbf{C}(t_f)\mathbf{\Phi}(t_f, \sigma)\mathbf{B}(\sigma) + \mathbf{H}(t_f)\delta(t_f - \sigma)\big]. \tag{5.5}$$

Now, the left-hand side is considered an *arbitrary* vector $\mathscr{y}$. It is not hard by construction to see that (5.5) is solved by

$$\mathbf{u}(\sigma) = \mathscr{G}^*(\sigma)\left[\int_{t_0}^{t_f}\mathscr{G}(\xi)\mathscr{G}^*(\xi)\, d\xi\right]^{-1}\mathscr{y}(t_f) \tag{5.6}$$

since, by substitution into (5.5),

$$\mathscr{y}(t_f) = \int_{t_0}^{t_f}\mathscr{G}(\sigma)\mathscr{G}^*(\sigma)\, d\sigma\left[\int_{t_0}^{t_f}\mathscr{G}(\xi)\mathscr{G}^*(\xi)\, d\xi\right]^{-1}\mathscr{y}(t_f) = \mathscr{y}(t_f).$$

Note that for the control (5.6) to exist, the matrix

$$\mathbf{Y}(t_f) \triangleq \int_{t_0}^{t_f}\mathscr{G}(\sigma)\mathscr{G}^*(\sigma)\, d\sigma \tag{5.7}$$

must be invertible.

However, the ad hoc approach of introducing (5.6) does not show the *necessary* condition for controllability. Can there be another control which will make the transfer without requiring $\mathbf{Y}(t_f)$ to be invertible? Rewrite (5.5) as follows:

$$\mathscr{Y} = \int_{t_0}^{t_f} \begin{pmatrix} \mathbf{g}_1^*(\sigma)\mathbf{u}(\sigma)\,d\sigma \\ \vdots \\ \mathbf{g}_{n_y}^*(\sigma)\mathbf{u}(\sigma)\,d\sigma \end{pmatrix}, \qquad \mathscr{G}^*(\sigma) = \begin{bmatrix}\mathbf{g}_1(\sigma) & \cdots & \mathbf{g}_{n_y}(\sigma)\end{bmatrix}. \tag{5.8}$$

Now, if $\mathscr{Y}$ is *arbitrary*, then each element $\mathscr{Y}_i$, $i = 1, 2, \ldots, n_y$, is arbitrary. Thus, some linear combination of the elements of the vector $\mathbf{g}_i(\sigma)$ is required to give a different value $\mathscr{Y}_i$ for each i. This requires that the vectors $\mathbf{g}_i(\sigma)$, $i = 1, 2, \ldots, n_y$, be *linearly independent* on the interval $\sigma \in [t_0, t_f]$. The test for linear independence of time-varying vectors on an interval is given by Theorem 2.6. Thus, the rows of $\mathscr{G}(t)$ are linearly independent on the interval $\sigma \in [t_0, t_f]$ iff

$$\int_{t_0}^{t_f} \mathscr{G}(\sigma)\mathscr{G}^*(\sigma)\,d\sigma > 0. \tag{5.9}$$

Define

$$\mathbf{X}(t_0, t_f) \triangleq \int_{t_0}^{t_f} \mathbf{\Phi}(t_f, \sigma)\mathbf{B}(\sigma)\mathbf{B}^*(\sigma)\mathbf{\Phi}^*(t_f, \sigma)\,d\sigma. \tag{5.10a}$$

Then, using the screening property of the Dirac delta, (5.7) becomes

$$\mathbf{Y}(t_f) = \mathbf{H}(t_f)\mathbf{H}^*(t_f)\delta + \mathbf{H}(t_f)\mathbf{B}^*(t_f)\mathbf{C}^*(t_f) + \mathbf{C}(t_f)\mathbf{B}(t_f)\mathbf{H}^*(t_f)$$
$$+ \mathbf{C}(t_f)\mathbf{X}(t_0, t_f)\mathbf{C}^*(t_f), \tag{5.10b}$$

where δ denotes an arbitrarily large positive scalar value. Note from Theorem 4.12 that (5.10a) may be computed by integrating from $t = t_0$ to $t = t_f$ the differential equation

$$\frac{d}{dt}\mathbf{X}(t_0, t) = \mathbf{A}(t)\mathbf{X}(t_0, t) + \mathbf{X}(t_0, t)\mathbf{A}^*(t) + \mathbf{B}(t)\mathbf{B}^*(t), \qquad \mathbf{X}(t_0, t_0) = 0. \tag{5.11}$$

This situation is now summarized below.

Theorem 5.1

The system (5.1) *is output controllable at* t_0 *if and only if* (5.10*b*) *has the property* $\mathbf{Y}(t_f) > 0$ *for some* $t_f > t_0$. *If the output is controllable, then one control that transfers* $\mathbf{y}(t_0)$ *to* $\mathbf{y}(t_f)$ *is given by* (5.6).

EXAMPLE 5.1

Show that $\dot{x} = tx + u$, $y = x$ is output controllable at every t_0.

Solution: In this case (5.11) yields

$$\dot{X} = 2tX + 1, \quad \text{or} \quad X(t_0, t_f) = \int_{t_0}^{t_f} e^{t^2 - \sigma^2}\, d\sigma,$$

which yields $Y(t_f) = X(t_0, t_f) > 0 \quad \forall t > t_0$. ■

EXAMPLE 5.2

Is this system output controllable?

$$\dot{\mathbf{x}} = \begin{bmatrix} a & 0 \\ 0 & a \end{bmatrix}\mathbf{x} + \begin{pmatrix} 1 \\ 1 \end{pmatrix} u = \mathbf{A}\mathbf{x} + \mathbf{B}u,$$

$$\mathbf{y} = \begin{bmatrix} 1 & 1 \\ 0 & 1 \end{bmatrix}\mathbf{x} + \begin{pmatrix} 0 \\ 1 \end{pmatrix} u = \mathbf{C}\mathbf{x} + \mathbf{H}u.$$

If so, find a control that will transfer $\mathbf{x}^T(0) = [x_1(0), x_2(0)]$ to $\mathbf{y}^T(t_f) = [y_1(t_f), y_2(t_f)]$.

Solution: We must determine if (5.10b) is positive definite. Equation (5.11) yields

$$\dot{\mathbf{X}}(t) = 2a\mathbf{X}(t) + \begin{bmatrix} 1 & 1 \\ 1 & 1 \end{bmatrix}, \qquad \mathbf{X}(0) = 0.$$

Hence, for $i, j = 1, 2$,

$$\dot{X}_{ij}(t) = 2aX_{ij}(t) + 1, \qquad X_{ij}(0) = 0,$$

which has solution

$$X_{ij}(t) = \frac{1}{-2a}(1 - e^{2at}) = \Delta(t), \qquad i, j = 1, 2.$$

Now, (5.10b) becomes

$$\mathbf{Y}(t_f) = \begin{pmatrix} 0 \\ 1 \end{pmatrix}(0 \quad 1)\delta + \begin{pmatrix} 0 \\ 1 \end{pmatrix}(1 \quad 1)\begin{bmatrix} 1 & 0 \\ 1 & 1 \end{bmatrix} + \begin{bmatrix} 1 & 1 \\ 0 & 1 \end{bmatrix}\begin{pmatrix} 1 \\ 1 \end{pmatrix}(0 \quad 1) + \begin{bmatrix} 4 & 2 \\ 2 & 1 \end{bmatrix}\Delta$$

$$= \begin{bmatrix} 4\Delta & 2(1 + \Delta) \\ 2(1 + \Delta) & \delta + 2 + \Delta \end{bmatrix},$$

$$\Delta \triangleq \frac{1 - e^{2at_f}}{-2a},$$

which is positive definite iff $\Delta\delta > 1$, $\Delta > 0$. Both of these circumstances are guaranteed for every $t_f > 0$ by definition of Δ and δ. Hence, the system is output controllable, but this example would *not* be output controllable if H were zero. Using (5.6), the reader may verify after some routine calculations that the transfer can be accomplished by

$$u(t) = \frac{e^{a(t_f - t)}}{2\Delta}\left[y_1(t_f) - \left(x_1(0) + x_2(0)\right)e^{at_f}\right], \qquad 0 \le t < t_f,$$

and that at $t = t_f$,

$$u(t_f) = -\left(\frac{1+\Delta}{4\Delta}\right)\left[y_1(t_f) - e^{at_f}\left(x_1(0) + x_2(0)\right)\right] + y_2(t_f) - x_2(0)e^{at_f}.$$

Time-Invariant Systems

If the matrices are constant, the test of Theorem 5.1 becomes

$$\mathbf{Y}(t_f) > \mathbf{0}, \qquad \mathbf{Y}(t_f) = \mathbf{HH}^*\delta + \mathbf{HB}^*\mathbf{C}^* + \mathbf{CBH}^* + \mathbf{CX}(t_0, t_f)\mathbf{C}^*.$$

Let us, therefore, look at the matrix $\mathbf{X}(t_0, t_f)$ for time-invariant systems. In this case, (5.10a) becomes

$$\mathbf{X}(t_0, t_f) = \int_{t_0}^{t_f} e^{\mathbf{A}(t_f - \zeta)}\mathbf{BB}^* e^{\mathbf{A}^*(t_f - \zeta)}\, d\zeta = \int_0^{t_f - t_0} e^{\mathbf{A}\sigma}\mathbf{BB}^* e^{\mathbf{A}^*\sigma}\, d\sigma. \qquad (5.12)$$

Note that the matrix integrand in (5.12) is positive semidefinite. Hence, the time integral of the matrix is at least positive semidefinite, and if $\mathbf{X}(t_0, t_f)$ is positive for any t_f, it is positive for every $t > t_f$. Hence, without loss of generality, we say that $\mathbf{X}(t_0, \infty) > \mathbf{0}$ iff a finite t_f exists such that $\mathbf{X}(t_0, t_f) > \mathbf{0}$. On the strength of this argument we take $t_f = \infty$ in (5.12) and write the more convenient integral

$$\mathbf{X} \triangleq \int_0^{\infty} e^{\mathbf{A}\sigma}\mathbf{BB}^* e^{\mathbf{A}^*\sigma}\, d\sigma, \qquad (5.13a)$$

which is exactly the state covariance matrix of (4.152) with unit impulses $\mathscr{U} = \mathbf{I}$. Then, from Theorem 4.12 we know that if $\mathbf{X}$ exists, it satisfies

$$\mathbf{0} = \mathbf{XA}^* + \mathbf{AX} + \mathbf{BB}^*. \qquad (5.13b)$$

Compare this with (4.152) to conclude that $\mathbf{X}$ in (5.13b) is the state covariance matrix if $u_i = 1$, $\forall i = 1, 2, \ldots, n_u$. Now, suppose $u_i \neq 0$, $\forall i = 1, 2, \ldots, n_u$. Then, the matrix $\mathscr{U}$ in (4.152) is positive definite and $\mathbf{X} > \mathbf{0}$ iff $\mathbf{X} > \mathbf{0}$ in (4.152). Matrix $\mathbf{X}$ in (5.13a) will be called the "controllability grammian" and Exercise 5.2 establishes the equivalence of a positive definite controllability grammian and a positive definite state covariance matrix. For the time-invariant system, output controllability requires, from (5.106), that

$$[\mathbf{HH}^*\delta + \mathbf{HB}^*\mathbf{C}^* + \mathbf{CBH}^* + \mathbf{CXC}^*] > \mathbf{0}.$$

Now, if $\mathbf{H} = \mathbf{0}$, then the necessary and sufficient condition for output controllability is $\mathbf{CXC}^* > \mathbf{0}$. Note that this is entirely equivalent to the test $\mathbf{R}_0 > 0$. Hence, the following conclusion.

Theorem 5.2
The system $\dot{\mathbf{x}} = \mathbf{Ax} + \mathbf{Bu}$, $\mathbf{y} = \mathbf{Cx}$ is output controllable iff $\mathbf{R}_0 > \mathbf{0}$, where $\mathbf{R}_0$ is the output covariance matrix defined in Chapter 4, $\mathbf{R}_0 = \mathbf{CXC}^*$, and $\mathbf{0} = \mathbf{XA}^* + \mathbf{AX} + \mathbf{B}\mathscr{U}\mathbf{B}^*$, where

$$\mathscr{U}_{ij} = u_i^2\delta_{ij}$$

and u_i is the nonzero strength of an impulse applied in the ith input channel.

Let us further examine the state covariance matrix $\mathbf{X}$ for unit input intensities. Use (4.91b) to write $e^{\mathbf{A}t} = \sum_{i=0}^{n_x-1}\mathbf{A}^i\alpha_i(t)$ and substitute into (5.13a):

$$\mathbf{X} = \int_0^\infty \sum_{i=0}^{n-1} \mathbf{A}^i\alpha_i(\sigma)\mathbf{BB}^* \sum_{i=0}^{n-1} \mathbf{A}^{i*}\alpha_i(\sigma)\, d\sigma. \tag{5.14}$$

Exercise 5.1
Verify that (5.14) can be written in the form

$$\mathbf{X} = \mathbf{W}_c\mathbf{\Theta}\mathbf{W}_c^*, \tag{5.15}$$

where

$$\mathbf{\Theta} \triangleq \int_0^\infty \begin{bmatrix} \alpha_0(\sigma)\mathbf{I}_{n_u} \\ \alpha_1(\sigma)\mathbf{I}_{n_u} \\ \vdots \\ \alpha_{n-1}(\sigma)\mathbf{I}_{n_u} \end{bmatrix} \begin{bmatrix} \alpha_0(\sigma)\mathbf{I}_{n_u} & \cdots & \alpha_{n_x-1}(\sigma)\mathbf{I}_{n_u} \end{bmatrix} d\sigma, \tag{5.16a}$$

$$\mathbf{W}_c \triangleq [\mathbf{B} \quad \mathbf{AB} \quad \cdots \quad \mathbf{A}^{n_x-1}\mathbf{B}]. \tag{5.16b}$$

Note that if input intensities are not unit, then the $\mathscr{U}$ appears in between the two matrices in brackets. Now, suppose $\mathbf{H} = \mathbf{0}$. Then, $\mathbf{R}_0 = \mathbf{CXC}^*[\mathbf{CW}_c]\mathbf{\Theta}[\mathbf{W}_c^*\mathbf{C}^*]$ cannot be positive definite unless the bracketed matrix has rank $n_y = \dim \mathbf{y}$. A stronger result is the following (due to $\mathbf{\Theta} > \mathbf{0}$):

Theorem 5.3
Let $\mathscr{U} > \mathbf{0}$. The necessary and sufficient condition under which the time-invariant system $\dot{\mathbf{x}} = \mathbf{Ax} + \mathbf{Bu}$, $\mathbf{y} = \mathbf{Cx}$ is output controllable is that the output covariance be positive definite,

$$\mathbf{R}_0 > \mathbf{0}, \quad \mathbf{X} \triangleq \int_0^\infty e^{\mathbf{A}t}\mathbf{B}\mathscr{U}\mathbf{B}^*e^{\mathbf{A}^*t}\, dt, \quad \mathbf{R}_0 = \mathbf{CXC}^* \tag{5.17a}$$

or, equivalently,

$$\text{rank}\,\mathbf{CW}_c = \text{rank}\,[\mathbf{CB}\quad \mathbf{CAB}\quad \cdots\quad \mathbf{CA}^{n_x-1}\mathbf{B}] = n_y. \tag{5.17b}$$

Exercise 5.2

Show that if $\mathbf{X}$ in (5.13) is positive definite, so is the matrix $\mathbf{X}$ defined by $\mathbf{0} = \mathbf{XA}^* + \mathbf{AX} + \mathbf{B}\mathscr{U}\mathbf{B}^*$ for any positive definite $\mathscr{U}$.

Exercise 5.3

Prove that condition (5.17) of Theorem 5.3 is sufficient by showing that $\boldsymbol{\Theta}$ is a positive definite matrix. (*Hint*: Theorem 2.4 is useful here.)

Exercise 5.4

Determine whether the flexible beam (3.78) is output controllable (in y), using models with only

(a) One mode of the beam, $i = 1$ in (3.78)
(b) Two modes of the beam, $i = 1, 2$ in (3.78)
(c) How many modes of the system may appear in the model before uncontrollability occurs?

Determine whether the system can reamin at this value of output $y(t_f) = 0$ after transfer from $y(0) = 1$. Explain in physical terms.

Exercise 5.5

Determine whether the space backpack (3.23) is output controllable.

Output Tracking

Nothing in the definition of output controllability states that the output will *stay* at $\mathbf{y}(t_f)$ or *track* a specified *function* $\bar{\mathbf{y}}(t)$. The definition is only concerned with *getting* to $\mathbf{y}(t_f)$. For $\mathbf{y}(t)$ to track an arbitrarily specified $\bar{\mathbf{y}}(t)$, the required $\mathbf{u}(t)$ might contain impulses. To synthesize a *well-behaved* function $\mathbf{u}(t)$, we shall restrict the class of $\bar{\mathbf{y}}(t)$ which we will track to have $\beta \le n_x - 1$ time derivatives. To have $\mathbf{y}(t) \equiv \bar{\mathbf{y}}(t)$ for all $t > t_0$, we must require all derivatives to be equal. Hence,

$$\begin{aligned} \mathbf{y}(t) \equiv \bar{\mathbf{y}}(t) \Rightarrow \mathbf{y}(t) &= \bar{\mathbf{y}}(t), \\ \dot{\mathbf{y}}(t) &= \dot{\bar{\mathbf{y}}}(t), \\ \ddot{\mathbf{y}}(t) &= \ddot{\bar{\mathbf{y}}}(t), \\ &\vdots \\ \overset{(\beta)}{\mathbf{y}}(t) &= \overset{(\beta)}{\bar{\mathbf{y}}}(t), \qquad \frac{d^\beta}{dt^\beta}y(t) \triangleq \overset{(\beta)}{y}(t) \end{aligned} \tag{5.18}$$

where

$$\frac{d^i}{dt^i}y(t) = CA^i x(t) + CA^{i-1}Bu + CA^{i-2}B\dot{u} + \cdots H\frac{d^i}{dt^i}u.$$

In matrix form, (5.18) leads to

$$\bar{\mathscr{y}} \triangleq \begin{pmatrix} \bar{\mathbf{y}}(t) \\ \dot{\bar{\mathbf{y}}}(t) \\ \vdots \\ \overset{(\beta)}{\bar{\mathbf{y}}}(t) \end{pmatrix} - \begin{bmatrix} \mathbf{C} \\ \mathbf{CA} \\ \vdots \\ \mathbf{CA}^{\beta} \end{bmatrix} x(t) = \begin{bmatrix} \mathbf{H} & & \mathbf{0} & \mathbf{0} & \cdots \\ \mathbf{CB} & & \mathbf{H} & \mathbf{0} & \cdots \\ \vdots & & \vdots & & \\ \mathbf{CA}^{\beta-1} & \mathbf{B} \cdots & \mathbf{CB} & \mathbf{H} & \end{bmatrix} \begin{pmatrix} \mathbf{u} \\ \dot{\mathbf{u}} \\ \vdots \\ \overset{(\beta-1)}{\mathbf{u}} \end{pmatrix}, \tag{5.19}$$

where $\bar{\mathscr{y}}$ is now an arbitrary vector of dimension $(\beta + 1)\mathbf{n}_y$. Recall from the Cayley–Hamilton theorem that $\mathbf{A}^{\beta}$ is not linearly independent of $\mathbf{A}^{\beta-k}$, $k = 1, 2, \ldots, \beta - 1$, if $\beta \geq n_x$. Hence, we cannot track a function with more than $n_x - 1$ independent derivatives. The solution for the vector of $\mathbf{u}$ derivatives is a *linear algebra* problem (2.68) and leads to the following results.

Theorem 5.4

The linear system (5.1) *can track the vector function* $\bar{y}(t)$ *up to its first* $\beta \leq n_x - 1$ *derivatives if* $\bar{\mathbf{y}}(t)$ *is sufficiently smooth to have* β *derivatives and if the matrix*

$$\mathscr{M}_{\beta} \triangleq \begin{bmatrix} \mathbf{H} & \mathbf{0} & \mathbf{0} & \cdots & \mathbf{0} \\ \mathbf{CB} & \mathbf{H} & \mathbf{0} & & \mathbf{0} \\ \mathbf{CAB} & \mathbf{CB} & \mathbf{H} & & \mathbf{0} \\ \vdots & \vdots & \vdots & & \vdots \\ \mathbf{CA}^{\beta-1}\mathbf{B} & \mathbf{CA}^{\beta-2}\mathbf{B} & \mathbf{CA}^{\beta-3}\mathbf{B} & & \mathbf{H} \end{bmatrix}$$

has full row $\text{rank}\,(\beta + 1)n_y$. *Conversely, if* $\mathbf{y}(t)$ *is the known output of the system* (5.1), *then an input* $u(t)$ *which generates that output can be determined if* $\mathscr{M}_{\beta}$ *has* $\text{rank}\,(\beta + 1)n_y$.

Note that the conditions of Theorem 5.4 can never be satisfied whenever $\mathbf{H} = \mathbf{0}$. In this case the arbitrary $\bar{\mathbf{y}}(t)$ cannot be matched exactly, but if rank $\mathscr{M}_{\beta} = \beta n_y$, then all derivatives $(d^i/dt^i)\bar{\mathbf{y}}(t)$, $i = 1, 2, \ldots, \beta$, can be tracked (note the omission of $i = 0$). This means that "tracking" occurs with a constant offset $\mathbf{y}(t) - \bar{\mathbf{y}}(t) =$ constant.

Output Disturbability

In many situations output controllability is *not* a desirable condition. These are conditions under which the inputs that are causing the output response are *undesirable* and unavoidable inputs. We shall distinguish these inputs as *disturbances*. Yet we will not attempt any formal definition delineating all possible types of inputs that may be characterized as disturbances. Indeed, whether an input is a "control" or a "disturbance" is often in the mind of the analyst. By *output disturbability* we shall mean the same thing as Definition 5.1, except that the "input" $\mathbf{w}(t)$ is considered now to be a "disturbance." This allows us to give meaning to the following type of question: "Under what circumstances can the system totally *reject* the disturbance $\mathbf{w}(t)$, in the sense that $\mathbf{y}(t)$ is governed by its undisturbed response regardless of the disturbance $\mathbf{w}(t)$?" The answer to this question is given below.

We assume $\mathbf{J} = \mathbf{0}$ in (5.1). For $\mathbf{y}(t)$ to follow only its undisturbed response, we require

$$\int_0^t \mathbf{C}e^{\mathbf{A}(t-\sigma)}\mathbf{D}\mathbf{w}(\sigma)\,d\sigma \equiv \mathbf{0}. \tag{5.20}$$

Inserting (4.91b) into (5.20) yields

$$\int_0^t \mathbf{C}\sum_{i=0}^{n_x-1} \mathbf{A}^i\alpha_i(t-\sigma)\mathbf{D}\mathbf{w}(\sigma)\,d\sigma \equiv \mathbf{0}$$

or

$$\sum_{i=0}^{n_x-1} \mathbf{C}\mathbf{A}^i\mathbf{D}\int_0^t \alpha_i(t-\sigma)\mathbf{w}(\sigma)\,d\sigma \equiv \mathbf{0}. \tag{5.21}$$

This leads to the following result.

Theorem 5.5

Complete disturbance rejection is accomplished in system (5.1) *with* $\mathbf{J} = \mathbf{0}$ *for arbitrary disturbance* $\mathbf{w}(t)$ *if and only if the Markov parameters* $\mathbf{M}_i \triangleq \mathbf{C}\mathbf{A}^i\mathbf{D}$ *are zero for* $i = 0, \ldots, n_x - 1$.

Proof: The vectors $\mathbf{v}_i \triangleq \int_0^t \alpha_i(t-\sigma)\mathbf{w}(\sigma)\,d\sigma$ in (5.21) are arbitrary, by assumption of $\mathbf{w}$. Thus, (5.21) states that

$$\sum_{i=0}^{n_x-1} \mathbf{M}_i v_i = \begin{bmatrix}\mathbf{M}_0 & \mathbf{M}_1 & \cdots & M_{n_x-1}\end{bmatrix}\mathbf{v} = \mathcal{M}\mathbf{v} = \mathbf{0} \tag{5.22}$$

is required for *arbitrary* vector $\mathbf{v}^* = (\mathbf{v}_0^*, \ldots, \mathbf{v}_{n_x-1}^*)$. From the linear algebra of Chapter 2, this is possible if and only if $\mathcal{M} = \mathbf{0}$. ■

A calculus of variations point of view is used in the very early work by Rozenoer [5.1] to prove the result in Theorem 5.5.

Exercise 5.6

Determine whether it is possible to locate the control force in the beam problem (3.78), (i.e., by selecting r_c) so that modes $i = 3, 4, \ldots, N$ will not be excited by a control $\mathbf{u}(t)$ designed to control modes $i = 1, 2$. (*Hint*: Assume $\mathbf{u}(t)$ is an arbitrary function and require complete disturbance rejection, taking the "output" to be the displacement of modes $i = 3, 4, \ldots N$.)

Exercise 5.7

As an application of the disturbance rejection properties in Theorem 5.5, consider the system described by

$$\begin{pmatrix} \dot{\mathbf{x}}_R \\ \dot{\mathbf{x}}_T \end{pmatrix} = \begin{bmatrix} \mathbf{A}_R & \mathbf{0} \\ \mathbf{0} & \mathbf{A}_T \end{bmatrix} \begin{pmatrix} \mathbf{x}_R \\ \mathbf{x}_T \end{pmatrix} + \begin{bmatrix} \mathbf{B}_R \\ \mathbf{B}_T \end{bmatrix} \mathbf{u} \qquad \begin{matrix} \mathbf{x}_R \in \mathscr{R}^n, \\ \mathbf{x}_T \in \mathscr{R}^{n_x - n}, \end{matrix}$$

$$\mathbf{x}_T = [0 \quad I] \begin{pmatrix} x_R \\ x_T \end{pmatrix}.$$

Show that the states $\mathbf{x}_T$ remain undisturbed by *any* control action $\mathbf{u}(t)$ if $\mathbf{A}^i_T \mathbf{B}_T = \mathbf{0}$ for $i = 0, 1, \ldots, n_x - n$.

Exercise 5.8

Treat the force f as a disturbance in the backpack problem (3.23). Using Theorem 5.5, determine whether this unknown force f will affect the attitude rate $\dot{\theta}(t)$.

5.2 State Controllability

It is much less obvious why one would require the *state* to be arbitrarily transferred since system performance specifications usually involve only the outputs. Nonetheless, we will find the requirement for state controllability in existence theorems of optimal control, and such notions also give additional insight into system behavior. Hence, we add the following definition.

Definition 5.2

The system (5.1) *is said to be "state controllable at t_0" if there exists a control* $\mathbf{u}(t)$, *and a* t_f, $t_0 \le t \le t_f < \infty$, *such that an arbitrarily specified* $\mathbf{x}(t_f)$ *can be achieved for an arbitrary* $\mathbf{x}(t_0)$.

The following theorems are immediate consequences of Theorems 5.1 and 5.3.

Theorem 5.6

The system (5.1) *is state controllable at t_0 iff* $\mathbf{X}(t_0, t_f) > 0$ *for some* $t_f > t_0$, *where* $\mathbf{X}(t_0, t_f)$ *is defined by* (5.10*a*).

The proof of this result is obtained by setting $\mathbf{C} = \mathbf{I}$, $\mathbf{H} = \mathbf{0}$ in Theorem 5.1.

Theorem 5.7

Let $\mathscr{U} > \mathbf{0}$. The time-invariant system $\dot{\mathbf{x}} = \mathbf{Ax} + \mathbf{Bu}$ is state controllable if and only if the state covariance matrix is positive definite,

$$\mathscr{X} > 0, \quad \mathscr{X} \triangleq \int_0^\infty e^{\mathbf{A}t}\mathbf{B}\mathscr{U}\mathbf{B}^* e^{\mathbf{A}^*t}\, dt, \qquad \mathscr{U} > \mathbf{0}, \tag{5.23a}$$

or, equivalently,

$$\operatorname{rank} \mathbf{W}_c = n_x, \qquad \mathbf{W}_c \triangleq [\mathbf{B}, \mathbf{AB}, \ldots, \mathbf{A}^{n_x - 1}\mathbf{B}]. \tag{5.23b}$$

The proof of this result is obtained by setting $\mathbf{C} = \mathbf{I}$, $\mathbf{H} = \mathbf{0}$ in Theorem 5.3, noting that $\mathscr{U}$ is a positive definite matrix here (and can include the identity matrix $\mathscr{U} = \mathbf{I}$).

Exercise 5.9

Prove that rank $\mathbf{W}_c = n_x$ if $\mathscr{X}(t) > \mathbf{0}$ where $\mathscr{X}(t)$ satisfies

$$\dot{\mathscr{X}} = \mathscr{X}\mathbf{A}^* + \mathbf{A}\mathscr{X} + \mathbf{B}\mathscr{U}\mathbf{B}^*, \qquad \mathscr{X}(0) = \mathbf{0}, \tag{5.24a}$$

and $t > 0$. Prove that if $\mathbf{A}$ has no eigenvalues in the closed right half-plane, then rank $\mathbf{W}_c = n_x$ iff $\mathscr{X} > \mathbf{0}$ where $\mathscr{X}$ satisfies

$$0 = \mathscr{X}\mathbf{A}^* + \mathbf{A}\mathscr{X} + \mathbf{B}\mathscr{U}\mathbf{B}^*. \tag{5.24b}$$

Exercise 5.10

Determine whether the backpack system (3.23) is state controllable.

Exercise 5.11

Determine whether the flexible beam (3.78) is state controllable for the two-mode model $i = 1, 2$. Can you choose r_c so that this two-mode system is not state controllable?

Exercise 5.12

Show that the columns of the matrix

$$\mathbf{W}_c = [\mathbf{B}, \mathbf{AB}, \ldots, \mathbf{A}^{n_x - 1}\mathbf{B}]$$

span the controllable subspace of the state space. Sketch the vectors $\mathbf{B}$ and $\mathbf{AB}$ for the system below and explain controllability in terms of the angle between the vectors $\mathbf{B}$ and $\mathbf{AB}$,

$$\dot{\mathbf{x}} = \begin{bmatrix} 0 & 1 \\ -1 & 0 \end{bmatrix}\mathbf{x} + \begin{pmatrix} 0 \\ 2 \end{pmatrix} u.$$

EXAMPLE 5.3

Determine whether (3.28) is state controllable.

Solution: Define

$$x_1 = r - R_0,$$

$$x_2 = \dot{r},$$

$$x_3 = R_0(\theta - \omega_0 t),$$

$$x_4 = R_0(\dot{\theta} - \omega_0),$$

and normalize R_0 and m to 1. Then, the linearized equations become

$$\begin{pmatrix} \dot{x}_1 \\ \dot{x}_2 \\ \dot{x}_3 \\ \dot{x}_4 \end{pmatrix} = \underbrace{\begin{bmatrix} 0 & 1 & 0 & 0 \\ 3\omega_0^2 & 0 & 0 & 2\omega_0 \\ 0 & 0 & 0 & 1 \\ 0 & -2\omega_0 & 0 & 0 \end{bmatrix}}_{\mathbf{A}} \begin{pmatrix} x_1 \\ x_2 \\ x_3 \\ x_4 \end{pmatrix} + \underbrace{\begin{bmatrix} 0 & 0 \\ 1 & 0 \\ 0 & 0 \\ 0 & 1 \end{bmatrix}}_{\mathbf{B}} \begin{pmatrix} u_1 \\ u_2 \end{pmatrix}. \tag{5.25}$$

Then,

$$\mathbf{W}_c = \begin{bmatrix} 0 & 0 & 1 & 0 & 0 & 2\omega_0 & -\omega_0^2 & 0 \\ 1 & 0 & 0 & 2\omega_0 & -\omega_0^2 & 0 & 0 & -2\omega_0^3 \\ 0 & 0 & 0 & 1 & -2\omega_0 & 0 & 0 & -4\omega_0^2 \\ 0 & 1 & -2\omega_0 & 0 & 0 & -4\omega_0 & 2\omega_0^3 & 0 \end{bmatrix} \text{ has rank 4,}$$

hence, (5.25) is controllable. ■

Exercise 5.13

Show in Example 5.3 that circular orbits are controllable from tangential thrusting alone (from u_2 in Fig. 3.3), but that circular orbits cannot be maintained by radial thrusting alone (from u_1 in Fig. 3.3).

5.2.1 CONTROLLABILITY OF TRANSFORMED COORDINATES

It is important to know whether the concept of controllability is a *system* property or a property of a particular *realization* of the system. The controllability tests of Theorem 5.1 and equations (5.23) do not require a *particular* state space realization (no particular basis is required). The results of the previous sections would not be very useful if it turned out that the system is controllable in one set of coordinates and not in another.

Consider the coordinate transformation $\mathbf{x} = \mathbf{T}\boldsymbol{x}$, where $\mathbf{T}$ is constant, nonsingular. In $\boldsymbol{x}$ coordinates, then, the system description (5.1) becomes

$$\dot{\boldsymbol{x}} = \mathbf{T}^{-1}\mathbf{A}\mathbf{T}\boldsymbol{x} + \mathbf{T}^{-1}\mathbf{B}\mathbf{u} = \mathscr{A}\boldsymbol{x} + \mathscr{B}\mathbf{u},$$
$$\mathbf{y} = \mathbf{C}\mathbf{T}\boldsymbol{x} + \mathbf{H}\mathbf{u} = \mathscr{C}\boldsymbol{x} + \mathbf{H}\mathbf{u}. \tag{5.26}$$

Now, the test of Theorem 5.1 is not affected by the change of coordinates since

$$\begin{aligned}\mathbf{Y}(t_f) &= \mathbf{H}(t_f)\mathbf{H}^*(t_f)\delta + \mathbf{H}(t_f)(\mathbf{T}^{-1}\mathbf{B}(t_f))^*(\mathbf{C}(t_f)\mathbf{T})^* + \mathbf{C}(t_f)\mathbf{T}\mathbf{T}^{-1}\mathbf{B}(t_f)\mathbf{H}^*(t_f)\\ &\quad + \mathbf{C}(t_f)\mathbf{T}\mathbf{T}^{-1}\mathbf{X}(t_f)\mathbf{T}^{-*}\mathbf{T}^*\mathbf{C}^*(t_f)\\ &= \mathbf{H}(t_f)\mathbf{H}^*(t_f)\delta + \mathbf{H}(t_f)\mathbf{B}^*(t_f)\mathbf{C}^*(t_f) + \mathbf{C}(t_f)\mathbf{B}(t_f)\mathbf{H}^*(t_f)\\ &\quad + \mathbf{C}(t_f)\mathbf{X}(t_f)\mathbf{C}^*(t_f)\end{aligned}$$

is obviously the *same* test because $\mathbf{Y}(t_f)$ is not altered by coordinate transformations. This comes as no surprise since Chapter 4 showed that coordinate transformations of the state do not change output functions $\mathbf{y}(t)$.

One may easily show that the *state* controllability tests (5.17) also remain invariant under coordinate transformation by setting $\mathbf{C} = \mathbf{I}$, $\mathbf{H} = \mathbf{0}$. This proves the following:

Theorem 5.8

Controllability of states or outputs of a linear dynamic system reamins invariant under coordinate transformation of the state.

Exercise 5.14

Show that controllability tests of Theorem 5.1 remain invariant even under time-varying transformations $\mathbf{T}(t)$.

Exercise 5.15

Prove directly that $\text{rank}\,[\mathbf{B}, \mathbf{AB}, \ldots, \mathbf{A}^{n-1}\mathbf{B}] = \text{rank}\,[\mathbf{T}^{-1}\mathbf{B}, (\mathbf{T}^{-1}\mathbf{A}\mathbf{T})\mathbf{T}^{-1}\mathbf{B}, \ldots, (\mathbf{T}^{-1}\mathbf{A}\mathbf{T})^{n-1}\mathbf{T}^{-1}\mathbf{B}]$.

The importance of these results is that controllability is a system property invariant under transformation. The advantage is that the tests (5.17) may be conducted in the most *convenient* realization of the system. Several examples of this advantage follow.

5.2.1.1 Controllability of Modal Coordinates

Consider the coordinate transformation to Jordan form $\mathbf{x} = \mathbf{E}\boldsymbol{x}$,

$$\mathbf{AE} = \mathbf{E}\Lambda, \qquad \Lambda = \text{block diag}\,\{\Lambda_1, \ldots, \Lambda_p\}. \tag{5.27}$$

Hence,

$$\dot{\boldsymbol{x}} = \Lambda\boldsymbol{x} + \mathbf{E}^{-1}\mathbf{B}\mathbf{u},$$
$$\mathbf{y} = \mathbf{C}\mathbf{E}\boldsymbol{x} + \mathbf{H}\mathbf{u}, \tag{5.28}$$

or

$$(\mathbf{E}^{-1}\mathbf{B})^* = [\mathbf{B}_1, \ldots, \mathbf{B}_p],$$
$$\dot{\boldsymbol{x}}_i = \Lambda_i\boldsymbol{x}_i + \mathbf{B}_i^*\mathbf{u}, \qquad i = 1, \ldots, p, \tag{5.29}$$
$$\mathbf{y} = \sum_{i=1}^{p} \mathbf{C}_i\boldsymbol{x}_i + \mathbf{H}\mathbf{u}, \qquad \mathbf{C}\mathbf{E} = [\mathbf{C}_1, \ldots, \mathbf{C}_p],$$

where p is the number of linearly independent eigenvectors of A, and Λ_i has the form (2.175). We ask now whether controllability tests are easier to conduct in the modal coordinates (5.29).

A straightforward substitution into (5.23b) yields

$$\mathbf{W}_c = [\mathbf{B}, \ldots, \mathbf{A}^{n-1}\mathbf{B}] = \begin{bmatrix} \mathbf{B}_1^* & \Lambda_1\mathbf{B}_1^* & \Lambda_1^2\mathbf{B}_1^* & \cdots & \Lambda_1^{n-1}\mathbf{B}_1^* \\ \mathbf{B}_2^* & \Lambda_2\mathbf{B}_2^* & \Lambda_2^2\mathbf{B}_2^* & \cdots & \Lambda_2^{n-1}\mathbf{B}_2^* \\ \vdots & \vdots & \vdots & & \vdots \end{bmatrix}, \tag{5.30}$$

where

$$\Lambda_i = \begin{bmatrix} \lambda_i & 1 & & \\ & \ddots & \ddots & \\ & & \lambda_i & 1 \\ & & & \ddots \\ & & & \lambda_i \end{bmatrix}, \qquad \Lambda_i^2 = \begin{bmatrix} \lambda_i^2 & 2\lambda_i & 1 & & \\ & \lambda_i^2 & \lambda_i & 1 & \\ & & \ddots & \ddots & \ddots \end{bmatrix}, \tag{5.31}$$

$$\Lambda_i^k = \begin{bmatrix} \lambda_i^k & k\lambda_i^{k-1} & \frac{1}{2!}k(k-1)\lambda_i^{k-2} & \cdots \\ & \lambda_i^k & k\lambda_i^{k-1} & \ddots \\ & & \lambda_i^k & \ddots \\ & & & \ddots \end{bmatrix}. \tag{5.32}$$

Hence,

$$\Lambda_i \mathbf{B}_i^* = \begin{bmatrix} \lambda_i \mathbf{b}_{i1}^* + \mathbf{b}_{i2}^* \\ \lambda_i \mathbf{b}_{i2}^* + \mathbf{b}_{i3}^* \\ \lambda_i \mathbf{b}_{i3}^* + \mathbf{b}_{i4}^* \\ \vdots \end{bmatrix}, \tag{5.33}$$

$$\Lambda_i^2 \mathbf{B}_i^2 = \begin{bmatrix} \lambda_i(\lambda_i \mathbf{b}_{i1}^* + \mathbf{b}_{i2}^*) + (\lambda_i \mathbf{b}_{i2}^* + \mathbf{b}_{i3}^*) \\ \lambda_i(\lambda_i \mathbf{b}_{i2}^* + \mathbf{b}_{i3}^*) + (\lambda_i \mathbf{b}_{i3}^* + \mathbf{b}_{i4}^*) \\ \lambda_i(\lambda_i \mathbf{b}_{i3}^* + \mathbf{b}_{i4}^*) + (\lambda_i \mathbf{b}_{i4}^* + \mathbf{b}_{i5}^*) \end{bmatrix}, \tag{5.34}$$

and so forth.

Now, define $\mathbf{b}_{il}^*$ as the *last* row of $\mathbf{B}_i^*$. Then, the following result can be proved [5.2].

Theorem 5.9

The system (5.29) *in Jordan form is state controllable if and only if*

$$\{\mathbf{b}_{il}, \mathbf{b}_{jl}, \ldots, \mathbf{b}_{kl}\}$$

is a linearly independent set of vectors, where Λ_i, $\Lambda_j, \ldots, \Lambda_k$ are the Jordan blocks associated with the same eigenvalue.*

Outline of Proof: The rank of the matrix $\mathbf{W}_c$ is not altered by multiplication by nonsingular matrices. It is possible to find nonsingular matrices $\mathbf{L}$ and $\mathbf{T}$ such that $\mathbf{W}_c$ is taken to the block triangular form

$$\mathbf{L}\mathbf{W}_c\mathbf{T} = \mathbf{L}[\mathbf{B}, \mathbf{A}\mathbf{B}, \ldots, \mathbf{A}^{n-1}\mathbf{B}]\mathbf{T}$$

$$= \left[\begin{array}{c|c|c|cc|c} \mathbf{b}_{13}^* & & & & & \\ \mathbf{b}_{22}^* & & & 0 & & \\ \hline \mathbf{b}_{12}^* & \mathbf{b}_{13}^* & & & & \\ \mathbf{b}_{21}^* & \mathbf{b}_{22}^* & & & & \\ \hline \mathbf{b}_{11}^* & \mathbf{b}_{12}^* & \mathbf{b}_{13}^* & & & \\ \hline \mathbf{B}_3^* & [\Lambda_3 - \lambda_1 I]\mathbf{B}_3^* & [\Lambda_3 - \lambda_1 I]^2\mathbf{B}_3^* & [\Lambda_3 - \lambda_1 I]^3\mathbf{B}_3^* & [\Lambda_3 - \lambda_1 I]^4\mathbf{B}_3^* & \cdots \end{array}\right] \tag{5.35}$$

*Suppose there is only one vector in this set. A linearly independent "set" of just *one* vector $\mathbf{b}_{i\ell}$ means $\mathbf{b}_{i\ell} \neq 0$.

shown here worked out for the example

$$\Lambda_1 = \begin{bmatrix} \lambda_1 & 1 & 0 \\ 0 & \lambda_1 & 1 \\ 0 & 0 & \lambda_1 \end{bmatrix}, \quad \Lambda_2 = \begin{bmatrix} \lambda_1 & 1 \\ 0 & \lambda_1 \end{bmatrix}, \quad \Lambda_3 = \begin{bmatrix} \lambda_3 & 1 \\ 0 & \lambda_3 \end{bmatrix},$$

$$\mathbf{B}_1^* = \begin{bmatrix} \mathbf{b}_{11}^* \\ \mathbf{b}_{12}^* \\ \mathbf{b}_{13}^* \end{bmatrix}, \quad \mathbf{B}_2^* = \begin{bmatrix} \mathbf{b}_{21}^* \\ \mathbf{b}_{22}^* \end{bmatrix}, \quad \mathbf{B}_3^* = \begin{bmatrix} \mathbf{b}_{31}^* \\ \mathbf{b}_{32}^* \end{bmatrix}.$$

According to the theorem, the necessary and sufficient conditions for controllability are (a) $\{\mathbf{b}_{13}, \mathbf{b}_{22}\}$ must be linearly independent,

$$\det \begin{bmatrix} \mathbf{b}_{13}^* \\ \mathbf{b}_{22}^* \end{bmatrix} [\mathbf{b}_{13} \quad \mathbf{b}_{22}] \neq 0,$$

and (b) $\{\mathbf{b}_{32}\}$ must be linearly independent ($\mathbf{b}_{32} \neq 0$).

If $\mathbf{A}$ has no repeated eigenvalues, then the Jordan blocks are all 1×1, and this situation simplifies to $\mathbf{b}_i^* \neq 0, \forall i = 1, \ldots, n_x$. ■

Corollary 1 to Theorem 5.9
The diagonal Jordan form simplifies (5.29) *to*

$$\begin{aligned} \dot{\boldsymbol{x}}_i &= \Lambda_i \boldsymbol{x}_i + \mathbf{B}_i^* \mathbf{u}, & \Lambda_i &= \lambda_i \mathbf{I}_{m_i}, \\ \mathbf{y} &= \sum_{i=1}^{n_x} \mathbf{C}_i \boldsymbol{x}_i + \mathbf{H}\mathbf{u}, & \mathbf{B}_i^* &\in \mathscr{C}^{m_i \times n_u}, \\ & & \mathbf{C}_i &\in \mathscr{C}^{n_y \times m_i}. \end{aligned} \tag{5.36}$$

In this case the system (5.36) *is controllable iff* rank $\mathbf{B}_i = m_i$ *for all* $i = 1, \ldots, p$. *If* rank $\mathbf{B}_i < m_i$, *the state* $\boldsymbol{x}_i \in \mathscr{C}^{m_i}$ *in* (5.36) *is uncontrollable. For distinct eigenvalues,*

$$\begin{aligned} \dot{\boldsymbol{x}}_i &= \lambda_i \boldsymbol{x}_i + \mathbf{B}_i^* u, & \boldsymbol{x}_i &\in \mathscr{C}^1, \\ \mathbf{y} &= \sum_{i=1} \mathbf{C}_i \boldsymbol{x}_i + \mathbf{H}\mathbf{u}. \end{aligned} \tag{5.37}$$

and $\mathbf{B}_i^*$ *is a row vector,* $\mathbf{B}_i^* = \mathbf{b}_i^*$, *and* rank $\mathbf{B}_i = m_i \Rightarrow \|\mathbf{b}_i\| > 0$ *is required for controllability.*

An important distinction between the controllability tests (5.23b) and the controllability test here is that (5.23b) only gives "yes or no" answers—the system is either controllable or it is not. If it is not, there is no information provided in the test which suggests *why* the system is uncontrollable. Singular value decomposition (Section 2.8.2) is the most reliable *calculation* of the rank of a matrix. In this section the modes of the system are tested for controllability one at a time. If the system is not controllable, then we can know *which* modes are not controllable.

Exercise 5.16

Determine whether the system is state controllable:

$$\dot{\mathbf{x}} = \begin{bmatrix} 1 & 0 & 0 \\ 0 & 1 & 0 \\ 0 & 0 & 2 \end{bmatrix} \mathbf{x} + \begin{bmatrix} 1 \\ 5 \\ 3 \end{bmatrix} u. \tag{5.38}$$

Check your answer three ways by applying Theorems 5.7 and 5.9 and the corollary to Theorem 5.9.

Exercise 5.16

Can systems be state controllable if some subsets of the states $\mathbf{x}_i$ are related to others $\mathbf{x}_j$ by $\mathbf{x}_i = \mathbf{L}\mathbf{x}_j$ for some constant matrix $\mathbf{L}$?

Taking the Laplace transform of (5.36) yields the transfer function $\mathbf{G}(s)$:

$$\mathbf{y}(s) = \left[\sum_{i=1}^{p} \mathbf{C}_i (s\mathbf{I} - \boldsymbol{\Lambda}_i)^{-1} \mathbf{B}_i^* \right] \mathbf{u}(s) = \mathbf{G}(s)\mathbf{u}(s). \tag{5.39}$$

Now, consider the special case of distinct λ_i ($m_i = 1$). Theorem 4.4 shows that the residue matrices are invariant under coordinate transformation. Thus, (5.39) illustrates that the ith residue is

$$\mathbf{J}_i = \mathbf{C}_i \mathbf{B}_i^*, \qquad \mathbf{G}(s) = \sum_{i=1}^{n_x} \frac{\mathbf{C}_i \mathbf{B}_i^*}{s - \lambda_i} = \sum_{i=1}^{n_x} \frac{\mathbf{J}_i}{s - \lambda_i}. \tag{5.40}$$

Note that *if* λ_i is distinct ($m_i = 1$), then

$$\mathbf{J}_i = \mathbf{c}_i \mathbf{b}_i^*, \tag{5.41}$$

which is a rank 1 matrix formed by the outer product of vectors $\mathbf{c}_i$, $\mathbf{b}_i$ (see Definition 2.8). This gives meaning to the decomposition used earlier in (4.38)–(4.40) and Fig. 4.8. Clearly, from Theorem 5.9 and (5.39), the ith residue $\mathbf{J}_i$ is zero if the ith mode is uncontrollable, so that *uncontrollable modes do not appear in the transfer from* $u(s)$ *to* $y(s)$ if this transfer function is in its simplest form (lowest order). Gilbert [5.3] showed that this lowest order is $n_{\min} = \Sigma_i(\operatorname{rank} \mathbf{J}_i)$.

Exercise 5.17

Can one determine whether the system described by

$$y(s) = \frac{s+1}{s^2+3s+2}u(s) \tag{5.42}$$

is controllable? Why or why not?

Exercise 5.18

For the system

$$G(s) = \begin{bmatrix} \dfrac{1}{s+1} & \dfrac{2}{s+1} \\ \dfrac{-1}{(s+1)(s+2)} & \dfrac{1}{s+2} \end{bmatrix},$$

the residues are

$$\mathbf{J}_1 = \begin{bmatrix} 1 & 2 \\ -1 & 0 \end{bmatrix}, \quad \mathbf{J}_2 = \begin{bmatrix} 0 & 0 \\ 1 & 1 \end{bmatrix}.$$

Find a state space realization.

EXAMPLE 5.4

Verify directly that rank $[\mathbf{B}, \mathbf{AB}, \ldots, \mathbf{A}^{n_x-1}\mathbf{B}] < n_x$ if $\mathbf{b}_i^* = 0$ for some i in (5.37). Assume distinct eigenvalues.

Solution: Without loss of generality, assume that the system is given in its Jordan form. Then,

$$\mathbf{B} = \begin{bmatrix} \mathbf{b}_1^* \\ \vdots \\ \mathbf{b}_{n_x}^* \end{bmatrix}, \quad \mathbf{A} = \begin{bmatrix} \lambda_1 & & \\ & \ddots & \\ & & \lambda_{n_x} \end{bmatrix}$$

and

$$\mathbf{W}_c \triangleq [\mathbf{B}, \mathbf{AB}, \ldots, \mathbf{A}^{n_x-1}\mathbf{B}] = \begin{bmatrix} \mathbf{b}_1^* & \lambda_1 \mathbf{b}_1^* & \cdots & \lambda_1^{n_x-1}\mathbf{b}_1^* \\ \vdots & \vdots & & \vdots \\ \mathbf{b}_i^* & \lambda_i \mathbf{b}_i^* & & \lambda_i^{n_x-1}\mathbf{b}_i^* \\ \vdots & \vdots & & \vdots \\ \mathbf{b}_{n_x}^* & \lambda_{n_x} \mathbf{b}_{n_x}^* & & \lambda_{n_x}^{n_x-1}\mathbf{b}_{n_x}^* \end{bmatrix}. \tag{5.43}$$

Thus, the ith row of $\mathbf{W}_c$ is zero if $\mathbf{b}_i = 0$ and in this event the rank of $\mathbf{W}_c$ cannot exceed $n_x - 1 < n_x$. ■

Exercise 5.19

Determine the range of values for the parameters α, β, Ω, and γ for which these systems are uncontrollable.

(a)

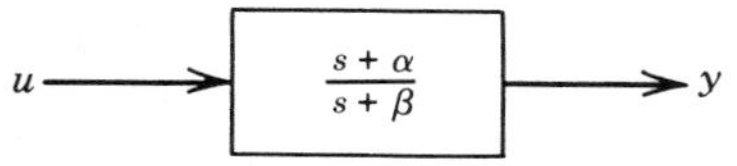

(b)

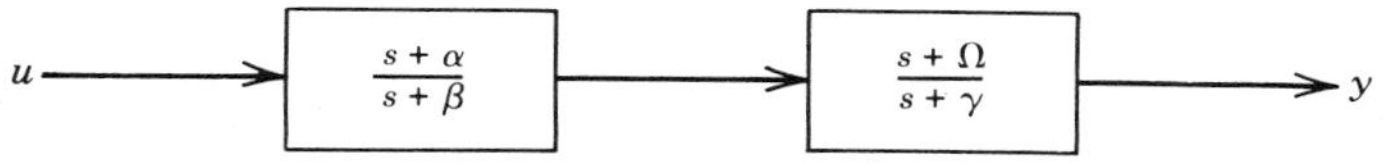

(c) In part (b), compute the angle between the vectors $\mathbf{B}$ and $\mathbf{AB}$. Give geometrical significance of this angle and controllability.

Explain why tandem connections of controllable subsystems might not yield a controllable system.

5.2.1.2 Controllability of Controllable Canonical Coordinates

From (5.24b), the system with $\mathcal{R}e\,\lambda_i[\mathbf{A}] < 0$ is controllable if $\mathbf{X} > \mathbf{0}$, where

$$\mathbf{0} = \mathbf{XA}^* + \mathbf{AX} + \mathbf{B}\mathcal{U}\mathbf{B}^* \tag{5.44}$$

and $\mathbf{X} = \int_0^\infty e^{\mathbf{A}t}\mathbf{B}\mathcal{U}\mathbf{B}^* e^{\mathbf{A}^*t}\,dt$ is called the state covariance matrix. Recall that the "controllability grammian" defined by $\mathbf{X} \triangleq \int_0^\infty e^{\mathbf{A}t}\mathbf{BB}^* e^{\mathbf{A}^*t}\,dt$ is actually the state covariance matrix with unit intensity impulse inputs $\mathcal{U} = \mathbf{I}$.

The positive definiteness of $\mathbf{X}$ is most readily determined when $\mathbf{X}$ is diagonal. To perform the controllability tests in these coordinates, proceed as follows: Without loss of generality, let $\mathcal{U} = \mathbf{I}$.

STEP 1: Solve for $\mathbf{X}$ ($\mathbf{X}$ exists if the controllable modes are stable):

$$\mathbf{0} = \mathbf{XA}^* + \mathbf{AX} + \mathbf{BB}^*. \tag{5.45}$$

STEP 2: Compute the singular value decomposition of the symmetric $\mathbf{X}$,

$$\mathbf{X} = \mathbf{L\Sigma L}^*, \qquad \mathbf{XL} = \mathbf{L\Sigma}, \tag{5.46}$$

to get the unitary matrix $\mathbf{L}$ ($\mathbf{L}^* = \mathbf{L}^{-1}$).

STEP 3: Apply the coordinate transformation $\mathbf{x} = \mathbf{L}x$ to get

$$\begin{aligned} \dot{x} &= \mathbf{L}^T\mathbf{A}\mathbf{L}x + \mathbf{L}^T\mathbf{B}\mathbf{u} = \mathscr{A}x + \mathscr{B}\mathbf{u}, \\ \mathbf{y} &= \mathbf{C}\mathbf{L}x + \mathbf{H}\mathbf{u} = \mathscr{C}x + \mathbf{H}\mathbf{u}, \end{aligned} \tag{5.47}$$

$$\mathbf{0} = \hat{\mathscr{X}}\mathscr{A}^* + \mathscr{A}\hat{\mathscr{X}} + \mathscr{B}\mathscr{B}^* \Rightarrow \hat{\mathscr{X}} = \Sigma = \begin{bmatrix} \sigma_1 & & \\ & \ddots & \\ & & \sigma_{n_x} \end{bmatrix}. \tag{5.48}$$

STEP 4: The test for controllability in the new coordinates is "$\hat{\mathscr{X}} > \mathbf{0}$?" Hence, the system is uncontrollable if and only if $\sigma_i = 0$ for some i. If $\sigma_i = 0$, then x_i in (5.47) is uncontrollable.

To show that $\hat{\mathscr{X}}$ is diagonal in these coordinates, write

$$\hat{\mathscr{X}} \triangleq \int_0^\infty e^{\mathscr{A}t}\mathscr{B}\mathscr{B}^* e^{\mathscr{A}^*t}\,dt = \int_0^\infty \mathbf{L}^* e^{\mathbf{A}t}\mathbf{L}\mathbf{L}^*\mathbf{B}\mathbf{B}^*\mathbf{L}\mathbf{L}^* e^{\mathbf{A}^*t}\mathbf{L}\,dt. \tag{5.49}$$

But since $\mathbf{L}$ is unitary ($\mathbf{L}\mathbf{L}^* = \mathbf{I}$),

$$\hat{\mathscr{X}} = \mathbf{L}^* \int_0^\infty e^{\mathbf{A}t}\mathbf{B}\mathbf{B}^* e^{\mathbf{A}^*t}\,dt\,\mathbf{L} = \mathbf{L}^*\mathbf{X}\mathbf{L}. \tag{5.50}$$

But from (5.49), $\mathbf{X} = \mathbf{L}\Sigma\mathbf{L}^*$. Hence,

$$\hat{\mathscr{X}} = \mathbf{L}^*\mathbf{X}\mathbf{L} = \mathbf{L}^*\mathbf{L}\Sigma\mathbf{L}^*\mathbf{L} = \Sigma. \tag{5.51}$$

Thus, we have the following result.

Theorem 5.10

The system (5.47) *is controllable iff* $\sigma_i^2 > 0, \forall i = 1, \ldots, n_x$.

The advantages of the computations required in Theorem 5.10 as opposed to the computations required in Theorem 5.9 are evident.

Exercise 5.20

Find a set of coordinates in which $\mathbf{W}_c\mathbf{W}_c^* =$ diagonal. These might also be called controllable canonical coordinates, but they require no assumptions on stability.

5.2.1.3 Other Controllable Canonical Forms

The coordinates in the previous Section are distinguished by the diagonal structure of the controllability grammian

$$\mathbf{X} \triangleq \int_0^\infty e^{\mathbf{A}t}\mathbf{B}\mathbf{B}^* e^{\mathbf{A}^*t}\,dt \tag{5.52}$$

and the uncontrollable states are distinguished by the zero entries $X_{ii} = 0$. Note that in these coordinates, **A** and **B** have no special structure.

Instead of a special structure for **X**, another set of coordinates is distinguished by a special structure of **A** and **B**, with no special structure for **X**:

$$\begin{pmatrix} \dot{\mathbf{x}}_1 \\ \dot{\mathbf{x}}_2 \end{pmatrix} = \begin{bmatrix} \mathbf{A}_{11} & \mathbf{A}_{12} \\ \mathbf{0} & \mathbf{A}_{22} \end{bmatrix} \begin{pmatrix} \mathbf{x}_1 \\ \mathbf{x}_2 \end{pmatrix} + \begin{bmatrix} \mathbf{B}_1 \\ \mathbf{0} \end{bmatrix} u = \mathbf{A}\mathbf{x} + \mathbf{B}\mathbf{u}, \tag{5.53}$$

where $(\mathbf{A}_1, \mathbf{B}_1)$ is a controllable pair. The controllability test matrix $\mathbf{W}_c$ in (5.23b) becomes

$$\mathbf{W}_c = \left\{ \begin{bmatrix} \mathbf{B}_1 \\ \mathbf{0} \end{bmatrix}, \begin{bmatrix} \mathbf{A}_{11}\mathbf{B}_1 \\ \mathbf{0} \end{bmatrix}, \ldots \begin{bmatrix} \mathbf{A}_{11}^{n_x - 1}\mathbf{B}_1 \\ \mathbf{0} \end{bmatrix} \right\}. \tag{5.54}$$

Now, since the matrix pair $(\mathbf{A}_{11}, \mathbf{B}_1)$ is controllable, the matrix $\mathbf{W}_c$ has its maximum rank = dimension of $\mathbf{x}_1$. Note that

1. $\mathbf{x}_1$ is controllable iff $(\mathbf{A}_{11}, \mathbf{B}_1)$ is controllable.
2. $\mathbf{x}_2$ is uncontrollable $\forall \mathbf{A}_{12}, \mathbf{A}_{22}$.

The coordinates of (5.53) are not unique since $\mathbf{A}_{12}$ and $\mathbf{A}_{22}$ are arbitrary. The special matrix structure of $(\mathbf{A}, \mathbf{B})$ in (5.53) is called "Kalman's controllable canonical form" [5.4].

Theorem 5.7 provides two basic tests for controllability. Another basic test is provided as follows.

Theorem 5.11

The matrix pair $(\mathbf{A}, \mathbf{B})$ *is controllable iff* $\operatorname{rank}[\lambda\mathbf{I} - \mathbf{A}, \mathbf{B}] = n_x$ *for all* λ.

Proof: First, prove that $\operatorname{rank}[\lambda\mathbf{I} - \mathbf{A}, \mathbf{B}] = n_x \Rightarrow (\mathbf{A}, \mathbf{B})$ controllable. If the rank condition holds, then

$$\begin{aligned} \ell^*[\lambda\mathbf{I} - \mathbf{A}, \mathbf{B}] &= \mathbf{0} \Rightarrow \ell = \mathbf{0}, \\ \lambda\ell^* = \ell^*\mathbf{A},\ \ell^*\mathbf{B} &= 0 \Rightarrow \ell = \mathbf{0}, \\ \lambda\ell^*\mathbf{B} = \ell^*\mathbf{A}\mathbf{B} &\Rightarrow \ell = \mathbf{0}, \end{aligned} \tag{5.55}$$

or

$$\{\lambda^2\ell^*\mathbf{B} = \lambda\ell^*\mathbf{A}\mathbf{B}\} \Rightarrow \{\lambda\ell^*\mathbf{A}\mathbf{B} = \lambda\ell^*\mathbf{A}^2\mathbf{B}\},$$

and so forth, until

$$\lambda\ell^*\mathbf{A}^{n_x-2}\mathbf{B} = \ell^*\mathbf{A}^{n_x-1}\mathbf{B} \Rightarrow \ell = \mathbf{0},$$

which all together imply

$$\ell^*[\mathbf{B}, \mathbf{AB}, \mathbf{A}^2\mathbf{B}, \ldots \mathbf{A}^{n_x-1}\mathbf{B}] = 0 \Rightarrow \ell = \mathbf{0}. \tag{5.56}$$

But this means rank $[\mathbf{B}, \mathbf{AB}, \ldots, \mathbf{A}^{n_x-1}\mathbf{B}] = n_x \Rightarrow (\mathbf{A}, \mathbf{B})$ controllable. This proves the first part. The geometrical significance of (5.56) is that the vectors $(\mathbf{A}^i\mathbf{B})$ represent those directions in state space in which we have some control authority. Hence, if there is some direction ℓ that is orthogonal to every direction of control authority, then there is no control in direction ℓ.

Now, prove that $(\mathbf{A}, \mathbf{B})$ uncontrollable $\Rightarrow$ a nonzero ℓ exists, and hence rank $[\lambda\mathbf{I} - \mathbf{A}, \mathbf{B}] < n_x$. Assume $(\mathbf{A}, \mathbf{B})$ in Kalman's controllable canonical form (5.53),

$$\mathbf{A} = \begin{bmatrix} \mathbf{A}_{11} & \mathbf{A}_{12} \\ \mathbf{0} & \mathbf{A}_{22} \end{bmatrix}, \qquad \mathbf{B} = \begin{bmatrix} \mathbf{B}_1 \\ \mathbf{0} \end{bmatrix},$$

where $(\mathbf{A}_{11}, \mathbf{B}_1)$ is controllable. Note that the directions of control authority are now $\mathbf{A}^i\mathbf{B} = [\mathbf{A}_{11}^i\mathbf{B}_i]$. Hence, we can choose $\ell^* = (\mathbf{0}, \ell_2^*)$, $\ell_2 \neq \mathbf{0}$, to satisfy (5.56) with $\ell \neq \mathbf{0}$. Hence, rank $[\lambda\mathbf{I} - \mathbf{A}, \mathbf{B}] < n_x$. ■

The next section considers a set of coordinates distinguished by the special form

$$\mathbf{A} = \begin{bmatrix} \mathbf{0} & \mathbf{I} \\ \mathbf{A}_{21} & \mathbf{A}_{22} \end{bmatrix}, \qquad \mathbf{B} = \begin{bmatrix} \mathbf{0} \\ \mathbf{B}_2 \end{bmatrix}. \tag{5.57}$$

5.2.1.4 Controllability of Vector Second-Order Systems

Consider the systems developed in Chapter 3 of the form

$$\begin{aligned} \mathcal{M}\ddot{\mathbf{q}} + \mathcal{D}\dot{\mathbf{q}} + \mathcal{K}\mathbf{q} &= \mathcal{B}\mathbf{u}, \\ \mathbf{y} &= \mathcal{P}\mathbf{q} + \mathcal{R}\dot{\mathbf{q}}. \end{aligned} \tag{5.58}$$

Such systems occur frequently in structural dynamics, fluid dynamics, circuits, and acoustics. A state space realization of (5.54) can be constructed where $\mathcal{M}$ is positive definite and $\mathcal{K}$ is at least positive semidefinite:

$$\begin{pmatrix} \dot{\mathbf{q}} \\ \ddot{\mathbf{q}} \end{pmatrix} = \begin{bmatrix} \mathbf{0} & \mathbf{I} \\ -\mathcal{M}^{-1}\mathcal{K} & -\mathcal{M}^{-1}\mathcal{D} \end{bmatrix} \begin{pmatrix} \mathbf{q} \\ \dot{\mathbf{q}} \end{pmatrix} + \begin{bmatrix} \mathbf{0} \\ \mathcal{M}^{-1}\mathcal{B} \end{bmatrix} \mathbf{u} = \mathbf{Ax} + \mathbf{Bu}, \tag{5.59}$$

$$\mathbf{y} = [\mathcal{P} \quad \mathcal{R}] \begin{pmatrix} \mathbf{q} \\ \dot{\mathbf{q}} \end{pmatrix} = \mathbf{Cx},$$

$$\mathcal{M} > \mathbf{0}, \qquad \mathcal{K} \geq \mathbf{0}. \tag{5.60}$$

Of course, many other choices of state variables are possible. The most straightforward approach to the controllability of (5.58) is to use the standard rank tests (5.17) and (5.23) applied to a state realization such as (5.59). The *disadvantage* of this standard approach is that no special properties of the matrices **A** and **B** have been utilized in (5.17) and (5.23). When special properties such as (5.60) are known, it is almost certain that the controllability tests can be simplified when these additional properties of **A** are utilized.

When mechanical systems of the type (5.58) have terms $-\mathbf{f}_G \triangleq \mathscr{D}\dot{\mathbf{q}}$ present with the property $\mathscr{D} = -\mathscr{D}^T$, they are called *gyroscopic* systems because the gyroscopic forces $\mathbf{f}_G$ arise from rotors on a structure, or they arise when $\mathbf{q}$ is defined with respect to a uniformly rotating frame of reference. Gyroscopic forces do not do work because $\int \mathbf{f}_G^T\, d\mathbf{q} = \int \dot{\mathbf{q}}^T \mathscr{D}\mathbf{q}\, dt \equiv 0$ along the motion. (The reader should verify that $\dot{\mathbf{q}}^T\mathscr{D}\mathbf{q} \equiv 0\ \forall \mathbf{q}(t)$ if $\mathscr{D} = -\mathscr{D}^T$.) The study of controllability and observability of gyroscopic systems may be found in refs. [5.5], [5.6], and [5.13]. Only the special case $\mathscr{D} = \mathbf{0}$ is treated here.

Nongyroscopic Undamped Systems

Now, we consider the following system:

$$\left.\begin{aligned} \mathscr{M}\ddot{\mathbf{q}} + \mathscr{K}\mathbf{q} &= \mathscr{B}\mathbf{u} \\ \mathbf{y} &= \mathscr{P}\mathbf{q} + \mathscr{R}\dot{\mathbf{q}} \end{aligned}\right\} \qquad \mathbf{q} \in \mathscr{R}^N, \tag{5.61}$$

where $\mathscr{M} > \mathbf{0}$, $\mathscr{K} \geq \mathbf{0}$. By employing Theorem 4.10, we know that $\mathscr{E}$ exists such that

$$\begin{aligned} \mathscr{E}^T\mathscr{M}\mathscr{E} &= \mathbf{I}, \\ \mathscr{E}^T\mathscr{K}\mathscr{E} &= \boldsymbol{\Omega}^2 \\ &= \operatorname{diag}\left[\boldsymbol{\Omega}_1^2(\text{multiplicity } m_1), \boldsymbol{\Omega}_2^2(\text{multiplicity } m_2) \ldots \boldsymbol{\Omega}_p^2(\text{multiplicity } m_p)\right] \end{aligned} \tag{5.62a}$$

where $\boldsymbol{\Omega}_i^2 = \operatorname{diag}[\omega_i^2, \omega_i^2, \ldots, \omega_i^2]$ is $m_i \times m_i$. Using this transformation $\mathbf{q} = \mathscr{E}\boldsymbol{\eta}$ with $\mathscr{D} = \mathbf{0}$, (5.59) becomes

$$\begin{aligned} \begin{pmatrix} \dot{\boldsymbol{\eta}} \\ \ddot{\boldsymbol{\eta}} \end{pmatrix} &= \begin{bmatrix} \mathbf{0} & \mathbf{I} \\ -\boldsymbol{\Omega}^2 & \mathbf{0} \end{bmatrix} \begin{pmatrix} \boldsymbol{\eta} \\ \dot{\boldsymbol{\eta}} \end{pmatrix} + \begin{bmatrix} \mathbf{0} \\ \mathscr{B}_1 \end{bmatrix} \mathbf{u}, \qquad \mathscr{B}_1 = \mathscr{E}^T\mathscr{B}, \\ \mathbf{y} &= [\mathscr{P}' \ \ \mathscr{R}'] \begin{pmatrix} \boldsymbol{\eta} \\ \dot{\boldsymbol{\eta}} \end{pmatrix}, \qquad [\mathscr{P}' \ \ \mathscr{R}'] = [\mathscr{P}\mathscr{E} \ \ \mathscr{R}\mathscr{E}], \end{aligned} \tag{5.62b}$$

and the controllability test (5.23b) becomes

$$\operatorname{rank} W_c = \operatorname{rank} \begin{bmatrix} \mathbf{0} & \mathscr{B}_1 & \mathbf{0} & -\boldsymbol{\Omega}^2\mathscr{B}_1 & \cdots & (-\boldsymbol{\Omega}^2)^{N-1}\mathscr{B}_1 \\ \mathscr{B}_1 & \mathbf{0} & -\boldsymbol{\Omega}^2\mathscr{B}_1 & \mathbf{0} & \cdots & \mathbf{0} \end{bmatrix} \overset{?}{=} 2N,$$

which reduces to the equivalent test

$$\text{rank}\left[\mathcal{B}_1, \boldsymbol{\Omega}^2\mathcal{B}_1, \dots, \boldsymbol{\Omega}^{2(N-1)}\mathcal{B}_1\right] \stackrel{?}{=} N. \tag{5.63}$$

Now, since $\boldsymbol{\Omega}^2$ is *diagonal*, the test (5.63) is just a *special case of the Jordan-form controllability corollary to Theorem* 5.9, *when the Jordan form is diagonal. Hence the following result*

Theorem 5.12

Label the partitions of $\mathcal{B}_1$ in (5.62) *as follows*:

$$\mathcal{B}_1 = \begin{bmatrix} \mathcal{B}_{m_1} \\ \vdots \\ \mathcal{B}_{m_p} \end{bmatrix}, \qquad \mathcal{B}_{m_i} \text{ is } m_i \times n_u$$

and m_i is the multiplicity of ω_i^2. The system described by (5.61) *and* (5.62) *is controllable iff*

$$\text{rank}\left[\mathcal{B}_{m_i}\right] = m_i, \qquad i = 1, \dots, p. \tag{5.64a}$$

When the eigenvalues ω_i^2 are distinct, (5.64*a*) *reduces to*

$$\mathscr{b}_i \neq 0, \qquad i = 1, \dots, N, \tag{5.64b}$$

where $\mathscr{b}_i^T$ is the ith row of $\mathcal{B}_1$.

Two differences between the computations of Theorem 5.12 and Theorem 5.9, are that (1) even for repeated frequencies, there is *always* a linearly independent set of eigenvectors for (5.62a) and the added complexity of Theorem 5.9 (due to the presence of generalized eigenvectors) is not necessary; (2) there are only *half* as many tests to conduct in (5.64) as in Theorem 5.9, or (5.36), or Theorem 5.10.

One final simplification is available as follows:

Theorem 5.13

The ith mode of the system (5.61) *is controllable iff for all complex values of* λ,

$$\text{rank}\left[\lambda^2\mathcal{M} + \mathcal{K}, \mathcal{B}\right] = N. \tag{5.65}$$

Proof: We have only to show that (5.65) is equivalent to the statement of Theorem 5.11. Note that the substitution of the (A, B) of (5.59) leads to the rank test

$$\text{rank}\begin{bmatrix} \lambda\mathbf{I} & -\mathbf{I} & 0 \\ \mathcal{M}^{-1}\mathcal{K} & \lambda\mathbf{I} & \mathcal{B} \end{bmatrix} = 2N, \qquad \forall\lambda. \tag{5.66}$$

This test is always satisfied when $\lambda \neq \lambda_i$ if λ_i is an eigenvalue of $\mathbf{A}$, since in this event $|\lambda\mathbf{I} - \mathbf{A}| \neq 0$ and $\text{rank}[\lambda\mathbf{I} - \mathbf{A}] = 2N$, leaving $\text{rank}[\lambda\mathbf{I} - \mathbf{A}, \mathbf{B}] = 2N$. Clearly, we have only to test (5.66) when $\lambda = \lambda_i$. Rank test (5.66) implies that

$$(\ell_1^* \ell^*)\begin{bmatrix} \lambda_1\mathbf{I} & -\mathbf{I} & \mathbf{0} \\ \mathscr{M}^{-1}\mathscr{K} & \lambda_i\mathbf{I} & \mathscr{B} \end{bmatrix} = 0 \Rightarrow (\ell_1^*, \ell^*) = \mathbf{0} \tag{5.67}$$

or

$$\begin{aligned} \lambda_i\ell_1^* + \ell^*\mathscr{M}^{-1}\mathscr{K} &= \mathbf{0}, \\ -\ell_1^* + \lambda_i\ell^* &= \mathbf{0}, \\ \ell^*\mathscr{B} &= \mathbf{0}. \end{aligned} \tag{5.68}$$

But since $\ell_1^* = \lambda_i\ell^*$, these tests can be rewritten in terms of ℓ:

$$\begin{aligned} \lambda_i^2\ell^* + \ell^*\mathscr{M}^{-1}\mathscr{K} = 0 = \ell^*\left[\lambda_i^2\mathbf{I} + \mathscr{M}^{-1}\mathscr{K}\right] &= 0, \\ \ell^*\mathscr{B} &= 0. \end{aligned} \tag{5.69}$$

Note that the rank of $[\lambda_i^2\mathbf{I} + \mathscr{M}^{-1}\mathscr{K}]$ is the same as the rank of $[\lambda_i^2\mathscr{M} + \mathscr{K}]$. Thus, the conditions

$$\ell^*\left[\lambda_i^2\mathscr{M} + \mathscr{K}, \mathscr{B}\right] = \mathbf{0} \Rightarrow \ell = \mathbf{0} \tag{5.70}$$

lead directly to (5.65). ■

Exercise 5.21

A symmetric spinning spacecraft has this description:

$$\mathbf{J}\dot{\omega} + \tilde{\omega}\mathbf{J}\omega = \mathscr{B}\mathbf{u}, \qquad \tilde{\omega} \triangleq \begin{bmatrix} 0 & \omega_3 & -\omega_2 \\ -\omega_3 & 0 & \omega_1 \\ \omega_2 & -\omega_1 & 0 \end{bmatrix}, \tag{5.71}$$

where $\mathbf{J}$ is the moment of inertia matrix. About principal axes,

$$\mathbf{J} \triangleq \begin{bmatrix} J_1 & 0 & 0 \\ 0 & J_2 & 0 \\ 0 & 0 & J_3 \end{bmatrix}, \qquad J_2 = J_3. \tag{5.72}$$

Linearize the equations of motion about

$$\bar{\omega} = \begin{pmatrix} \bar{\omega}_1 \\ 0 \\ 0 \end{pmatrix}$$

to get

$$\mathbf{J}\dot{\boldsymbol{\omega}} + \mathscr{D}\boldsymbol{\omega} = \mathscr{B}\mathbf{u}, \qquad \mathscr{D} = -\mathscr{D}^T. \tag{5.73}$$

(a) Find $\mathscr{D}$ and determine whether this system is controllable from only one torque applied about the 2-axis, hence $\mathscr{B}^T = (0 \quad 1 \quad 0)$. Explain what is and is not controllable.

(b) Determine whether ω_2 and ω_3 are controllable from u.

Exercise 5.22

Using both Theorems 5.12 and 5.13, determine whether (3.23) is state controllable from T_c or f alone. Assume $\overline{\mathscr{C}}_{22} = 0$.

Exercise 5.23

Using both Theorems 5.12 and 5.13, determine whether the launch vehicle (3.15) is state controllable (at launch $V = 0$).

Exercise 5.24

Using both Theorems 5.12 and 5.13, determine whether (3.78) is state controllable. Let $i = 1, 2, 3$.

5.3 *Observability*

Having a reliable model and knowing its controllability properties gives insight about what is possible and not possible to accomplish with control $\mathbf{u}(t)$. Now, we wish to know what is possible and not possible to know about the internal properties of the system (the states), given only the outputs $\mathbf{y}(t)$. The solution to

$$\begin{aligned} \dot{\mathbf{x}} &= \mathbf{A}(t)\mathbf{x} + \mathbf{B}(t)\mathbf{u}, \\ \mathbf{y} &= \mathbf{C}(t)\mathbf{x} + \mathbf{H}(t)\mathbf{u} \end{aligned} \tag{5.74}$$

is given by

$$\mathbf{y}(t) = \mathbf{C}(t)\boldsymbol{\Phi}(t, t_0)\mathbf{x}(t_0) + \int_{t_0}^{t}[\mathbf{C}(t)\boldsymbol{\Phi}(t, \sigma)\mathbf{B}(\sigma) + \mathbf{H}(t)\delta(t - \sigma)]\mathbf{u}(\sigma)\, d\sigma. \tag{5.75}$$

The question of observability involves the following:

Definition 5.3

The system (5.74) *is "observable at time t" if* $\mathbf{x}(t_0)$ *can be uniquely determined from knowledge of* $\mathbf{u}(\tau), \mathbf{y}(\tau)$, $0 \le \tau \le t$.

It should be noted that if $\mathbf{x}$ at some *different* time $\mathbf{t}_1$ is desired, then $\mathbf{x}(t_1) = \mathbf{\Phi}(t_1, t_0)\mathbf{x}(t_0) + \int_{t_0}^{t_1}[\mathbf{\Phi}(t_1, \sigma)\mathbf{B}(\sigma) + \mathbf{H}(t_1)\delta(t - \sigma)]\mathbf{u}(\sigma)\, d\sigma$ can be computed once $\mathbf{x}(t_0)$ is known. There is, therefore, no loss of generality in restricting our attention in Definition 5.3 to computation of $\mathbf{x}(t_0)$, since this will give us the capability to know any $\mathbf{x}(t)$ for $t \neq t_0$.

To pursue this study of observability of (5.74), we collect all known quantities (from Definition 5.3) on the left-hand side of (5.75) and call this quantity $\mathscr{y}(t)$:

$$\mathscr{y}(t) \triangleq \mathbf{y}(t) - \int_{t_0}^{t}[\mathbf{C}(t)\mathbf{\Phi}(t, \sigma)\mathbf{B}(\sigma) + \mathbf{H}(t)\delta(t - \sigma)]\mathbf{u}(\sigma)\, d\sigma$$

$$= \mathbf{C}(t)\mathbf{\Phi}(t, t_0)\mathbf{x}(t_0). \tag{5.76}$$

See that this is *equivalent* to the solution of the unforced system:

$$\begin{aligned} \dot{\mathbf{x}} &= \mathbf{A}(t)\mathbf{x}, \\ \mathscr{y} &= \mathbf{C}(t)\mathbf{x}. \end{aligned} \tag{5.77}$$

Hence, we lose no generality in observability studies by examining only the *unforced* system, since observability obviously deals only with the matrix pair $\{\mathbf{A}(t), \mathbf{C}(t)\}$. Now, multiply (5.76) from the left by $[\mathbf{C}(t)\mathbf{\Phi}(t, t_0)]^*$ and integrate:

$$\int_{t_0}^{t_f}[\mathbf{\Phi}^*(t, t_0)\mathbf{C}^*(t)y(t) = \mathbf{\Phi}^*(t, t_0)\mathbf{C}^*(t)\mathbf{C}(t)\mathbf{\Phi}(t, t_0)\mathbf{x}(t_0)]\, dt. \tag{5.78}$$

Changing the dummy variable of integration to σ and integrating up to t_f yields for $\mathbf{x}(t_0)$

$$\mathbf{x}(t_0) = \left[\int_{t_0}^{t_f}\mathbf{\Phi}^*(\sigma, t_0)\mathbf{C}^*(\sigma)\mathbf{C}(\sigma)\mathbf{\Phi}(\sigma, t_0)\, d\sigma\right]^{-1}\int_{t_0}^{t_f}\mathbf{\Phi}^*(\sigma, t_0)\mathbf{C}^*(\sigma)\mathscr{y}(\sigma)\, d\sigma. \tag{5.79}$$

Note that for the solution (5.79) to exist, the matrix

$$\mathbf{K}(t_0, t_f) \triangleq \int_{t_0}^{t_f}\mathbf{\Phi}^*(\sigma, t_0)\mathbf{C}^*(\sigma)\mathbf{C}(\sigma)\mathbf{\Phi}(\sigma, t_0)\, d\sigma \tag{5.80}$$

must be invertible. By using Theorem 4.12, we note that the matrix $\mathbf{K}(t_0, t_f)$ is the solution to the differential equation

$$\frac{d}{dt}\mathbf{K}(t, t_f) = -\mathbf{K}(t, t_f)\mathbf{A}(t) - \mathbf{A}^*(t)\mathbf{K}(t, t_f) - \mathbf{C}^*(t)\mathbf{C}(t), \qquad \mathbf{K}(t_f, t_f) = \mathbf{0}. \tag{5.81}$$

Note also from the structure of (5.80) that $\mathbf{K}(t, t_f)$ cannot be negative. Hence, if $\mathbf{K}$ is nonsingular, it is positive definite. The conclusion then is that the system (5.77) is observable if $\mathbf{K}(t_0, t_f) > \mathbf{0}$. But this *sufficient* condition does not establish necessity.

From Chapter 2 we know that (5.76) has a solution $\mathbf{x}(t_0)$ iff $\mathscr{y}(t)$ lies in the space spanned by the columns of the matrix $[\mathbf{C}(t)\mathbf{\Phi}(t, t_0)]$ on the interval $[t, t_0]$. Since $\mathscr{y}(t)$ is arbitrary, the columns must span the *entire* space. Hence, the columns must be linearly independent. From Theorem 2.6, the columns of the matrix $[\mathbf{C}(t)\mathbf{\Phi}(t, t_0)]$ are linearly independent on the interval $[t, t_0]$ iff (5.80) is nonsingular. This establishes the *necessity* of the above-stated $\mathbf{K}(t_0, t_1) > \mathbf{0}$ requirement, and these results are now summarized.

Theorem 5.14

The linear system (5.74) *is observable at time* t_f *iff the solution* $\mathbf{K}(t_0, t_f)$ *of* (5.81) *is positive definite.*

Note that these results hold also if $\mathbf{C^*C}$ is replaced by $\mathbf{C^*QC}$ for any $\mathbf{Q} > \mathbf{0}$.

Time-Invariant Systems

For constant $(\mathbf{A}, \mathbf{C})$, the observability test is greatly simplified; (5.80) becomes

$$\mathbf{K}(t_0, t_f) = \int_{t_0}^{t_f} e^{\mathbf{A}^*(\sigma - t_0)}\mathbf{C}^*\mathbf{C}e^{\mathbf{A}(\sigma - t_0)}\,d\sigma = \int_0^{t_f - t_0} e^{\mathbf{A}^*\tau}\mathbf{C}^*\mathbf{C}e^{\mathbf{A}\tau}\,d\tau, \tag{5.82}$$

which is the matrix $\mathscr{K}$ in (4.165) when $\mathbf{Q} = \mathbf{I}$ and $t_f \to \infty$. Substitute

$$e^{\mathbf{A}\tau} = \sum_{i=0}^{n_x - 1} \alpha_i(\tau)\mathbf{A}^i \tag{5.83}$$

to get

$$\mathbf{K}(t_0, t_f) = \int_0^{t_f - t_0} \sum_{i=0}^{n_x - 1} \alpha_i(\tau)\mathbf{A}^{i*}\mathbf{C}^*\mathbf{C} \sum_{k=0}^{n_x - 1} \mathbf{A}^k\alpha_k(\tau)\,d\tau \tag{5.84}$$

$$= [\mathbf{C}^*, \mathbf{A}^*\mathbf{C}^*, \ldots, \mathbf{A}^{n_x - 1*}\mathbf{C}^*]\mathbf{\Omega}(t_f, t_0)\begin{bmatrix} \mathbf{C} \\ \mathbf{CA} \\ \vdots \\ \mathbf{CA}^{n_x - 1} \end{bmatrix} \tag{5.85}$$

$$= \mathbf{W}_0^*\mathbf{\Omega}(t_f, t_0)W_0, \qquad \mathbf{W}_0^* \triangleq [\mathbf{C}^*, \mathbf{A}^*\mathbf{C}^*, \ldots, \mathbf{A}^{n_x - 1*}\mathbf{C}^*],$$

where $\mathbf{\Omega}$ is defined by

$$\mathbf{\Omega}(t_f, t_0) \triangleq \int_0^{t_f - t_0} \begin{bmatrix} \alpha_0(\tau)\mathbf{I}_{n_y} \\ \vdots \\ \alpha_{n_x - 1}(\tau)\mathbf{L}_{n_y} \end{bmatrix} \left[\alpha_0(\tau)\mathbf{I}_{n_y}, \ldots, \alpha_{n_x - 1}(\tau)\mathbf{I}_{n_y}\right] d\tau. \tag{5.86}$$

Clearly, $\mathbf{K}(t_0, t_f)$ cannot have rank n_x if $\mathbf{W}_0$ does not rank n_x. It is left as an exercise to the reader to show that $\boldsymbol{\Omega}(t_f, t_0)$, $t_f > t_0$, is positive definite (see Exercise 5.3). This shows that $\mathbf{K}(t_0, t_f) > \mathbf{0}$ iff $\mathbf{W}_0$ has rank n_x. These results also hold iff $\mathbf{C}^*\mathbf{C}$ is replaced by $\mathbf{C}^*\mathbf{QC}$ for any $\mathbf{Q} > \mathbf{0}$. This leads to the following results, using definition (4.165).

Theorem 5.15

The time-invariant system

$$\dot{\mathbf{x}} = \mathbf{Ax}, \qquad \mathbf{y} = \mathbf{Cx} \tag{5.87}$$

is observable iff

$$\operatorname{rank} \mathbf{W}_0 = n_x, \qquad \mathbf{W}_0^* \triangleq [\mathbf{C}^*, \mathbf{A}^*\mathbf{C}^*, \ldots, \mathbf{A}^{n_x - 1*}\mathbf{C}^*]. \tag{5.88}$$

If $\mathcal{R}e\, \lambda_i[\mathbf{A}] < 0$, $\forall i = 1, \ldots, n$, *then* $\mathscr{K} \triangleq \int_0^\infty e^{\mathbf{A}^*\sigma}\mathbf{C}^*\mathbf{QC}e^{\mathbf{A}\sigma}\, d\sigma$ *exists and satisfies, for any* $\mathbf{Q} > \mathbf{0}$,

$$\mathbf{0} = \mathscr{K}\mathbf{A} + \mathbf{A}^*\mathscr{K} + \mathbf{C}^*\mathbf{QC}, \tag{5.89}$$

and $\mathscr{K} > 0$ *iff* $(\mathbf{A}, \mathbf{C})$ *is an observable pair.*

EXAMPLE 5.5

Show that without the above qualification the following statement is *untrue*: "The solution of $\mathscr{K}\mathbf{A} + \mathbf{A}^*\mathscr{K} + \mathbf{C}^*\mathbf{QC} = \mathbf{0}$ is $\mathscr{K} > \mathbf{0}$ iff $(\mathbf{A}, \mathbf{C})$ is an observable pair."

Solution: The counterexample

$$(\mathscr{K} = \mathbf{I}, \mathbf{C} = \mathbf{0}, \mathbf{A} = -\mathbf{A}^*) \tag{5.90}$$

shows that $\mathscr{K} > \mathbf{0}$ is a possible solution to (5.89) even if $(\mathbf{A}, \mathbf{C})$ is not observable. In this case, $\mathscr{K}$ is not unique since $\lambda_i + \lambda_j = 0$ for a skew-symmetric $\mathbf{A}$. Thus, a positive definite $\mathscr{K}$ satisfies (5.89), but the $\mathscr{K}$ that satisfies $\mathscr{K} = \int_0^\infty e^{\mathbf{A}^*t}\mathbf{C}^*\mathbf{QC}e^{\mathbf{A}t}\, dt$ is *not* the positive definite solution of (5.89). In fact, the integral solution is $\mathscr{K} = \mathbf{0}$. ■

EXAMPLE 5.6

Determine whether the orbit of Fig. 3.3 can be determined from range measurements $r(t)$ alone. That is, is (3.28) observable with output $r - R_0$? Can the orbit be determined from $\theta(t) - \omega_0 t$ measurements alone?

Solution: The state space realization of (3.28) is given by (5.25), which yields the $\mathbf{W}_0$ of (5.88) with $\mathbf{C} = [1 \quad 0 \quad 0 \quad 0]$ for $y = r$:

$$\mathbf{W}_0 = \begin{bmatrix} 1 & 0 & 0 & 0 \\ 0 & 1 & 0 & 0 \\ 3\omega_0^2 & 0 & 0 & 2\omega_0 \\ 0 & -\omega_0^2 & 0 & 0 \end{bmatrix} \text{ has rank } 3 < 4 \Rightarrow \left\{ \begin{array}{l} \text{orbit can't be determined} \\ \text{from range measurements} \\ r(t). \end{array} \right.$$

Now, for $y = \theta - \omega_0 t$, $\mathbf{C} = [0 \quad 0 \quad 1 \quad 0]$ and

$$\mathbf{W}_0 = \begin{bmatrix} 0 & 0 & 1 & 0 \\ 0 & 0 & 0 & 1 \\ 0 & -2\omega_0 & 0 & 0 \\ -6\omega_0^2 & 0 & 0 & -4\omega_0^2 \end{bmatrix} \text{has rank } 4 \Rightarrow \begin{cases} \text{orbit can be determined} \\ \text{from azimuth} \\ \text{measurements.} \end{cases}$$

■

5.4 Observability of Transformed Coordinates

This section shows that observability is a system property invariant under coordinate transformation.

Consider the coordinate transformation $\mathbf{x} = \mathbf{T}\boldsymbol{x}$, where $\mathbf{T}$ is constant, nonsingular. In $\boldsymbol{x}$-coordinates, the time-invariant system description then becomes

$$\begin{aligned} \dot{\boldsymbol{x}} &= \mathbf{T}^{-1}\mathbf{A}\mathbf{T}\boldsymbol{x}, \\ \boldsymbol{y} &= \mathbf{C}\mathbf{T}\boldsymbol{x}. \end{aligned} \tag{5.91}$$

Now, for observability we must test the rank of

$$\begin{bmatrix} \mathbf{CT} \\ (\mathbf{CT})(\mathbf{T}^{-1}\mathbf{AT}) \\ \vdots \\ (\mathbf{CT})(\mathbf{T}^{-1}\mathbf{AT})^{n_x-1} \end{bmatrix} = \begin{bmatrix} \mathbf{C} \\ \mathbf{CA} \\ \vdots \\ \mathbf{CA}^{n_x-1} \end{bmatrix} \mathbf{T} = \mathbf{W}_0\mathbf{T}. \tag{5.92}$$

But $\operatorname{rank} \mathbf{W}_0 = \operatorname{rank} \mathbf{W}_0\mathbf{T}$ for any nonsingular $\mathbf{T}$. Hence, observability remains invariant under coordinate transformation.

Exercise 5.25

Show that the observability of time-varying systems remain invariant under time-varying coordinate transformation $\mathbf{x} = \mathbf{T}(t)\boldsymbol{x}$.

5.4.1 OBSERVABILITY OF MODAL COORDINATES

Consider the modal coordinates (5.29). Now, the matrix $\mathbf{W}_0$ defined by (5.88) is

$$\mathbf{W}_0 = \begin{bmatrix} \mathbf{C}_1 & \mathbf{C}_2 & \cdots & \mathbf{C}_p \\ \mathbf{C}_1\mathbf{\Lambda}_1 & \mathbf{C}_2\mathbf{\Lambda}_2 & & \mathbf{C}_p\mathbf{\Lambda}_p \\ \mathbf{C}_1\mathbf{\Lambda}_1^2 & \mathbf{C}_2\mathbf{\Lambda}_2^2 & & \mathbf{C}_p\mathbf{\Lambda}_p^2 \\ \vdots & & & \vdots \\ \mathbf{C}_1\mathbf{\Lambda}_1^{n_x-1} & \mathbf{C}_2\mathbf{\Lambda}_2^{n_x-1} & & \mathbf{C}_p\mathbf{\Lambda}_p^{n_x-1} \end{bmatrix}, \tag{5.93}$$

where Λ_i is described by (5.35) and

$$\mathbf{C}_i\Lambda_i = [\mathbf{C}_{i1}\lambda_i, \mathbf{C}_{i1} + \lambda_i\mathbf{C}_{i2}, \mathbf{C}_{i2} + \lambda_i\mathbf{C}_{i3}, \ldots], \qquad \mathbf{C}_i \triangleq [\mathbf{C}_{i1}, \mathbf{C}_{i2}, \mathbf{C}_{i3}, \ldots], \tag{5.94}$$

$$\mathbf{C}_i\Lambda_i^2 = [\mathbf{C}_{i1}\lambda_i^2, \mathbf{C}_i\lambda_i + \lambda_i(\mathbf{C}_{i1} + \lambda_i\mathbf{C}_{i2}), \mathbf{C}_{i1} + \lambda_i\mathbf{C}_{i2} + \lambda_i(\mathbf{C}_{i2} + \lambda_i\mathbf{C}_{i3}), \ldots]. \tag{5.95}$$

$$\vdots$$

etc.

Now, define $\mathbf{C}_{i1}$ as the *first* column of $\mathbf{C}_i$. Then, the following result can be proved [5.2].

Theorem 5.16

The system (5.29) *in Jordan form is observable if and only if*

$$[\mathbf{C}_{i1}, \mathbf{C}_{j1}, \ldots, \mathbf{C}_{k1}] \tag{5.96}$$

is a linearly independent set of vectors, where $\Lambda_i, \Lambda_j, \ldots, \Lambda_k$ *are Jordan blocks that are associated with the same eigenvalue.*

The proof follows similar steps as the proof of Theorem 5.10 and is omitted.

EXAMPLE 5.7

The system described by (5.29) with

$$\Lambda_1 = \begin{bmatrix} \lambda_1 & 1 & 0 \\ 0 & \lambda_1 & 1 \\ 0 & 0 & \lambda_1 \end{bmatrix}, \quad \Lambda_2 = \begin{bmatrix} \lambda_1 & 1 \\ 0 & \lambda_1 \end{bmatrix}, \quad \Lambda_3 = \begin{bmatrix} \lambda_3 & 1 \\ 0 & \lambda_3 \end{bmatrix}, \tag{5.97}$$

$$\mathbf{C}_1 = [\mathbf{C}_{11} \quad \mathbf{C}_{12} \quad \mathbf{C}_{13}], \quad \mathbf{C}_2 = [\mathbf{C}_{21} \quad \mathbf{C}_{22}], \quad C_3 = [\mathbf{C}_{31} \quad \mathbf{C}_{32}]$$

is observable iff (a) $[\mathbf{C}_{11}, \mathbf{C}_{21}]$ are linearly independent vectors and (b) $\mathbf{C}_{31}$ is linearly independent ($\mathbf{C}_{31} \neq 0$).

Corollary to Theorem 5.15

In cases where the Jordan form of $\mathbf{A}$ *is diagonal,* (5.29) *reduces to* (5.36), *which is observable iff* rank $\mathbf{C}_i = m_i$.

Exercise 5.26

Determine whether the system

$$\dot{\mathbf{x}} = \begin{bmatrix} 1 & 0 & 0 \\ 0 & 1 & 0 \\ 0 & 0 & 2 \end{bmatrix}\mathbf{x}, \qquad y = [1 \quad 2 \quad 3]\mathbf{x} \tag{5.98}$$

is observable.

See from Fig. 4.8 and equations (5.39) through (5.41) that the ith residue of the transfer function between $u(s)$ and $y(s)$ is zero if mode i is unobservable (Theorem 5.15). Hence, unobservable modes do not appear in the transfer function from $u(s)$ to $y(s)$ if this transfer function is in its simplest form.

Exercise 5.27

Can one determine whether the system (5.43) is observable? Explain.

EXAMPLE 5.8

Show that rank $\mathbf{W}_0 < n_x$ if $\mathbf{C}_i = 0$ for any i. Assume distinct eigenvalues.

Solution: Since $\mathbf{C} = [\mathbf{C}_1 \ldots \mathbf{C}_{nx}]$, $\mathbf{A} = \text{diag}[\ldots \lambda_i \ldots]$, then the matrix

$$\mathbf{W}_0 = \begin{bmatrix} \mathbf{C}_1 & \cdots & \mathbf{C}_{n_x} \\ \mathbf{C}_1\lambda_1 & \cdots & \mathbf{C}_{n_x}\lambda_{n_x} \\ \vdots & & \\ \mathbf{C}_1\lambda_1^{n_x-1} & \cdots & \mathbf{C}_{n_x}\lambda_{n_x}^{n_x-1} \end{bmatrix} \tag{5.99}$$

has a zero column if $\mathbf{C}_i = \mathbf{0}$ for any i. Hence, rank $\mathbf{W}_0 < n_x$. ■

Exercise 5.28

Determine the range of values for α, β, Ω, and γ that ensure observability for the following systems:

(a)

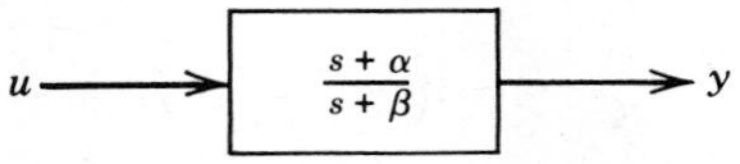

(b)

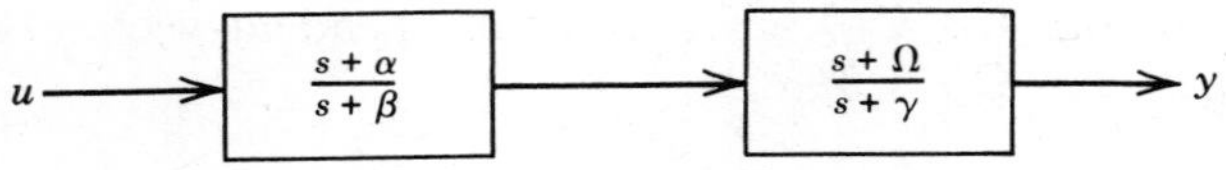

(c) Determine the angle between the vectors **C** and **CA**. Relate this angle to observability in (a) and (b).

5.4.2 OBSERVABILITY OF OBSERVABLE CANONICAL COORDINATES

From (5.89), the system with $\mathscr{Re}\, \lambda_i[\mathbf{A}] < 0$ is observable if $\mathbf{K} > \mathbf{0}$ where $\mathbf{K}$ satisfies

$$\mathbf{0} = \mathbf{KA} + \mathbf{A}^*\mathbf{K} + \mathbf{C}^*\mathbf{C}. \tag{5.100}$$

The choice of coordinates in this section is that which makes $\mathbf{K}$ diagonal. The observability tests in these coordinates proceed as follows:

STEP 1: Solve (5.100) for $\mathbf{K}$.

STEP 2: Compute the singular value decomposition of the symmetric matrix $\mathbf{K}$:

$$\mathbf{K} = \mathbf{N}\boldsymbol{\Sigma}\mathbf{N}^*, \qquad \mathbf{N}^*\mathbf{N} = \mathbf{I}. \tag{5.101}$$

STEP 3: Apply the coordinate transformation $\mathbf{x} = \mathbf{N}\boldsymbol{x}$ to get

$$\begin{aligned} \dot{\boldsymbol{x}} &= \mathbf{N}^*\mathbf{A}\mathbf{N}\boldsymbol{x}, \\ \mathbf{y} &= \mathbf{C}\mathbf{N}\boldsymbol{x} \end{aligned} \tag{5.102}$$

for which the observability grammian $\mathscr{K}$ satisfies

$$\begin{aligned} \mathscr{K} &= \int_0^\infty e^{\mathbf{N}^*\mathbf{A}^*\mathbf{N}t}\mathbf{N}^*\mathbf{C}^*\mathbf{C}\mathbf{N}e^{\mathbf{N}^*\mathbf{A}\mathbf{N}t}\,dt \\ &= \mathbf{N}^*\left[\int_0^\infty e^{\mathbf{A}^*t}\mathbf{C}^*\mathbf{C}e^{\mathbf{A}t}\,dt\right]\mathbf{N} = \mathbf{N}^*\mathbf{K}\mathbf{N}. \end{aligned} \tag{5.103}$$

Hence, from (5.101) and (5.103),

$$\mathscr{K} = \mathbf{N}^*(\mathbf{K})\mathbf{N} = \mathbf{N}^*(\mathbf{N}\boldsymbol{\Sigma}\mathbf{N}^*)\mathbf{N} = \Sigma, \tag{5.104}$$

which is a diagonal matrix. The observability test in the new coordinates is as follows:

Theorem 5.17
System (5.102) *is observable iff* $\sigma_i^2 > 0$ *for all* $i = 1, \ldots, n_x$ *where*

$$\boldsymbol{\Sigma} = \begin{bmatrix} \sigma_1 & & \\ & \ddots & \\ & & \sigma_{n_x} \end{bmatrix}. \tag{5.105}$$

5.4.3 OTHER OBSERVABLE CANONICAL COORDINATES

Instead of requiring special properties of the matrix **K**, one may require special properties of **A** and **C** as follows:

$$\begin{pmatrix} \dot{\mathbf{x}}_1 \\ \dot{\mathbf{x}}_2 \end{pmatrix} = \begin{bmatrix} \mathbf{A}_{11} & \mathbf{0} \\ \mathbf{A}_{21} & \mathbf{A}_{22} \end{bmatrix} \begin{pmatrix} \mathbf{x}_1 \\ \mathbf{x}_2 \end{pmatrix},$$

$$\mathbf{y} = [\mathbf{C}_1 \quad \mathbf{0}] \begin{pmatrix} \mathbf{x}_1 \\ \mathbf{x}_2 \end{pmatrix}, \tag{5.106}$$

whereupon the observability test matrix W_0 in (5.97) becomes

$$\mathbf{W}_0^* = \begin{bmatrix} \mathbf{C}_1^* & \mathbf{A}_{11}^*\mathbf{C}_1^* & \cdots & \mathbf{A}_{11}^{*n_x-1}\mathbf{C}_1^* \\ \mathbf{0} & \mathbf{0} & & \mathbf{0} \end{bmatrix} \tag{5.107}$$

The special matrix pair $(\mathbf{A}, \mathbf{C})$ in (5.106) is called Kalman's observable form if (A_{11}, C_1) is an observable pair. Note that (1) $\mathbf{x}_1$ is observable iff $(\mathbf{A}_{11}, \mathbf{C}_{11})$ is an observable pair; (2) $\mathbf{x}_2$ is unobservable $\forall \mathbf{A}_{21}, \mathbf{A}_{22}$.

A final simplification is as follows.

Theorem 5.18

The matrix pair $(\mathbf{A}, \mathbf{C})$ is observable iff

$$\text{rank} \begin{bmatrix} \mathbf{C} \\ \lambda\mathbf{I} - \mathbf{A} \end{bmatrix} = n_x \qquad \text{for all } \lambda. \tag{5.108}$$

Proof: First prove that the rank condition $\Rightarrow$ $(\mathbf{A}, \mathbf{C})$ observable. If the rank condition holds, then

$$\begin{bmatrix} \mathbf{C} \\ \lambda\mathbf{I} - \mathbf{A} \end{bmatrix} \mathbf{e} = \mathbf{0} \Rightarrow e = 0 \tag{5.109}$$

or

$$\mathbf{Ce}, \qquad \lambda\mathbf{e} = \mathbf{Ae} \Rightarrow \mathbf{e} = \mathbf{0}. \tag{5.110}$$

Hence,

$$\lambda\mathbf{Ce} = \mathbf{CAe} \tag{5.111}$$

and

$$\lambda\mathbf{C}\lambda\mathbf{e} = \lambda\mathbf{CAe} \Rightarrow \lambda\mathbf{CAe} = \mathbf{CA}^2\mathbf{e}, \tag{5.112}$$

and so forth,

$$\lambda \mathbf{C}\mathbf{A}^{n_x-2}\mathbf{e} = \mathbf{C}\mathbf{A}^{n_x-1}\mathbf{e} \Rightarrow \mathbf{e} = \mathbf{0}. \tag{5.113}$$

All together this leads to

$$\begin{bmatrix} \mathbf{C} \\ \mathbf{CA} \\ \vdots \\ \mathbf{CA}^{n_x-1} \end{bmatrix} \mathbf{e} = \mathbf{0} \Rightarrow \mathbf{e} = \mathbf{0}, \tag{5.114}$$

which implies rank $[\mathbf{C}^T, \mathbf{A}^T\mathbf{C}^T, \ldots, \mathbf{A}^{n_x-1^T}\mathbf{C}^T] = n_x$, which is the observability condition.

Now, prove that $(\mathbf{A}, \mathbf{C})$ unobservable $\Rightarrow$ a nonzero $\mathbf{e}$ exists above. Assume $(\mathbf{A}, \mathbf{C})$ in Kalman's observable canonical form,

$$\mathbf{A} = \begin{bmatrix} \mathbf{A}_{11} & \mathbf{0} \\ \mathbf{A}_{21} & \mathbf{A}_{22} \end{bmatrix}, \qquad \mathbf{C} = [\mathbf{C}_1 \quad \mathbf{0}],$$

where $(\mathbf{A}_{11}, \mathbf{C}_1)$ is observable. Choose $\mathbf{e}^T = (\mathbf{0}, \mathbf{e}_2^T)$ to see that

$$\mathbf{CA}^i = \left[\mathbf{CA}_{11}^i, \mathbf{0}\right].$$

Hence, (5.114) is satisfied for arbitrary $\mathbf{e}_2 \neq \mathbf{0}$, $\mathbf{e}^T = (\mathbf{0}, \mathbf{e}_2^T)$. This proves $(\mathbf{A}, \mathbf{C})$ observable $\Rightarrow$ (5.108). ■

The next section considers special structures of $(\mathbf{A}, \mathbf{C})$ of the form

$$\mathbf{A} = \begin{bmatrix} \mathbf{0} & \mathbf{I} \\ \mathbf{A}_{21} & \mathbf{A}_{22} \end{bmatrix}, \qquad \mathbf{C} = [\mathbf{C}_1 \quad \mathbf{C}_2]. \tag{5.115}$$

5.4.4 OBSERVABILITY OF VECTOR SECOND-ORDER SYSTEMS

The special system (5.58) is now of interest.

Nongyroscopic Systems

Now, consider the observability of

$$\mathscr{M}\ddot{\mathbf{q}} + \mathscr{K}\mathbf{q} = \mathbf{0}, \quad \mathbf{y} = \mathscr{P}\mathbf{q} + \mathscr{R}\dot{\mathbf{q}}, \tag{5.116a}$$

$$\mathscr{M} > \mathbf{0}, \quad \mathscr{K} > \mathbf{0}, \tag{5.116b}$$

where the inputs are suppressed with no loss in generality. The state equations (5.62b) apply here with the addition of the output equation yielding the observability test

$$\operatorname{rank} \mathbf{W}_0 = \operatorname{rank} \begin{bmatrix} \mathscr{P}' & \mathscr{R}' \\ -\mathscr{R}'\Omega^2 & \mathscr{P}' \\ -\mathscr{P}'\Omega^2 & -\mathscr{R}'\Omega^2 \\ \mathscr{R}'\Omega^4 & -\mathscr{P}'\Omega^2 \end{bmatrix} \overset{?}{=} 2N. \tag{5.117}$$

Note that by defining

$$\mathbf{F} = \begin{bmatrix} \mathscr{P}' & \mathscr{R}' \\ -\mathscr{R}'\Omega^2 & \mathscr{P}' \end{bmatrix}, \qquad \tilde{\Omega} \triangleq \begin{bmatrix} \Omega^2 & 0 \\ 0 & \Omega^2 \end{bmatrix}, \tag{5.118}$$

the matrix W_0 in (5.117) can be put into the form

$$\mathbf{W}_0 = \begin{bmatrix} \mathbf{F} \\ \mathbf{F}\tilde{\Omega} \\ \vdots \\ \mathbf{F}\tilde{\Omega}^{N-1} \end{bmatrix}. \tag{5.119}$$

Our goal is to use the simplified version of Theorem 5.16 which assumes that the Jordan form is diagonal. Since $\tilde{\Omega}$ in (5.118) is diagonal, we are in a position to state our result. However, the simplest form of the results require the repeated frequencies in $\tilde{\Omega}$ to be adjacent. To remedy this defect in $\tilde{\Omega}$, apply the orthonormal similarity transformation

$$\mathbf{T}_U = \left[\begin{array}{cccccc} \mathbf{I}_{m_1} & 0 & 0 & \cdot & \cdots & 0 \\ 0 & 0 & \mathbf{I}_{m_2} & \cdot & \cdots & 0 \\ 0 & 0 & 0 & \cdot & \cdots & \mathbf{I}_{m_p} \\ \hline 0 & \mathbf{I}_{m_1} & 0 & 0 & \cdots & 0 \\ 0 & 0 & 0 & \mathbf{I}_{m_2} & \cdots & 0 \\ & & & \cdot & & \\ 0 & 0 & 0 & 0 & \cdots & 0 \end{array}\right] \tag{5.120}$$

to get

$$\hat{\tilde{\Omega}} = \mathbf{T}_U^T\tilde{\Omega}\mathbf{T}_U = \begin{bmatrix} \omega_1^2\mathbf{I}_{2m_1} & & & \\ & \omega_2^2\mathbf{I}_{2m_2} & & \\ & & \ddots & \\ & & & \omega_{N_p}^2\mathbf{I}_{2m_p} \end{bmatrix}, \qquad \mathbf{F}\mathbf{T}_U = \hat{\mathbf{F}} = \left[\hat{\mathbf{F}}_1, \ldots, \hat{\mathbf{F}}_p\right], \tag{5.121}$$

where $\hat{\mathbf{F}}_i$ has $2m_i$ columns. In the same manner, $\mathscr{P}'$ and $\mathscr{R}'$ are partitioned:

$$\mathscr{P}' = \left[\mathscr{P}'_1, \ldots, \mathscr{P}'_p\right], \qquad \mathscr{R}' = \left[\mathscr{R}'_1, \ldots, \mathscr{R}'_p\right]. \tag{5.122}$$

Then, the $\hat{\mathbf{F}}_i$ in (5.121) has the form

$$\hat{\mathbf{F}}_i = \left[\begin{array}{c|c} \mathscr{P}'_i & \mathscr{R}'_i \\ \hline -\mathscr{R}'_i\omega_i^2 & \mathscr{P}'_i \end{array}\right]. \tag{5.123}$$

Now, the test

$$\operatorname{rank} \mathbf{W}_0 \mathbf{T}_U = \operatorname{rank} \begin{bmatrix} \hat{\mathbf{F}} \\ \hat{\mathbf{F}}\hat{\hat{\Omega}} \\ \vdots \\ \hat{\mathbf{F}}\hat{\hat{\Omega}}^{N-1} \end{bmatrix} \tag{5.124}$$

yields these results that follow directly from the corollary to Theorem 5.9.

Theorem 5.19

The system (5.116) *is observable iff*

$$\operatorname{rank} \hat{\mathbf{F}}_i = 2m_i, \qquad i = 1, \ldots, p. \tag{5.125}$$

If $m_i = 1$ for all i, then the matrices $\mathscr{P}'_i, \mathscr{R}'_i$ in (5.123) become vectors $\not{p}_i$, $\not{r}_i$, respectively. Hence, rank $\hat{\mathbf{F}}_i = 2$ iff $|\hat{\mathbf{F}}_i^T \hat{\mathbf{F}}_i| \neq 0$, and

$$\hat{\mathbf{F}}_i^T \hat{\mathbf{F}}_i = \begin{bmatrix} \not{p}_i^T & -\not{r}_i^T\omega_i^2 \\ \not{r}_i^T & \not{p}_i^T \end{bmatrix} \begin{bmatrix} \not{p}_i & \not{r}_i \\ -\not{r}_i\omega_i^2 & \not{p}_i \end{bmatrix} \tag{5.126}$$

gives

$$\left|\hat{\mathbf{F}}_i^T \hat{\mathbf{F}}_i\right| = \left(\|\not{p}_i\|^2 + \|\not{r}_i\|^2\omega_i^4\right)\left(\|\not{p}_i\|^2 + \|\not{r}_i\|^2\right) - \left(\not{p}_i^T\not{r}_i\right)^2\left(1 - \omega_i^2\right)^2. \tag{5.127}$$

Hence, the following holds:

Corollary to Theorem 5.19

If ω_i^2 *is distinct for all* i, *then* (5.116) *is observable iff*

$$\left(\|\not{p}_i\|^2 + \|\not{r}_i\|^2\omega_i^2\right)^2 + \left(1 - \omega_i^2\right)^2 \sin^2 \beta_i \|\not{p}_i\|^2 \|\not{r}_i\|^2 > 0, \tag{5.128}$$

for

$$i = 1, \ldots, N,$$

where β_i is defined by

$$\mathscr{p}_i^T \mathscr{r}_i = \|\mathscr{p}_i\| \, \|\mathscr{r}_i\| \cos \beta_i . \tag{5.129}$$

If, in addition, the ω_i^2 are all nonzero, then (5.128) *reduces to*

$$\|\mathscr{p}_i\|^2 + \|\mathscr{r}_i\|^2 > 0, \qquad i = 1, \ldots, N. \tag{5.130}$$

Exercise 5.29

Prove the corollary to Theorem 5.19. *Hint*: Substitute (5.129) into (5.127).

Theorem 5.18 provides yet another convenient condition for observability. The following results from substitution of

$$\mathbf{A} = \begin{bmatrix} \mathbf{0} & \mathbf{I} \\ -\mathscr{M}^{-1}\mathscr{K} & \mathbf{0} \end{bmatrix}, \qquad \mathbf{C} = [\mathscr{P} \quad \mathscr{R}], \tag{5.131}$$

into (5.108).

Theorem 5.20

The ith mode of the system (5.116) *is observable iff rank*

$$\begin{bmatrix} \mathscr{M}\lambda_i^2 + \mathscr{K} \\ \mathscr{P} + \lambda_i \mathscr{R} \end{bmatrix} = N. \tag{5.132}$$

The potential advantage of Theorems 5.13, 5.18, 5.19, and 5.20 is that the observability or controllability of any mode may be determined without use of the eigenvectors. Only the eigenvalues and the system matrices are needed.

5.5 *Adjoint Systems and Duality*

The linear system is state controllable if $\mathbf{X}(t_0, t_f)$ in (5.10a) is positive definite and observable if $\mathbf{K}(t_0, t_f)$ in (5.80) is positive definite. By comparison of these two equations it is clear that if one substitutes column I for column II

I	II
$\mathbf{\Phi}^*(t, \cdot)$	$\mathbf{\Phi}(\cdot, t)$
$\mathbf{C}^*(t)$	$\mathbf{B}(t)$
t_0	t_f

(5.133)

to convert matrix $\mathbf{K}(t_0, t_f)$ in (5.80) into the matrix $\mathbf{X}(t_0, t_f)$. Now, if $\mathbf{\Phi}^*(t, \cdot)$

replaces $\Phi(\cdot, t)$ what replaces $\mathbf{A}(t)$? That is, if $\Phi(t, \cdot)$ satisfies (4.83),

$$\dot{\Phi}^*(t, \cdot) = \Phi^*(t, \cdot)\mathbf{A}^*(t), \tag{5.134}$$

what differential equation does $\Phi(\cdot, t)$ obey? The inversion property 2 of Theorem 4.6 shows that

$$\Phi^{-1}(t, \cdot) = \Phi(\cdot, t) \tag{5.135}$$

and

$$\frac{d}{dt}\Phi(\cdot, t) = -\Phi(\cdot, t)\mathbf{A}(t). \tag{5.136}$$

Hence, the comparison of (5.134) and (5.136) shows that the table (5.133) is completed by substituting $-\mathbf{A}^*(t)$ for $\mathbf{A}(t)$. To summarize:

Theorem 5.21 (duality)
The system described by

$$\begin{aligned} \dot{\mathbf{x}} &= \mathbf{A}(t)\mathbf{x} + \mathbf{B}(t)\mathbf{u}, \\ \mathbf{y} &= \mathbf{C}(t)\mathbf{x} \end{aligned} \tag{5.137}$$

is state controllable (observable) if and only if the system

$$\begin{aligned} \dot{\mathscr{x}} &= -\mathbf{A}^*(t)\mathscr{x} + \mathbf{C}^*(t)\mathscr{u}, \\ \mathscr{y} &= \mathbf{B}^*(t)\mathscr{x} \end{aligned} \tag{5.138}$$

is observable (controllable).

Exercise 5.30
Show that the unforced responses of (5.137) and (5.138) have the property that

$$\mathbf{x}^*(t)\mathscr{x}(t) = \text{constant } \forall t.$$

Equation (5.138) is called the "adjoint" system associated with (5.137).

5.6 *On Relative Controllability, Observability*

Recall in (4.62) through and in Fig. 4.8 and in (5.42b) that the residue was factored into the product of two terms $\mathbf{c}_i$ and $\mathbf{b}_i^*$. Such factorizations are *arbitrary* to within a scalar; see for single-input/single-output systems that for the given unique residue

$\mathbf{J}_i$ that c_i and b_i can be scaled,

$$c_i = \frac{J_i}{b_i^*} \qquad \text{for } \textit{arbitrary } b_i \neq 0. \tag{5.139}$$

Now, the corollary to Theorem 5.9 gives the controllability test $\|\mathbf{b}_i\|^2 \triangleq \mathbf{b}_i^* \mathbf{b}_i \overset{?}{>} 0$. Hence, the reader might be tempted to interpret the magnitude of $\|\mathbf{b}_i\|^2$ and $\|\mathbf{c}_i\|^2$ as a measure of the *relative degree of* controllability and observability of mode i. This would be a mistake since $\mathbf{c}_i$ and $\mathbf{b}_i$ have arbitrary lengths. This point is also clear from the definitions of $\mathbf{c}_i$ and $\mathbf{b}_i^*$ in (4.59), $\mathbf{c}_i = \mathbf{C}\mathbf{e}_i$, $\mathbf{b}_i^* = \boldsymbol{\ell}_i^* \mathbf{B}$, where $\mathbf{e}_i$ and $\boldsymbol{\ell}_i^*$ are the right and left eigenvectors, respectively. Since eigenvectors may have any length, the norm of $\mathbf{c}_i$ or $\mathbf{b}_i$ has no physical significance. It may be noted, however, that the residue is unique and hence so is its norm

$$\begin{aligned}
\|\mathbf{J}_i\|^2 &= \operatorname{tr} \mathbf{J}_i^* \mathbf{J}_i = \operatorname{tr} (\mathbf{c}_i \mathbf{b}_i^*)^* (\mathbf{c}_i \mathbf{b}_i^*) \\
&= \operatorname{tr} \mathbf{b}_i \mathbf{c}_i^* \mathbf{c}_i \mathbf{b}_i^* \\
&= \operatorname{tr} \mathbf{c}_i^* \mathbf{c}_i \mathbf{b}_i^* \mathbf{b}_i \\
&= \|\mathbf{c}_i\|^2 \|\mathbf{b}_i\|^2.
\end{aligned} \tag{5.140}$$

Thus, even though *neither* $\|\mathbf{c}_i\|^2$ nor $\|\mathbf{b}_i\|^2$ is unique, their *product* is unique.

5.6.1 MODAL COST ANALYSIS

The residue $\mathbf{J}_i$ determines the contribution of mode i in the transfer function. Sometimes, it is important to determine the contribution of mode i in the scalar cost function. This contribution will be called the "modal cost." We shall compute the modal costs for the system

$$\begin{aligned}
\dot{x}_1 &= \lambda_i x_i + \mathbf{b}_i^* \mathbf{u}, \qquad i = 1, \ldots, n_x, \quad u_\alpha(t) = \mu_\alpha \delta(t), \\
\mathbf{y} &= \sum_{i=1}^{n_x} \mathbf{c}_i x_i, \qquad x_i(0) = 0,
\end{aligned} \tag{5.141}$$

where the cost function is

$$\mathscr{V} = \sum_{\alpha=1}^{n_u} \int_0^\infty \mathbf{y}^{\alpha*}(t) \mathbf{Q} \mathbf{y}^\alpha(t)\, dt, \tag{5.142}$$

where $\mathbf{y}^\alpha(t)$ is the response to $u_\alpha(t) = \mu_\alpha \delta(t)$. We assume, $\mathscr{Re}\, \lambda_i < 0\ \forall i$, $\lambda_i \neq \lambda_j$, throughout this section, even though the underlying concepts are not so limited.

The "modal cost" $\mathscr{V}_{x_i}$ computation is given by (4.175) when x_i is a "modal" coordinate,

$$\mathscr{V}_{x_i} = [\mathbf{XC}^*\mathbf{QC}]_{ii}, \qquad \mathbf{0} = \mathbf{XA}^* + \mathbf{AX} + \mathbf{B}\mathscr{U}\mathbf{B}^*. \tag{5.143}$$

Now, substitute $\mathbf{A} = \text{diag}[\ldots \lambda_i \ldots]$ and solve for the ij element of the equation for $\mathbf{X}$:

$$X_{ij} = \frac{-\mathbf{b}_i^* \mathscr{U} \mathbf{b}_j}{\bar{\lambda}_j + \lambda_i}, \qquad X_{ii} = \frac{-\mathbf{b}_i^* \mathscr{U} \mathbf{b}_i}{2\,\mathscr{Re}\,\lambda_i}. \tag{5.144}$$

Hence,

$$\mathscr{V}_{x_i} = \sum_{j=1}^{n_x} \frac{-\mathbf{b}_i^* \mathscr{U} \mathbf{b}_j \mathbf{c}_j^* \mathbf{Q} \mathbf{c}_i}{\bar{\lambda}_j + \lambda_i} \tag{5.145a}$$

or, equivalently,

$$\mathscr{V}_{x_i} = -\sum_{j=1}^{n_x} \frac{\operatorname{tr} \mathbf{b}_i^* \mathscr{U} \mathbf{b}_j \mathbf{c}_j^* \mathbf{Q} \mathbf{c}_i}{\bar{\lambda}_j + \lambda_i} = -\sum_{j=1}^{n_x} \frac{\operatorname{tr} \mathbf{c}_i \mathbf{b}_i^* \mathscr{U} (\mathbf{c}_j \mathbf{b}_j^*)^* \mathbf{Q}}{\bar{\lambda}_j + \lambda_i}.$$

But since the residue is related to the parameters $\mathbf{c}_i$ and $\mathbf{b}_i$ by

$$\mathbf{J}_i = \mathbf{c}_i \mathbf{b}_i^*,$$

then the modal cost in terms of the residue matrices is

$$\mathscr{V}_{x_i} = -\sum_{j=1}^{n_x} \frac{\operatorname{tr} \mathbf{J}_i \mathscr{U} \mathbf{J}_j^* \mathbf{Q}}{\lambda_i + \bar{\lambda}_j}. \tag{5.145b}$$

Exercise 5.31

Show that $\mathscr{V}_{x(i+1)} = \bar{\mathscr{V}}_{x_i}$ if $\lambda_{i+1} = \bar{\lambda}_i$.

Now, let $\lambda_{i+1} = \bar{\lambda}_i$. For this complex set of eigenvalues, we shall compute only one cost for the entire second-order mode. This cost is $\mathscr{V}_{x_i} + \mathscr{V}_{x(i+1)} = 2\,\mathscr{Re}\,\mathscr{V}_{x_i}$ by Exercise 5.31. This expression is especially simple when the mode is very lightly damped, as shown by the following.

Theorem 5.22

For the asymptotically stable, nondefective system

$$\dot{x}_i = \lambda_i x_i + \mathbf{b}_i^* \mathbf{u}, \qquad \mathbf{y} = \sum_{i=1}^{n_x} \mathbf{c}_i x_i, \tag{5.146a}$$

$i = 1, 2, \ldots, n_x$, define ζ_i, ω_i by

$$\lambda_i = -\zeta_i\omega_i + j\omega_i\sqrt{1-\zeta_i^2}, \qquad \zeta_i > 0, \quad \omega_i > 0, \tag{5.146b}$$

and let $\lambda_{i+1} = \bar{\lambda}_i$ for some i. Suppose $\mathbf{b}_i \neq \mathbf{0}$, $\mathbf{c}_i \neq \mathbf{0}$. Then, if either (i), (ii) or (iii) below holds,

(i) $$\mathbf{b}_i^* \mathscr{U} \mathbf{b}_k = 0, \qquad k \neq i,$$

(ii) $$\mathbf{c}_k^* \mathbf{Q} \mathbf{c}_i = 0, \qquad k \neq i,$$

(iii) $$\zeta_i \ll \frac{|\omega_i - \omega_k|}{2\omega_i} \qquad \forall k \neq i,$$

the ith modal cost is given by

$$\mathscr{V}_{x_i} = \mathscr{V}_{x(i+1)} = \frac{\|\mathbf{b}_i\|_{\mathscr{U}}^2 \|\mathbf{c}_i\|_{\mathbf{Q}}^2}{2\zeta_i\omega_i} + 0(\zeta_i), \tag{5.147}$$

where $0(\zeta_i)$ is zero under conditions (i) or (ii) and is of order ζ_i otherwise.

Proof: Write (5.145a):

$$\mathscr{V}_{x_i} = \frac{-\mathbf{b}_i^* \mathscr{U} \mathbf{b}_i \mathbf{c}_i^* \mathbf{Q} \mathbf{c}_i}{\lambda_i + \bar{\lambda}_i} - \sum_{k=1 \neq i}^{n_x} \frac{\mathbf{b}_i^* \mathscr{U} \mathbf{b}_k \mathbf{c}_k^* \mathbf{Q} \mathbf{c}_i}{\lambda_i + \bar{\lambda}_k}, \tag{5.148}$$

where, from (5.146),

$$\lambda_i + \bar{\lambda}_i = -2\zeta_i\omega_i, \qquad \lambda_i + \bar{\lambda}_k = -(\zeta_i\omega_i + \zeta_k\omega_k) + j(\omega_i - \omega_k). \tag{5.149}$$

The magnitude of the first term in $\mathscr{V}_{x_i}$ is much larger than the second iff

$$\frac{\|\mathbf{b}_i\|_{\mathscr{U}}^2 \|\mathbf{c}_i\|_{\mathscr{Q}}^2}{2\zeta_i\omega_i} \gg \frac{|\mathbf{b}_i^* \mathscr{U} \mathbf{b}_k \mathbf{c}_k^* \mathbf{Q} \mathbf{c}_i|}{\left[(\zeta_i\omega_i + \zeta_k\omega_k)^2 + (\omega_i - \omega_k)^2\right]^{1/2}} \qquad \forall k \neq i. \tag{5.150}$$

This is obviously true if (i) or (ii) holds, assuming mode i is both controllable and observable ($b_i \neq 0$, $c_i \neq 0$). The inequality is also true if

$$(2\zeta_i\omega_i)^2 \ll (\zeta_i\omega_i + \zeta_k\omega_k)^2 + (\omega_i - \omega_k)^2. \tag{5.151}$$

This is satisfied if

$$(2\zeta_i\omega_i)^2 \ll (\omega_i - \omega_k)^2 \tag{5.152}$$

or

$$\zeta_i \ll \frac{|\omega_i - \omega_k|}{2\omega_i} \qquad \forall k \neq i, \tag{5.153}$$

which verifies (iii) and $\mathscr{V}_{x_i}$ in (5.147). To verify the $\mathscr{V}_{x(i+1)}$ expression, substitute $\bar{\lambda}_i$ for λ_i in the above proof, since $\lambda_{i+1} = \bar{\lambda}_i$. ■

Note that even though the magnitudes $\|\mathbf{b}_i\|$ and $\|\mathbf{c}_i\|$ *separately* have no meaning (as discussed earlier), their *product* is unique (and the residues $\mathbf{J}_i$ are unique). Hence, the modal cost (5.147) is unique.

Now, we apply Theorem 5.22 to vector second-order systems of the form

$$\mathscr{M}\ddot{\mathbf{q}} + \mathscr{D}\dot{\mathbf{q}} + \mathscr{K}\mathbf{q} + \mathscr{B}\mathbf{u}, \quad \mathbf{y} = \mathscr{P}\mathbf{q} + \mathscr{R}\dot{\mathbf{q}}, \qquad \mathbf{q} \in \mathscr{R}^N, \tag{5.154}$$

where

$$\mathscr{M} = \mathscr{M}^T > \mathbf{0}, \quad \mathscr{K} = \mathscr{K}^T \geq \mathbf{0}, \quad \mathscr{D} = \alpha\mathscr{M} + \beta\mathscr{K}$$

for some scalars α, β. The Jordan form of the state space model of this system was developed in (4.115) through (4.119), using $\mathbf{q} = \mathscr{E}\eta$, yielding

$$\dot{\mathbf{x}} = \begin{bmatrix} \Lambda_c & \mathbf{0} \\ \mathbf{0} & \bar{\Lambda}_c \end{bmatrix}\mathbf{x} + \begin{bmatrix} \mathbf{I} & \mathbf{I} \\ \Lambda_c & \bar{\Lambda}_c \end{bmatrix}^{-1}\begin{bmatrix} \mathbf{0} \\ \mathscr{E}^T\mathscr{B} \end{bmatrix}u,$$
$$\mathbf{y} = [\mathscr{P}\mathscr{E} \quad \mathscr{R}\mathscr{E}]\begin{bmatrix} \mathbf{I} & \mathbf{I} \\ \Lambda_c & \bar{\Lambda}_c \end{bmatrix}\mathbf{x}, \quad \begin{pmatrix} \eta \\ \dot{\eta} \end{pmatrix} = \begin{bmatrix} \mathbf{I} & \mathbf{I} \\ \Lambda_c & \bar{\Lambda}_c \end{bmatrix}\mathbf{x}, \tag{5.155}$$

where

$$\mathscr{E}^T\mathscr{M}\mathscr{E} = \mathbf{I}, \quad \mathscr{E}^T\mathscr{K}\mathscr{E} = \Omega^2, \quad \mathscr{E}^T\mathscr{D}\mathscr{E} = 2\zeta\Omega$$

and

$$\Lambda_c \triangleq -\zeta\Omega + j\Omega[\mathbf{I} - \zeta^2]^{1/2}, \qquad \zeta = \operatorname{diag}[\ldots\zeta_i\ldots],$$
$$\bar{\Lambda}_c \triangleq -\zeta\Omega - j\Omega[\mathbf{I} - \zeta^2]^{1/2}, \qquad \Omega = \operatorname{diag}[\ldots\Omega_i\ldots].$$

Note that any transformation which diagonalizes $\mathscr{M}$ and $\mathscr{K}$ will also diagonalize $\mathscr{D} = \alpha\mathscr{M} + \beta\mathscr{K}$. This assumed structure of $\mathscr{D}$ is called Rayleigh damping [5.17]. We have assumed that there are no rigid body modes ($\omega_i > 0 \ \forall i$), and we shall maintain this assumption throughout this section. Otherwise, generalized eigenvectors are needed which complicate our discussion of the vector second-order case. More details may be found in ref. [5.5]. The inverse required in (5.155) is given by

(3.85). Using this result, the Jordan form becomes

$$\dot{\mathbf{x}} = \begin{bmatrix} \Lambda_c & \mathbf{0} \\ \mathbf{0} & \overline{\Lambda}_c \end{bmatrix} \mathbf{x} + \begin{bmatrix} -\mathbf{I} \\ \mathbf{I} \end{bmatrix} [\overline{\Lambda}_c - \Lambda_c]^{-1} \mathscr{E}^T \mathscr{B} u,$$

$$\mathbf{y} = [\mathscr{P}\mathscr{E} + \mathscr{R}\mathscr{E}\Lambda_c, \mathscr{P}\mathscr{E} + \mathscr{R}\mathscr{E}\overline{\Lambda}_c] x. \tag{5.156}$$

Exercise 5.32

Show that the transfer matrix of (5.156) may be written

$$\mathbf{G}(s) = \left\{ -(\mathscr{P}\mathscr{E} + \mathscr{R}\mathscr{E}\Lambda_c)(s\mathbf{I} - \Lambda_c)^{-1} + (\mathscr{P}\mathscr{E} + \mathscr{R}\mathscr{E}\overline{\Lambda}_c)(s\mathbf{I} - \overline{\Lambda}_c)^{-1} \right\}$$

$$\times (\overline{\Lambda}_c - \Lambda_c)^{-1} \mathscr{E}^T \mathscr{B} \tag{5.157a}$$

and show that if $\not{p}_i$, $\not{r}_i$, $\not{b}_i$ denote the ith column of $\mathscr{P}\mathscr{E}, \mathscr{R}\mathscr{E}, \mathscr{B}^T\mathscr{E}$, respectively, then $\mathbf{G}(s)$ may be written

$$\mathbf{G}(s) = \sum_{i=1}^{N} \frac{(s\not{r}_i + \not{p}_i)\not{b}_i^*}{s^2 + 2\zeta_i \omega_i s + \omega_i^2}. \tag{5.157b}$$

Now, we may apply (5.147) to obtain the modal cost of any lightly damped mode. The modal cost $\mathscr{V}_{x_i}$ and $\mathscr{V}_{x(i+1)}$ when $\lambda_{i+1} = \overline{\lambda}_i$ and $\zeta_i \ll |\omega_i - \omega_k|/2\omega_i$ is given by

$$\mathscr{V}_{x_i} = \frac{\mathbf{b}_i^* \mathscr{U} \mathbf{b}_i \mathbf{c}_i^* \mathbf{Q} \mathbf{c}_i}{2\zeta_i \omega_i}. \tag{5.158}$$

Now, the relationships between the $\mathbf{b}_i$, $\mathbf{c}_i$ in (4.146) and the $\not{b}_i$, $\not{p}_i$, $\not{r}_i$ in (5.156, 5.157) are

$$\mathbf{b}_i^* = \frac{\not{b}_i^*}{2j\omega_i\sqrt{1-\zeta_i^2}}, \qquad \text{hence, } \mathbf{b}_i = \frac{\not{b}_i}{-2j\omega_i\sqrt{1-\zeta_i^2}},$$

$$\mathbf{c}_i = \not{p}_i + \not{r}_i\left(-\zeta_i\omega_i + j\omega_i\sqrt{1-\zeta_i^2}\right), \qquad \text{hence, } \mathbf{c}_i^* = \not{p}_i^* + \not{r}_i^*\left(-\zeta_i\omega_i - j\omega_i\sqrt{1-\zeta_i^2}\right).$$

After these substitutions, (5.158) becomes,

$$\mathscr{V}_{x_i} = \frac{\not{b}_i^* \mathscr{U} \not{b}_i [\not{p}_i^* \mathbf{Q} \not{p}_i + \not{r}_i^* \mathbf{Q} \not{r}_i \omega_i^2 - 2\zeta_i\omega_i \not{p}_i^* \mathbf{Q} \not{r}_i]}{8\zeta_i\omega_i^3(1-\zeta_i^2)}. \tag{5.159}$$

Likewise for $i = N+1, N+2, \ldots, 2N$, (5.156) yields

$$\mathbf{b}_i^* = \frac{\mathcal{b}_i^*}{-2j\omega_i\sqrt{1-\zeta_i^2}}, \qquad \text{hence, } \mathbf{b}_i = \frac{\mathcal{b}_i}{2j\omega_i\sqrt{1-\zeta_i^2}},$$

$$\mathbf{c}_i = \mathcal{p}_i + \mathcal{r}_i\left(-\zeta_i\omega_i - j\omega_i\sqrt{1-\zeta_i^2}\right), \qquad \text{hence, } \mathbf{c}_i^* = \mathcal{p}_i^* + \mathcal{r}_i^*\left(-\zeta_i\omega_i + j\omega_i\sqrt{1-\zeta_i^2}\right),$$

leading to $\mathscr{V}_{x(i+1)} = \mathscr{V}_{x_i}$ as given by (5.159). Now, the conditions of Theorem 5.22 lead to the following.

Corollary to Theorem 5.22

For the asymptotically stable system described by

$$\ddot{\eta}_i + 2\zeta_i\omega_i\dot{\eta}_i + \omega_i^2\eta_i = \mathcal{b}_i^*\mathbf{u}, \qquad i = 1, 2, \ldots, N,$$

$$\mathbf{y}_p = \sum_{i=1}^{N}(p_i\eta_i), \tag{5.160}$$

$$\mathbf{y}_r = \sum_{i=1}^{N}(r_i\dot{\eta}_i),$$

the modal cost associated with $\eta_i(t)$ for the cost function

$$\mathscr{V} = \sum_{i=1}^{n_u}\int_0^\infty \mathbf{y}^{\alpha*}(t)\mathbf{Q}\mathbf{y}^{\alpha}(t)\,dt, \quad \mathbf{y}^* \triangleq (\mathbf{y}_p^*, \mathbf{y}_r^*), \quad \mathbf{Q} = \begin{bmatrix}\mathbf{Q}_p & \mathbf{0}\\ \mathbf{0} & \mathbf{Q}_r\end{bmatrix}, \quad \mathcal{p}_i = \begin{pmatrix}\mathbf{p}_i\\ \mathbf{0}\end{pmatrix}, \quad \mathcal{r}_i = \begin{pmatrix}\mathbf{0}\\ \mathbf{r}_i\end{pmatrix}$$

with impulsive inputs $u_\alpha(t) = \mathcal{u}_\alpha\delta(t)$, is given by

$$\mathscr{V}_{\eta i} \cong \frac{\mathcal{b}_i^*\,\mathscr{U}\,\mathcal{b}_i\left(\mathbf{p}_i^*\mathbf{Q}_p\mathbf{p}_i + \mathbf{r}_i^*\mathbf{Q}_r\mathbf{r}_i\omega_i^2\right)}{4\zeta_i\omega_i^3}, \tag{5.161}$$

where $\mathscr{U}_{\alpha\beta} = 0$, $\alpha \neq \beta$, $\mathscr{U}_{\alpha\alpha} = \mathcal{u}_\alpha^2$.

Proof: Coordinate $\eta_i(t)$ of the system (5.160) is associated with system eigenvalues $-\zeta_i\omega_i \pm j\omega_i\sqrt{1-\zeta_i^2}$. The modal cost associated with the complex eigenvalue is given by (5.154). Hence, the sum of the costs associated with each complex pair of eigenvalues is given by the sum of $\mathscr{V}_{x_i}$ and $\mathscr{V}_{x(i+1)}$ in (5.154), where $\mathscr{V}_{x_i}$ simplifies to (5.159) under the conditions of the corollary. Furthermore, the last term in (5.159) is zero since

$$\mathcal{p}_i^*\mathbf{Q}\mathcal{r}_i = [\mathbf{p}_i^* \;\; \mathbf{0}]\begin{bmatrix}\mathbf{Q}_p & \mathbf{0}\\ \mathbf{0} & \mathbf{Q}_r\end{bmatrix}\begin{pmatrix}\mathbf{0}\\ \mathbf{r}_i\end{pmatrix} = 0.$$

Hence, the facts $\mathscr{V}_{\eta_i} = \mathscr{V}_{x_i} + \mathscr{X}_{x(i+1)}$ and $\mathscr{V}_{x_i} = \mathscr{V}_{x(i+1)}$ and (5.159) lead immediately to result (5.161). It should be noted that the relationship between the

Jordan modal coordinates x_i in (5.155) and the second-order modes of (5.160) is given by

$$\begin{pmatrix} \boldsymbol{\eta} \\ \dot{\boldsymbol{\eta}} \end{pmatrix} = \begin{bmatrix} \mathbf{I} & \mathbf{I} \\ \Lambda_c & \bar{\Lambda}_c \end{bmatrix} \mathbf{x}. \tag{5.162}$$

Hence $\eta_i = x_i + x_{i+1}$, where x_i is the ith element of x and x_{i+1} is the $(N + i)$th element of x. ■

Exercise 5.33

For the system (5.160), show that the contribution of the αth input in the modal cost (5.161) is

$$\mathscr{V}_{\eta_i}^{u_\alpha} = b_{i\alpha} u_\alpha^2 \frac{p_i^* Q_p p_i + r_i^* Q_r r_i \omega_i^2}{4\zeta_i \omega_i^3}, \tag{5.163}$$

where

$$b_{i\alpha} \triangleq [\mathscr{E}^T \mathscr{B}]_{i\alpha} = \mathbf{e}_i^T \mathscr{B}_{\alpha\,\mathrm{col}}.$$

Note that (5.163) gives the effect of the αth input on the ith mode.

EXAMPLE 5.9

Do a modal cost analysis of the simply supported Euler–Bernoulli beam of length L and uniform mass density ρ, with an impulsive force applied at position r_c and an output deflection at r_0. Let $Q_p = 1$.

Solution: The beam is modeled in (3.53) through (3.79) and summarized here (modal damping added):

$$\ddot{\eta}_i + 2\zeta_i \omega_i \dot{\eta}_i + \omega_i^2 \eta_i = b_i^* \mathbf{u},$$
$$\mathbf{y} = \sum_{i=1}^{N} \mathbf{p}_i \eta_i, \tag{5.164}$$

where, for $i = 1, 2, \ldots, \infty$,

$$b_i^* \triangleq \Psi_i(r_c) = \sqrt{\frac{2}{\rho L}} \sin\left(r_c i \frac{\pi}{L}\right) \quad \text{if } u \text{ is a force input,}$$

$$b_i^* \triangleq \Phi_i(r_c) = \frac{\pi i}{L} \sqrt{\frac{2}{\rho L}} \cos\left(r_c i \frac{\pi}{L}\right) \quad \text{if } u \text{ is a torque input,}$$

$$p_i \triangleq \Psi_i(r_0) = \sqrt{\frac{2}{\rho L}} \sin\left(r_0 i \frac{\pi}{L}\right),$$

$$\omega_i = \sqrt{\frac{EI}{\rho}} \left(\frac{i\pi}{L}\right)^2.$$

Now, the use of (5.161) yields the modal costs

$$\mathscr{V}_{\eta_i} = u^2 \frac{2}{\rho L} \frac{\sin^2(r_c i\pi/L)}{4\zeta_i \left[\sqrt{\frac{EI}{\rho}}\left(\frac{i\pi}{L}\right)^2\right]^3} \left(\frac{2}{\rho L}\sin^2\left(r_0 i\frac{\pi}{L}\right)\right)$$

$$= \left[\frac{u^2 L^4}{\rho^{1/2}(EI)^{3/2}\pi^6}\right]\left[\frac{\sin^2(r_c i\pi/L)\sin^2\left(r_0 i\frac{\pi}{L}\right)}{\zeta_i i^6}\right]. \tag{5.165}$$

Now, the use of the Rayleigh damping model (4.114) yields

$$\zeta_i = \frac{\bar{\zeta}}{\omega_i(\omega_1+\omega_2)}\left[\omega_1\omega_2 + \omega_i^2\right]$$

$$= \frac{\bar{\zeta}(EI/\rho)(\pi/L)^4[2^2+i^4]}{i^2(1+2^2)(EI/\rho)(\pi/L)^4} = \frac{\bar{\zeta}(4+i^4)}{5i^2}. \tag{5.166}$$

This yields the modal cost

$$\mathscr{V}_{\eta_i} = \frac{5u^2L^4}{\sqrt{\rho(EI)^3}\,\pi^6}\left[\frac{\sin^2(r_c i\pi/L)\sin^2(r_0 i\pi/L)}{\bar{\zeta}_i^4(4+i^4)}\right]. \tag{5.167}$$

■

Exercise 5.34

(a) Determine the five most critical modes of the Euler–Bernoulli beam (3.80) using a damping model $\zeta_i = \bar{\zeta} = \text{const} = 0.005$. [Use (5.165).]

(b) Determine the five most critical modes of the beam (3.80) using the Rayleigh damping model. [Use (5.167).] Compare your two results and discuss the sensitivity of the modal costs to the damping model.

Since $|\sin(\cdot)| \le 1$, the modal costs in (5.165) may easily be bounded:

$$|\mathscr{V}_{\eta_i}| \le \frac{u^2L^4}{\sqrt{\rho(EI)^3}\,\pi^6 i^6 \zeta_i}. \tag{5.168}$$

Also note that

$$\sum_{i=1}^{\infty}\mathscr{V}_{\eta_i} \le \sum_{i=1}^{\infty}|\mathscr{V}_{\eta_i}| \le \sum_{i=1}^{\infty}\frac{u^2L^4}{\sqrt{\rho(EI)^3}\,(\pi i)^6\zeta_i}.$$

We shall now determine the number of modes N to be retained in the model (5.164) to guarantee that the error in the finite sum $\Sigma_{i=1}^{N}|\mathscr{V}_{\eta_i}|$ is smaller than a specified error, $\varepsilon > 0$, with respect to the infinite sum. First, let $\zeta_i = \bar{\zeta} =$ constant. Find N such that

$$\frac{\displaystyle\sum_{i=1}^{N} \frac{u^2L^4}{\sqrt{\rho(EI)^3}}(\pi i)^6\bar{\zeta}}{\displaystyle\sum_{i=1}^{\infty} \frac{u^2L^4}{\sqrt{\rho(EI)^3}}(\pi i)^6\bar{\zeta}} \geq (1-\varepsilon). \tag{5.169a}$$

From (5.170), this reduces to the determination of N such that,

$$\sum_{i=1}^{N} \frac{1}{i^6} \leq (1-\varepsilon) \sum_{i=1}^{\infty} \frac{1}{i^6}. \tag{5.169b}$$

Now, the value of N can be determined as follows. From algebra,

$$\beta \triangleq \sum_{i=1}^{\infty} i^{-6} = \frac{\pi^6}{945}.$$

Write (5.169b) as

$$0 \geq -\sum_{i=1}^{N} \frac{1}{i^6} + \sum_{i=1}^{N} \frac{1}{i^6} + \sum_{i=N+1}^{\infty} \frac{1}{i^6} - \varepsilon \sum_{i=1}^{\infty} \frac{1}{i^6}$$

$$= \sum_{N+1}^{\infty} \frac{1}{i^6} - \varepsilon\beta. \tag{5.170}$$

But

$$\sum_{i=N+1}^{\infty} \frac{1}{i^6} \leq \int_{N+1}^{\infty} \frac{dx}{x^6} = \frac{(N+1)^{-5}}{5}. \tag{5.171}$$

Hence, from (5.170) and (5.171),

$$0 \geq \frac{(N+1)^{-5}}{5} - \varepsilon\beta$$

or

$$N \geq -1 + (5\varepsilon\beta)^{-1/5} = -1 + \left(\frac{5\varepsilon\pi^6}{945}\right)^{-1/5}$$

TABLE 5.1 Number of Modes Required in Beam Model to Satisfy Equation (5.169)

ε	N
0.1	1
0.01	1
0.001	2
0.0001	4
0.00001	7

Thus, N is the smallest integer equal to or greater than the value on the right-hand side. A table of values of N versus ε is given in Table 5.1, taken from ref. [5.21].

More sophisiticated methods of model reduction are given in Chapter 6.

EXAMPLE 5.10

Find the modal costs for the system described by (3.80).

Solution: From (3.80),

$$\omega_i = i^2, \quad p_i = \Psi_i(r_0) = \sin(0.45\pi i), \quad r_i = 0, \quad \ell_i = i. \qquad \blacksquare$$

From (5.161),

$$\mathscr{V}_{\eta_i} = \frac{u^2 i \sin^2(0.45\pi i)}{4\zeta_i i^6} = \frac{50}{i^4}\sin^2(0.45\pi i)$$

Note that this is the product of two functions of i.

EXAMPLE 5.11

Plot $\mathscr{V}_{\eta_i}$ as a function of i for the beam in (3.80), except let the input be a torque and the output be an angular displacement.

Solution: The modal cost is

$$\mathscr{V}_{\eta_i} = \frac{50}{i^2}\cos^2(0.45\pi i),$$

as plotted in Fig. 5.1. Note that for torque inputs the leading term is $50/i^2$, as opposed to $50/i^4$ for force inputs. The change in the type of output (from rectilinear displacement to angular displacement) changes the sin function to cos. The ranking of the modes in Fig. 5.1 by modal costs is 2, 4, 1, 3, 5, 6. $\blacksquare$

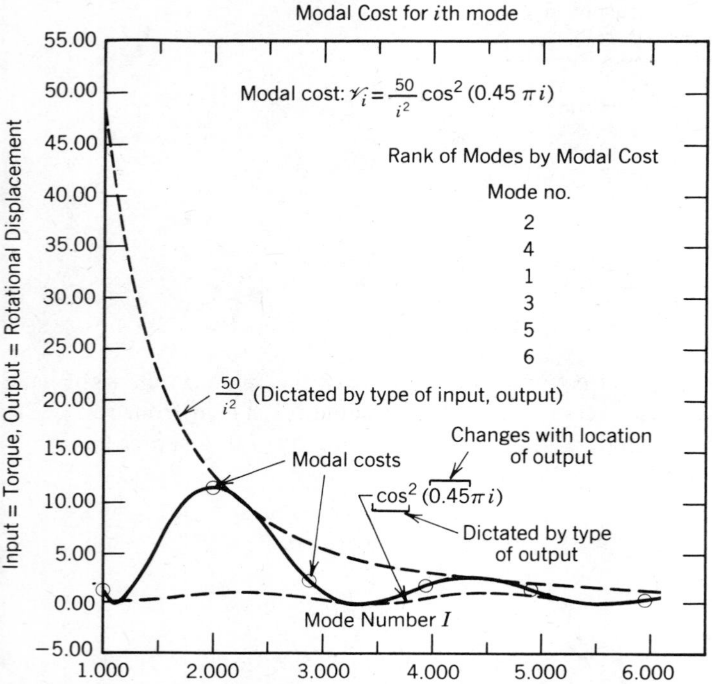

***Figure 5.1** Modal Cost Analysis for the Euler– Bernoulli Beam*

5.6.2 BALANCED COORDINATES

Earlier, it was established that the *magnitudes* of neither the observability indicators $\|\mathbf{c}_i\|$, $\|\mathbf{W}_0\|$, $\|\mathbf{K}\|$ nor the controllability indicators $\|\mathbf{b}_i\|$, $\|\mathbf{W}_c\|$, $\|\mathbf{X}\|$ have meaning in an absolute sense. In other words, controllability and observability tests have only "yes" or "no" answers. One may always normalize a controllable and observable system such that it is *equally* controllable and observable in some sense. If $\mathbf{T}$ is a coordinate transformation matrix of the states $\mathbf{x} = \mathbf{T}\boldsymbol{x}$, then one can always find a $\mathbf{T}$ such that the controllability and observability grammians are equal and diagonal,

$$\mathbf{X} = \mathbf{K} = \text{diag}\left[\ldots \sigma_i^2 \ldots\right], \tag{5.172}$$

so that *each* state variable is equally controllable and observable. The construction of these internally "balanced" coordinates were first discovered by Mullis and Roberts [5.19] and Moore [5.11]. Let $(\mathbf{A}, \mathbf{B}, \mathbf{C})$ be any state representation of the system. We first solve for $\mathbf{X}$ and $\mathbf{K}$ from

$$\mathbf{0} = \mathbf{X}\mathbf{A}^* + \mathbf{A}\mathbf{X} + \mathbf{B}\mathbf{B}^*, \tag{5.173a}$$

$$\mathbf{0} = \mathbf{K}\mathbf{A} + \mathbf{A}^*\mathbf{K} + \mathbf{C}^*\mathbf{C}. \tag{5.173b}$$

Then, compute the spectral decomposition of the matrix **XK**,

$$\mathbf{XK} = \mathbf{E}_b \Lambda_b \mathbf{E}_b^{-1}, \qquad \Lambda_b = \text{diag}, \tag{5.174}$$

and use $\mathbf{E}_b$ as a coordinate transformation $\mathbf{x} = \mathbf{E}_b \mathbf{x}_b$ of

$$\begin{aligned} \dot{\mathbf{x}} &= \mathbf{Ax} + \mathbf{Bu}, \\ \mathbf{y} &= \mathbf{CX}, \end{aligned} \tag{5.175}$$

so that

$$\begin{aligned} \dot{\mathbf{x}}_b &= \mathbf{E}_b^{-1}\mathbf{AE}_b\mathbf{x}_b + \mathbf{E}_b^{-1}\mathbf{Bu}, \\ \mathbf{y} &= \mathbf{CE}_b\mathbf{x}_b \end{aligned} \tag{5.176}$$

is balanced. To show this property, compute the matrices $\mathbf{X}_b = \mathbf{E}_b^{-1}\mathbf{XE}_b^{-*}$, $\mathbf{K}_b = \mathbf{E}_b^{*}\mathbf{KE}_b$. Now, from (5.174),

$$\begin{aligned} \mathbf{X}_b &= \mathbf{E}_b^{-1}\mathbf{XE}_b^{-*} = \mathbf{E}_b^{-1}\mathbf{E}_b\Lambda_b\mathbf{E}_b^{-1}\mathbf{K}^{-1}\mathbf{E}_b^{-*} = \Lambda_b(\mathbf{E}_b^{*}\mathbf{KE}_b)^{-1} \\ &= \Lambda_b\mathbf{K}_b^{-1}. \end{aligned}$$

Also using (5.174),

$$\mathbf{K}_b = \mathbf{E}_b^{*}\mathbf{KE}_b = \mathbf{E}_b^{*}\mathbf{X}^{-1}\mathbf{E}_b\Lambda_b\mathbf{E}_b^{-1}\mathbf{E}_b = \mathbf{X}_b^{-1}\Lambda_b.$$

Now it is clear that $\mathbf{X}_b\mathbf{K}_b = \Lambda_b$. To show that $\mathbf{X}_b = \mathbf{K}_b = \Lambda_b^{1/2}$, we ask

$$\mathbf{X}_b = \mathbf{E}_b^{-1}\mathbf{XE}_b^{-*} \stackrel{?}{=} \mathbf{E}_b^{*}\mathbf{KE}_b = \mathbf{K}_b.$$

But since $\mathbf{K} = \mathbf{X}^{-1}\mathbf{E}_b\Lambda_b\mathbf{E}_b^{-1}$, this yields

$$\mathbf{X}_b = \mathbf{E}_b^{*}\mathbf{KE}_b = \mathbf{E}_b^{*}\mathbf{X}^{-1}\mathbf{E}_b\Lambda_b\mathbf{E}_b^{-1}\mathbf{E}_b = \mathbf{X}_b^{-1}\Lambda_b$$

or

$$\mathbf{X}_b^2 = \Lambda_b.$$

(Similarly, one can show that $\mathbf{K}_b^2 = \Lambda_b$.) Since we have established $\mathbf{X}_b\mathbf{K}_b = \Lambda_b$ and $\mathbf{X}_b = \Lambda_b^{1/2}$, this implies that $\mathbf{K}_b = \Lambda_b^{1/2}$ also. We now have only to prove that the Λ_b in (5.174) is diagonal and has only positive diagonal elements. Otherwise, we cannot compute $\Lambda_b^{1/2}$. The second deficiency in the above calculation of balanced coordinates (5.176) is the assumption of controllability and observability of the initial system (5.175). To remove this deficiency and also to prove the existence of $\Lambda_b^{1/2}$ in a constructive way, we offer the following algorithm which begins with three assumptions: (1) The controllable modes (of 5.175) are asymptotically stable; (2) the observable modes of (5.175) are asymptotically stable; (3) the matrix **A** in (5.175) has no eigenvalues with the property $\lambda_i[\mathbf{A}] + \lambda_j[\mathbf{A}] = 0$ for any i, j.

The Balanced Realization Algorithm

STEP 1: Solve for $\mathbf{X}$ from

$$\mathbf{0} = \mathbf{XA}^* + \mathbf{AX} + \mathbf{BB}^*. \qquad (5.177)$$

This is possible to do uniquely iff $\mathbf{A}$ has no eigenvalues that are symmetric about the $j\omega$ axis ($\lambda_j + \lambda_i \neq 0\ \forall i, j$) and if the controllable modes are stable.

STEP 2: Find the singular value decomposition of $\mathbf{X}$:

$$\mathbf{X} = [\mathbf{U}_{11} \quad \mathbf{U}_{12}]\begin{bmatrix} \boldsymbol{\Sigma}_1 & \mathbf{0} \\ \mathbf{0} & \mathbf{0} \end{bmatrix}\begin{bmatrix} \mathbf{U}_{11}^* \\ \mathbf{U}_{12}^* \end{bmatrix} = \mathbf{U}_{11}\boldsymbol{\Sigma}_1\mathbf{U}_{11}^*. \qquad (5.178a)$$

Now, the columns of $\mathbf{U}_{11}$ span the controllable subspace and the columns of $\mathbf{U}_{12}$ span the uncontrollable subspace:

$$\boldsymbol{\Sigma}_1 = \operatorname{diag}\{\sigma_{11}, \ldots, \sigma_{1n_c}\}, \qquad \mathbf{U}_{11} \in \mathscr{R}^{n_x \times n_c}. \qquad (5.178b)$$

STEP 3: Solve for $\mathbf{K}$:

$$\mathbf{0} = \mathbf{KA} + \mathbf{A}^*\mathbf{K} + \mathbf{C}^*\mathbf{C}. \qquad (5.179)$$

This is possible if $\lambda_i + \lambda_j \neq \forall i, j$ holds and if the observable modes are stable.

STEP 4: Find the singular value decomposition

$$\mathbf{T}_1^*\mathbf{K}\mathbf{T}_1 = [\mathbf{U}_{21} \quad \mathbf{U}_{22}]\begin{bmatrix} \boldsymbol{\Sigma}_2 & \mathbf{0} \\ \mathbf{0} & \mathbf{0} \end{bmatrix}\begin{bmatrix} \mathbf{U}_{21}^* \\ \mathbf{U}_{22}^* \end{bmatrix} = \mathbf{U}_{21}\boldsymbol{\Sigma}_2\mathbf{U}_{21}^*, \qquad (5.180a)$$

where $T_1 = U_{11}\Sigma_1^{1/2}$. Now, U_{21} columns span the subspace that is both controllable and observable; U_{22} columns span the controllable unobservable subspace; and

$$\boldsymbol{\Sigma}_2 = \operatorname{diag}\{\sigma_{21}, \ldots, \sigma_{2n_{c0}}\}, \qquad U_{21} \in \mathscr{R}^{n_c \times n_{c0}}. \qquad (5.180b)$$

The parameters of the new system are

$$\begin{aligned} \dot{\mathbf{x}}_b &= \mathbf{A}_b\mathbf{x}_b + \mathbf{B}_b\mathbf{u}, \qquad \mathbf{x}_b \in \mathscr{R}^{n_{c0}}, \\ \mathbf{y} &= \mathbf{C}_b\mathbf{x}_b, \\ \mathbf{A}_b &= \left(\boldsymbol{\Sigma}_2^{1/4}\mathbf{U}_{21}^*\boldsymbol{\Sigma}_1^{-1/2}\mathbf{U}_{11}^*\right)\mathbf{A}\left(\mathbf{U}_{11}\boldsymbol{\Sigma}_1^{1/2}\mathbf{U}_{21}\boldsymbol{\Sigma}_2^{-1/4}\right), \qquad (5.181) \\ \mathbf{B}_b &= \left(\boldsymbol{\Sigma}_2^{1/4}\mathbf{U}_{21}^*\boldsymbol{\Sigma}_1^{-1/2}\mathbf{U}_{11}^*\right)\mathbf{B}, \\ \mathbf{C}_b &= \mathbf{C}\left(\mathbf{U}_{11}\boldsymbol{\Sigma}_1^{1/2}\mathbf{U}_{21}\boldsymbol{\Sigma}_2^{-1/4}\right). \end{aligned}$$

Note from Step 2 that $\mathbf{\Sigma}_1$ contains all of the nonzero singular values of $\mathbf{X}$ and n_c is the number of controllable states (i.e., the controllable subspace is spanned by $\mathbf{U}_{11}$). Note from Step 4 that $\mathbf{\Sigma}_2$ contains all of the nonzero singular values of $\mathbf{T}_1^*\mathbf{K}\mathbf{T}_1$, and hence n_{c0} is the dimension of the controllable states that are also observable (i.e., the controllable observable subspace is spanned by $\mathbf{U}_{21}$).

Exercise 5.35

If the original system $(\mathbf{A}, \mathbf{B}, \mathbf{C})$ is observable and controllable (that is, if $n_{c0} = n_x$), show that $\mathbf{A}_b = \mathbf{E}_b^{-1}\mathbf{A}\mathbf{E}_b$, $\mathbf{B}_b = \mathbf{E}_b^{-1}\mathbf{B}$, $\mathbf{C}_b = \mathbf{C}\mathbf{E}_b$, where $\mathbf{E}_b = \mathbf{T}_1\mathbf{T}_2$, $\mathbf{T}_1 = \mathbf{U}_{11}\mathbf{\Sigma}_1^{1/2}$, and $\mathbf{T}_2 = \mathbf{U}_{21}\mathbf{\Sigma}_2^{-1/4}$.

Exercise 5.36

Verify that (5.181) is both controllable and observable even if the triple $(\mathbf{A}, \mathbf{B}, \mathbf{C})$ is not controllable and observable.

Exercise 5.37

Verify that the states of (5.181) are "equally" controllable and observable in the sense that the controllability and observability grammians are both equal to the diagonal matrix $\mathbf{\Sigma}_2^{1/2}$. That is, $\mathbf{\Sigma}_2^{1/2}$ obeys both

$$\mathbf{\Sigma}_2^{1/2}\mathbf{A}_b^* + \mathbf{A}_b\mathbf{\Sigma}_2^{1/2} + \mathbf{B}_b\mathbf{B}_b^* = \mathbf{0} \tag{5.182a}$$

and

$$\mathbf{\Sigma}_2^{1/2}\mathbf{A}_b + \mathbf{A}_b^*\mathbf{\Sigma}_2^{1/2} + \mathbf{C}_b^*\mathbf{C}_b = \mathbf{0}. \tag{5.182b}$$

It is for this reason, (5.182), that (5.181) is called a "balanced" realization.

Exercise 5.38

Find the balanced realization of the model of a ship in deep water (taken from ref. [5.12]). See Fig. 5.2.

The reader is reminded that the balanced realization algorithm as given by (5.177) through (5.181) does not apply if the assumptions 1 through 3 (page 251) do not hold. (For example, the algorithm cannot be applied to the ship in shallow water.) The remedy for this deficiency is postponed until after the next section. Meanwhile, an alternate algorithm is given that does *not* presume conditions assumptions 1 through 3 on $(\mathbf{A}, \mathbf{B}, \mathbf{C})$, since solutions for $\mathbf{X}$ and $\mathbf{K}$ will not be required.

Modified Balanced Realization Algorithm

STEP 1: Compute

$$\mathbf{W}_c \triangleq [\mathbf{B}, \mathbf{A}\mathbf{B}, \ldots, \mathbf{A}^{n_x-1}\mathbf{B}],$$

$$\mathbf{W}_0^* \triangleq [\mathbf{C}^*, \mathbf{A}^*\mathbf{C}^*, \ldots, \mathbf{A}^{*n_x-1}\mathbf{C}^*].$$

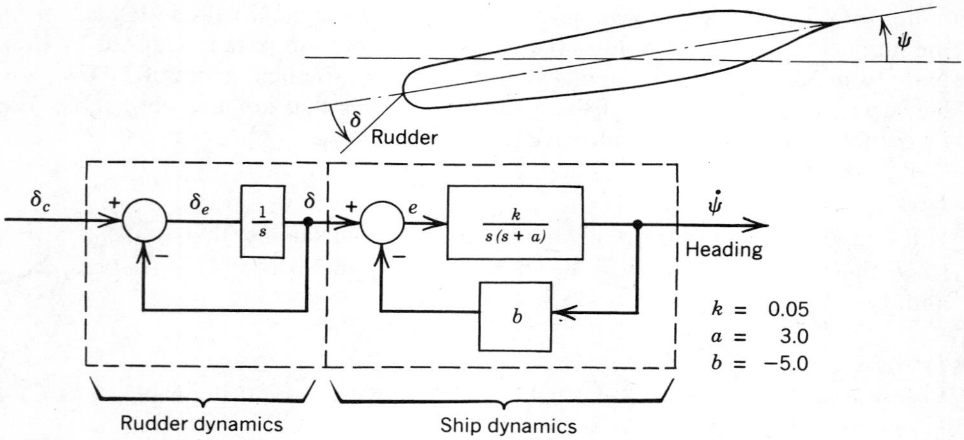

Figure 5.2 *Ship Dynamics*

STEP 2: Find the singular value decomposition of $\mathbf{W}_c\mathbf{W}_c^*$:

$$\mathbf{W}_c\mathbf{W}_c^* = [\mathbf{U}_{11} \quad \mathbf{U}_{12}]\begin{bmatrix}\boldsymbol{\Sigma}_1 & \mathbf{0}\\ \mathbf{0} & \mathbf{0}\end{bmatrix}\begin{bmatrix}\mathbf{U}_{11}^*\\ \mathbf{U}_{12}^*\end{bmatrix} = \mathbf{U}_{11}\boldsymbol{\Sigma}_1\mathbf{U}_{11}^* = \mathbf{T}_1\mathbf{T}_1^*. \quad (5.183)$$

STEP 3: Find the singular value decomposition of

$$\mathbf{T}_1^*\mathbf{W}_0^*\mathbf{W}_0\mathbf{T}_1 = [\mathbf{U}_{21} \quad \mathbf{U}_{22}]\begin{bmatrix}\boldsymbol{\Sigma}_2 & \mathbf{0}\\ \mathbf{0} & \mathbf{0}\end{bmatrix}\begin{bmatrix}\mathbf{U}_{21}^*\\ \mathbf{U}_{22}^*\end{bmatrix} = \mathbf{U}_{21}\boldsymbol{\Sigma}_2\mathbf{U}_{21}^* = \mathbf{T}_2\boldsymbol{\Sigma}_2^{3/2}\mathbf{T}_2^*, \quad (5.184)$$

where $\mathbf{T}_1 = \mathbf{U}_{11}\boldsymbol{\Sigma}_1^{1/2}$, $\mathbf{T}_2 = \mathbf{U}_{21}\boldsymbol{\Sigma}_2^{-1/4}$. The balanced realization is now given by (5.181) using these new definitions of $\mathbf{U}_{11}$, $\boldsymbol{\Sigma}_1$, $\mathbf{U}_{21}$, and $\boldsymbol{\Sigma}_2$.

Exercise 5.39

Show that the modified balanced realization (5.181) using data from (5.183) has the property

$$\mathscr{W}_c\mathscr{W}_c^* = \mathscr{W}_0^*\mathscr{W}_0 = \boldsymbol{\Sigma}_2^{1/2}, \quad (5.185a)$$

where the realization (A_b, B_b, C_b) so modified is labeled $(\mathscr{A}_b, \mathscr{B}_b, \mathscr{C}_b)$ and

$$\mathscr{W}_c \triangleq [\mathscr{B}_b, \mathscr{A}_b\mathscr{B}_b, \ldots, \mathscr{A}_b^{n_{c0}-1}\mathscr{B}_b], \quad (5.185b)$$

$$\mathscr{W}_0^* \triangleq [\mathscr{C}_b^*, \mathscr{A}_b^*\mathscr{C}_b^*, \ldots, \mathscr{A}_b^{*n_{c0}-1}\mathscr{C}_b^*]. \quad (5.185c)$$

Exercise 5.40

Compute the balanced realization of

$$\dot{\mathbf{x}} = \begin{bmatrix} -1 & 0 \\ 0 & -10 \end{bmatrix}\mathbf{x} + \begin{pmatrix} 1 \\ 70 \end{pmatrix} u,$$

$$y = [1 \; -0.2]\mathbf{x},$$

using both the balanced realization algorithm and the modified balanced realization algorithm.

Exercise 5.41

Use the modified balanced realization algorithm to compute balanced coordinates for the ship in shallow water in Exercise 5.38.

5.6.3 HESSENBERG COORDINATES

Cost-decoupled coordinates, modal coordinates, and balanced coordinates all have special advantages. Our final set of coordinates which we introduce for special advantages is called Hessenberg coordinates, and two special cases will be introduced in Sections 5.6.3.1 and 5.6.3.2.

5.6.3.1 Controllable Hessenberg Coordinates

Definition 5.4

Coordinates with these properties are called normalized controllable Hessenberg coordinates:

$$\mathbf{0} = \mathbf{A} + \mathbf{A}^* + \mathbf{C}^*\mathbf{C}, \tag{5.186a}$$

$$\begin{bmatrix} \mathbf{B} & \mathbf{A} \\ & \mathbf{C} \end{bmatrix} = \left[\begin{array}{c|cccccc} \mathbf{B}_1 & \mathbf{A}_{11} & \mathbf{A}_{12} & \cdots & \cdot & \cdot & \mathbf{A}_{1q} \\ \mathbf{0} & \mathbf{A}_{21} & \mathbf{A}_{22} & & & & \cdot \\ \mathbf{0} & \mathbf{0} & \mathbf{A}_{32} & & & & \cdot \\ & \cdot & & \ddots & & & \cdot \\ & \cdot & & & \ddots & & \cdot \\ & \cdot & & & & & \mathbf{A}_{q-1,q} \\ \mathbf{0} & \mathbf{0} & \cdot & & 0 & \mathbf{A}_{q,q-1} & \mathbf{A}_{qq} \\ \hline & \mathbf{C}_1 & \mathbf{C}_2 & \cdots & \cdot & \cdot & \mathbf{C}_q \end{array}\right], \tag{5.186b}$$

where $\mathbf{A}_{ij} \in \mathscr{R}^{n_i \times n_j}$ *and rank* $[\mathbf{A}_{i+1,j}] = n_{i+1} \leq n_i$, $n_x = \Sigma_{i=1}^{q} n_i$, *and rank* $\mathbf{B}_1 = n_u$. *Also,* $\prod_{i=0}^{k} \mathbf{A}_{i+1,i}\mathbf{B}_i$, $k = 0, 1, \ldots, q-1$, *are all upper triangular or upper echelon matrices with the first nonzero entry in each row positive.*

Suppose $\mathbf{A}_{i+1,i} \in \mathscr{R}^{n_{i+1} \times n_i}$, *where* $n_{i+1} = 2$, $n_i = 4$. *Then, the possible forms of* $\mathbf{A}_{i+1,i}$ *are*

$$\begin{bmatrix} + & * & * & * \\ 0 & + & * & * \end{bmatrix}, \begin{bmatrix} + & * & * & * \\ 0 & 0 & + & * \end{bmatrix}, \begin{bmatrix} + & * & * & * \\ 0 & 0 & 0 & + \end{bmatrix},$$

$$\begin{bmatrix} 0 & + & * & * \\ 0 & 0 & + & * \end{bmatrix}, \begin{bmatrix} 0 & + & * & * \\ 0 & 0 & 0 & + \end{bmatrix}, \begin{bmatrix} 0 & 0 & + & * \\ 0 & 0 & 0 & + \end{bmatrix},$$

*where + denotes a positive number and * denotes an unspecified number. We shall now show that every controllable, observable, asymptotically stable, proper system has a unique normalized controllable Hessenberg form and that cotnrollability is established by inspection of the* $\mathbf{A}_{i+1,i}$ *blocks. Any* $\mathbf{A}_{i+1,i}$ *which is zero implies* x_k, $k = i+1, \ldots, q$, *are all uncontrollable states variables.*

Lemma 5.1

Given any controllable, observable, asymptotically stable triple $(\mathbf{A}, \mathbf{B}, \mathbf{C})$, *one can construct a unique normalized controllable Hessenberg realization.*

Proof: We shall prove only the scalar input case. (In this case, $n_i = 1$ for all i.) Uniqueness for the multiple-input case is discussed in ref. [5.20].

Let $(\mathbf{A}_\alpha, \mathbf{B}_\alpha, \mathbf{C}_\alpha)$ and $(\mathbf{A}_\beta, \mathbf{B}_\beta, \mathbf{C}_\beta)$ be two controllable Hessenberg realizations of a given transfer function $\mathbf{G}(s)$. Any two realizations of $\mathbf{G}(s)$ that are both controllable and observable are related by a coordinate transformation. (For proof of this, we await Theorem 6.1.) Hence, the observability grammians $\mathbf{K}_\beta = \mathbf{K}_\alpha = \mathbf{I}$ and due to controllability and observability, the two realizations are related by

$$\mathbf{A}_\beta = \mathbf{T}^{-1}\mathbf{A}_\alpha\mathbf{T}, \quad \mathbf{B}_\beta = \mathbf{T}^{-1}\mathbf{B}_\alpha, \quad \mathbf{C}_\beta = \mathbf{C}_\alpha\mathbf{T}.$$

Now, $\mathbf{K}_\alpha$ transforms according to $\mathbf{K}_\beta = \mathbf{T}^*\mathbf{K}_\alpha\mathbf{T} = \mathbf{T}^*\mathbf{T} = \mathbf{I}$. Hence, $\mathbf{T}$ must be unitary. Now, $\mathbf{W}_{c\beta} = \mathbf{T}^{-1}\mathbf{W}_{c\alpha}$, where

$$\mathbf{W}_{c\alpha} = \left[\mathbf{B}_\alpha, \mathbf{A}_\alpha\mathbf{B}_\alpha, \ldots, \mathbf{A}_\alpha^{q-1}\mathbf{B}_\alpha\right]$$

is upper triangular with positive diagonal elements, and $[W_{c\alpha}]_{kk} = \prod_{i=0}^{k-1}\mathbf{A}_{\alpha_{i+1,i}}\mathbf{B}_{\alpha_1}$, where $\mathbf{A}_{1,0} \triangleq \mathbf{I}$. Hence,

$$\mathbf{W}_{c\beta} = \mathbf{T}^{-1}\mathbf{W}_{c\alpha},$$

with both $W_{c\beta}$ and $W_{c\alpha}$ upper triangular. This implies that T^{-1} must be upper triangular. Likewise, from $W_{c\alpha} = TW_{c\beta}$, T must be lower triangular. The only unitary matrix for which $\mathbf{T}$ and $\mathbf{T}^*$ are both upper triangular is $\mathbf{T} = \mathbf{I}$. This proves the uniqueness of the normalized controllable Hessenberg coordinates. ■

Lemma 5.2

Let $\dot{\mathbf{x}} = \mathbf{A}\mathbf{x} + \mathbf{B}\mathbf{u}$ *have the normalized controllable Hessenberg form* (5.186), *and define the substate* $\mathbf{x}_i \in \mathscr{R}^{n_i}$ *consistent with the block sizes* n_i *in* (5.186). *Then, substates* x_α, $\alpha = k+1, \ldots, q$, *are all uncontrollable iff* $\mathbf{A}_{k+1,k}$ *is zero.*

Proof: The controllability matrix W_c has the structure

$$\mathbf{W}_c = \begin{bmatrix} \mathbf{B}_1 & * & * & * & * \\ \mathbf{0} & \mathbf{A}_{21}\mathbf{B}_1 & * & * & * \\ \mathbf{0} & \mathbf{0} & \mathbf{A}_{32}\mathbf{A}_{21}\mathbf{B}_1 & * & * \\ \mathbf{0} & \mathbf{0} & \mathbf{0} & \mathbf{A}_{43}\mathbf{A}_{32}\mathbf{A}_{21}\mathbf{B}_1 & * \\ \mathbf{0} & \mathbf{0} & \mathbf{0} & \mathbf{0} & \mathbf{A}_{54}\mathbf{A}_{43}\mathbf{A}_{32}\mathbf{A}_{21}\mathbf{B}_1 \end{bmatrix} \tag{5.187}$$

shown for $q = 5$. Also, $\mathbf{A}_{21}$ multiplies all nonzero blocks of the second-row block, and $\mathbf{A}_{32}\mathbf{A}_{21}$ multiplies all nonzero blocks of the third-row block, and so on. Hence, $\mathbf{A}_{i+1,i} = \mathbf{0}$ implies that $\mathbf{W}_c$ has only zero rows below, including the $i+1$ block of rows. (Recall that the $\mathbf{A}_{i+1,i}$ has $n_{i+1} \le n_i$ rows.) ■

The Normalized Controllable Hessenberg Algorithm

Given $(\mathbf{A}, \mathbf{B}, \mathbf{C})$ asymptotically stable, observable, and controllable. Find the unique normalized controllable Hessenberg form $(\mathbf{A}_{CH}, \mathbf{B}_{CH}, \mathbf{C}_{CH})$.

STEP 1: Compute $\mathbf{K}$ from $\mathbf{0} = \mathbf{KA} = +\mathbf{A}^*\mathbf{K} + \mathbf{C}^*\mathbf{C}$.

STEP 2: Compute the upper-triangular factor $\mathbf{T}_0$ of $\mathbf{K}$, $\mathbf{K} = \mathbf{T}_0\mathbf{T}_0^*$. This step yields $(\mathbf{T}_0^{-1}\mathbf{A}\mathbf{T}_0, \mathbf{T}_0^{-1}\mathbf{B}, \mathbf{C}\mathbf{T}_0, \mathbf{T}_0^*\mathbf{K}\mathbf{T}_0) = (\mathbf{A}_0, \mathbf{B}_0, \mathbf{C}_0, \mathbf{I})$.

STEP 3: Compute the singular value decomposition of $\mathbf{B}_0$,

$$\mathbf{B}_0 = \mathbf{U}_0\begin{bmatrix}\boldsymbol{\Sigma}_0\\ \mathbf{0}\end{bmatrix}\mathbf{V}_0^*, \qquad \boldsymbol{\Sigma}_0 > \mathbf{0}.$$

This step yields $(\mathbf{U}_0^*\mathbf{A}_0\mathbf{U}_0, \mathbf{U}_0^*\mathbf{B}_0, \mathbf{C}_0\mathbf{U}_0, \mathbf{K}_1 = \mathbf{U}_0^*\mathbf{U}_0) = (\mathbf{A}_1, \mathbf{B}_1, \mathbf{C}_1, \mathbf{K}_1 = \mathbf{I})$, where

$$\mathbf{B}_1 = \begin{bmatrix}\boldsymbol{\Sigma}_0\mathbf{V}_0^*\\ \mathbf{0}\end{bmatrix}, \qquad n_1 \triangleq \dim \boldsymbol{\Sigma}_0.$$

STEP 4: Set $i = 1$ on the first iteration and compute the singular value decomposition $(\mathbf{U}_i, \boldsymbol{\Sigma}_i, \mathbf{V}_i)$ of the $(n_x - r_i) \times r_i$ matrix $(r_i \triangleq \sum_{\alpha=1}^{i} n_\alpha)$,

$$[\mathbf{A}_i]_{21} = \mathbf{U}_i\begin{bmatrix}\boldsymbol{\Sigma}_i & \mathbf{0}\\ \mathbf{0} & \mathbf{0}\end{bmatrix}\mathbf{V}_i^*, \quad \mathbf{U}_i = \left[\mathbf{U}_{i_1}, \mathbf{U}_{i_2}\right], \quad n_{i+1} \triangleq \dim \boldsymbol{\Sigma}_i.$$

Now, define

$$\mathbf{T}_i \triangleq \begin{bmatrix}\mathbf{I}_i & \mathbf{0}\\ \mathbf{0} & \mathbf{U}_i\end{bmatrix}, \qquad \mathbf{I}_i = \text{identity of dimension } r_i;$$

$$\mathbf{A}_{i+1} \triangleq \mathbf{T}_i^*\mathbf{A}_i\mathbf{T}_i, \quad \mathbf{B}_{i+1} = \mathbf{T}_i^*\mathbf{B}_i, \quad \mathbf{C}_{i+1} = \mathbf{C}_i\mathbf{T}_i.$$

STEP 5: Set $i = i + 1$ and repeat step 4. This process continues until $r_i = n_x$. At this point the matrices $(\mathbf{A}_i, \mathbf{B}_i, \mathbf{C}_i, \mathbf{K}_i) = (\mathbf{A}_{CH}, \mathbf{B}_{CH}, \mathbf{C}_{CH}, \mathbf{I})$ have the correct block structure; but without an additional constraint on $\mathbf{V}_i$, $i = 1, 2, \ldots,$ the leading blocks are not upper triangular with the first nonzero entry positive as required in the normalized controllable Hessenberg form. To accomplish this, choose the unit vectors in $\mathbf{V}_i$ (the freedom here is $\pm\mathbf{v}_\alpha$, where $\mathbf{v}_\alpha$ is the α column of $\mathbf{V}_i$) so that $[\mathbf{A}_i]_{21}$ is a positive number (in the scalar case, $n_1 = 1$). In the matrix case, the first nonzero row elements of $[\mathbf{A}_i]_{21}$ must be positive.

Note that all $(\mathbf{A}, \mathbf{C})$ which satisfy $\mathbf{A} + \mathbf{A}^* + \mathbf{C}^*\mathbf{C} = \mathbf{0}$ imply certain constraints between $\mathbf{A}$ and $\mathbf{C}$ which serve to reduce the number of independent parameters in the normalized controllable Hessenberg form. This dependence is illustrated by the structure

$$\left[\begin{array}{c|c} \mathbf{B} & \mathbf{A} \\ \hline & \mathbf{C} \end{array}\right] = \left[\begin{array}{c|cccc} \mathbf{B}_1 & -\frac{1}{2}\mathbf{C}_1^*\mathbf{C}_1 - \mathbf{S}_1 & -\mathbf{C}_1^*\mathbf{C}_2 - \mathbf{A}_{21}^* & -\mathbf{C}_1^*\mathbf{C}_3 & -\mathbf{C}_1^*\mathbf{C}_4 \\ 0 & \mathbf{A}_{21} & -\frac{1}{2}\mathbf{C}_2^*\mathbf{C}_2 - \mathbf{S}_2 & -\mathbf{C}_2^*\mathbf{C}_3 - \mathbf{A}_{32}^* & -\mathbf{C}_2^*\mathbf{C}_4 \\ 0 & \mathbf{0} & \mathbf{A}_{32} & -\frac{1}{2}\mathbf{C}_3^*\mathbf{C}_3 - \mathbf{S}_3 & -\mathbf{C}_3\mathbf{C}_4 - \mathbf{A}_{43}^* \\ 0 & \mathbf{0} & \mathbf{0} & \mathbf{A}_{43} & -\frac{1}{2}\mathbf{C}_4^*\mathbf{C}_4 - \mathbf{S}_4 \\ \hline & \mathbf{C}_1 & \mathbf{C}_2 & \mathbf{C}_3 & \mathbf{C}_4 \end{array}\right], \tag{5.188}$$

where $\mathbf{S}_i$ represents a skew Hermitian matrices of size $n_i \times n_i$. Thus, the free parameters in a normalized controllable form are: $\mathbf{B}_1, \mathbf{C}_1, \mathbf{C}_2, \ldots \mathbf{C}_q$, $\mathbf{A}_{21}, \mathbf{A}_{32}, \ldots \mathbf{A}_{q,q-1}$, $\mathbf{S}_1, \mathbf{S}_2, \ldots \mathbf{S}_q$.

EXAMPLE 5.12

Describe all single-input/output third-order systems in normalized controllable Hessenberg form.

Solution:

$$\left[\begin{array}{c|c} \mathbf{B}_{\mathrm{CH}} & \mathbf{A}_{\mathrm{CH}} \\ \hline & \mathbf{C}_{\mathrm{CH}} \end{array}\right] = \left[\begin{array}{c|ccc} \mathbf{b} & -c_1^2/2 & -c_1c_2 - a_{21} & -c_1c_3 \\ 0 & a_{21} & -c_2^2/2 & -c_2c_3 - a_{32} \\ 0 & 0 & a_{32} & -c_3^2/2 \\ \hline & c_1 & c_2 & c_3 \end{array}\right], \tag{5.189}$$

where $b > 0$, $a_{21} > 0$, $a_{32} > 0$ if $(\mathbf{A}, \mathbf{B})$ is controllable. Note that the number of free parameters (required to describe all third-order systems) is correct at six $(b_1, a_{21}, a_{32}, c_1, c_2, c_3)$. The transfer function of (A_{CH}, B_{CH}, C_{CH}) is

$$\mathbf{C}_{\mathrm{CH}}(s\mathbf{I} - \mathbf{A}_{\mathrm{CH}})^{-1}\mathbf{B}_{\mathrm{CH}} = \frac{b}{\Delta}\left\{c_1\left[\left(s + \frac{c_2^2}{2}\right)\left(s + \frac{c_3^2}{2}\right) + a_{32}(c_2c_3 + a_3D2)\right]\right.$$
$$\left. + c_2a_{21}\left(s + \frac{c_3^2}{2}\right) + c_3a_{21}a_{32}\right\}, \tag{5.190}$$

where

$$\Delta \triangleq \left(s + \frac{c_1^2}{2}\right)\left(s + \frac{c_2^2}{2}\right)\left(s + \frac{c_3^2}{2}\right) + \left(s + \frac{c_1^2}{2}\right)(c_2 c_3 + a_{32}) a_{32}$$

$$+ \left(s + \frac{c_3^2}{2}\right)(c_1 c_2 + a_{21}) a_{21}. \qquad \blacksquare$$

5.6.3.2 Observable Hessenberg Coordinates

Definition 5.5

The observable controllable, asymptotically stable triple $(\mathbf{A}, \mathbf{B}, \mathbf{C})$ *is in normalized observable Hessenberg form when*

$$\left[\begin{array}{c|c} \mathbf{C} & \\ \hline \mathbf{A} & \mathbf{B} \end{array}\right] = \left[\begin{array}{cccccc|c} \mathbf{C}_1 & \mathbf{0} & \mathbf{0} & \mathbf{0} & \cdots & \mathbf{0} & \\ \hline \mathbf{A}_{11} & \mathbf{A}_{12} & \mathbf{0} & \mathbf{0} & \cdots & \mathbf{0} & \mathbf{B}_1 \\ \mathbf{A}_{21} & \mathbf{A}_{22} & \mathbf{A}_{23} & \mathbf{0} & \cdots & \mathbf{0} & \mathbf{B}_2 \\ \mathbf{A}_{31} & \mathbf{A}_{32} & \mathbf{A}_{33} & \mathbf{A}_{34} & \ddots & \cdot & \mathbf{B}_3 \\ \mathbf{A}_{41} & \mathbf{A}_{42} & \mathbf{A}_{43} & \mathbf{A}_{44} & \ddots & \cdot & \mathbf{B}_4 \\ \cdot & \cdot & \cdot & \cdot & & \cdot & \cdot \\ \cdot & \cdot & \cdot & \cdot & & \ddots & \cdot \\ \cdot & \cdot & \cdot & \cdot & & \mathbf{A}_{q-1,q} & \cdot \\ \mathbf{A}_{q1} & \mathbf{A}_{q2} & \mathbf{A}_{q3} & \mathbf{A}_{q4} & & \mathbf{A}_{qq} & \mathbf{B}_q \end{array}\right], \qquad (5.191)$$

$$\mathbf{A} + \mathbf{A}^* + \mathbf{B}\mathbf{B}^* = \mathbf{0},$$

where $\mathbf{A}_{ij} \in \mathscr{R}^{n_i \times n_j}$ *and*

$$\operatorname{rank} \mathbf{A}_{i,i+1} = n_{i+1} \le n_i, \qquad \sum_{i=1}^{q} n_i = n_x,$$

$$\operatorname{rank} \mathbf{C}_1 = n_1 = n_y,$$

and $\prod_{i=1}^{k} \mathbf{C}_1 \mathbf{A}_{i,i+1}$, $k = 1, 2, 3, \ldots, q-1$, *are lower-echelon matrices with the first nonzero entry in each column positive.*

Suppose $n_i = 4$ *and* $n_{i+1} = 2$. *Then, the possible forms of* $\mathbf{A}_{i,i+1}$ *are*

$$A_{i,i+1} = \begin{bmatrix} +0 \\ *+ \\ ** \\ ** \end{bmatrix}, \begin{bmatrix} +0 \\ *0 \\ *+ \\ ** \end{bmatrix}, \begin{bmatrix} +0 \\ *0 \\ *0 \\ *+ \end{bmatrix}, \begin{bmatrix} 00 \\ +0 \\ *+ \\ ** \end{bmatrix}, \begin{bmatrix} 00 \\ +0 \\ *0 \\ *+ \end{bmatrix}, \begin{bmatrix} 00 \\ 00 \\ 0+ \\ *+ \end{bmatrix},$$

where $+$ *denotes a positive number and* $*$ *denotes an unspecified number.*

Lemma 5.3

There is a unique normalized observable Hessenberg representation for any controllable, observable, asymptotically stable, strictly proper linear time-invariant system.

Proof: We shall prove this only for the single-output case. In this case, $1 = n_y = n_1 = n_2 = \cdots = n_q$. Suppose two minimal normalized observable Hessenberg realizations exist, $(A_\alpha, B_\alpha, C_\alpha)$ and $(\mathbf{A}_\beta, \mathbf{B}_\beta, \mathbf{C}_\beta)$. Then, $\mathbf{X}_\alpha = \mathbf{I}$ and $\mathbf{X}_\beta = \mathbf{I}$. Theorem 6.1 will show that any two n_xth-order state space representations of an n_xth-order controllable, observable system must be related by a coordinate transformation. Hence,

$$\mathbf{A}_\beta = \mathbf{T}^{-1}\mathbf{A}_\alpha\mathbf{T}, \quad \mathbf{B}_\beta = \mathbf{T}^{-1}\mathbf{B}_\alpha, \quad \mathbf{C}_\beta = \mathbf{C}_\alpha\mathbf{T}, \quad \mathbf{X}_\beta = \mathbf{T}^{-1}\mathbf{X}_\alpha\mathbf{T}^{-*}.$$

Thus, the given facts $\mathbf{X}_\beta = \mathbf{I}$, $\mathbf{X}_\alpha = \mathbf{I}$ imply that $\mathbf{TT}^* = \mathbf{I}$. Hence, $\mathbf{T}$ must be unitary. Now, since $n_i = 1$ for all i, the observability matrix is lower-triangular:

$$\mathbf{W}_{0\alpha} = \begin{bmatrix} \mathbf{C}_\alpha \\ \mathbf{C}_\alpha\mathbf{A}_\alpha \\ \vdots \\ \mathbf{C}_\alpha\mathbf{A}_\alpha^{q-1} \end{bmatrix} = \begin{bmatrix} C_1 & & & \\ * & C_1A_{12} & & \\ * & * & C_1A_{12}A_{23} & \\ * & * & * & \ddots \end{bmatrix}, \tag{5.192}$$

and the first nonzero entry in each column is positive $(+)$. $\mathbf{W}_{0\beta}$ is also lower-trinagular with this structure. Now, since $\mathbf{W}_{0\beta} = \mathbf{W}_{0\alpha}T$, we conclude that both $\mathbf{T}$ and $\mathbf{T}^{-1}$ must be lower-triangular with positive numbers on the diagonal to preserve the lower-triangular structure of $\mathbf{W}_{0\alpha}, \mathbf{W}_{0\beta}$). To summarize, we must find $\mathbf{T}$ which is unitary, with both $\mathbf{T}$ and $\mathbf{T}^{-1} = \mathbf{T}^*$ lower-triangular. The only matrix with this property is $\mathbf{T} = \mathbf{I}$. Hence, $(\mathbf{A}_\alpha, \mathbf{B}_\alpha, \mathbf{C}_\alpha) = (\mathbf{A}_\beta, \mathbf{B}_\beta, \mathbf{C}_\beta)$ is unique. ■

The construction of the normalized observable Hessenberg coordinates follows.

The Normalized Observable Hessenberg Algorithm:

Given a controllable, observable, asymptotically stable system $(\mathbf{A}, \mathbf{B}, \mathbf{C})$ with $|\mathbf{CC}^*| \neq 0$, find the unique $\mathbf{A}_{OH}, \mathbf{B}_{OH}, \mathbf{C}_{OH})$ observable Hessenberg form.

STEP 1: Compute $\mathbf{X}$ from $\mathbf{0} = \mathbf{XA}^* + \mathbf{AX} + \mathbf{BB}^*$.

STEP 2: Compute the lower triangular factor of

$$\mathbf{X} = \mathbf{T}_0\mathbf{T}_0^*.$$

This step yields $\mathbf{X}_0 = \mathbf{I} = \mathbf{T}_0^{-1}\mathbf{X}\mathbf{T}_0^{-*}$ for the coordinates $\mathbf{A}_0 = \mathbf{T}_0^{-1}\mathbf{A}\mathbf{T}_0$, $\mathbf{B}_0 = \mathbf{T}_0^{-1}\mathbf{B}$, $\mathbf{C}_0 = \mathbf{C}\mathbf{T}_0$.

STEP 3: Compute the singular value decomposition

$$\mathbf{C}_0 = \mathbf{C}\mathbf{T}_0 = \mathbf{U}_0[\boldsymbol{\Sigma}_0\mathbf{0}]\begin{bmatrix} \mathbf{V}_{01}^* \\ \mathbf{V}_{02}^* \end{bmatrix} = [\mathbf{U}_0\boldsymbol{\Sigma}_0, \ \ \mathbf{0}]\mathbf{V}_0^*.$$

This step yields

$$\mathbf{A}_1 = \mathbf{V}_0^*\mathbf{A}_0\mathbf{V}_0, \quad \mathbf{B}_1 = \mathbf{V}_0^*\mathbf{B}_0, \quad \mathbf{C}_1 = \mathbf{C}_0\mathbf{V}_0, \quad \mathbf{X}_1 = \mathbf{V}_0\mathbf{X}_0\mathbf{V}_0^* = \mathbf{I},$$

where

$$\mathbf{C}_1 = [\mathbf{U}_0\boldsymbol{\Sigma}_0, \mathbf{0}], \qquad n_1 \triangleq \dim \Sigma_0.$$

STEP 4: Set $i = 1$ on the first iteration and compute the singular value decomposition $(\mathbf{U}_i, \boldsymbol{\Sigma}_i, \mathbf{V}_i)$ of the $r_i \times (n_x - r_i)$ matrix $(r_i \triangleq \Sigma_{\alpha=1}^{i} n_\alpha)$,

$$[\mathbf{A}_i]_{12} = \mathbf{U}_i\begin{bmatrix} \boldsymbol{\Sigma}_i & \mathbf{0} \\ \mathbf{0} & \mathbf{0} \end{bmatrix}\mathbf{V}_i^*, \quad \mathbf{V}_i = [\mathbf{V}_{i1}, \mathbf{V}_{i2}], \quad n_{i+1} \triangleq \dim \Sigma_i.$$

Now, define

$$\mathbf{T}_i \triangleq \begin{bmatrix} \mathbf{I}_i & \mathbf{0} \\ \mathbf{0} & \mathbf{V}_i \end{bmatrix}, \qquad \mathbf{I}_i = \text{identity of dim } r_i,$$

$$\mathbf{A}_{i+1} = \mathbf{T}_i^*\mathbf{A}_i\mathbf{T}_i, \quad \mathbf{B}_{i+1} = \mathbf{T}_i^*\mathbf{B}_i, \quad \mathbf{C}_{i+1} = \mathbf{C}_i\mathbf{T}_i.$$

STEP 5: Set $i = i + 1$ and repeat step 4. This process continues until $r_i = n_x$. At this point the matrices $(\mathbf{A}_i, \mathbf{B}_i, \mathbf{C}_i, \mathbf{X}_i) = (\mathbf{A}_{OH}, \mathbf{B}_{OH}, \mathbf{C}_{OH}, \mathbf{I})$ have the correct block structure, but without additional constraints on $\mathbf{U}_i$, $i = 1, 2, \ldots,$ the leading blocks are not lower-triangular with the first nonzero entry positive as required in the normalized observable Hessenberg form. To accomplish this, choose the unit vectors in $\mathbf{U}_i$ (the freedom here is $\pm u_\alpha$, where u_α is the α column of $\mathbf{U}_i$), so that $[\mathbf{A}_i]_{12}$ is a positive number (in the scalar case, $n_1 = 1$). In the matrix case, the first nonzero elements of the columns of $[A_i]_{12}$ must be positive.

Note that all $(\mathbf{A}, \mathbf{B})$ which satisfy $\mathbf{A} + \mathbf{A}^* + \mathbf{B}\mathbf{B}^* = \mathbf{0}$ imply certain constraints between $\mathbf{A}$ and $\mathbf{B}$ which serve to reduce the number of independent parameters in the normalized observable Hessenberg form. This dependence is illustrated by the structure

$$\left[\begin{array}{c|c} \mathbf{C} & \\ \hline \mathbf{A} & \mathbf{B} \end{array}\right]$$

$$= \left[\begin{array}{cccc|c} \mathbf{C}_1 & \mathbf{0} & \mathbf{0} & \mathbf{0} & \\ \hline -\frac{1}{2}\mathbf{B}_1\mathbf{B}_1^* - \mathbf{S}_1 & \mathbf{A}_{12} & \mathbf{0} & \mathbf{0} & \mathbf{B}_1 \\ -\mathbf{B}_2\mathbf{B}_1^* - \mathbf{A}_{12}^* & -\frac{1}{2}\mathbf{B}_2\mathbf{B}_2^* - \mathbf{S}_2 & \mathbf{A}_{23} & \mathbf{0} & \mathbf{B}_2 \\ -\mathbf{B}_3\mathbf{B}_1^* & -\mathbf{B}_3\mathbf{B}_2^* - \mathbf{A}_{23}^* & -\frac{1}{2}\mathbf{B}_3\mathbf{B}_3^* - \mathbf{S}_3 & \mathbf{A}_{34} & \mathbf{B}_3 \\ -\mathbf{B}_4\mathbf{B}_1^* & -\mathbf{B}_4\mathbf{B}_2^* & -\mathbf{B}_4\mathbf{B}_3^* - \mathbf{A}_{34}^* & -\frac{1}{2}\mathbf{B}_4\mathbf{B}_4^* - \mathbf{S}_4 & \mathbf{B}_4 \end{array}\right], \tag{5.193}$$

where $\mathbf{S}_i \in \mathscr{R}^{n_i \times n_i}$ represents a skew-Hermitian matrix. Thus the free parameters in a normalized observable Hessenberg form are $\mathbf{C}_1, \mathbf{B}_1, \mathbf{B}_2, \ldots, \mathbf{B}_q, \mathbf{A}_{12}, \mathbf{A}_{23}, \ldots \mathbf{A}_{q-1,q}$, $\mathbf{S}_1, \mathbf{S}_2, \ldots, \mathbf{S}_q$.

EXAMPLE 5.13

Describe all single-input/output third-order systems in their normalized observable Hessenberg form.

Solution:

$$\left[\begin{array}{c|c} \mathbf{C}_{\mathrm{OH}} & \\ \hline \mathbf{A}_{\mathrm{OH}} & \mathbf{B}_{\mathrm{OH}} \end{array}\right] = \left[\begin{array}{ccc|c} c_1 & 0 & 0 & \\ \hline -\dfrac{b_1^2}{2} & A_{12} & 0 & b_1 \\ -b_1 b_2 - A_{12} & -\dfrac{b_2^2}{2} & A_{23} & b_2 \\ -b_1 b_3 & -b_2 b_3 - A_{23} & -\dfrac{b_3^2}{2} & b_3 \end{array}\right]$$

with transfer function

$$\mathbf{C}_{\mathrm{OH}}(s\mathbf{I} - \mathbf{A}_{\mathrm{OH}})^{-1}\mathbf{B}_{\mathrm{OH}} = \frac{C_1}{\Delta}\left\{ b_1\left[\left(s + \frac{b_2^2}{2}\right)\left(s + \frac{b_3^2}{2}\right) + A_{23}(A_{23} + b_2 b_3)\right]\right.$$

$$\left. + b_2 A_{12}\left(s + \frac{b_3^2}{2}\right) + b_3\left[(b_1 b_2 + A_{12})(b_2 b_3 + A_{23}) - b_1 b_3\left(s + \frac{b_2^2}{2}\right)\right]\right\}, \quad (5.195)$$

where

$$\Delta = \left(s + \frac{b_1^2}{2}\right)\left(s + \frac{b_2^2}{2}\right)\left(s + \frac{b_3^2}{2}\right) + A_{12}(b_1 b_2 + A_{12})\left(s + \frac{b_3^2}{2}\right)$$

$$+ A_{23}(b_2 b_3 + A_{23})\left(s + \frac{b_1^2}{2}\right) + A_{12} A_{23} b_1 b_3.$$

The normalized observable Hessenberg form requires the choices $c_1 > 0$, $A_{12} > 0$, $A_{23} > 0$.

The next chapter will exploit certain advantages of the observable Hessenberg form.

Closure

This chapter introduces the fundamental concepts of observability and controllability. If a system is controllable and observable, then it will be so in any transformed coordinates. Several special coordinates are introduced for simple examination of observability and controllability. These special coordinates included modal, balanced, and Hessenberg coordinates. These special coordinates included modal, balanced, and Hessenberg coordinates. Chapter 10 will warn us that all physical systems are neither observable nor controllable. Hence, we must understand in subsequent chapters what control design methods depend upon this property for existence. The concept of modal cost analysis is introduced to assign a value to each mode when the system performance is measured by the norm of the impulse response.

References

5.1 L. I. Rozenoer, "A Variational Approach to the Problem of Invariance of Automatic Control Systems," translated from *Automatika i Telemekhanika*, 24(6), 744–756, June 1964. Part II followed in *ibid.*, 24(7), 861–870, July 1964.

5.2 W. L. Brogan, *Modern Control Theory*. Quantum Publishers, New York, 1974, p. 304.

5.3 E. G. Gilbert, "Controllability and Observability in Multivariable Control Systems," *J. Siam Control*, Ser. A, 1(2), 128–151, 1963.

5.4 R. E. Kalman, "Mathematical Description of Linear Dynamical Systems," *J. Soc. Ind. Appl. Math.-Control Series*, Ser. A, 1(2), 1964, 152–192.

5.5 P. C. Hughes and R. E. Skelton, "Controllability and Observability of Linear Matrix Second Order Systems," *J. Applied Mechanics*, 47(2), 415–420, June 1980.

5.6 P. C. Hughes and R. E. Skelton, "Controllability and Observability for Flexible Spacecraft," *J. Guidance and Control*, 3(5), 452–459, Sept. 1980.

5.7 R. E. Skelton and C. Gregory, "Measurement Feedback and Model Reduction by Modal Cost Analysis," *1979 Joint Automatic Control Conf. Proceedings*, 211–218, 1979.

5.8 R. E. Skelton and A. Yousuff, "Component Cost Analysis of Large Scale Systems," *Int. J. Control*, 37(2), 285–302, 1983.

5.9 R. E. Skelton and P. C. Hughes, "Modal Cost Analysis for Linear Matrix Second Order Systems," *J. Dynamic Systems, Measurement and Control*, 102, 151–158, Sept. 1980.

5.10 R. E. Skelton, P. C. Hughes, and H. B. Hablani, "Order Reduction for Models of Space Structures Using Modal Cost Analysis," *J. Guidance and Control*, 5(4), 351–357, July 1982.

5.11 B. C. Moore, "Principal Component Analysis in Linear Systems; Controllability, Observability and Model Reduction," *IEEE Trans. Autom. Control*, AC-26, 17–32, 1981.

5.12 J. Van Amerongen and A. J. Udink Ten Cate, "Model Reference Adaptive Autopilots for Ships," *Automatica*, 11, 441–449, 1975.

5.13 A. J. Laub and W. F. Arnold, "Controllability and Observability criteria for multivariable linear second order models," *IEEE Trans. Autom. Control*, AC-29, 163–165, 1984.

5.14 V. M. Popov, "Hyperstability and Optimality of Automatic Systems with Several Control Functions," *Rev. Roum. Sci.-Electrotechn. et Energ.*, 9, 629–690, 1964.

5.15 B. D. O. Anderson and D. G. Luenberger, "Design of Multivariable Feedback System," *Proc. IEE*, 114, 395–399, 1967.

5.16 W. M. Wonham, "On Pole Assignment in Multi-Input Controllable Linear Systems," *IEEE Trans. Autom. Control*, AC-12, 660–665, 1967.

5.17 J. W. S. Rayleigh, *The Theory of Sound*. (Macmillan 1877, 1894, 1926, 1929; Dover, 1945, New York.

5.18 R. W. Clough and J. Penzien, *Dynamics of Structures*, McGraw-Hill, New York, 1975, pp. 194–199.

5.19 C. T. Mullis and R. A. Roberts, "The Use of Second-Order Information in the Approximation of Discrete Time Linear Systems," *IEEE Trans. Acoustics, Speech, and Signal Processing*, ASSP-24(3), June 1976.

5.20 B. D. O. Anderson and R. E. Skelton, "The Generation of all q-Markov Covers," *IEEE J. Circuits and Systems*, to appear, 1987.

5.21 A. Hu and R. E. Skelton, "Modeling and Control of Beam-Like Structures," *J. Sound and Vibration*, to appear, 1987.

CHAPTER 6

Equivalent Realizations and Model Reduction

Up to this point we have discussed various properties of linear systems: (a) transfer functions, (b) output correlations, and (c) cost functions; and in Section 4.5 we have determined that coordinate transformations on a given realization will not alter these properties. However, there may exist many other realizations that preserve these properties besides those obtained by similarity transformations. Specifically, we wish now to determine the *minimal-order* realizations which preserve the exact properties of interest. From Table 4.2 we see that this is equivalent to asking for the minimal realization to match the functions

(a) $\mathbf{C}e^{\mathbf{A}t}\mathbf{B}$
(b) $\mathbf{C}e^{\mathbf{A}t}\mathbf{X}\mathbf{C}^*$
(c) $\operatorname{tr}\mathbf{C}\mathbf{X}\mathbf{C}^*$

respectively. We may incorporate a weighting matrix $\mathbf{Q}$ into each of these (a), (b), (c) by redefining the output $\mathbf{y}$ by $\sqrt{\mathbf{Q}}\,\mathbf{y}$, where $\sqrt{\mathbf{Q}^*}\sqrt{\mathbf{Q}} = \mathbf{Q}$. See in this case that $\mathscr{y} = \sqrt{\mathbf{Q}}\,\mathbf{y} = \sqrt{\mathbf{Q}}\,\mathbf{C}\mathbf{x}$, $\mathscr{y}^*\mathscr{y} = \mathbf{y}^*\mathbf{Q}\mathbf{y}$. So, one may replace $\mathbf{C}$ by $\sqrt{\mathbf{Q}}\,\mathbf{C}$ to include a weight $\mathbf{Q}$ in the problem. A physical motivation for choices of $\mathbf{Q}$ will appear in Chapter 8.

There are two reasons why the minimal realization is an important goal: simplification of *synthesis* and *analysis*. If electronic circuits or other dynamic devices must be constructed to yield the desired property (this is the *synthesis* problem), then we wish to use the fewest number of components possible. On the other hand, if we are engaged in the analysis of a given complicated system, we save on hand and computer work (and also on numerical *accuracy!*) by using the simplest possible model of that system.

6.1 Transfer Equivalent Realizations (TERs)

It has already been proved that similarity transformations do not alter the transfer function (Section 4.3.3). Hence, we may use any convenient set of coordinates to perform our study. Here, we choose modal coordinates. Since now A is in Jordan form

$$\mathbf{C}e^{\mathbf{A}t}\mathbf{B} = [\mathbf{C}_1 \dots \mathbf{C}_p]\begin{bmatrix} e^{\Lambda_1 t} & & \\ & \ddots & \\ & & e^{\Lambda_p t}\end{bmatrix}\begin{bmatrix}\mathbf{B}_1 \\ \vdots \\ \mathbf{B}_p\end{bmatrix}$$

$$= \sum_{i=1}^{p} \mathbf{C}_i e^{\Lambda_i t}\mathbf{B}_i,$$

where Λ_i, $i = 1, \dots, p$, are the Jordan blocks of $\mathbf{A}$ and the transfer function is

$$\mathbf{G}(s) = \mathbf{C}(s\mathbf{I} - \mathbf{A})^{-1}\mathbf{B} = \sum_{i=1}^{p} \mathbf{C}_i[sI - \Lambda_i]^{-1}\mathbf{B}_i.$$

Definition 6.1
A minimal transfer equivalent realization (TER) of (A, B, C) is a realization of lowest order that matches the transfer function $\mathbf{C}(s\mathbf{I} - \mathbf{A})^{-1}\mathbf{B}$.

Gilbert [6.1] and Kalman [6.2] proved the following.

Theorem 6.1
Of all TERs of

$$\dot{\mathbf{x}} = \mathbf{A}\mathbf{x} + \mathbf{B}\mathbf{u}, \qquad \mathbf{x} \in \mathscr{C}^{n_x}, \qquad\qquad (6.1)$$
$$\mathbf{y} = \mathbf{C}\mathbf{x},$$

minimal TERs are observable and controllable, and these minimal TERs are related by a similarity transformation. Furthermore, if any realization is not controllable and observable, it can be reduced in order without changing its transfer function.

Proof: We shall prove only the simplest case for distinct eigenvalues. In this case,

$$\mathbf{G}(s) = \mathbf{C}(s\mathbf{I} - \mathbf{A})^{-1}\mathbf{B} = \sum_{i=1}^{n_x} \frac{\mathbf{J}_i}{s - \lambda_i}, \qquad \mathbf{J}_i = \mathbf{c}_i\mathbf{b}_i^*. \qquad (6.2)$$

Hence, the degree of denominator of $\mathbf{G}(s)$ can be reduced in order below n_x iff $\mathbf{J}_i = 0$ for some i. But from the corollaries to Theorems 5.9 and 5.15, $\mathbf{J}_i = 0$ is possible iff mode i is either uncontrollable ($\mathbf{b}_i^* = 0$) or unobservable ($\mathbf{c}_i = 0$). Even without requiring modal coordinates for the discussion, it is clear from (6.2) that in the single-input/output case, if the polynomials $\mathbf{C}[\text{adj}\,(s\mathbf{I} - \mathbf{A})]\mathbf{B}$ and $|s\mathbf{I} - \mathbf{A}|$ have a common factor, then the same transfer function can be realized by a lower-order realization. ■

Theorem 6.1 proves that *the minimal transfer equivalent realization (TER) is controllable and observable.*

EXAMPLE 6.1

Find a minimal TER for (5.42).

Solution: The rectangular method of programming of (5.42) yields the block diagram and the state equations

$$\dot{\mathbf{x}} = \begin{bmatrix} -3 & 1 \\ -2 & 0 \end{bmatrix}\mathbf{x} + \begin{bmatrix} 1 \\ 1 \end{bmatrix}u, \qquad y = [1 \quad 0]\mathbf{x}, \tag{6.3}$$

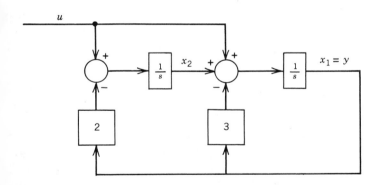

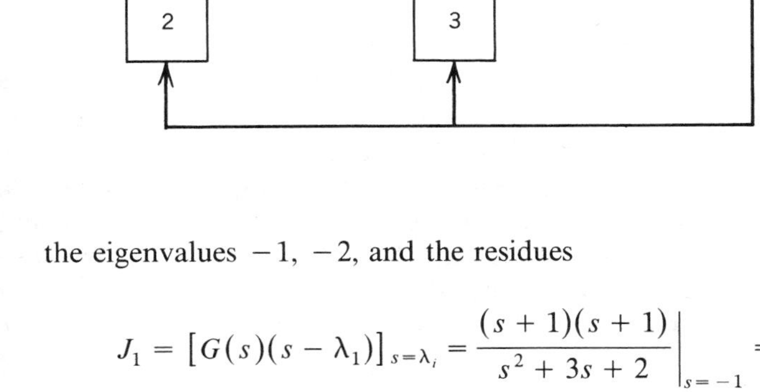

the eigenvalues -1, -2, and the residues

$$J_1 = [G(s)(s - \lambda_1)]_{s=\lambda_i} = \left.\frac{(s+1)(s+1)}{s^2 + 3s + 2}\right|_{s=-1} = \left.\frac{s+1}{s+2}\right|_{s=-1} = 0,$$

$$J_2 = [G(s)(s - \lambda_2)]_{s=\lambda_2} = \left.\frac{(s+1)(s+2)}{s^2 + 3s + 2}\right|_{s=-2} = 1.$$

Thus, according to Theorem 6.1, the system can be reduced to first order. Indeed, had we noted (5.42) more carefully, we could have seen that

$$\frac{s+1}{s^2+3s+2} = \frac{s+1}{(s+1)(s+2)} = \frac{1}{s+2}$$

yields the *first*-order controllable observable system

$$\begin{aligned} \dot{x} &= -2x + u, \\ y &= x. \end{aligned} \tag{6.4}$$

See that (6.3) and (6.4) have the same transfer function, as promised by Theorem 6.1. Note also that a different state space realization of (5.42) is the phase variable form with the state equations

$$\begin{aligned} \dot{\mathbf{x}} &= \begin{bmatrix} 0 & 1 \\ -2 & -3 \end{bmatrix}\mathbf{x} + \begin{pmatrix} 0 \\ 1 \end{pmatrix}u, \\ \mathbf{y} &= [1 \quad 1]\mathbf{x}. \end{aligned} \tag{6.5}$$

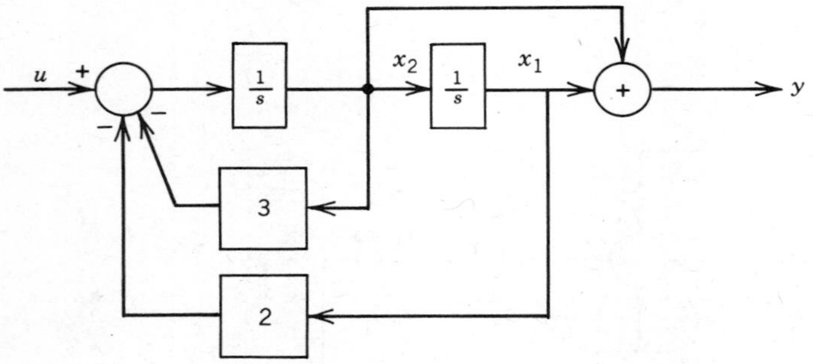

Verify that (6.3) is observable but uncontrollable, while (6.5) is unobservable but controllable. See also that (6.3) and (6.5) have the same transfer function *but that they are not similar*. That is, no coordinate transformation will transform (6.3) into (6.5) or vice versa, for the existence of such a transformation would contradict Theorem 5.8 and Section 5.3.1. The conclusion from this discussion is that uncontrollable subsystems or states may always be deleted from the state space realization without altering the transfer function. The balanced coordiantes of Section 5.6.2 also allows the construction of the minimal realization with respect to the transfer function. ■

Theorem 6.2

The realization $(\mathbf{A}_b, \mathbf{B}_b, \mathbf{C}_b)$ *given by* (5.181) *is a minimal transfer equivalent realization* (*TER*) *of* $(\mathbf{A}, \mathbf{B}, \mathbf{C})$.

Proof: See that the controllability and observability grammians are equal, and positive definite, from (5.182). ■

What Your Transfer Function Never Told You

Transfer equivalent realizations (TERs) can be very useful, but they can also lead to disaster without proper formulation of the problem. Practically every physical system is not completely controllable (note that several of the systems in Chapter 3 are uncontrollable). Hence, a hasty search for the TER will delete he uncontrollable part of the system even if this part is observable. The analyst must decide if this is OK. Quite often it is not OK. Consider, for example, the need to control a system ($\dot{\mathbf{x}} = \mathbf{Ax} + \mathbf{Bu} + \mathbf{Dw}$, $\mathbf{y} = \mathbf{Cx}$) in the presence of troublesome disturbances satisfying some model ($\dot{\mathbf{x}}_w = \mathbf{A}_w\mathbf{x}_w + \mathbf{u}_w$, $\mathbf{w} = \mathbf{C}_w\mathbf{x}_w$). The complete system

$$\begin{pmatrix}\dot{\mathbf{x}}\\ \dot{\mathbf{x}}_w\end{pmatrix} = \begin{bmatrix}\mathbf{A} & \mathbf{DC}_w\\ \mathbf{0} & \mathbf{A}_w\end{bmatrix}\begin{pmatrix}\mathbf{x}\\ \mathbf{x}_w\end{pmatrix}\mathbf{u} + \begin{bmatrix}\mathbf{0}\\ \mathbf{I}\end{bmatrix}\mathbf{u}_w + \begin{bmatrix}\mathbf{B}\\ \mathbf{0}\end{bmatrix}\mathbf{u},$$
$$\mathbf{y} = [\mathbf{C} \quad \mathbf{0}]\begin{pmatrix}\mathbf{x}\\ \mathbf{x}_w\end{pmatrix}, \tag{6.6}$$

is not controllable from $\mathbf{u}$, but it can easily be observable (suppose it is). A TER of this system from input $\mathbf{u}$ to output $\mathbf{y}$ is useless for control design because it ignores the disturbance. The uncontrollable part (the disturbance $\mathbf{x}_w$), generates the need to control in the first place, yet would be eliminated from the model in such a TER. However, a TER from the *total* set of inputs $\begin{pmatrix}\mathbf{u}\\ \mathbf{u}_w\end{pmatrix}$ to $\mathbf{y}$ is useful for design and analysis.

Even without disturbances, caution is required in the use of transfer functions. Recall from Chapter 4 that for $\dot{\mathbf{x}} = \mathbf{Ax} + \mathbf{Bu}$, $\mathbf{y} = \mathbf{Cx}$, Laplace transform of the output $\mathbf{y}(s)$ is

$$\mathbf{y}(s) = \mathbf{C}(s\mathbf{I} - \mathbf{A})^{-1}\mathbf{x}(0) + \mathbf{C}(s\mathbf{I} - \mathbf{A})^{-1}\mathbf{Bu}(s).$$

Hence, the actual response of the system $\mathbf{y}(t)$ excited by some nonzero $\mathbf{x}(0)$ could yield properties (even instabilities!) which cannot be explained by the transfer function $\mathbf{C}(s\mathbf{I} - \mathbf{A})^{-1}\mathbf{B}$. For example, observable modes can be excited by $\mathbf{x}(0)$ but may not appear in $\mathbf{C}(s\mathbf{I} - \mathbf{A})^{-1}\mathbf{B}$ due to pole-zero cancellation (uncontrollability). However, it is safe to seek the minimal TER between *all* inputs $\begin{pmatrix}\mathbf{u}\\ \mathbf{x}(0)\end{pmatrix}$ and $\mathbf{y}$,

$$\mathbf{y}(s) = \mathbf{C}(s\mathbf{I} - \mathbf{A})^{-1}[\mathbf{I}, \mathbf{B}]\begin{bmatrix}\mathbf{x}(0)\\ \mathbf{u}(s)\end{bmatrix}.$$

EXAMPLE 6.2

Prove that two different controllable realizations of a system are related by the coordinate transformation

$$\mathbf{x}_1 = \mathbf{Tx}_2, \qquad \mathbf{T} \triangleq \mathbf{W}_{C1}\mathbf{W}_{C2}^*(\mathbf{W}_{C2}\mathbf{W}_{C2}^*)^{-1}. \tag{6.7}$$

Solution: The system $(\mathbf{A}_1, \mathbf{B}_1, \mathbf{C}_1)$ with state $\mathbf{x}_1$ has

$$\mathbf{W}_{C1} = \left[\mathbf{B}_1, \mathbf{A}_1\mathbf{B}_1, \ldots, \mathbf{A}_1^{n_x-1}\mathbf{B}_1\right],$$

and any coordinate transformation $(\mathbf{T}^{-1}\mathbf{A}_1\mathbf{T}, \mathbf{T}^{-1}\mathbf{B}_1, \mathbf{C}_1\mathbf{T})$ yields

$$\mathbf{W}_{C2} = \mathbf{T}^{-1}\left[\mathbf{B}_1, \mathbf{A}_1\mathbf{B}_1, \ldots, \mathbf{A}_1^{n_x-1}\mathbf{B}_1\right] = \mathbf{T}^{-1}\mathbf{W}_{C1}. \tag{6.8}$$

Since $(\mathbf{A}_1, \mathbf{B}_1)$ is controllable by assumption, then any TER of $(\mathbf{A}_1, \mathbf{B}_1)$ will be controllable. Hence, $\mathbf{W}_{C2}\mathbf{W}_{C2}^*$ has an inverse, and (6.8) yields

$$\mathbf{T} = \mathbf{W}_{C1}\mathbf{W}_{C2}^*(\mathbf{W}_{C2}\mathbf{W}_{C2}^*)^{-1}.$$

■

EXAMPLE 6.3

Prove that any two observable realizations of a system are related by the coordinate transformation

$$\mathbf{x}_1 = \mathbf{T}\mathbf{x}_2, \qquad \mathbf{T} = (\mathbf{W}_{01}^*\mathbf{W}_{01})^{-1}\mathbf{W}_{01}^*\mathbf{W}_{02} \tag{6.9}$$

Solution: The system $(\mathbf{A}_1, \mathbf{B}_1, \mathbf{C}_1)$ with state $\mathbf{x}_1$ has

$$\mathbf{W}_{01}^* = \left[\mathbf{C}_1^*, \mathbf{A}_1^*\mathbf{C}_1^*, \ldots, \mathbf{A}_1^{*n_x-1}\mathbf{C}_1^*\right],$$

and any similarity transformation $(\mathbf{T}^{-1}\mathbf{A}_1\mathbf{T}, \mathbf{T}^{-1}\mathbf{B}_1, \mathbf{C}_1\mathbf{T})'$ yields

$$\mathbf{W}_{02}^* = \mathbf{T}^*\mathbf{W}_{01}^*. \tag{6.10}$$

Since $(\mathbf{A}_1, \mathbf{C}_1)$ is observable by assumption, then any TER of $(\mathbf{A}_1, \mathbf{B}_1, \mathbf{C}_1)$ will be observable (and coordinate transformations do not alter transfer functions, Theorem 4.3). Hence, $\mathbf{W}_{01}^*\mathbf{W}_{01}$ has an inverse, and (6.10) leads to

$$\mathbf{T}^* = \mathbf{W}_{02}^*\mathbf{W}_{01}(\mathbf{W}_{01}^*\mathbf{W}_{01})^{-1},$$

which verifies (6.9). ■

EXAMPLE 6.4

Verify that two controllable, observable realizations of a system have the properties

$$\mathbf{W}_{01}^*\mathbf{W}_{01}\mathbf{W}_{C1}\mathbf{W}_{C2}^* = \mathbf{W}_{01}^*\mathbf{W}_{02}\mathbf{W}_{C2}\mathbf{W}_{C2}^*. \tag{6.11}$$

Solution: From Example 6.2 we must conclude that the transformation is

$$\mathbf{T} = \mathbf{W}_{C1}\mathbf{W}_{C2}^*(\mathbf{W}_{C2}\mathbf{W}_{C2}^*)^{-1}, \tag{6.12a}$$

and from Example 6.3 we must conclude that the transformation is

$$\mathbf{T} = (\mathbf{W}_{01}^* \mathbf{W}_{01})^{-1} \mathbf{W}_{01}^* \mathbf{W}_{02}. \tag{6.12b}$$

Hence, if these **T**'s are unique, they must be equal. The linear algebra problems (6.8) and (6.10) each have unique solutions iff $\mathbf{W}_{C1}^*$ has full column rank and $\mathbf{W}_{01}$ has full column rank. These conditions are both satisfied by assumption of controllability and observability. Hence, **T** is unique, and the **T** of (6.7) and (6.9) can be set equal to give (6.11) immediately. ∎

6.2 *Output Correlation Equivalent Realization (COVER)*

Note that matching the transfer function allows one to match $\mathbf{y}(t)$ for *every* possible choice of $\mathbf{u}(t)$. If the inputs are restricted to a special class, then it might *not* be necessary to match the transfer function in order to match $\mathbf{y}(t)$. Also, it may not be required to match $\mathbf{y}(t)$ exactly, but to match special properties of $\mathbf{y}(t)$. In this section we focus on the output correlation

$$\mathbf{R}(t) = \sum_{i=1}^{n_u} \int_0^\infty \mathbf{y}^i(t+\tau)\mathbf{y}^{i*}(\tau)\, d\tau \tag{6.13}$$

for impulsive inputs (all input channels should be included in the number n_u). From (4.122c), the Laplace transform of $\mathbf{R}(t)$ has the power series in $1/s$,

$$\mathbf{R}(s) = \mathbf{C}[s\mathbf{I} - \mathbf{A}]^{-1}\mathbf{X}\mathbf{C}^* = \sum_{i=0}^{\infty} \frac{\mathbf{R}_i}{s^{i+1}},$$

where **X** has been called a state *covariance* matrix in Chapter 4 (p. 000) and $\mathbf{R}_i$ the *covariance parameters*, where

$$\mathbf{R}_i = \mathbf{C}\mathbf{A}^i\mathbf{X}\mathbf{C}^*, \qquad \mathbf{0} = \mathbf{X}\mathbf{A}^* + \mathbf{A}\mathbf{X} + \mathbf{B}\mathscr{U}\mathbf{B}^*. \tag{6.14}$$

By contrast, the transfer function has the power series (4.160d) in $1/s$,

$$\mathbf{G}(s) = C[s\mathbf{I} - \mathbf{A}]^{-1}\mathbf{B} = \sum_{i=0}^{\infty} \frac{\mathbf{M}_i}{s^{i+1}}, \qquad \mathbf{M}_i = \mathbf{C}\mathbf{A}^i\mathbf{B}. \tag{6.15}$$

Definition 6.2

A covariance equivalent realization (COVER) of a linear system $(\mathbf{A}, \mathbf{B}, \mathbf{C})$ *is one which matches* $\mathbf{R}_i$, $i = 0, 1, \ldots, \infty$, *where* $\mathbf{R}_i$ *is defined by* (6.14). *A COVER obviously matches* $\mathbf{R}(t)$ *in* (6.13) *for all* t.

Definition 6.3
A q-COVER of a linear system $(\mathbf{A}, \mathbf{B}, \mathbf{C})$ *is one which matches* $\mathbf{R}_i$, $i = 0, 1, \ldots, q-1$.

Definition 6.4
A q-Markov COVER of a linear system $(\mathbf{A}, \mathbf{B}, \mathbf{C})$ *is one which matches* $\mathbf{M}_i$ *and* $\mathbf{R}_i$, $i = 0, 1, \ldots, q-1$.

We shall give an algorithm to construct a q-Markov COVER for a controllable but not necessarily observable system. But first some motivation.

The parameters $\mathbf{M}_i$ and $\mathbf{R}_i$ have the following physical significance. Let $\mathbf{y}^i(t)$ be the output due to an impulsive input in the ith input $u_i(t) = \mathscr{u}_i \delta(t)$, $\mathscr{u}_i^2 > 0$, $u_j(t) = 0$, $i \neq j$. Then $\mathscr{U}^{1/2} = \text{diag}[\mathscr{u}_i \ldots]$ and

$$\left[\mathbf{y}^1(t), \mathbf{y}^2(t), \ldots, \mathbf{y}^{n_u}(t)\right] = \mathbf{C}e^{\mathbf{A}t}\mathbf{B}\mathscr{U}^{1/2} = \left[\sum_{i=0}^{\infty} \frac{\mathbf{M}_i t^i}{i!}\right]\mathscr{U}^{1/2}. \tag{6.16}$$

Hence,

$$\int_0^{\infty} \mathbf{y}^1(t+\tau)\mathbf{y}^{1*}(\tau)\,d\tau + \cdots + \int_0^{\infty} \mathbf{y}^{n_u}(t+\tau)\mathbf{y}^{n_u^*}(\tau)\,d\tau = \mathbf{C}e^{\mathbf{A}t}\mathbf{X}\mathbf{C}^* = \sum_{i=0}^{\infty} \frac{\mathbf{R}_i t^i}{i!}. \tag{6.17}$$

Hence, a q-Markov COVER matches the first q derivatives of the impulse response $\mathbf{C}e^{\mathbf{A}t}\mathbf{B}\mathscr{U}^{1/2}$ at $t = 0$ and also the first q derivatives of the output correlation $\mathbf{C}e^{\mathbf{A}t}\mathbf{X}\mathbf{C}^*$ at $t = 0$. Now, in the limit as $q \to \infty$, the entire impulse response $\mathbf{C}e^{\mathbf{A}t}\mathbf{B}\mathscr{U}^{1/2}$ (and its Laplace transform, the transfer function $\mathbf{C}(s\mathbf{I} - \mathbf{A})^{-1}\mathbf{B}$) and the entire output correlation $\mathbf{C}e^{\mathbf{A}t}\mathbf{X}\mathbf{C}^*$ are matched. However, the interesting cases are for $0 < q < n_x$ since these give reduced order models. In these cases the *transient* properties will not exactly match that of the original system but will start off at $t = 0$ in the "right direction" up to q derivatives of $\mathbf{y}(t)$, and the *steady-state* properties (6.17) will not exactly match those of the original system but will match the steady-state quantity

$$\mathbf{R}_0 = \sum_{i=1}^{n_u} \int_0^{\infty} \mathbf{y}^i(\tau)\mathbf{y}^{i*}(\tau)\,d\tau$$

plus $q - 1$ derivatives of the output correlation. For $q \geq 1$, this means that for a multioutput system the mean squared value of *each* output, y_α, $\alpha = 1, 2, \ldots, n_y$, defined by

$$[y_{\alpha_{\text{RMS}}}]^2 \triangleq \left[\sum_{i=1}^{n_u} \int_0^{\infty} y_\alpha^2(\tau)\,d\tau\right] = \left[\sum_{i=1}^{n_u} \int_0^{\infty} \mathbf{y}^i(\tau)\mathbf{y}^{i*}(\tau)\,d\tau\right]_{\alpha\alpha} \tag{6.18}$$

is preserved in the q-Markov COVER. This can be important since the usual purpose of simulation or control design is to monitor (or control) the mean squared value of multiple outputs (antenna tracking, flexible structure deflections at multiple points on the structure, etc.). The q-Markov COVER method of model reduction which we will discuss has this property and does not compromise information about the RMS values of each of the multiple outputs in the model reduction process.

Theorem 6.3

Given any asymptotically stable, controllable system $(\mathbf{A}, \mathbf{B}, \mathbf{C}, \mathbf{X})$ *where*

$$\mathbf{X} \triangleq \int_0^\infty e^{\mathbf{A}t}\mathbf{B}\mathscr{U}\mathbf{B}^* e^{\mathbf{A}^*t}\, dt > 0, \tag{6.19}$$

then $(\mathbf{A}_R, \mathbf{B}_R, \mathbf{C}_R, \mathbf{X}_R)$ *is a q-Markov COVER of* $(\mathbf{A}, \mathbf{B}, \mathbf{C}, \mathbf{X})$ *if*

$$\mathbf{A}_R = \mathbf{T}_R^* \mathbf{V}_1^* \mathbf{T}^{-1}\mathbf{A}\mathbf{T}\mathbf{V}_1\mathbf{T}_R, \tag{6.20a}$$

$$\mathbf{B}_R = \mathbf{T}_R^* \mathbf{V}_1^* \mathbf{T}^{-1}\mathbf{B}, \tag{6.20b}$$

$$\mathbf{C}_R = \mathbf{C}\mathbf{T}\mathbf{V}_1\mathbf{T}_R, \tag{6.20c}$$

provided $(\mathbf{A}_R, \mathbf{B}_R)$ *is a controllable pair,* $\mathbf{T}_R$ *is an arbitrary unitary matrix, and* $\mathbf{T}$ *is any matrix factor of* $\mathbf{X}$,

$$\mathbf{X} = \mathbf{T}\mathbf{T}^*, \tag{6.20d}$$

where $\mathbf{V}_1$ *is defined by the singular value decomposition*

$$\mathbf{W}_{0q}\mathbf{T} = [\mathbf{U}_1 \quad \mathbf{U}_2]\begin{bmatrix} \Sigma & \mathbf{0} \\ \mathbf{0} & \mathbf{0} \end{bmatrix}\begin{bmatrix} \mathbf{V}_1^* \\ \mathbf{V}_2^* \end{bmatrix}, \qquad \mathbf{W}_{0q} \triangleq \begin{bmatrix} \mathbf{C} \\ \mathbf{C}\mathbf{A} \\ \vdots \\ \mathbf{C}\mathbf{A}^{q-1} \end{bmatrix}. \tag{6.20e}$$

For the proof, we need the following result.

Lemma 6.1

If $\mathbf{A}_R + \mathbf{A}_R^* + \mathbf{B}_R\mathscr{U}\mathbf{B}_R^* = \mathbf{0}$, *then* $\mathbf{A}_R$ *is asymptotically stable iff* $(\mathbf{A}_R, \mathbf{B}_R)$ *is a controllable pair. If* $(\mathbf{A}_R, \mathbf{B}_R)$ *is not a controllable pair, then* $\mathbf{A}_R$ *has at least one eigenvalue on the* $j\omega$ *axis.* $\mathbf{A}_R$ *cannot have any eigenvalues in the right half-plane.*

Proof of Lemma in the Case of Nondefective $\mathbf{A}_R$: Replace $\mathbf{A}_R$ by its spectral decomposition $\mathbf{A}_R = \mathbf{E}\Lambda\mathbf{E}^{-1}$ and multiply $\mathbf{A}_R + \mathbf{A}_R^* + \mathbf{B}_R\mathscr{U}\mathbf{B}_R^* = 0$ from the left by $\mathbf{E}^{-1}$ and from the right by $\mathbf{E}^{-*}$. This yields

$$\begin{aligned} &\mathbf{E}^{-1}[\mathbf{0} = (\mathbf{E}\Lambda\mathbf{E}^{-1})^* + \mathbf{E}\Lambda\mathbf{E}^{-1} + \mathbf{B}_R\mathscr{U}\mathbf{B}_R^*]\mathbf{E}^{-*}, \\ &\mathbf{0} = \mathbf{E}^{-1}\mathbf{E}^{-*}\Lambda^* + \Lambda\mathbf{E}^{-1}\mathbf{E}^{-*} + \mathbf{E}^{-1}\mathbf{B}_R\mathscr{U}\mathbf{B}_R^*\mathbf{E}^{-*}. \end{aligned} \tag{6.21}$$

The left eigenvectors of $\mathbf{A}$ are $\mathbf{l}_i^*$:

$$\mathbf{E}^{-1} = \begin{bmatrix} \vdots \\ \mathbf{l}_i^* \\ \vdots \end{bmatrix}.$$

Then, the iith element of (6.21) is

$$0 = \mathbf{l}_i^* \mathbf{l}_i \bar{\lambda}_i + \lambda_i \mathbf{l}_i^* \mathbf{l}_i + \mathbf{l}_i^* \mathbf{B}_R \mathscr{U} \mathbf{B}_R^* \mathbf{l}_i$$

or, equivalently,

$$\bar{\lambda}_i + \lambda_i = 2\,\mathscr{Re}\,\lambda_i = -\frac{\mathbf{l}_i^* \mathbf{B}_R \mathscr{U} \mathbf{B}_R^* \mathbf{l}_i}{\mathbf{l}_i^* \mathbf{l}_i} = -\frac{\|\mathbf{l}_i\|^2 \mathbf{B}_R \mathscr{U} \mathbf{B}_R^*}{\|\mathbf{l}_i\|^2},$$

which cannot be a positive number. This number can be zero iff $\mathbf{l}_i^* \mathbf{B}_R = \mathbf{0}$, but this is precisely the uncontrollability condition for mode i. Hence, $\mathbf{A}_R$ is asymptotically stable iff $(\mathbf{A}_R, \mathbf{B}_R)$ is controllable. ■

Proof of Theorem 6.3: Define

$$\hat{\mathbf{A}} \triangleq \mathbf{T}^{-1}\mathbf{A}\mathbf{T}, \quad \hat{\mathbf{B}} \triangleq \mathbf{T}^{-1}\mathbf{B}, \quad \hat{\mathbf{C}} = \mathbf{C}\mathbf{T}, \quad \hat{\mathbf{X}} = \mathbf{T}^{-1}\mathbf{X}\mathbf{T}^{-*} = \mathbf{I},$$

$$\hat{\mathbf{W}}_{0q} = \mathbf{W}_{0q}\mathbf{T} = \mathbf{U}\begin{bmatrix} \boldsymbol{\Sigma} & \mathbf{0} \\ \mathbf{0} & \mathbf{0} \end{bmatrix}\begin{bmatrix} \mathbf{V}_1^* \\ \mathbf{V}_2^* \end{bmatrix}. \tag{6.22}$$

We must prove that

$$\hat{\mathbf{W}}_{0q}[\hat{\mathbf{B}}, \hat{\mathbf{C}}^*] = \mathbf{W}_{0qR}[\mathbf{B}_R, \mathbf{C}_R^*], \qquad \mathbf{W}_{0qR} \triangleq \begin{bmatrix} \mathbf{C}_R \\ \mathbf{C}_R \mathbf{A}_R \\ \vdots \\ \mathbf{C}_R \mathbf{A}_R^{q-1} \end{bmatrix} \tag{6.23}$$

and that

$$\mathbf{X}_R \triangleq \int_0^\infty e^{\mathbf{A}_R t} \mathbf{B}_R \mathscr{U} \mathbf{B}_R^* e^{\mathbf{A}_R^* t}\, dt = \mathbf{I}.$$

(Note that $\mathbf{M}_i = [\hat{\mathbf{W}}_{0q}\hat{\mathbf{B}}]_{i \text{ block of rows}}$ and $\mathbf{R}_i = [\hat{\mathbf{W}}_{0q}\hat{\mathbf{C}}^*]_{i \text{ block of rows}}$.) First, we shall prove that $\mathbf{X}_R = \mathbf{I}$. $(\mathbf{A}_R, \mathbf{B}_R)$ is assumed to be controllable, in which case (by Lemma 6.1) $\mathbf{X}_R$ exists and satisfies

$$\mathbf{0} = \mathbf{X}_R \mathbf{A}_R^* + \mathbf{A}_R \mathbf{X}_R + \mathbf{B}_R \mathscr{U} \mathbf{B}_R^*$$

or

$$\mathbf{0} = \mathbf{X}_R\left[\mathbf{T}_R^*\mathbf{V}_1^*\hat{\mathbf{A}}\mathbf{V}_1\mathbf{T}_R\right]^* + \left[\mathbf{T}_R^*\mathbf{V}_1^*\hat{\mathbf{A}}\mathbf{V}_1\mathbf{T}_R\right]\mathbf{X}_R + \mathbf{T}_R^*\mathbf{V}_1^*\hat{\mathbf{B}}\mathscr{U}\hat{\mathbf{B}}_*\mathbf{V}_1\mathbf{T}_R.$$

$\mathbf{X}_R = \mathbf{I}$ is clearly one solution since

$$\mathbf{0} = \mathbf{T}_R^*\mathbf{V}_1^*[\hat{\mathbf{A}}^* + \hat{\mathbf{A}} + \hat{\mathbf{B}}\mathscr{U}\hat{\mathbf{B}}^*]\mathbf{V}_1\mathbf{T}_R$$

is satisfied due to the fact that $\hat{\mathbf{X}} = \mathbf{I}$ implies that $\hat{\mathbf{A}}^* + \hat{\mathbf{A}} + \hat{\mathbf{B}}\mathscr{U}\hat{\mathbf{B}}^* = \mathbf{0}$. But by Lemma 6.1, $\mathbf{A}_R$ is asymptotically stable, by virtue of the controllability assumption in Theorem 6.3. Hence, $\lambda_i[\mathbf{A}_R] + \lambda_j[\mathbf{A}_R] \neq 0$ for all i, j guarantees that $\mathbf{X}_R = \mathbf{I}$ is unique.

Now, to prove (6.23), write, from (6.20),

$$\mathbf{W}_{0qR}[\mathbf{B}_R, \mathbf{C}_R^*] = \begin{bmatrix} \hat{\mathbf{C}}\mathbf{V}_1\mathbf{T}_R \\ \vdots \\ \hat{\mathbf{C}}\mathbf{V}_1\mathbf{T}_R\left[\mathbf{T}_R^*\mathbf{V}_1^*\hat{\mathbf{A}}\mathbf{V}_1\mathbf{T}_R\right]^{q-1} \end{bmatrix}\left[\mathbf{T}_R^*\mathbf{V}_1^*\hat{\mathbf{B}}, \mathbf{T}_R^*\mathbf{V}_1^*\hat{\mathbf{C}}^*\right].$$

Noting that $\mathbf{T}_R\mathbf{T}_R^* = \mathbf{I}$, this becomes

$$\mathbf{W}_{0qR}[\mathbf{B}_R, \mathbf{C}_R^*] = \begin{bmatrix} \hat{\mathbf{C}} \\ \vdots \\ \hat{\mathbf{C}}\hat{\mathbf{A}}^{q-1} \end{bmatrix}\mathbf{V}_1\mathbf{V}_1^*[\hat{\mathbf{B}}, \hat{\mathbf{C}}^*] = \hat{\mathbf{W}}_{0q}\mathbf{V}_1\mathbf{V}_1^*[\hat{\mathbf{B}}, \hat{\mathbf{C}}^*].$$

But $\hat{\mathbf{W}}_{0q}\mathbf{V}_2 = \mathbf{0}$ and $\mathbf{V}_1\mathbf{V}_1^* = \mathbf{I} - \mathbf{V}_2\mathbf{V}_2^*$ from (6.22) yield $\mathbf{W}_{0qR}[\mathbf{B}_R, \mathbf{C}_R^*] = \hat{\mathbf{W}}_{0q}[\hat{\mathbf{B}}, \hat{\mathbf{C}}^*]$. ■

Note that coordinate transformations do not change the Markov parameters $\mathbf{M}_i$ or the covariance parameters $\mathbf{R}_i$. Hence, any transformation of the result of (6.20), $(\mathbf{P}^{-1}\mathbf{A}_R\mathbf{P}, \mathbf{P}^{-1}\mathbf{B}_R, \mathbf{C}_R\mathbf{P})$, is also a q-Markov COVER.

Theorem 6.4

A q-Markov COVER of $(\mathbf{A}, \mathbf{B}, \mathbf{C})$ *is obtained by truncation of observable Hessenberg coordinates* (5.191) *to retain the first* $n_r = \sum_{i=1}^{q} n_i$ *states.*

Proof: The theorem states that $(\mathbf{A}_R, \mathbf{B}_R, \mathbf{C}_R)$ is a q-Markov COVER of $(\mathbf{A}_{\mathrm{OH}}, \mathbf{B}_{\mathrm{OH}}, \mathbf{C}_{\mathrm{OH}})$ if

$$\mathbf{A}_R = [\mathbf{I} \quad \mathbf{0}]\mathbf{A}_{\mathrm{OH}}\begin{bmatrix}\mathbf{I}\\ \mathbf{0}\end{bmatrix}, \quad \mathbf{B}_R = [\mathbf{I} \quad \mathbf{0}]\mathbf{B}_{\mathrm{OH}}, \quad \mathbf{C}_R = \mathbf{C}_{\mathrm{OH}}\begin{bmatrix}\mathbf{I}\\ \mathbf{0}\end{bmatrix}, \qquad (6.24)$$

where $(\mathbf{A}_{OH}, \mathbf{B}_{OH}, \mathbf{C}_{OH})$ is the observable Hessenberg form

$$\mathbf{A}_{OH} = \begin{bmatrix} \mathbf{A}_{11} & \mathbf{A}_{12} & \mathbf{0} & \mathbf{0} & \cdots & \mathbf{0} \\ \mathbf{A}_{21} & \mathbf{A}_{22} & \mathbf{A}_{23} & \mathbf{0} & \cdots & \cdot \\ \mathbf{A}_{31} & & & \ddots & \cdots & \cdot \\ \cdot & & & & \cdots & \cdot \\ \cdot & & & & \cdots & \mathbf{0} \\ \cdot & & & & \cdots & \mathbf{A}_{p-1,p} \\ \mathbf{A}_{p1} & \cdot & \mathbf{A}_{p,p-2} & \mathbf{A}_{p,p-1} & \cdots & \mathbf{A}_{p,p} \end{bmatrix},$$

$$\mathbf{B}_{OH} = \begin{bmatrix} \mathbf{B}_1 \\ \mathbf{B}_2 \\ \vdots \\ \mathbf{B}_p \end{bmatrix}, \tag{6.25}$$

$$\mathbf{C}_{OH} = [\mathbf{C}_1 \quad \mathbf{0} \quad \mathbf{0} \quad \mathbf{0} \quad \cdots \quad \mathbf{0}], \qquad \mathscr{X}_{OH} = \mathbf{I},$$

$$n_x = \sum_{i=1}^{p} n_i, \qquad \operatorname{rank} \mathbf{A}_{i,i+1} = n_{i+1} \le n_i.$$

Our proof relies only on this structure, and the *unique* observable Hessenberg form of Definition 5.5 is not required. That is, $\mathbf{A}_{i,i+1}$ need not be triangular. To prove the theorem, we must show that

$$\mathbf{W}_{0q}[\mathbf{B}_{OH}, \mathbf{C}_{OH}{}^*] = \mathbf{W}_{0qR}[\mathbf{B}_R, \mathbf{C}_R^*], \quad \mathscr{X}_{OH} = \mathbf{I}, \quad \mathscr{X}_R = \mathbf{I}. \tag{6.26}$$

In observable Hessenberg coordinates the observability matrix has the structure given in (5.192),

$$\mathbf{W}_{0q} = \begin{bmatrix} \mathbf{C}_1 & \mathbf{0} & \mathbf{0} & \mathbf{0} & \cdots & \mathbf{0} \\ * & \mathbf{C}_1\mathbf{A}_{12} & \mathbf{0} & \mathbf{0} & \cdots & \mathbf{0} \\ * & * & \mathbf{C}_1\mathbf{A}_{12}\mathbf{A}_{23} & \mathbf{0} & \cdots & \mathbf{0} \\ \vdots & & & \ddots & & \ddots \\ * & \cdot & & \mathbf{C}_1\mathbf{A}_{12}\mathbf{A}_{23}\ldots\mathbf{A}_{q-1,q} & & \mathbf{0} \end{bmatrix}$$

$$= [\mathbf{W}_{0qR}, \mathbf{0}], \tag{6.27}$$

where $\mathbf{W}_{0qR} \triangleq \mathbf{W}_{0q}\begin{bmatrix} \mathbf{I} \\ \mathbf{0} \end{bmatrix}$ and the dimension of the zero matrix is $(n_1 q) \times (\Sigma_{i=1}^p n_i - \Sigma_{i=1}^q n_i) = n_1 q \times (\Sigma_{q+1}^p n_i)$. Hence, any choice of $q < p$ yields a *reduced*-order model of dimension $r_q = \Sigma_{i=1}^q n_i$, and the zero matrix appears in

(6.27). Now, use (6.24) and (6.27) to rewrite (6.26):

$$\mathbf{W}_{0q}[\mathbf{B}, \mathbf{C}^*] = [\mathbf{W}_{0qR}, \mathbf{0}][\mathbf{B}, \mathbf{C}^*] = \mathbf{W}_{0qR}[\mathbf{I}, \mathbf{0}][\mathbf{B}, \mathbf{C}^*] = \mathbf{W}_{0qR}[\mathbf{B}_R, \mathbf{C}_R^*].$$

Now, we have only to prove that $\mathscr{X}_R = \mathbf{I}$. Under the assumption of controllability of $(\mathbf{A}_R, \mathbf{B}_R)$, this implies that

$$\mathbf{0} = \mathbf{A}_R^* + \mathbf{A}_R + \mathbf{B}_R \mathscr{U} \mathbf{B}_R^*.$$

From (6.24),

$$\mathbf{0} = [\mathbf{I} \quad \mathbf{0}]\mathbf{A}_{\mathrm{OH}}^* \begin{bmatrix} \mathbf{I} \\ \mathbf{0} \end{bmatrix} + [\mathbf{I} \quad \mathbf{0}]\mathbf{A}_{\mathrm{OH}} \begin{bmatrix} \mathbf{I} \\ \mathbf{0} \end{bmatrix} + [\mathbf{I} \quad \mathbf{0}]\mathbf{B}_{\mathrm{OH}} \mathscr{U} \mathbf{B}_{\mathrm{OH}}^* \begin{bmatrix} \mathbf{I} \\ \mathbf{0} \end{bmatrix}$$

or

$$\mathbf{0} = [\mathbf{I} \quad \mathbf{0}][\mathbf{A}_{\mathrm{OH}}^* + \mathbf{A}_{\mathrm{OH}} + \mathbf{B}_{\mathrm{OH}} \mathscr{U} \mathbf{B}_{\mathrm{OH}}^*] \begin{bmatrix} \mathbf{I} \\ \mathbf{0} \end{bmatrix},$$

which holds by virtue of the observable Hessenberg property $\mathbf{A}_{\mathrm{OH}}^* + \mathbf{A}_{\mathrm{OH}} + \mathbf{B}_{\mathrm{OH}} \mathscr{U} \mathbf{B}_{\mathrm{OH}}^* = \mathbf{0}$. ■

EXAMPLE 6.5

Find a 2-Markov COVER of all stable third-order systems. Third-order systems can be described in observable Hessenberg form as follows:

$$\begin{bmatrix} \mathbf{C} & \\ \mathbf{A} & \mathbf{B} \end{bmatrix} = \begin{bmatrix} c_1 & 0 & 0 & \\ -b_1^2/2 & A_{12} & 0 & b_1 \\ -b_1 b_2 - A_{12} & -b_2^2/2 & A_{23} & b_2 \\ -b_1 b_3 & -b_2 b_3 - A_{23} & -b_3^2/2 & b_3 \end{bmatrix}, \qquad \mathscr{X} = \mathbf{I}.$$

Solution: From Theorem 6.4,

$$\begin{bmatrix} \mathbf{C}_R & \\ \mathbf{A}_R & \mathbf{B}_R \end{bmatrix} = \begin{bmatrix} c_1 & 0 & \\ -b_1^2/2 & A_{12} & b_1 \\ -b_1 b_2 - A_{12} & -b_2^2/2 & b_2 \end{bmatrix}$$

is a 2-Markov COVER of $(\mathbf{A}, \mathbf{B}, \mathbf{C})$, provided that $(b_1^2 + b_2^2)(A_{12} + b_1 b_2/2) \neq 0$ [for controllability of $(\mathbf{A}, \mathbf{B})$], and $(A_R, B_R, C_R) = (-b_1^2/2, b_1, c_1)$ is a 1-Markov COVER of $(\mathbf{A}, \mathbf{B}, \mathbf{C})$, provided that $b_1 \neq 0$ [for controllability of (A_R, B_R)]. ■

Exercise 6.1
Show that a 1-Markov COVER of the system described by

$$G(s) = \frac{\alpha s + 1}{(s+\rho)(s+\gamma)}$$

is given by

$$G_R(s) = \frac{\alpha}{s + \dfrac{\alpha^2\rho\gamma(\rho+\gamma)}{1+\alpha^2\rho\gamma}}$$

and, hence, the 1-Markov COVER of the system

$$G(s) = \frac{0.9(s+1.1)}{(s+1)(s+10)}$$

(which has a pole and zero near cancellation) is

$$G_R(s) = \frac{0.9}{s+9.8}.$$

Verify that the 1-Markov COVER gives the correct TER [that is, $G_R(s) = G(s)$] when either $\alpha = 1/\rho$ or $\alpha = 1/\gamma$.

Theorem 6.5
If $(\mathbf{A}, \mathbf{B}, \mathbf{C})$ *is any asymptotically stable and controllable system, then*

$$\mathbf{A}_R \triangleq \mathscr{L}\mathbf{A}\mathscr{R}, \quad \mathbf{B}_R \triangleq \mathscr{L}\mathbf{B}, \quad \mathbf{C}_R \triangleq \mathbf{C}\mathscr{R}, \tag{6.28a}$$

$$\mathscr{L} \triangleq \mathbf{U}_1^* \mathbf{W}_{0q}, \qquad \mathscr{R} \triangleq \mathbf{X}\mathscr{L}^*(\mathscr{L}\mathbf{X}\mathscr{L}^*)^{-1}, \tag{6.28b}$$

and

$$\mathbf{W}_{0q} = [\mathbf{U}_1 \quad \mathbf{U}_2]\begin{bmatrix} \boldsymbol{\Sigma} & \mathbf{0} \\ \mathbf{0} & \mathbf{0} \end{bmatrix}\begin{bmatrix} \mathbf{V}_1^* \\ \mathbf{V}_2^* \end{bmatrix}, \qquad \mathbf{W}_{0q} \triangleq \begin{bmatrix} \mathbf{C} \\ \vdots \\ \mathbf{CA}^{q-1} \end{bmatrix}, \tag{6.28c}$$

is a q-Markov COVER.

Proof: First, we illustrate that these results are *independent* of the choice of coordinates for $(\mathbf{A}, \mathbf{B}, \mathbf{C})$. Then, we shall show that $(\mathbf{A}_R, \mathbf{B}_R, \mathbf{C}_R)$ is a q-Markov COVER when $(\mathbf{A}, \mathbf{B}, \mathbf{C})$ is in observable Hessenberg coordinates.

Note that the matrices $\mathbf{M}_i \triangleq \mathbf{CA}^i\mathbf{B} = (\mathbf{CT})(\mathbf{T}^{-1}\mathbf{AT})^i(\mathbf{T}^{-1}\mathbf{B})$ and $\mathbf{R}_i = \mathbf{CA}^i\mathbf{XC}^* = (\mathbf{CT})(\mathbf{T}^{-1}\mathbf{AT})^i(\mathbf{T}^{-1}\mathbf{XT}^{-*})(\mathbf{T}^*\mathbf{C}^*)$ are not influenced by a change in coordinates. Hence, if $(\mathbf{A}_R, \mathbf{B}_R, \mathbf{C}_R)$ is a q-Markov COVER of $(\mathbf{A}, \mathbf{B}, \mathbf{C})$, then it is also a q-Markov cover of $(\mathbf{T}^{-1}\mathbf{A}\mathbf{T}, \mathbf{T}^{-1}\mathbf{B}, \mathbf{C}\mathbf{T})$ for any nonsingular $\mathbf{T}$. Now, let $\mathbf{T}$ be the transformation which takes $(\mathbf{A}, \mathbf{B}, \mathbf{C})$ to observable Hessenberg form, $(\mathbf{A}_{\mathrm{OH}} = \mathbf{T}^{-1}\mathbf{A}\mathbf{T},\ \mathbf{B}_{\mathrm{OH}} = \mathbf{T}^{-1}\mathbf{B},\ \mathbf{C}_{\mathrm{OH}} = \mathbf{C}\mathbf{T})$. Now, apply (6.28) to $(\mathbf{A}_{\mathrm{OH}}, \mathbf{B}_{\mathrm{OH}}, \mathscr{X}_{\mathrm{OH}})$ to obtain, using (6.27) and $\mathscr{X}_{\mathrm{OH}} = \mathbf{I}$,

$$\mathbf{W}_{0q} = \left[\mathbf{W}_{0qR}, \mathbf{0}\right],$$

$$\mathscr{L} = \mathbf{U}_1^*\mathbf{W}_{0q} = \mathbf{U}_1^*\left[\mathbf{W}_{0qR}, \mathbf{0}\right] = \mathbf{U}_1^*\mathbf{W}_{0qR}[\mathbf{I}, \mathbf{0}] = \mathbf{P}^{-1}[\mathbf{I} \quad \mathbf{0}],$$

$$\mathscr{R} = \begin{bmatrix}\mathbf{I}\\ \mathbf{0}\end{bmatrix}\mathbf{P}^{-*}\left[\mathbf{P}^{-1}[\mathbf{I} \quad \mathbf{0}]\begin{bmatrix}\mathbf{I}\\ \mathbf{0}\end{bmatrix}\mathbf{P}^{-*}\right]^{-1} = \begin{bmatrix}\mathbf{I}\\ \mathbf{0}\end{bmatrix}\mathbf{P}^{-*}\mathbf{P}^*\mathbf{P} = \begin{bmatrix}\mathbf{I}\\ \mathbf{0}\end{bmatrix}\mathbf{P}.$$

Hence, (6.28) yields

$$\begin{aligned}\mathbf{A}_R &= \mathbf{P}^{-1}[\mathbf{I} \quad \mathbf{0}]\mathbf{A}_{\mathrm{OH}}\begin{bmatrix}\mathbf{I}\\ \mathbf{0}\end{bmatrix}\mathbf{P},\\ \mathbf{B}_R &= \mathbf{P}^{-1}[\mathbf{I} \quad \mathbf{0}]\mathbf{B}_{\mathrm{OH}}, \qquad \mathbf{C}_R = \mathbf{C}_{\mathrm{OH}}\begin{bmatrix}\mathbf{I}\\ \mathbf{0}\end{bmatrix}\mathbf{P},\end{aligned} \tag{6.29}$$

which is simply a coordinate transformation from the result of Theorem 6.4 given in (6.24). The coordinate transformation matrix is $\mathbf{P}^{-1} = \mathbf{U}_1^*\mathbf{W}_{0qR}$, where $\mathbf{W}_{0qR}$ is defined in (6.27). Since a nonsingular coordinate transformation does not alter Markov or covariance parameters the proof is complete. ■

6.3 Cost-Equivalent Realizations

Consider the system

$$\begin{aligned}\dot{\mathbf{x}} &= \mathbf{A}\mathbf{x} + \mathbf{B}\mathbf{u},\\ \mathbf{y} &= \mathbf{C}\mathbf{x}.\end{aligned} \tag{6.30}$$

We now ask if there is a coordinate system

$$\begin{aligned}\dot{\mathscr{x}} &= \mathbf{T}^{-1}\mathbf{A}\mathbf{T}\mathscr{x} + \mathbf{T}^{-1}\mathbf{B}\mathbf{u},\\ \mathbf{y} &= \mathbf{C}\mathbf{T}\mathscr{x}\end{aligned}$$

such that the truncation of the last state (or the last block of states) $\mathscr{x}_n$ from the state vector $\mathbf{x}$ allows the reduced model

$$\begin{aligned}\dot{\mathscr{x}}_R &= [\mathbf{I} \quad \mathbf{0}]\mathbf{T}^{-1}\mathbf{A}\mathbf{T}\begin{bmatrix}\mathbf{I}\\ \mathbf{0}\end{bmatrix}\mathscr{x}_R + [\mathbf{I} \quad \mathbf{0}]\mathbf{T}^{-1}\mathbf{B}\mathbf{u},\\ \mathbf{y}_R &= \mathbf{C}\mathbf{T}\begin{bmatrix}\mathbf{I}\\ \mathbf{0}\end{bmatrix}\mathscr{x}_R\end{aligned}$$

to perturb the cost function

$$\mathscr{V} = \operatorname{tr} \mathbf{X}\mathbf{C}^*\mathbf{C} = \sum_{i=1}^{n_u} \int_0^\infty \mathbf{y}^i(t+\tau)\mathbf{y}^{i*}(\tau)\, d\tau$$

by the smallest possible amount. These will be "optimal" coordinates in the sense mentioned. We now show that the observable Hessenberg coordinates have this property if the reduced model has dimension $r \geq n_y = n_1$. Note that in observable Hessenberg coordinates,

$$\mathscr{V} = \operatorname{tr} \mathscr{X}_{\mathrm{OH}} \mathbf{C}_{\mathrm{OH}}^* \mathbf{C}_{\mathrm{OH}} = \operatorname{tr} \mathbf{C}_{\mathrm{OH}}^* \mathbf{C}_{\mathrm{OH}} = \operatorname{tr} \begin{bmatrix} \mathbf{C}_1^*\mathbf{C}_1 & \mathbf{0} \\ \mathbf{0} & \mathbf{0} \end{bmatrix} = \operatorname{tr} \mathbf{C}_1^*\mathbf{C}_1. \qquad (6.31)$$

Now, delete the last block of $(n_x - n_1)$ states to obtain the reduced model

$$\dot{\mathbf{x}}_R = \mathbf{A}_R\mathbf{x}_R + \mathbf{B}_R\mathbf{u}, \qquad \mathbf{y}_R = \mathbf{C}_R\mathbf{x}_R,$$

$$\mathbf{A}_R = \begin{bmatrix} \mathbf{I}_{n_1} & \mathbf{0} \end{bmatrix} \mathbf{A}_{\mathrm{OH}} \begin{bmatrix} \mathbf{I}_{n_1} \\ \mathbf{0} \end{bmatrix}, \quad \mathbf{B}_R = \begin{bmatrix} \mathbf{I}_{n_1} & \mathbf{0} \end{bmatrix} \mathbf{B}_{\mathrm{OH}}, \quad \mathbf{C}_R = \mathbf{C} \begin{bmatrix} \mathbf{I}_{n_1} \\ \mathbf{0} \end{bmatrix} = \mathbf{C}_1 \qquad (6.32)$$

to see that the cost function

$$\mathscr{V}_R = \operatorname{tr} \mathscr{X}_R \mathbf{C}_R^* \mathbf{C}_R = \operatorname{tr} \mathbf{C}_R^* \mathbf{C}_R = \operatorname{tr} \mathbf{C}_1^* \mathbf{C}_1 \qquad (6.33)$$

is identical to that for the original system, $\mathscr{V}$. Hence, the cost perturbation is

$$\Delta\mathscr{V} \triangleq \mathscr{V} - \mathscr{V}_R = 0,$$

the smallest possible perturbation magnitude. This $(\mathbf{A}_R, \mathbf{B}_R, \mathbf{C}_R)$ is, therefore, called a "cost-equivalent realization." Theorem 6.5 shows that this cost-equivalent realization (6.32) is also a 1-Markov COVER. Hence, the q-Markov COVER is a cost equivalent realization for any $q \geq 1$. Of course, the q-Markov COVER has many other properties besides matching the cost function.

Now, suppose the number of states to be deleted exceeds $n_x - n_1$. In this case, $\Delta\mathscr{V} \neq 0$, but we wish to show that $|\Delta\mathscr{V}|$ is smallest by truncating states in the following "cost-decoupled" coordinates.

Definition 6.5
Let any realization $(\mathbf{A}, \mathbf{B}, \mathbf{C})$ *with this property*

$$\mathbf{X}\mathbf{C}\mathbf{Q}^*\mathbf{C} = \text{diag.}, \qquad \mathbf{0} = \mathbf{X}\mathbf{A}^* + \mathbf{A}\mathbf{X} + \mathbf{B}\mathscr{U}\mathbf{B}^* \qquad (6.34)$$

be called a realization in "cost-decoupled" coordinates.

Now, we shall show an algorithm which will construct cost-decoupled coordinates, but first consider "approximately" cost-decoupled coordinates as follows.

Lemma 6.2

The coordinates (5.160) *are cost decoupled if* (*i*) *or* (*ii*) *of Theorem* 5.22 *holds true, and are approximately cost decoupled if condition* (*iii*) *of Theorem* 5.22 *holds true.*

Proof: The proof requires that the off-diagonal elements be small or zero compared to the diagonal elements of [**XC*QC**]. The ikth element of this matrix is, from (5.144),

$$[\mathbf{XC^*QC}]_{ik} = \sum_{\alpha=1}^{n_x} \frac{\mathbf{b}_i^* \mathcal{U} \mathbf{b}_\alpha \mathbf{c}_\alpha^* \mathbf{Q} \mathbf{c}_k}{\lambda_i + \bar{\lambda}_k}, \qquad i \neq k, \tag{6.35a}$$

and the magnitude of this is "small" compared to the diagonal elements

$$[\mathbf{XC^*QC}]_{ii} = \frac{\mathbf{b}_i^* \mathcal{U} \mathbf{b}_i \mathbf{c}_i^* \mathbf{Q} c_i}{-2\,\mathcal{Re}\,\lambda_i} + 0(\zeta_i) \tag{6.35b}$$

if either (i), (ii), or (iii) of Theorem 5.22 holds true. The proof of this follows exactly as the steps in the proof of Theorem 5.22. ■

The following algorithm will construct coordinates that are exactly cost decoupled.

Cost-Decoupled Coordinates

STEP 1: Compute $\mathbf{X}$ from $\mathbf{0} = \mathbf{XA}^* + \mathbf{AX} + \mathbf{B}\mathcal{U}\mathbf{B}^*$.

STEP 2: Factor $\mathbf{X} = \mathbf{T}_1\mathbf{T}_1^*$.

STEP 3: Find the singular value decomposition of $\mathbf{T}_1^*\mathbf{C}^*\mathbf{CT}_1$,

$$\mathbf{T}_1^*\mathbf{C}^*\mathbf{CT}_1 = \mathbf{T}_2\begin{bmatrix} \Sigma & \mathbf{0} \\ \mathbf{0} & \mathbf{0} \end{bmatrix}\mathbf{T}_2^*, \qquad \mathbf{T}_2\mathbf{T}_2^* = \mathbf{I}.$$

STEP 4: The cost-decoupled coordinates of $(\mathbf{A}, \mathbf{B}, \mathbf{C}, \mathbf{X})$ are given by $(\mathbf{T}^{-1}\mathbf{AT}, \mathbf{T}^{-1}\mathbf{B}, \mathbf{CT}, \mathbf{T}^{-1}\mathbf{XT}^{-*})$, where $\mathbf{T} = \mathbf{T}_1\mathbf{T}_2$.

In these coordinates it follows immediately from steps 2 and 3 that

$$(\mathbf{T}^{-1}\mathbf{XT}^{-*})(\mathbf{CT})^*(\mathbf{CT}) = \begin{bmatrix} \Sigma & \mathbf{0} \\ \mathbf{0} & \mathbf{0} \end{bmatrix},$$

where the singular values are $\sigma_i > 0$ and

$$\Sigma = \begin{bmatrix} \sigma_1 & & & \\ & \sigma_2 & & \\ & & \ddots & \\ & & & \sigma_{n_y} \end{bmatrix}, \qquad \sigma_i \geq \sigma_{i+1}.$$

This algorithm provides a constructive proof that $\mathbf{XC^*QC}$ is nondefective with nonnegative eigenvalues, since

$$\mathbf{T}^{-1}[\mathbf{XC^*QC}]\mathbf{T} = \begin{bmatrix} \boldsymbol{\Sigma} & \mathbf{0} \\ \mathbf{0} & \mathbf{0} \end{bmatrix} = \text{diag},$$

where $\boldsymbol{\Sigma} > \mathbf{0}$ always exists from the singular value decompositions of step 3. Now, consider model reduction by "component cost analysis." [Recall the definitions from (4.174).] The component cost $\mathscr{V}_{x_i}$ is the contribution of the ith state to the cost function. The total cost function satisfies

$$\mathscr{V} = \sum_{i=1}^{n_x} \mathscr{V}_{x_i},$$

where, from (4.174),

$$\mathscr{V}_{x_i} = [\mathbf{XC^*QC}]_{ii}. \tag{6.36}$$

Now, in cost-decoupled coordinates this gives

$$\mathscr{V}_{x_i} = \sigma_i, \qquad i = 1, 2, \ldots, n_1,$$

$$\mathscr{V}_{x_i} = 0 \qquad i = n_{i+1}, \ldots, n_x$$

Now, if we arrange the states so that $\sigma_i \geq \sigma_{i+1}$, then deleting states from the bottom of the state vector (in cost-decoupled coordinates) causes the smallest perturbation in $\Delta\mathscr{V}$ since

$$\Delta\mathscr{V} = \sum_{i=1}^{n_1} \sigma_i - \sum_{i=1}^{r} \sigma_i = \sum_{r+1}^{n_1} \sigma_i$$

is smaller than any other sum of $[n_1 - (r+1)]$ values of σ_i (by virtue of $\sigma_i \geq \sigma_{i+1}$).

This proves the following. Let $(\mathbf{A}_R, \mathbf{B}_R, \mathbf{C}_R)$ be the truncation of $(\mathbf{A}, \mathbf{B}, \mathbf{C})$ in the sense that

$$(\mathbf{A}_R, \mathbf{B}_R, \mathbf{C}_R) = \left([\mathbf{I} \quad \mathbf{0}]\mathbf{A}\begin{bmatrix} \mathbf{I} \\ \mathbf{0} \end{bmatrix}, (\mathbf{I} \quad \mathbf{0})\mathbf{B}, \mathbf{C}\begin{bmatrix} \mathbf{I} \\ \mathbf{0} \end{bmatrix} \right).$$

This is equivalent to the deletion of the bottom part of the state vector of $\dot{\mathbf{x}} = \mathbf{Ax} + \mathbf{Bu}$, $\mathbf{y} = \mathbf{Cx}$ to form the reduced system $\dot{\mathbf{x}}_R = \mathbf{A}_R\mathbf{x}_R + \mathbf{B}_R\mathbf{u}$, $\mathbf{y}_R = \mathbf{C}_R\mathbf{x}_R$. Let $\mathscr{V} = \operatorname{tr} \mathscr{X}\mathbf{C^*QC}$ and $\mathscr{V}_R = \operatorname{tr} \mathscr{X}_R\mathbf{C}_R^*\mathbf{QC}_R$ be the cost function associated with $(\mathbf{A}, \mathbf{B}, \mathbf{C})$ and $(\mathbf{A}_R, \mathbf{B}_R, \mathbf{C}_R)$, respectively.

Theorem 6.6

Let the dimension of the reduced model $(\mathbf{A}_R, \mathbf{B}_R, \mathbf{C}_R)$ be $\mathbf{x}_R \in \mathscr{R}^r$, obtained by truncating the bottom $(n_x - r)$ states from $\dot{\mathbf{x}} = \mathbf{Ax} + \mathbf{Bu}$, $\mathbf{y} = \mathbf{Cx}$ as above. The magnitude of $\Delta\mathscr{V} =$

$\operatorname{tr} \mathbf{XC^*QC} - \operatorname{tr} \mathbf{X}_R \mathbf{C}_R^* \mathbf{QC}_R$ *is smallest for any given* $0 \leq r \leq n$ *if the truncation occurs in cost-decoupled coordinates (defined by Definition 6), where*

$$\mathbf{XC^*QC} = \begin{bmatrix} \mathbf{\Sigma} & \mathbf{0} \\ \mathbf{0} & \mathbf{0} \end{bmatrix}, \quad \mathbf{\Sigma} = \operatorname{diag}[\ldots \sigma_i \ldots], \qquad \sigma_i \geq \sigma_{i+1}.$$

Now, cost-decoupled coordinates have another property of significance. In cost-decoupled coordinates the component cost $\mathscr{V}_{xi} = [\mathbf{XC^*QC}]_{ii}$ has a dual meaning.

Corollary to Theorem 6.6

By definition, $\mathscr{V}_{x_i} = [\mathbf{XC^*QC}]_{ii}$ *represents the contribution of the ith state in the cost function* $\mathscr{V}$ *of the full order system; and in cost-decoupled coordinates* $\mathscr{V}_{x_i}$ *also represents the difference between* $\mathscr{V}$ *and the cost function* $\mathscr{V}_R$ *of the system after* x_i *is deleted.*

Hence, the component costs of cost-decoupled coordinates allow one to predict exactly the perturbation of the cost function if a specified state or set of these states were deleted from the model. (Thus, a cost-equivalent realization follows by deleting states with zero component costs). Take note, however, that the component costs $\mathscr{V}_{x_i} = [\mathbf{XC^*QC}]_{ii}$, $i = 1, 2, \ldots,$ are useful as a criterion for model reduction (i.e., retaining states in the order of their component costs $\mathscr{V}_{x1} \geq \mathscr{V}_{x2} \geq \mathscr{V}_{x3} \geq \cdots$) only for cost-decoupled coordinates. These comments and restrictions do not apply to input or output costs analysis, which are both invariant with a change of basis (see Chapter 4).

Frequency Domain Interpretations

Recall from Theorem 4.15 that the Fourier transform of the output correlation is

$$\begin{aligned} \mathscr{F}\{\rangle \mathbf{y}(t+\tau), \mathbf{y}(\tau)\langle\} &= [\mathscr{F}\{\mathbf{y}(t)\}][\mathscr{F}\{y(t)\}]^* \\ &= \mathbf{Y}(j\omega)\mathbf{Y}^*(j\omega) \\ &= \mathbf{G}(j\omega)\mathbf{U}(j\omega)\mathbf{U}^*(j\omega)\mathbf{G}^*(j\omega), \end{aligned} \tag{6.37a}$$

where $[\mathbf{Y}(j\omega)\mathbf{Y}^*(j\omega)]$ and $[\mathbf{U}(j\omega)\mathbf{U}^*(j\omega)]$ are the power spectral densities of the output $\mathbf{y}(t)$ and input $\mathbf{u}(t)$, respectively, and

$$\mathbf{G}(j\omega) = \mathbf{C}(j\omega\mathbf{I} - \mathbf{A})^{-1}\mathbf{B}. \tag{6.37b}$$

Now, let us define a transfer function $\mathbf{W}(j\omega)$ as an "all pass network" if it satisfies

$$\mathbf{W}(j\omega)[\mathbf{U}(j\omega)\mathbf{U}^*(j\omega)]\mathbf{W}^*(j\omega) = [\mathbf{U}(j\omega)\mathbf{U}^*(j\omega)]. \tag{6.37c}$$

Note that if a transfer function $\mathbf{G}(s)$ has an all-pass factor

$$\mathbf{G}(s) = \mathbf{G}_R(s)\mathbf{W}(s), \tag{6.38}$$

then, from (6.37a),

$$\mathscr{F}\{\rangle \mathbf{y}(t+\tau), \mathbf{y}(\tau)\langle\} = \mathbf{G}_R(j\omega)\mathbf{W}(j\omega)\mathbf{U}(j\omega)\mathbf{U}^*(j\omega)\mathbf{W}^*(j\omega)\mathbf{G}_R^*(j\omega)$$
$$= \mathbf{G}_R(j\omega)\mathbf{U}(j\omega)\mathbf{U}^*(j\omega)\mathbf{G}_R^*(j\omega). \tag{6.39}$$

This proves the following:

Theorem 6.7
Any linear system with a transfer function $\mathbf{G}(s)$ *that contains an all-pass factor* $\mathbf{W}(s)$ *defined by* (6.37c) *has a COVER whose transfer function is* $\mathbf{G}_R(s)$ *defined from* (6.38), *and the minimal order COVER is obtained by factoring all all-pass networks out of* $\mathbf{G}(s)$.

EXAMPLE 6.6
Given

$$\mathbf{G}(s) = \frac{(s-\alpha_1)(s-\alpha_2)}{(s+\alpha_1)(s+\alpha_2)(s+\alpha_3)(s+\alpha_4)}.$$

Find the minimal COVER.

$$\mathbf{G}(j\omega)[U(j\omega)U^*(j\omega)]G^*(j\omega)$$
$$= \frac{(j\omega-\alpha_1)(j\omega-\alpha_2)}{(j\omega+\alpha_1)(j\omega+\alpha_2)(j\omega+\alpha_3)(j\omega+\alpha_4)}$$
$$\times [U(j\omega)U^*(j\omega)]\frac{\overline{(j\omega-\alpha_1)}\,\overline{(j\omega-\alpha_2)}}{\overline{(j\omega+\alpha_1)}\,\overline{(j\omega+\alpha_2)}\cdot\overline{(j\omega+\alpha_3)}\,\overline{(j\omega+\alpha_4)}}$$
$$= \frac{|j\omega-\alpha_i|^2|j\omega-\alpha_2|^2}{|j\omega+\alpha_1|^2|j\omega+\alpha_2|^2|j\omega+\alpha_3|^2|j\omega+\alpha_4|^2}[U(j\omega)U^*(j\omega)]$$
$$= \frac{1}{|j\omega+\alpha_3|^2|j\omega+\alpha_4|^2}U(j\omega)U^*(j\omega).$$

Solution: Hence, the minimal COVER is described by

$$G_R(j\omega) = \frac{1}{(j\omega+\alpha_3)(j\omega+\alpha_4)},$$

and the all-pass network is

$$W(j\omega) = \frac{(j\omega-\alpha_1)(j\omega-\alpha_2)}{(j\omega+\alpha_1)(j\omega+\alpha_2)}.$$ ■

EXAMPLE 6.7

Find the minimal COVER for

$$\begin{pmatrix} \dot{x}_1 \\ \dot{x}_2 \\ \dot{x}_3 \end{pmatrix} = \begin{bmatrix} -3 & -4 & -2 \\ 0 & -2 & -2 \\ 0 & 0 & -1 \end{bmatrix} \begin{pmatrix} x_1 \\ x_2 \\ x_3 \end{pmatrix} + \begin{pmatrix} 1 \\ 1 \\ 1 \end{pmatrix} u$$

$$y = [1 \quad 0 \quad 0]\mathbf{x}$$

that has the transfer function $G(s) = \dfrac{(s-1)(s-2)}{(s+1)(s+2)(s+3)}$.

Solution: A solution of (6.34) for $\mathbf{X}$ gives

$$\mathbf{X} = \begin{bmatrix} \frac{1}{6} & 0 & 0 \\ 0 & \frac{1}{4} & 0 \\ 0 & 0 & \frac{1}{2} \end{bmatrix}$$

Hence, the first 3-Markov and covariance parameters are

$$R_0 = \mathbf{CXC}^* = \tfrac{1}{6}, \qquad M_0 = \mathbf{CB} = 1,$$

$$R_1 = \mathbf{CAXC}^* = -\tfrac{1}{2}, \qquad M_1 = \mathbf{CAB} = -9,$$

$$R_2 = \mathbf{CA}^2\mathbf{XC}^* = \tfrac{3}{2}, \qquad M_2 = \mathbf{CA}^2\mathbf{B} = 45,$$

and, from (6.28), a 1-Markov COVER is

$$(A_R, B_R, C_R) = (-3, 1, 1).$$

Or, from (6.36), noting that the realization $(\mathbf{A}, \mathbf{B}, \mathbf{C})$ is in cost-decoupled coordinates, component cost analysis yields $\mathscr{V}_{x1} = \frac{1}{6}$, $\mathscr{V}_{x2} = \mathscr{V}_{x3} = 0$, leading to deletion of x_2 and x_3 to give the cost-equivalent realization

$$(A_R, B_R, C_R) = (-3, 1, 1),$$

all of which have the transfer function

$$C_R(sI - A_R)^{-1}B_R = \frac{1}{s+3},$$

which is a COVER of $(\mathbf{A}, \mathbf{B}, \mathbf{C})$ by Theorem 6.7. Note that the 1-Markov COVER did not contain any of the all-pass network that was present in $G(s)$. ■

Exercise 6.2

For the (**A**, **B**, **C**) in Example 6.7 find a 1-Markov COVER and a 2-Markov COVER by applying Theorem 6.4.

Exercise 6.3

For the (**A**, **B**, **C**) in Example 6.7, find a 1-Markov COVER and a 2-Markov COVER by applying Theorem 6.3. Plot the impulse response of these three systems and compare.

Exercise 6.4

Let $\dot{\theta}$ be the output of the space backpack (3.23) and let the inputs be T_c and f. Find a 1-Markov COVER of (3.23). Find a 2-Markov COVER.

6.4 Balanced Model Reduction

The balanced coordinates developed in (5.177) through (5.181) suggest a simple model reduction procedure. By deleting those states associated with the smallest singular values σ_i of the balanced configuration

$$\begin{aligned} \dot{\mathbf{x}}_b &= \mathbf{A}_b\mathbf{x}_b + \mathbf{B}_b\mathbf{u}, \qquad \mathbf{x}_b \in \mathscr{R}^{n_x}, \\ \mathbf{y} &= \mathbf{C}_b\mathbf{x}_b, \end{aligned} \tag{6.40a}$$

$$\mathbf{X}_b = \mathbf{K}_b = \mathbf{\Sigma}_b = \begin{bmatrix} \sigma_1 & & & \\ & \sigma_2 & & \\ & & \ddots & \\ & & & \sigma_{n_x} \end{bmatrix}, \qquad \sigma_i \geq \sigma_{i+1}, \tag{6.40b}$$

$$\mathbf{0} = \mathbf{X}_b\mathbf{A}_b^* + \mathbf{A}_b\mathbf{X}_b + \mathbf{B}_b\mathscr{U}\mathbf{B}_b^*, \qquad \mathbf{0} = \mathbf{K}_b\mathbf{A}_b + \mathbf{A}_b^*\mathbf{K}_b + \mathbf{C}_b^*\mathbf{Q}\mathbf{C}_b,$$

the reduced model is

$$\dot{\mathbf{x}}_R = \mathbf{A}_R\mathbf{x}_R + \mathbf{B}_R\mathbf{u}, \quad \mathbf{y}_R = \mathbf{C}_R\mathbf{x}_R, \quad \mathbf{x}_R \in \mathscr{R}^r, \tag{6.41a}$$

where

$$\begin{aligned} \mathbf{X}_R &= \mathbf{K}_R = \mathbf{\Sigma}_R = [\mathbf{I}_r \quad \mathbf{0}]\mathbf{\Sigma}_b\begin{bmatrix} \mathbf{I}_r \\ \mathbf{0} \end{bmatrix}, \\ \mathbf{A}_R &= [\mathbf{I}_r \quad \mathbf{0}]\mathbf{A}_b\begin{bmatrix} \mathbf{I}_r \\ \mathbf{0} \end{bmatrix}, \\ \mathbf{B}_R &= [\mathbf{I}_r \quad \mathbf{0}]\mathbf{B}_b, \\ \mathbf{C}_R &= \mathbf{C}_b\begin{bmatrix} \mathbf{I}_r \\ \mathbf{0} \end{bmatrix}, \end{aligned} \tag{6.41b}$$

TABLE 6.1 Matched Parameters in Model Reduction

METHOD	NUMBER OF PARAMETERS MATCHED	ORDER OF REDUCED MODEL r	PARAMETERS MATCHED
Cost-equivalent	1	$r = n_y$	$\mathscr{V} = \mathscr{V}_R$
Component cost analysis (in cost-decoupled coordinates)	r	r selected	$\mathscr{V}_{x_i} = \mathscr{V}_{x_{Ri}}$ $i = 1 \to r$
Balancing	r	r selected	$\sigma_i = \sigma_{R_i}$, $i = 1 \to r$
q-Markov COVER	$qn_y(n_y + n_u + 1)$	$r = \text{rank}\,[\mathbf{W}_{0q}]$	$\mathbf{M}_i = \mathbf{M}_{R_i}$, $i = 0, \to q-1$ $\mathbf{R}_i = \mathbf{R}_{R_i}$

and *each* state of the reduced model preserves the same degree of controllability/observability product σ_i as in the full-order model. Hence, in the spirit of *equivalences*, we should think of the balancing method of model reduction as a "component controllability/observability equivalence" method. By comparison, the component cost method preserves component costs, and the q-Markov COVER method preserves q-Markov and q-covariance parameters. No one method will give the best approximation in all circumstances, unless, of course, one defines the criteria for "best" to be precisely the criterion that one of the methods is based upon. The analyst must decide whether it is more reasonable in the application to match the cost function, or the Markov and covariance parameters, or the controllability/observability of each state, or the component cost of each state. For those methods which match parameters of these types, Table 6.1 shows the number of parameters matched between the reduced and full models for each of these methods. The q-Markov COVER method matches far more parameters. However, the more important measure of performance is the error in the responses $y(t) - y_R(t)$. Working a number of examples with each method will provide valuable insight.

An upper bound of the transfer function error can be provided for the balancing method of model reduction. Define

$$\Delta\mathbf{G}(s) = \mathbf{G}(s) - \mathbf{G}_R(s) = \mathbf{C}[s\mathbf{I} - \mathbf{A}]^{-1}\mathbf{B} - \mathbf{C}_R[s\mathbf{I} - \mathbf{A}_R]^{-1}\mathbf{B}_R.$$

Then ref. [6.19] proves that the maximum eigenvalue over all ω of $[\Delta G(j\omega)]^*[\Delta G(j\omega)]$ is bounded by

$$\lambda_{\max}^{1/2}\{[\Delta\mathbf{G}(j\omega)]^*[\Delta\mathbf{G}(j\omega)]\} \le \sum_{i=r+1}^{n} \sigma_i. \tag{6.42}$$

However, even though the errors associated with the balancing method has a known bound, the balancing method does not necessarily yield the smallest errors. Consider the following exercises to illustrate this point.

Exercise 6.6

Use balancing methods to reduce this second order balanced system to first order,

$$\dot{\mathbf{x}} = \begin{bmatrix} -0.005 & -0.990 \\ -0.990 & -5000 \end{bmatrix}\mathbf{x} + \begin{bmatrix} 1 \\ 100 \end{bmatrix}u = \mathbf{Ax} + \mathbf{Bu},$$

$$\mathbf{y} = [1 \quad 100]\mathbf{x} = \mathbf{Cx}.$$

Show that the singular values are

$$\mathbf{X} = \mathbf{K} = \begin{bmatrix} 100 & 0 \\ 0 & 1 \end{bmatrix},$$

indicating that state x_1 should be retained, yielding reduced-order model

$$\dot{x}_R = -0.005x_R + u, \qquad y_R = x_R.$$

Now, show that the unit impulse response of the reduced system is

$$y_R(t) = e^{-0.005t}$$

and that the unit impulse response of the original system is

$$y(t) = 0.961e^{-0.00485t} + 10^4e^{-5000t},$$

and that the normalized error is

$$\frac{\int_0^\infty [y(t) - y_R(t)]^2\, dt}{\int_0^\infty y^2(t)\, dt} = \frac{\operatorname{tr} \mathbf{X}_e\mathbf{C}_e^*\mathbf{C}_e}{\operatorname{tr} \mathbf{XC^*C}} = 0.995,$$

indicating a very poor reduced-order model, measured by the unit impulse response errors $\|y(t) - y_R(t)\|^2$.

Hint: The error system $(\mathbf{A}_e, \mathbf{B}_e, \mathbf{C}_e, \mathbf{X}_e)$ is obviously characterized by

$$\dot{\mathbf{x}}_e = \begin{pmatrix} \dot{\mathbf{x}} \\ \dot{\mathbf{x}}_R \end{pmatrix} = \begin{bmatrix} \mathbf{A} & \mathbf{0} \\ \mathbf{0} & A_R \end{bmatrix}\begin{pmatrix} \mathbf{x} \\ x_R \end{pmatrix} + \begin{bmatrix} \mathbf{B} \\ B_R \end{bmatrix}u = \mathbf{A}_e\mathbf{x}_e + \mathbf{B}_e u,$$

$$y_e = (y - y_R) = [\mathbf{C} \ -C_R]\begin{pmatrix} \mathbf{x} \\ x_R \end{pmatrix},$$

$$\mathbf{0} = \mathbf{X}_e\mathbf{A}_e^* + \mathbf{A}_e\mathbf{X}_e + \mathbf{B}_e\mathbf{B}_e^*.$$

Exercise 6.7

Given the same $(\mathbf{A}, \mathbf{B}, \mathbf{C})$ in the above exercise, find a first-order model using component cost analysis of the balanced coordinates. Show from (4.175) that the component costs are

$$\mathscr{V}_{x1} = [\mathbf{XC^*C}]_{11} = 10^2, \qquad \mathscr{V}_{x2} = [\mathbf{XC^*C}]_{22} = 10^4,$$

indicating that the second state should be retained to yield the first-order model

$$\dot{x}_R = -5000x_R + 100u, \qquad y_R = 100x_R,$$

with impulse response

$$y_R(t) = 10^4 e^{-5000t}.$$

Show that the error is now

$$\frac{\operatorname{tr} \mathbf{X}_e \mathbf{C}_e^* \mathbf{C}_e}{\operatorname{tr} \mathbf{XC^*C}} = 0.098,$$

indicating a better result than in Exercise 6.6.

Exercise 6.8

Transform the system in Exercise 6.6 to normalized observable Hessenberg coordinates and truncate to first order. Compare with results in the above examples.

Exercise 6.9

Find a 1-Markov COVER of the $(\mathbf{A}, \mathbf{B}, \mathbf{C})$ in Exercise 6.6. Compute the same error measure as in Exercises 6.6 and 6.7. *Hint*: Use Theorem 6.5 with $U_1 = 1$, and show that

$$\mathscr{L} = (1 \quad 100), \quad \mathscr{R} = \begin{pmatrix} 1 \\ 1 \end{pmatrix} \frac{1}{101}, \quad A_R = -5000, \quad B_R = 10{,}001, \quad C_R = 1,$$

yielding the same results as in Exercise 6.7.

The conclusion from these exercises is that methods which have no known bounds on modeling error are not precluded from providing smaller errors.

6.5 Model Reduction by Singular Perturbation

Up to this point we have restricted our attention to realizations that are *equivalent* with respect to some criterion. Of course, one could reduce the order of the system *below* the "minimal" dimension and actually degrade the performance instead of demanding exact equivalence. There are many schemes for reducing the order of models which do not require equivalence in any sense. This is the general subject of model reduction. However, the control problem and the model reduction problem cannot be truly separated. Thus, it serves little purpose to dwell on the subject in detail prior to discussions of the control design problems. We shall return to this subject later—now, only a few examples.

Sometimes, the system contains a large spectral separation among the eigenvalues of the matrix **A**. This happens when the states can be separated into "slow" and "fast" states x_1 and x_2, respectively,

$$\begin{pmatrix} \dot{\mathbf{x}}_1 \\ \varepsilon\dot{\mathbf{x}}_2 \end{pmatrix} = \begin{bmatrix} \mathbf{A}_{11} & \mathbf{A}_{12} \\ \mathbf{A}_{21} & \mathbf{A}_{22} \end{bmatrix} \begin{pmatrix} \mathbf{x}_1 \\ \mathbf{x}_2 \end{pmatrix} + \begin{bmatrix} \mathbf{B}_1 \\ \mathbf{B}_2 \end{bmatrix} \mathbf{u}, \qquad \mathbf{y} = [\mathbf{C}_1 \quad \mathbf{C}_2] \begin{pmatrix} \mathbf{x}_1 \\ \mathbf{x}_2 \end{pmatrix}, \tag{6.43}$$

where ε is a small positive scalar. As a first approximation of this system, the small parameter ε is set equal to zero:

$$\begin{pmatrix} \dot{\mathbf{x}}_1 \\ \mathbf{0} \end{pmatrix} = \begin{bmatrix} \mathbf{A}_{11} & \mathbf{A}_{12} \\ \mathbf{A}_{21} & \mathbf{A}_{22} \end{bmatrix} \begin{pmatrix} \mathbf{x}_1 \\ \mathbf{x}_{2_{ss}} \end{pmatrix} + \begin{bmatrix} \mathbf{B}_1 \\ \mathbf{B}_2 \end{bmatrix} \mathbf{u}, \qquad \mathbf{y} = [\mathbf{C}_1 \quad \mathbf{C}_2] \mathbf{x}, \tag{6.44}$$

where $\mathbf{x}_{2_{ss}} \triangleq \lim_{t \to \infty} \mathbf{x}_2(t)$ represents the steady-state value of $\mathbf{x}_2(t)$, if it exists. Thus, under the assumption that $\mathbf{x}_{2_{ss}}$ exists,* (6.44) indicates that $\mathbf{x}_{2_{ss}}$ satisfies

$$\mathbf{A}_{22}\mathbf{x}_{2_{ss}} = -\mathbf{A}_{21}\mathbf{x}_1 - \mathbf{B}_2\mathbf{u}$$

or

$$\mathbf{x}_{2_{ss}} = -\mathbf{A}_{22}^{-1}(\mathbf{A}_{21}x_1 + \mathbf{B}_2 u), \tag{6.45}$$

where $\mathbf{A}_{22}$ is presumed to have an inverse; otherwise, a pseudo-inverse is used. Substituting (6.45) into (6.44) gives

$$\dot{\mathbf{x}}_1 = \left[\mathbf{A}_{11} - \mathbf{A}_{12}\mathbf{A}_{22}^{-1}\mathbf{A}_{21}\right]\mathbf{x}_1 + \left[\mathbf{B}_1 - \mathbf{A}_{12}\mathbf{A}_{22}^{-1}\mathbf{B}_2\right]\mathbf{u}, \qquad \mathbf{y} = \left[\mathbf{C}_1 - \mathbf{C}_2\mathbf{A}_{22}^{-1}\mathbf{A}_{21}\right]\mathbf{x}_1 - \left[\mathbf{C}_2\mathbf{A}_{22}^{-1}\mathbf{B}_2\right]\mathbf{u}. \tag{6.46}$$

This method of approximation is called "singular perturbation" and has been studied extensively, [6.7] and [6.8].

*$\mathbf{x}_{2_{ss}}$ exists if the controllable, observable modes of $(\mathbf{A}, \mathbf{B}, \mathbf{0} \quad \mathbf{I})$ are stable, where $\mathbf{x}_2 = [\mathbf{0} \quad \mathbf{I}]\mathbf{x}$.

EXAMPLE 6.8

For the system

$$\mathbf{A} = \begin{bmatrix} -1 & 0 \\ 0 & -10 \end{bmatrix}, \quad \mathbf{B} = \begin{pmatrix} 1 \\ 70 \end{pmatrix}, \quad \mathbf{C} = [1, -0.2],$$

let the small parameter $\varepsilon = 0.1$ be used to write the system in the form

$$\begin{pmatrix} \dot{x}_1 \\ \varepsilon\dot{x}_2 \end{pmatrix} = \begin{bmatrix} -1 & 0 \\ 0 & -1 \end{bmatrix}\begin{pmatrix} x_1 \\ x_2 \end{pmatrix} + \begin{bmatrix} 1 \\ 7 \end{bmatrix} u,$$

$$y = [1 \quad -0.2]\begin{pmatrix} x_1 \\ x_2 \end{pmatrix}.$$

The singular perturbation (6.46) gives

$$\dot{x}_1 = -x_1 + u,$$

$$y = x_1 - 1.4u, \tag{6.47}$$

$$G(s) = \frac{1}{s+1} - 1.4 = \frac{-1.4s - 0.4}{s+1}.$$

EXAMPLE 6.9

Compare (6.47) with the reduced model obtained by truncating balanced coordinates with the smallest σ_i, where $\mathbf{X} = \mathbf{K} = \operatorname{diag}[\ldots \sigma_i^2 \ldots]$.

Solution: From equation (5.181) and Exercise (5.40),

$$\mathbf{X} = \begin{bmatrix} 1/2 & 70/11 \\ 70/11 & 245 \end{bmatrix}, \quad \mathbf{K} = \begin{bmatrix} 1/2 & -0.2/11 \\ -0.2/11 & 0.002 \end{bmatrix},$$

$$\mathbf{X} = \mathbf{T}_1\mathbf{T}_1^* \Rightarrow \mathbf{T}_1 = \begin{bmatrix} -0.41 & -0.58 \\ -15.65 & -0.02 \end{bmatrix},$$

$$\mathbf{T}_1^*\mathbf{K}\mathbf{T}_1 = \begin{bmatrix} 0.34 & -0.05 \\ -0.05 & 0.17 \end{bmatrix} = \mathbf{U}_2\mathbf{\Sigma}_2\mathbf{U}_2^*,$$

$$\mathbf{T}_2 = \mathbf{U}_2\mathbf{\Sigma}_2^{-1/4} = \begin{bmatrix} -0.97 & -0.25 \\ 0.25 & -0.97 \end{bmatrix}\begin{bmatrix} 1.30 & 0 \\ 0 & 1.59 \end{bmatrix} = \begin{bmatrix} -1.26 & -0.39 \\ 0.32 & -1.5 \end{bmatrix},$$

$$\mathbf{T} = \mathbf{T}_1\mathbf{T}_2 = \begin{bmatrix} 0.327 & 1.05 \\ 19.68 & 6.12 \end{bmatrix}.$$

The balanced realization is

$$\mathbf{T}^{-1}\mathbf{AT} = \begin{bmatrix} -10.96 & -3.10 \\ 3.10 & -0.04 \end{bmatrix},$$

$$\mathbf{T}^{-1}\mathbf{B} = \begin{bmatrix} 3.61 \\ -0.17 \end{bmatrix}, \qquad \mathbf{CT} = [-3.61 \quad -0.17],$$

$$\mathbf{T}^{-1}\mathbf{XT}^{-*} = \mathbf{T}^{*}\mathbf{KT} = \begin{bmatrix} 0.59 & 0 \\ 0 & 0.39 \end{bmatrix}.$$

Since $\sigma_2^2 = 0.39 < \sigma_1^2 = 0.59$, state 2 should be deleted as the least observable, controllable, giving

$$\dot{x}_R = -10.96x_R + 3.61u,$$

$$y_R = -3.61x_R, \tag{6.48}$$

$$G(s) = \frac{-3.61^2}{s + 10.96}. \qquad \blacksquare$$

Note that the balanced truncation (6.48) produced an eigenvalue very close to the *fast* mode of the original system, whereas the singular perturbation (6.47) retains the *slow* mode. Actually, the singular perturbation method *presumes* that the slow mode is more important *a priori*. ignoring the input/output matrices. All the methods of Sections 6.1 through 6.4 rely explicitly on the input/output matrices. The advantage of the singular perturbation method is its simplicity. It requires no coordinate transformations nor complex caluclations.

EXAMPLE 6.10

Compare the above results with model reduction using component cost analysis and cost-decoupled coordinates.

Solution: Using the previous calculation of **X** and $\mathbf{T}_1$,

$$\mathbf{T}_1^{*}\mathbf{C}^{*}\mathbf{QCT}_1 = \begin{bmatrix} 7.42 & -1.58 \\ -1.58 & 0.34 \end{bmatrix} = \mathbf{U}_2\mathbf{\Sigma}_2\mathbf{U}_2^{*} = \begin{bmatrix} -0.98 & 0.21 \\ 0.21 & 0.98 \end{bmatrix} \begin{bmatrix} 7.75 & 0 \\ 0 & 0 \end{bmatrix},$$

where

$$\mathbf{T}_2 = \begin{bmatrix} -0.98 & 0.21 \\ 0.21 & 0.98 \end{bmatrix},$$

$$\mathbf{T} = \mathbf{T}_1\mathbf{T}_2,$$

$$\mathbf{T}^{-1}\mathbf{AT} = \begin{bmatrix} -10.90 & 2.10 \\ -4.22 & -0.10 \end{bmatrix}, \quad \mathbf{T}^{-1}\mathbf{B} = \begin{bmatrix} 4.67 \\ 0.45 \end{bmatrix}, \quad \mathbf{CT} = [-2.78 \quad 0].$$

Hence, the reduced-order model is obtained by retaining the state with the largest component cost, x_1 in this case, since $\mathscr{V}_1 = [\mathbf{T}^{-1}\mathbf{XC^*QCT}]_{11} = 7.7$, $\mathscr{V}_2 = [\mathbf{T}^{-1}\mathbf{XC^*QCT}]_{22} = 0$. Hence,

$$\begin{aligned} \dot{x}_R &= -10.90x_R + 4.67u, \\ y &= -2.78x_R, \\ G(s) &= \frac{-(2.78)(4.67)}{s + 10.90}. \end{aligned} \tag{6.49}$$

Note from (6.47), (6.48), and (6.49) that the methods that focus on input/output structure produced similar results, (6.48) and (6.49). ■

The reader is referred to many other methods of model reduction in refs. [6.3] through [6.18].

6.6 Realizing Models from Output Data

Often, occasions arise when the system under study is too complicated to apply known laws of physics to every component of the complex system. Even when the physical system is quite simple, there may be uncertainties about the parameters of the model. In either event, it is desired to have an alternate way to develop a state space description of the system, one that operates directly on the output data $\mathbf{y}(t)$. We shall call this procedure *identification*.

6.6.1 IDENTIFICATION THEORY — A NAIVE APPROACH

Chapter 2 provided our first application to identification theory. Theorem 2.10 states that if $\mathscr{x}(t)$ and $\mathscr{y}(t)$ are given vector functions of time, then the $\mathscr{A}$ which minimizes

$$\min_{\mathscr{A}} \| \mathscr{A}\mathscr{x}(t) - \mathscr{y}(t) \|^2$$

is

$$\begin{aligned} \mathscr{A} &= (\rangle \mathscr{y}(t), \mathscr{x}(t) \langle)(\rangle \mathscr{x}(t), \mathscr{x}(t) \langle)^{-1} \\ &= \left(\int \mathscr{y}(t)\mathscr{x}^*(t)\, dt \right)\left(\int \mathscr{x}(t)\mathscr{x}^*(t)\, dt \right)^{-1}. \end{aligned} \tag{6.50}$$

Now, apply this result to

$$\dot{\mathbf{x}} = \mathbf{Ax} + \mathbf{Bu},$$

$$\mathbf{y} = \mathbf{Cx} + \mathbf{Du}$$

to get

$$\mathscr{y} \triangleq \begin{pmatrix} \dot{\mathbf{x}} \\ \mathbf{y} \end{pmatrix} = \begin{bmatrix} \mathbf{A} & \mathbf{B} \\ \mathbf{C} & \mathbf{D} \end{bmatrix} \begin{pmatrix} \mathbf{x} \\ \mathbf{u} \end{pmatrix} = \mathscr{A} \mathscr{x}.$$

Hence, applying (6.50) yields

$$\begin{bmatrix} \mathbf{A} & \mathbf{B} \\ \mathbf{C} & \mathbf{D} \end{bmatrix} = \left[\int_0^\infty \begin{pmatrix} \dot{\mathbf{x}}(t) \\ \mathbf{y}(t) \end{pmatrix} (\mathbf{x}^T(t)\mathbf{u}^T(t))\, dt \right] \left[\int_0^\infty \begin{pmatrix} \mathbf{x}(t) \\ \mathbf{u}(t) \end{pmatrix} (\mathbf{x}^T(t)\mathbf{u}^T(t))\, dt \right]^{-1} \tag{6.51a}$$

or

$$\mathbf{A} = \int_0^\infty \dot{\mathbf{x}}(t)\mathbf{x}^T(t)\, dt\, \mathbf{L}_{xx} + \int_0^\infty \dot{\mathbf{x}}(t)\mathbf{u}^T(t)\, dt\, \mathbf{L}_{ux}, \tag{6.51b}$$

$$\mathbf{B} = \int_0^\infty \dot{\mathbf{x}}(t)\mathbf{x}^T(t)\, dt\, L_{xu} + \int_0^\infty \dot{\mathbf{x}}(t)\mathbf{u}^T(t)\, dt\, \mathbf{L}_{uu}, \tag{6.51c}$$

$$\mathbf{C} = \int_0^\infty \mathbf{y}(t)\mathbf{x}^T(t)\, dt\, \mathbf{L}_{xx} + \int_0^\infty \mathbf{y}(t)\mathbf{u}^T(t)\, dt\, \mathbf{L}_{ux}, \tag{6.51d}$$

$$\mathbf{D} = \int_0^\infty \mathbf{y}(t)\mathbf{x}^T(t)\, dt\, \mathbf{L}_{xu} + \int_0^\infty \mathbf{y}(t)\mathbf{u}^T(t)\, dt\, \mathbf{L}_{uu}, \tag{6.51e}$$

where

$$\begin{bmatrix} \mathbf{L}_{xx} & \mathbf{L}_{xu} \\ \mathbf{L}_{ux} & \mathbf{L}_{uu} \end{bmatrix} \triangleq \left[\int_0^\infty \begin{pmatrix} \mathbf{x}(t) \\ \mathbf{u}(t) \end{pmatrix} [\mathbf{x}^T(t)\mathbf{u}^T(t)]\, dt \right]^{-1}. \tag{6.51f}$$

Of course, to synthesize this approach, one needs $\dot{\mathbf{x}}(t)$, $\mathbf{x}(t)$, $\mathbf{u}(t)$, and $\mathbf{y}(t)$, *not* just $\mathbf{y}(t)$! This is why the approach is called naive. This much information is not usually available. On the other hand, some of the realizations we have discussed in Sections 6.2 and 6.3 have square invertible $\mathbf{C}$ matrices. In this case, $\mathbf{y}$ and $\mathbf{x}$ have the same dimension. So, (6.51) might be useful in such cases when $\mathbf{x}$ can be recovered from $\mathbf{y}$.

EXAMPLE 6.11

Find the unknown plant generating the data

$$y(t) = 2e^{-t}, \qquad u(t) = \delta(t). \tag{6.52}$$

Solution: To use (6.51), we presume that $C = 1$, since $x(t)$ is not available. Then, from (6.51),

$$\begin{bmatrix} L_{xx} & L_{xu} \\ L_{ux} & L_{uu} \end{bmatrix} = \begin{bmatrix} \int_0^\infty (2e^{-t})(2e^{-t})\,dt & \int_0^\infty 2e^{-t}\delta(t)\,dt \\ \int_0^\infty 2e^{-t}\delta(t)\,dt & \int_0^\infty \delta^2(t)\,dt \end{bmatrix}^{-1} = \begin{bmatrix} 2 & 2 \\ 2 & \delta(0) \end{bmatrix}^{-1},$$

$$A = \int_0^\infty \frac{(-2e^{-t})(2e^{-t})\,dt\,\delta(0)}{2\delta(0) - 4} + \int_0^\infty \frac{(-2e^{-t})\delta(t)\,dt\,(-2)}{2\delta(0) - 4}$$

$$= \frac{-2\delta(0) + 4}{2\delta(0) - 4} = -1,$$

$$B = \int_0^\infty \frac{(-2e^{-t})(2e^{-t})}{2\delta(0) - 4}\,dt\,(-2) + \int_0^\infty \frac{(-2e^{-t})\delta(t)}{2\delta(0) - 4}\,dt\,2$$

$$= \frac{4 - 4}{2\delta(0) - 4} = 0,$$

$$C = \int_0^\infty \frac{(2e^{-t})(2e^{-t})}{2\delta(0) - 4}\,dt\,\delta(0) + \int_0^\infty \frac{(2e^{-t})\delta(t)}{2\delta(0) - 4}\,dt\,(-2)$$

$$= \frac{2\delta(0) - 4}{2\delta(0) - 4} = 1,$$

$$D = \int_0^\infty \frac{(2e^{-t})^2\,dt\,(-2)}{2\delta(0) - 4} + \int_0^\infty \frac{2e^{-t}\delta(t)\,dt\,2}{2\delta(0) - 4} = 0.$$

Now, as a check, ask if the identified system

$$\dot{x} = -x,$$

$$y = x$$

can generate the data $y = 2e^{-t}$. It can, if $x(0) = 2$. But the original problem stated that $u(t) = \delta(t)$. Hence, the algorithm (6.51) did not identify this system exactly but produced instead an *equivalent* result. The correct answer is

$$[A, B, C, D] = \left[-1, \frac{1}{\alpha}, 2\alpha, 0\right]. \tag{6.53}$$

for any $\alpha \neq 0$.

■

6.6.2 1-COVER IDENTIFICATION

Now, we wish to apply a version of the 1-Markov COVER realization algorithm which allows its construction from output data $\mathbf{R}_0$ and $\mathbf{R}_1$:

$$\mathbf{R}_0 = \int_0^\infty \sum_{i=1}^{n_u} \mathbf{y}^i(\tau)\mathbf{y}^{i*}(\tau)\,d\tau,$$
$$\mathbf{R}_1 = \frac{d}{dt}\left[\int_0^\infty \sum_{i=1}^{n_u} \mathbf{y}^i(t+\tau)\mathbf{y}^{i*}(\tau)\,d\tau\right]_{t=0}. \tag{6.54}$$

Suppose the outputs $y_\alpha(t)$, $\alpha = 1, 2, \ldots, n_y$, are linearly independent. Then, $\mathbf{C}$ has linearly independent rows and the 1-Markov COVER given by (6.28)

$$\mathbf{A}_R = \mathscr{L}\mathbf{A}\mathscr{R}, \quad \mathbf{B}_R = \mathscr{L}\mathbf{B}, \quad \mathbf{C}_R = \mathbf{C}\mathscr{R}$$
$$\mathscr{L} = \mathbf{U}_1^*\mathbf{W}_{0q}, \qquad \mathscr{R} = \mathbf{X}\mathscr{L}^*(\mathscr{L}\mathbf{X}\mathscr{L}^*)^{-1} \tag{6.55}$$

simplifies to

$$\mathscr{L} = \mathbf{C}, \qquad \mathscr{R} = \mathbf{XC}^*(\mathbf{CXC}^*)^{-1},$$
$$\mathbf{A}_R = \mathbf{CAXC}^*(\mathbf{CXC}^*)^{-1} = \mathbf{R}_1\mathbf{R}_0^{-1},$$
$$\mathbf{B}_R = \mathbf{CB}, \tag{6.56}$$
$$\mathbf{C}_R = \mathbf{CXC}^*(\mathbf{C}\mathscr{X}\mathbf{C}^*)^{-1} = \mathbf{I}.$$

Note that this value of $\mathbf{B}_R$ yields

$$\mathbf{B}_R\mathscr{U}\mathbf{B}_R^* = \mathbf{CB}\mathscr{U}\mathbf{B}^*\mathbf{C}^* = \mathbf{C}[-\mathbf{XA}^* - \mathbf{AX}]\mathbf{C}^* = -(\mathbf{R}_1^* + \mathbf{R}_1). \tag{6.57}$$

Hence, the following adaptation of (6.28) allows $(\mathbf{A}_R, \mathbf{B}_R, \mathbf{C}_R)$ to be constructed only from data $(\mathbf{R}_0, \mathbf{R}_1)$.

Theorem 6.8 (1-COVER Identification)
Given the data $\mathbf{R}_0$ and $\mathbf{R}_1$ from a linear system's impulse responses (6.54), *a realization $(\mathbf{A}_R, \mathbf{B}_R, \mathbf{C}_R)$ which matches $\mathbf{R}_0$ and $\mathbf{R}_1$ is given by*

$$\mathbf{C}_R = \mathbf{I},$$
$$\mathbf{A}_R = \mathbf{R}_1\mathbf{R}_0^{-1}, \tag{6.58}$$
$$\mathbf{B}_R\mathscr{U}\mathbf{B}_R^* = -(\mathbf{R}_1 + \mathbf{R}_1^*).$$

Note that this result does not promise the matching of Markov parameters. Hence, any matrix factor of $-(\mathbf{R}_1 + \mathbf{R}_1^*)$ yields the 1-COVER (which matches $\mathbf{R}_0$ and $\mathbf{R}_1$), but generally this factor will not promise $\mathbf{C}_R\mathbf{B}_R = \mathbf{CB} = \mathbf{M}_0$. Hence, it is not a 1-Markov COVER.

The reader may also be suspicious of the claim that the realization (6.58) matches $\mathbf{R}_0$ and $\mathbf{R}_1$ since Theorem 6.3 only promised to match $\mathbf{R}_0, \mathbf{R}_1, \ldots, \mathbf{R}_{q-1}$ covariance parameters, and $q = 1$ in Theorem 6.6. This dilemma is resolved as follows.

Corollary

For any linear stable system whose covariances are defined, the relationship between Markov and covariance parameters is given by

$$\mathbf{R}_{2k+1} + \mathbf{R}_{2k+1}^* = -\mathbf{M}_0\mathscr{U}\mathbf{M}_{2k}^* + \mathbf{M}_1\mathscr{U}\mathbf{M}_{2k-1}^*$$

$$-\mathbf{M}_2\mathscr{U}\mathbf{M}_{2k-2}^* \cdots -\mathbf{M}_{2k}\mathscr{U}\mathbf{M}_0^*, \tag{6.59a}$$

$$\mathbf{R}_{2k} - \mathbf{R}_{2k}^* = \mathbf{M}_0\mathscr{U}\mathbf{M}_{2k-1}^* - \mathbf{M}_1\mathscr{U}\mathbf{M}_{2k-2}^*$$

$$+\mathbf{M}_2\mathscr{U}\mathbf{M}_{2k-3}^* \cdots -\mathbf{M}_{2k-1}\mathscr{U}\mathbf{M}_0^*, \tag{6.59b}$$

for $k = 0, 1, 2, \ldots$.

Proof: First, observe that

$$\mathbf{Z}(s) + \mathbf{Z}^T(-s) = \mathbf{G}(s)\mathscr{U}\mathbf{G}^T(-s), \tag{6.60}$$

where

$$\mathbf{Z}(s) = \mathbf{C}(s\mathbf{I} - \mathbf{A})^{-1}\mathbf{X}\mathbf{C}^T, \qquad \mathbf{G}(s) = \mathbf{C}(s\mathbf{I} - \mathbf{A})^{-1}\mathbf{B}.$$

To see this, write

$$\mathbf{C}(s\mathbf{I} - \mathbf{A})^{-1}\mathbf{X}\mathbf{C}^T + \mathbf{C}\mathbf{X}(-s\mathbf{I} - \mathbf{A}^T)^{-1}\mathbf{C}^T$$

$$= \mathbf{C}(s\mathbf{I} - \mathbf{A})^{-1}\mathbf{B}\mathscr{U}\mathbf{B}^T(-s\mathbf{I} - \mathbf{A}^T)^{-1}\mathbf{C}^T,$$

where

$$\mathbf{B}\mathscr{U}\mathbf{B}^T = -\mathbf{X}\mathbf{A}^T - \mathbf{A}\mathbf{X} \tag{6.61}$$

yields

$$\mathbf{C}(s\mathbf{I} - \mathbf{A})^{-1}[\mathbf{X}(-s\mathbf{I} - \mathbf{A}^T) + (s\mathbf{I} - \mathbf{A})\mathbf{X}](-s\mathbf{I} - \mathbf{A}^T)^{-1}\mathbf{C}^T$$

$$= \mathbf{C}(s\mathbf{I} - \mathbf{A})^{-1}[-\mathbf{X}\mathbf{A}^T - \mathbf{A}\mathbf{X}](-s\mathbf{I} - \mathbf{A}^T)^{-1}\mathbf{C}^T$$

$$= \mathbf{C}(s\mathbf{I} - \mathbf{A})^{-1}\mathbf{B}\mathscr{U}\mathbf{B}^T(-s\mathbf{I} - \mathbf{A}^T)^{-1}\mathbf{C}^T$$

$$= \mathbf{G}(s)\mathscr{U}\mathbf{G}^T(-s).$$

Now, write

$$\mathbf{Z}(s) = \sum_{i=0}^{\infty} \frac{\mathbf{R}_i}{s^{i+1}}, \qquad \mathbf{G}(s) = \sum_{i=0}^{\infty} \frac{\mathbf{M}_i}{s^{i+1}}$$

and set like powers of s on each side of (6.60) equal. This produces (6.59) ■

Now, use the corollary to see that for $k = 0$, (6.59a) yields

$$\mathbf{R}_1 + \mathbf{R}_1^* = -\mathbf{M}_0 \mathscr{U} \mathbf{M}_0^*,$$

which follows also from (6.57). Hence, any realization which matches $\mathbf{M}_0$ (as the 1-Markov COVER does) *also* matches the symmetric part of $\mathbf{R}_1$. However, the particular realizations $(\mathbf{A}_R, \mathbf{B}_R, \mathbf{C}_R)$ given by (6.58) match the entire $\mathbf{R}_1$ and not just the symmetric part. This follows from

$$\mathbf{R}_0 = \mathbf{C}_R \mathbf{X}_R \mathbf{C}_R^* = \mathbf{X}_R,$$

$$\mathbf{R}_1 = \mathbf{C}_R \mathbf{A}_R \mathbf{X}_R \mathbf{C}_R^* = \mathbf{A}_R \mathbf{X}_R = \mathbf{R}_1 \mathbf{R}_0^{-1} \mathbf{R}_0 = \mathbf{R}_1.$$

EXAMPLE 6.12

Solve Example 6.11 using Theorem 6.8.

Solution: From the given data,

$$\mathbf{R}(t) = \int_0^{\infty} y(t+\tau) y(\tau)\, d\tau = \int_0^{\infty} 2e^{-(t+\tau)} 2e^{-\tau}\, d\tau$$

$$= 2e^{-t}.$$

Now, $R_0 = R(0) = 2$ and $R_1 = \dot{R}(0) = -2$. From (6.58), $A_R = R_1 R_0^{-1} = -1$, $B_R^2 = -2R_1 = 4$, $B_R = \pm 2$, $C_R = 1$. This yields the transfer function $C_R(sI - A_R)^{-1} B_R = \pm 2/(s+1)$, which agrees with the correct answer given by (6.53), with $\alpha = \frac{1}{2}$, $B_R = 2$. The transfer of the correct realization is

$$G(s) = C_R(sI - A_R)^{-1} B_R = \frac{2}{s+1}.$$

However, the 1-COVER algorithm of Theorem 6.8 may miss the first Markov parameter by a sign ± 2. ■

Exercise 6.5

Use the 1-COVER identification algorithm (6.58) to find a state space realization to fit the data

$$\mathbf{y}(t) = \begin{pmatrix} \sin\left(t + \frac{\pi}{4}\right) \\ \cos\left(2t + \frac{\pi}{4}\right) \end{pmatrix}, \qquad \mathbf{u}(t) = \begin{pmatrix} \delta(t) \\ 2\delta(t) \end{pmatrix}.$$

6.6.3 FREQUENCY DOMAIN IDENTIFICATION

Theorem 4.15 can be used for identification in the frequency domain. If the Fourier transform of the input/output correlation

$$\mathscr{F}\{\rangle \mathbf{y}_{ss}(t+\tau), \mathbf{u}(\tau)\langle\}$$

is available together with the Fourier transform of the power spectral density of the input $\mathscr{F}\{\rangle \mathbf{u}(t+\tau), \mathbf{u}(\tau)\langle\}$, then (4.191) immediately suggests that

$$\mathbf{G}(j\omega) = \mathscr{F}\{\rangle \mathbf{y}_{ss}(t+\tau), \mathbf{u}(\tau)\langle\}[\mathscr{F}\{\rangle \mathbf{u}(t+\tau), \mathbf{u}(\tau)\langle\}]^{-1}. \qquad (6.62)$$

EXAMPLE 6.13

Find the transfer function for the linear system whose input is described by

$$U(j\omega) = \frac{1}{1-\omega^2}$$

and whose output is described by

$$Y(j\omega) = \frac{j\omega + 1}{-\omega^2 + 1}.$$

Solution: From (4.191),

$$G(j\omega) = [Y(j\omega)U^*(j\omega)][U(j\omega)U^*(j\omega)]^{-1},$$

$$G(j\omega) = \frac{j\omega+1}{(1-\omega^2)^2}\left[\frac{1}{(1-\omega^2)^2}\right]^{-1} = j\omega + 1. \qquad \blacksquare$$

Exercise 6.10

Repeat the above example, except suppose white noise inputs $U(j\omega)U^*(j\omega) = 1$.

Closure

This chapter describes all models that match the transfer function matrix, the output correlation matrix, and a specified number of coefficients in a series expansion of both the impulse response and the output correlation. Since these models are all of different order and perhaps of lower order than the original system, these "model reductions" match a specified number of parameters of the original system.

Other coordinates are introduced which preserve the component cost of each retained state (these are called cost-decoupled coordinates) and the products of controllability and observability of each retained state (these are called balanced coordinates).

References

6.1 E. G. Gilbert, "Controllability and Observability in Multivariable Conrol Systems," *J. SIAM Control*, Ser. A, 1(2), 128–151, 1964.

6.2 R. E. Kalman, "Mathematical Description of Linear Dynamical Systems," *J. Soc. Ind. Appl. Math.-Control Series*, Ser. A, 1(2), 152–192, 1964.

6.3 R. E. Skelton and A. Yousuff, "Component Cost Analysis of Large Scale Systems," *Int. J. Control*, 37(2), 285–304, 1983.

6.4 A. Yousuff, D. A. Wagie, and R. E. Skelton, "Linear Systems Approximations Via Covariance Equivalent Realizations," *J. Math. Anal. Applications*, 106(1), 91–114, 1985.

6.5 A. Yousuff and R. E. Skelton, "Controller Reduction by Component Cost Analysis," *IEEE Trans. Autom. Control*, AC-29(6), 520–530, 1984.

6.6 D. Wagie and R. E. Skelton, "A Projection Approach to Covariance Equivalent Realizations of Discrete Systems," *IEEE Trans. Autom. Control*, AC-31(12), 1114–1120, 1986.

6.7 P. V. Kokotovic, R. E. O'Malley, Jr., and P. Sannuti, "Singular Perturbations and Order Reduction in Control Theory—An Overview," *Automatica*, 12, 123–132, 1976.

6.8 W. Eckhaus, "Formal Approximations and Singular Perturbation Techniques," *SIAM Rev.*, 19, 593, 1977.

6.9 B. C. Moore, "Principal Component Analysis in Linear Systems: Controllability, Observability and Model Reduction," *IEEE Trans. Autom. Control*, AC-26(1), 17–32, 1981.

6.10 M. F. Hutton and GB. Friedland, "Routh Approximations for Reducing Order of Linear, Time-Invariant Systems," *IEEE Trans. Autom. Control*, AC-20(3), 329–337, 1975.

6.11 B. D. O. Anderson, "The Inverse Problem of Stationary Covariance Generation," *J. Statistical Phys.*, 1(1), 133–141, 1969.

6.12 E. I. Verriest, "Low Sensitivity Design and Optimal Order Reduction for the LQG Problem," *Proc. 24th Symp. Circuits Systems*, Chicago, June 1981, pp. 365–369.

6.13 E. C. Y. Tse, J. V. Medanic, and W. R. Perkins, "Generalized Hessenberg Transformations for Reduced-Order Modeling of Large Scale Systems," *Int. J. Control*, 27, 493–512, 1978.

6.14 R. N. Mishra and D. A. Wilson, "A New Algorithm for Optimal Reduction of Multivariable Systems," *Int. J. Control*, 31(3), 443–466, 1980.

6.15 D. C. Hyland and D. S. Bernstein, "The Optimal Projection Approach to Model Reduction and the Relationship Between the Methods of Wilson and Moore," *Proc. 23rd IEEE CDC*, Las Vegas, December 1984.

6.16 S. Y. Kung and D. W. Lin, "Optimal Hankel-Norm Model Reductions; Multivariable Systems," *IEEE Trans. Autom. Control*, AC-26, 832–852, 1981.

6.17 L. Parnebo and L. M. Silverman, "Model Reduction Via Balanced State Space Representation," *IEEE Trans. Autom. Control*, AC-27, 1982.

6.18 J. Rissanen and T. Kailath, "Partial Realization of Random Systems," *Automatica*, 8, 389–396, 1972.

6.19 K. Glover, "All Optimal Hankel-Norm Approximations of Linear Multivariable Systems and Their L^{∞}-Error Bounds, *Int. J. Control*, 39(6), 1115–1193, 1984.

CHAPTER 7

Stability

Chapter 4 characterizes the solution $\mathbf{y}(t)$ for a linear system, and Chapter 5 provides tests to determine when $\mathbf{y}(t)$ can be driven to any value (controllability) and also when $\mathbf{y}(t)$ contains information about all the states (observability). There, we limit our concept of stability to "eigenvalues of $\mathbf{A}$ in the left half-plane." A different property of $\dot{\mathbf{x}} = \mathbf{A}\mathbf{x}$ that is not addressed in these chapters is whether $\lim_{t\to\infty} \mathbf{x}(t)$ exists. The general (oversimplified) idea of "stability" is that this limit exists. For nonlinear systems

$$\begin{aligned} \dot{x} &= \mathbf{f}(x, \mathbf{u}, t), \\ \mathbf{y} &= \mathbf{g}(x, \mathbf{u}, t), \end{aligned} \tag{7.1}$$

it may happen that $\lim_{t\to\infty} \mathbf{y}(t)$ exists for certain $x(0)$ and not for others. Hence, "stability" is generally not a *system* property but a property of a *particular solution* $\bar{x}(t)$ corresponding to a particular $\bar{x}(0)$. This forces us to talk about the behavior "in the vicinity of a solution."

7.1 Liapunov Stability

Now, consider an arbitrarily small perturbation from $\bar{x}(t)$ beginning at $x(0)$. Figure 7.1 illustrates an unforced solution $\bar{x}(t)$ corresponding to the initial state $x(0)$. This perturbation might result from a sudden disturbance or a slightly different initial condition $x(0) = \bar{x}(0) + \varepsilon\eta$. The stability notion made precise by Liapunov is that if the "neighboring" trajectory $x(t)$ eminating from $x(0)$ *remains* neighboring to $\bar{x}(t)$, $(\|x(t) - \bar{x}(t)\| < \delta)$ for small ε and δ, then the trajectory $\bar{x}(t)$ is stable in the sense of Liapunov.

Definition 7.1

The solution $\bar{x}(t)$ is stable in the sense of Liapunov iff for any $\delta > \varepsilon$ there exists an $\varepsilon > 0$ such that

$$\|x(0) - \bar{x}(0)\| < \varepsilon \Rightarrow \|x(t) - \bar{x}(t)\| < \delta \qquad \textit{for all } t \geq 0.$$

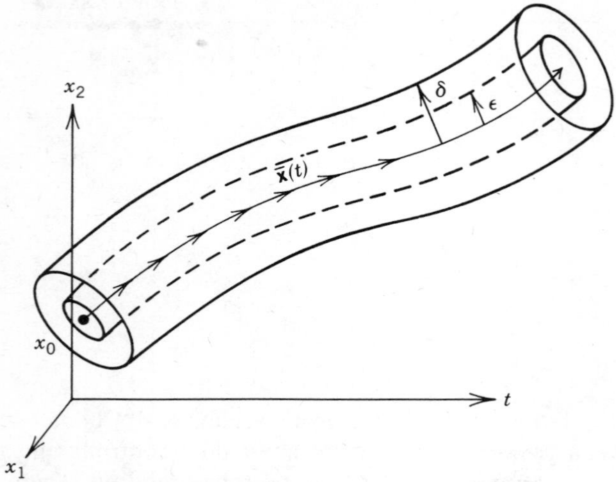

Figure 7.1 *Liapunov Stability*

Now, suppose $\mathscr{V}(\mathscr{x}(t))$ is some scalar such that $\|\mathscr{x}(0) - \bar{\mathscr{x}}(0)\| = 0$ implies $\mathscr{V}(\mathscr{x}(0)) = 0$. Furthermore, suppose $\mathscr{V}(\mathscr{x}(t)) = \infty$ iff $\|\mathscr{x}(t) - \bar{\mathscr{x}}(t)\| = \infty$. Verify that one such function $\mathscr{V}(\mathscr{x}(t))$ is $\mathscr{V}(\mathscr{x}(t)) \triangleq (\mathscr{x}(t) - \bar{\mathscr{x}}(t))^* \mathbf{Q}(\mathscr{x}(t) - \bar{\mathscr{x}}(t))$ for any $\mathbf{Q} = \mathbf{Q}^* > \mathbf{0}$. Hence, Definition 7.1 can be written with any $\mathscr{V}(\mathscr{x}(t))$ replacing $\|\mathscr{x}(t) - \bar{\mathscr{x}}(t)\|$, such that $\mathscr{V}(\mathscr{x}(t)) = 0, \infty$ iff $\|\mathbf{x}(t) - \bar{\mathbf{x}}(t)\| = 0, \infty$, respectively. Such a scalar function $\mathscr{V}(\mathscr{x}(t))$ is called a "Liapunov function" if it can be used to prove stability or instability by Liapunov's definitions.

Definition 7.2

The solution $\bar{\mathscr{x}}(t)$ is asymptotically stable iff it is stable in the sense of Liapunov and in addition $\lim_{t \to \infty} \|\mathscr{x}(t) - \bar{\mathscr{x}}(t)\| = 0$.

Two simplifications will first be made before continuing. Define a new state

$$\mathbf{x}(t) \triangleq \mathscr{x}(t) - \bar{\mathscr{x}}(t) \tag{7.2}$$

so that (7.1) becomes

$$\begin{aligned} \dot{\mathbf{x}} &= \mathbf{f}(\mathbf{x}, \mathbf{u}, t), \\ \mathbf{y} &= \mathbf{g}(\mathbf{x}, \mathbf{u}, t) \end{aligned} \tag{7.3}$$

and the particular solution of interest now is the *null solution* ($\mathbf{x}(t) = \mathbf{0}$) of (7.3). Since the discussions of "neighboring" trajectories involve only small perturbations from $\mathbf{x}(t) = \mathbf{0}$, we may as well *linearize* the equations about the null solution. There is only one circumstance for which stability of the linearized equations does not dictate the stability properties of the nonlinear system, and we shall later point out this circumstance. The linearized equation is

$$\dot{\mathbf{x}} = \mathbf{Ax} + \mathbf{Bu},$$
$$\mathbf{y} = \mathbf{Cx}, \tag{7.4a}$$

where

$$\mathbf{A} = \left[\frac{\partial \mathbf{f}}{\partial \mathbf{x}}\right]^T_{\mathbf{x}=\mathbf{0},\,\mathbf{u}=\mathbf{0}}, \qquad \mathbf{B} = \left[\frac{\partial \mathbf{f}}{\partial \mathbf{u}}\right]^T_{\mathbf{x}=\mathbf{0},\,\mathbf{u}=\mathbf{0}}. \tag{7.4b}$$

Now, if $\mathscr{V}$ is defined as in Chapters 4 through 6, then the cost function we associated with the linear system (7.4) is

$$\mathscr{V} = \sum_{i=1}^{n_u} \int_0^\infty \mathbf{y}^{i*}(t)\mathbf{Q}\mathbf{y}^i(t)\,dt = \operatorname{tr}\mathbf{XC^*QC},$$
$$\mathbf{0} = \mathbf{XA^*} + \mathbf{AX} + \mathbf{B}\mathscr{U}\mathbf{B^*}. \tag{7.5}$$

But keep in mind that a *different* $\mathscr{V}$ may be also defined by excitation only from initial conditions, *applied one at a time* to yield

$$\mathscr{V} = \sum_{i=1}^{n_x} \int_0^\infty \mathbf{y}^{i*}(t)\mathbf{Q}\mathbf{y}^i(t)\,dt = \operatorname{tr}\mathbf{XC^*QC},$$
$$\mathbf{0} = \mathbf{XA^*} + \mathbf{AX} + \mathbf{X}_0, \qquad X_{0_{ij}} = x_i^2(0)\,\delta_{ij}, \tag{7.6}$$

or by a *simultaneous* application of initial conditions $x_i(0)$ to yield

$$\begin{aligned}
\mathscr{V} &= \int_0^\infty \mathbf{y}^*(t)\mathbf{Q}\mathbf{y}(t)\,dt \\
&= \int_0^\infty (\mathbf{C}e^{\mathbf{A}t}\mathbf{x}(0))^*\mathbf{QC}e^{\mathbf{A}t}\mathbf{x}(0)\,dt \\
&= \mathbf{x}^*(0)\left[\int_0^\infty e^{\mathbf{A}^*t}\mathbf{C^*QC}e^{\mathbf{A}t}\,dt\right]\mathbf{x}(0) \\
&= \mathbf{x}^*(0)\mathbf{Kx}(0), \qquad \mathbf{0} = \mathbf{KA} + \mathbf{A^*K} + \mathbf{C^*QC}.
\end{aligned} \tag{7.7}$$

Now, the theory in this chapter will not be significantly altered by the use of any of these functions $\mathscr{V}$. However, to maintain a close identity with the particular initial conditions that generate the special trajectory about which we seek stability information, we shall use (7.7) exclusively in this chapter. We seek a Liapunov function which is useful for linear systems in the proof of stability by Definition 7.1.

Equation (7.4) retains only the first term in the Taylor's series expansion. Recall from Chapter 4 that by *increasing* the dimension of the state space, higher-order terms of the Taylor's series can be accommodated with linear or bilinear models. This was accomplished by introducing new states which were *products* of the original states (x_1, x_1^2, x_1x_2, etc.; see Section 3.2). The step we now introduce involves products of all the original states, but it is different in the sense that we shall *reduce* the system to a cost-equivalent (more specifically a "Liapunov function" equivalent) realization and study the stability properties of this cost-equivalent realization.

Theorem 7.1

The system (7.4) is stable in the sense of Liapunov if and only if its cost-equivalent realization

$$\dot{\mathbf{x}}_R = \mathbf{A}_R \mathbf{x}_R, \qquad \mathscr{V}_R(\mathbf{x}_R) = \mathscr{V}(\mathbf{x}) \tag{7.8}$$

is stable in the sense of Liapunov, provided $\mathscr{V}_R(\mathbf{x}_R)$ and $\mathscr{V}(\mathbf{x})$ are positive definite functions of $\mathbf{x}_R$ and $\mathbf{x}$, respectively.

Proof: First, note that if $\mathscr{V}(\mathbf{x}) > 0\ \forall \mathbf{x} \neq \mathbf{0}$, then $\mathscr{V}(\mathbf{x})$ serves as a metric of distance from the origin of $\mathbf{x}$, and $\|\mathbf{x}(t)\|$ in Definitions 7.1 and 7.2 can be replaced by $\mathscr{V}(\mathbf{x})$. Now, suppose (7.4) is stable and (7.8) is not. Then, from Definition 7.1, there exists an ε such that

$$\mathscr{V}(\mathbf{x}(0)) < \varepsilon \Rightarrow \mathscr{V}(\mathbf{x}(t)) < \delta \tag{7.9}$$

for every δ; and since (7.8) is unstable,

$$\mathscr{V}_R(\mathbf{x}_R(0)) < \varepsilon \not\Rightarrow \mathscr{V}_R(\mathbf{x}_R) < \delta. \tag{7.10}$$

But since $\mathscr{V}_R(\mathbf{x}_R) = \mathscr{V}(\mathbf{x})$, statements (7.9) and (7.10) are contradictory. Now, suppose (7.8) is stable and (7.4) is not. Then there exists an ε such that

$$\mathscr{V}_R(\mathbf{x}_R(0)) < \varepsilon \Rightarrow \mathscr{V}_R(\mathbf{x}_R) < \delta \tag{7.11}$$

for every δ, but

$$\mathscr{V}(\mathbf{x}(0)) < \varepsilon \not\Rightarrow \mathscr{V}(\mathbf{x}) < \delta. \tag{7.12}$$

Again, since $\mathscr{V}_R(\mathbf{x}_R) = \mathscr{V}(\mathbf{x})$ statements (7.11) and (7.12) are contradictory. Hence, the theorem is proved. ■

The cost-equivalent realizations of Chapter 6 are not scalar models (unless $n_y = 1$). Hence, they are not used here. Now, it is always possible to construct a *scalar* cost-equivalent realization of (7.4). To show this, define

$$\mathbf{x}_R(t) \triangleq \int_t^{\infty} \mathbf{y}^*(\sigma)\mathbf{Q}\mathbf{y}(\sigma)\,d\sigma = \int_t^{\infty} \mathbf{x}^*(t)e^{\mathbf{A}^*(\sigma-t)}\mathbf{C}^*\mathbf{Q}\mathbf{C}e^{\mathbf{A}(\sigma-t)}\mathbf{x}(t)\,d\sigma$$

$$= \mathbf{x}^*(t)\mathbf{K}\mathbf{x}(t), \qquad \mathbf{K} \triangleq \int_0^{\infty} e^{\mathbf{A}^*\tau}\mathbf{C}^*\mathbf{Q}\mathbf{C}e^{\mathbf{A}\tau}\,d\tau, \tag{7.13}$$

where we have substituted $\tau = \sigma - t$, $d\tau = d\sigma$. Also note that

$$\mathscr{V} = x_R(0) \triangleq \mathscr{V}_R \tag{7.14a}$$

and

$$\dot{x}_R = -\mathbf{y}^*\mathbf{Q}\mathbf{y} = A_R x_R, \tag{7.14b}$$

where

$$A_R = \frac{-\mathbf{x}^*\mathbf{C}^*\mathbf{Q}\mathbf{C}\mathbf{x}}{\mathbf{x}^*\mathbf{K}\mathbf{x}}. \tag{7.14c}$$

It is clear from (7.14a) that $\mathscr{V}_R$ is a positive function if $x_R(0) \neq 0$, and $\mathscr{V}$ is a positive function of $\mathbf{x}(0)$ if $\mathbf{x}(0) \neq \mathbf{0}$. In this latter case, this requires $\mathbf{K}$ to be a positive definite matrix. From (7.11) and Chapter 5, we conclude that, for any $\mathbf{Q} > \mathbf{0}$, $\mathbf{K} > \mathbf{0}$ iff $(\mathbf{A}, \mathbf{C})$ is an observable pair. Since all the conditions of Theorem 7.1 are met, the system (7.5) is Liapunov stable if and only if the cost-equivalent realization (7.13a) is Liapunov stable. To study (7.13) more carefully and to understand the range of values that A_R might have, we examine the extrema of A_R.

Theorem 7.2

If $(\mathbf{A}, \mathbf{C})$ *is an observable pair, then* $\mathbf{K}^{-1}$ *exists and*

$$\lambda_{\min}[\mathbf{K}^{-1}\mathbf{C}^*\mathbf{Q}\mathbf{C}] \leqq -A_R \leqq \lambda_{\max}[\mathbf{K}^{-1}\mathbf{C}^*\mathbf{Q}\mathbf{C}], \tag{7.14d}$$

where $\lambda_{\min}, \lambda_{\max}$ *denote minimum and maximum eigenvalues.*

Proof: We shall prove only the case for real variables. An extremum of (7.14c) satisfies

$$0 = \frac{\partial}{\partial \mathbf{x}}\left\{-(\mathbf{x}^T\mathbf{K}\mathbf{x})^{-1}(\mathbf{x}^T\mathbf{C}^T\mathbf{Q}\mathbf{C}\mathbf{x})\right\}$$

$$= (\mathbf{x}^T\mathbf{K}\mathbf{x})^{-2}(\mathbf{K}\mathbf{x})(\mathbf{x}^T\mathbf{C}^T\mathbf{Q}\mathbf{C}\mathbf{x}) - (\mathbf{x}^T\mathbf{K}\mathbf{x})^{-1}(\mathbf{C}^T\mathbf{Q}\mathbf{C}\mathbf{x}) = 0.$$

Multiply by $\mathbf{x}^T\mathbf{K}\mathbf{x}$ to obtain

$$\frac{\mathbf{x}^T\mathbf{C}^T\mathbf{Q}\mathbf{C}\mathbf{x}}{\mathbf{x}^T\mathbf{K}\mathbf{x}}\mathbf{K}\mathbf{x} - \mathbf{C}^T\mathbf{Q}\mathbf{C}\mathbf{x} = \mathbf{0}.$$

Multiply from the left by $\mathbf{K}^{-1}$ (but $|\mathbf{K}| \neq 0$ requires observability of the pair $(\mathbf{A}, \mathbf{C})$ in (7.13), if $\mathbf{Q} > \mathbf{0}$):

$$-A_R\mathbf{x} = \mathbf{K}^{-1}\mathbf{C}^T\mathbf{Q}\mathbf{C}\mathbf{x}, \qquad -A_R = \frac{\mathbf{x}^T\mathbf{C}^T\mathbf{Q}\mathbf{C}\mathbf{x}}{\mathbf{x}^T\mathbf{K}\mathbf{x}}. \tag{7.15}$$

The scalar $-A_R$ and the vector $\mathbf{x}$ in (7.15) are obviously eigenvalues and eigenvectors of the matrix $[\mathbf{K}^{-1}\mathbf{C}^T\mathbf{Q}\mathbf{C}]$. Hence, the extremum values of A_R are the extremum values of the eigenvalues of $[\mathbf{K}^{-1}\mathbf{C}^T\mathbf{Q}\mathbf{C}]$. ■

As a consequence of this theorem, it follows that

$$\dot{x}_R(t) \leqq A_{R_{\min}} x_R(t), \qquad A_{R_{\min}} = -\lambda_{\max}[\mathbf{K}^{-1}\mathbf{C}^T\mathbf{Q}\mathbf{C}]. \tag{7.16}$$

Hence,

$$x_R(t) \leqq e^{A_{R_{\min}}t} x_R(0). \tag{7.17}$$

From (7.17) and Definitions 7.1 and 7.2, it is clear that (7.8) is stable in the sense of Liapunov if $A_{R_{\max}} \leqq 0$ (to see this, pick $\varepsilon = \delta$). From Theorem 7.2, this requires all eigenvalues of $[\mathbf{K}^{-1}\mathbf{C}^T\mathbf{Q}\mathbf{C}]$ to be zero or positive. This is guaranteed if $\mathbf{K} > \mathbf{0}$ (since $\mathbf{K} > \mathbf{0} \Rightarrow \mathbf{K}^{-1} > \mathbf{0}$). The conditions for existence of a $\mathbf{K} > \mathbf{0}$ is now the only remaining issue.

Theorem 7.3

If the matrix pair $(\mathbf{A}, \mathbf{C})$ *is observable, then there exists a* $\mathbf{K} > \mathbf{0}$ *solution of*

$$\mathbf{0} = \mathbf{K}\mathbf{A} + \mathbf{A}^T\mathbf{K} + \mathbf{C}^T\mathbf{Q}\mathbf{C}, \qquad \mathbf{Q} > \mathbf{0}, \tag{7.18}$$

iff $\mathbf{A}$ *is asymptotically stable. Such a solution will also be unique.*

Proof: The uniqueness question first: Liapunov equations (7.18) have unique solutions iff $\lambda_i + \lambda_j \neq 0 \ \forall i, j$ (Chapter 2). If $\mathbf{A}$ is asymptotically stable, there will be no positive real parts of λ_j to add to the negative λ_i to give $\lambda_i + \lambda_j = 0$. Hence, $\mathbf{K}$ is unique if $\mathbf{A}$ is asymptotically stable. The observability (weighted) grammian given by (7.13) exists iff the observable modes are asymptotically stable. Otherwise, the integral blows up. (Visualize (7.13) in modal coordinates to see this clearly). But since observability is a *given*, we conclude that $\mathbf{K}$ exists iff $\mathbf{A}$ is asymptotically stable. Thus, the upper bound on $\mathbf{K} < \infty$ is established by asymptotic stability, whereas the lower bound on $\mathbf{K} > \mathbf{0}$ is established by observability. Finally, if the grammian $\mathbf{K}$ in (7.13) exists, it is known to satisfy (7.18). ■

Now, we may conclude that $A_{R_{\max}} < 0$ *iff* (7.18) *has a solution* $\mathbf{K} > 0$. Hence, the following:

Corollary 7.1
The cost-equivalent realization (7.8) *is asymptotically stable iff* (7.18) *has a solution* $\mathbf{K} > \mathbf{0}$.

Perhaps it is disappointing that the stability of the *scalar* cost-equivalent realization (7.8) in the final analysis requires the calculation of an $n_x \times n_x$ matrix **K**, which in turn establishes the stability of (7.4) directly. However, the point of view is important, allowing one to relate the stability of a high-order and a lower-order system. Besides, the primary advantage in Liapunov's point of view holds for nonlinear systems, which we do not explore. The use of states of the form (7.13) has been used to great advantage by Siljak [7.1] in the study of stability of large-scale systems.

EXAMPLE 7.1
Use Theorem 7.3 to determine whether the null solution of

$$\dot{\mathbf{x}} = \mathbf{A}\mathbf{x}, \qquad \mathbf{A} = \begin{bmatrix} 1 & 0 \\ 0 & -1 \end{bmatrix}$$

is stable.

Solution: The answer is obvious. The eigenvalues are 1, -1. But the technique of Thoerem 7.3 is to be illustrated. Choose

$$\mathbf{Q} = \mathbf{I}, \qquad \mathbf{C} = \begin{bmatrix} 1 & 0 \\ 0 & 1 \end{bmatrix}.$$

This choice makes $(\mathbf{A}, \mathbf{C})$ observable, and the solution for **K** gives

$$\mathbf{K} = \begin{bmatrix} -\frac{1}{2} & \Delta \\ \Delta & \frac{1}{2} \end{bmatrix},$$

which is not positive definite. Hence, **A** is unstable. Note that the Δ terms in **K** are *arbitrary!* Why is **K** not unique? Now, choose $Q = 1$, $\mathbf{C} = (1, 1)$. This choice makes $(\mathbf{A}, \mathbf{C})$ observable, and (7.18) gives

$$2K_{11} + 1 = 0, \quad -K_{12} + K_{12} + 1 = 0, \quad -2K_{22} + 1 = 0,$$

which has no solution. Hence, **A** is unstable. ∎

Theorem 7.4
The null solution of a system is asymptotically stable iff there exists a positive definite function of **x**,

$$x_R(\mathbf{x}) > 0, \tag{7.19a}$$

which satisfies either

$$\dot{x}_R(x) < 0 \tag{7.19b}$$

or

$$\{\dot{x}_R(\mathbf{x}) \leqq 0 \text{ and } (\dot{x}_R \equiv 0 \Rightarrow \mathbf{x} = \mathbf{0})\}. \tag{7.19c}$$

The null solution is stable in the sense of Liapunov iff there exists $x_R(\mathbf{x}) > 0$ *such that* $\dot{x}_R(\mathbf{x}) \leqq 0$.

Proof: The necessary and sufficient conditions are given by Theorem 7.3 for linear systems, which is our focus here. Hence, we have only to show that the conditions (7.19a), (7.19b) or (7.19a), (7.19c) are equivalent to the conditions of Theorem 7.3. Let $x_R(\mathbf{x})$ be the positive definite function of $\mathbf{x}$,

$$x_R(\mathbf{x}) = \mathbf{x}^*\mathbf{K}\mathbf{x}, \qquad \mathbf{K} > \mathbf{0}. \tag{7.20}$$

Then,

$$\begin{aligned} \dot{x}_R(\mathbf{x}) &= \dot{\mathbf{x}}^*\mathbf{K}\mathbf{x} + \mathbf{x}^*\mathbf{K}\dot{\mathbf{x}} \\ &= \mathbf{x}^*[\mathbf{A}^*\mathbf{K} + \mathbf{K}\mathbf{A}]\mathbf{x}. \end{aligned}$$

Define $\mathbf{C}^*\mathbf{Q}\mathbf{C}$ by

$$\mathbf{A}^*\mathbf{K} + \mathbf{K}\mathbf{A} = -\mathbf{C}^*\mathbf{Q}\mathbf{C}, \qquad \mathbf{Q} > \mathbf{0}, \tag{7.21}$$

and $\mathbf{y} \triangleq \mathbf{C}\mathbf{x}$. Then,

$$\dot{x}_R(\mathbf{x}) = -\mathbf{x}^*\mathbf{C}^*\mathbf{Q}\mathbf{C}\mathbf{x} = -\mathbf{y}^*\mathbf{Q}\mathbf{y}. \tag{7.22}$$

Now, if there exists a $\mathbf{C}$ such that (7.20) and (7.21) hold, then $\dot{x}_R(\mathbf{x}) \leqq 0$ from (7.22). Also, from (7.22), see that $\dot{x}_R \equiv 0 \Leftrightarrow \mathbf{y}(t) \equiv \mathbf{0}$, so that (7.19c) means the condition

$$\{\mathbf{y}(t) \equiv 0 \Rightarrow \mathbf{x} = \mathbf{0}\},$$

which is just the observability condition (Chapter 5) on the pair $(\mathbf{A}, \mathbf{C})$. Hence, the existence of a $\mathbf{K} > \mathbf{0}$ which satisfies (7.21) for any $\mathbf{C}$ such that $(\mathbf{A}, \mathbf{C})$ is observable, and $\mathbf{Q} > \mathbf{0}$, establishes that (7.19a, c) hold, and this is in fact the statement of Theorem 7.3. Now, suppose $\dot{x}_R(\mathbf{x}) < 0$ as in (7.19b). This is accomplished by any square *nonsingular* matrix $\mathbf{C}$, in which case the pair $(\mathbf{A}, \mathbf{C})$ is observable regardless of $\mathbf{A}$, and Theorem 7.3 again applies. The final sentence of the theorem follows immediately by setting $\varepsilon = \delta$ in Definition 7.1 and substituting $x_R(\mathbf{x})$ for $\|\mathbf{x}\|^2$. ■

Theorem 7.4 was first proved by Liapunov for both linear and nonlinear systems, even though we have restricted our attention to linear systems.

EXAMPLE 7.2

Use Theorem 7.3 to examine stability of $m\ddot{y} + ky = 0$.

Solution:

$$\mathbf{x}^T = (y, \dot{y}), \qquad \mathbf{A} = \begin{bmatrix} 0 & 1 \\ -\dfrac{k}{m} & 0 \end{bmatrix}.$$

Choose $C = (1 \quad 0)$ so that $(\mathbf{A}, \mathbf{C})$ is observable. The solution of (7.18) is

$$\mathbf{K} = \text{no solution}.$$

Hence, the system is not asymptotically stable. It is stable in the sense of Liapunov from Theorem 7.4 by showing that $x_R(\mathbf{x}) > 0$, $\dot{x}_R(\mathbf{x}) \leqq 0$, with $x_R(\mathbf{x}) \triangleq \frac{1}{2}ky^2 + \frac{1}{2}m\dot{y}^2$. ■

Theorem 7.5

The null solution of the nonlinear system $\dot{\mathbf{x}} = \mathbf{f}(\mathbf{x}, t)$ is asymptotically stable iff its linearized model $\dot{\mathbf{x}} = \mathbf{A}\mathbf{x}$ is asymptotically stable, provided

$$\mathbf{A}^T = \left.\frac{\partial \mathbf{f}}{\partial \mathbf{x}}\right|_{\mathbf{x}=\mathbf{0}} \quad \text{exists} \tag{7.23}$$

and

$$\lim_{\mathbf{x}\to 0} \frac{\| \mathbf{h}(\mathbf{x}, t) \|}{\|\mathbf{x}\|} = 0, \tag{7.24}$$

where

$$\dot{\mathbf{x}} = \mathbf{f}(\mathbf{x}, t) = \mathbf{A}\mathbf{x} + \mathbf{h}(\mathbf{x}, t). \tag{7.25}$$

Proof: From Theorem 7.4, we wish to show that $\dot{x}_R(\mathbf{x}) < 0$ for some real $x_R(\mathbf{x}) > 0$. Choose

$$x_R(\mathbf{x}) = \mathbf{x}^T\mathbf{K}\mathbf{x}, \tag{7.26}$$

$$\begin{aligned} \dot{x}_R(\mathbf{x}) &= \dot{\mathbf{x}}^T\mathbf{K}\mathbf{x} + \mathbf{x}^T\mathbf{K}\dot{\mathbf{x}} \\ &= (\mathbf{h}^T(\mathbf{x}, t) + \mathbf{x}^T\mathbf{A}^T)\mathbf{K}\mathbf{x} + \mathbf{x}^T\mathbf{K}(\mathbf{A}\mathbf{x} + \mathbf{h}(\mathbf{x}, t)) \\ &= \mathbf{x}^T(\mathbf{A}^T\mathbf{K} + \mathbf{K}\mathbf{A})\mathbf{x} + 2\mathbf{x}^T\mathbf{K}\mathbf{h}. \end{aligned} \tag{7.27}$$

Now, the linear part of the system is asymptotically stable iff $\mathbf{A}^T\mathbf{K} + \mathbf{K}\mathbf{A} = -\mathbf{C}^T\mathbf{Q}\mathbf{C}$ for *any* choice of $\mathbf{C}, \mathbf{Q}$ subject to $(\mathbf{A}, \mathbf{C})$ observable, $\mathbf{Q} > \mathbf{0}$, $\mathbf{K} > \mathbf{0}$. Choose $\mathbf{C} = \mathbf{I}$, $\mathbf{Q} = \mathbf{I}$. Then, $\dot{x}_R(\mathbf{x})$ becomes

$$\dot{x}_R(\mathbf{x}) = -\mathbf{x}^T\mathbf{x} + 2\mathbf{x}^T\mathbf{K}\mathbf{h}$$

$$= -\mathbf{x}^T\mathbf{x}\left(1 - \frac{2\mathbf{x}^T\mathbf{K}\mathbf{h}}{\mathbf{x}^T\mathbf{x}}\right). \tag{7.28}$$

But by the Cauchy–Schwartz inequality,

$$\frac{2\mathbf{x}^T\mathbf{K}\mathbf{h}}{\mathbf{x}^T\mathbf{x}} \leqq \left(\frac{2\|\mathbf{x}^T\mathbf{K}\|}{\|\mathbf{x}\|}\right)\left(\frac{\|\mathbf{h}\|}{\|\mathbf{x}\|}\right) \tag{7.29}$$

and the second term in parenthesis approaches zero in the vicinity of the origin, by (7.24). Hence, the second inequality $(2\mathbf{x}^T\mathbf{K}\mathbf{h}/\mathbf{x}^T\mathbf{x} = \varepsilon < 1)$ and (7.28) becomes $\dot{x}_R(\mathbf{x}) = -\mathbf{x}^T\mathbf{x}(1 - \varepsilon) < 0$ and the proof is complete. ■

Exercise 7.1 [Likins 7.9]

Determine whether the null solution of

$$m\ddot{y} - my\Omega^2 + k\frac{\ell - \ell_0}{\ell}y + c\dot{y} = 0 \tag{7.30}$$

is asymptotically stable, where

$$\ell_0 < a\left(1 - \frac{m\Omega^2}{k}\right), \qquad \ell \triangleq \sqrt{y^2 + a^2}. \tag{7.31}$$

The physical entities are described in Fig. 7.2. Repeat with the inequality in (7.31) reversed. Give physical interpretations to your results.

The *only* time that linear analysis fails to determine the stability or instability of a solution of the underlying nonlinear system is when the linearized model is only stable and not asymptotically stable. In this case the discarded nonlinear terms dictate stability or instability of the nonlinear systems.

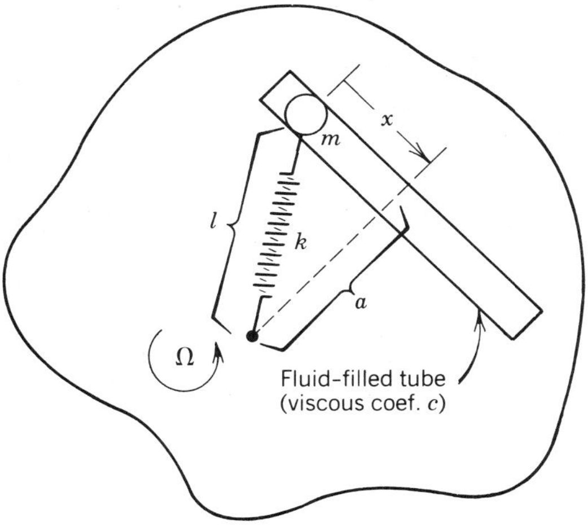

Ω: angular speed

l_0: Undeformed length of spring

$$m\ddot{x} - mx\Omega^2 + k\left(\frac{l - l_0}{l}\right)x + c\dot{x} = 0$$

$$l_0 < a\left(1 - \frac{m\Omega^2}{k}\right) \qquad l = \sqrt{x^2 + a^2}$$

Linearized about $x = 0$:

$$m\ddot{x} + c\dot{x} + \left(k - \Omega^2 m - \frac{kl_0}{a}\right)x = 0$$

$$\text{asymptotic stability} \Leftrightarrow \left(k - \Omega^2 m - \frac{kl_0}{a}\right) > 0$$

Figure 7.2 *Spinning Spacecraft*

7.2 *Bounded-Input / Bounded-Output Stability*

Note that the stability concepts of Section 7.1 involved the initial states and not the inputs. In this section we involve the inputs and not the initial state.

Definition 7.3
The system

$$\begin{aligned}\dot{\mathbf{x}} &= \mathbf{Ax} + \mathbf{Bu},\\ \mathbf{y} &= \mathbf{Cx}\end{aligned} \tag{7.32}$$

is bounded-input/bounded-output (BIBO) stable iff

$$\left\{\begin{matrix} \|\mathbf{u}(t)\| < \infty \\ \mathbf{x}(0) = \mathbf{0} \end{matrix}\right\} \Rightarrow \|\mathbf{y}(t)\| < \infty. \tag{7.33}$$

Proof: To develop conditions for BIBO stability, solve (7.32),

$$\mathbf{y}(t) = \int_0^t \mathbf{C}\boldsymbol{\Phi}(t, \sigma)\mathbf{B}(\sigma)\mathbf{u}(\sigma)\, d\sigma. \tag{7.34}$$

Now, let $\|\mathbf{u}(\sigma)\| < \mathscr{L}$ for some arbitrarily large number $\mathscr{L}$. Note that

$$\|\mathbf{y}(t)\| = \left\|\int_0^t \mathbf{C}(t)\boldsymbol{\Phi}(t, \sigma)\mathbf{B}(\sigma)\mathbf{u}(\sigma)\, d\sigma\right\| \tag{7.35}$$

and

$$\|\mathbf{y}(t)\| \leqq \int_0^t \|\mathbf{C}(t)\mathbf{\Phi}(t,\sigma)\mathbf{B}(\sigma)\mathbf{u}(\sigma)\|\,d\sigma,$$

and, by the Cauchy–Schwartz inequality,

$$\begin{aligned}\|\mathbf{y}(t)\| &\leqq \int_0^t \|\mathbf{C}(t)\mathbf{\Phi}(t,\sigma)\mathbf{B}(\sigma)\|\,\|\mathbf{u}(\sigma)\|\,d\sigma \\ &\leqq \mathscr{L}\int_0^t \|\mathbf{C}(t)\mathbf{\Phi}(t,\sigma)\mathbf{B}(\sigma)\|\,d\sigma. \qquad (7.36)\end{aligned}$$

Hence, $\|\mathbf{y}\| < \infty$ if

$$\int_0^t \|\mathbf{C}(t)\mathbf{\Phi}(t,\sigma)\mathbf{B}(\sigma)\|\,d\sigma < \infty. \qquad (7.37)$$

Now, to prove that (7.37) is necessary, note that, if for every finite k

$$\int_{t_0}^{t_k} \|\mathbf{C}(t)\mathbf{\Phi}(t,\sigma)\mathbf{B}(\sigma)\|\,d\sigma \geqq k, \qquad \|u(t)\| \leqq L,$$

for some t_0, then there exists no bound for $\|\mathbf{y}(t)\|$. To verify that this is true, use the norm

$$\|\mathbf{y}\| = \max_{\|\mathbf{v}\|=1} \mathbf{v}^*\mathbf{y}$$

(note $\mathbf{v}^*\mathbf{y} = \|\mathbf{v}\|\,\|\mathbf{y}\|\cos\theta = \|\mathbf{y}\|\cos\theta$, which is max when $\cos\theta = 1$, hence $\mathbf{v} = \mathbf{y}/\|\mathbf{y}\|$) to show that, for any given t_k,

$$\begin{aligned}\|\mathbf{y}(t_k)\| &= \int_{t_0}^{t_k} \max_{\|\mathbf{v}(t_k)\|=1}\ \max_{\mathbf{u}(\sigma)} \mathbf{v}^*(t_k)\mathbf{C}(t_k)\mathbf{\Phi}(t_k,\sigma)\mathbf{B}(\sigma)\mathbf{u}(\sigma)\,d\sigma \\ &= \int_{t_0}^{t_k} \|\mathbf{C}(t_k)\mathbf{\Phi}(t_k,\sigma)\mathbf{B}(\sigma)\|\,d\sigma\,\mathscr{L}.\end{aligned}$$

Thus, $\|\mathbf{y}(t_k)\|$ cannot be bounded if no bound exists on the norm of $\mathbf{C}(t_k)\mathbf{\Phi}(t_k,\sigma)\mathbf{B}(\sigma)$. ■

This proof follows that of ref. [7.2], where more detailed discussion on stability may be found. In summary of the above results,

Theorem 7.6
The linear system has BIBO stability iff (7.37) *holds for all* $t > 0$.

Exercise 7.2
Determine whether

$$\dot{x} = e^{t}x + e^{-2t}u,$$
$$y = x$$

are BIBO stable.

For time-invariant systems, Theorem 7.6 simplifies; (7.37) becomes

$$\int_0^t \|\mathbf{C}e^{\mathbf{A}(t-\sigma)}\mathbf{B}\|\,d\sigma \leqq \|\mathbf{C}\|\left\|\int_0^t e^{\mathbf{A}(t-\sigma)}\,d\sigma\right\|\|\mathbf{B}\| < \infty$$

or, for finite $\mathbf{C}, \mathbf{B}$,

$$\left\|\int_0^t e^{\mathbf{A}(t-\sigma)}\,d\sigma\right\| < \infty.$$

Now, substitute the spectral decomposition of $e^{\mathbf{A}(t-\sigma)}$,

$$e^{\mathbf{A}(t-\sigma)} = \mathbf{E}e^{\Lambda(t-\sigma)}\mathbf{E}^{-1},$$

where $\mathbf{AE} = \mathbf{E}\Lambda$, to get

$$\left\|\int_0^t \mathbf{E}e^{\Lambda(t-\sigma)}\mathbf{E}^{-1}\,d\sigma\right\| \leqq \|\mathbf{E}\|\left\|\int_0^t e^{\Lambda(t-\sigma)}\,d\sigma\right\|\|\mathbf{E}^{-1}\|$$
$$= \left\|\int_0^t e^{\Lambda(t-\sigma)}\,d\sigma\right\| < \infty.$$

But since

$$e^{\Lambda(t-\sigma)} = \text{block diag}\,[\ldots e^{\Lambda_i(t-\sigma)}\ldots],$$

where

$$e^{\Lambda_i(t-\sigma)} = e^{\lambda_i(t-\sigma)}\begin{bmatrix} 1 & t & \frac{t^2}{2} & \cdots \\ 0 & 1 & t & \cdots \\ 0 & 0 & 1 & \cdots \\ 0 & & & \cdots \end{bmatrix},$$

then

$$\left\|\int_0^t e^{\Lambda(t-\sigma)}\,d\sigma\right\| < \infty$$

iff

$$\sum_i \left[\left(1 + t + \frac{t^2}{2} + \cdots \right) e^{\lambda_i t} \right]^2 < \infty,$$

which holds iff $\mathscr{Re}\, \lambda_i < 0\ \forall i$. In summary,

Theorem 7.7
Time-invariant linear systems are BIBO stable if all eigenvalues of **A** *lie in the open left half-plane. If* (**A**, **B**, **C**) *is a minimal transfer-equivalent realization (TER) of the system, then the system is BIBO stable iff all eigenvalues of* **A** *lie in the open left half-plane, except that distinct roots may appear on the* $j\omega$ *axis.*

The only new point to verify is the difference between the first and second sentences of Theorem 7.7. The first sentence cannot be written with "if and only if" because uncontrollable or unobservable modes might be unstable ($\lambda_i[\mathbf{A}] > 0$) even though $\mathbf{y}(t)$ is well behaved.

In conclusion, if (**A**, **B**, **C**) is a minimal TER, then the Liapunov equation

$$\mathbf{0} = \mathbf{KA} + \mathbf{A}^T\mathbf{K} + \mathbf{C}^T\mathbf{QC}$$

has solution **K** with the property

$$\{\mathbf{0} < \mathbf{K} < \infty \text{ iff } \mathscr{Re}\, \lambda_i[\mathbf{A}] < 0\},$$

which is equivalent to asymptotic stability in the sense of Liapunov. For minimal TERs, there is no difference between Liapunov stability and BIBO stability, since both allow distinct eigenvalues on the $j\omega$ axis.

7.3 Exponential and Uniform Stability

Liapunov stability definitions involved the initial state and not the inputs. BIBO stability concepts involved the inputs and not the initial state. The definition of exponential stability involves neither the initial states nor the inputs.

Definition 7.4
The system

$$\dot{\mathbf{x}} = \mathbf{A}(t)\mathbf{x},$$

is exponentially stable iff there exists scalars $\gamma > 0$, $\lambda < 0$, *such that*

$$\|\mathbf{\Phi}(t, t_0)\| \leqq \gamma e^{\lambda(t-t_0)}, \qquad \dot{\mathbf{\Phi}}(t, t_0) = \mathbf{A}(t)\mathbf{\Phi}(t, t_0),$$

$$\mathbf{\Phi}(t_0, t_0) = \mathbf{I}.$$

for all $t > t_0$.

Definition 7.5

The system $\dot{\mathbf{x}} = \mathbf{A}(t)\mathbf{x}$ *is said to be uniformly stable if there exists a* $\gamma > 0$ *such that* $\|\boldsymbol{\Phi}(t, t_0)\| \leqq \gamma$ *for all* $t \geqq t_0$.

Definition 7.6

A transformation $\mathbf{x}(t) = \mathbf{P}^{-1}(t)\mathbf{z}(t)$ *is called a Liapunov transformation if*

(i) $\mathbf{P}(t)$ *and* $\dot{\mathbf{P}}(t)$ *are bounded on the interval* $-\infty < t < \infty$.
(ii) $0 < p \leqq |\mathbf{P}(t)|$ *for all* t, *for some constant scalar* p.

Theorem 7.8

Let $\mathbf{P}(t)$ *be the Liapunov transformation. The system* $\dot{\mathbf{x}}(t) = \mathbf{A}(t)\mathbf{x}(t)$ *is uniformly (exponentially) stable iff the transformed system,* $\mathbf{x} = \mathbf{P}^{-1}(t)\boldsymbol{\eta}$,

$$\dot{\boldsymbol{\eta}} = \left[\mathbf{P}(t)\mathbf{A}(t)\mathbf{P}^{-1}(t) - \mathbf{P}(t)\dot{\mathbf{P}}^{-1}(t)\right]\boldsymbol{\eta}$$

is uniformly (exponentially) stable.

Theorem 7.9

If $\mathbf{A}$ *is constant or periodic (as in Section* 4.4.3), *the system* $\dot{\mathbf{x}} = \mathbf{A}(t)\mathbf{x}$ *is exponentially stable iff* $\lim_{t\to\infty} \|\boldsymbol{\Phi}(t, t_0)\| = 0$.

Proof: Necessity is obvious, so we shall concentrate on sufficiency. If $\mathbf{A}$ is periodic, then Theorem 4.11 guarantees the existence of a Liapunov transformation $\mathbf{P}(t)$ such that $\mathbf{P}(t)\mathbf{A}(t)\mathbf{P}^{-1}(t) - \mathbf{P}(t)\dot{\mathbf{P}}^{-1}(t) = \hat{\mathbf{A}} =$ constant. Now, either $\mathbf{A}$ is constant or the transformed $\hat{\mathbf{A}}$ is constant. In either case the exponential stability is determined by (using either $\mathbf{A}$ or $\hat{\mathbf{A}}$) $\lim_{t\to\infty}\|e^{\hat{\mathbf{A}}t}\| = 0$. But

$$\|e^{\hat{\mathbf{A}}t}\| = \|\mathbf{E}e^{\boldsymbol{\Lambda}t}\mathbf{E}^{-1}\| \leqq \|\mathbf{E}\|\,\|\mathbf{E}^{-1}\|\,\|e^{\boldsymbol{\Lambda}t}\| = \|e^{\boldsymbol{\Lambda}t}\|,$$

where $\boldsymbol{\Lambda}$ is the Jordan form of $\hat{\mathbf{A}}$ and $e^{\boldsymbol{\Lambda}t}$ is a block-diagonal matrix composed of terms like $t^k e^{\alpha t}$ or $t^k e^{\alpha t} \sin \omega t$, or $t^k e^{\alpha t} \cos \omega t$. In either case, $\|e^{\boldsymbol{\Lambda}t}\|$ goes to zero iff there exists $\gamma > 0$ such that $t^k e^{\alpha t} \leqq \gamma e^{\lambda t}$. This completes the proof. ■

7.4 Stabilizability and Detectability

With some understanding of stability, we are now prepared to discuss various methods of modifying the stability properties by feedback control. In this section we consider the controllability and observability of linear time-invariant systems under the influence of the state feedback policy

$$\mathbf{u} = \mathbf{G}\mathbf{x} + \mathbf{v}.$$

The block diagram of the closed-loop system is given in Fig. 7.3. The concept of

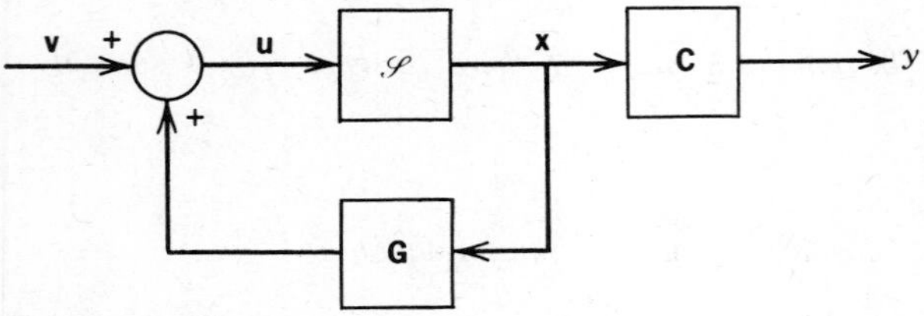

Figure 7.3 *State Feedback*

stabilizability deals with (and is equivalent to) the existence of a state feedback **G** such that the closed-loop system is asymptotically stable.

Definition 7.7
The system $\dot{\mathbf{x}} = \mathbf{A}\mathbf{x} + \mathbf{B}\mathbf{u}$ *is said to be stabilizable iff the uncontrollable modes of* $(\mathbf{A}, \mathbf{B})$ *are asymptotically stable.*

Note from this definition that there exists no **G** that will stabilize the matrix $[\mathbf{A} + \mathbf{B}\mathbf{G}]$ unless $(\mathbf{A}, \mathbf{B})$ is stabilizable.

The closed-loop system is described by

$$\begin{aligned} \dot{\mathbf{x}} &= (\mathbf{A} + \mathbf{B}\mathbf{G})\mathbf{x} + \mathbf{B}\mathbf{v}, \\ \mathbf{u} &= \mathbf{G}\mathbf{x} + \mathbf{v}, \\ \mathbf{y} &= \mathbf{C}\mathbf{x}. \end{aligned} \tag{7.38}$$

We ask now if the system is controllable from **v** and observable from **y**. That is, we must determine whether the matrix pair $(\mathbf{A} + \mathbf{B}\mathbf{G}, \mathbf{B})$ is controllable and whether the matrix pair $(\mathbf{A} + \mathbf{B}\mathbf{G}, \mathbf{C})$ is observable. Multiply the following two matrices:

$$[\mathbf{B}, \mathbf{A}\mathbf{B}, \ldots, \mathbf{A}^{n_x - 1}\mathbf{B}] \begin{bmatrix} \mathbf{I} & \mathbf{G}\mathbf{B} & \mathbf{G}\hat{\mathbf{A}}\mathbf{B} & \cdots & \mathbf{G}\hat{\mathbf{A}}^{n_x-2}\mathbf{B} \\ \mathbf{0} & \mathbf{I} & \mathbf{G}\mathbf{B} & \cdots & \mathbf{G}\hat{\mathbf{A}}^{n_x-3}\mathbf{B} \\ \mathbf{0} & \mathbf{0} & \mathbf{I} & & \mathbf{G}\hat{\mathbf{A}}^{n_x-4}\mathbf{B} \\ \vdots & \vdots & \vdots & & \vdots \\ \mathbf{0} & \mathbf{0} & \mathbf{0} & & \mathbf{G}\mathbf{B} \\ \mathbf{0} & \mathbf{0} & \mathbf{0} & \cdots & \mathbf{I} \end{bmatrix} = \mathbf{W}_c \mathscr{I}$$

to see that

$$\mathbf{W}_c \mathscr{I} = [\mathbf{B}, \hat{\mathbf{A}}\mathbf{B}, \ldots, \hat{\mathbf{A}}^{n_x - 1}\mathbf{B}] \triangleq \hat{\mathbf{W}}_c, \qquad \hat{\mathbf{A}} \triangleq \mathbf{A} + \mathbf{B}\mathbf{G}. \tag{7.39}$$

Hence the rank of the closed loop controllability matrix $\hat{\mathbf{W}}_c$ is always equal to the rank of the open loop controllability matrix $\mathbf{W}_c$ if and only if the matrix $\mathscr{I}$ is always nonsingular. This is obvious by inspection of $\mathscr{I}$. Hence, the following result.

Theorem 7.10
The matrix pair $(\mathbf{A}, \mathbf{B})$ *is controllable iff the matrix pair* $(\mathbf{A} + \mathbf{BG}, \mathbf{B})$ *is controllable.*

From the duality results of Theorem 5.16, we know that $(\mathbf{A}, \mathbf{B})$ is a controllable pair if and only if $(-\mathbf{A}^*, \mathbf{B}^*)$ is observable (the sign of $-\mathbf{A}^*$ doesn't matter here). The "dual" of Theorem 7.10 is, therefore, as follows:

Theorem 7.11
The matrix pair $(\mathbf{A}, \mathbf{C})$ *is observable iff the matrix pair* $(\mathbf{A} - \mathbf{FC}, \mathbf{C})$ *is observable.*

Proof: One may prove this directly from Theorem 7.10 by noting that controllability of the pair $(\mathbf{A} + \mathbf{BG}, \mathbf{B})$ is equivalent to observability of $(-(\mathbf{A} + \mathbf{BG})^*, \mathbf{B}^*)$. Now, rename matrices so that $-\mathbf{A}^* \to \mathbf{A}$, $\mathbf{G}^* \to \mathbf{F}$, $\mathbf{B}^* \to \mathbf{C}$. Then, $(-(\mathbf{A} + \mathbf{BG})^*, \mathbf{B}^*) \to (\mathbf{A} - \mathbf{FC}, \mathbf{C})$, and Theorem 7.10 leads immediately to Theorem 7.11. ■

We will later find use of the following definition.

Definition 7.8
The system $\dot{\mathbf{x}} = \mathbf{Ax}$, $\mathbf{y} = \mathbf{Cx}$ *is said to be detectable iff the unobservable modes of* $(\mathbf{A}, \mathbf{C})$ *are asymptotically stable.*

From this definition it follows that $(\mathbf{A} - \mathbf{FC})$ is not stable for any choice of $\mathbf{F}$ unless $(\mathbf{A}, \mathbf{C})$ is detectable. This result will be needed in Chapters 8 and 9.

Now, we wish to show that the output feedback control policy

$$\mathbf{u} = \mathscr{G}\mathbf{y} + \mathbf{v}$$

destroys neither controllability nor observability. The closed-loop system matrices in this case are $(\mathbf{A} + \mathbf{B}\mathscr{G}\mathbf{C}, \mathbf{B}, \mathbf{C})$.

Theorem 7.12
The matrix pair $(\mathbf{A} + \mathbf{B}\mathscr{G}\mathbf{C}, \mathbf{B})$ *is controllable iff* $(\mathbf{A}, \mathbf{B})$ *is controllable. The matrix pair* $(\mathbf{A} + \mathbf{B}\mathscr{G}\mathbf{C}, C)$ *is observable iff* $(\mathbf{A}, \mathbf{C})$ *is observable.*

Proof: Theorem 7.10 holds for any $\mathbf{G}$. Let $\mathbf{G} = \mathscr{G}\mathbf{C}$, then Theorem 7.10 proves the first part of Theorem 7.12. Now, Theorem 7.11 holds for any $\mathbf{F}$. Choose $\mathbf{F} = -\mathbf{B}\mathscr{G}$ to see that Theorem 7.11 proves the second part of Theorem 7.12. ■

Exercise 7.3
Show by example that observability is not always preserved by state feedback. *Hint*: Choose $\mathbf{G}$ such that $\mathbf{A} + \mathbf{BG} = \alpha\mathbf{I}$ for any α.

Exercise 7.4

Verify that observability of the closed-loop system is preserved for arbitrary state feedback gain $\mathbf{G}$ if rank $\mathbf{C} = n_x$.

The matrix $(\mathbf{A} + \mathbf{BG})$ is said to have an arbitrary spectrum if there exists a $\mathbf{G}$ to place the eigenvalues of $(\mathbf{A} + \mathbf{BG})$ arbitrarily, subject to the constraint of complex conjugate pairs.

Theorem 7.13

The matrix $(\mathbf{A} + \mathbf{BG})$ *has an arbitrary spectrum iff* $(\mathbf{A}, \mathbf{B})$ *is a controllable pair.*

By duality, we also have the following result:

Theorem 7.14

The matrix $(\mathbf{A} - \mathbf{FC})$ *has an arbitrary spectrum iff* $(\mathbf{A}, \mathbf{C})$ *is an observable pair.*

These theorems are proved in refs. [7.3], [7.4], and [7.5], and because of their extended length for the multiple-input (output) case, the proofs are omitted here. A constructive proof for the single-input (output) case will appear under control design by pole assignment in the next section.

7.5 Pole Assignment

If only the eigenvalues are of interest, then the control design task is easier since the parameters to be assigned are only n_x in number. The reader is cautioned, however, that the eigenvalue location is rarely an adequate statement of control objectives, since the eigenvectors, the zeros, the cost function, and the output correlation are all ignored in this control objective. Also, it is not clear where the poles *should* be, since the physical objectives must first be translated into the language of pole locations. In this translation, there is always some abuse of the original objectives. Nonetheless, pole assignment has been a popular control technique and deserves some mention. We shall limit our attention, however, to single-input/single-output systems with state feedback.

Let the desired characteristic polynomial be denoted by $\hat{\Delta}(s)$. Then, we wish to find $\mathbf{G}$ such that

$$\begin{aligned} \hat{\Delta}(s) &= |s\mathbf{I} - (\mathbf{A} + \mathbf{BG})| \\ &= \left|(s\mathbf{I} - \mathbf{A})\left(\mathbf{I} - (s\mathbf{I} - \mathbf{A})^{-1}\mathbf{BG}\right)\right| \\ &= |s\mathbf{I} - \mathbf{A}|\left|\mathbf{I} - (s\mathbf{I} - \mathbf{A})^{-1}\mathbf{BG}\right| \\ &= \Delta(s)\left|\mathbf{I} - (s\mathbf{I} - \mathbf{A})^{-1}\mathbf{BG}\right|, \end{aligned} \tag{7.40}$$

where $\Delta(s)$ is the characteristic polynomial of $\mathbf{A}$. Now, use the result from linear algebra.

Theorem 7.14
Let $|\mathbf{A}|$ denote the determinant of $\mathbf{A}$. Then,

$$|\mathbf{I}_n + \mathbf{MN}| = |\mathbf{I}_m + \mathbf{NM}|, \qquad \mathbf{M} \in \mathscr{R}^{n\times m}, \quad \mathbf{N} \in \mathscr{R}^{m\times n}. \tag{7.41}$$

The proof of Theorem 7.14 is left as an exercise to the reader.
The characteristic polynomial $\hat{\Delta}(s)$ may now be written

$$\hat{\Delta}(s) = \Delta(s)\left|\mathbf{I} - \mathbf{G}(s\mathbf{I} - \mathbf{A})^{-1}\mathbf{B}\right|. \tag{7.42}$$

Using (4.88) and the fact that $\mathbf{I}_{n_u} = 1$ for single-input/single-output systems,

$$\hat{\Delta}(s) = \Delta(s)\left|1 - \frac{1}{\Delta(s)}\mathbf{G}\left[\mathbf{S}_{n-1}s^{n_x-1} + \mathbf{S}_{n-2}s^{n_x-2} + \cdots \mathbf{S}_1 s + \mathbf{S}_0\right]\mathbf{B}\right|, \tag{7.43}$$

where $\mathbf{S}_{n-1} = \mathbf{I}$ using the resolvent algorithm (4.90). Now, (7.43) leads to

$$\Delta(s) - \hat{\Delta}(s) = \mathbf{G}\left[s^{n_x-1}\mathbf{I} + \mathbf{S}_{n-2}s^{n_x-2} + \cdots \mathbf{S}_1 s + \mathbf{S}_0\right]\mathbf{B}. \tag{7.44}$$

But from (4.90), the right-hand side becomes

$$\Delta(s) - \hat{\Delta}(s) = \mathbf{G}\Big[s^{n_x-1}\mathbf{I} + \left(\mathbf{A} + a_{n_x-1}\mathbf{I}\right)s^{n_x-2} + \left(\mathbf{A}^2 + a_{n_x-1}\mathbf{A} + a_{n_x-2}\mathbf{I}\right)s^{n-3} + \cdots\Big]\mathbf{B}. \tag{7.45}$$

Write the left-hand side as

$$\Delta(s) - \hat{\Delta}(s) = \sum_{i=0}^{n_x-1} a_i s^i - \sum_{i=0}^{n_x-1} \hat{a}_i s^i = \sum_{i=0}^{n_x-1} (a_i - \hat{a}_i)s^i = \sum_{i=0}^{n_x-1} \tilde{a}_i s^i. \tag{7.46}$$

Comparison of (7.45) and (7.46) leads to

$$\begin{aligned} \tilde{a}_{n_x-1} &= \mathbf{GB}, \\ \tilde{a}_{n_x-2} &= \mathbf{G}\left(\mathbf{AB} + a_{n_x-1}\mathbf{B}\right), \\ \tilde{a}_{n_x-3} &= \mathbf{G}\left(\mathbf{A}^2\mathbf{B} + a_{n_x-1}\mathbf{AB} + a_{n_x-2}\mathbf{B}\right), \end{aligned} \tag{7.47}$$

or, in vector form, with $\mathbf{b} \triangleq \mathbf{B}$, to

$$\tilde{\mathbf{a}}^T = \mathbf{G}[\mathbf{b}, \mathbf{A}\mathbf{b}, \mathbf{A}^2\mathbf{b}, \ldots \mathbf{A}^{n_x-1}\mathbf{b}] \begin{bmatrix} 1 & a_{n_x-1} & a_{n_x-2} & & \\ 0 & 1 & a_{n_x-1} & & \\ 0 & 0 & 1 & & \\ & & & \ddots & \\ 0 & 0 & 0 & & 1 \end{bmatrix} = \mathbf{G}\mathbf{W}_c\mathbf{T}_\Delta. \quad (7.48)$$

Since $\mathbf{T}_\Delta$ is always invertible, we have the following conclusion for *arbitrary* $\tilde{\mathbf{a}}$:

$$\mathbf{G} = \tilde{\mathbf{a}}^T \mathbf{T}_\Delta^{-1} \mathbf{W}_c^{-1}, \quad (7.49)$$

where the requirement for controllability, $|\mathbf{W}_c| \neq 0$, is obvious for *arbitrary* $\tilde{\mathbf{a}}$. For a *specific* $\tilde{\mathbf{a}}$, a solution of (7.48) might exist without the controllability conditions. See Section 2.6 for necessary and sufficient conditions for the solution of (7.48) for a *specific* $\tilde{\mathbf{a}}$.

Exercise 7.5

Show that a necessary and sufficient condition for a solution of (7.48) is $(\mathbf{W}_c\mathbf{T}_\Delta)^+(\mathbf{W}_c\mathbf{T}_\Delta)\tilde{\mathbf{a}} = \tilde{\mathbf{a}}$, where $+$ denotes a pseudo inverse.

EXAMPLE 7.3

Given

$$\mathbf{A} = \begin{bmatrix} 10 & 22 \\ -8 & -17 \end{bmatrix}, \quad \mathbf{B} = \begin{pmatrix} 2 \\ -1 \end{pmatrix}. \quad (7.50)$$

Find a state feedback control that gives the closed-loop system the pole locations $(-6, -p)$. Can p be placed arbitrarily?

Solution: The closed-loop system must have the characteristic polynomial

$$(\lambda + 6)(\lambda + p) = \lambda^2 + (p + 6)\lambda + 6p. \quad (7.51)$$

The open-loop system has the characteristic polynomial

$$\lambda^2 - \operatorname{tr}\mathbf{A}\lambda + |\mathbf{A}| = \lambda^2 + 7\lambda + 6. \quad (7.52)$$

Hence, the $\tilde{\mathbf{a}}^T$ vector in (7.48) is

$$(\tilde{a}_1, \tilde{a}_0) = (7 - p - 6, 6 - 6p),$$

and the other required matrices in (7.48) are

$$\mathbf{W}_c = \begin{bmatrix} 2 & -2 \\ -1 & 1 \end{bmatrix} = \text{rank } 1, \quad \mathbf{T}_\Delta = \begin{bmatrix} 1 & 7 \\ 0 & 1 \end{bmatrix}. \quad (7.53)$$

Hence, from (7.48),

$$(1-p, 6(1-p)) = (g_1 g_2)\begin{bmatrix} 2 & -2 \\ -1 & 1 \end{bmatrix}\begin{bmatrix} 1 & 7 \\ 0 & 1 \end{bmatrix} = (g_1 g_2)\begin{bmatrix} 2 & 12 \\ -1 & -6 \end{bmatrix}, \tag{7.54}$$

which has a solution iff (see Theorem 2.7 and its corollary)

$$\text{rank}\begin{bmatrix} 2 & -1 \\ 12 & -6 \end{bmatrix} = \text{rank}\begin{bmatrix} 2 & -1 & 1-p \\ 12 & -6 & 6(1-p) \end{bmatrix}. \tag{7.55}$$

Since the left-hand side has rank 1, the two rows of the right-hand side must be linearly dependent. That is, the second row must be 6 times the first row, and it is. Thus, there is a solution to (7.48) or, equivalently, to

$$1 - p = 2g_1 - g_2,$$

$$6(1-p) = 6(2g_1 - g_2).$$

Any choice of p can be obtained using

$$g_2 = 2g_1 + p - 1, \qquad (g_1 \text{ arbitrary}). \tag{7.56}$$

■

The point of this example is to illustrate that the controllability required in (7.49) is sufficient but not necessary. But when controllability is not present, there are some poles that cannot be moved by state feedback. In this example the reader should show that (-6) is an uncontrollable mode.

Exercise 7.6

Find the state feedback controller that places poles of equation (3.28) at $\omega_{n_1} = \omega_{n_2} = 0.02$ rad/sec, $\lambda_1 = 0.4$, $\lambda_2 = 0.8$,

$$\lambda_{1,2} = -\zeta_1 \omega_{n_1} \pm j\omega_{n_1}\sqrt{1-\zeta_1^2},$$

$$\lambda_{3,4} = -\zeta_2 \omega_{n_2} \pm j\omega_{n_2}\sqrt{1-\zeta_2^2}.$$

Let $R_0 = 10^6$, $m = 100$, $\omega_0 = 1/900$.

7.6 *Covariance Assignment*

For almost thirty years, control by eigenvalue assignment (the subject of the previous section) and quadratic optimization (the subject of Chapter 8) have been the most popular controller design methods for multi-input, multi-output systems.

However, they can guarantee only that the control system state vector as a whole behaves well. Very little is said about the transient behavior of individual state variables. In practice, this means that a system controlled by such methods can have a transient behavior where some state variables becomes so large that the linear model is no longer valid or some components fail (i.e., springs break, etc.). The idea of *covariance control* introduced in this section is to ensure good behavior of each state variable separately.

In this section we would like to find a state feedback gain $\mathbf{G}$ such that the system

$$\dot{\mathbf{x}} = \mathbf{A}\mathbf{x} + \mathbf{B}\mathbf{u} + \mathbf{B}\mathbf{w}, \quad w_i = \omega_i \delta(t), \qquad i = 1, 2, \ldots, n_u,$$
$$\mathbf{y} = \mathbf{C}\mathbf{x}, \tag{7.57}$$
$$\mathbf{u} = \mathbf{G}\mathbf{x}$$

has a specified output covariance matrix

$$\mathbf{R}_0 = \sum_{i=1}^{n_u} \int_0^\infty \mathbf{y}^i(t)\mathbf{y}^{i*}(t)\, dt. \tag{7.58}$$

In this way one can assign specified root mean squared values to *each* output simultaneously, since, for multiple inputs, $[R_0]_{\alpha\alpha} = \Sigma_{i=1}^{n_u} \int_0^\infty [y_\alpha^i(t)]^2\, dt$ is the mean squared value to be assigned to the output $y_\alpha(t)$. For systems that require specific RMS behavior of multiple outputs, this is an important design approach.

First, we consider $\mathbf{C} = \mathbf{I}$ and solve the state covariance assignment problem. Our task is to find all values of $\mathbf{G}$ such that

$$\mathbf{0} = \hat{\mathbf{X}}(\mathbf{A} + \mathbf{B}\mathbf{G})^* + (\mathbf{A} + \mathbf{B}\mathbf{G})\hat{\mathbf{X}} + \mathbf{B}\mathscr{U}\mathbf{B}^*, \tag{7.59}$$

where $\hat{\mathbf{X}}$ is the assigned value of the state covariance, where

$$\dot{\mathbf{x}} = \mathbf{A}\mathbf{x} + \mathbf{B}(\mathbf{u} + \mathbf{w}), \qquad w_i(t) = \omega_i \delta(t),$$
$$\mathbf{u} = \mathbf{G}\mathbf{x},$$
$$\mathscr{U}_{ij} = \omega_i^2 \delta_{ij}.$$

From (7.59),

$$\mathbf{0} = (\hat{\mathbf{X}}\mathbf{A}^* + \mathbf{A}\hat{\mathbf{X}} + \mathbf{B}\mathscr{U}\mathbf{B}^*) + \hat{\mathbf{X}}(\mathbf{B}\mathbf{G})^* + \mathbf{B}\mathbf{G}\hat{\mathbf{X}}. \tag{7.60}$$

This is satisfied by all $\mathbf{G}$ which yield

$$\mathbf{B}\mathbf{G}\hat{\mathbf{X}} = -\tfrac{1}{2}(\hat{\mathbf{X}}\mathbf{A}^* + \mathbf{A}\hat{\mathbf{X}} + \mathbf{B}\mathscr{U}\mathbf{B}^* + \mathbf{S}), \tag{7.61}$$

where $\mathbf{S} = -\mathbf{S}^*$ is an arbitrary skew-Hermitian matrix [prove by substitution of

(7.61) into (7.60)]. Now, the solution of the linear equation (7.61) for **G** forms the essence of our conclusions. To facilitate these developments, define $(\mathbf{U}_B, \mathscr{Q}, \hat{\mathscr{Q}}, \mathscr{S})$ by

$$\begin{bmatrix} \mathbf{U}_{B1}^* \\ \mathbf{U}_{B2}^* \end{bmatrix} \mathbf{B}\mathbf{B}^+ [\mathbf{U}_{B1}, \mathbf{U}_{B2}] = \begin{bmatrix} \mathbf{I}_b & \mathbf{0} \\ \mathbf{0} & \mathbf{0} \end{bmatrix}, \qquad \mathbf{B}^+ = \text{Moore–Penrose inverse}, \quad b \triangleq \text{rank } B, \tag{7.62a}$$

$$\hat{\mathscr{Q}} \triangleq \hat{\mathbf{X}}A^* + \mathbf{A}\hat{\mathbf{X}} + \mathbf{B}\mathscr{U}\mathbf{B}^*, \tag{7.62b}$$

$$\mathscr{Q} \triangleq \mathbf{U}_B^* \hat{\mathscr{Q}} \mathbf{U}_B = \begin{bmatrix} \mathbf{U}_{B1}^* \\ \mathbf{U}_{B2}^* \end{bmatrix} \hat{\mathscr{Q}} [\mathbf{U}_{B1}, \mathbf{U}_{B2}] = \begin{bmatrix} \mathscr{Q}_{11} & \mathscr{Q}_{12} \\ \mathscr{Q}_{12}^* & \mathscr{Q}_{22} \end{bmatrix}, \tag{7.62c}$$

$$\mathscr{S} \triangleq \mathbf{U}_B^* \mathbf{S} \mathbf{U}_B = -\mathscr{S}^*, \tag{7.62d}$$

where $\mathbf{U}_{B1}$ is $n_x \times b$ and $\mathbf{U}_{B2}$ is $n_x \times (n_x - b)$. Now, premultiply (7.61) from the left by $\mathbf{U}_B^*$ and from the right by $\mathbf{U}_B$. This yields

$$\mathbf{U}_B^* \mathbf{B}\mathbf{G}\hat{\mathbf{X}}\mathbf{U}_B = -\tfrac{1}{2}(\mathscr{Q} + \mathscr{S})$$

or

$$\mathbf{U}_B^* \mathbf{B}\mathbf{G} = -\tfrac{1}{2}(\mathscr{Q} + \mathscr{S})\mathbf{U}_B^* \hat{\mathbf{X}}^{-1}. \tag{7.63}$$

This linear equation has a solution [see (2.62)] iff

$$\mathbf{U}_B^* \mathbf{B}(\mathbf{U}_B^* \mathbf{B})^+ (\mathscr{Q} + \mathscr{S})\mathbf{U}_B^* \hat{\mathbf{X}}^{-1} = (\mathscr{Q} + \mathscr{S})\mathbf{U}_B^* \hat{\mathbf{X}}^{-1}$$

or, equivalently,

$$\mathbf{U}_B^* \mathbf{B}\mathbf{B}^+ \mathbf{U}_B (\mathscr{Q} + \mathscr{S}) = (\mathscr{Q} + \mathscr{S}).$$

But (7.62a) reduces this to

$$\begin{bmatrix} \mathbf{I}_b & \mathbf{0} \\ \mathbf{0} & \mathbf{0} \end{bmatrix} \begin{bmatrix} \mathscr{Q}_{11} + \mathscr{S}_{11} & \mathscr{Q}_{12} + \mathscr{S}_{12} \\ \mathscr{Q}_{21} + \mathscr{S}_{21} & \mathscr{Q}_{22} + \mathscr{S}_{22} \end{bmatrix} = \begin{bmatrix} \mathscr{Q}_{11} + \mathscr{S}_{11} & \mathscr{Q}_{12} + \mathscr{S}_{12} \\ \mathscr{Q}_{21} + \mathscr{S}_{21} & \mathscr{Q}_{22} + \mathscr{S}_{22} \end{bmatrix},$$

which requires

$$\mathbf{0} = \mathscr{Q}_{21} + \mathscr{S}_{21}, \qquad \mathbf{0} = \mathscr{Q}_{22} + \mathscr{S}_{22}.$$

Now, since $\mathscr{Q}_{21} = \mathscr{Q}_{12}^*$, $\mathscr{Q}_{22} = \mathscr{Q}_{22}^*$, and $\mathscr{S}_{22} = -\mathscr{S}_{22}^*$, we have these conclusions:

$$\mathscr{S}_{21} = -\mathscr{Q}_{21}, \quad \mathscr{Q}_{22} = \mathbf{0}, \quad \mathscr{S}_{22} = \mathbf{0}. \tag{7.64}$$

Hence,

$$\mathcal{Q} + \mathcal{S} = \begin{bmatrix} \mathcal{Q}_{11} + \mathcal{S}_{11} & 2\mathcal{Q}_{12} \\ \mathbf{0} & \mathbf{0} \end{bmatrix} \tag{7.65}$$

This constraint is summarized as follows. The result gives the assignability condition for state covariances.

Theorem 7.15

A linear system (7.59) *may possess the state covariance value* $\hat{\mathbf{X}}$ (*for some choice of* $\mathbf{G}$) *iff*

$$\mathcal{Q}_{22} = \mathbf{U}_{B2}^*[\hat{\mathbf{X}}\mathbf{A}^* + \mathbf{A}\hat{\mathbf{X}} + \mathbf{B}\mathcal{U}\mathbf{B}^*]\mathbf{U}_{B2} = \mathbf{0} \tag{7.66}$$

where $\mathbf{U}_{B2}$ *is defined by* (7.62a).

Exercise 7.7

Show from the definition of $\mathbf{U}_{B2}$ in (7.62a) that $\mathbf{U}_{B2}^*\mathbf{B}\mathcal{U}\mathbf{B}^*\mathbf{U}_{B2} = \mathbf{0}$ and hence that (7.66) reduces to $\mathbf{U}_{B2}^*[\hat{\mathbf{X}}\mathbf{A}^* + \mathbf{A}\hat{\mathbf{X}}]\mathbf{U}_{B2} = \mathbf{0}$.

Now, suppose (7.66) is satisfied for the selected $\hat{\mathbf{X}}$. The set of all $\mathbf{G}$ matrices that yield $\hat{\mathbf{X}}$ is computed as follows. Use (2.65) to solve the linear algebra equation (7.63),

$$\mathbf{G} = -\tfrac{1}{2}\mathbf{B}^+\mathbf{U}_B(\mathcal{Q} + \mathcal{S})\mathbf{U}_B^*\hat{\mathbf{X}}^{-1} - \mathbf{B}^+\mathbf{U}_B\mathbf{U}_B^*\mathbf{B}\mathbf{Z} + \mathbf{Z},$$

where $\mathbf{Z}$ is arbitrary and where the use of (7.62) and (7.65) gives

$$\mathbf{G} = -\tfrac{1}{2}\mathbf{B}^+[\mathbf{U}_{B1}(\mathcal{Q}_{11} + \mathcal{S}_{11})\mathbf{U}_{B1}^* + 2U_{B1}\mathcal{Q}_{12}\mathbf{U}_{B2}^*]\hat{\mathbf{X}}^{-1} - \mathbf{B}^+\mathbf{B}\mathbf{Z} + \mathbf{Z}. \tag{7.67}$$

Note, however, that the last two terms in (7.67) do not contribute to the control term $\mathbf{Bu}$ in (7.59) since

$$\mathbf{Bu} = \mathbf{BGx} = \mathbf{B}(\text{first term}) - \mathbf{BB}^+\mathbf{BZ} + \mathbf{BZ} = \mathbf{B}(\text{first term}).$$

Thus, the last two terms in (7.67) are called trivial since they do not contribute to the state trajectories or to $\mathbf{X}$.

Theorem 7.16

If there exists a $\mathbf{G}$ *such that* (7.59) *takes on the state covariance value* $\hat{\mathbf{X}}$, *then all nontrivial* $\mathbf{G}$'s *that give* $\mathbf{X}$ *this value* $\mathbf{X} = \hat{\mathbf{X}}$ *are given by*

$$\mathbf{G} = -\tfrac{1}{2}\mathbf{B}^+\mathbf{U}_{B1}[(\mathcal{Q}_{11} + \mathcal{S}_{11})\mathbf{U}_{B1}^* + 2\mathcal{Q}_{12}\mathbf{U}_{B2}^*]\hat{\mathbf{X}}^{-1}, \tag{7.68}$$

where $\mathcal{S}_{11}$ *is an arbitrary skew-Hermitian matrix and* $\mathbf{U}_{B1}, \mathbf{U}_{B2}, \mathcal{Q}_{12}, \mathcal{Q}_{11}$ *are defined by equation* (7.62).

State covariance assignment is introduced for stochastic problems in refs. [7.6] through [7.8].

EXAMPLE 7.4

Show that there does not exist a state feedback gain $\mathbf{G}$ which assigns the value

$$\mathbf{X} \triangleq \int_0^{\infty} \mathbf{x}\mathbf{x}^* \, dt = \hat{\mathbf{X}} = \mathbf{I}$$

to the system

$$\dot{\mathbf{x}} = \begin{bmatrix} -1 & 0 \\ 0 & -10 \end{bmatrix} \mathbf{x} + \begin{pmatrix} 1 \\ 70 \end{pmatrix} (\mathbf{u} + \mathbf{w}), \qquad u(t) = \delta(t).$$

Solution: Since $\mathbf{B}$ has linearly independent columns,

$$\mathbf{B}^+ = (\mathbf{B}^*\mathbf{B})^{-1}\mathbf{B}^*,$$

then

$$\mathbf{B}\mathbf{B}^+ = \mathbf{B}(\mathbf{B}^*\mathbf{B})^{-1}\mathbf{B}^* = \begin{bmatrix} 1 & 70 \\ 70 & 4900 \end{bmatrix} \frac{1}{4901}$$

and

$$[\mathbf{U}_{B1} \quad \mathbf{U}_{B2}] = \begin{bmatrix} \dfrac{4901}{\sqrt{4900 + 4901^2}} & \dfrac{70}{\sqrt{4901}} \\ \dfrac{70}{\sqrt{4900 + 4901^2}} & -\dfrac{1}{\sqrt{4901}} \end{bmatrix}.$$

Hence, from (7.66),

$$\mathbf{U}_{B2}^*[\mathbf{A}^* + \mathbf{A} + \mathbf{B}\mathbf{B}^*\mathscr{U}^2]\mathbf{U}_{B2} = -9820 \neq 0,$$

proving that $\hat{\mathbf{X}} = I$ is not assignable. ■

EXAMPLE 7.5

Suppose we wish to design a roll attitude regulator for a missle disturbed by impulsive roll torques. The control objective is to keep roll attitude small while staying within the physical limits of aileron deflection and aileron deflection rate. This example is taken from ref. [7.6].

We estimate the RMS value of the disturbing torque to be equivalent to 5 degrees of aileron deflection. The system is modeled as

$$\begin{bmatrix} \dot{\delta} \\ \dot{\omega} \\ \dot{\phi} \end{bmatrix} = \begin{bmatrix} 0 & 0 & 0 \\ 10 & -1 & 0 \\ 0 & 1 & 0 \end{bmatrix} \begin{bmatrix} \delta \\ \omega \\ \phi \end{bmatrix} + \begin{bmatrix} 1 \\ 0 \\ 0 \end{bmatrix} u(t) + \begin{bmatrix} 0 \\ 1 \\ 0 \end{bmatrix} w(t),$$

where δ = aileron deflection, ω = roll angular velocity, ϕ = roll angle, u = command signal to the aileron actuators, and $w(t)$ is an impulse disturbance source with intensity $W = 1150\ \text{deg}^2/\text{sec}^3$.

By specifying the state covariance, we can assign the desired RMS values for δ, ω, and ϕ. Also, we wish to keep the RMS value of u small. For multiple-input systems, this is achieved by proper choice of a skew-symmetric matrix in (7.68). However, for single-input systems, this freedom is not available as only one gain matrix will achieve the specified covariance (that is, $\mathscr{S}_{11} = -\mathscr{S}_{11}^* = 0$ if $\mathscr{S}_{11}$ is a scalar). Instead, since the RMS value of u is implicit in the specifications of the RMS values of the system states δ, ω, and ϕ, a large RMS value for u can be avoided by reasonable but acceptable specifications of the RMS values of ϕ, ω, and δ. For example, if we demand a very small RMS value for roll angle, then the missle dynamics must be very fast resulting in large RMS roll velocities and aileron deflection rates.

With these considerations, we assign the RMS values for δ, ω, and ϕ as,

$$\delta_{\text{RMS}} \triangleq \left[\int_0^\infty \delta^2\, dt\right]^{1/2} = 11\ \text{deg}, \quad \phi_{\text{RMS}} = 1\ \text{deg}, \quad \omega_{\text{RMS}} = 9\ \text{deg/sec}.$$

Assigning the desired cross correlations may be motivated as follows. Since the angle ϕ cannot react instantaneously to changes in the angular velocity ω, there is no time correlation at $t = 0$ between ω and ϕ at any given time in the steady state. Hence, we can assign $\mathscr{C}_{\phi\omega}(0) = 0$. Assigning $\mathscr{C}_{\delta\phi}(0)$ and $\mathscr{C}_{\omega\delta}(0)$ requires a little more work, employing the existence conditions (7.66), where

$$[\mathbf{U}_{B1}, \mathbf{U}_{B2}] = \begin{bmatrix} 1 & 0 & 0 \\ 0 & 1 & 0 \\ 0 & 0 & 1 \end{bmatrix}.$$

The assigned state covariance is then chosen to satisfy (7.66) with the desired diagonal elements. $\mathbf{B} = \begin{pmatrix} 0 \\ 1 \\ 0 \end{pmatrix}$ replaces $\mathbf{B} = \begin{pmatrix} 1 \\ 0 \\ 0 \end{pmatrix}$ in (7.66) since $\mathbf{D} \neq \mathbf{B}$ in the example. Also, $\mathscr{U} = 1150$ in (7.66). This gives uniquely

$$\hat{\mathbf{X}} = \begin{bmatrix} 121 & -49.4 & -8.1 \\ -49.4 & 81 & 0 \\ -8.1 & 0 & 1 \end{bmatrix}.$$

Substituting $\mathbf{A}$, $\mathbf{B}$, $\mathbf{U}_B$, and $\hat{\mathbf{X}}$ into equation (7.68) yields the desired gain

$$\mathbf{G} = [-14.56, -24.43, -68.57].$$

The resulting input covariance is

$$\mathscr{C}_{uu}(0) = \mathbf{G}\hat{\mathbf{X}}\mathbf{G}^T = 27{,}380\ (\text{deg/sec})^2$$

Closure

This chapter presents five notions of stability for liner systems: (i) eigenvalues, (ii) Liapunov stability, (iii) bounded-input/bounded-output (BIBO) stability, (iv) exponential stability, and (v) uniform stability.

Liapunov stability is shown to be related to an observability condition (and this explains the location of this chapter following observability concepts). BIBO stability is shown to be equivalent to Liapunov stability for minimal transfer equivalent representations (TER) of systems. Otherwise, they are not equivalent, since Liapunov theory can determine the stability of uncontrollable, unobservable states and BIBO stability cannot. Uniform stability keeps the norm of the state transition matrix below a finite value, and exponential stability bounds the norm of the state transition matrix from above by an exponentially decreasing function of time.

Stabilizability and detectability conditions respectively guarantee the existence of a $\mathbf{G}$ or $\mathbf{F}$ which will stabilize $(\mathbf{A} + \mathbf{BG})$ or $(\mathbf{A} - \mathbf{FC})$, respectively. The former conditions are useful in state feedback control, whereas the latter conditions will be useful in state estimation (Chapter 9). Both conditions are needed in the optimal control of Chapter 8.

In anticipation of the optimization of feedback controls in Chapter 8, Chapter 7 presents two control design methods that have simple linear algebra solutions. The first is pole assignment, and the second is covariance assignment. In the first, a state feedback gain is found which places the eigenvalues of $[\mathbf{A} + \mathbf{BG}]$ at specified locations. There is no restriction on the choice of these locations if $(\mathbf{A}, \mathbf{B})$ is a controllable pair. In the second method, all state feedback gains are found which assign the state covariance to a specified matrix value. This is extremely useful when it is desired to assign all the RMS values of the states $\sqrt{\int x_i^2\, dt}$ to specified values. In many engineering problems this is desired because stability alone does not guarantee this kind of bound on each state.

References

7.1 D. D. Siljak, *Large-Scale Dynamic Systems, Stability and Structure*. North-Holland, Amsterdam, 1978.

7.2 Roger W. Brockett, *Finite Dimensional Linear Systems*. John Wiley, New York, 1970.

7.3 V. M. Popov, "Hyperstability and Optimality of Automatic Systems with Several Control Functions," *Rev. Roum. Sci.-Electrotechn. et Energ.* 9, 629–690.

7.4 B. D. O. Anderson and D. G. Luenberger, "Design of Multivariable Feedback System," *Proc. IEE*, 114, 395–399, March 1967.

7.5 W. M. Wonham, "On Pole Assignment in Multi-input Controllable Linear Systems," *IEEE Trans. Autom. Control*, AC-12, 660–665, 1967.

7.6 A. Hotz and R. E. Skelton, "A Covariance Control Theory," *Int. J. Control*, 46, 1987.

7.7 E. Collins and R. E. Skelton, "A Theory of State Covariance Assignment for Discrete Systems," *IEEE Trans. Autom. Control*, AC-32, 1, 1987.

7.8 A Hotz and R. E. Skelton, "Controller Design for Robust Stability and Performance," *ACC Proceedings*, Seattle, Wash., June 1986.

7.9 P. W. Likins, Lecture Notes UCLA, 1975.

CHAPTER 8

Optimal Control of Time-Invariant Systems

*F*rom the concepts of Liapunov stability we know for an asymptotically stable solution of the system that some function exists that is decreased during the natural (unforced) trajectory of $\mathbf{x}(t)$. This function (the Liapunov function) might represent a physical entity such as system energy. We now ask if it might be possible to decrease this function *even faster* with some judicious choice of control inputs $\mathbf{u}(t)$. But the particular Liapunov function which the open-loop system trajectory is decreasing might not be the function the designer wishes to decrease. Therefore, the task we pose in this chapter is: Given a function of the designer's choice, find a control input $\mathbf{u}(\cdot)$ that will keep this function as small as possible. This is the philosophy of optimal control. This chapter focuses on linear systems with quadratic functions to be minimized. This is the simplest optimal control problem, and more fundamental approaches require a separate course dedicated to optimal control.

As shown in Sections 7.5 and 7.6, control by *assignment* of system properties (e.g., pole or covariance values) does not require optimal control as long as the existence conditions are satisfied (controllability for the case of arbitrary pole assignment and $\mathscr{Q}_{22} = \mathbf{0}$ in equation (7.66) in the case of covariance assignment). However, if these conditions are not satisfied, then one will naturally wonder *how close* to their desired values is it possible to get the poles or the covariances. This leads to optimization problems. To illustrate difficulties in posing such problems, we shall write down the necessary conditions to solve an optimal covariance problem. Then, we shall go on to more tractable problems, keeping in mind that to compromise with mathematical simplicity, control theory almost always provides answers to slightly *different* questions than those we directly seek to answer.

8.1 Optimal Covariance Control

Suppose we desire the output covariance matrix $\mathbf{R}_0$ to be as small as possible using some choice of state feedback gain $\mathbf{G}$, subject to some compromise on the size of the input covariance matrix $\mathbf{U}$. This suggests the optimization problem

$$\min_{\mathbf{G}}\left(\|\mathbf{R}_0\|_{\mathbf{Q}}^2 + \|\mathbf{U}\|_{\mathbf{R}}^2\right), \qquad \mathbf{Q} > \mathbf{0}, \quad \mathbf{R} > \mathbf{0}, \tag{8.1}$$

subject to

$$\mathbf{R}_0 \triangleq \mathbf{C}\mathscr{X}\mathbf{C}^*, \qquad \mathbf{0} = \mathscr{X}\hat{\mathbf{A}}^* + \hat{\mathbf{A}}\mathscr{X} + \mathbf{DWD}^*, \tag{8.2a}$$

$$\mathbf{U} \triangleq \mathbf{G}\mathscr{X}\mathbf{G}^*, \qquad \hat{\mathbf{A}} \triangleq \mathbf{A} + \mathbf{BG}. \tag{8.2b}$$

For convenience, all matrices in this chapter are assumed real unless stated otherwise. The underlying system we have in mind is

$$\begin{aligned} &\dot{\mathbf{x}} = \mathbf{Ax} + \mathbf{Bu} + \mathbf{Dw}, \qquad w_i(t) = \omega_i\delta(t), \\ &\mathbf{u} = \mathbf{Gx}, \qquad \mathbf{x}(t_0) = \mathbf{0}, \\ &\mathbf{y} = \mathbf{Cx}, \end{aligned} \tag{8.3}$$

where $\mathbf{w}(t)$ is a disturbance vector of impulses of strength ω_i, applied one at a time so that (from Chapter 4)

$$\mathbf{W} = \begin{bmatrix} \omega_1^2 & & \\ & \omega_2^2 & \\ & & \ddots \end{bmatrix}. \tag{8.4}$$

If the actual disturbances are not impulses, we remind the reader of the disturbance modeling techniques in Chapter 4 leading to an equation of the form (8.3) but of larger dimension.

Now, the constraints (8.2) are augmented to the objective function (8.1) by the use of Lagrange multipliers (recall the use of Lagrange multipliers in Section 2.7.3). There must be one Lagrange multiplier Λ_{ij} if for each constraint, $N_{ji} = 0$ in the matrix of constraints $\mathbf{N} \triangleq \mathscr{X}\hat{\mathbf{A}}^* + \hat{\mathbf{A}}\mathscr{X} + \mathbf{DWD}^*$. There are n_x^2 constraints ($N_{ji} = 0$, $i, j = 1, 2, \ldots, n_x$), although they are not independent constraints since $\mathbf{N}$ is symmetric. (The Lagrange multiplier rule does not require independent constraints.)

Note also that the identity

$$\sum_{i,j=1}^{n_x} \Lambda_{ij} N_{ji} = \operatorname{tr} \boldsymbol{\Lambda}\mathbf{N} \tag{8.5}$$

allows the constrained optimization problems (8.1) and (8.2) to be written simply

$$\min_{\mathbf{G},\,\Lambda,\,\mathscr{X}} \left\{ \operatorname{tr} \mathbf{R}_0^* \mathbf{Q} \mathbf{R}_0 + \operatorname{tr} \mathbf{U}^* \mathbf{R} \mathbf{U} + \operatorname{tr} \Lambda(\mathscr{X}\hat{\mathbf{A}}^* + \hat{\mathbf{A}}\mathscr{X} + \mathbf{DWD}^*) \right\}.$$

Define

$$\begin{aligned}\hat{\mathscr{V}} = {} & \operatorname{tr} \mathbf{C}\mathscr{X}\mathbf{C}^*\mathbf{QC}\mathscr{X}\mathbf{C}^* + \operatorname{tr} \mathbf{G}\mathscr{X}\mathbf{G}^*\mathbf{RG}\mathscr{X}\mathbf{G}^* \\ & + \operatorname{tr} \Lambda(\mathscr{X}(\mathbf{A} + \mathbf{BG})^* + (\mathbf{A} + \mathbf{BG})\mathscr{X} + \mathbf{DWD}^*).\end{aligned}$$

This leads to the necessary conditions (the reader should review Section 2.7.1.3)

$$\frac{\partial \hat{\mathscr{V}}}{\partial \mathbf{G}} = \mathbf{0} = 2\{[\mathbf{G}\mathscr{X}\mathbf{G}^*\mathbf{R} + \mathbf{RG}\mathscr{X}\mathbf{G}^*]\mathbf{G} + \mathbf{B}^*\Lambda\}\mathscr{X}, \tag{8.6}$$

$$\frac{\partial \hat{\mathscr{V}}}{\partial \Lambda} = \mathbf{0} = \mathscr{X}(\mathbf{A} + \mathbf{BG})^* + (\mathbf{A} + \mathbf{BG})\mathscr{X} + \mathbf{DWD}^*, \tag{8.7}$$

$$\begin{aligned}\frac{\partial \hat{\mathscr{V}}}{\partial \mathscr{X}} = \mathbf{0} = {} & (\mathbf{A} + \mathbf{BG})^*\Lambda + \Lambda(\mathbf{A} + \mathbf{BG}) + \mathbf{G}^*\mathbf{RG}\mathscr{X}\mathbf{G}^*\mathbf{G} + \mathbf{C}^*\mathbf{QC}\mathscr{X}\mathbf{C}^*\mathbf{C} \\ & + \mathbf{G}^*\mathbf{G}\mathscr{X}\mathbf{G}^*\mathbf{RG} + \mathbf{C}^*\mathbf{C}\mathscr{X}\mathbf{C}^*\mathbf{QC},\end{aligned} \tag{8.8}$$

where it is apparent that both Λ and $\mathscr{X}$ will be symmetric matrices. By defining

$$\mathscr{R} \triangleq \mathbf{GXG}^*\mathbf{R} + \mathbf{RGXG}^* = \mathbf{UR} + \mathbf{RU}, \tag{8.9}$$

we may write the first equation (8.6) in the form (this is not a *solution* of the equation since $\mathscr{R}$ contains $\mathbf{G}$)

$$\mathbf{G} = -\mathscr{R}^{-1}\mathbf{B}^*\Lambda, \tag{8.10}$$

and this substitution into (8.8) leads to (after some manipulation)

$$\mathbf{0} = \Lambda\mathbf{A} + \mathbf{A}^*\Lambda - \Lambda\mathbf{B}\mathscr{R}^{-1}\mathbf{B}^*\Lambda + \mathscr{Q}, \tag{8.11}$$

where $\mathscr{Q} = \mathbf{C}^*\mathbf{QC}\mathscr{X}\mathbf{C}^*\mathbf{C} + \mathbf{C}^*\mathbf{C}\mathscr{X}\mathbf{C}^*\mathbf{QC} = \mathbf{C}^*[\mathbf{QR}_0 + \mathbf{R}_0\mathbf{Q}]\mathbf{C}$. In summary, we cite these results.

Theorem 8.1

The necessary conditions for the solution of problem (8.1), (8.2) *are given by*

$$\mathbf{G} = -\mathscr{R}^{-1}\mathbf{B}^*\Lambda, \qquad \mathscr{R} \triangleq \mathbf{UR} + \mathbf{RU}, \tag{8.12a}$$

$$\mathbf{0} = \Lambda\mathbf{A} + \mathbf{A}^*\Lambda - \Lambda\mathbf{B}\mathscr{R}^{-1}\mathbf{B}^*\Lambda + \mathbf{C}^*\mathscr{Q}\mathbf{C}, \qquad \mathscr{Q} \triangleq \mathbf{QR}_0 + \mathbf{R}_0\mathbf{Q}, \tag{8.12b}$$

where $\mathbf{U}$, $\mathbf{R}_0$, *and* $\mathscr{X}$ *are defined by*

$$\mathbf{U} = \mathbf{G}\mathscr{X}\mathbf{G}^*, \qquad \mathbf{R}_0 = \mathbf{C}\mathscr{X}\mathbf{C}^*, \tag{8.12c}$$

$$\mathbf{0} = \mathscr{X}(\mathbf{A} + \mathbf{BG})^* + (\mathbf{A} + \mathbf{BG})\mathscr{X} + \mathbf{DWD}^*. \tag{8.12d}$$

The analytical solution to these equations is not known, and iterative approaches are required, except in simple cases as illustrated in this example.

Exercise 8.1

Find the solution to problem (8.1), (8.2) for the system

$$\dot{x} = ax + b(u + w), \quad y = cx, \qquad W = 1, \quad R = \rho Q$$

and show that the optimal covariance control gain must satisfy

$$G^4 + \frac{2a}{b}G^3 - \frac{c^4}{\rho} = 0$$

and that the optimal gain for $a = b = c = \rho = 1$ *is* $G = -2.107$.

8.2 Linear Quadratic Impulse (LQI) Optimal Control

In this section we assume as usual that the disturbances $\mathbf{w}(t)$ and $\mathbf{v}(t)$ are impulses, that the objective function to be minimized is a quadratic function of the outputs $\mathbf{y}(t)$ and the controls $\mathbf{u}(t)$, and that the order of the controller is equal to the order of the plant. We will call this the linear quadratic impulse (LQI) optimal control problem. To begin the problem formulation, we write the form of the plant

$$\begin{aligned} \dot{\mathbf{x}} &= \mathbf{Ax} + \mathbf{Bu} + \mathbf{Dw}, \qquad w_i(t) = \omega_i\delta(t), \\ \mathbf{y} &= \mathbf{Cx}, \\ \mathbf{z} &= \mathbf{Mx} + \mathbf{v}, \qquad v_i(t) = \nu_i\delta(t), \end{aligned} \tag{8.13}$$

and the controller

$$\begin{aligned} \mathbf{u} &= \mathbf{Gx}_c, \\ \dot{\mathbf{x}}_c &= \mathbf{A}_c\mathbf{x}_c + \mathbf{Fz}, \end{aligned} \tag{8.14}$$

where $\mathbf{w}(t)$ and $\mathbf{v}(t)$ represent plant and measurement impulsive disturbances, respectively, with intensity ω_i for the ith element of $\mathbf{w}$ and ν_i for the ith element of

v. The given matrices are (**A**, **B**, **C**, **M**, **D**), and those to be determined are (**G**, $\mathbf{A}_c$, **F**). The closed-loop system will be described in terms of coordinates

$$\mathscr{x} \triangleq \begin{pmatrix} \tilde{\mathbf{x}} \\ \mathbf{x}_c \end{pmatrix} \triangleq \begin{pmatrix} \mathbf{x} - \mathbf{x}_c \\ \mathbf{x}_c \end{pmatrix}. \tag{8.15a}$$

That is, the original coordinates are related to the new coordinates by the transformation

$$\begin{pmatrix} \mathbf{x} \\ \mathbf{x}_c \end{pmatrix} = \begin{bmatrix} \mathbf{I} & \mathbf{I} \\ \mathbf{0} & \mathbf{I} \end{bmatrix} \begin{pmatrix} \tilde{\mathbf{x}} \\ \mathbf{x}_c \end{pmatrix} = \begin{bmatrix} \mathbf{I} & \mathbf{I} \\ \mathbf{0} & \mathbf{I} \end{bmatrix} \mathscr{x}. \tag{8.15b}$$

From (8.13) through (8.15), we have

$$\dot{\mathscr{x}} = \mathscr{A}\mathscr{x} + \mathscr{D}\mathscr{w},$$

$$\mathscr{y} = \mathscr{C}\mathscr{x},$$

$$\begin{bmatrix} \mathscr{A} & \mathscr{D} \\ \mathscr{C} & \end{bmatrix} = \begin{bmatrix} \mathbf{A} - \mathbf{FM} & \mathbf{A} - \mathbf{FM} + \mathbf{BG} - \mathbf{A}_c & \mathbf{D} & -\mathbf{F} \\ \mathbf{FM} & \mathbf{FM} + \mathbf{A}_c & \mathbf{0} & \mathbf{F} \\ \mathbf{C} & \mathbf{C} & & \\ \mathbf{0} & \mathbf{G} & & \end{bmatrix}, \tag{8.16}$$

where $\mathscr{w}^* = (\mathbf{w}^*, \mathbf{v}^*)$. The function we wish to minimize is

$$\mathscr{V} = \sum_{i=1}^{n_w + n_v} \int_0^\infty \mathscr{y}^{i*}(t)\mathscr{Q}\mathscr{y}^i(t)\,dt, \qquad \mathscr{Q} = \begin{bmatrix} \mathbf{Q} & \mathbf{0} \\ \mathbf{0} & \mathbf{R} \end{bmatrix} > \mathbf{0}, \tag{8.17}$$

where $\mathscr{y}^i(t)$ is the impulse response when only the ith element from the vector $(\mathbf{w}^*, \mathbf{v}^*)$ is acting and $\mathscr{x}(0) = \mathbf{0}$. Now, the cost function $\mathscr{V}$ has value (from Chapter 4)

$$\mathscr{V} = \operatorname{tr} \mathscr{X}\mathscr{C}^*\mathscr{Q}\mathscr{C}, \tag{8.18}$$

where $\mathscr{X}$ satisfies

$$\mathbf{0} = \mathscr{X}\mathscr{A}^* + \mathscr{A}\mathscr{X} + \mathscr{D}\mathscr{W}\mathscr{D}^*, \tag{8.19}$$

where the intensities of the impulses are described by

$$\mathscr{W} = \begin{bmatrix} \mathbf{W} & \mathbf{0} \\ \mathbf{0} & \mathbf{V} \end{bmatrix}, \quad \mathbf{W} = \begin{bmatrix} w_1^2 & & & \\ & w_2^2 & & \\ & & \ddots & \\ & & & w_{n_w}^2 \end{bmatrix}, \quad \mathscr{V} = \begin{bmatrix} v_1^2 & & & \\ & v_2^2 & & \\ & & \ddots & \\ & & & v_{nz}^2 \end{bmatrix}.$$

The use of Lagrange multipliers requires the minimization of

$$\hat{\mathscr{V}} = \operatorname{tr} \mathscr{X}\mathscr{C}^*\mathscr{Q}\mathscr{C} + \operatorname{tr} \mathscr{K}[\mathscr{X}\mathscr{A}^* + \mathscr{A}\mathscr{X} + \mathscr{D}\mathscr{W}\mathscr{D}^*], \tag{8.20}$$

leading to the necessary conditions [the reader should review Section 2.7.1.3 leading up to equation (2.99)]

$$\frac{\partial \hat{\mathscr{V}}}{\partial \mathscr{K}} = \mathbf{0} = \mathscr{X}\mathscr{A}^* + \mathscr{A}\mathscr{X} + \mathscr{D}\mathscr{W}\mathscr{D}^*, \tag{8.21}$$

$$\frac{\partial \hat{\mathscr{V}}}{\partial \mathscr{X}} = \mathbf{0} = \mathscr{K}\mathscr{A} + \mathscr{A}^*\mathscr{K} + \mathscr{C}^*\mathscr{Q}\mathscr{C}, \tag{8.22}$$

$$\frac{\partial \hat{\mathscr{V}}}{\partial \mathbf{G}} = \mathbf{0}, \quad \frac{\partial \hat{\mathscr{V}}}{\partial \mathbf{A}_c} = \mathbf{0}, \quad \frac{\partial \hat{\mathscr{V}}}{\partial \mathbf{F}} = \mathbf{0}, \tag{8.23}$$

where for convenience we have assumed all real parameters (hence, in this section $\mathbf{M}^* = \mathbf{M}^T$). Let the $2n_x \times 2n_x$ matrices $\mathscr{K}$ and $\mathscr{X}$ be partitioned into four $n_x \times n_x$ partitions labeled with appropriate subscripts $\mathscr{K}_{11}$, $\mathscr{K}_{12}$, $\mathscr{K}_{22}$, $\mathscr{X}_{11}$, $\mathscr{X}_{12}$, $\mathscr{X}_{22}$, and so on, noting that $\mathscr{K}_{21} = \mathscr{K}_{12}^*$, $\mathscr{X}_{21} = \mathscr{X}_{12}^*$, $\mathscr{K}_{11} = \mathscr{K}_{11}^*$, $\mathscr{K}_{22} = \mathscr{K}_{22}^*$, $\mathscr{X}_{11} = \mathscr{X}_{11}^*$, $\mathscr{X}_{22} = \mathscr{X}_{22}^*$. Then, the structure of $\mathscr{A}, \mathscr{D}, \mathscr{C}$ in (8.16) yields

$$\mathscr{A}\mathscr{X} = \begin{bmatrix} (\mathbf{A}-\mathbf{FM})\mathscr{X}_{11} + (\mathbf{A}+\mathbf{BG}-\mathbf{FM}-\mathbf{A}_c)\mathscr{X}_{12}^* & (\mathbf{A}-\mathbf{FM})\mathscr{X}_{12} + (\mathbf{A}+\mathbf{BG}+\mathbf{FM}-\mathbf{A}_c)\mathscr{X}_{22} \\ \mathbf{FM}\mathscr{X}_{11} + (\mathbf{FM}+\mathbf{A}_c)\mathscr{X}_{12}^* & \mathbf{FM}\mathscr{X}_{12} + (\mathbf{FM}+\mathbf{A}_c)\mathscr{X}_{22} \end{bmatrix},$$

$$\mathscr{K}\mathscr{A} = \begin{bmatrix} \mathscr{K}_{11}(\mathbf{A}-\mathbf{FM}) + \mathscr{K}_{12}\mathbf{FM} & \mathscr{K}_{11}(\mathbf{A}+\mathbf{BG}-\mathbf{FM}-\mathbf{A}_c) + \mathscr{K}_{12}(\mathbf{FM}+\mathbf{A}_c) \\ \mathscr{K}_{12}^*(\mathbf{A}-\mathbf{FM}) + \mathscr{K}_{22}\mathbf{FM} & \mathscr{K}_{12}^*(\mathbf{A}+\mathbf{BG}-\mathbf{FM}-\mathbf{A}_c) + \mathscr{K}_{22}(\mathbf{FM}+\mathbf{A}_c) \end{bmatrix},$$

$$\mathscr{C}^*\mathscr{Q}\mathscr{C} = \begin{bmatrix} \mathbf{C^*QC} & \mathbf{C^*QC} \\ \mathbf{C^*QC} & \mathbf{C^*QC} + \mathbf{G^*RG} \end{bmatrix}, \quad \mathscr{D}\mathscr{W}\mathscr{D}^* = \begin{bmatrix} \mathbf{DWD}^* + \mathbf{FVF}^* & -\mathbf{FVF}^* \\ -\mathbf{FVF}^* & \mathbf{FVF}^* \end{bmatrix}.$$

Now, the $2n_x \times 2n_x$ equation (8.21) leads to these three $n_x \times n_x$ equations:

$$(\mathbf{A}-\mathbf{FM})\mathscr{X}_{11} + (\mathbf{A}+\mathbf{BG}-\mathbf{FM}-\mathbf{A}_c)\mathscr{X}_{12}^* + \mathscr{X}_{12}(\mathbf{A}+\mathbf{BG}-\mathbf{FM}-\mathbf{A}_c)^* + \mathscr{X}_{11}(\mathbf{A}-\mathbf{FM})^* + \mathbf{DWD}^* + \mathbf{FVF}^* = 0, \tag{8.24a}$$

$$(\mathbf{A}-\mathbf{FM})\mathscr{X}_{12} + (\mathbf{A}+\mathbf{BG}-\mathbf{FM}-\mathbf{A}_c)\mathscr{X}_{22} + \mathscr{X}_{12}(\mathbf{FM}+\mathbf{A}_c)^* + \mathscr{X}_{11}(\mathbf{FM})^* - \mathbf{FVF}^* = \mathbf{0}, \tag{8.24b}$$

$$\mathbf{FM}\mathscr{X}_{12} + \mathscr{X}_{12}^*(\mathbf{FM})^* + (\mathbf{FM}+\mathbf{A}_c)\mathscr{X}_{22} + \mathscr{X}_{22}(\mathbf{FM}+\mathbf{A}_c)^* + \mathbf{FVF}^* = \mathbf{0}. \tag{8.24c}$$

The $2n_x \times 2n_x$ equation (8.22) leads to these three $n_x \times n_x$ equations:

$$\mathscr{K}_{11}(\mathbf{A} - \mathbf{FM}) + (\mathbf{A} - \mathbf{FM})^*\mathscr{K}_{11} + \mathscr{K}_{12}\mathbf{FM} + (\mathbf{FM})^*\mathscr{K}_{12}^* + \mathbf{C}^*\mathbf{QC} = \mathbf{0}, \tag{8.25a}$$

$$\mathscr{K}_{11}(\mathbf{A} + \mathbf{BG} - \mathbf{FM} - \mathbf{A}_c) + \mathscr{K}_{12}(\mathbf{FM} + \mathbf{A}_c) + (\mathbf{FM})^*\mathscr{K}_{22} + (\mathbf{A} - \mathbf{FM})^*\mathscr{K}_{12} + \mathbf{C}^*\mathbf{QC} = \mathbf{0}, \tag{8.25b}$$

$$\mathscr{K}_{12}^*(\mathbf{A} + \mathbf{BG} - \mathbf{FM} - \mathbf{A}_c) + (\mathbf{A} + \mathbf{BG} - \mathbf{FM} - \mathbf{A}_c)^*\mathscr{K}_{12} + \mathscr{K}_{22}(\mathbf{FM} + \mathbf{A}_c) + (\mathbf{FM} + \mathbf{A}_c)^*\mathscr{K}_{22} + \mathbf{C}^*\mathbf{QC} + \mathbf{G}^*\mathbf{RG} = \mathbf{0}. \tag{8.25c}$$

Equations (8.23) are developed as follows. In terms of the partitioned matrices, (8.20) becomes

$$\begin{aligned}\hat{\mathscr{V}} = {} & \operatorname{tr}[\mathscr{X}_{11} + 2\mathscr{X}_{12} + \mathscr{X}_{22}]\mathbf{C}^*\mathbf{QC} + \operatorname{tr}\mathscr{X}_{22}\mathbf{G}^*\mathbf{RG} \\ & + \operatorname{tr} 2\mathscr{K}_{11}\mathscr{X}_{12}(\mathbf{A} + \mathbf{BG} - \mathbf{FM} - \mathbf{A}_c)^* + \operatorname{tr} 2\mathscr{K}_{11}\mathscr{X}_{11}(\mathbf{A} - \mathbf{FM})^* \\ & + \operatorname{tr}\mathscr{K}_{11}(\mathbf{DWD}^* + \mathbf{FVF}^*) \\ & + \operatorname{tr}\mathscr{K}_{12}[\mathscr{X}_{12}^*(\mathbf{A} - \mathbf{FM})^* + \mathscr{X}_{22}(\mathbf{A} + \mathbf{BG} - \mathbf{FM} - \mathbf{A}_c)^* \\ & \qquad + (\mathbf{FM} + \mathbf{A}_c)\mathscr{X}_{12}^* + \mathbf{FM}\mathscr{X}_{11} - \mathbf{FVF}^*] \\ & + \operatorname{tr}\mathscr{K}_{12}^*[(\mathbf{A} - \mathbf{FM})\mathscr{X}_{12} + (\mathbf{A} + \mathbf{BG} - \mathbf{FM} - \mathbf{A}_c)\mathscr{X}_{22} \\ & \qquad + \mathscr{X}_{12}(\mathbf{FM} + \mathbf{A}_c)^* + \mathscr{X}_{11}(\mathbf{FM})^* - \mathbf{FVF}^*] \\ & + \operatorname{tr}\mathscr{K}_{22}[2\mathscr{X}_{22}(\mathbf{FM} + \mathbf{A}_c)^* + 2\mathscr{X}_{12}^*(\mathbf{FM})^* + \mathbf{FVF}^*],\end{aligned} \tag{8.26}$$

where we have used the identities (2.83) and (2.84)

$$\operatorname{tr}\mathbf{AB} = \operatorname{tr}\mathbf{BA} = \operatorname{tr}\mathbf{A}^*\mathbf{B}^*, \qquad (\mathbf{A}, \mathbf{B} \text{ real}),$$

for convenient collections of terms. Now, the differentiation of (8.26) yields for (8.23)

$$\frac{\partial\hat{\mathscr{V}}}{\partial\mathbf{G}} = \mathbf{0} = 2\mathbf{RG}\mathscr{X}_{22} + 2\mathbf{B}^*\mathscr{K}_{11}\mathscr{X}_{12} + 2\mathbf{B}^*\mathscr{K}_{12}\mathscr{X}_{22}, \tag{8.27}$$

$$\frac{\partial\hat{\mathscr{V}}}{\partial\mathbf{A}_c} = \mathbf{0} = -2\mathscr{K}_{11}\mathscr{X}_{12} - 2\mathscr{K}_{12}\mathscr{X}_{22} + 2\mathscr{K}_{12}^*\mathscr{X}_{12} + 2\mathscr{K}_{22}\mathscr{X}_{22}, \tag{8.28}$$

$$\begin{aligned}\frac{\partial\hat{\mathscr{V}}}{\partial\mathbf{F}} = \mathbf{0} = {} & 2[\mathscr{K}_{11} + \mathscr{K}_{22} - \mathscr{K}_{12} - \mathscr{K}_{12}^*]\mathbf{FV} \\ & + 2[(\mathscr{K}_{12}^* - \mathscr{K}_{11})\mathscr{X}_{11} + (\mathscr{K}_{12}^* - \mathscr{K}_{11})\mathscr{X}_{12} \\ & \qquad + (\mathscr{K}_{22} - \mathscr{K}_{12})\mathscr{X}_{12}^* + (\mathscr{K}_{22} - \mathscr{K}_{12})\mathscr{X}_{22}]\mathbf{M}^*.\end{aligned} \tag{8.29}$$

Now, (8.28) allows (8.27) to be rewritten

$$\mathbf{G}\mathscr{X}_{22} = -\mathbf{R}^{-1}\mathbf{B}^*[\mathscr{K}_{22}\mathscr{X}_{22} + \mathscr{K}_{12}^*\mathscr{X}_{12}], \tag{8.30}$$

and (8.28) allows (8.29) to be rewritten

$$\begin{aligned}[\mathscr{K}_{11} + \mathscr{K}_{22} - \mathscr{K}_{12} - \mathscr{K}_{12}^*]\mathbf{F} \\ = [(\mathscr{K}_{11} - \mathscr{K}_{12}^*)\mathscr{X}_{11} + (\mathscr{K}_{12} - \mathscr{K}_{22})\mathscr{X}_{12}^*]\mathbf{M}^*\mathbf{V}^{-1},\end{aligned} \tag{8.31}$$

assuming that the indicated inverses exist. Our task is to find a solution of the set of equations (8.24), (8.25), (8.30), and (8.31). We begin by showing that (8.24b) is satisfied by the choice

$$\mathscr{X}_{12} = \mathbf{0}, \quad \mathbf{A} + \mathbf{BG} - \mathbf{FM} - \mathbf{A}_c = \mathbf{0}, \quad [\mathscr{X}_{11}\mathbf{M}^* - \mathbf{FV}]\mathbf{F}^* = \mathbf{0}. \tag{8.32}$$

This leads to

$$\mathbf{A}_c = \mathbf{A} + \mathbf{BG} - \mathbf{FM}, \tag{8.33}$$

$$\mathbf{F} = \mathscr{X}_{11}\mathbf{M}^*\mathbf{V}^{-1}. \tag{8.34}$$

Now, (8.31) and (8.34) are consistent if $\mathscr{X}_{12} = \mathbf{0}$ and

$$\mathscr{K}_{12} = \mathscr{K}_{22}, \tag{8.35}$$

where

$$|\mathscr{K}_{11} - \mathscr{K}_{22}| \neq 0 \text{ yields a unique } \mathbf{F}. \tag{8.36}$$

We are now prepared to summarize our results.

Theorem 8.2

Let the linear system

$$\begin{aligned}\dot{\mathbf{x}} &= \mathbf{Ax} + \mathbf{Bu} + \mathbf{Dw}, && \mathbf{x} \in \mathscr{R}^{n_x}, \quad \mathbf{w} \in \mathscr{R}^{n_w}, \quad \mathbf{x}(0) = \mathbf{0}, \\ \mathbf{y} &= \mathbf{Cx}, && \mathbf{y} \in \mathscr{R}^{n_y}, \\ \mathbf{z} &= \mathbf{Mx} + \mathbf{v}, && \mathbf{z} \in \mathscr{R}^{n_z}, \quad \mathbf{v} \in \mathscr{R}^{n_z},\end{aligned} \tag{8.37}$$

be subject to impulsive disturbances $w_i = \omega_i\delta(t)$, $i = 1, 2, \ldots, n_w$, *and* $v_j = \upsilon_j\delta(t)$, $j = 1, 2, \ldots, n_v$, *applied one at a time where* $\mathbf{y}^i(t)$ *and* $\mathbf{u}^i(t)$ *represent the output of the plant and the output of the* n_x*th order controller*

$$\begin{aligned}\mathbf{u} &= \mathbf{Gx}_c, && \mathbf{u} \in \mathscr{R}^{n_u} \\ \dot{\mathbf{x}}_c &= \mathbf{A}_c\mathbf{x}_c + \mathbf{Fz}, && \mathbf{x}_c \in \mathscr{R}^{n_x}, \quad \mathbf{x}_c(0) = \mathbf{0},\end{aligned} \tag{8.38}$$

when only the ith disturbance element in the vector $(\mathbf{w}^*, \mathbf{v}^*)$ *is acting. Then, the scalar function*

$$\mathscr{V} = \sum_{i=1}^{n_w+n_z} \int_0^\infty \left[\mathbf{y}^{i*}(t)\mathbf{Q}\mathbf{y}^i(t) + \mathbf{u}^{i*}(t)\mathbf{R}\mathbf{u}^i(t)\right] dt, \qquad \mathbf{Q} > \mathbf{0}, \quad \mathbf{R} > \mathbf{0} \tag{8.39}$$

is minimized by the choice of controller parameters $\mathbf{G}, \mathbf{A}_c, \mathbf{F}$ *which satisfy*

$$\mathbf{A}_c = \mathbf{A} + \mathbf{B}\mathbf{G} - \mathbf{F}\mathbf{M}, \tag{8.40a}$$

$$\mathbf{G} = -\mathbf{R}^{-1}\mathbf{B}^*\mathscr{K}_{22}, \qquad \mathbf{0} = \mathscr{K}_{22}\mathbf{A} + \mathbf{A}^*\mathscr{K}_{22} - \mathscr{K}_{22}\mathbf{B}\mathbf{R}^{-1}\mathbf{B}^*\mathscr{K}_{22} + \mathbf{C}^*\mathbf{Q}\mathbf{C}, \tag{8.40b}$$

$$\mathbf{F} = \mathscr{X}_{11}\mathbf{M}^*\mathbf{V}^{-1}, \qquad \mathbf{0} = \mathscr{X}_{11}\mathbf{A}^* + \mathbf{A}\mathscr{X}_{11} - \mathscr{X}_{11}\mathbf{M}^*\mathbf{V}^{-1}\mathbf{M}\mathscr{X}_{11} + \mathbf{D}\mathbf{W}\mathbf{D}^*. \tag{8.40c}$$

The minimum value of the cost function $\mathscr{V}$ *is*

$$\mathscr{V} = \operatorname{tr}(\mathscr{X}_{11} + \mathscr{X}_{22})\mathbf{C}^*\mathbf{Q}\mathbf{C} + \operatorname{tr}\mathscr{X}_{22}\mathbf{G}^*\mathbf{R}\mathbf{G} \tag{8.41a}$$

or

$$\mathscr{V} = \operatorname{tr}\mathscr{K}_{11}\mathbf{D}\mathbf{W}\mathbf{D}^* + \operatorname{tr}\tilde{\mathscr{K}}\mathbf{F}\mathbf{V}\mathbf{F}^*, \tag{8.41b}$$

where $\mathscr{X}_{22}$ *is computed from*

$$\mathbf{0} = \mathscr{X}_{22}(\mathbf{A} + \mathbf{B}\mathbf{G})^* + (\mathbf{A} + \mathbf{B}\mathbf{G})\mathscr{X}_{22} + \mathbf{F}\mathbf{V}\mathbf{F}^*. \tag{8.42a}$$

and $\mathscr{K}_{11}$ *from*

$$\mathbf{0} = \tilde{\mathscr{K}}(\mathbf{A} - \mathbf{F}\mathbf{M}) + (\mathbf{A} - \mathbf{F}\mathbf{M})^*\tilde{\mathscr{K}} + \mathbf{G}^*\mathbf{R}\mathbf{G}, \qquad \mathscr{K}_{11} = \tilde{\mathscr{K}} + \mathscr{K}_{22}. \tag{8.42b}$$

$\mathbf{F}$ *is unique if* $(\mathbf{A} - \mathbf{F}\mathbf{M}, \mathbf{G})$ *is observable, and* $\mathbf{G}$ *is unique if* $(\mathbf{A} + \mathbf{B}\mathbf{G}, \mathbf{F})$ *is controllable. We assume that* $\mathscr{K}_{22}$ *and* $\mathscr{X}_{11}$ *are unique positive definite solutions of* (8.40b) *and* (8.40c).

Proof: The necessary condition for this optimization problem is recorded by (8.24), (8.25), (8.30), and (8.31). First, we will show one solution to these equations, and then we must show that the solution is unique. In the above listed equation (8.24b), set $\mathscr{X}_{12} = \mathbf{0}$, $\mathbf{A}_c = \mathbf{A} + \mathbf{B}\mathbf{G} - \mathbf{F}\mathbf{M}$, $\mathscr{K}_{12} = \mathscr{K}_{22}$, $\mathbf{G} = -\mathbf{R}^{-1}\mathbf{B}^*\mathscr{K}_{22}$, $\mathbf{F} = \mathscr{X}_{11}\mathbf{M}^*\mathbf{V}^{-1}$ and see that (8.24a) then reduces to the (8.40c) calculation for $\mathscr{X}_{11}$, and that (8.25c) reduces to the (8.40b) calculation for $\mathscr{K}_{22}$, and that (8.30) reduces to the $\mathbf{G}$ calculation in (8.40b), and that (8.31) reduces to the $\mathbf{F}$ calculation in (8.40c), and that (8.24c) reduces to (8.42a). Equation (8.24b) is identically zero on each side. Equation (8.25b) reduces to the $\mathscr{K}_{22}$ equation in (8.40b), and hence is redundant to (8.25c). This leaves only (8.25a) to be reconciled. The equation serves to define $\mathscr{K}_{11}$, but let us subtract (8.25b) from (8.25a). This yields (8.42b) or, equivalently,

$$(\mathscr{K}_{11} - \mathscr{K}_{22})(\mathbf{A} - \mathbf{F}\mathbf{M}) + (\mathbf{A} - \mathbf{F}\mathbf{M})^*(\mathscr{K}_{11} - \mathscr{K}_{22}) + \mathbf{G}^*\mathbf{R}\mathbf{G} = \mathbf{0},$$

where it is clear from Theorem 4.12 that if $(\mathscr{K}_{11} - \mathscr{K}_{22})$ exists, it is given by

$$\mathscr{K}_{11} - \mathscr{K}_{22} = \int_0^\infty e^{(\mathbf{A}-\mathbf{FM})^*t}\mathbf{G}^*\mathbf{RG}e^{(\mathbf{A}-\mathbf{FM})t}\,dt, \tag{8.43}$$

which is positive definite iff $(\mathbf{A} - \mathbf{FM}, \mathbf{G})$ is an observable pair. The linear equation (8.31) has a unique solution $\mathbf{F}$ if $\mathscr{K}_{11} - \mathscr{K}_{22}$ is nonsingular, and this will be the case whenever $(\mathbf{A} - \mathbf{FM}, \mathbf{G})$ turns out to be an observable pair. Note from (8.30) and (8.42) that $\mathbf{G}$ is unique if $\mathscr{X}_{22}$ is nonsingular. This will be the case whenever $(\mathbf{A} + \mathbf{BG}, \mathbf{F})$ is a controllable pair. $\mathscr{K}_{22}$ and $\mathscr{X}_{11}$ are unique positive definite matrices by assumption. This completes the proof. ■

This theorem will be strengthened later by the conditions under which $\mathscr{K}_{22}$ and $\mathscr{X}_{11}$ are positive definite and by closed-loop stability conditions. For the moment, let us analyze the impact of Theorem 8.2. The $\mathscr{X}$ and the $\mathscr{K}$ matrices of (8.21) and (8.22) have the structure

$$\mathscr{K} = \begin{bmatrix} \mathscr{K}_{11} & \mathscr{K}_{22} \\ \mathscr{K}_{22} & \mathscr{K}_{22} \end{bmatrix}, \qquad \mathscr{X} = \begin{bmatrix} \mathscr{X}_{11} & \mathbf{0} \\ \mathbf{0} & \mathscr{X}_{22} \end{bmatrix}, \tag{8.44}$$

where $\mathscr{X}_{12} = \mathbf{0}$ implies that

$$\mathscr{X}_{12} \triangleq \sum_{i=1}^{n_w+n_z} \int_0^\infty \tilde{\mathbf{x}}^i(t)\mathbf{x}_c^{i*}(t)\,dt = 0. \tag{8.45}$$

Hence, according to Definition 4.5, we say that the plant error $\tilde{\mathbf{x}}(t)$ and the controller state $\mathbf{x}_c(t)$ in the LQI optimal system are not time correlated at $t = 0$. Note also that due to (8.33), equation (8.16) reduces to

$$\begin{aligned}\begin{pmatrix} \dot{\tilde{\mathbf{x}}} \\ \dot{\mathbf{x}}_c \end{pmatrix} &= \begin{bmatrix} \mathbf{A} - \mathbf{FM} & \mathbf{0} \\ \mathbf{FM} & \mathbf{A} + \mathbf{BG} \end{bmatrix}\begin{pmatrix} \tilde{\mathbf{x}} \\ \mathbf{x}_c \end{pmatrix} + \begin{bmatrix} \mathbf{D} & -\mathbf{F} \\ \mathbf{0} & \mathbf{F} \end{bmatrix}\begin{pmatrix} \mathbf{w} \\ \mathbf{v} \end{pmatrix}, \\ \begin{pmatrix} \mathbf{y} \\ \mathbf{u} \end{pmatrix} &= \begin{bmatrix} \mathbf{C} & \mathbf{C} \\ \mathbf{0} & \mathbf{G} \end{bmatrix}\begin{pmatrix} \tilde{\mathbf{x}} \\ \mathbf{x}_c \end{pmatrix}. \end{aligned} \tag{8.46}$$

Three observations are immediate:

The first observation is that the eigenvalues of the closed-loop optimal LQI system are $\lambda_i[\mathbf{A} - \mathbf{FM}]$, $i = 1, 2, \ldots, n_x$, and $\lambda_j[\mathbf{A} + \mathbf{BG}]$, $j = 1, 2, \ldots, n_x$. Hence, the closed-loop system is stable iff $[\mathbf{A} - \mathbf{FM}]$ and $[\mathbf{A} + \mathbf{BG}]$ are stable.

The second observation is that $[\mathbf{A} + \mathbf{BG}]$ would be the closed-loop system matrix if *state* feedback were used $(\mathbf{u} = \mathbf{Gx})$ for control. Hence, half of the eigenvalues of the optimal LQI system are those which would have been obtained by state feedback control. Note that $\mathbf{G}$ is determined by the choice of weights $\mathbf{Q}$ and $\mathbf{R}$. See (8.40b).

The third observation contains a real surprise. See that

$$\dot{\tilde{\mathbf{x}}} = [\mathbf{A} - \mathbf{FM}]\tilde{\mathbf{x}} + [\mathbf{D} \quad -\mathbf{F}]\begin{pmatrix}\mathbf{w}\\ \mathbf{v}\end{pmatrix}, \tag{8.47}$$

together with the assumption that $\mathbf{A} - \mathbf{FM}$ is stable, implies that $\lim_{t\to\infty} \tilde{\mathbf{x}} = \mathbf{0}$ or, equivalently, that *controller state* $\mathbf{x}_c(t)$ *converges to the plant state* $\mathbf{x}(t)$ regardless of the strength of the disturbance impulses in $\mathbf{w}, \mathbf{v}$ (and regardless of the initial state $\mathbf{x}(0)$). This convergence occurs with the time constants associated with $[\mathbf{A} - \mathbf{FM}]$. In other words, without any *a priori* request to do so, the LQI optimal controller consists of a controller state $\mathbf{x}_c(t)$ which seeks to *mimic* the plant trajectory $\mathbf{x}(t)$. In fact, the LQI controller has two parts: (i) $\mathbf{u} = \mathbf{G}\mathbf{x}_c$, the *control law*; and (ii) $\dot{\mathbf{x}}_c = \mathbf{A}\mathbf{x}_c + \mathbf{B}\mathbf{u} + \mathbf{F}(\mathbf{z} - \mathbf{M}\mathbf{x}_c)$, the *state estimator* [given this name by virtue of the property $\mathbf{x}_c(t) \to \mathbf{x}(t)$]. The speed at which $\mathbf{x}_c(t)$ converges to $\mathbf{x}(t)$ is governed by $\mathbf{F}$, and $\mathbf{F}$ is completely dictated by the intensities of the plant and measurement impulse disturbances. See (8.40c). More importantly, $\mathbf{F}$ is *independent* of $\mathbf{G}$ or the $\mathbf{Q}, \mathbf{R}$ parameters that dictate $\mathbf{G}$.

It is appealing logic to argue that the LQI controller has provided a state estimator to quickly track the plant trajectory and to apply a state feedback control law to control the system as though the state were available for feedback. In the next section we will indeed find that *the* $\mathbf{G}$ in (8.40b) *is the same gain that would be LQI optimal if the state were available for feedback*. However, the above logic fails for a different reason. Note that the calculations (8.40b, c) for $\mathbf{F}$ and $\mathbf{G}$ are quite independent, being dictated by the separate parameters $(\mathbf{W}, \mathbf{V})$ and $(\mathbf{Q}, \mathbf{R})$, respectively. Hence, the optimal LQI controller is just as happy (i.e., just as optimal) to have the state estimator (whose time constants are those of $[\mathbf{A} - \mathbf{FM}]$) *slower* than the time constants of $[\mathbf{A} + \mathbf{BG}]$ for state feedback. That is, the controller is still optimal when $\mathbf{F}$ is small [from small $\mathbf{W}$, large $\mathbf{V}$, see (8.40c)], and $\mathbf{G}$ is large [from large $\mathbf{Q}$, small $\mathbf{R}$, see (8.40b)], yielding much faster time constants in $[\mathbf{A} + \mathbf{BG}]$ than in $[\mathbf{A} - \mathbf{FM}]$. Hence, one cannot rationalize the presence of a state estimator in the LQI controller by the logic of a relatively fast estimator (even though such logic has been used in ad hoc juxtaposition of a state estimator and an optimal control law for state feedback for deterministic systems [8.1]). The appearance of a state estimator in the LQI controller remains a surprise. More will be said in Chapter 9 about this separation of the estimation and control problems, after developing more properties of the estimator. The structure of the optimal LQI controller is in Fig. 8.1.

EXAMPLE 8.1

Solve the LQI problem if

$$\mathbf{A} = \begin{bmatrix} 0 & -1 \\ 1 & 0 \end{bmatrix}, \quad \mathbf{B} = \begin{bmatrix} 1 \\ 0 \end{bmatrix}, \quad \mathbf{D} = \begin{bmatrix} 1 & 1 \\ -(15 + \varepsilon) & \varepsilon \end{bmatrix},$$

$$\mathbf{C} = [2 \quad 0], \quad \mathbf{M} = [1 \quad -1], \quad V = Q = R = 1, \quad \mathbf{W} = \mathbf{I}, \quad \varepsilon = 0.0663725.$$

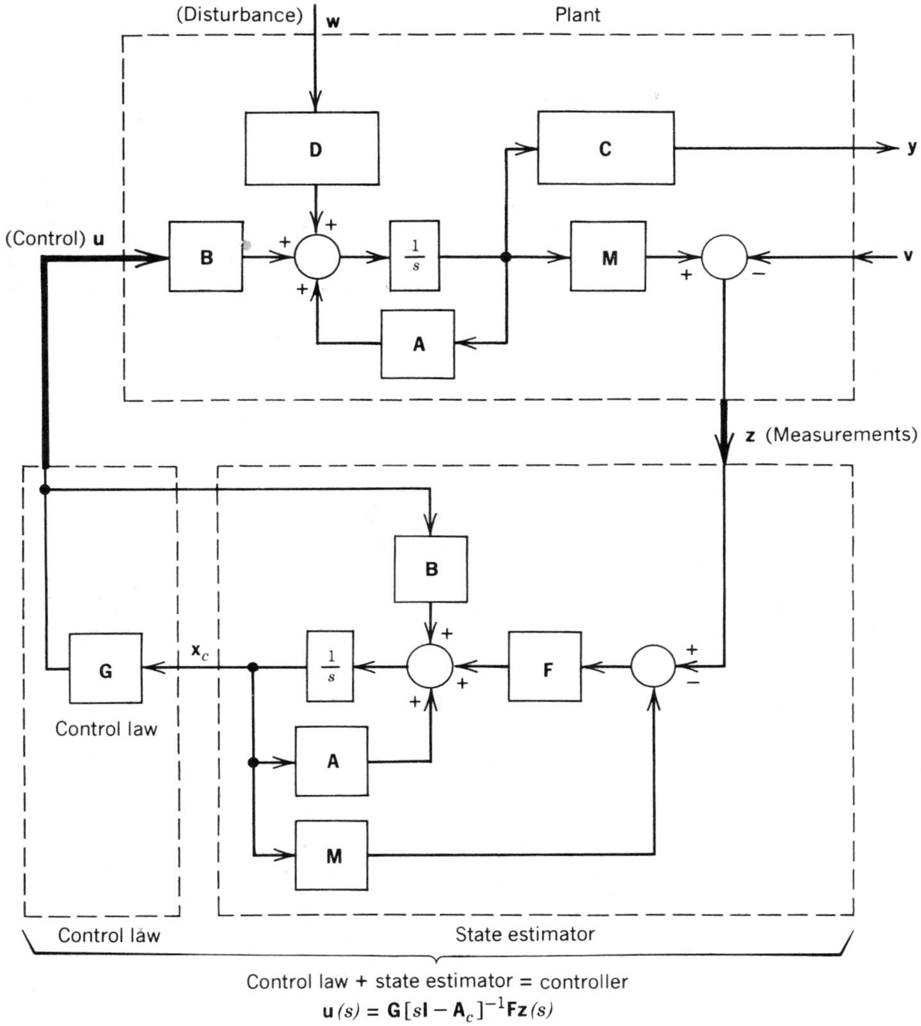

Figure 8.1 *Optimal LQI Controller*

Solution: The solution from Theorem 8.2 is

$$\mathbf{A}_c = \begin{bmatrix} -3 & 0 \\ 5 & -4 \end{bmatrix}, \quad \mathbf{F} = \begin{bmatrix} 1 \\ -4 \end{bmatrix}, \quad \mathbf{G} = [-2 \quad 0],$$

which is clearly not a minimal TER since

$$\mathbf{G}(s\mathbf{I} - \mathbf{A}_c)^{-1}\mathbf{F} = \frac{-2}{s+3}.$$

Hence, the *first*-order controller given by

$$A_{CR} = -3, \quad F_R = 1, \quad G_R = -2$$

is an optimal LQI controller since it yields the same value of $\mathscr{V}$ as does the second-order controller $(\mathbf{A}_c, \mathbf{F}, \mathbf{G})$. The point of this example is to show that even if the plant is a minimal TER, the LQI controller might not be minimal. That is, even if $(\mathbf{A}, \mathbf{B}, \mathbf{C})$ and $(\mathbf{A}, \mathbf{D}, \mathbf{M})$ are both observable, controllable triples, the triple $(\mathbf{A}_c, \mathbf{F}, \mathbf{G})$ might not be *controllable*, *observable* (or even *stable*, as the next example shows). ■

EXAMPLE 8.2

Find the optimal LQI controller for the system whose transfer function is

$$\mathbf{C}(s\mathbf{I} - \mathbf{A})^{-1}\mathbf{B} = \frac{s-1}{s(s-2)} = \mathbf{M}(s\mathbf{I} - \mathbf{A})^{-1}\mathbf{D},$$

with unit strength impulses entering the system through $w(t)$ and $v(t)$ in the state space realization

$$\dot{\mathbf{x}} = \begin{bmatrix} 0 & 1 \\ 0 & 2 \end{bmatrix}\mathbf{x} + \begin{bmatrix} 0 \\ 1 \end{bmatrix}(u + w),$$

$$y = [-1 \quad 1]\mathbf{x}, \qquad z = y + v.$$

Show that the closed-loop system minimizing

$$\mathscr{V} = \sum_{i=1}^{n_w + n_v} \int_0^\infty \left[y^{i^2}(t) + \rho u^{i^2}(t) \right] dt$$

is asymptotically stable for all $\rho > 0$ but that the controller itself (with transfer function $\mathbf{H}(s) = \mathbf{G}(s\mathbf{I} - \mathbf{A}_c)^{-1}\mathbf{F}$) is never stable for any $\rho > 0$.

Solution: Table 8.1 lists the eigenvalues of the closed-loop system $\{\lambda_i[\mathbf{A} + \mathbf{BG}]$ and $\lambda_i[\mathbf{A} - \mathbf{FM}]\}$ and the eigenvalues of the controller $\lambda_i[A_c]$, as ρ is varied from 10^{-6} to 10^4. Note that the closed-loop system is always stable and that the controller itself is never stable. The explanation for this phenomenon is best illustrated on the root locus. In Fig. 8.2 see that the open-loop poles and zeros are located by $\times$ and $\bigcirc$. The controller poles and zeros are located by dotted $\times$ and $\bigcirc$. Note that there is an unstable branch of the root locus *for any* choice of controller poles (the reader should try some possible combinations to see this)

TABLE 8.1 LQI Controller for Example 8.2

$$\lambda_i[A - FM] = (-0.457, -2.189), \qquad H(s) = \frac{k(s - z_1)}{(s - p_1)(s - p_2)}$$

ρ	$-\mathbf{G}$	$\lambda_i[\mathbf{A} + \mathbf{BG}]$	k	z_1	p_1	p_2
10^{-6}	[1000, 1003]	−1, −1000	1.596×10^4	6.262×10^{-2}	10.15	-1.016×10^3
10^{-4}	[100, 103]	−0.9999, −100	1.625×10^3	6.155×10^{-2}	9.107	-1.148×10^2
10^{-2}	[10, 13.14]	−0.9852, −10.15	1.916×10^2	5.218×10^{-2}	5.596	−21.38
1	[1, 4.646]	−0.4569, −2.189	53.46	1.87×10^{-2}	2.924	−10.22
10^2	[0.1, 4.052]	−0.05, −2.002	42.26	2.366×10^{-3}	2.444	09.141
10^4	[0.01, 4.005]	−0.005, −2	41.27	2.424×10^{-4}	2.394	−9.045

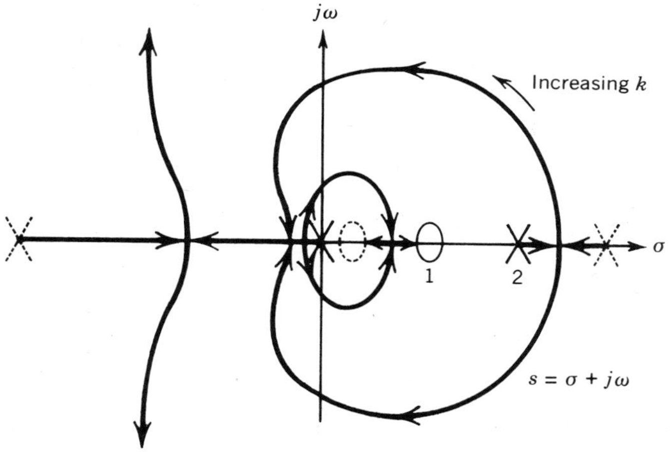

Figure 8.2 *Root Locus Using Pole/Zero Configurations from LQI Controllers*

unless one controller pole is in the right half-plane (as shown in Fig. 8.2). It is also interesting to note that the four optimal closed-loop poles are always real. ■

Example 8.2 illustrates a class of systems that *cannot* be stabilized by a stable controller. Such systems are always unstable (in the open loop), but not all unstable plants require unstable controllers. Johnson [8.2] uses the adjective "homeopathic" to describe the class of plants requiring unstable controllers, since a *dose of instability* (the controller) is required to cure the *disease of instability* (in the plant).

Exercise 8.2

Define the generalized LQI problem by

$$\mathscr{V} = \sum_{i=1}^{n_w+n_z+2n_x} \int_0^\infty [\mathbf{y}^{i*}(t)\mathbf{Q}\mathbf{y}^i(t) + \mathbf{u}^{i*}(t)\mathbf{R}\mathbf{u}^i(t)]\,dt, \tag{8.48a}$$

$$\dot{\mathbf{x}} = \mathbf{A}\mathbf{x} + \mathbf{B}\mathbf{u} + \mathbf{D}\mathbf{w}, \tag{8.48b}$$

$$\mathbf{y} = \mathbf{C}\mathbf{x},$$

$$\mathbf{z} = \mathbf{M}\mathbf{x} + \mathbf{v},$$

$$\mathbf{u} = \mathbf{G}\mathbf{x}_c, \tag{8.48c}$$

$$\dot{\mathbf{x}}_c = \mathbf{A}_c\mathbf{x}_c + \mathbf{F}\mathbf{z},$$

where $\mathbf{y}^i(t), \mathbf{u}^i(t)$ are the responses of the closed-loop system when subjected to an impulse in the ith channel of the vector $(\mathbf{w}^*, \mathbf{v}^*)$ corresponding to $(1 \le i \le n_w + n_v)$, or an initial condition in the ith element of the state vector $(\mathbf{x}^*(0), \mathbf{x}_c^*(0))$ corresponding to $(n_w + n_v < i \le n_w + n_v + 2n_x)$. Show that if a solution exists, then the solution to the generalized LQI problem is

$$\mathbf{G} = -\mathbf{R}^{-1}\mathbf{B}^*\mathscr{K}_{22}, \quad \mathbf{0} = \mathscr{K}_{22}\mathbf{A} + \mathbf{A}^*\mathscr{K}_{22} - \mathscr{K}_{22}\mathbf{B}\mathbf{R}^{-1}\mathbf{B}^*\mathscr{K}_{22} + \mathbf{C}^*\mathbf{Q}\mathbf{C}, \tag{8.49a}$$

$$\mathbf{F} = \mathscr{X}_{11}\mathbf{M}^*\mathbf{V}^{-1}, \quad 0 = \mathscr{X}_{11}\mathbf{A}^* + \mathbf{A}\mathscr{X}_{11} - \mathscr{X}_{11}\mathbf{M}^*\mathbf{V}^{-1}\mathbf{M}\mathscr{X}_{11} + \mathbf{D}\mathbf{W}\mathbf{D}^* + \mathbf{X}_0, \tag{8.49b}$$

$$\begin{aligned}\mathscr{V} &= \operatorname{tr}(\mathscr{X}_{11} + \mathscr{X}_{22})\mathbf{C}^*\mathbf{Q}\mathbf{C} + \operatorname{tr}\mathscr{X}_{22}\mathbf{G}^*\mathbf{R}\mathbf{G} \\ &= \operatorname{tr}\left(\tilde{\mathscr{K}} + \mathscr{K}_{22}\right)(\mathbf{D}\mathbf{W}\mathbf{D}^* + \mathbf{X}_0) + \operatorname{tr}\tilde{\mathscr{K}}\mathbf{F}\mathbf{V}\mathbf{F}^* + \operatorname{tr}\mathscr{K}_{22}\mathbf{X}_{c0},\end{aligned} \tag{8.49c}$$

where

$$\mathbf{0} = \mathscr{X}_{22}(\mathbf{A} + \mathbf{B}\mathbf{G})^* + (\mathbf{A} + \mathbf{B}\mathbf{G})\mathscr{X}_{22} + \mathbf{F}\mathbf{V}\mathbf{F}^* + \mathbf{X}_{c0}, \tag{8.49d}$$

$$\mathbf{0} = \tilde{\mathscr{K}}(\mathbf{A} - \mathbf{F}\mathbf{M}) + (\mathbf{A} - \mathbf{F}\mathbf{M})^*\tilde{\mathscr{K}} + \mathbf{G}^*\mathbf{R}\mathbf{G}, \qquad \mathscr{K}_{11} = \tilde{\mathscr{K}} + \mathscr{K}_{22}, \tag{8.49e}$$

$$\mathbf{X}_0 \triangleq \begin{bmatrix} x_1^2(0) & & & \\ & x_2^2(0) & & \\ & & \ddots & \\ & & & x_{n_x}^2(0) \end{bmatrix}, \quad X_{c0} \triangleq \begin{bmatrix} x_{c_1}^2(0) & & & \\ & x_{c_2}^2(0) & & \\ & & \ddots & \\ & & & x_{c_{n_x}}^2(0) \end{bmatrix}. \tag{8.50}$$

As in Theorem 8.2, assume observability of $(\mathbf{A}-\mathbf{FM},\mathbf{G})$ and controllability of $(\mathbf{A}+\mathbf{BG},\mathbf{F})$ for uniqueness of $\mathbf{F}$ and $\mathbf{G}$.

Exercise 8.3

Show that the transfer function of the controller in the LQI problem is

$$\mathbf{H}(s)=\mathbf{G}[s\mathbf{I}-(\mathbf{A}+\mathbf{BG}-\mathbf{FM})]^{-1}\mathbf{F}, \tag{8.51}$$

where

$$\mathbf{u}(s)=\mathbf{H}(s)\mathbf{z}(s).$$

As a special case of Theorem 8.2, it is easy to show that the following is true.

Corollary to Theorem 8.2 (Initial conditions applied simultaneously in lieu of one at a time.) *Let the system*

$$\begin{aligned}\dot{\mathbf{x}}&=\mathbf{Ax}+\mathbf{Bu}, \qquad \mathbf{x}(0)=\mathbf{x}_0\\ \mathbf{y}&=\mathbf{Cx},\\ \mathbf{z}&=\mathbf{Mx}+\mathbf{v}\end{aligned} \tag{8.52}$$

be subject to impulsive disturbances $v_i=\nu_i\delta(t)$, $w_i=\omega_i\delta(t)$ one at a time, where $\mathbf{y}^i(t)$ and $\mathbf{u}^i(t)$ represent the output of the plant and the output of the n_xth-order controller

$$\begin{aligned}\mathbf{u}&=\mathbf{G}\mathbf{x}_c,\\ \dot{\mathbf{x}}_c&=\mathbf{A}_c\mathbf{x}_c+\mathbf{Fz}, \qquad \mathbf{x}_c(0)=\mathbf{x}_{c0},\end{aligned} \tag{8.53}$$

when only the ith disturbance element in $[\mathbf{w}^*(t),\mathbf{v}^*(t)]$ is acting, or $(\mathbf{x}_0,\mathbf{x}_{c0})$ is applied (this is denoted as the $i=1$ event). Then, the scalar function

$$\mathscr{V}=\sum_{i=1}^{1+n_w+n_v}\int^{\infty}[\mathbf{y}^{i*}(t)\mathbf{Q}\mathbf{y}^i(t)+\mathbf{u}^{i*}(t)\mathbf{R}\mathbf{u}^i(t)]\,dt, \qquad \mathbf{Q}>\mathbf{0}, \quad \mathbf{R}>\mathbf{0}, \tag{8.54}$$

is minimized by the choice of controller parameters $\mathbf{G},\mathbf{A}_c,\mathbf{F}$ satisfying

$$\mathbf{A}_c=\mathbf{A}+\mathbf{BG}-\mathbf{FM}, \tag{8.55a}$$

$$\mathbf{G}=-\mathbf{R}^{-1}\mathbf{B}^*\mathscr{K}_{22}, \qquad \mathbf{0}=\mathscr{K}_{22}\mathbf{A}+\mathbf{A}^*\mathscr{K}_{22}-\mathscr{K}_{22}\mathbf{BR}^{-1}\mathbf{B}^*\mathscr{K}_{22}+\mathbf{C}^*\mathbf{QC}, \tag{8.55b}$$

$$\mathbf{F}=\mathscr{X}_{11}\mathbf{M}^*\mathbf{V}^{-1}, \qquad \mathbf{0}=\mathscr{X}_{11}\mathbf{A}^*+\mathbf{A}\mathscr{X}_{11}-\mathscr{X}_{11}\mathbf{M}^*\mathbf{V}^{-1}\mathbf{M}\mathscr{X}_{11}+\mathbf{x}_0\mathbf{x}_0^*+\mathbf{DWD}^*, \tag{8.55c}$$

and the minimum value of the cost function $\mathscr{V}$ is

$$\begin{aligned}\mathscr{V}&=\operatorname{tr}(\mathscr{X}_{11}+\mathscr{X}_{22})\mathbf{C}^*\mathbf{QC}+\operatorname{tr}\mathscr{X}_{22}\mathbf{G}^*\mathbf{RG} &(8.56a)\\ &=\operatorname{tr}\tilde{\mathscr{K}}\mathbf{FVF}^*+\mathbf{x}_{c0}^*\mathscr{K}_{22}\mathbf{x}_{c0}+\operatorname{tr}[\mathscr{K}_{22}+\tilde{\mathscr{K}}][\mathbf{DWD}^*+\mathbf{x}_0\mathbf{x}_0^*], &(8.56b)\end{aligned}$$

where $\mathscr{X}_{22}$ is computed from

$$\mathbf{0} = \mathscr{X}_{22}(\mathbf{A} + \mathbf{BG})^* + (\mathbf{A} + \mathbf{BG})\mathscr{X}_{22} + \mathbf{FVF}^* + \mathbf{x}_{c0}\mathbf{x}_{c0}^* \tag{8.57}$$

and $\mathscr{K}_{11}$ from

$$\mathbf{0} = \tilde{\mathscr{K}}(\mathbf{A} - \mathbf{FM}) + (\mathbf{A} - \mathbf{FM})^*\tilde{\mathscr{K}} + \mathbf{G}^*\mathbf{RG}, \qquad \mathscr{K}_{11} = \tilde{\mathscr{K}} + \mathscr{K}_{22}. \tag{8.58}$$

$\mathbf{F}$ *is unique if* $(\mathbf{A} - \mathbf{FM}, \mathbf{G})$ *is observable and* $\mathbf{G}$ *is unique if* $(\mathbf{A} + \mathbf{BG}, \mathbf{F})$ *is controllable. We assume that $\mathscr{K}_{22}$ and $\mathscr{X}_{11}$ are unique positive definite solutions of* (8.55).

Proof: It was shown in Chapter 4 that the trajectories $\mathbf{x}(t)$ of $\{\dot{\mathbf{x}} = \mathbf{Ax}, \mathbf{x}(0) = \mathbf{x}_0\}$ and $\{\dot{\mathbf{x}} = \mathbf{Ax} + \mathbf{Dw}, \mathbf{x}(0) = \mathbf{0}\}$ are the same if $\mathbf{Dw}(t) = \mathbf{x}_0\delta(t)$. Hence, in Theorem 8.2 let $\mathbf{D} = \mathbf{x}_0$, $\mathbf{w}(t) = \delta(t)$. This yields $\mathbf{DWD}^* = \mathbf{x}_0\mathbf{x}_0^*$, reducing Theorem 8.2 to the results of the Corollary immediately when $\mathbf{x}_0$ *replaces* the excitation $w_i(t) = \omega_i\delta(t)$ applied one at a time. Now, for the sum over both types of excitation, merely replace $\mathbf{X}_0$ and $\mathbf{X}_{c0}$ in (8.49) by $\mathbf{x}_0\mathbf{x}_0^*$ and $\mathbf{x}_{c0}\mathbf{x}_{c0}^*$, respectively. It is clearly seen from (8.55) and (8.56) that $\mathbf{x}_{c0}$ does not influence the optimal control gains $\mathbf{F}, \mathbf{G}$, and that $\mathscr{V}$ is decreased by the choice $\mathbf{x}_{c0} = \mathbf{0}$. Hence, for these steady-state (infinite time integral) cost functions, there is no advantage in choosing a nonzero initial condition on the controller, and the practical choice of $\mathbf{x}_c(0)$ in both (8.48c) and (8.53) is $\mathbf{x}_c(0) = \mathbf{0}$.

8.3 The Linear Quadratic (LQ) Optimal Measurement Feedback Control

Our next task is to solve the LQI optimal control problem in the case of measurement feedback. That is, minimize

$$\mathscr{V} = \sum_{i=1}^{n_x+n_w} \int_0^\infty \left[\mathbf{y}^{i*}(t)\mathbf{Q}\mathbf{y}^i(t) + \mathbf{u}^{i*}(t)\mathbf{R}\mathbf{u}^i(t)\right] dt, \qquad \mathbf{Q} > \mathbf{0}, \quad \mathbf{R} > \mathbf{0}, \tag{8.59}$$

subject to

$$\dot{\mathbf{x}} = \mathbf{Ax} + \mathbf{Bu} + \mathbf{Dw}, \qquad \mathbf{x}_i(0) = \mathbf{x}_{i0},$$

$$\mathbf{z} = \mathbf{Mx},$$

$$\mathbf{u} = \mathbf{Gz},$$

where the excitations $w_i(t) = \omega_i\delta(t)$, $x_i(0) = x_{i0}$ are applied one at a time as usual.

Then, we have

$$\dot{\mathbf{x}} = \mathbf{Ax} + \mathbf{BG}(\mathbf{Mx}) + \mathbf{Dw}, \tag{8.60}$$

$$\begin{pmatrix} \mathbf{y} \\ \mathbf{u} \end{pmatrix} = \begin{bmatrix} \mathbf{C} \\ \mathbf{GM} \end{bmatrix} \mathbf{x} = \mathscr{C}\mathbf{x},$$

$$\mathscr{V} = \operatorname{tr} \mathscr{X}\mathscr{C}^*\mathscr{Q}\mathscr{C} = \operatorname{tr} \mathscr{X}(\mathbf{C}^*\mathbf{QC} + \mathbf{M}^*\mathbf{G}^*\mathbf{RGM}), \tag{8.61}$$

$$\mathbf{0} = \mathscr{X}(\mathbf{A} + \mathbf{BGM})^* + (\mathbf{A} + \mathbf{BGM})\mathscr{X} + \mathbf{DWD}^* + \mathbf{X}_0. \tag{8.62}$$

This leads to the augmented function

$$\hat{\mathscr{V}} = \operatorname{tr} \mathscr{X}(\mathbf{C}^*\mathbf{QC} + \mathbf{M}^*\mathbf{G}^*\mathbf{RGM})$$
$$+ \operatorname{tr} \mathscr{K}\left[\mathscr{X}(\mathbf{A} + \mathbf{BGM})^* + (\mathbf{A} + \mathbf{BGM})\mathscr{X} + \mathbf{DWD}^* + \mathbf{X}_0\right] \tag{8.62}$$

and the necessary conditions

$$\frac{\partial \hat{\mathscr{V}}}{\partial \mathbf{G}} = \mathbf{0} = 2\mathbf{RGM}\mathscr{X}\mathbf{M}^* + 2\mathbf{B}^*\mathscr{K}\mathscr{X}\mathbf{M}^*, \tag{8.63}$$

$$\frac{\partial \hat{\mathscr{V}}}{\partial \mathscr{X}} = \mathbf{0} = \mathscr{K}(\mathbf{A} + \mathbf{BGM}) + (\mathbf{A} + \mathbf{BGM})^*\mathscr{K} + \mathbf{C}^*\mathbf{QC} + \mathbf{M}^*\mathbf{G}^*\mathbf{RGM}, \tag{8.64}$$

$$\frac{\partial \hat{\mathscr{V}}}{\partial \mathscr{K}} = \mathbf{0} = \mathscr{X}(\mathbf{A} + \mathbf{BGM})^* + (\mathbf{A} + \mathbf{BGM})\mathscr{X} + \mathbf{DWD}^* + \mathbf{X}_0. \tag{8.65}$$

This leads to the following summary.

Theorem 8.3

The measurement feedback LQ problem described by (8.59) *and* (8.60) *has a unique solution if*

(a) $\mathbf{M}$ *has linearly independent rows,* $\mathbf{R} > \mathbf{0}$,

(b) *The controllable modes of* $(\mathbf{A} + \mathbf{BGM}, \mathbf{D})$ *and the observable modes of* $(\mathbf{A} + \mathbf{BGM}, \mathbf{C})$ *are asymptotically stable,*

(c) $[\mathbf{A} + \mathbf{BGM}]$ *has no eigenvalues symmetric about the imaginary axis.*

The solution must satisfy

$$\mathbf{G} = -\mathbf{R}^{-1}\mathbf{B}^*\mathscr{K}\mathscr{X}\mathbf{M}^*(\mathbf{M}\mathscr{X}\mathbf{M}^*)^{-1}, \tag{8.66}$$

$$\mathbf{0} = \mathscr{K}(\mathbf{A} + \mathbf{BGM}) + (\mathbf{A} + \mathbf{BGM})^*\mathscr{K} + \mathbf{M}^*\mathbf{G}^*\mathbf{RGM} + \mathbf{C}^*\mathbf{QC}, \tag{8.67}$$

$$\mathbf{0} = \mathscr{X}(\mathbf{A} + \mathbf{BGM})^* + (\mathbf{A} + \mathbf{BGM})\mathscr{X} + \mathbf{DWD}^* + \mathbf{X}_0, \tag{8.68}$$

and the minimum value of the cost function is

$$\mathscr{V} = \operatorname{tr} \mathscr{X}(\mathbf{C}^*\mathbf{Q}\mathbf{C} + \mathbf{M}^*\mathbf{G}^*\mathbf{R}\mathbf{G}\mathbf{M})$$

$$= \operatorname{tr} \mathscr{K}(\mathbf{D}\mathbf{W}\mathbf{D}^* + \mathbf{X}_0). \tag{8.69}$$

If $w(t) = \mathbf{0}$, set $\mathbf{W} = \mathbf{0}$. If $\mathbf{x}(0) = \mathbf{0}$, then set $\mathbf{X}_0 = \mathbf{0}$. If $x_i(0)$ are all applied simultaneously, then set $\mathbf{X}_0 = \mathbf{x}_0\mathbf{x}_0^*$.

The uniqueness follows from conditions for unique solutions of Liapunov equations (b) and (c), and condition (a) implies that the measurements $z_1(t), z_2(t), \ldots$ are linearly independent. Unfortunately, (8.66) through (8.68) have no analytical solution and numerical iteration is required. The reader may refer to refs. [8.3] and [8.4] for iterative schemes and approximate methods for solving (8.66) through (8.68).

A special case of Theorem 8.3 is much simpler to solve. We define this special case when $\mathbf{u} = \mathbf{G}\mathbf{x}$ as the "state feedback control."

Corollary to Theorem 8.3

The optimal LQ state feedback control is

$$\mathbf{u} = \mathbf{G}\mathbf{x}, \qquad \mathbf{G} = -\mathbf{R}^{-1}\mathbf{B}^*\mathbf{K}, \tag{8.70a}$$

$$\mathbf{0} = \mathbf{K}\mathbf{A} + \mathbf{A}^*\mathbf{K} - \mathbf{K}\mathbf{B}\mathbf{R}^{-1}\mathbf{B}^*\mathbf{K} + \mathbf{C}^*\mathbf{Q}\mathbf{C}, \tag{8.70b}$$

$$\mathscr{V} = \operatorname{tr} \mathbf{K}(\mathbf{D}\mathbf{W}\mathbf{D}^* + \mathbf{X}_0) = \operatorname{tr} \mathbf{X}(\mathbf{C}^*\mathbf{Q}\mathbf{C} + \mathbf{G}^*\mathbf{R}\mathbf{G}), \tag{8.70c}$$

where

$$\mathbf{0} = \mathbf{X}(\mathbf{A} + \mathbf{B}\mathbf{G})^* + (\mathbf{A} + \mathbf{B}\mathbf{G})\mathbf{X} + \mathbf{D}\mathbf{W}\mathbf{D}^* + \mathbf{X}_0 \tag{8.70d}$$

and the cost function is

$$\mathscr{V} = \sum_{i=1}^{n_x + n_w} \int_0^\infty [\mathbf{y}^*(t)\mathbf{Q}\mathbf{y}(t) + \mathbf{u}^*(t)\mathbf{R}\mathbf{u}(t)]^{(i)}\, dt, \tag{8.70e}$$

where $x_i(0)$, $w_i(t) = \omega_i\delta(t)$ are applied one at a time. If the initial states are applied simultaneously, then replace $\mathbf{X}_0$ by $\mathbf{x}_0\mathbf{x}_0^*$. If $\mathbf{w}(t) = \mathbf{0}$, then set $\mathbf{W} = \mathbf{0}$. If $\mathbf{x}(0) = \mathbf{0}$, then set $\mathbf{X}_0 = \mathbf{0}$.

Proof: Setting $\mathbf{M} = \mathbf{I}$ reduces (8.66) to (8.70a) and reduces (8.67) to (8.70b). Equation (8.68) is not needed in the optimal control but may be useful in cost function calculations. The proof of the corollary has verified the replacement of $\mathbf{D}\mathbf{W}\mathbf{D}$ by $\mathbf{x}_0\mathbf{x}_0^*$ when initial conditions rather than impulsive disturbances excite the system. ■

The optimal control problems of Theorem 8.3 and its corollary are called LQ rather than LQI problems, since they have meaning whether the excitation comes from initial state or impulsive disturbances.

Note that equations of the form (8.70b) have appeared numerous times in (8.12b), (8.40b), (8.49a), and (8.55b). These are called Riccati equations after Count Riccati, who first studied their properties in the scalar case. In the matrix case, equation (8.70b) is sometimes called the matrix Riccati equation, but we shall not add the matrix adjective. It should be clear whether the equation is a scalar or a matrix. We shall call (8.70b) the Riccati equation, and its properties deserve careful study.

Theorem 8.4

If

$$\mathbf{K} = \int_0^\infty e^{(\mathbf{A}+\mathbf{BG})^* t}(\mathbf{C}^*\mathbf{QC} + \mathbf{G}^*\mathbf{RG})\, e^{(\mathbf{A}+\mathbf{BG})t}\, dt, \qquad \mathbf{G} = -\mathbf{R}^{-1}\mathbf{B}^*\mathbf{K}, \tag{8.71}$$

exists, then $\mathbf{K}$ *satisfies*

$$\mathbf{0} = \mathbf{KA} + \mathbf{A}^*\mathbf{K} - \mathbf{KBR}^{-1}\mathbf{B}^*\mathbf{K} + \mathbf{C}^*\mathbf{QC}. \tag{8.72}$$

(a) $\mathbf{K}$ *exists iff there are no modes of* $(\mathbf{A}, \mathbf{B}, \mathbf{C})$ *that are simultaneously*

- *Observable*
- *Uncontrollable*
- *Unstable*

(b) *If* $\mathbf{K}$ *exists, then* $\mathbf{K} > \mathbf{0}$ *iff* $(\mathbf{A}, \mathbf{C})$ *is an observable pair.*

(c) *If* $(\mathbf{A}, \mathbf{C})$ *is detectable and* $(\mathbf{A}, \mathbf{B})$ *stabilizable, then* $[\mathbf{A} + \mathbf{BG}]$ *is asymptotically stable, where*

$$\mathbf{G} = -\mathbf{R}^{-1}\mathbf{B}^*\mathbf{K}. \tag{8.73}$$

(d) *Under the condition of* (c), $\mathbf{K}$ *is the unique positive semidefinite solution of* (8.72).

Proof:

(a) Suppose that $\mathbf{K}$ exists but there are modes that are simultaneously observable, uncontrollable, unstable. Then,

$$\mathscr{V} = \int_0^\infty (\mathbf{y}^*\mathbf{Qy} + \mathbf{u}^*\mathbf{Ru})\, dt = \mathbf{x}_0^*\mathbf{Kx}_0 < \infty, \tag{8.74}$$

and some unstable mode is observable in $\mathbf{y}$. Furthermore, this mode is uncontrollable. Hence, for *any* control $\mathbf{y}(t)^*\mathbf{Qy}(t) \to \infty$. But this contradicts $\mathscr{V} < \infty$. This proves that $\{\infty > \mathbf{K}\}$ implies {no observable, unstable uncontrollable modes}. Now, suppose the converse—suppose there are now modes that are observable, unstable, uncontrollable, but $\mathbf{K}$ does not exist. Then,

$$\mathscr{V} = \mathbf{x}_0^* \mathbf{K} \mathbf{x}_0 = \infty \quad (\mathbf{x}_0 \text{ finite}). \tag{8.75}$$

But since all unstable modes are controllable by assumption, there exists some control such that $\mathbf{x}(t) \to \mathbf{0}$. This control obviously yields $\mathscr{V} < \infty$, which contradicts the given $\mathscr{V} = \infty$ for the optimal control. This proves that {no observable, unstable, uncontrollable modes} implies $\{\mathbf{K} < \infty\}$.

(b) Suppose $\{\mathbf{x}_0^* \mathbf{K} \mathbf{x}_0 > 0\}$ but $(\mathbf{A}, \mathbf{C})$ is not observable. Then, $\{\mathbf{x}_0^* \mathbf{K} \mathbf{x}_0 = \int_0^\infty (\mathbf{y}^* \mathbf{Q} \mathbf{y} + \mathbf{u}^* \mathbf{R} \mathbf{u})\, dt > 0\}$ but $\{\mathbf{y}(t) \equiv \mathbf{0}$ for some $\mathbf{x}_0 \neq \mathbf{0}\}$. Hence, for this $\mathbf{x}_0$, $\mathbf{x}_0^* \mathbf{K} \mathbf{x}_0 = \int_0^\infty \mathbf{u}^* \mathbf{R} \mathbf{u}\, dt$, since $\mathbf{y}(t) \equiv \mathbf{0}$. But this is minimized by choosing $\mathbf{u}(t) \equiv \mathbf{0}$, which implies that $\mathscr{V} = \int_0^\infty \mathbf{u}^* \mathbf{R} \mathbf{u}\, dt = 0$ which contradicts $\mathscr{V} > 0$. This proves that $\{\mathbf{K} > \mathbf{0}\}$ implies $\{(A, \mathbf{C})$ observable$\}$. Now, suppose the converse. Suppose $(\mathbf{A}, \mathbf{C})$ is observable but $\mathbf{x}_0^* \mathbf{K} \mathbf{x}_0 = 0$ for some $\mathbf{x}_0 \not\equiv \mathbf{0}$. Since $(\mathbf{A}, \mathbf{C})$ is observable, $\mathbf{x}_0 \not\equiv \mathbf{0} \to \mathbf{y}(t) \neq \mathbf{0}$ for $t > 0$. Now, since $\mathbf{y}(t) \not\equiv \mathbf{0}$, then $\int_0^\infty (\mathbf{y}^* \mathbf{Q} \mathbf{y} + \mathbf{u}^* \mathbf{R} \mathbf{u})\, dt = \mathbf{x}_0^* \mathbf{K} \mathbf{x}_0 > 0$, which contradicts the assumption $\mathbf{x}_0^* \mathbf{K} \mathbf{x}_0 = 0$. This proves that $\{(\mathbf{A}, \mathbf{C})$ observable$\}$ implies $\{\mathbf{K} > \mathbf{0}\}$.

(c) Choose

$$\mathscr{V}(t) \triangleq \int_t^\infty (\mathbf{y}^* \mathbf{Q} \mathbf{y} + \mathbf{u}^* \mathbf{R} \mathbf{u})\, dt = \mathbf{x}^*(t) \mathbf{K} \mathbf{x}(t) \tag{8.76}$$

as a Liapunov function. Then, for asymptotic stability, Theorem 7.4 requires $\{\mathscr{V} > 0,\ \dot{\mathscr{V}} \leq 0$, and $\dot{\mathscr{V}} \equiv 0$ implies $x = 0\}$. Clearly, from (b), $\mathscr{V} > 0$ is established if $(\mathbf{A}, \mathbf{C})$ is observable, and

$$\dot{\mathscr{V}} = -\mathscr{y}^* \mathscr{Q} \mathscr{y}, \quad \mathscr{y}^* \triangleq (\mathbf{y}^* \mathbf{u}^*),$$

$$\dot{\mathbf{x}} = \mathscr{A} \mathbf{x}, \qquad \mathscr{A} = \mathbf{A} + \mathbf{B}\mathbf{G}, \qquad \mathscr{Q} = \begin{bmatrix} \mathbf{Q} & \mathbf{0} \\ \mathbf{0} & \mathbf{R} \end{bmatrix}. \tag{8.77}$$

$$\mathscr{y} = \mathscr{C} \mathbf{x}, \qquad \mathscr{C}^* = (\mathbf{C}^* \quad \mathbf{G}^*),$$

which is at least a negative semidefinite function of $\mathbf{x}$, and $\dot{\mathscr{V}} \equiv 0$ is equivalent to $\mathscr{y}(t) \equiv \mathbf{0}$. Hence, asymptotic stability is established if $\{\mathscr{y}(t) \equiv \mathbf{0}$ implies $\mathbf{x} = \mathbf{0}\}$, but this is just the observability condition on $(\mathscr{A}, \mathscr{C})$ in (8.77). Therefore, $\mathscr{A}$ is asymptotically stable if $(\mathbf{A}, \mathbf{C})$ and $(\mathscr{A}, \mathscr{C})$ are both observable.

The observability grammians for $(\mathbf{A}, \mathbf{C})$ and $(\mathscr{A}, \mathscr{C})$ are

$$\mathbf{K}_0 = \int_0^\infty e^{\mathbf{A}^* t} \mathbf{C}^* \mathbf{Q} \mathbf{C} e^{\mathbf{A} t}\, dt, \tag{8.78}$$

$$\mathbf{K} = \int_0^\infty e^{\mathscr{A}^* t} \mathscr{C}^* \mathscr{Q} \mathscr{C} e^{\mathscr{A} t}\, dt, \tag{8.79}$$

where $\mathbf{K}_0 > \mathbf{0}$ iff $(\mathbf{A}, \mathbf{C})$ is observable and $\mathbf{K} > \mathbf{0}$ iff $(\mathscr{A}, \mathscr{C})$ is observable. But from (b), $\mathbf{K} > \mathbf{0}$ iff $(\mathbf{A}, \mathbf{C})$ is observable. Hence $\mathbf{K}_0 > \mathbf{0}$ iff $\mathbf{K} > \mathbf{0}$, and therefore $(\mathscr{A}, \mathscr{C})$ is observable iff $(\mathbf{A}, \mathbf{C})$ is observable. This shows that the above conditioning of stability of $\mathscr{A}$ on $(\mathbf{A}, \mathbf{C})$ and $(\mathscr{A}, \mathscr{C})$ observability can now be reduced to just observability of $(\mathbf{A}, \mathbf{C})$. Of course, the optimal control (through $\mathbf{K}$) is only influenced by the modes observable in $\mathbf{y}$. Suppose there are modes which are unobservable in $\mathbf{y}$. For $\mathscr{A}$ to be asymptotically stable, these unobservable modes must be asymptotically stable. This is the definition of "detectability." Hence, we conclude that $\mathscr{A}$ is asymptotically stable if $(\mathbf{A}, \mathbf{C})$ is detectable. The condition for the existence of $\mathbf{K}$ in all of the above discussions is $(\mathbf{A}, \mathbf{B})$ stabilizable (no unstable, uncontrollable modes). Thus, part (c) is proved.

(d) Part (d) follows immediately from Theorem D.1 by substituting into (D6), $\mathbf{A} \to (\mathbf{A} + \mathbf{BG})^*$, $\mathbf{X} \to \mathbf{K}$, $\mathbf{B} \to (\mathbf{A} + \mathbf{BG})$, $\mathbf{Q} \to \mathbf{C}^*\mathbf{QC} + \mathbf{G}^*\mathbf{RG}$, and noting that $\lambda_i \neq -\lambda_j$ for any i, j due to the stability of $(\mathbf{A} + \mathbf{BG})$ from part (c). ■

Exercise 8.4

Show that the control that minimizes

$$\mathscr{V} = \int_0^\infty (\mathbf{y}^*\mathbf{u}^*) \begin{bmatrix} \mathbf{Q} & \mathbf{N} \\ \mathbf{N}^* & \mathbf{R} \end{bmatrix} \begin{pmatrix} \mathbf{y} \\ \mathbf{u} \end{pmatrix} dt, \qquad \begin{bmatrix} \mathbf{Q} & \mathbf{N} \\ \mathbf{N}^* & \mathbf{R} \end{bmatrix} > \mathbf{0},$$

subject to the constraint

$$\dot{\mathbf{x}} = \mathbf{Ax} + \mathbf{Bu}, \qquad \mathbf{x}(0) = \mathbf{x}_0,$$

$$\mathbf{y} = \mathbf{Cx}$$

satisfies

$$\mathbf{u} = -\mathbf{R}^{-1}(\mathbf{B}^*\mathbf{K} + \mathbf{N}^*\mathbf{C})\mathbf{x},$$

where $\mathbf{K}$ satisfies

$$\mathbf{0} = \mathbf{K}(\mathbf{A} - \mathbf{BR}^{-1}\mathbf{N}^*\mathbf{C}) + (\mathbf{A} - \mathbf{BR}^{-1}\mathbf{N}^*\mathbf{C})^*\mathbf{K} - \mathbf{KBR}^{-1}\mathbf{B}^*\mathbf{K} + \mathbf{C}^*[\mathbf{Q} - \mathbf{NR}^{-1}\mathbf{N}^*]\mathbf{C}.$$

8.4 *Modal Methods for Solving Riccati Equations*

Rearrange (8.72) in the form $-\mathbf{C}^*\mathbf{QC} - \mathbf{A}^*\mathbf{K} = \mathbf{K}(\mathbf{A} - \mathbf{BR}^{-1}\mathbf{B}^*\mathbf{K})$ and add the identity $\mathbf{A} - \mathbf{BR}^{-1}\mathbf{B}^*\mathbf{K} = \mathbf{A} + \mathbf{BG}$ to obtain

$$\begin{bmatrix} \mathbf{A} - \mathbf{BR}^{-1}\mathbf{B}^*\mathbf{K} \\ -\mathbf{C}^*\mathbf{QC} - \mathbf{A}^*\mathbf{K} \end{bmatrix} = \begin{bmatrix} \mathbf{A} + \mathbf{BG} \\ \mathbf{K}(\mathbf{A} + \mathbf{BG}) \end{bmatrix} = \begin{bmatrix} \mathbf{I} \\ \mathbf{K} \end{bmatrix} (\mathbf{A} + \mathbf{BG}). \tag{8.80}$$

The left-hand side further factors into

$$\begin{bmatrix} \mathbf{A} & -\mathbf{BR}^{-1}\mathbf{B}^* \\ -\mathbf{C}^*\mathbf{QC} & -\mathbf{A}^* \end{bmatrix}\begin{bmatrix} \mathbf{I} \\ \mathbf{K} \end{bmatrix} = \begin{bmatrix} \mathbf{I} \\ \mathbf{K} \end{bmatrix}(\mathbf{A} + \mathbf{BG}). \tag{8.81}$$

The spectral decomposition of $\mathbf{A} + \mathbf{BG} = \mathbf{E}_1\boldsymbol{\Lambda}_1\mathbf{E}_1^{-1}$ allows

$$\begin{bmatrix} \mathbf{A} & -\mathbf{BR}^{-1}\mathbf{B}^* \\ -\mathbf{C}^*\mathbf{QC} & -\mathbf{A}^* \end{bmatrix}\begin{bmatrix} \mathbf{I} \\ \mathbf{K} \end{bmatrix} = \begin{bmatrix} \mathbf{I} \\ \mathbf{K} \end{bmatrix}\mathbf{E}_1\boldsymbol{\Lambda}_1\mathbf{E}_1^{-1} \tag{8.82}$$

or

$$\mathscr{H}\mathbf{E} = \begin{bmatrix} \mathbf{A} & -\mathbf{BR}^{-1}\mathbf{B}^* \\ -\mathbf{C}^*\mathbf{QC} & -\mathbf{A}^* \end{bmatrix}\begin{bmatrix} \mathbf{E}_1 \\ \mathbf{KE}_1 \end{bmatrix} = \begin{bmatrix} \mathbf{E}_1 \\ \mathbf{KE}_1 \end{bmatrix}\boldsymbol{\Lambda}_1 = \mathbf{E}\boldsymbol{\Lambda}_1. \tag{8.83}$$

Now, since $\boldsymbol{\Lambda}_1$ is a Jordan form, the columns of $\mathbf{E}$ must be eigenvectors of $\mathscr{H}$. In fact, the columns of $\mathbf{E}$ are eigenvectors associated with the n_x eigenvalues of $(\mathbf{A} + \mathbf{BG})$. But it is not clear *which* eigenvalues of $\mathscr{H}$ are associated with $(\mathbf{A} + \mathbf{BG})$. Part (c) of Theorem 8.4 provides a clue. If $(\mathbf{A}, \mathbf{B}, \mathbf{C})$ is a stabilizable, detectable triple, then n_x of the *stable* eigenvalues of $\mathscr{H}$ are eigenvalues of $(\mathbf{A} + \mathbf{BG})$. Theorem 8.6 will show us that there can *only* be n_x stable eigenvalues of $\mathscr{H}$. In anticipation of that result, we state the following.

Theorem 8.5

If $(\mathbf{A}, \mathbf{B})$ is stabilizable and $(\mathbf{A}, \mathbf{C})$ is detectable, then the solution of the nonlinear equation

$$\mathbf{0} = \mathbf{KA} + \mathbf{A}^*\mathbf{K} - \mathbf{KBR}^{-1}\mathbf{B}^*\mathbf{K} + \mathbf{C}^*\mathbf{QC} \tag{8.84}$$

is

$$\mathbf{K} = \mathbf{E}_2\mathbf{E}_1^{-1}, \tag{8.85}$$

where $\mathbf{E}_1, \mathbf{E}_2$ are obtained from the eigenvector calculation

$$\mathscr{H}\mathbf{E} = \begin{bmatrix} \mathbf{A} & -\mathbf{BR}^{-1}\mathbf{B}^* \\ -\mathbf{C}^*\mathbf{QC} & -\mathbf{A}^* \end{bmatrix}\begin{bmatrix} \mathbf{E}_1 \\ \mathbf{E}_2 \end{bmatrix} = \begin{bmatrix} \mathbf{E}_1 \\ \mathbf{E}_2 \end{bmatrix}\boldsymbol{\Lambda}_1 = \mathbf{E}\boldsymbol{\Lambda}_1, \tag{8.86}$$

where $\boldsymbol{\Lambda}_1$ contains the left half-plane eigenvalues of the $2n_x \times 2n_x$ matrix $\mathscr{H}$.

Now, we must show that only n_x eigenvalues of $\mathscr{H}$ are in the left half-plane.

Theorem 8.6

The $2n_x$ eigenvalues of $\mathscr{H}$ are

$$\begin{aligned} &\lambda_i[\mathbf{A} + \mathbf{BG}], \quad && i = 1, \ldots, n_x, \\ &\lambda_{i+n_x} = -\lambda_i, \quad && i = 1, \ldots, n_x. \end{aligned} \tag{8.87}$$

Proof: The similarity transform of $\mathscr{H}$ given by $\mathbf{T}^{-1}\mathscr{H}\mathbf{T}$, where

$$\mathbf{T} = \begin{bmatrix} \mathbf{I} & \mathbf{0} \\ \mathbf{K} & \mathbf{I} \end{bmatrix}, \qquad \mathbf{T}^{-1} = \begin{bmatrix} \mathbf{I} & \mathbf{0} \\ -\mathbf{K} & \mathbf{I} \end{bmatrix}, \tag{8.88}$$

is

$$\mathbf{T}^{-1}\mathscr{H}\mathbf{T} = \begin{bmatrix} \mathbf{A} - \mathbf{B}\mathbf{R}^{-1}\mathbf{B}^*\mathbf{K} & -\mathbf{B}\mathbf{R}^{-1}\mathbf{B}^* \\ -\mathbf{K}\mathbf{A} - \mathbf{C}^*\mathbf{Q}\mathbf{C} + \mathbf{K}\mathbf{B}\mathbf{R}^{-1}\mathbf{B}^*\mathbf{K} - \mathbf{A}^*\mathbf{K} & \mathbf{K}\mathbf{B}\mathbf{R}^{-1}\mathbf{B}^* - \mathbf{A}^* \end{bmatrix}. \tag{8.89}$$

But the lower left-hand corner matrix is zero by (8.84). Hence, the eigenvalues of $\mathscr{H}$ are those of $[\mathbf{A} - \mathbf{B}\mathbf{R}^{-1}\mathbf{B}^*\mathbf{K}]$ and $[\mathbf{K}\mathbf{B}\mathbf{R}^{-1}\mathbf{B}^* - \mathbf{A}^*]$, and the theorem is proved since for any matrix, $\lambda_i[\mathscr{A}] = \bar{\lambda}_i[\mathscr{A}^*]$ and $\lambda_i[-\mathscr{A}] = -\lambda_i[\mathscr{A}]$, and λ_i occur in complex conjugate pairs. ■

The main conclusion from Theorems 8.5 and 8.6 is that the stable eigenvalues of $\mathscr{H}$ are associated with the matrix $\mathbf{A} + \mathbf{B}\mathbf{G}$ if $(\mathbf{A}, \mathbf{B}, \mathbf{C})$ is stabilizable and detectable. The modal calculations of Theorem 8.5 are presented for insight into the modal behavior of optimal systems. Also, the method (8.85) requires great care when the Jordan form of $[\mathbf{A} + \mathbf{B}\mathbf{G}]$ is not diagonal since generalized eigenvectors have to be calculated in this case. There are other approaches to the numerical solution of Riccati equations which are more efficient and accurate. See refs. [8.5] and [8.6].

8.5 Root Locus of the LQ Optimal State Feedback Controller

The characteristic polynomial of the matrix $\mathscr{H}$ is also of interest. We shall need the identities

1. $\begin{vmatrix} \mathbf{A} & \mathbf{B} \\ \mathbf{C} & \mathbf{D} \end{vmatrix} = |\mathbf{A}|\,|\mathbf{D} - \mathbf{C}\mathbf{A}^{-1}\mathbf{B}|$
2. $|\mathbf{I} + \mathbf{M}\mathbf{N}| = |\mathbf{I} + \mathbf{N}\mathbf{M}|$
3. $|s\mathbf{I} + \mathbf{A}^T| = (-1)^{n_x}|-s\mathbf{I} - \mathbf{A}^T| = (-1)^{n_x}|-s\mathbf{I} - \mathbf{A}|$

Now, label the characteristic polynomial of $\mathscr{H}$ as $\Delta(s)$,

$$\Delta(s) = |s\mathbf{I} - \mathscr{H}| = \begin{vmatrix} s\mathbf{I} - \mathbf{A} & \mathbf{B}\mathbf{R}^{-1}\mathbf{B}^T \\ \mathbf{C}^T\mathbf{Q}\mathbf{C} & s\mathbf{I} + \mathbf{A}^T \end{vmatrix}. \tag{8.90}$$

But from identity **1**,

$$\begin{aligned} \Delta(s) &= |s\mathbf{I} - \mathbf{A}|\,\left|s\mathbf{I} + \mathbf{A}^T - \mathbf{C}^T\mathbf{Q}\mathbf{C}(s\mathbf{I} - \mathbf{A})^{-1}\mathbf{B}\mathbf{R}^{-1}\mathbf{B}^T\right| \\ &= |s\mathbf{I} - \mathbf{A}|\,\left|\mathbf{I} - \mathbf{C}^T\mathbf{Q}\mathbf{C}(s\mathbf{I} - \mathbf{A})^{-1}\mathbf{B}\mathbf{R}^{-1}\mathbf{B}^T(s\mathbf{I} + \mathbf{A}^T)^{-1}\right|\,|s\mathbf{I} + \mathbf{A}^T|. \end{aligned}$$

Now, use identity **3**:

$$\Delta(s) = |s\mathbf{I} - \mathbf{A}|\,| - s\mathbf{I} - \mathbf{A}|(-1)^{n_x}\left|\mathbf{I} - \mathbf{C}^T\mathbf{Q}\mathbf{C}(s\mathbf{I} - \mathbf{A})^{-1}\mathbf{B}\mathbf{R}^{-1}\mathbf{B}^T(s\mathbf{I} + \mathbf{A}^T)^{-1}\right|.$$

Now, use identity **2**:

$$\Delta(s) = (-1)^{n_x}|s\mathbf{I} - \mathbf{A}|\,| - s\mathbf{I} - \mathbf{A}|\left|\mathbf{I} - \mathbf{Q}\mathbf{C}(s\mathbf{I} - \mathbf{A})^{-1}\mathbf{B}\mathbf{R}^{-1}\mathbf{B}^T(s\mathbf{I} + \mathbf{A}^T)^{-1}\mathbf{C}^T\right|.$$

Noting that

$$\mathscr{G}(s) \triangleq \mathbf{C}(s\mathbf{I} - \mathbf{A})^{-1}\mathbf{B}, \qquad \Delta_A(s) \triangleq |s\mathbf{I} - \mathbf{A}|$$

and, from identity **3**,

$$\mathscr{G}^T(-s) = \mathbf{B}^T(-s\mathbf{I} - \mathbf{A})^{-T}\mathbf{C}^T = \frac{\mathbf{B}^T\left[\operatorname{adj}(-s\mathbf{I} - \mathbf{A})^T\right]\mathbf{C}^T}{|-s\mathbf{I} - \mathbf{A}|}$$

$$= \frac{(-1)^{n_x-1}\mathbf{B}^T\left[\operatorname{adj}(s\mathbf{I} + \mathbf{A})^T\right]\mathbf{C}^T}{(-1)^{n_x}|s\mathbf{I} + \mathbf{A}^T|} = -\mathbf{B}^T(s\mathbf{I} + \mathbf{A}^T)^{-1}\mathbf{C}^T,$$

$\Delta(s)$ can be written in the form

$$\Delta(s) = (-1)^{n_x}\Delta_A(s)\Delta_A(-s)\left|\mathbf{I} + \mathbf{Q}\mathscr{G}(s)\mathbf{R}^{-1}\mathscr{G}^T(-s)\right|. \tag{8.91}$$

Note that the characteristic polynomial $\Delta(s)$ is dictated by choice of $\mathbf{Q}, \mathbf{R}$.

Scalar Input / Scalar Output (SISO) Systems

We press the result (8.91) a bit further for scalar input/scalar output (SISO) systems. In this case (8.91) becomes

$$\Delta(s) = (-1)^{n_x}\Delta_A(s)\Delta_A(-s)\left|1 + \frac{Q}{R}\mathscr{G}(s)\mathscr{G}^T(-s)\right|. \tag{8.92}$$

But

$$\mathscr{G}(s) = \frac{Z(s)}{\Delta_A(s)}, \qquad Z(s) = \mathbf{C}[\operatorname{adj}(s\mathbf{I} - \mathbf{A})]\mathbf{B}. \tag{8.93}$$

Note that the degree of the polynomial $Z(s)$ is equal to or less than $n_x - 1$, since the adjoint matrix is composed of determinants with one less column and row. Hence, (8.92) becomes

$$(-1)^{n_x}\Delta(s) = \Delta_A(s)\Delta_A(-s) + \frac{Q}{R}Z(s)Z(-s), \tag{8.94}$$

where the first term is of degree $2n_x$ and the second term is of degree $\leq 2(n_x - 1)$. The question of interest now is the location of the eigenvalues of $[\mathbf{A} + \mathbf{BG}]$ as $Q/R = (0 \to \infty)$.

The Q / R → 0 Case:
As $Q/R \to 0$, (8.94) becomes

$$\Delta_0(s) = (-1)^{n_x}\Delta_A(s)\Delta_A(-s). \tag{8.95}$$

Hence, the optimal poles are the *stable* roots of (8.95).

EXAMPLE 8.3
Determine the location of the optimal poles for

$$\mathscr{V} = \int_0^\infty (y^2 + \rho u^2)\, dt,$$

$$\dot{\mathbf{x}} = \begin{bmatrix} -1 & & \\ & 2 & \\ & & -3 \end{bmatrix}\mathbf{x} + \begin{pmatrix} 1 \\ 1 \\ 1 \end{pmatrix} u, \qquad \mathbf{x}(0) = \mathbf{x}_0,$$

$$y = [1 \quad 1 \quad 1]\mathbf{x},$$

in the limit as $\rho \to \infty$. This approaches a "minimal energy control" because by comparison the weight on $\mathbf{y}$ is arbitrarily small.

Solution: Since the system is stabilizable and detectable, (8.95) gives all eigenvalues of $\mathscr{H}$ and the eigenvalues of $\mathbf{A} + \mathbf{BG}$ are the left half-plane eigenvalues of $\mathscr{H}$. Hence, from (8.95),

$$\Delta(s) = (-1)^3(s+1)(s-2)(s+3)(s-1)(s+2)(s-3).$$

Therefore, the eigenvalues of $\mathbf{A} + \mathbf{BG}$ are $-1, -2, -3$, but only two of these are the same as the open-loop eigenvalues. (The solid X's in Fig. 8.3 are the

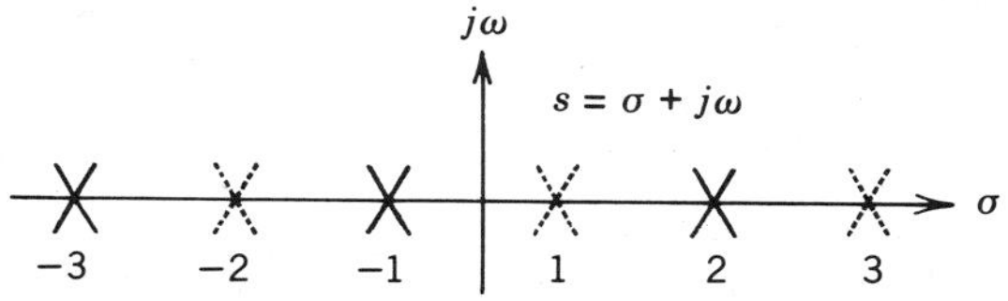

Figure 8.3 *Eigenvalues of* $\mathscr{H}$

open-loop eigenvalues). The other (-2) is the *mirror image* of the unstable open-loop pole. The mirror images are shown dotted in Fig. 8.3. ■

The $Q/R \to \infty$ Case:
Multiply the characteristic polynomial (8.94) by R/Q to obtain

$$(-1)^{n_x}\frac{R}{Q}\Delta(s) = 0 = \frac{R}{Q}\Delta_A(s)\Delta_A(-s) + Z(s)Z(-s) \tag{8.96}$$

and let $R/Q \to 0$. Now, for *finite* values of s, this limit yields

$$0 = Z(s)Z(-s).$$

Now, see that

$$\begin{aligned}\Delta_A(s)\Delta_A(-s)\big|_{|s|\to\infty} &= \left(s^{n_x} + \cdots\ a_1 s + a_0\right)\left((-s)^{n_x} + \cdots\ a_1(-s) + a_0\right)\Big|_{\infty}\\ &= (-1)^{n_x}s^{2n_x},\\ Z(s)Z(-s)\big|_{\infty} &= \left(\alpha_m s^m + \cdots\ \alpha_1 s + \alpha_0\right)\left(\alpha_m(-s)^m + \cdots\ \alpha_1(-s) + \alpha_0\right)\Big|_{\infty}\\ &= (-1)^m s^{2m}\alpha_m^2.\end{aligned}$$

Then, for *infinite* values of the complex number s, (8.96) becomes

$$0 = \frac{R}{Q}(-1)^{n_x}s^{2n_x} + s^{2m}(-1)^m\alpha_m^2. \tag{8.97}$$

Recall from the definition (8.93) that $m \le n_x - 1$. Multiplying (8.97) by s^{-2m} leads to

$$s^{2(n_x-m)} = -\frac{Q}{R}\alpha_m^2(-1)^{n_x-m} = \frac{Q}{R}\alpha_m^2(-1)^{n_x-m+1} = \frac{Q}{R}\alpha_m^2 e^{j180k(r+1)} \tag{8.98}$$

whose LHP roots are listed in Table 8.2, with $r = n_x - m$, $k =$ odd integer. The plots of these roots in Table 8.2 yield the patterns in Fig. 8.4, which are called Butterworth configurations [8.7].

We now summarize these results.

Theorem 8.7
Let $\mathbf{B}$ *and* $\mathbf{C}^*$ *both be* $n_x \times 1$ *matrices with* $(\mathbf{A}, \mathbf{B})$ *stabilizable and* $(\mathbf{A}, \mathbf{C})$ *detectable. Let* $\mathbf{K}$ *be the positive semidefinite or positive definite solution of* $\mathbf{0} = \mathbf{KA} + \mathbf{A}^*\mathbf{K} - \mathbf{KBR}^{-1}\mathbf{B}^*\mathbf{K} + \mathbf{C}^*Q\mathbf{C}$,

TABLE 8.2 Roots of Equation (8.98)

$r = n_x - m$	$\|s\|$	ANGLE OF s (LHP)
1	$\sqrt{\frac{Q}{R}}\,\alpha_m$	$+180°$
2	$\left[\frac{Q}{R}\alpha_m^2\right]^{1/4}$	$\pm 135°$
3	$\left[\frac{Q}{R}\alpha_m^2\right]^{1/6}$	$\pm 120°,\ +180°$
4	$\left[\frac{Q}{R}\alpha_m^2\right]^{1/8}$	$\pm 112.5°,\ \pm 157.5°$
r	$\left[\frac{Q}{R}\alpha_m^2\right]^{1/2r}$	$180°\left(\frac{r+1}{2r}\right)k$, k odd

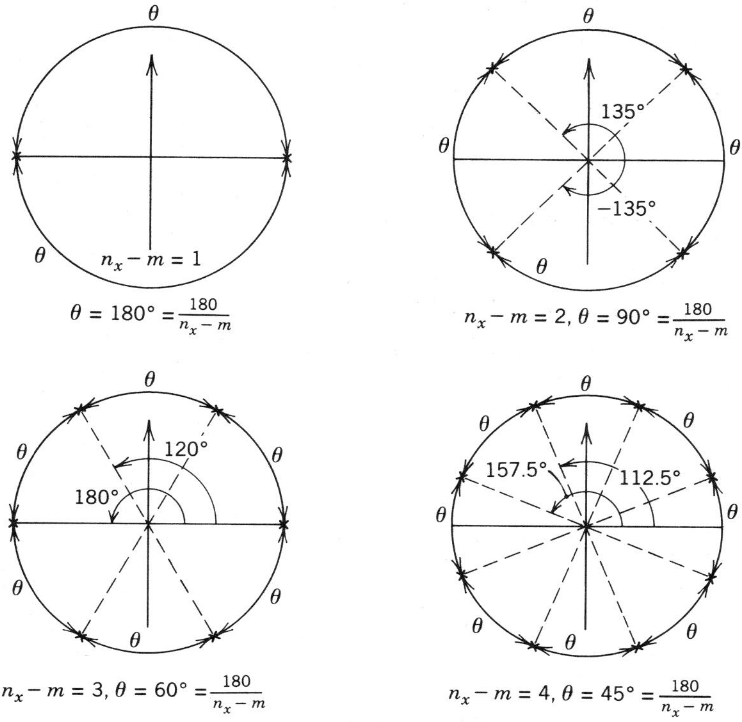

Figure 8.4 *Butterworth Pole Configurations*

and $\mathbf{G} = -\mathbf{R}^{-1}\mathbf{B}^*\mathbf{K}$. *Define*

$$\mathcal{G}(s) = \mathbf{C}[s\mathbf{I} - \mathbf{A}]^{-1}\mathbf{B} = \frac{Z(s)}{\Delta_A(s)},$$

where $m \triangleq$ *degree of* $Z(s) \triangleq \alpha_m s^m + \cdots + \alpha_1 s + \alpha_0$, *and* $n_x \triangleq$ *degree of* $\Delta_A(s)$. *Then, the eigenvalues of* $(\mathbf{A} + \mathbf{BG})$ *go to the left half-plane roots of* $\Delta_\Delta(s)\Delta_A(-s)$ *in the limit as* $Q/R \to 0$. *Alternately, in the limit as* $R/Q \to 0$, *then* m *of the eigenvalues of* $(\mathbf{A} + \mathbf{BG})$ *go to the left half-plane roots of* $Z(s)Z(-s)$, *whereas the remaining* $(n_x - m)$ *eigenvalues of* $\mathbf{A} + \mathbf{BG}$ *go to infinity in the manner*

$$s \simeq \left[\frac{Q}{R}\alpha_m^2\right]^{1/2r} e^{j90k(r+1)r^{-1}}, \qquad k = odd, \quad r = n_x - m.$$

EXAMPLE 8.4

Determine the location of the optimal poles for

$$\mathscr{V} = \int_0^\infty (y^2 + \rho u^2)\, dt,$$

$$\mathcal{G}(s) = \frac{(s+1)(s-1)}{(s+2)^2}$$

in the limit as $\rho \to 0$.

Solution: In this problem, $Z(s) = (s+1)(s-1)$. Hence, the poles of the closed-loop optimal state feedback system are the stable roots of

$$\Delta_\infty(s) = Z(s)Z(-s) = (s+1)(s-1)(s-1)(s+1),$$

or $s = -1, -1$, which are not exactly at the location of the open zeros, but at the mirror image of any right half-plane zeros. ■

8.6 Nyquist Plot and Stability Margin of the LQ Controller

Here, we continue our study of the state feedback system $\dot{\mathbf{x}} = [\mathbf{A} + \mathbf{BG}]\mathbf{x}$, where $\mathbf{G} = -\mathbf{R}^{-1}\mathbf{B}^*\mathbf{K}$, $\mathbf{0} = \mathbf{KA} + \mathbf{A}^*\mathbf{K} - \mathbf{KBR}^{-1}\mathbf{B}^*\mathbf{K} + \mathbf{C}^*\mathbf{QC}$. We wish to show that such control designs have an infinite gain margin and a phase margin $\geq 60°$. Toward this end, write the Riccati equation in the form

$$\mathbf{K}[-\mathbf{A} + \lambda\mathbf{I}] + [-\mathbf{A}^* - \lambda\mathbf{I}]\mathbf{K} + \mathbf{KBR}^{-1}\mathbf{B}^*\mathbf{K} = \mathbf{C}^*\mathbf{QC}.$$

Multiply from the left by $\mathbf{R}^{-1/2}\mathbf{B}^*[-\mathbf{A}^* - \lambda\mathbf{I}]^{-1}$ and from the right by $[-\mathbf{A} +$

$\lambda \mathbf{I}]^{-1}\mathbf{B}\mathbf{R}^{-1/2}$ to obtain

$$\mathbf{R}^{-1/2}\mathbf{B}^*[-\mathbf{A}^* - \lambda\mathbf{I}]^{-1}\mathbf{K}\mathbf{B}\mathbf{R}^{-1/2} + \mathbf{R}^{-1/2}\mathbf{B}^*\mathbf{K}[-\mathbf{A} + \lambda\mathbf{I}]^{-1}\mathbf{B}\mathbf{R}^{-1/2}$$

$$+\mathbf{R}^{-1/2}\mathbf{B}^*[-\mathbf{A}^* - \lambda\mathbf{I}]^{-1}\mathbf{K}\mathbf{B}\mathbf{R}^{-1}\mathbf{B}^*\mathbf{K}[-\mathbf{A} + \lambda\mathbf{I}]^{-1}\mathbf{B}\mathbf{R}^{-1/2}$$

$$= \mathbf{R}^{-1/2}\mathbf{B}^*[-\mathbf{A}^* - \lambda\mathbf{I}]^{-1}\mathbf{C}^*\mathbf{Q}\mathbf{C}[-\mathbf{A} + \lambda\mathbf{I}]^{-1}\mathbf{B}\mathbf{R}^{-1/2}.$$

Note that $-\mathbf{K}\mathbf{B}\mathbf{R}^{-1/2} = \mathbf{G}^*\mathbf{R}^{1/2}$. Add $\mathbf{I}$ to each side of the equation. Then,

$$\mathbf{I} - \mathbf{R}^{-1/2}\mathbf{B}^*[-\mathbf{A}^* - \lambda\mathbf{I}]^{-1}\mathbf{G}^*\mathbf{R}^{1/2} - \mathbf{R}^{1/2}\mathbf{G}[-\mathbf{A} + \lambda\mathbf{I}]^{-1}\mathbf{B}\mathbf{R}^{-1/2}$$

$$+\mathbf{R}^{-1/2}\mathbf{B}^*[-\mathbf{A}^* - \lambda\mathbf{I}]^{-1}\mathbf{G}^*\mathbf{R}\mathbf{G}[-\mathbf{A} + \lambda\mathbf{I}]^{-1}\mathbf{B}\mathbf{R}^{-1/2}$$

$$= \mathbf{I} + \mathbf{R}^{-1/2}\mathbf{B}^*[-\mathbf{A}^* - \lambda\mathbf{I}]^{-1}\mathbf{C}^*\mathbf{Q}\mathbf{C}[-\mathbf{A} + \lambda\mathbf{I}]^{-1}\mathbf{B}\mathbf{R}^{-1/2},$$

which factors into the form

$$\left[\mathbf{I} - \mathbf{R}^{-1/2}\mathbf{B}^*[-\mathbf{A}^* - \lambda\mathbf{I}]^{-1}\mathbf{G}^*\mathbf{R}^{1/2}\right]\left[\mathbf{I} - \mathbf{R}^{1/2}\mathbf{G}[-\mathbf{A} + \lambda\mathbf{I}]^{-1}\mathbf{B}\mathbf{R}^{-1/2}\right]$$

$$= \mathbf{I} + \mathbf{R}^{-1/2}\mathbf{B}^*[-\mathbf{A} - \lambda\mathbf{I}]^{-1}\mathbf{C}^*\mathbf{Q}\mathbf{C}[-\mathbf{A} + \lambda\mathbf{I}]^{-1}\mathbf{B}\mathbf{R}^{-1/2}.$$

Let $\lambda = j\omega$:

$$\left[\mathbf{I} - \mathbf{R}^{-1/2}\mathbf{B}^*[-\mathbf{A}^* - j\omega\mathbf{I}]^{-1}\mathbf{G}^*\mathbf{R}^{1/2}\right]\left[\mathbf{I} - \mathbf{R}^{1/2}\mathbf{G}[-\mathbf{A} + j\omega\mathbf{I}]^{-1}\mathbf{B}\mathbf{R}^{-1/2}\right]$$

$$= \mathbf{I} + \mathbf{R}^{-1/2}\mathbf{B}^*[-\mathbf{A} - j\omega\mathbf{I}]^{-1}\mathbf{C}^*\mathbf{Q}\mathbf{C}[-\mathbf{A} + j\omega\mathbf{I}]^{-1}\mathbf{B}\mathbf{R}^{-1/2}$$

$$= [\mathbf{I} + \mathbf{H}^*(j\omega)\mathbf{H}(j\omega)],$$

where $\mathbf{H}^*(j\omega) \triangleq \mathbf{R}^{-1/2}\mathbf{B}^*[-\mathbf{A} - j\omega\mathbf{I}]^{-1}\mathbf{C}^*\mathbf{Q}^{1/2}$. Now, since $\mathbf{H}^*(j\omega)\mathbf{H}(j\omega) \geq \mathbf{0}$, we have the inequality

$$[\mathbf{I} + \mathbf{L}^*(j\omega)][\mathbf{I} + \mathbf{L}(j\omega)] \geq \mathbf{I}, \qquad \mathbf{L}^*(j\omega) \triangleq -\mathbf{R}^{-1/2}\mathbf{B}^*[-\mathbf{A}^* - j\omega\mathbf{I}]^{-1}\mathbf{G}^*\mathbf{R}^{1/2}.$$

Now, for the single-input/output case, we must have

$$|1 + L(j\omega)| \geq 1. \tag{8.99}$$

To interpret this inequality for the feedback system of Fig. 8.5, plot these vectors $1 + L(j\omega)$ and $L(j\omega)$ on the Nyquist plot Fig. 8.6 to see that (8.99) implies that the Nyquist plot for the optimal state feedback system never penetrates the unit circle centered around the $-1 + j0$ point.

Note that the phase margin in Fig. 8.6 is defined at the point where the unit circle (dotted and centered at the origin) intersects the polar plot of the open-loop transfer

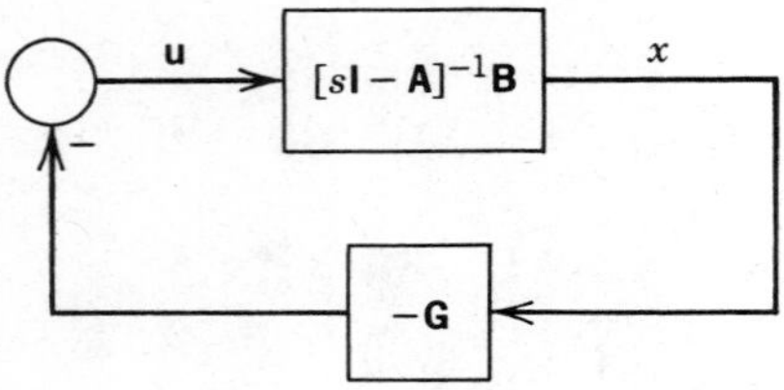

Figure 8.5 State Feedback Optimal Control Loop

function $L(j\omega)$. Let $x_1 + jy_1$ describe points on the unit circle centered at -1. Hence, $(x_1 + 1)^2 + y_1^2 = 1$. Let $x_2 + jy_2$ describe points on the unit circle centered at the origin. Hence, $x_2^2 + y_2^2 = 1$. At the intersection of these two circles, $x_1 = x_2$, $y_1 = y_2$, satisfying

$$\left(x_1^2 + 1\right)^2 + y_1^2 = x_2^2 + y_2^2 = x_1^2 + y_1^2,$$

which leads to $x_1 = x_2 = -1/2$, $y_1 = y_2 = \sqrt{3}/2$. The minimum possible phase margin is therefore described by the angle $\phi_m = \tan^{-1}(\sqrt{3/2}/1/2) = 60°$; see Fig. 8.7. Such phase margins are not guaranteed for *any* of the other controllers of Chapter 8. However parameter sensitivity is equally as important as phase margin, and this property can be minimized by LQI controllers, as will be shown in Chapter 10.

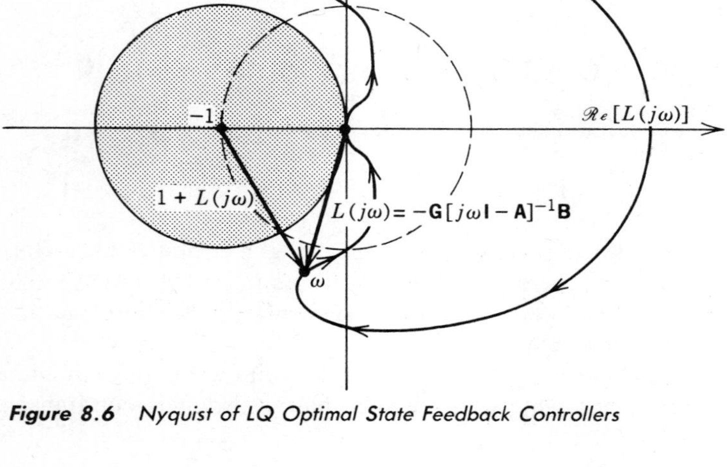

Figure 8.6 Nyquist of LQ Optimal State Feedback Controllers

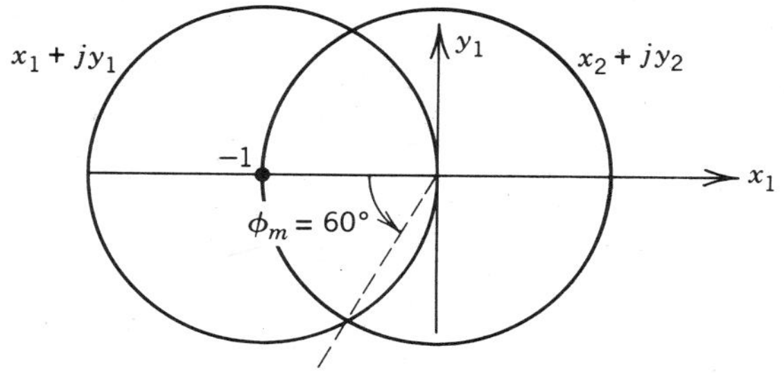

Figure 8.7 *Minimum Phase Margin of State Feedback Controllers*

EXAMPLE 8.5

Consider the space backpack system described by (3.23) and Fig. 3.2. The transfer function from the input $T_c(s)$ to the output $\dot{\theta}(s) \triangleq \mathscr{L}[\dot{\theta}(t)]$ is

$$\dot{\theta}(s) = \left[\frac{-s^3\overline{\mathscr{M}}_{13}\left(s^2\overline{\mathscr{M}}_{22} + s\overline{\mathscr{C}}_{22} + \overline{\mathscr{K}}_{22}\right)}{s^4\overline{\mathscr{M}}_{13}\left[\left(s^2\overline{\mathscr{M}}_{22} + s\overline{\mathscr{C}}_{22} + \overline{\mathscr{K}}_{22}\right)\left(\overline{\mathscr{M}}_{11} - \overline{\mathscr{M}}_{33}\right) - s^2\overline{\mathscr{M}}_{12}^2\right]}\right]T_c(s),$$

which clearly displays three pole-zero cancellations at the origin. This means that three rigid body modes are either uncontrollable or unobservable or both. These correspond to ψ, $\dot{\psi}$, and θ, as may be seen by inspection of (3.23). Note that ψ, $\dot{\psi}$, and θ do not appear in the equations of motion (except with zero coefficients), and their deletion will not alter the remaining variables. The state equations for the remaining variables $(\gamma, \dot{\theta}, \dot{\gamma})$ are (using the numerical values given in Fig. 3.2)

$$\dot{\mathbf{x}} = \begin{pmatrix} \dot{\gamma} \\ \ddot{\theta} \\ \ddot{\gamma} \end{pmatrix} = \begin{bmatrix} 0 & 0 & 1.000 \\ 28.564 & 0 & 0.041 \\ -79.285 & 0 & -0.113 \end{bmatrix}\begin{pmatrix} \gamma \\ \dot{\theta} \\ \dot{\gamma} \end{pmatrix}$$

$$+ \begin{bmatrix} 0.000 & 0.000 \\ -0.030 & -0.0016 \\ 0.041 & -0.0047 \end{bmatrix}\begin{pmatrix} T_c + T_w \\ f + f_w \end{pmatrix} = \mathbf{Ax} + \mathbf{B(u + w)},$$

where the measurements are

$$\mathbf{z} = \begin{pmatrix} \dot{\theta} + \dot{\theta}_w \\ \gamma + \gamma_w \end{pmatrix} = \begin{bmatrix} 0 & 1 & 0 \\ 1 & 0 & 0 \end{bmatrix}\begin{pmatrix} \gamma \\ \dot{\theta} \\ \dot{\gamma} \end{pmatrix} + \begin{pmatrix} \dot{\theta}_w \\ \gamma_w \end{pmatrix} = \mathbf{Mx} + \mathbf{v}.$$

The outputs we wish to control are $\dot{\theta}$ and γ,

$$\mathbf{y} = \begin{pmatrix} \dot{\theta} \\ \gamma \end{pmatrix}. \tag{8.100}$$

The cost function to be minimized is

$$\mathscr{V} = \sum_{i=1}^{4} \int_0^\infty (\mathbf{y}^{i*}\mathbf{Q}\mathbf{y}^i + \mathbf{u}^{i*}\mathbf{R}\mathbf{u}^i)\,dt, \qquad \mathbf{Q} = \mathbf{I}, \quad \mathbf{R} = \rho\mathbf{I}, \tag{8.101}$$

where the four impulsive inputs are T_w, f_w, $\dot{\theta}_w$, and γ_w. From Theorem 8.2, the optimal control is

$$\begin{aligned}
\mathbf{u} &= \begin{pmatrix} T_c \\ f \end{pmatrix} = -\mathbf{R}^{-1}\mathbf{B}^*\mathbf{K}\mathbf{x}_c, & \mathbf{0} &= \mathbf{K}\mathbf{A} + \mathbf{A}^*\mathbf{K} - \mathbf{K}\mathbf{B}\mathbf{R}^{-1}\mathbf{B}^*\mathbf{K} + \mathbf{C}^*\mathbf{Q}\mathbf{C}, \\
\dot{\mathbf{x}}_c &= \mathbf{A}_c\mathbf{x}_c + \mathbf{F}\mathbf{z}, & \mathbf{A}_c &= \mathbf{A} + \mathbf{B}\mathbf{G} - \mathbf{F}\mathbf{M}, \\
\mathbf{F} &= \mathbf{P}\mathbf{M}^*\mathbf{V}^{-1}, & \mathbf{0} &= \mathbf{P}\mathbf{A}^* + \mathbf{A}\mathbf{P} - \mathbf{P}\mathbf{M}^*\mathbf{V}^{-1}\mathbf{M}\mathbf{P} + \mathbf{D}\mathbf{W}\mathbf{D}^*,
\end{aligned} \tag{8.102}$$

where, in this example, $\mathbf{D} = \mathbf{B}$, $\mathbf{Q} = \mathbf{I}$, $\mathbf{R} = \rho\mathbf{I}$, $\mathbf{W} = \mathbf{I}$, $\mathbf{V} = \mathbf{I}$. This yields, for this example,

$$\mathscr{V}_y = \sum_{i=1}^{4} \int_0^\infty \mathbf{y}^{i*}\mathbf{y}^i\,dt, \qquad \mathscr{V}_u = \sum_{i=1}^{4} \int_0^\infty \mathbf{u}^{i*}\mathbf{u}^i\,dt, \tag{8.103}$$

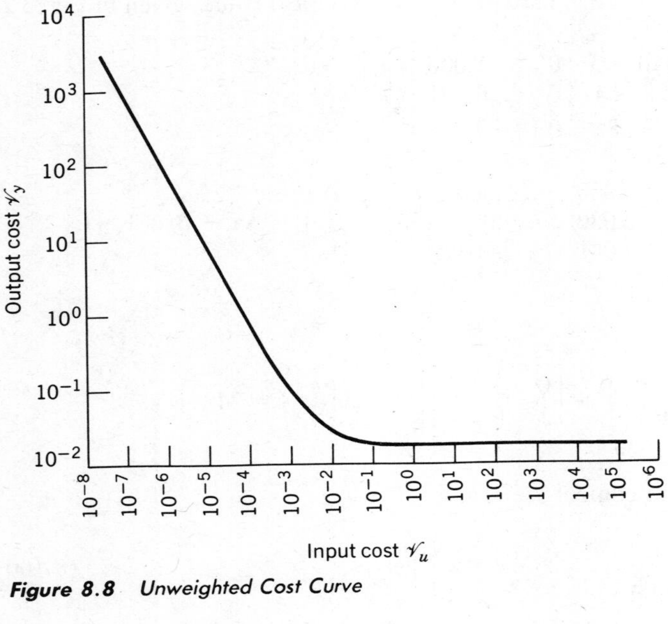

Figure 8.8 *Unweighted Cost Curve*

TABLE 8.3 LQI Design for $\rho = 1.0000d - 10$

$$\mathbf{K} = \begin{bmatrix} 5.2250d-02 & 1.8049d-03 & 1.3620d-03 \\ 1.8049d-03 & 4.6471d-04 & 1.0050d-04 \\ 1.3620d-03 & 1.0050d-04 & 7.5457d-05 \end{bmatrix}$$

$$\mathbf{G} = \begin{bmatrix} 1.1061d+04 & 9.9240d+04 & -4.6078d+02 \\ 9.3536d+04 & 1.2302d+04 & 5.1903d+03 \end{bmatrix}$$

$$\mathbf{P} = \begin{bmatrix} 9.3510d-05 & -7.2371d-06 & 4.3982d-09 \\ -7.2371d-06 & 1.6794d-02 & -2.6711d-03 \\ 4.3982d-09 & -2.6711d-03 & 7.4139d-03 \end{bmatrix}$$

$$\mathbf{F} = \begin{bmatrix} -7.2371d-06 & 9.3510d-05 \\ 1.6794d-02 & -7.2371d-06 \\ -2.6711d-03 & 4.3982d-09 \end{bmatrix}$$

$$\mathbf{A} + \mathbf{BG} = \begin{bmatrix} 0.\quad d+00 & 0.\quad d+00 & 1.0000d+00 \\ 2.1000d+02 & -3.0151d+03 & 5.4922d+00 \\ -9.7107d+02 & 3.9917d+03 & -4.3356d+01 \end{bmatrix}$$

Eigenvalues of $\mathbf{A} + \mathbf{BG}$:

$-1.8045d+01 \quad +1.9125d+01i$
$-1.8045d+01 \quad -1.9125d+01i$
$-3.0224d+03 \quad +0.\quad d+00i$

$$\mathbf{A} - \mathbf{FM} = \begin{bmatrix} -9.3510d-05 & 7.2371d-06 & 1.0000d+00 \\ 2.8564d+01 & -1.6794d-02 & 4.0806d-02 \\ -7.9285d+01 & 2.6711d-03 & -1.1326d-01 \end{bmatrix}$$

Eigenvalues of $\mathbf{A} - \mathbf{FM}$:

$-5.7160d-02 \quad +8.9040d+00i$
$-5.7160d-02 \quad -8.9040d+00i$
$-1.5832d-02 \quad +0.\quad d+00i$

$$\mathbf{A} + \mathbf{BG} - \mathbf{FM} = \begin{bmatrix} -9.3510d-05 & 7.2371d-06 & 1.0000d+00 \\ 2.1000d+02 & -3.0151d+03 & 5.4922d+00 \\ -9.7107d+02 & 3.9917d+03 & -4.3356d+01 \end{bmatrix}$$

Eigenvalues of $\mathbf{A} + \mathbf{BG} - \mathbf{FM}$:

$-1.8045d+01 \quad +1.9125d+01i$
$-1.8045d+01 \quad -1.9125d+01i$
$-3.0224d+03 \quad +0.\quad d+00i$

$\mathcal{V}_y = 1.6888d-02$ (output cost)
$\mathcal{V}_u = 6.7977d+02$ (input cost)
$\mathcal{V}^{\text{act 1}} = -8.6041d-03$ (torquer cost)
$\mathcal{V}^{\text{act 2}} = -3.6305d-04$ (thruster cost)
$\mathcal{V}^{\text{sens 1}} = 7.9205d-03$ (pitch rate sensor cost)
$\mathcal{V}^{\text{sens 2}} = 4.3277d-07$ (leg angle sensor cost)

which are related by $\mathscr{V} = \mathscr{V}_y + \rho\mathscr{V}_u$ and are computed from

$$\mathscr{V}_y = \mathrm{tr}\,(\mathbf{P} + \mathbf{X}_c)\mathbf{C}^*\mathbf{C}, \qquad \mathscr{V}_u = \mathrm{tr}\,\mathbf{X}_c\mathbf{G}^*\mathbf{G}, \tag{8.104}$$

where $\mathbf{X}_c$ satisfies

$$\mathbf{0} = \mathbf{X}_c(\mathbf{A} + \mathbf{BG})^* + (\mathbf{A} + \mathbf{BG})\mathbf{X}_c + \mathbf{FVF}^*. \tag{8.105}$$

To simplify notation from Theorem 8.2, we have defined $\mathbf{X}_c \triangleq \mathscr{X}_{22}$, $\mathbf{P} \triangleq \mathscr{X}_{11}$, $\mathbf{K} \triangleq \mathscr{K}_{22}$.

For this example, the plot of $\mathscr{V}_y$ versus $\mathscr{V}_u$ appears in Fig. 8.8 on a log–log scale. Note that the curve is approximated by two straight lines on this log–log scale. The small $\mathscr{V}_u$ portion of the curve is approximated by a straight line with a slope of -1 (decade of $\mathscr{V}_y$ per decade of $\mathscr{V}_u$), and the large $\mathscr{V}_u$ portion of the curve is governed by

$$\min_{\rho} \mathscr{V}_y = \mathrm{tr}\,\mathbf{PC}^*\mathbf{C}. \tag{8.106}$$

TABLE 8.4 LQI Design for $\rho = 1.000d-02$

$$\mathbf{K} = \begin{bmatrix} 3.6980d+01 & 1.7870d-01 & 7.0681d-02 \\ 1.7870d-01 & 6.3155d+00 & 2.2750d+00 \\ 7.0681d-02 & 2.2750d+00 & 1.2860d+00 \end{bmatrix}$$

$$\mathbf{G} = \begin{bmatrix} 2.5088d-01 & 9.7770d+00 & 1.6184d+00 \\ 6.2393d-02 & 2.1000d+00 & 9.7613d-01 \end{bmatrix}$$

$$\mathbf{A} + \mathbf{BG} = \begin{bmatrix} 0.\quad d+00 & 0.\quad d+00 & 1.0000d+00 \\ 2.8556d+01 & -2.9849d-01 & -9.6285d-03 \\ -7.9275d+01 & 3.8907d-01 & -5.1819d-02 \end{bmatrix}$$

Eigenvalues of $\mathbf{A} + \mathbf{BG}$:

$-9.5974d-02 \quad +8.9025d+00i$
$-9.5974d-02 \quad -8.9025d+00i$
$-1.5836d-01 \quad +0.\quad d+00i$

$$\mathbf{A} + \mathbf{BG} - \mathbf{FM} = \begin{bmatrix} -9.3510d-05 & 7.2371d-06 & 1.0000d+00 \\ 2.8556d+01 & -3.1529d-01 & -9.6285d-03 \\ -7.9275d+01 & 3.9174d-01 & -5.1819d-02 \end{bmatrix}$$

Eigenvalues of $\mathbf{A} + \mathbf{BG} - \mathbf{FM}$:

$-9.6498d-02 \quad +8.9023d+00i$
$-9.6498d-02 \quad -8.9023d+00i$
$-1.7420d-01 \quad +0\quad d+00i$

$\mathscr{V}_y = 1.7682d-02$ (output cost)
$\mathscr{V}_u = 7.9237d-02$ (input cost)
$\mathscr{V}^{\text{act 1}} = -9.3669d-03$ (torquer cost)
$\mathscr{V}^{\text{act 2}} = -3.9807d-04$ (thruster cost)
$\mathscr{V}^{\text{sens 1}} = 7.9172d-03$ (pitch rate sensor cost)
$\mathscr{V}^{\text{sens 2}} = 1.1311d-07$ (leg angle sensor cost)

See from (8.102) that **P** and **C** are independent of ρ and that X_c is a positive semidefinite or positive definite matrix. Hence, the lowest bound on $\mathscr{V}_y$ is given by the value

$$\min_{\rho} \operatorname{tr} [\mathbf{P} + \mathbf{X}_c(\rho)] \mathbf{C}^* \mathbf{C} = \operatorname{tr} \mathbf{P} \mathbf{C}^* \mathbf{C}, \tag{8.107}$$

which represents the best output performance available from the optimal control theory. Using the notation $i = \sqrt{-1}$, $d - 03 \triangleq \times 10^{-3}$, Table 8.3 lists the controller design data for $\rho = 10^{-10}$, Table 8.4 lists the data for $\rho = 10^{-2}$, Table 8.5 lists the data for $\rho = 1.0$, Table 8.6 lists the data for $\rho = 10^2$, and Table 8.7 lists the data for $\rho = 10^{10}$. Included in these tables are the input costs (review Section 4.5.2) for inputs $\dot{\theta}_w$ and γ_w (denoted as "pitch rate sensor cost" and "leg angle sensor cost"). In every case, the pitch rate sensor is a more critical sensor since its effectiveness (input cost) is at least four orders of magnitude greater than the effectiveness of the leg angle sensor. This suggests that the γ sensor could be deleted

TABLE 8.5 LQI Design for $\rho = 1.0000d + 00$

$$\mathbf{K} = \begin{bmatrix} 4.9612d+01 & 3.3592d-01 & 1.2733d-01 \\ 3.3592d-01 & 6.3163d+01 & 2.2756d+01 \\ 1.2733d-01 & 2.2756d+01 & 8.8240d+00 \end{bmatrix}$$

$$\mathbf{G} = \begin{bmatrix} 4.9423d-03 & 9.7769d-01 & 3.2670d-01 \\ 1.1468d-03 & 2.1005d-01 & 7.8620d-02 \end{bmatrix}$$

$$\mathbf{A} + \mathbf{BG} = \begin{bmatrix} 0. \quad d+00 & 0. \quad d+00 & 1.0000d+00 \\ 2.8564d+01 & -2.9849d-02 & 3.0818d-02 \\ -7.9285d+01 & 3.8906d-02 & -1.0030d-01 \end{bmatrix}$$

Eigenvalues of **A** + **BG**:

$-5.7160d-02 \quad +8.9040d+00i$
$-5.7160d-02 \quad -8.9040d+00i$
$-1.5832d-02 \quad +0. \quad d+00i$

$$\mathbf{A} + \mathbf{BG} - \mathbf{FM} = \begin{bmatrix} -9.3510d-05 & 7.2371d-06 & 1.0000d+00 \\ 2.8564d+01 & -4.6644d-02 & 3.0818d-02 \\ -7.9285d+01 & 4.1578d-02 & -1.0030d-01 \end{bmatrix}$$

Eigenvalues of **A** + **BG** − **FM**:

$-5.7688d-02 \quad +8.9040d+00i$
$-5.7688d-02 \quad -8.9040d+00i$
$-3.1664d-02 \quad +0. \quad d+00i$

$\mathscr{V}_y = 2.4809d-02$ (output cost)
$\mathscr{V}_u = 7.9161d-03$ (input cost)
$\mathscr{V}^{\text{act }1} = -1.6180d-02$ (torquer cost)
$\mathscr{V}^{\text{act }2} = -7.1245d-04$ (thruster cost)
$\mathscr{V}^{\text{sens }1} = 7.9161d-03$ (pitch rate sensor cost)
$\mathscr{V}^{\text{sens }2} = 3.8807d-09$ (leg angle sensor cost)

TABLE 8.6 LQI Design for $\rho = 1.0000d + 02$

$$\mathbf{K} = \begin{bmatrix} 4.9842d+01 & 3.5783d-01 & 1.3522d-01 \\ 3.5783d-01 & 6.3163d+02 & 2.2756d+02 \\ 1.3522d-01 & 2.2756d+02 & 8.2611d+01 \end{bmatrix}$$

$$\mathbf{G} = \begin{bmatrix} 5.2814d-05 & 9.7769d-02 & 3.4967d-02 \\ 1.2196d-05 & 2.1005d-02 & 7.5970d-03 \end{bmatrix}$$

$$\mathbf{A} + \mathbf{BG} = \begin{bmatrix} 0. \quad d+00 & 0. \quad d+00 & 1.0000d+00 \\ 2.8564d+01 & -2.9849d-03 & 3.9738d-02 \\ -7.9285d+01 & 3.8906d-03 & -1.1187d-01 \end{bmatrix}$$

Eigenvalues of **A** + **BG**:

$-5.6637d-02 \quad +8.9040d+00i$
$-5.6637d-02 \quad -8.9040d+00i$
$-1.5832d-03 \quad +0. \quad d+00i$

$$\mathbf{A} + \mathbf{BG} - \mathbf{FM} = \begin{bmatrix} -9.3510d-05 & 7.2371d-06 & 1.0000d+00 \\ 2.8564d+01 & -1.9779d-02 & 3.9738d-02 \\ -7.9285d+01 & 6.5618d-03 & -1.1187d-01 \end{bmatrix}$$

Eigenvalues of **A** + **BG** − **FM**:

$-5.7165d-02 \quad +8.9040d+00i$
$-5.7165d-02 \quad -8.9040d+00i$
$-1.7415d-02 \quad +0. \quad d+00i$

$\mathscr{V}_y = 9.6053d-02$ (output cost)
$\mathscr{V}_u = 7.9161d-04$ (input cost)
$\mathscr{V}^{\text{act 1}} = -8.4282d-02$ (torquer cost)
$\mathscr{V}^{\text{act 2}} = -3.8557d-03$ (thruster cost)
$\mathscr{V}^{\text{sens 1}} = 7.9161d-03$ (pitch rate sensor cost)
$\mathscr{V}^{\text{sens 2}} = 1.9244d-09$ (leg angle sensor cost)

without significant change in performance (if the system remains observable with a deletion of the γ sensor). The reader can verify that this is the case.

The reader should plot the eigenvalues of **A** + **BG** to see that they approach a Butterworth pattern as ρ decreases. Verify that as ρ decreases, the complex eigenvalue pair (of **A** + **BG**) converges to a damping of $\zeta = 0.707$. The values in Tables 8.3 through 8.7 labeled $\mathscr{V}^{\text{act 1}}, \mathscr{V}^{\text{act 2}}$ are computed by $\mathscr{V}^{\text{act } i} = \mathscr{V}_{u_i} - \mathscr{V}_{w_i}$, $i = 1, 2$, where $\mathscr{V}_{u_i}$ is the output cost of (8.46) for the output u_i and $\mathscr{V}_{w_i}$ is the input cost of (8.46) for the input w_i. The calculation $\mathscr{V}^{\text{act } i}$ serves as a measure of effectiveness for the actuator, with control signal $u_i(t)$ and disturbance signal (often called a "noise" signal) $w_i(t)$ emerging from the ith actuator. The total signal coming from the ith actuator is $u_i + w_i$, where $\mathscr{V}_{u_i}/\mathscr{V}_{w_i}$ is effectively the "signal-to-noise" ratio for the actuator. Hence, subtracting the "bad" contribution $\mathscr{V}_{w_i}$ from the "good" contribution yields the actuator effectiveness $\mathscr{V}^{\text{act } i} = \mathscr{V}_{u_i} - \mathscr{V}_{w_i}$, which utilizes the input/output cost analysis of Section 4.5.2. According to Tables 8.3 through 8.7 the torquer actuator has the largest noise contribution (note the larger negative value for thruster cost), and a redesign would suggest improving the

TABLE 8.7 LQI Design for $\rho = 1.0000d + 10$

$$\mathbf{K} = \begin{bmatrix} 4.9844d+01 & 3.6027d-01 & 1.3610d-01 \\ 3.6027d-01 & 6.3386d+06 & 2.2836d+06 \\ 1.3610d-01 & 2.2836d+06 & 8.2272d+05 \end{bmatrix}$$

$$\mathbf{G} = \begin{bmatrix} 5.3193d-13 & 9.8114d-06 & 3.5348d-06 \\ 1.2278d-13 & 2.1079d-06 & 7.5941d-07 \end{bmatrix}$$

$$\mathbf{A} + \mathbf{BG} = \begin{bmatrix} 0.\quad d+00 & 0.\quad d+00 & 1.0000d+00 \\ 2.8564d+01 & -2.9954d-07 & 4.0806d-02 \\ -7.9285d+01 & 3.9044d-07 & -1.1326d-01 \end{bmatrix}$$

Eigenvalues of $\mathbf{A} + \mathbf{BG}$:

$-5.6632d-02 \quad +8.9040d+00i$

$-5.6632d-02 \quad -8.9040d+00i$

$-1.5888d-07 \quad +0.\quad d+00i$

$$\mathbf{A} + \mathbf{BG} - \mathbf{FM} = \begin{bmatrix} -9.3510d-05 & 7.2371d-06 & 1.0000d+00 \\ 2.8564d+01 & -1.6795d-02 & 4.0806d-02 \\ -7.9285d+01 & 2.6715d-03 & -1.1326d-01 \end{bmatrix}$$

Eigenvalues of $\mathbf{A} + \mathbf{BG} - \mathbf{FM}$:

$-5.7160d-02 \quad +8.9040d+00i$

$-5.7160d-02 \quad -8.9040d+00i$

$-1.5832d-02 \quad +0.\quad d+00i$

$\mathscr{V}_y = 7.8884d+02$ (output cost)

$\mathscr{V}_u = 7.9440d-08$ (input cost)

$\mathscr{V}^{\text{act 1}} = -7.5936d+02$ (torquer cost)

$\mathscr{V}^{\text{act 2}} = -3.5049d+01$ (thruster cost)

$\mathscr{V}^{\text{sens 1}} = 7.9721d-03$ (pitch rate sensor cost)

$\mathscr{V}^{\text{sens 2}} = 1.9209d-09$ (leg angle sensor cost)

quality of this torquer so that it contains smaller disturbances, and would suggest that reducing the disturbances in the torquer actuator is more effective than a redesign that reduces the disturbances in the thruster. Reference [8.9] describes the sensor and actuator costs more completely.

Exercise 8.5

Rather than using the plant matrix $\mathbf{A}$, suppose the optimal LQ or LQI problem is solved for a "biased" plant matrix $\hat{\mathbf{A}} = \mathbf{A} + \eta\mathbf{I}$ for some constant $\eta > 0$. The optimal theory places the eigenvalues

$$\lambda_i[\hat{\mathbf{A}} + \mathbf{BG}], \quad \lambda_i[\hat{\mathbf{A}} - \mathbf{FM}], \qquad i = 1, 2, \ldots, n_x,$$

in the LHP if $(\mathbf{A}, \mathbf{B}, \mathbf{C})$ and $(\mathbf{A}, \mathbf{D}, \mathbf{M})$ are both stabilizable, detectable triples. Show that the controller

$$\mathbf{H}(s) = \mathbf{G}\left[s\mathbf{I} - (\hat{\mathbf{A}} + \mathbf{BG} - \mathbf{FM})\right]^{-1}\mathbf{F}$$

applied to the system

$$\dot{\mathbf{x}} = \mathbf{Ax} + \mathbf{Bu} + \mathbf{Dw}, \quad \mathbf{y} = \mathbf{Cx}, \quad \mathbf{z} = \mathbf{Mx} + \mathbf{v}$$

will guarantee that all $2n_x$ eigenvalues of the closed-loop systems will lie to the left of the vertical line $-\eta + j\omega$, $-\infty \leq \omega \leq \infty$. Also, show in this LQ problem that the Liapunov function $\mathscr{V}(\mathbf{x}(t)) = \mathbf{x}^*(t)\mathbf{Kx}(t)$ has the property $\dot{\mathscr{V}}/\mathscr{V} \leq -2\eta$.

8.7 Optimal Control of Vector Second-Order Systems

Consider the task of determining the optimal state feedback control to minimize

$$\mathscr{V} = \int_0^\infty (\mathbf{y}^*\mathbf{Qy} + \mathbf{u}^*\mathbf{Ru})\, dt \tag{8.108}$$

subject to

$$\begin{aligned} \mathscr{M}\ddot{\mathbf{q}} + \mathscr{D}\dot{\mathbf{q}} + \mathscr{K}\mathbf{q} &= \mathscr{B}\mathbf{u}, \qquad \mathbf{q}(0) = \mathbf{q}_0, \quad \dot{\mathbf{q}}(0) = \dot{\mathbf{q}}_0, \\ \mathbf{y} &= \mathscr{P}\mathbf{q} + \mathscr{R}\dot{\mathbf{q}}, \\ \mathbf{u} &= \mathbf{G}_1\mathbf{q} + \mathbf{G}_2\dot{\mathbf{q}}. \end{aligned} \tag{8.109}$$

This class of problems occurs so frequently in acoustics, electronic circuits, and mechanical systems that it is worth special attention. It is, of course, a special case of the general theory of the previous sections, but these sections did not take advantage of any special properties of the matrices (**A**, **B**, **C**, **D**, **M**). We shall look at nongyroscopic systems ($\mathscr{D} = \mathscr{D}^T$). We assume $\mathscr{M} = \mathscr{M}^T > \mathbf{0}$, $\mathscr{K} = \mathscr{K}^T \geq \mathbf{0}$.

See that (8.109) may be put into the state form

$$\begin{aligned} \dot{\mathbf{x}} &= \mathbf{Ax} + \mathbf{Bu}, \qquad \mathbf{x}(0) = \mathbf{x}_0, \\ \mathbf{y} &= \mathbf{Cx}, \\ \mathbf{u} &= \mathbf{Gx} \end{aligned} \tag{8.110a}$$

if

$$\mathbf{A} = \begin{bmatrix} \mathbf{0} & \mathbf{I} \\ -\Omega^2 & -\mathscr{D}' \end{bmatrix}, \qquad \mathbf{B} = \begin{bmatrix} \mathbf{0} \\ \mathscr{B}_0 \end{bmatrix},$$
$$\mathbf{C} = [\mathscr{P}\mathscr{E} \quad \mathscr{R}\mathscr{E}] \triangleq [\mathscr{P}_0 \quad \mathscr{R}_0], \qquad \mathscr{B}_0 \triangleq \mathscr{E}^*\mathscr{B}, \quad \mathscr{D}' \triangleq \mathscr{E}^*\mathscr{D}\mathscr{E}, \tag{8.110b}$$

where (5.58) holds. That is, $\mathscr{E}^*\mathscr{M}\mathscr{E} = \mathbf{I}$, $\mathscr{E}^*\mathscr{K}\mathscr{E} = \Omega^2$; and, using (8.110b), see that the necessary conditions for optimality of the state feedback linear quadratic (LQ) problem (8.70) reduced to

$$\mathbf{u} = \mathbf{G}_1\boldsymbol{\eta} + \mathbf{G}_2\dot{\boldsymbol{\eta}}, \quad \mathbf{G}_1 = -\mathbf{R}^{-1}\mathscr{B}_0^*\mathbf{K}_{12}^*, \quad \mathbf{G}_2 = -\mathbf{R}^{-1}\mathscr{B}_0^*\mathbf{K}_{22}, \quad \mathbf{q} = \mathscr{E}\boldsymbol{\eta}, \tag{8.111a}$$

$$\mathbf{0} = -\mathbf{K}_{12}\boldsymbol{\Omega}^2 - \boldsymbol{\Omega}^2\mathbf{K}_{12} - \mathbf{K}_{12}\mathscr{B}_0\mathbf{R}^{-1}\mathscr{B}_0^*\mathbf{K}_{12} + \mathscr{P}_0^*\mathbf{Q}\mathscr{P}_0, \tag{8.111b}$$

$$\mathbf{0} = \mathbf{K}_{11} - \mathbf{K}_{12}\mathscr{D}' - \boldsymbol{\Omega}^2\mathbf{K}_{22} - \mathbf{K}_{12}\mathscr{B}_0\mathbf{R}^{-1}\mathscr{B}_0^*\mathbf{K}_{22} + \mathscr{P}_0\mathbf{Q}\mathscr{R}_0, \tag{8.111c}$$

$$\mathbf{0} = \mathbf{K}_{12}^* + \mathbf{K}_{12} - \mathbf{K}_{22}\mathscr{D}' - \mathscr{D}'^*\mathbf{K}_{22} - \mathbf{K}_{22}\mathscr{B}_0\mathbf{R}^{-1}\mathscr{B}_0^*\mathbf{K}_{22} + \mathscr{R}_0^*\mathbf{Q}\mathscr{R}_0. \tag{8.111d}$$

Note that these $n_q \times n_q$ equations are uncoupled to the extent that they may be solved one at a time in the order (8.111b), (8.111d), (8.111c). Hence, the $2n_q$th-order state realization (8.110b) leads to a $2n_q \times 2n_q$ Riccati equation which can be solved in smaller pieces; three sequential $n_q \times n_q$ equations, one of which, (8.111c), is linear! This is a substantial computational savings. Note also that *each* of the Riccati equations, (8.111b) and (8.111d), may be solved by the methods of Sections 8.4 and 8.5. $\mathbf{K}_{11}$ is not needed except to verify $\mathbf{K} > \mathbf{0}$ and to compute the value of the cost function. Note from (8.111b) that $\mathbf{G}_1$ is the optimal control gain for the hypothetical system

$$\{\dot{\mathbf{x}} = -\boldsymbol{\Omega}^2\mathbf{x} + \mathbf{B}_0\mathbf{u},\ \mathbf{y} = \mathbf{P}_0\mathbf{x}\}$$

and that G_2 is the optimal gain for the hypothetical system

$$\{\dot{\mathbf{x}} = -\mathscr{D}'\mathbf{x} + \mathbf{B}_0\mathbf{u},\ \mathbf{y} = \hat{\mathbf{R}}\mathbf{x}\},$$

which is optimal for $\mathscr{V} = \int_0^\infty ((\mathbf{y}^*\mathbf{y} + \mathbf{u}^*\mathbf{R}\mathbf{u})\,dt$, and where $\hat{\mathbf{R}}^*\hat{\mathbf{R}} = 2\mathbf{K}_{12} + \mathbf{R}_0^*\mathbf{Q}\mathbf{R}_0$.

From these results it is possible to state an important special case. Suppose

$$\mathscr{P} = 0, \quad \mathscr{R} = \mathscr{B}^*, \quad \mathscr{D} = \mathbf{0}, \quad \mathbf{R} = \mathbf{Q}^{-1}\alpha^2.$$

In flexible structural dynamics the condition $\mathscr{R}_0 = \mathscr{B}_0^*$ will be called a "collocated actuators and rate sensors" condition, and this arises whenever rectilinear displacement rate measurements occur at the same location (and direction) as a control force, or whenever angular displacement rate measurements occur at the same location (and direction) as a control torque.

Theorem 8.8

Given the stabilizable, detectable system

$$\mathscr{M}\ddot{\mathbf{q}} + \mathscr{K}\mathbf{q} = \mathscr{B}\mathbf{u}, \qquad \mathbf{q}(0) = \mathbf{q}_0, \quad \dot{\mathbf{q}}(0) = \dot{\mathbf{q}}_0,$$

$$\mathbf{y} = \mathscr{B}^*\dot{\mathbf{q}}. \tag{8.112}$$

The control that minimizes

$$\mathscr{V} = \int_0^\infty \left(\alpha^2\mathbf{y}^*\mathscr{G}\mathbf{y} + \mathbf{u}^*\mathscr{G}^{-1}\mathbf{u}\right) dt, \qquad \mathscr{G} = \mathscr{G}^* > \mathbf{0}, \tag{8.113}$$

is

$$\mathbf{u} = -\alpha\mathscr{G}\mathbf{y}, \tag{8.114}$$

where α *is any positive scalar and* $\mathscr{G}$ *is any positive definite matrix.*

Proof: One may show that

$$\{\mathscr{P}_0 = \mathbf{0},\ \ \mathscr{R}_0 = \mathscr{B}_0^T,\ \ \mathbf{G} = -\alpha\mathscr{G},\ \ \mathbf{Q} = \alpha^2\mathscr{G},\ \ \mathbf{R} = \mathscr{G}^{-1},\ \ \mathbf{K}_{12} = \mathbf{0},$$

$$\mathbf{K}_{22} = \alpha\mathbf{I},\ \ \mathbf{K}_{11} = \alpha\Omega^2\} \tag{8.115}$$

solves the measurement feedback LQ problem and that the answer is unique and that the minimal value of the cost function is $V = \alpha[\eta_0^*\Omega^2\eta_0 + \dot{\eta}_0^*\dot{\eta}_0]$. However, it is easier to show that (8.114) is actually the optimal *state* feedback solution (8.111). This proves the optimality of the measurement feedback solution (8.114), since measurement feedback is a more restrictive constraint than state feedback; hence, optimality of the control (8.114) as a state feedback solution implies optimality of the control (8.114) as a measurement feedback solution.

For state feedback, we must satisfy (8.111), which becomes

$$\mathbf{0} = \mathbf{0}, \tag{8.116a}$$

$$\mathbf{0} = \mathbf{K}_{11} - \Omega^2\alpha, \tag{8.116b}$$

$$\mathbf{0} = -\alpha^2\mathscr{B}_0\mathscr{G}\mathscr{B}_0^* + \alpha^2\mathscr{B}_0\mathscr{G}\mathscr{B}_0^*, \tag{8.116c}$$

and (8.111a) becomes

$$\begin{aligned}\mathbf{G} = [\mathbf{G}_1, \mathbf{G}_2] = -\mathbf{R}^{-1}\mathbf{B}_0^*\mathbf{K} &= -\mathscr{G}\mathscr{B}_0^*[\mathbf{K}_{12}, \mathbf{K}_{22}]\\ &= -\mathscr{G}\mathscr{B}_0^*[\mathbf{0}, \alpha\mathbf{I}]\\ &= [\mathbf{0}, -\alpha\mathscr{G}\mathscr{B}_0^*].\end{aligned}$$

Hence,

$$\mathbf{u} = \mathbf{G}\mathbf{x} = -\alpha\mathscr{G}\mathscr{B}_0^*\dot{\eta} = -\alpha\mathscr{G}\mathscr{B}^*\mathscr{E}\mathscr{E}^{-1}\dot{\mathbf{q}} = -\alpha\mathscr{G}\mathscr{B}^*\dot{\mathbf{q}}, \tag{8.117}$$

which is the result (8.114). Now, to prove uniqueness, we rely on part (d) of Theorem 8.4 and the stabilizability and detectability of (8.111). ■

Exercise 8.6

Determine what special properties are associated with the optimal control problem with $\{\mathscr{R} = \mathbf{0},\ \mathscr{P} = \mathscr{B}_0^*,\ \mathbf{Q} = \alpha^2\mathscr{G},\ \mathbf{R} = \mathscr{G}^{-1}\}$.

Exercise 8.7

Find the optimal control for the Euler–Bernoulli beam in Section 3.7.2 and equation (3.78) if the controls are: u_1 = force at location $r_1 = 0.45L$, u_2 = force at location $r_2 = 0.30L$; the outputs of interest are: y_1 = rectilinear velocity at r_1, y_2 = rectilinear velocity at r_2; the measurements are: $z_1 = y_1 + v_1$, $z_2 = y_2 + v_2$; and the actuator disturbances $w_1 = w_2 = \delta(t)$ are unit strength impulses.

Hence, the equations of motion are

$$\ddot{\eta}_i + 2\zeta_i \omega_i \dot{\eta}_i + \omega_i^2 \eta_i = \mathscr{b}_i^* (\mathbf{u} + \mathbf{w}), \qquad i = 1, 2, \ldots, N,$$

$$\mathbf{y} = \sum_{i=1}^{N} \mathscr{c}_i \dot{\eta}_i, \quad \mathbf{z} = \sum_{i=1}^{N} \mathscr{c}_i \dot{\eta}_i + \mathbf{v}, \quad \zeta_i = 0.005,$$

$$\omega_i = \sqrt{\frac{EI}{\rho}} \left(\frac{i\pi}{L} \right)^2 = i^2,$$

where

$$\psi_i(r_1) = \sqrt{\frac{2}{\rho L}} \sin\left(\frac{i\pi r_1}{L} \right) = \sin(0.45\pi i)$$

and

$$\mathscr{b}_i^* = [\psi_i(r_1), \psi_i(r_2)],$$

$$\mathscr{c}_i^* = [\psi_i(r_1), \psi_i(r_2)], \qquad \psi_i(r_2) = \sin(0.3\pi i),$$

and the cost function to be minimized is

$$\mathscr{V} = \sum_{i=1}^{4} \int_0^\infty [\mathbf{y}^{i*}(t)\mathbf{y}^i(t) + \mathbf{u}^{i*}(t)\mathbf{u}^i(t)]\, dt, \tag{8.118}$$

where $u^i(t)$, $y^i(t)$ are responses in the closed-loop system due to impulses applied in the ith element of the vector (w_1, w_2, v_1, v_2). Design an optimal LQI controller that receives $\mathbf{z}(t)$ as an input and produces $\mathbf{u}(t)$ as an output.

(i) Show that the optimal controller (*Hint*: Theorem 8.2) is given by

$$\mathbf{u}(s) = -s\mathscr{B}^*[s^2\mathbf{I} + 2\mathscr{B}\mathscr{B}^* s + \Omega^2]^{-1}\mathscr{B}\mathbf{z}(s).$$

(ii) Show that if $\mathbf{v}(t) \equiv 0$, the optimal controller is given by

$$\mathbf{u}(s) = -\mathbf{z}(s),$$

where for $i = 1, 2, \ldots, N$,

$$\mathscr{B}_{i\text{th row}} = \ell_i^* = [\psi_i(r_1), \psi_i(r_2)] = [\sin(0.45\pi i), \sin(0.3\pi i)].$$

(iii) Show that the closed-loop eigenvalues of the system in (ii) are roots of the polynomial

$$|s^2\mathbf{I} + \mathscr{B}\mathscr{B}^* s + \Omega^2|.$$

(iv) Show the closed-loop eigenvalues of the system in (i) are the same as in (iii), except repeated with multiplicity 2. Show that

$$\mathbf{A} + \mathbf{B}\mathbf{G} = \mathbf{A} - \mathbf{F}\mathbf{M} = \begin{bmatrix} \mathbf{0} & \mathbf{I} \\ \Omega^2 & -\mathscr{B}\mathscr{B}^* \end{bmatrix}.$$

(v) Show that the closed-loop system in (ii) is asymptotically stable iff the matrix pair $(\Omega^2, \mathscr{B})$ is observable. Otherwise, the unobservable modes (eigenvalues) remain at their open-loop location.

(vi) Compute the value of the optimal cost in (i) and (ii). Compare to find the performance cost associated with imperfect measurements. Show that the cost in (i) is $\mathscr{V} = 2\,\mathrm{tr}\,\mathscr{B}^*\mathscr{B}$ and the cost in (ii) is $\mathscr{V} = \mathrm{tr}\,\mathscr{B}^*\mathscr{B}$. Hence, the presence of disturbances in the measurements doubles the cost.

Exercise 8.8

Evaluate your design in Exercise 8.7 by computing the cost functions

(i) $\mathscr{V} = \sum_{i=1}^{2} \int_0^\infty \mathbf{y}^{i*}(t)\mathbf{Q}\mathbf{y}^i(t)\,dt$, $\mathbf{u}(t) \equiv \mathbf{0}$.

(ii) $\mathscr{V}$ as in (8.118) with "plant" = your controller design model in Exercise 8.7.

(iii) $\mathscr{V}$ as in (8.118) with your controller from Exercise 8.6, but with a "plant" = the three modes with the largest modal cost. Note that in this case, the impact of some modeling errors is assessed since the controller is optimal for a different model than the more correct higher-order plant. This mismatch between the actual plant and the control design model is *always* the case in practice, since the controller is always based upon a model which is only an approximation of the actual plant.

Solution Hints: The modal costs are computed from (5.161):

$$\mathscr{V}_{\eta_i} = \frac{\ell_i^* \mathscr{U} \ell_i \left(p_i^* \mathbf{Q}_p p_i + r_i^* \mathbf{Q}_r r_i \omega_i^2 \right)}{4\zeta_i \omega_i^3} = \frac{\ell_i^* \ell_i r_i^* r_i}{4\zeta_i \omega_i}$$

$$= \frac{\left[\psi_i^2(r_1) + \psi_i^2(r_2)\right]^2}{4(0.005)i^2}$$

$$= 50\frac{\left[\sin^2(0.45\pi i) + \sin^2(0.3\pi i)\right]^2}{i^2},$$

yielding the modal costs

MODE i	MODAL COST
1	132.851
2	12.500
3	4.394
4	1.492
5	4.5
6	1.389
7	0.0928
8	2.557
9	0.306
10	0.5

Hence, because we are restricted to three modes by the problem statement, we take modes 1, 2, and 5 for our control design model and

$$\Omega^2 = \begin{bmatrix} 1 & 0 & 0 \\ 0 & 4 & 0 \\ 0 & 0 & 25 \end{bmatrix}, \qquad \mathscr{R}_0^* = \mathscr{B}_0 = \begin{bmatrix} \sin(0.45\pi) & \sin(0.3\pi) \\ \sin(0.45\pi 2) & \sin(0.3\pi 2) \\ \sin(0.45\pi 5) & \sin(0.3\pi 5) \end{bmatrix}$$

$$\mathbf{K} = 2\rho^{1/4}\begin{bmatrix} 2\rho^{1/2} & \rho^{1/4} \\ \rho^{1/4} & 1 \\ & \end{bmatrix}.$$

Note however, that a better (bigger) gap exists between the modal costs of modes 3 and 4. Hence, a better model for design would be modes 1, 2, 5, 3.

Exercise 8.9

(a) Show that the system

$$\dot{\mathbf{x}} = \begin{bmatrix} 0 & 1 \\ 0 & 0 \end{bmatrix}\mathbf{x} + \begin{pmatrix} 0 \\ 1 \end{pmatrix}u, \qquad \mathbf{x}(0) = \mathbf{x}_0,$$

with the state feedback control $\mathbf{u} = \mathbf{G}\mathbf{x}$, $\mathbf{G} = [-0.1, -0.1]$, is asymptotically stable.

(b) Show that there does not exist any $\mathbf{Q} \geq \mathbf{0}$ or $\mathbf{Q} > \mathbf{0}$ such that the $\mathbf{G}$ in (a) minimizes

$$\mathscr{V} = \int_0^\infty \left(\mathbf{x}^*\mathbf{Q}\mathbf{x} + u^2\rho\right) dt.$$

Explain why.

Exercise 8.10
Show that the control that minimizes

$$\mathscr{V} = \int_0^\infty \left(Q_{11}\eta^2 + Q_{22}\dot{\eta}^2 + \rho u^2\right) dt$$

for the second-order system

$$\ddot{\eta} + 2\zeta\omega\dot{\eta} + \omega^2\eta = \mathbf{b}^*\mathbf{u}$$

yields the closed-loop system

$$\ddot{\eta} + 2\zeta_c\omega_c\dot{\eta} + \omega_c^2\eta = 0,$$

where the relationships between the weights Q_{11}, Q_{22}, ρ, and the modal data are

$$Q_{11} = \frac{\left(\omega_c^4 - \omega^4\right)\rho}{\mathbf{b}^*\mathbf{b}}$$

$$Q_{22} = \frac{\left\{2\left(\omega^2 - \omega_c^2\right) + 4\left(\zeta_c^2\omega_c^2 - \zeta^2\omega^2\right)\right\}\rho}{\mathbf{b}^*\mathbf{b}}.$$

EXAMPLE 8.6
It is desired to regulate the spin speed ω of the spinning body in Fig. 8.9 so that ω remains close to the value $\omega = \Omega$. Let the angular moment of inertia of the

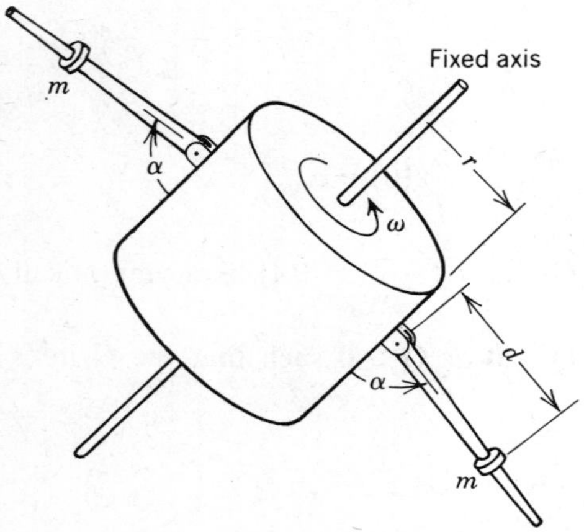

Figure 8.9 *A Spinning Govenor System*

body (mounted on the fixed spin axis, a drum of radius $r = 1$ m) be J. Let the control be the rate at which masses m move in or out along the massless arms (making angle α with the main body). Hence, $u \triangleq \dot{d}$. The coefficient of viscous friction at the hinge joints is c, and the movable masses have mass m. Linearize the nonlinear equations of motion

$$\dot{\omega} = \frac{-4m\omega(r + d\sin\alpha)(d\sin\alpha - \dot{\alpha}\, d\cos\alpha)}{J + 2m(r + d\sin\alpha)^2}, \tag{8.119}$$

$$\ddot{\alpha} = -\left[\frac{c}{m} + \frac{2\dot{d}}{d}\right]\dot{\alpha} + \omega^2\left[\frac{r}{d} + \sin\alpha\right]\cos\alpha$$

about the desired operating conditions

$$\omega(t) = \Omega,$$

$$\alpha(t) = \frac{\pi}{2}\ \text{rad},$$

$$d(t) = d_0$$

to obtain

$$\begin{bmatrix} \dot{x}_1 \\ \dot{x}_2 \\ \dot{x}_3 \\ \dot{x}_4 \end{bmatrix} = \begin{bmatrix} 0 & 0 & 0 & 0 \\ 0 & 0 & 1 & 0 \\ 0 & -\gamma^2 & -\dfrac{c}{m} & 0 \\ 0 & 0 & 0 & 0 \end{bmatrix} \begin{bmatrix} x_1 \\ x_2 \\ x_3 \\ x_4 \end{bmatrix} + \begin{bmatrix} -\beta \\ 0 \\ 0 \\ 1 \end{bmatrix} u, \tag{8.120}$$

$$y = [1 \quad 0 \quad 0 \quad 0]\mathbf{x},$$

where

$$\beta \triangleq \frac{4m\Omega(r + d_0)}{\left[J + 2m(r + d_0)^2\right]}, \qquad \gamma^2 \triangleq \Omega^2\left(1 + \frac{r}{d_0}\right),$$

$$u \triangleq \dot{d}, \quad x_1 \triangleq \omega - \Omega, \quad x_2 = \alpha - \frac{\pi}{2}, \quad x_3 = \dot{\alpha}, \quad x_4 = d - d_0.$$

Find the control that minimizes

$$\mathscr{V} = \int_0^\infty (y^2 + \rho u^2)\, dt, \qquad x(0) = x_0. \tag{8.121}$$

Solution: The reader may verify that states x_2, x_3, x_4 in (8.120) are not observable in y and therefore may be deleted. Without this step, the available theory does not apply, since the system (8.120) is neither detectable nor stabilizable (states x_1 and x_4 are not asymptotically stable nor controllable; states x_2, x_3, x_4 are unobservable and not asymptotically stable). Yet the problem has a solution as we shall show. The observable part of the system is

$$\dot{x}_1 = -\beta u, \qquad y = x_1, \tag{8.122}$$

and this is the only part of the system that influences the value of $\mathscr{V}$ in (8.121). The reader can readily verify that the Riccati equation for (8.122) yields $K_1 = \rho^{1/2}\beta^{-2}$ and $G_1 = -R^{-1}B^*K = \beta^{-1}\rho^{-1/2}$ yielding $\dot{x}_1 = -\rho^{1/2}x_1$. Hence, the optimal control for (8.120) is

$$u = \dot{d} = Gx = G_1x_1 = \beta^{-1}\rho^{-1/2}(\omega - \Omega).$$

Note that optimality does not imply stability unless (A, B, C) is stabilizable and detectable, since state x_4 may go to infinity (i.e., the mass m may run off the arm). ■

EXAMPLE 8.7

For a particular optimal state feedback control problem we must solve $\mathbf{0} = \mathbf{KA} + \mathbf{A}^*\mathbf{K} - \mathbf{KBR}^{-1}\mathbf{B}^*\mathbf{K} + \mathbf{C}^*\mathbf{C}$ with

$$\mathbf{A} = \begin{bmatrix} 1 & -1 \\ 0 & 0 \end{bmatrix}, \quad \mathbf{B} = \begin{bmatrix} 0 \\ 1 \end{bmatrix}, \quad \mathbf{C} = [0 \quad 1].$$

There are two solutions,

$$\mathbf{K}_1 = \begin{bmatrix} 0 & 0 \\ 0 & 1 \end{bmatrix} \geq \mathbf{0} \quad \text{and} \quad \mathbf{K}_2 = \begin{bmatrix} 8 & 4 \\ 4 & 3 \end{bmatrix} > \mathbf{0},$$

one of which is positive definite. Which is the correct answer, K_1 or K_2?

Solution: Theorem 8.4, part (b), provides an immediate answer. The optimal $\mathbf{K}$ is positive definite iff $(\mathbf{A}, \mathbf{C})$ is observable. The above $(\mathbf{A}, \mathbf{C})$ is *not* observable. Hence, the positive semidefinite solution K_1 is the correct answer even though there exists a positive definite solution to the Riccati equation. The closed-loop optimal system is unstable,

$$\mathbf{A} + \mathbf{BG} = \begin{bmatrix} 1 & -1 \\ 0 & -1 \end{bmatrix}, \qquad \mathbf{G} = -\mathbf{R}^{-1}\mathbf{B}^*\mathbf{K}_1,$$

illustrating that optimality does not guarantee stability. See that $(\mathbf{A}, \mathbf{C})$ is not detectable in this problem and hence part c of Theorem 8.4 does not apply. Now, in modal coordinates,

$$\Lambda = \mathbf{E}^{-1}\mathbf{A}\mathbf{E} = \begin{bmatrix} 1 & 0 \\ 0 & 0 \end{bmatrix}, \quad \mathbf{C}\mathbf{E} = [0 \quad 1], \quad \mathbf{E}^{-1}\mathbf{B} = \begin{pmatrix} -1 \\ 1 \end{pmatrix}, \quad \mathbf{E}^{-1} = \begin{bmatrix} 1 & -1 \\ 0 & 1 \end{bmatrix},$$

one may of course delete the unobservable mode and proceed to get the same answer as above. Deleting the unobservable mode yields the reduced-order model

$$A_R = 0, \quad C_R = I, \quad B_R = 1,$$

leading to

$$0 = K_R A_R + A_R^* K_R - K_R B_R^{-1} B_R^* K_R + C_R^* Q C_R$$

or

$$K_R = 1, \qquad G_R = R^{-1} B_R^* K_R = -1.$$

In the original coordinates, this yields

$$u = G_R x_R = [0 \quad G_R]\begin{pmatrix} x_T \\ x_R \end{pmatrix} = [0 \quad G_R]\mathbf{E}^{-1}\mathbf{x} = \mathbf{G}\mathbf{x} = [0 \quad -1]\begin{bmatrix} 1 & -1 \\ 0 & 1 \end{bmatrix}\mathbf{x},$$

yielding the same result as before,

$$\mathbf{A} + \mathbf{B}\mathbf{G} = \begin{bmatrix} 1 & -1 \\ 0 & -1 \end{bmatrix}.$$ ■

Exercise 8.11

Consider a spacecraft attitude control problem (pitch axis only). The equation of motion is $\ddot{\theta} = T/J = u$, where J is the pitch axis moment of inertia, $u \triangleq T/J$ is the control acceleration, and $\theta(t)$ is pitch attitude. For the initial conditions $\theta(0) = 10°$, $\dot{\theta}(0) = 0$ and the cost function $\mathscr{V} = \int_0^\infty (4\theta^2 + \rho u^2)\, dt$ one may plot the optimal responses as in Fig. 8.10. Show that the open-loop transfer function is

$$GH(s) = \frac{2}{s^2 \rho^{1/4}}(s + \rho^{-1/4})$$

and that the closed-loop system has damping $\zeta = 0.707$ and undamped natural frequency $\omega_n = \sqrt{2}\,\rho^{-1/4}$.

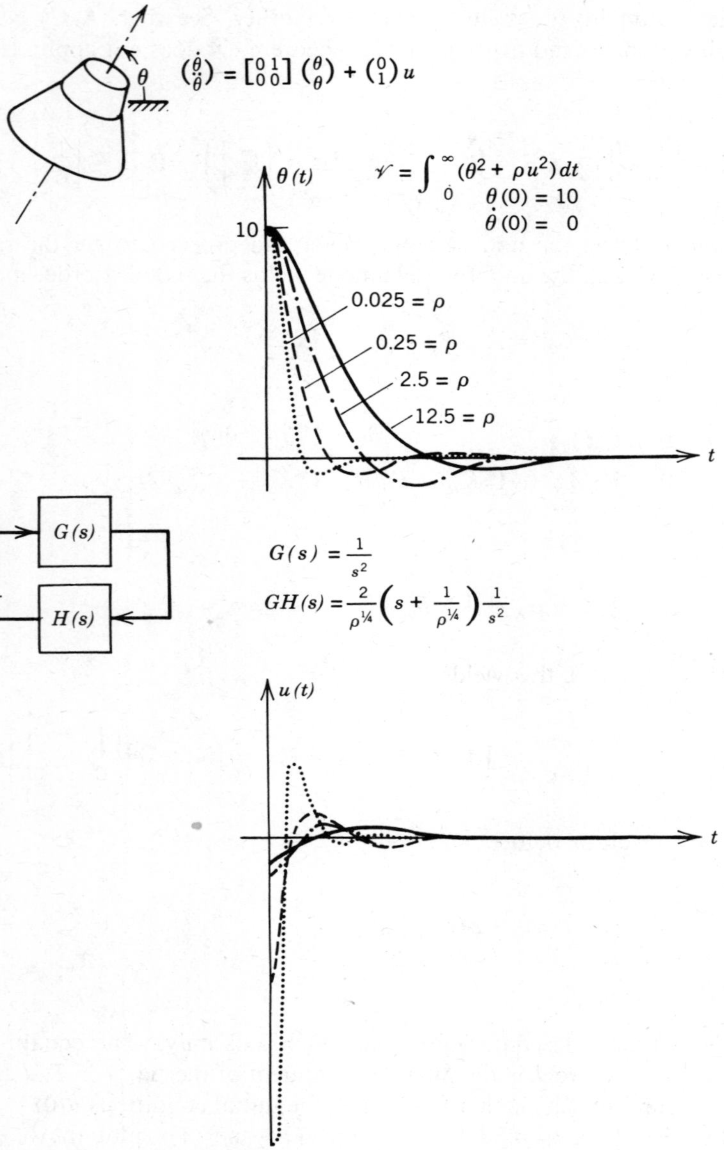

Figure 8.10 *Optimal Attitude Control Responses*

8.8 Disturbance Accommodation

It is the rule rather than the exception that the system of interest does *not* satisfy the conditions of Theorem 8.4. Consider the important problem of controlling in the presence of a disturbance $\mathbf{w}(t)$ that satisfies

$$\mathbf{w} = \mathbf{C}_w \mathbf{x}_w, \qquad \dot{\mathbf{x}}_w = \mathbf{A}_w \mathbf{x}_w + \mathbf{w}_w. \tag{8.123}$$

Augmenting this disturbance model to the plant

$$\dot{\mathbf{x}} = \mathbf{A}\mathbf{x} + \mathbf{B}\mathbf{u} + \mathbf{D}\mathbf{w}, \qquad \mathbf{y} = \mathbf{C}\mathbf{x}, \quad \mathbf{z} = \mathbf{M}\mathbf{x} + \mathbf{v} \tag{8.124}$$

yields the system uncontrollable from $\mathbf{u}$,

$$\begin{pmatrix} \dot{\mathbf{x}} \\ \dot{\mathbf{x}}_w \end{pmatrix} = \begin{bmatrix} \mathbf{A} & \mathbf{D}\mathbf{C}_w \\ \mathbf{0} & \mathbf{A}_w \end{bmatrix} \begin{pmatrix} \mathbf{x} \\ \mathbf{x}_w \end{pmatrix} + \begin{bmatrix} \mathbf{B} \\ \mathbf{0} \end{bmatrix} \mathbf{u} + \begin{bmatrix} \mathbf{0} \\ \mathbf{I} \end{bmatrix} \mathbf{w}_w, \qquad \mathbf{y} = [\mathbf{C} \;\; \mathbf{0}] \begin{pmatrix} \mathbf{x} \\ \mathbf{x}_w \end{pmatrix}, \quad \mathbf{z} = [\mathbf{M} \;\; \mathbf{0}] \begin{pmatrix} \mathbf{x} \\ \mathbf{x}_w \end{pmatrix} + \mathbf{v}. \tag{8.125}$$

Since $\mathbf{A}_w$ may be unstable (disturbances are not usually well behaved), the system may not be stabilizable. A straightforward attempt to solve the LQI problem for the system (8.125) (which we shall write as $\dot{\mathbf{x}}_d = \mathbf{A}_d \mathbf{x}_d + \mathbf{B}_d \mathbf{u} + \mathbf{D}_d \mathbf{w}_w$, $\mathbf{y} = \mathbf{C}_d \mathbf{x}_d$, $\mathbf{z} = \mathbf{M}_d \mathbf{x}_d + \mathbf{v}$) leads to two Riccati equations of the form

$$\mathbf{0} = \mathbf{K}_d \mathbf{A}_d + \mathbf{A}_d^* \mathbf{K}_d - \mathbf{K}_d \mathbf{B}_d \mathbf{R}^{-1} \mathbf{B}_d^* \mathbf{K}_d + \mathbf{C}_d^* \mathbf{Q} \mathbf{C}_d, \tag{8.126}$$

$$\mathbf{0} = \mathbf{P}_d \mathbf{A}_d^* + \mathbf{A}_d \mathbf{P}_d - \mathbf{P}_d \mathbf{M}_d^* \mathbf{V}^{-1} \mathbf{M}_d \mathbf{P}_d + \mathbf{D}_d \mathbf{W} \mathbf{D}_d^* + \mathbf{X}_0, \tag{8.127}$$

where the $\mathbf{X}_0$ term is added only if initial states $\mathbf{x}_d(0)$ are to be considered in the problem [see (8.49b)]. From Theorem 8.4, we know that $\mathbf{K}_d$ exists iff there are no modes of $(\mathbf{A}_d, \mathbf{B}_d, \mathbf{C}_d)$ which are observable, uncontrollable, unstable. This condition could easily be violated by (8.125). Hence, $\mathbf{K}_d$ might not exist in our straightforward approach to the LQI problem. Similar difficulties are not expected with the $\mathbf{P}_d$ matrix since $(\mathbf{A}_d, \mathbf{D}_d, \mathbf{M}_d)$ is usually an observable and controllable triple. If there are states which are unobservable in both $(\mathbf{A}_d, \mathbf{C}_d)$ and $(\mathbf{A}_d, \mathbf{M}_d)$, which can be determined by checking the observability of the pair $\left(\mathbf{A}_d, \begin{bmatrix} \mathbf{C}_d \\ \mathbf{M}_d \end{bmatrix}\right)$, they should be deleted since they do not affect the cost function. Hence, in the "state estimator" part of the optimal LQI controller, the generation of the state estimate $\mathbf{x}_c(t) \to \mathbf{x}_d(t)$ usually occurs without incident. That is, $\mathbf{x}_c(t)$ contains a converging estimate of

$\mathbf{x}_d(t)$ which also contains the disturbance state $\mathbf{x}_w(t)$. In this way the state of LQI controller plays the role of a "disturbance estimator" and can asymptotically *generate the disturbances without actually measuring them.*

Thus, the difficulties in this disturbance accommodation problem (8.125) is the calculation of control gains $\mathbf{G}_d = -\mathbf{R}^{-1}\mathbf{B}_d^*\mathbf{K}_d$ rather than the estimator gains $\mathbf{F}_d = \mathbf{P}_d\mathbf{M}_d^*\mathbf{V}^{-1}$. We offer two "tricks" to approximate the optimal solution $\mathbf{G}_d$ when we cannot calculate it directly due to unstabilizability of $(\mathbf{A}_d, \mathbf{B}_d)$.

8.8.1 DISTURBANCE UTILIZATION CONTROL

We assume here that $(\mathbf{A}, \mathbf{B})$ is stabilizable but that $(\mathbf{A}_d, \mathbf{B}_d)$ is not. The trick in this section is to simply force (8.125) to be stabilizable by pole-shifting $\mathbf{A}_w$. This may be accomplished by replacing $\mathbf{A}_w$ in (8.125) by $\hat{\mathbf{A}}_w \triangleq \mathbf{A}_w - \varepsilon\mathbf{I}$, where $\varepsilon > 0$ is chosen large enough to make $\hat{\mathbf{A}}_w(\varepsilon)$ asymptotically stable. In this modified situation, the optimal controller is

$$\begin{aligned} \mathbf{u} &= \mathbf{G}_d(\varepsilon)\mathbf{x}_c, \qquad & \mathbf{G}_d(\varepsilon) &= -\mathbf{R}^{-1}\mathbf{B}_d^*\mathbf{K}_d(\varepsilon), \\ \dot{\mathbf{x}}_c &= \mathbf{A}_c\mathbf{x}_c + \mathbf{F}_d\mathbf{z}, \qquad & \mathbf{F}_d(0) &= \mathbf{P}_d\mathbf{M}_d^*\mathbf{V}^{-1} = \mathbf{F}_d, \end{aligned} \tag{8.128}$$

where $\mathbf{K}_d(\varepsilon)$ solves

$$\mathbf{0} = \mathbf{K}_d(\varepsilon)\mathbf{A}_d(\varepsilon) + \mathbf{A}_d^*(\varepsilon)\mathbf{K}_d(\varepsilon) - \mathbf{K}_d(\varepsilon)\mathbf{B}_d\mathbf{R}^{-1}\mathbf{B}_d^*\mathbf{K}_d(\varepsilon) + \mathbf{C}^*\mathbf{Q}\mathbf{C} \tag{8.129}$$

and

$$\mathbf{A}_d(\varepsilon) \triangleq \begin{bmatrix} \mathbf{A} & \mathbf{D}\mathbf{C}_w \\ \mathbf{0} & \hat{\mathbf{A}}_w(\varepsilon) \end{bmatrix}. \tag{8.130}$$

The optimal control gain is $\mathbf{G}_d(0)$, and in some cases one can set $\varepsilon = 0$ in the final expression for $\mathbf{G}_d(\varepsilon)$ to get the optimal result (see Example 8.8). In other cases the analyst should compute $\mathbf{G}_d(\varepsilon)$ for smaller and smaller values of ε until either (i) $\hat{\mathbf{A}}_w$ ceases to be stable or (ii) $\mathbf{G}_d(\varepsilon)$ ceases to change with decreases in ε.

EXAMPLE 8.8

A vertically fired sounding rocket has mass m, thrust f, weight mg, velocity v, and momentum $h = mv$. The equation of motion is

$$\dot{h} = f - mg. \tag{8.131}$$

Suppose the desired momentum is $\bar{h}$; then, a reasonable cost function for evaluating performance is

$$\mathscr{V} = \int_0^\infty \left[(h - \bar{h})^2 + \rho f^2\right] dt. \tag{8.132}$$

We wish to regulate the thrust $f(t)$ to minimize $\mathscr{V}$ in the presence of *unknown* but constant weight $w \triangleq mg$. The constant is modeled by $\dot{w} = 0$, and the augmented model expands (8.131) to

$$\begin{pmatrix} \dot{\tilde{h}} \\ \dot{w} \end{pmatrix} = \begin{bmatrix} 0 & -1 \\ 0 & 0 \end{bmatrix} \begin{pmatrix} \tilde{h} \\ w \end{pmatrix} + \begin{bmatrix} 1 \\ 0 \end{bmatrix} u, \qquad \tilde{h} \triangleq h - \bar{h},$$

$$y = \tilde{h} = [1 \quad 0] \begin{bmatrix} \tilde{h} \\ w \end{bmatrix}, \qquad \begin{pmatrix} \tilde{h}(0) \\ w(0) \end{pmatrix} = \begin{pmatrix} -\bar{h} \\ \bar{w} \end{pmatrix}. \tag{8.133}$$

Now, this system is not stabilizable, but the addition of $-\varepsilon$ in the 22 element of the $\mathbf{A}_d$ matrix yields the stabilizable system parameters

$$\mathbf{A}_d(\varepsilon) = \begin{bmatrix} 0 & -1 \\ 0 & -\varepsilon \end{bmatrix}, \quad \mathbf{B}_d = \begin{bmatrix} 1 \\ 0 \end{bmatrix}, \quad \mathbf{C}_d = [1 \quad 0], \quad \mathbf{M}_d = \mathbf{C}_d,$$

where we presume the measurement $z = y + v$, $v(t) = \delta(t)$. The optimal controller is given by solving (8.127), (8.128), and (8.129) with $\mathbf{W} = \mathbf{0}$,

$$u = \mathbf{G}_d \mathbf{x}_c, \qquad \dot{\mathbf{x}}_c = \mathbf{A}_c \mathbf{x}_c + \mathbf{F}_d z,$$

where

$$\mathbf{K}_d(\varepsilon) = \begin{bmatrix} \rho^{1/2} & -\rho\left[1 + \varepsilon\rho^{1/2}\right]^{-1} \\ -\rho\left[1 + \varepsilon\rho^{1/2}\right]^{-1} & \rho(1 + \rho^{1/2})\left[1 + \varepsilon\rho^{1/2}\right]^{-2} \end{bmatrix}$$

and by direct calculation, since $(\mathbf{A}_d, \mathbf{M}_d)$ is observable and $(\mathbf{A}_d, \mathbf{x}_d(0))$ is controllable,

$$\mathbf{P}_d(0) = \begin{bmatrix} [\bar{h}^2 + 2\bar{w}]^{1/2} & -\bar{w} \\ -\bar{w} & \bar{w}\left[\bar{h} + (\bar{h}^2 + 2w)^{1/2}\right] \end{bmatrix},$$

leading to

$$\mathbf{F}_d = \mathbf{P}_d \mathbf{M}_d^* V^{-1} = \begin{bmatrix} [\bar{h}^2 + 2\bar{w}]^{1/2} \\ -\bar{w} \end{bmatrix},$$

$$\mathbf{G}_d(\varepsilon) = -\mathbf{R}^{-1}\mathbf{B}_d^*\mathbf{K}_d(\varepsilon) = -\rho^{-1}\left[\rho^{1/2}, -\rho\left[1 + \varepsilon\rho^{1/2}\right]^{-1}\right] \tag{8.134a}$$

$$= \left[-\rho^{-1/2}, \left[1 + \varepsilon\rho^{1/2}\right]^{-1}\right]$$

which immediately yields the optimal answer

$$\mathbf{G}_d(0) = \left[-\rho^{-1/2}, 1\right] = -\mathbf{R}^{-1}\mathbf{B}_d^*\mathbf{K}_d(0), \tag{8.134b}$$

where the optimal value

$$\mathbf{K}_d(0) = \begin{bmatrix} \rho^{1/2} & -\rho \\ -\rho & \rho(1 + \rho^{1/2}) \end{bmatrix}$$

can *not* be computed directly using $\mathbf{A}_d(0)$. (Try it.) Thus, the optimal control is

$$f = -\rho^{-1/2}\hat{h} + \hat{w},$$

where the controller state is $\mathbf{x}_c^* = (\hat{h}, \hat{w})$, satisfying $\dot{\mathbf{x}}_c = \mathbf{A}_c\mathbf{x}_c + \mathbf{F}_d z$, $u = \mathbf{G}_d\mathbf{x}_c$, or, specifically,

$$\begin{pmatrix} \dot{\hat{h}} \\ \dot{\hat{w}} \end{pmatrix} = \begin{bmatrix} -\rho^{-1/2} - (\bar{h}^2 + 2\bar{w})^{1/2} & 0 \\ \bar{w} & 0 \end{bmatrix} \begin{pmatrix} \hat{h} \\ \hat{w} \end{pmatrix} + \begin{bmatrix} (\bar{h}^2 + 2\bar{w})^{1/2} \\ -\bar{w} \end{bmatrix} z$$

$$f = \begin{bmatrix} -\rho^{-1/2} & 1 \end{bmatrix} \begin{pmatrix} \hat{h} \\ \hat{w} \end{pmatrix}$$

with the block diagram in Fig. 8.11 and transfer function

$$\frac{f(s)}{z(s)} = \mathbf{G}_d[s\mathbf{I} - \mathbf{A}_c]^{-1}\mathbf{F}_d = -\frac{as + b}{s(s + c)},$$

where

$$a = \bar{w} + \left[\frac{\bar{h}^2 + 2\bar{w}}{\rho}\right]^{1/2},$$

$$b = \bar{w}\rho^{-1/2},$$

$$c = \rho^{-1/2} + (\bar{h}^2 + 2\bar{w})^{1/2}.$$

The reader should verify that

$$\lim_{t \to \infty} y(t) = 0,$$

regardless of the constant value of w or of the assumed value of $\bar{w}$ in (8.133). This is the result of the *integral* nature of the controller. The integral controller will always appear using the procedure in this section whenever $\mathbf{A}_w = \mathbf{0}$ (constant disturbances). Thus, the controllers of this (and the next) section provide a direct response to the design requirements stated in the "disturbance modeling principle" of Section 4.3.

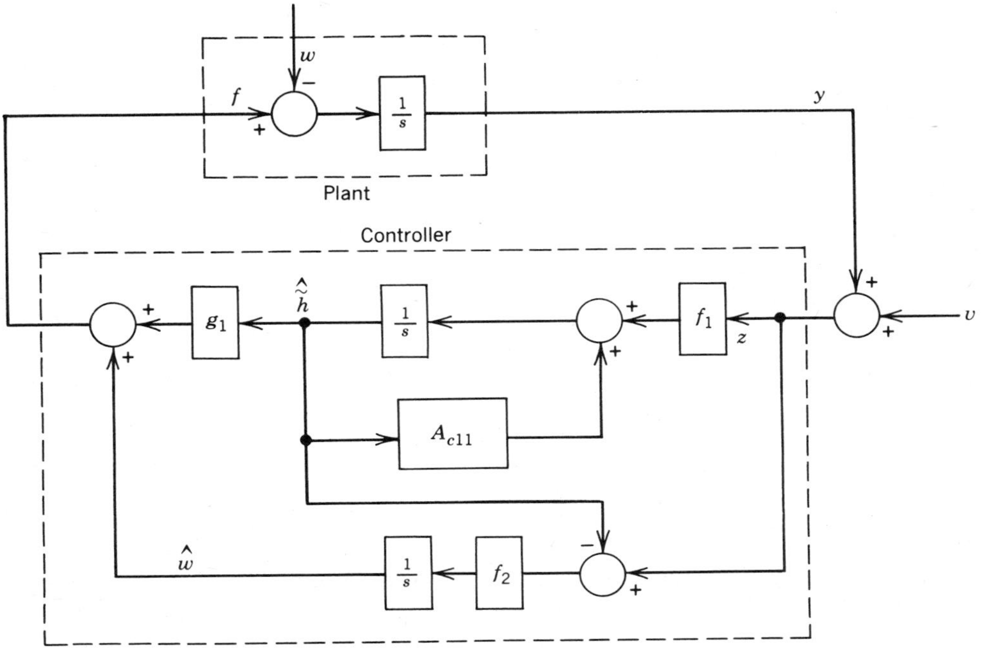

$$A_{c11} = -\rho^{-1/2} - (\bar{h}^2 + 2\bar{w})^{1/2},\ g_1 = -\rho^{-1/2},\ f_1 = (\bar{h}^2 + 2\bar{w})^{1/2}$$
$$f_2 = -\bar{w}$$

Figure 8.11 *An Integral Controller for Accommodating A Constant Disturbance*

8.8.2 DISTURBANCE CANCELLATION CONTROL

The controller in Section 8.8.1 is optimal for the cost function

$$\mathscr{V} = \int_0^\infty (\mathbf{y}^*\mathbf{Q}\mathbf{y} + \mathbf{u}^*\mathbf{R}\mathbf{u})\,dt,$$

subject to $\dot{\mathbf{x}} = \mathbf{A}\mathbf{x} + \mathbf{B}\mathbf{u} + \mathbf{D}\mathbf{w}$. This is called disturbance utilization because the controller optimally "utilizes" the disturbance if the disturbance serves to help reduce $\mathscr{V}$ and only "cancels" the disturbance when the disturbance serves as a bad effect, increasing $\mathscr{V}$. In this section we *presume* that the disturbance has a bad effect, and we choose to cancel the disturbance. This is clearly not always the proper thing to do since such a controller will expend energy fighting a "friendly" disturbance.

The strategy is to decompose the control $\mathbf{u}$ into two parts, $\mathbf{u} = \mathbf{u}_c + \mathbf{u}_w$, where $\mathbf{u}_w$ is chosen to cancel the disturbance. Hence,

$$\mathbf{B}\mathbf{u}_w + \mathbf{D}\mathbf{w} = \mathbf{0}, \tag{8.135}$$

and $\mathbf{u}_c$ is chosen to provide the optimal control for the *undisturbed* system

$$\mathscr{V} = \int_0^\infty (\mathbf{y}^*\mathbf{Q}\mathbf{y} + \mathbf{u}_c^*\mathbf{R}\mathbf{u}_c)\,dt, \quad \dot{\mathbf{x}} = \mathbf{A}\mathbf{x} + \mathbf{B}\mathbf{u}_c, \quad \mathbf{y} = \mathbf{C}\mathbf{x}. \tag{8.136}$$

Now, a solution $\mathbf{u}_w$ which solves (8.135) exists if $\mathbf{D} = \mathbf{BL}$ for some matrix $\mathbf{L}$ (see Section 2.6). In this case one solution of (8.135) is

$$\mathbf{u}_w = -\mathbf{B}^+\mathbf{D}\mathbf{w}, \tag{8.137}$$

where $\mathbf{B}^+$ is the Moore–Penrose inverse (see Section 2.6). The optimal control for (8.136) is

$$\mathbf{u}_c = -\mathbf{R}^{-1}\mathbf{B}^*\mathbf{K}\mathbf{x}, \tag{8.138}$$

yielding the total control

$$\mathbf{u} = \mathbf{u}_c + \mathbf{u}_w = [-\mathbf{R}^{-1}\mathbf{B}^*\mathbf{K}, -\mathbf{B}^+\mathbf{D}]\begin{pmatrix}\mathbf{x}\\ \mathbf{w}\end{pmatrix}, \tag{8.139}$$

assuming the availability of the state $\mathbf{x}_d^* = (\mathbf{x}^*, \mathbf{w}^*)$. Of course, the theorems of Chapter 8 show that when the state is not available for feedback, then the optimal controller is obtained by replacing the state by an estimate of the state (provided by the controller dynamics (8.128)). Hence, the only degree of suboptimality in the disturbance cancellation controller,

$$\begin{aligned} \mathbf{u} &= -[\mathbf{R}^{-1}\mathbf{B}^*\mathbf{K}, \mathbf{B}^+\mathbf{D}]\mathbf{x}_c = \hat{\mathbf{G}}_d\mathbf{x}_c, \\ \dot{\mathbf{x}}_c &= \mathbf{A}_c\mathbf{x}_c + \mathbf{F}_d\mathbf{z}, \end{aligned} \tag{8.140}$$

is due to the error $\mathbf{G}_d - \hat{\mathbf{G}}_d$, where $\mathbf{G}_d$ is given in (8.128) in the limit as $\varepsilon \to 0$, and $\hat{\mathbf{G}}_d \triangleq -[\mathbf{R}^{-1}\mathbf{B}^*\mathbf{K}, \mathbf{B}^+\mathbf{D}]$. The controller (8.140) is said to employ an "orthogonal filter" whenever the disturbance models of Section 4.3 have been employed, using orthogonal basis functions; see equations (4.74) through (4.76), for example.

It is interesting to note that disturbance cancellation (8.140) is sometimes the optimal controller, as shown in the following example.

EXAMPLE 8.9

For the system (8.131), it follows that

$$K = \rho^{1/2}, \quad A = 0, \quad B = 1, \quad D = -1, \quad C = 1.$$

Hence, the disturbance-cancelling controller (8.140) is given by

$$\hat{\mathbf{G}}_d = [-\rho^{-1/2}, 1],$$

which is identical to the disturbance utilization control gain (8.134).

8.9 Optimal Tracking and Servomechanisms

In this section we assume that it is desired to find a control $\mathbf{u}(t)$ that will cause the output $\mathbf{y}(t)$ to track a *specified* trajectory $\bar{\mathbf{y}}(t)$ very closely. Suppose that the vector function $\bar{\mathbf{y}}(t)$ can be described as the solution of the differential equation

$$\begin{aligned} \bar{\mathbf{y}} &= \mathbf{C}_y \mathbf{x}_y, \\ \dot{\mathbf{x}}_y &= \mathbf{A}_y \mathbf{x}_y + \mathbf{w}_y \end{aligned} \tag{8.141}$$

for some impulsive excitation $\mathbf{w}_y$. A natural cost function to minimize in this case is

$$\mathscr{V} = \int_0^\infty [(\mathbf{y} - \bar{\mathbf{y}})^* \mathbf{Q} (\mathbf{y} - \bar{\mathbf{y}}) + \mathbf{u}^* \mathbf{R} \mathbf{u}] \, dt, \tag{8.142}$$

subject to the constraints

$$\begin{aligned} \begin{pmatrix} \dot{\mathbf{x}} \\ \dot{\mathbf{x}}_y \end{pmatrix} &= \begin{bmatrix} \mathbf{A} & \mathbf{0} \\ \mathbf{0} & \mathbf{A}_y \end{bmatrix} \begin{pmatrix} \mathbf{x} \\ \mathbf{x}_y \end{pmatrix} + \begin{bmatrix} \mathbf{B} \\ \mathbf{0} \end{bmatrix} \mathbf{u} + \begin{bmatrix} \mathbf{D} & \mathbf{0} \\ \mathbf{0} & \mathbf{I} \end{bmatrix} \begin{pmatrix} \mathbf{w} \\ \mathbf{w}_y \end{pmatrix}, \\ \mathbf{y} - \bar{\mathbf{y}} = \tilde{\mathbf{y}} &= [\mathbf{C} \ \ -\mathbf{C}_y] \begin{pmatrix} \mathbf{x} \\ \mathbf{x}_y \end{pmatrix}, \\ \mathbf{z} &= [\mathbf{M} \ \ \mathbf{0}] \begin{pmatrix} \mathbf{x} \\ \mathbf{x}_y \end{pmatrix} + \mathbf{v}, \end{aligned} \tag{8.143}$$

which is written

$$\begin{aligned} \dot{\mathbf{x}}_t &= \mathbf{A}_t \mathbf{x}_t + \mathbf{B}_t \mathbf{u} + \mathbf{D}_t \mathbf{w}_t, \\ \tilde{\mathbf{y}} &= \mathbf{C}_t \mathbf{x}_t, \\ \mathbf{z} &= \mathbf{M}_t \mathbf{x}_t + \mathbf{v}, \end{aligned} \tag{8.144}$$

Now, $(\mathbf{A}_t, \mathbf{M}_t)$ is not observable and will not be detectable if $\mathbf{A}_y$ is unstable. Such tricks as offered in Section 8.8 may be used to approximate an optimal solution.

EXAMPLE 8.10

We wish the output of the system $\dot{x} = x + u$, $y = x$, $z = x + v$ to follow a sine function $\bar{y}(t) = \sin t$. Find a controller driven by $z(t)$ which minimizes the tracking error in the sense that

$$\mathscr{V} = \int_0^\infty \left[(y - \bar{y})^2 + \rho u^2 \right] dt$$

is minimized, where $v = \delta(t)$, $x(0) = 1$.

Solution: Let $x_{y1} = \sin t$, $x_{y2} = \cos t$. Then,

$$\bar{y} = [1 \quad 0]\begin{pmatrix} x_{y1} \\ x_{y2} \end{pmatrix} = \mathbf{C}_y \mathbf{x}_y,$$

$$\begin{pmatrix} \dot{x}_{y1} \\ \dot{x}_{y2} \end{pmatrix} = \begin{bmatrix} 0 & 1 \\ -1 & 0 \end{bmatrix}\begin{pmatrix} x_{y1} \\ x_{y2} \end{pmatrix} = \mathbf{A}_y \mathbf{x}_y, \qquad \begin{pmatrix} x_{y1}(0) \\ x_{y2}(0) \end{pmatrix} = \begin{pmatrix} 0 \\ 1 \end{pmatrix},$$

and the augmented system

$$\begin{pmatrix} \dot{x} \\ \dot{x}_{y1} \\ \dot{x}_{y2} \end{pmatrix} = \begin{bmatrix} 1 & 0 & 0 \\ 0 & 0 & 1 \\ 0 & -1 & 0 \end{bmatrix}\begin{pmatrix} x \\ x_{y1} \\ x_{y2} \end{pmatrix} + \begin{pmatrix} 1 \\ 0 \\ 0 \end{pmatrix} u = \mathbf{A}_t \mathbf{x}_t + \mathbf{B}_t u,$$

$$\begin{pmatrix} \tilde{y} \\ z \end{pmatrix} = \begin{bmatrix} 1 & -1 & 0 \\ 1 & 0 & 0 \end{bmatrix}\begin{pmatrix} x \\ x_{y1} \\ x_{y2} \end{pmatrix} + \begin{pmatrix} 0 \\ v \end{pmatrix} = \begin{bmatrix} \mathbf{C}_t \\ \mathbf{M}_t \end{bmatrix}\mathbf{x}_t + \begin{pmatrix} 0 \\ v \end{pmatrix},$$

$$\mathbf{x}_t^*(0) = (1 \quad 0 \quad 1)$$

is not detectable in $(\mathbf{A}_t, \mathbf{M}_t)$. Detectability is artificially induced by changing $\mathbf{A}_y$ to

$$\hat{\mathbf{A}}_y(\varepsilon) \triangleq \mathbf{A}_y - \varepsilon \mathbf{I} = \begin{bmatrix} -\varepsilon & 1 \\ -1 & -\varepsilon \end{bmatrix}.$$

Solving the modified problem yields

$$\mathbf{F}_t(\varepsilon) = \mathbf{P}_t(\varepsilon)\mathbf{M}_t^* \mathbf{V}^{-1} = \begin{bmatrix} P_{t_{11}} \\ P_{t_{12}} \\ P_{t_{13}} \end{bmatrix} = \begin{bmatrix} 1 + \sqrt{2} \\ \left[1 + (\varepsilon + \sqrt{2})^2\right]^{-1} \\ (\varepsilon + \sqrt{2})\left[1 + (\varepsilon + \sqrt{2})^2\right]^{-1} \end{bmatrix} = \begin{bmatrix} f_1 \\ f_2 \\ f_3 \end{bmatrix},$$

and the gain we seek is

$$\mathbf{F}_t(0) = \begin{pmatrix} 1 + \sqrt{2} \\ \dfrac{1}{3} \\ \dfrac{\sqrt{2}}{3} \end{pmatrix}.$$

The control gains are

$$\mathbf{G}_t(\varepsilon) = -\mathbf{R}^{-1}\mathbf{B}_t^*\mathbf{K}_t(\varepsilon) = -\frac{1}{\rho}(\mathbf{K}_{11} \quad \mathbf{K}_{12} \quad \mathbf{K}_{13})$$

$$= \left[-1 - \sqrt{1+\rho^{-1}}, \frac{\varepsilon + \sqrt{1+\rho^{-1}}}{\rho\left[1 + \left(\varepsilon + \sqrt{1+\rho^{-1}}\right)^2\right]}, \frac{1}{\rho\left[1 + \left(\varepsilon + \sqrt{1+\rho^{-1}}\right)^2\right]} \right],$$

whereupon the optimal gain is

$$\mathbf{G}_t(0) = \left[-1 - \sqrt{1+\rho^{-1}}, \frac{\sqrt{1+\rho^{-1}}}{\rho[2+\rho^{-1}]}, \frac{1}{\rho[2+\rho^{-1}]} \right]$$

$$= [g_1, g_2, g_3],$$

$$\mathbf{A}_c = \mathbf{A}_t + \mathbf{B}_t\mathbf{G}_t(0) - \mathbf{F}_t(0)\mathbf{M}_t$$

$$= \begin{bmatrix} -\left[1 + \sqrt{2} + \sqrt{1+\rho^{-1}}\right] & \dfrac{\sqrt{1+\rho^{-1}}}{2\rho+1} & \dfrac{1}{2\rho+1} \\ -\dfrac{1}{3} & 0 & 1 \\ -\dfrac{\sqrt{2}}{3} & -1 & 0 \end{bmatrix},$$

and the transfer function of the controller is

$$\mathscr{H}(s) = \mathbf{G}_t(0)[s\mathbf{I} - \mathbf{A}_c]^{-1}\mathbf{F}_t(0).$$

In the steady state, this controller will drive the output y in a sinusoidal manner, $\lim_{t\to\infty} y(t) = \sin t$, achieving the desired tracking. ■

8.10 Weight Selections in the LQ and LQI Problems

Thus far, we have not offered much motivation on the choice of **Q** and **R**. In this section a methodology is given for selection of these weights.

The following possibilities exist on the choice of weights **Q** and **R**:

1. Weights **Q**, **R** may be specified by the chosen *function* (e.g., if one wishes to minimize the kinetic energy T of a rigid body $2T = \boldsymbol{\omega}^T\mathbf{J}\boldsymbol{\omega}$, then the choice of **y**

and $\mathbf{Q}$ is fixed by $\mathbf{y} = \boldsymbol{\omega}$ = the angular velocity vector about body fixed axes, and $\mathbf{Q} = \mathbf{J}$ = the moment of inertia matrix about body fixed axes and the center of mass).

2. Weights may be chosen so as to achieve a specified pole assignment (i.e., specified values of $\lambda_i[\mathbf{A} + \mathbf{BGM}]$). The choice of weights that accomplishes this follows from equation (8.91) and is discussed in ref. [8.8].

3. Weights may be chosen so as to achieve specified performance bounds [8.9],

$$\int_0^\infty y_i^2 \, dt \le \sigma_i^2, \qquad i = 1, 2, \ldots, n_y,$$

$$\int_0^\infty u_i^2 \, dt \le \mu_i^2, \qquad i = 1, 2, \ldots, n_u.$$

In this section we develop a means to accomplish possibility 3, or rather special cases of 3. The root mean squared (RMS) values of the (multiple) outputs

$$y_{i_{\text{RMS}}} \triangleq \langle y_i(t), y_i(t)\rangle^{1/2} = \left[\int_0^\infty y_i^2(t)\, dt\right]^{1/2}, \qquad i = 1, 2, \ldots, n_y,$$

are usually of interest in engineering problems. For example, it may be required in the space backpack problem (3.23) to simultaneously limit the RMS value of torso rate $\dot{\theta}$ to $\dot{\theta}_{\text{RMS}} \le 10°/\text{sec}$ and the RMS leg deflections $\gamma_{\text{RMS}} \le 10°$. Failure of a control design to meet either of these objectives may lead to dismissal of the controller from further consideration. In Fig. 3.1, it may be required to constrain the RMS values of r_1, β, and θ during flight. The elastic beam of Fig. 3.6 may require RMS deflections at 10 critical points in the structure $\mu(r_1, t), \mu(r_2, t), \mu(r_3, t), \ldots, \mu(r_{10}, t)$ to satisfy inequality constraints $\mu(r_i, t)_{\text{RMS}} \le 1$ millimeter, $i = 1, 2, 3, \ldots, 10$. Such constraints may be imposed to satisfy such performance requirements as mirror imaging (extending the beam example to a plate), antenna pointing, simultaneous aiming or tracking of multiple targets, and so on. Or, such constraints may also reflect the safety limits of the structure, since deflections exceeding a certain value will exceed the stress limitations of the structure or exceed the linear elastic region causing the dynamic structure to behave in a nonlinear manner. We suggest, therefore, that the following *multiobjective* control problem has much practical significance:

The Output Constrained (OC) Controller

Minimize $\mathscr{V}_u = \sum_{i=1}^{n_w} \int_0^\infty [\mathbf{u}^{*i}\mathbf{R}\mathbf{u}^i]\, dt$ subject to the constraints

$$\dot{\mathbf{x}} = \mathbf{Ax} + \mathbf{Bu} + \mathbf{Dw}, \qquad w_i = \omega_i \delta(t),$$
$$\mathbf{y} = \mathbf{Cx}, \qquad \mathbf{u} = \mathbf{Gx}, \tag{8.145}$$

$$y_{\alpha_{\text{RMS}}} \triangleq \left\{\sum_{i=1}^{n_w} \int_0^\infty \left[y_\alpha^{i^2}(t)\right] dt\right\}^{1/2} \le \sigma_\alpha, \qquad \alpha = 1, 2, \ldots, n_y, \tag{8.146}$$

where $\{\mathbf{A}, \mathbf{B}, \mathbf{C}, \mathbf{D}, \mathbf{W}, \mathbf{V}, \mathbf{M}, \mu_1, \mu_2, \ldots, \mu_{n_u}, \sigma_1, \sigma_2, \ldots, \sigma_{n_y}\}$ are given, and $R_{ii}^{-1} = \mu_i^2$ = desired limit on control MS $(u_{i_{\mathrm{RMS}}})^2$.

The OC controller is developed as follows. The expressions

$$(y_{\alpha_{\mathrm{RMS}}})^2 \triangleq \sum_{i=1}^{n_u} \int_0^{\infty} y_\alpha^{i^2}(t)\, dt = \mathbf{c}_\alpha^* \mathbf{X} \mathbf{c}_\alpha, \qquad \mathbf{X} \triangleq \int_0^{\infty} e^{\hat{\mathbf{A}}t} \mathbf{DWD}^* e^{\hat{\mathbf{A}}^* t}\, dt, \tag{8.147}$$

$$\mathscr{V}_u = \operatorname{tr} \mathbf{RG^*XG}, \qquad \hat{\mathbf{A}} \triangleq \mathbf{A} + \mathbf{BG}, \tag{8.148}$$

allows the OC problem to be restated as follows: $\min \operatorname{tr} \mathbf{RG^*XG}$ subject to constraints

$$\mathbf{0} = \mathbf{X}\hat{\mathbf{A}}^* + \hat{\mathbf{A}}\mathbf{X} + \mathbf{DWD}^*, \qquad \mathbf{W} = \begin{bmatrix} \omega_1^2 & & \\ & \omega_2^2 & \\ & & \ddots \end{bmatrix}, \tag{8.149}$$

$$\mathbf{c}_\alpha^* \mathbf{X} \mathbf{c}_\alpha - \sigma_\alpha^2 \le 0, \qquad \mathbf{C}^* = [\mathbf{c}_1, \mathbf{c}_2, \ldots, \mathbf{c}_{n_y}]. \tag{8.150}$$

If we change the *inequality* to *equality* constraints, then the straightforward use of Lagrange multipliers leads to

$$\hat{\mathscr{V}} \triangleq \min_G \mathscr{V}$$

$$\hat{\mathscr{V}} \triangleq \Bigg\{ \operatorname{tr} \mathbf{RGXG}^* + \operatorname{tr} \mathbf{K}[\mathbf{X}(\mathbf{A} + \mathbf{BG})^* + (\mathbf{A} + \mathbf{BG})\mathbf{X} + \mathbf{DWD}^*] + \sum_{\alpha=1}^{n_y} \lambda_\alpha \left[\operatorname{tr} \mathbf{X} \mathbf{c}_\alpha \mathbf{c}_\alpha^* - \sigma_\alpha^2\right] \Bigg\}. \tag{8.151}$$

Note that $\Sigma_\alpha \mathbf{c}_\alpha \lambda_\alpha \mathbf{c}_\alpha^* = \mathbf{C^*QC}$ if $\mathbf{Q} \triangleq \operatorname{diag}[\ldots \lambda_\alpha \ldots]$, yielding the necessary conditions

$$\frac{\partial \mathscr{V}}{\partial \mathbf{K}} = \mathbf{0} = \mathbf{X}(\mathbf{A} + \mathbf{BG})^* + (\mathbf{A} + \mathbf{BG})\mathbf{X} + \mathbf{DWD}^*, \tag{8.152}$$

$$\frac{\partial \mathscr{V}}{\partial \mathbf{X}} = \mathbf{0} = \mathbf{K}(\mathbf{A} + \mathbf{BG}) + (\mathbf{A} + \mathbf{BG})^*\mathbf{K} + \mathbf{G^*RG} + \mathbf{C^*QC}, \tag{8.153}$$

$$\frac{\partial \mathscr{V}}{\partial \lambda_\alpha} = 0 = \mathbf{c}_\alpha^* \mathbf{X} \mathbf{c}_\alpha - \sigma_\alpha^2, \qquad \alpha = 1, 2, \ldots, n_y, \tag{8.154}$$

$$\frac{\partial \mathscr{V}}{\partial \mathbf{G}} = \mathbf{0} = 2\mathbf{RGX} + 2\mathbf{B^*KX} \quad \text{or} \quad \mathbf{G} = -\mathbf{R}^{-1}\mathbf{B^*K}. \tag{8.155}$$

Note that (8.155) together with (8.153) leads to (8.70b). This leads to the following conclusion.

Theorem 8.9

If there exists a state feedback controller which minimizes

$$\mathscr{V}_u = \sum_{i=1}^{n_w} \int_0^\infty \mathbf{u}^{i*} \mathbf{R} \mathbf{u}^i \, dt \tag{8.156}$$

subject to the equality constraints

$$\sum_{i=1}^{n_w} \int_0^\infty y_\alpha^{i2}(t) \, dt = \sigma_\alpha^2, \qquad \alpha = 1, 2, \ldots, n_y, \tag{8.157}$$

$$\dot{\mathbf{x}} = \mathbf{A}\mathbf{x} + \mathbf{B}\mathbf{u} + \mathbf{D}\mathbf{w}, \quad \mathbf{y} = \mathbf{C}\mathbf{x}, \quad w_i = \omega_i \delta(t), \tag{8.158}$$

then the solution is given by an LQ control (8.70*a*, *b*) *for some choice of diagonal weights* $Q_{ij} = Q_{ii}\delta_{ij}$.

Now, this theorem establishes the existence of diagonal weights in the solution, but the weights are not specified. This theorem also fails to find the OC controller since the inequality constraints were replaced by equality constraints. Next, we offer an iterative algorithm which attempts to overcome both of these limitations of Theorem 8.9.

The Weight Selection Algorithm for the OC Controller

STEP **1**: Solve the LQ problem with $Q_{ij} = \sigma_i^{-2}\delta_{ij}$, $R_{ij} = \mu_i^{-2}\delta_{ij}$:

$$\mathbf{0} = \mathbf{K}\mathbf{A} + \mathbf{A}^*\mathbf{K} - \mathbf{K}\mathbf{B}\mathbf{R}^{-1}\mathbf{B}^*\mathbf{K} + \mathbf{C}^*\mathbf{Q}\mathbf{C}, \qquad \mathbf{G} = -\mathbf{R}^{-1}\mathbf{B}^*\mathbf{K}.$$

STEP **2**: Replace $Q_{\alpha\alpha}$ by $[\mathbf{c}_\alpha^* \mathbf{X} \mathbf{c}_\alpha / \sigma_\alpha^2] Q_{\alpha\alpha}$ and return to Step 1. Repeat until $Q_{\alpha\alpha}$ stops changing (for all $\alpha = 1, 2, \ldots, n_y$).

This algorithm decreases the weight $Q_{\alpha\alpha}$ on the output $y_\alpha(t)$ in the standard LQ cost function (8.59) whenever the current RMS value of y_α is below its allowed limit σ_α, and it increases the weight otherwise. This will always force the RMS value of $y_\alpha(t)$ in the proper direction on the next iteration. Note that whenever the RMS value of the output y_α *remains* below its allowed upper limit σ_α (certain outputs may be well behaved owing to the natural dynamics of the system), the weight selection algorithm drives the weight $Q_{\alpha\alpha}$ to zero. In this event, $y_{\alpha_{\mathrm{RMS}}}$ remains below its constraint $y_{\alpha_{\mathrm{RMS}}} < \sigma_\alpha$. If the weight selection algorithm converges, it cannot converge to a value exceeding the constraint $y_{\alpha_{\mathrm{RMS}}} > \sigma_\alpha$, since such an event would drive the weight $Q_{\alpha\alpha}$ toward infinity and the term $Q_{\alpha\alpha} y_\alpha^2$ would dominate the cost function causing a minimization which reduces $y_{\alpha_{\mathrm{RMS}}}$. Experience suggests rapid convergence of the above weight selection algorithm.

Now, consider the input constrained (IC) controller defined as follows.

The Input Constrained (IC) Controller

Minimize $\mathscr{V}_y = \sum_{i=1}^{n_w} \int_0^\infty \mathbf{y}^{i*}\mathbf{Q}\mathbf{y}^i\, dt$ subject to the constraints

$$\dot{\mathbf{x}} = \mathbf{A}\mathbf{x} + \mathbf{B}\mathbf{u} + \mathbf{D}\mathbf{w}, \quad w_i = \omega_i \delta(t), \quad \mathbf{y} = \mathbf{C}\mathbf{x},$$

$$\mathbf{u} = \mathbf{G}\mathbf{x}, \mathbf{G}^* = [\mathbf{g}_1, \mathbf{g}_2, \ldots, \mathbf{g}_{n_u}]$$

$$u_{\alpha_{\mathrm{RMS}}} \triangleq \left[\sum_{i=1}^{n_w} \int_0^\infty u_\alpha^{i^2}(t)\, dt \right]^{1/2} = \mathbf{g}_\alpha^* \mathbf{X} \mathbf{g}_\alpha \le \mu_\alpha, \qquad \alpha = 1, 2, \ldots, n_u.$$

We summarize the IC controller as follows. First, an existence result.

Theorem 8.10

The state feedback controller which minimizes

$$\mathscr{V}_y = \sum_{i=1}^{n_w} \int_0^\infty \mathbf{y}^{i*}\mathbf{Q}\mathbf{y}^i\, dt, \qquad \mathbf{Q} \ge \mathbf{0}$$

subject to the equality constraints

$$u_{\alpha_{\mathrm{RMS}}} \triangleq \mu_\alpha, \qquad \alpha = 1, 2, \ldots, n_u,$$
$$\dot{\mathbf{x}} = \mathbf{A}\mathbf{x} + \mathbf{B}\mathbf{u} + \mathbf{D}\mathbf{w}, \quad \mathbf{y} = \mathbf{C}\mathbf{x}, \quad \mathbf{u} = \mathbf{G}\mathbf{x}, \quad w_i = \omega_i \delta(t), \tag{8.159}$$

is given by an LQ control (8.70*a*, *b*) *for some choice of diagonal weight* $R_{ij} = R_{ii}\delta_{ij}$.

The Weight Selection Algorithm for the IC Controller

STEP 1: Solve the LQ problem with $R_{ij} = \mu_i^{-2}\delta_{ij}$, $Q_{ij} = \sigma_i^{-2}\delta_{ij}$,

$$\mathbf{0} = \mathbf{K}\mathbf{A} + \mathbf{A}^*\mathbf{K} - \mathbf{K}\mathbf{B}\mathbf{R}^{-1}\mathbf{B}^*\mathbf{K} + \mathbf{C}^*\mathbf{Q}\mathbf{C}, \qquad \mathbf{G} = -\mathbf{R}^{-1}\mathbf{B}^*\mathbf{K}. \tag{8.160}$$

STEP 2: Replace $R_{\alpha\alpha}$ by $[\mathbf{g}_\alpha^* \mathbf{X} \mathbf{g}_\alpha / \mu_\alpha^2] R_{\alpha\alpha}$, $\alpha = 1, 2, \ldots, n_u$, and return to Step 1. Repeat until $R_{\alpha\alpha}$ stops changing for all α.

EXAMPLE 8.11

For the spacecraft of Fig. 3.4 and eq. (3.34) with $\omega_1(0) = \omega_2(0) = 1$ rad/sec, find a state feedback controller which will provide nutation control to the extent

$$\omega_{1_{\mathrm{RMS}}} \le 1 \frac{\mathrm{rad}}{\mathrm{sec}} \qquad \omega_{2_{\mathrm{RMS}}} \le \sqrt{10} \frac{\mathrm{rad}}{\mathrm{sec}} \tag{8.161}$$

Let $\dfrac{(J_{11} - J_{33})}{J_{22}} \bar{\omega}_3 = -1$, $R = 1$. Use only one control u_1 = torque about axis 1.

Solution:

STEP **1:**

$$\mathbf{A} = \begin{bmatrix} 0 & -1 \\ 1 & 0 \end{bmatrix}, \quad \mathbf{B} = \begin{bmatrix} 1 \\ 0 \end{bmatrix}, \quad \mathbf{C} = \begin{bmatrix} 1 & 0 \\ 0 & 1 \end{bmatrix}, \quad R = 1, \quad \mathbf{x}(0) = \begin{pmatrix} 1 \\ 1 \end{pmatrix},$$

$$W = 0, \qquad \mathbf{Q}(\text{initial}) = \begin{bmatrix} 1 & 0 \\ 0 & 1/10 \end{bmatrix} = \mathbf{Q}(0),$$

$$\mathbf{K}(0) = \begin{bmatrix} 1.098 & 0.049 \\ 0.049 & 1.151 \end{bmatrix},$$

$$\mathbf{G}(0) = -[1.098, 0.049], \qquad \mathbf{A} + \mathbf{BG}(0) = \begin{bmatrix} -1.098 & -1.049 \\ 1 & 0 \end{bmatrix}$$

$$\mathbf{X}(0) = \begin{bmatrix} 0.933 & -0.5 \\ -0.5 & 2.193 \end{bmatrix}, \qquad \begin{aligned} \omega_{1_{\text{RMS}}} &= 0.966, \\ \omega_{2_{\text{RMS}}} &= 1.481, \\ \mathscr{V}_u(0) &= \mathbf{GXG}^* = 1.076. \end{aligned}$$

Now, (8.161) is satisfied on the first iteration, but the algorithm continues in order to demonstrate (for this example) convergence of the weight selection algorithm to the minimal $\mathscr{V}_u$ soltuion.

STEP **2:**

$$Q_{11}(1) = \left[\frac{1 + \sqrt{1.1}/(4\sqrt{1.1} - 2)}{1} \right] Q_{11}(0) = \frac{1 + \sqrt{1.1}}{4\sqrt{1.1} - 2} = 0.933,$$

$$Q_{22}(1) = \left[\frac{4 + \sqrt{1.1}}{\sqrt{1.1}\,(4\sqrt{1.1} - 2)} \Big/ 10 \right] Q_{22}(0) = 0.022$$

$$\mathbf{K}(1) = \begin{bmatrix} 0.955 & 0.011 \\ 0.011 & 0.965 \end{bmatrix}, \qquad \mathbf{G}(1) = [-0.955, -0.011],$$

$$\mathbf{A} + \mathbf{BG}(1) = \begin{bmatrix} -0.955 & -1.0109 \\ 1 & 0 \end{bmatrix},$$

$$\mathbf{X}(1) = \begin{bmatrix} 1.050 & -0.500 \\ -0.500 & 2.500 \end{bmatrix} \Rightarrow \begin{aligned} \omega_{1_{\text{RMS}}} &= 1.025, \\ \omega_{2_{\text{RMS}}} &= 1.581, \\ \mathscr{V}_u(1) &= 0.947, \end{aligned}$$

Third Iteration:

$$Q_{11}(2) = \left[\frac{1.05}{1}\right] Q_{11}(1) = 0.980,$$

$$Q_{22}(2) = \left[\frac{2.5}{10}\right] Q_{22}(1) = 0.005$$

$$\mathbf{K}(2) = \begin{bmatrix} 0.980 & 0 \\ 0 & 0.980 \end{bmatrix}, \quad \mathbf{G}(2) = [-0.980, 0],$$

$$\mathbf{A} + \mathbf{BG}(2) = \begin{bmatrix} -0.980 & -1 \\ 1 & 0 \end{bmatrix},$$

$$\mathbf{X}(2) = \begin{bmatrix} 1.020 & -0.500 \\ -0.500 & 2.510 \end{bmatrix}, \qquad \begin{aligned} \omega_{1_{\mathrm{RMS}}} &= 1.010, \\ \omega_{2_{\mathrm{RMS}}} &= 1.580, \\ \mathscr{V}_u(2) &= 0.980. \end{aligned}$$

Fourth Iteration:

$$Q_{11}(3) = \left[\frac{1.02}{1}\right] Q_{11}(2) = 0.990,$$

$$Q_{22}(3) = \left[\frac{2.51}{10}\right] Q_{22}(2) = 0.001.$$

This algorithm converges to

$$Q_{11} = 1, \qquad Q_{22} = 0,$$

$$\mathbf{K} = \begin{bmatrix} 1 & 0 \\ 0 & 1 \end{bmatrix}, \quad \mathbf{A} + \mathbf{BG} = \begin{bmatrix} -1 & -1 \\ 1 & 0 \end{bmatrix}, \quad \mathbf{G} = [-1 \quad 0],$$

$$\mathbf{X} = \begin{bmatrix} 1 & -.5 \\ -.5 & 2.5 \end{bmatrix}, \quad \begin{aligned} \omega_{1_{\mathrm{RMS}}} &= 1.000, \\ \omega_{2_{\mathrm{RMS}}} &= 1.581, \\ \mathscr{V}_u &= 1.000. \end{aligned}$$

These results satisfy the inequality constraints (8.161), using the control

$$u_1 = -\omega_1.$$

To synthesize this control policy, only the spacecraft angular velocity about axis 1 is required. Hence, there is no need for sensors measuring ω_2, nor is there a need to penalize ω_2 in the cost function even though the specifications (8.161) impose constraints on ω_2. ■

Exercise 8.12

Repeat Example 8.11 using torques about both axes u_1 and u_2.

The example illustrates a general trend in practical control problems. Only the critical outputs need to be weighted in the LQ and LQI problems. The critical outputs can be identified as those which produce nonzero weights in the weight selection algorithm. Note that **Q** and **R** affect only **G** and **K** in the LQI problem. Hence, the weight selection algorithms above may be extended to the case of dynamic controllers driven by disturbed measurements $\mathbf{z} = \mathbf{M}\mathbf{x} + \mathbf{v}$, by correcting only the calculation $y_{\alpha_{\mathrm{RMS}}}$ or $u_{\alpha_{\mathrm{RMS}}}$, as follows.

The Weight Selection Algorithm for the Dynamic Output Constrained Controller (DOC)

Given $(\mathbf{A}, \mathbf{B}, \mathbf{C}, \mathbf{M}, \mathbf{W}, \mathbf{V}, \mathbf{R}, \sigma_i,\ i = 1, 2, \ldots, n_y,\ \mu_j,\ j = 1, 2, \ldots, n_u)$.

STEP 1: Compute the LQI controller gain **F**:

$$\mathbf{0} = \mathbf{P}\mathbf{A}^* + \mathbf{A}\mathbf{P} - \mathbf{P}\mathbf{M}^*\mathbf{V}^{-1}\mathbf{M}\mathbf{P} + \mathbf{D}\mathbf{W}\mathbf{D}^* + \mathbf{X}_0, \qquad \mathbf{F} = \mathbf{P}\mathbf{M}^*\mathbf{V}^{-1}. \tag{8.162}$$

Set $Q_{ij} = \sigma_i^{-2}\delta_{ij}$, $i = 1, 2, \ldots, n_y$, and $R_{ij} = \mu_i^{-2}\delta_{ij}$, $i = 1, 2, \ldots, n_u$.

STEP 2: Compute

$$\mathbf{0} = \mathbf{K}\mathbf{A} + \mathbf{A}^*\mathbf{K} - \mathbf{K}\mathbf{B}\mathbf{R}^{-1}\mathbf{B}^*\mathbf{K} + \mathbf{C}^*\mathbf{Q}\mathbf{C}, \qquad \mathbf{G} = -\mathbf{R}^{-1}\mathbf{B}^*\mathbf{K}, \tag{8.163}$$

$$\mathbf{0} = \mathbf{X}_c(\mathbf{A} + \mathbf{B}\mathbf{G})^* + (\mathbf{A} + \mathbf{B}\mathbf{G})\mathbf{X}_c + \mathbf{F}\mathbf{V}\mathbf{F}^* + \mathbf{X}_{co} \tag{8.164}$$

$$(y_{i_{\mathrm{RMS}}})^2 \triangleq \mathbf{c}_i^*[\mathbf{P} + \mathbf{X}_c]\mathbf{c}_i, \qquad \mathbf{C}^* = [\mathbf{c}_1, \mathbf{c}_2, \ldots, \mathbf{c}_{n_y}]. \tag{8.165}$$

STEP 3: Replace Q_{ii} with $[y_{i_{\mathrm{RMS}}}/\sigma_i]^2 Q_{ii}$ for $i = 1, 2, \ldots, n_y$. Return to Step 2. Repeat until Q_{ii} stops changing.

The Weight Selection Algorithm for the Dynamic Input Constrained Controller (DIC)

Given $(\mathbf{A}, \mathbf{B}, \mathbf{C}, \mathbf{M}, \mathbf{W}, \mathbf{V}, \mathbf{Q}, \mu_i,\ i = 1, 2, \ldots, n_u,\ \mathbf{x}(0), \mathbf{x}_c(0))$.

STEP 1: Compute the LQI controller gain **F**,

$$\mathbf{0} = \mathbf{P}\mathbf{A}^* + \mathbf{A}\mathbf{P} - \mathbf{P}\mathbf{M}^*\mathbf{V}^{-1}\mathbf{M}\mathbf{P} + \mathbf{D}\mathbf{W}\mathbf{D}^* + \mathbf{X}_0, \qquad \mathbf{F} = \mathbf{P}\mathbf{M}^*\mathbf{V}^{-1}. \tag{8.166}$$

Set $R_{ij} = \mu_i^{-2}\delta_{ij}$, $i = 1, 2, \ldots, n_u$.

STEP 2: Compute

$$0 = \mathbf{KA} + \mathbf{A}^*\mathbf{K} - \mathbf{KBR}^{-1}\mathbf{B}^*\mathbf{K} + \mathbf{C}^*\mathbf{QC},$$

$$\mathbf{G} = -\mathbf{R}^{-1}\mathbf{B}^*\mathbf{K}, \tag{8.167}$$

$$0 = \mathbf{X}_c(\mathbf{A} + \mathbf{BG})^* + (\mathbf{A} + \mathbf{BG})\mathbf{X}_c + \mathbf{FVF}^* + \mathbf{X}_{c0}, \tag{8.168}$$

$$(u_{i_{\mathrm{RMS}}})^2 \triangleq \mathbf{g}_i^*\mathbf{X}_c\mathbf{g}_i, \qquad \mathbf{G}^* = [\mathbf{g}_1, \mathbf{g}_2, \ldots, \mathbf{g}_{n_u}]. \tag{8.169}$$

STEP 3: Replace R_{ii} with $[u_{i_{\mathrm{RMS}}}/\mu_i]^2 R_{ii}$, $i = 1, 2, \ldots, n_u$. Return to Step 2. Repeat until R_{ii}, $i = 1, 2, \ldots, n_u$, stops changing.

Exercise 8.13

Find the DOC controller for Example 8.11 that satisfies $\omega_{1_{\mathrm{RMS}}} \leq 1$ rad/sec, $\omega_{2_{\mathrm{RMS}}} \leq \sqrt{10}$ rad/sec. Let $z = \omega_2 + v$.

8.11 *Optimal Reduced Order Controllers*

This chapter has produced two types of controllers which are tractable for relatively easy design: (i) controllers of order n_x (equal to plant order), and (ii) state feedback controllers. These two controllers are given in Theorem 8.2 and the corollary to Theorem 8.3, respectively. Theorem 8.3 describes a controller of order zero that feeds back only the measurement vector. This problem is much more difficult and no closed form solution exists. Section 8.11 fills the gap between theories that produce n_xth order and zero order controllers. In this section the order of the controller is specified *a priori* by the designer. The solution to this problem will be of similar difficulty to the measurement feedback control problem of Theorem 8.3. Let n_c be the chosen order of the controller and let $\mathbf{x}_c$ be the state of the controller,

$$\begin{aligned} \mathbf{u} &= \mathbf{Gx}_c + \mathbf{Hz} \\ \dot{\mathbf{x}}_c &= \mathbf{A}_c\mathbf{x}_c + \mathbf{Fz} \end{aligned} \tag{8.170}$$

where the disturbance-free measurements $\mathbf{z}(t)$ emanate from the plant

$$\begin{aligned} \dot{\mathbf{x}} &= \mathbf{Ax} + \mathbf{Bu} + \mathbf{Dw} \\ \mathbf{y} &= \mathbf{Cx} \\ \mathbf{z} &= \mathbf{Mx}. \end{aligned} \tag{8.171}$$

The closed loop system is described by

$$\dot{\mathscr{x}} \triangleq \begin{pmatrix} \dot{\mathbf{x}} \\ \dot{\mathbf{x}}_c \end{pmatrix} = \begin{bmatrix} \mathbf{A} + \mathbf{BHM} & \mathbf{BG} \\ \mathbf{FM} & \mathbf{A}_c \end{bmatrix} \begin{pmatrix} \mathbf{x} \\ \mathbf{x}_c \end{pmatrix} + \begin{bmatrix} \mathbf{D} \\ \mathbf{0} \end{bmatrix} \mathbf{w} = \mathscr{A}\mathscr{x} + \mathscr{D}\mathbf{w}$$

$$\mathscr{y} \triangleq \begin{pmatrix} \mathbf{y} \\ \mathbf{u} \end{pmatrix} = \begin{bmatrix} \mathbf{C} & \mathbf{0} \\ \mathbf{HM} & \mathbf{G} \end{bmatrix} \begin{pmatrix} \mathbf{x} \\ \mathbf{x}_c \end{pmatrix} = \mathscr{C}\mathscr{x} \tag{8.172}$$

It is helpful to note that the matrices $(\mathscr{A}, \mathscr{B}, \mathscr{C}, \mathscr{D})$ have the structure

$$\mathscr{A} \triangleq \begin{bmatrix} \mathbf{A} & \mathbf{0} \\ \mathbf{0} & \mathbf{0} \end{bmatrix} + \begin{bmatrix} \mathbf{B} & \mathbf{0} \\ \mathbf{0} & \mathbf{I} \end{bmatrix} \begin{bmatrix} \mathbf{G} & \mathbf{H} \\ \mathbf{A}_c & \mathbf{F} \end{bmatrix} \begin{bmatrix} \mathbf{0} & \mathbf{I} \\ \mathbf{M} & \mathbf{0} \end{bmatrix} = \mathbf{A}_0 + \mathbf{B}_0\mathbf{G}_0\mathbf{M}_0 \tag{8.173a}$$

$$\mathscr{D} \triangleq \begin{bmatrix} \mathbf{D} \\ \mathbf{0} \end{bmatrix}, \tag{8.173b}$$

$$\mathscr{C} \triangleq \begin{bmatrix} \mathbf{C} & \mathbf{0} \\ \mathbf{0} & \mathbf{0} \end{bmatrix} + \begin{bmatrix} \mathbf{0} & \mathbf{0} \\ \mathbf{I} & \mathbf{0} \end{bmatrix} \begin{bmatrix} \mathbf{G} & \mathbf{H} \\ \mathbf{A}_c & \mathbf{F} \end{bmatrix} \begin{bmatrix} \mathbf{0} & \mathbf{I} \\ \mathbf{M} & \mathbf{0} \end{bmatrix} = \mathbf{C}_0 + \mathbf{I}_0\mathbf{G}_0\mathbf{M}_0. \tag{8.173c}$$

This structure allows the controller design parameters to be lumped conveniently into one matrix denoted by $\mathbf{G}_0$. The definitions of $(\mathbf{G}_0, \mathbf{A}_0, \mathbf{B}_0, \mathbf{M}_0, \mathbf{C}_0, \mathbf{I}_0)$, are implied in the above equations for $(\mathscr{A}, \mathscr{B}, \mathscr{C})$. Following the LQI theory of this chapter the closed loop system performance is described by

$$\mathscr{V} = \operatorname{tr} \mathbf{X}_0 \mathscr{C}^* \mathscr{Q} \mathscr{C}, \qquad \mathscr{Q} = \begin{bmatrix} \mathbf{Q} & \mathbf{0} \\ \mathbf{0} & \mathbf{R} \end{bmatrix}, \tag{8.174a}$$

$$\mathbf{0} = \mathbf{X}_0\mathscr{A}^* + \mathscr{A}\mathbf{X}_0 + \mathscr{D}\mathbf{W}\mathscr{D}^*, \tag{8.174b}$$

and the controller parameters are sought to solve

$$\min_{\mathbf{G}_0\mathbf{X}_0\mathbf{K}_0} \hat{\mathscr{V}}, \hat{\mathscr{V}} \triangleq \{\operatorname{tr} \mathbf{X}_0\mathscr{C}^*\mathscr{Q}\mathscr{C} + \operatorname{tr} \mathbf{K}_0(\mathbf{X}_0\mathscr{A}^* + \mathscr{A}\mathbf{X}_0 + \mathscr{D}\mathbf{W}\mathscr{D}^*)\}, \tag{8.175}$$

$$\frac{\partial \hat{\mathscr{V}}}{\partial \mathbf{K}_0} = \mathbf{X}_0\mathscr{A}^* + \mathscr{A}\mathbf{X}_0 + \mathscr{D}\mathbf{W}\mathscr{D}^* = \mathbf{0}, \tag{8.176a}$$

$$\frac{\partial \hat{\mathscr{V}}}{\partial \mathbf{X}_0} = \mathbf{K}_0\mathscr{A} + \mathscr{A}^*\mathbf{K}_0 + \mathscr{C}^*\mathscr{Q}\mathscr{C} = \mathbf{0}, \tag{8.176b}$$

$$\frac{\partial \hat{\mathscr{V}}}{\partial \mathbf{G}_0} = 2\mathbf{I}_0^*\mathscr{Q}(\mathbf{C}_0 + \mathbf{I}_0\mathbf{G}_0\mathbf{M}_0)\mathbf{X}_0\mathbf{M}_0^* + 2\mathbf{B}_0^*\mathbf{K}_0\mathbf{X}_0\mathbf{M}_0^* = \mathbf{0}. \tag{8.176c}$$

The first two of these equations are straightforward applications of section 2.7.1.3.

The third equation follows after substituting the definitions of $(\mathscr{A}, \mathscr{C})$ into (8.175). Using (8.173) and noting that

$$\mathbf{I}_0^* \mathscr{Q} \mathbf{C}_0 = \mathbf{0}, \tag{8.177a}$$

$$\mathscr{C}^* \mathscr{Q} \mathscr{C} = \mathbf{C}_0^* \mathscr{Q} \mathbf{C}_0 + \mathbf{M}_0^* \mathbf{G}_0^* \mathbf{R}_0 \mathbf{G}_0 \mathbf{M}_0, \tag{8.177b}$$

$$\mathbf{R}_0 \triangleq \mathbf{I}_0^* \mathscr{Q} \mathbf{I}_0 = \begin{bmatrix} \mathbf{R} & \mathbf{0} \\ \mathbf{0} & \mathbf{0} \end{bmatrix}, \tag{8.177c}$$

(8.175) becomes

$$\mathbf{0} = \mathbf{X}_0(\mathbf{A}_0 + \mathbf{B}_0\mathbf{G}_0\mathbf{M}_0)^* + (\mathbf{A}_0 + \mathbf{B}_0\mathbf{G}_0\mathbf{M}_0)\mathbf{X}_0 + \mathscr{D}\mathbf{W}\mathscr{D}^*, \tag{8.178a}$$

$$\mathbf{0} = \mathbf{K}_0(\mathbf{A}_0 + \mathbf{B}_0\mathbf{G}_0\mathbf{M}_0) + (\mathbf{A}_0 + \mathbf{B}_0\mathbf{G}_0\mathbf{M}_0)^*\mathbf{K}_0 + \mathbf{C}_0^* \mathscr{Q} \mathbf{C}_0 + \mathbf{M}_0^* \mathbf{G}_0^* \mathbf{R}_0 \mathbf{G}_0 \mathbf{M}_0, \tag{8.178b}$$

$$\mathbf{0} = 2\mathbf{R}_0\mathbf{G}_0\mathbf{M}_0\mathbf{X}_0\mathbf{M}_0^* + 2\mathbf{B}_0^*\mathbf{K}_0\mathbf{X}_0\mathbf{M}_0^*. \tag{8.178c}$$

A comparison of (8.178) with (8.63) through (8.65) confirms that the optimal dynamic controller of specified order n_c leads to the same type of mathematical problem as in the optimal measurement feedback control. Indeed, the measurement feedback equations (8.63) through (8.65) are a special case of the dynamic controller equations (8.178). Also, the state feedback controller (corollary to Theorem 8.3) is a special case of the measurement feedback controller (Theorem 8.3). Hence, equations (8.178) are more fundamental since they generate previous results as special cases. Thus, the optimal *dynamic controller* of order n_c is obtained by solving the optimal *measurement feedback problem* for the augmented system $(\mathbf{A}_0, \mathbf{B}_0, \mathbf{M}_0, \mathbf{C}_0)$. Furthermore, note from the structure of the $\mathbf{A}_0$ matrix that this augmented plant is composed of the original plant plus a set of n_c integrators (implying n_c zero eigenvalues). These integrators are controllable ($(\mathbf{A}_0, \mathbf{B}_0)$ is a controllable pair) but not observable [$(\mathbf{A}_0, \mathbf{C}_0)$ is not is an observable pair]. This is yet another reminder that a minimal transfer equivalent realization (TER) of a system will not generally provide enough information for good control policies. In the present situation we can certainly see that if the unobservable, uncontrollable parts of $(\mathbf{A}_0, \mathbf{B}_0, \mathbf{C}_0, \mathbf{M}_0)$ are discarded, the dynamic controller for $(\mathbf{A}, \mathbf{B}, \mathbf{C}, \mathbf{M})$ reduces to the measurement feedback controller $\mathbf{u} = \mathbf{G}\mathbf{z}$. That is, the order of the dynamic controller is equal to the number of unobservable, uncontrollable states in the system $(\mathbf{A}_0, \mathbf{B}_0, \mathbf{C}_0)$. Equations (8.178) are somewhat more difficult to treat than equations (8.63) through (8.65) due to the singular $\mathbf{R}_0$ matrix. Assuming $\mathbf{R} > \mathbf{0}$, the matrix $\mathbf{R}_0$ has rank equal to n_u.

Exercise 8.14

Prove that $(\mathbf{A}_0, \mathbf{B}_0)$ is controllable pair if $(\mathbf{A}, \mathbf{B})$ is. Prove that $(\mathbf{A}_0, \mathbf{C}_0)$ is not an observable pair.

TABLE 8.8 Optimal Controllers

COST FUNCTION $\mathscr{V}$	EXCITATION	COST VALUE	SPECIAL COMPUTATIONS	OPTIMAL CONTROLLER
$\sum_{i=1}^{n_w+n_v+2n_x}\mathscr{V}^{(i)}$	One at a time: $w_i(t) = \omega_i\delta(t)$ $v_i(t) = v_i\delta(t)$ $x_i(0) = x_{i0}$ $x_{c_i}(0) = x_{c_{i0}}$	$\mathscr{V} = \mathrm{tr}\,(\mathbf{P} + \mathbf{X}_c)\mathbf{C}^*\mathbf{QC} + \mathrm{tr}\,\mathbf{X}_c\mathbf{G}^*\mathbf{RG}$ $= \mathrm{tr}\,(\tilde{\mathbf{K}} + \mathbf{K})(\mathbf{DWD}^* + \mathbf{X}_0)$ $+ \mathrm{tr}\,\tilde{\mathbf{K}}\mathbf{FVF}^* + \mathrm{tr}\,\mathbf{KX}_{c0}$	$\mathbf{0} = \mathbf{PA}^* + \mathbf{AP} - \mathbf{PM}^*\mathbf{V}^{-1}\mathbf{MP} + \mathbf{DWD}^* + \mathbf{X}_0$ $\mathbf{0} = \mathbf{X}_c(\mathbf{A} + \mathbf{BG})^* + (\mathbf{A} + \mathbf{BG})\mathbf{X}_c + \mathbf{FVF}^* + \mathbf{X}_{c0}$	GLQI Controller: $\mathbf{u} = \mathbf{Gx}_c$ [eq. (8.55)]
$\sum_{i=1}^{n_x+n_w+n_v}\mathscr{V}^{(i)}$	One at a time: $v_i(t) = v_i\delta(t)$ simultaneously: $x_i(0) = x_{i0}$	$\mathscr{V} = \mathrm{tr}\,(\mathbf{P} + \mathbf{X}_c)\mathbf{C}^*\mathbf{QC} + \mathrm{tr}\,\mathbf{X}_c\mathbf{G}^*\mathbf{RG}$ $= \mathrm{tr}\,(\tilde{\mathbf{K}} + \mathbf{K})(\mathbf{DWD}^* + \mathbf{x}_0\mathbf{x}_0^*)$ $+ \mathrm{tr}\,\tilde{\mathbf{K}}\mathbf{FVF}^*$	$\mathbf{0} = \mathbf{PA}^* + \mathbf{AP} - \mathbf{PM}^*\mathbf{V}^{-1}\mathbf{MP} + \mathbf{x}_0\mathbf{x}_0^* + \mathbf{DWD}^*$ $\mathbf{0} = \mathbf{X}_c(\mathbf{A} + \mathbf{BG})^* + (\mathbf{A} + \mathbf{BG})\mathbf{X}_c + \mathbf{FVF}^*$	LQI Controller: $\mathbf{u} = \mathbf{Gx}_c$ $\mathbf{x}_c(0) = \mathbf{0}$
$\sum_{i=1}^{n_w+n_x}\mathscr{V}^{(i)}$	One at a time: $\mathrm{w}_i(t) = \omega_i\delta(t)$ $\mathrm{x}_i(0) = \mathrm{x}_{i0}$	$\mathscr{V} = \mathrm{tr}\,\mathbf{X}(\mathbf{C}^*\mathbf{QC} + \mathbf{G}^*\mathbf{RG})$ $= \mathrm{tr}\,\mathbf{K}(\mathbf{DWD}^* + \mathbf{X}_0)$	$\mathbf{0} = \mathbf{X}(\mathbf{A} + \mathbf{BG})^* + (\mathbf{A} + \mathbf{BG})\mathbf{X} + \mathbf{DWD}^* + \mathbf{X}_0$	*LQ Controller*: $\mathbf{u} = \mathbf{Gx}$ [*eq.* (8.70)]
Common Data	$\dot{\mathbf{x}} = \mathbf{Ax} + \mathbf{Bu} + \mathbf{Dw}$, $\mathbf{0} = \mathbf{KA} + \mathbf{A}^*\mathbf{K} - \mathbf{KBR}^{-1}\mathbf{B}^*\mathbf{K} + \mathbf{C}^*\mathbf{QC}$, $\mathbf{0} = \tilde{\mathbf{K}}(\mathbf{A} - \mathbf{FM}) + (\mathbf{A} - \mathbf{FM})^*\tilde{\mathbf{K}} + \mathbf{G}^*\mathbf{RG}$, $\mathbf{y} = \mathbf{Cx}$, $\mathbf{z} = \mathbf{Mx} + \mathbf{v}$, $\dot{\mathbf{x}}_c = \mathbf{A}_c\mathbf{x}_c + \mathbf{Fz}$, $\mathbf{A}_c = \mathbf{A} + \mathbf{BG} - \mathbf{FM}$, $\mathbf{F} = \mathbf{PM}^*\mathbf{V}^{-1}$, $\mathbf{G} = -\mathbf{R}^{-1}\mathbf{B}^*\mathbf{K}$, $\mathbf{W} \triangleq \mathrm{diag}[\ldots \omega_i^2 \ldots]$, $\mathbf{V} \triangleq \mathrm{diag}[\ldots v_i^2 \ldots]$, $\mathbf{X}_0 \triangleq \mathrm{diag}[\ldots x_{i0}^2 \ldots]$, $\mathbf{X}_{c0} \triangleq \mathrm{diag}[\ldots \mathbf{x}_{c_{i0}}^2 \ldots]$, $\mathscr{V} \triangleq \int_0^\infty [\mathbf{y}^*(t)\mathbf{Q}\mathbf{y}(t) + \mathbf{u}^*(t)\mathbf{R}\mathbf{u}(t)]\,dt$, $\mathscr{V}^{(i)} \triangleq$ value of $\mathscr{V}$ when only the ith excitation is applied [from $\mathbf{w}(t), \mathbf{v}(t), \mathbf{x}(0), \mathbf{x}_c(0)$].			

Closure

Chapter 8 provides a completely deterministic theory of optimal control and estimation which includes impulsive disturbances and initial conditions on the plant and controller states. Table 8.3 summarizes the solutions for different optimization problems. These problems are distinguished by the definition of the system excitation. When the excitations $w_i(t) = \omega_i \delta(t)$, $v_i(t) = v_i \delta(t)$, $x_i(0) = x_{i0}$, $x_{c_i}(0) = x_{c_{i0}}$ are applied one at a time, the optimal controller is GLQI in Table 8.8. When the initial conditions $x_i(0)$, $x_{c_i}(0)$ are all applied simultaneously (as opposed to one at a time), then $\mathbf{X}_0$ and $\mathbf{X}_{c0}$ in the GLQI controller changes to $\mathbf{x}_0 \mathbf{x}_0^*$ and $\mathbf{x}_{c0} \mathbf{x}_{c0}^*$, respectively. In addition, if we set $\mathbf{x}_{c0} = \mathbf{0}$, then we obtain the optimal controller labeled LQI in Table 8.8. These same variations exist for the state feedback controller labeled LQ.

References

8.1 T. E. Fortmann and K. L. Hitz, *An Introduction to Linear Control Systems*. Marcel Dekker, New York, 1977.

8.2 C. D. Johnson, "The Phenomenon of Homeopathic Instability in Dynamical Systems," *Int. J. Control*, 33(1), 159–173, 1981.

8.3 W. S. Levine, T. L. Johnson, and M. Athens, "Optimal Limited State Variable Feedback Controllers for Linear Systems," *IEEE Trans. Autom. Control*, AC-15, 557–563, 1971. See also R. L. Kosut, "Suboptimal Control of Linear Time-Invariant Systems Subject to Control Structure Constraints," *IEEE Trans. Autom. Control*, AC-15, 557–563, 1970.

8.4 D. C. Hyland and D. S. Bernstein, "The Optimal Projection Equations for Fixed-Order Dynamic Compensation," *IEEE Trans. Autom. Control*, AC-29, 1034–1037, 1984.

8.5 B. D. O. Anderson and J. B. Moore, *Linear Optimal Control*. Prentice-Hall, Englewood Cliffs, N.J., 1971.

8.6 A. J. Laub, "A Schur Method for Solving Algebraic Riccati Equations," *IEEE Trans. Autom. Control*, AC-24(6), 913–921, 1979.

8.7 L. Weinberg, *Network Analysis and Synthesis*. McGraw-Hill, New York, 1962.

8.8 A. H. Harvey and G. Stein, "Quadratic Weights for Asymptotic Regulator Properties," *IEEE Trans. Autom. Control*, AC-23, 378–387, June 1978.

8.9 M. Delorenzo and R. Skelton, "Space Structure Control Design by Variance Assignment," *AIAA J. Guidance, Control and Dynamics*, 8(4), 454–462, 1985.

CHAPTER 9

State Estimation

In Chapter 5 we considered the question, "Under what conditions can we uniquely recover the state vector $\mathbf{x}(0)$, given only the output response $\mathbf{y}(t)$ and knowledge of the inputs $\mathbf{u}(t)$?" See Definition 5.3. The general answer to this question was provided by Theorems 5.14 and 5.15. That is, for the time invariant systems,

$$\dot{\mathbf{x}} = \mathbf{A}\mathbf{x} + \mathbf{B}\mathbf{u}, \qquad \mathbf{y} = \mathbf{C}\mathbf{x} + \mathbf{H}\mathbf{u}.$$

The initial state

$$\mathbf{x}(0) = \mathbf{K}^{-1}\int_0^{\infty} e^{\mathbf{A}^*\sigma}\mathbf{C}^*\left\{\mathbf{y}(t) - \int_0^t \left(\mathbf{C}^{A(t-\tau)}\mathbf{B} + \mathbf{H}\delta(t-\tau)\right)\mathbf{u}(\tau)\,d\tau\right\}d\sigma,$$

$$\mathbf{K} \triangleq \int_0^{\infty} e^{\mathbf{A}^*\sigma}\mathbf{C}^*\mathbf{C}e^{\mathbf{A}\sigma}\,d\sigma$$

can be computed if the indicated inverse exists and theorem 5.15 provides the necessary and sufficient conditions for this inverse to exist. Now, since

$$\mathbf{x}(t) = e^{\mathbf{A}t}\mathbf{x}(0) + \int_0^t e^{\mathbf{A}(t-\sigma)}\mathbf{B}\mathbf{u}(\sigma)\,d\sigma,$$

we have a construction of $\mathbf{x}(t)$ from the data $\mathbf{y}(t), \mathbf{u}(t)$ by combining the above eqs.,

$$\mathbf{x}(t) = e^{\mathbf{A}t}\mathbf{K}^{-1}\int_0^{\infty} e^{\mathbf{A}^*\sigma}\mathbf{C}^*\left\{\mathbf{y}(\sigma) - \int_0^{\sigma}\left[\mathbf{C}e^{\mathbf{A}(\sigma-\tau)}\mathbf{B}\mathbf{u}(\tau) + \mathbf{H}\delta(\sigma-\tau)\right]d\tau\right\}d\sigma$$

$$+ \int_0^t e^{\mathbf{A}(t-\sigma)}\mathbf{B}\mathbf{u}(\sigma)\,d\sigma,$$

where **K** satisfies $\mathbf{0} = \mathbf{KA} + \mathbf{A^*K} + \mathbf{C^*C}$ if the observable modes are stable. The calculation of $\mathbf{x}(t)$ from $\mathbf{y}(t), \mathbf{u}(t)$ is a formidable task since the above equation requires integrals on an infinite interval. The theory of observability (see Definition 5.3) allows infinity to be replaced by a finite time t_f, and we say that "the state is observable at time t_f" if the integral is positive definite,

$$\int_0^{t_f} e^{\mathbf{A}^*\sigma}\mathbf{C}^*\mathbf{C}e^{\mathbf{A}\sigma}\, d\sigma,$$

but this calculation is also difficult. Hence, we shall rely on the theory of Chapter 5 to tell us when it is possible to construct the state (that is, when the system is observable), but we shall not use the above direct approach to construct this state. We shall seek another, more tractable approach.

The pole assignment and covariance assignment of Chapter 7 presume that the state is available for feedback. This is usually not the case in practice. This chapter will provide practical means to estimate the state $\mathbf{x}(t)$ in real time, given the past measurement data and the inputs up to the present time. In this problem formulation, we also add disturbances to the measurement equations $\mathbf{z} = \mathbf{Mx} + \mathbf{v}$, where $z_i(t)$, $i = 1, 2, \ldots, n_z$, represent the outputs of the n_z measurement devices and $v_i = \mathscr{v}_i\delta(t)$ represents an impulsive disturbance of strength $\mathscr{v}_i$. Such real-time state estimators are useful in signal processing and tracking problems, in addition to feedback control purposes.

This chapter seeks to develop "state estimation" devices which receive as inputs the system measurements $\mathbf{z}(t)$ and all known inputs $\mathbf{u}(t)$ and produce as outputs a function we shall call "$\hat{\mathbf{x}}(t)$," which has the property

$$\lim_{t\to\infty} [\mathbf{x}(t) - \hat{\mathbf{x}}(t)] = 0. \tag{9.1}$$

The time constants that govern the speed of convergence of $\hat{\mathbf{x}}(t)$ to $\mathbf{x}(t)$ must be selectable by the designer if the estimate is to be useful. We shall call $\hat{\mathbf{x}}(t)$ with property (9.1) an "estimate" of the state $\mathbf{x}(t)$, and the device which generates $\hat{\mathbf{x}}(t)$ is called the "state estimator." We seek the best *linear* state esimator of the linear system

$$\begin{aligned} \dot{\mathbf{x}} &= \mathbf{Ax} + \mathbf{Bu} + \mathbf{Dw}, \qquad \mathbf{x}(0) = \mathbf{x}_0, \\ \mathbf{z} &= \mathbf{Mx} + \mathbf{v}, \end{aligned} \tag{9.2}$$

where $v_i(t) = \mathscr{v}_i\delta(t)$ is an impulsive measurement disturbance of intensity $\mathscr{v}_i$ and $w_i(t) = \mathscr{w}_i\delta(t)$ is an impulsive disturbance of intensity $\mathscr{w}_i$. Now, if the estimator is linear, it must have the linear form

$$\dot{\hat{\mathbf{x}}} = \mathbf{F}_1\hat{\mathbf{x}} + \mathbf{F}_2\mathbf{u} + \mathbf{F}_3\mathbf{z}, \tag{9.3}$$

for some choice of $\mathbf{F}_i$, $i = 1, 2, 3$. To test for the property (9.1), define $\tilde{\mathbf{x}} \triangleq \mathbf{x} - \hat{\mathbf{x}}$

and subtract (9.3) from (9.2) to get

$$\begin{aligned}\dot{\tilde{\mathbf{x}}} &= \mathbf{Ax} + \mathbf{Bu} + \mathbf{Dw} - \mathbf{F}_1\hat{x} - \mathbf{F}_2 u - \mathbf{F}_3 z \\ &= [\mathbf{A} - \mathbf{F}_3\mathbf{M}]\mathbf{x} + [\mathbf{B} - \mathbf{F}_2]\mathbf{u} + \mathbf{Dw} - \mathbf{F}_1(\mathbf{x} - \tilde{\mathbf{x}}) - \mathbf{F}_3\mathbf{v} \\ &= [\mathbf{A} - \mathbf{F}_3\mathbf{M} - \mathbf{F}_1]\mathbf{x} + \mathbf{F}_1\tilde{\mathbf{x}} + [\mathbf{B} - \mathbf{F}_2]\mathbf{u} + \mathbf{Dw} - \mathbf{F}_3\mathbf{v}. \end{aligned} \tag{9.4}$$

Since (9.1) must be achieved for *every* trajectory $\mathbf{x}(t)$ and for *every* input $\mathbf{u}(t)$, the coefficients of these variables must be zero in (9.4), as a necessary condition for state estimation

$$\begin{aligned}\mathbf{A} - \mathbf{F}_3\mathbf{M} - \mathbf{F}_1 &= \mathbf{0}, \\ \mathbf{B} - \mathbf{F}_2 &= \mathbf{0}.\end{aligned} \tag{9.5}$$

This leaves (9.4) in the form $\dot{\tilde{\mathbf{x}}} = \mathbf{F}_1\tilde{\mathbf{x}} + \mathbf{Dw} - \mathbf{F}_3\mathbf{v}$ for arbitrary $\tilde{\mathbf{x}}(0)$. Hence, the sufficient condition to achieve (9.1) is (9.5) and a choice of $\mathbf{F}_1$ such that

$$\mathscr{Re}\,\lambda_i[\mathbf{F}_1] < 0, \qquad i = 1, 2, \ldots, n_x. \tag{9.6}$$

The results (9.5) and (9.6) are summarized as follows, defining $\mathbf{F} \triangleq \mathbf{F}_3$.

Theorem 9.1

A linear state estimator for (9.2) *of order* n_x *must satisfy*

$$\dot{\hat{\mathbf{x}}} = \mathbf{A}\hat{\mathbf{x}} + \mathbf{Bu} + \mathbf{F}(\mathbf{z} - \mathbf{M}\hat{\mathbf{x}}), \tag{9.7}$$

where $\mathbf{u}$ *is the known input, and* $\mathbf{F}$ *must be chosen so that* $[\mathbf{A} - \mathbf{FM}]$ *is asymptotically stable.*

One way to stabilize $\mathbf{A} - \mathbf{FM}$ is to use pole assignment to find $\mathbf{F}$ such that the eigenvalues of $[\mathbf{A} - \mathbf{FM}]$ take on assigned values as in Chapter 7. Another approach is to use the dual of the optimal control result [Theorem 8.4, (8.73)] which stabilizes $\mathbf{A} + \mathbf{BG}$. Recall from Theorem 8.4, part (c), and

$$\mathscr{Re}\,\lambda_i[\mathbf{A} + \mathbf{BG}] < 0 \text{ if}$$

1. $(\mathbf{A}, \mathbf{B})$ stabilizable, $(\mathbf{A}, \mathbf{C})$ detectable
2. $\mathbf{G} = -\mathbf{R}^{-1}\mathbf{B}^T\mathbf{K}, \qquad \mathbf{0} = \mathbf{KA} + \mathbf{A}^*\mathbf{K} - \mathbf{KBR}^{-1}\mathbf{B}^*\mathbf{K} + \mathbf{C}^*\mathbf{QC}$ (9.8)

for any $\mathbf{Q} > \mathbf{0}$, $\mathbf{R} > \mathbf{0}$.

Rename the matrices of this mathematical result

$$\mathbf{A} \to \mathbf{A}^*$$
$$\mathbf{B} \to \mathbf{M}^*$$
$$\mathbf{G} \to -\mathbf{F}^*$$
$$\mathbf{K} \to \mathbf{P}$$
$$\mathbf{C} \to \mathbf{D}^*$$
$$\mathbf{R} \to \mathbf{V}$$
$$\mathbf{Q} \to \mathbf{W}$$

to obtain the restatement of (9.8),

$$\mathscr{Re}\,\lambda_i[\mathbf{A}^* - \mathbf{M}^*\mathbf{F}^*] < 0 \text{ if}$$

1. $(\mathbf{A}^*, \mathbf{M}^*)$ stabilizable, $(\mathbf{A}^*, \mathbf{D}^*)$ detectable

2. $$-\mathbf{F}^* = -\mathbf{V}^{-1}\mathbf{M}\mathbf{P}, \qquad \mathbf{0} = \mathbf{P}\mathbf{A}^* + \mathbf{A}\mathbf{P} - \mathbf{P}\mathbf{M}^*\mathbf{V}^{-1}\mathbf{M}\mathbf{P} + \mathbf{D}\mathbf{W}\mathbf{D}^* \tag{9.9}$$

for any $\mathbf{V} > \mathbf{0}$, $\mathbf{W} > \mathbf{0}$.

But since the eigenvalues of $\mathscr{A}$ and $\mathscr{A}^*$ are the same and since $\{(\mathbf{A}^*, \mathbf{M}^*)$ detectable$\} \Leftrightarrow \{(\mathbf{A}, \mathbf{M})$ stabilizable$\}$, (9.9) reduces to the following.

Theorem 9.2
The matrix $\mathbf{A} - \mathbf{F}\mathbf{M}$ *is asymptotically stable,*

$$\mathscr{Re}\,\lambda_i[\mathbf{A} - \mathbf{F}\mathbf{M}] < 0 \text{ if}$$

1. $(\mathbf{A}, \mathbf{M})$ *detectable*, $(\mathbf{A}, \mathbf{D})$ *stabilizable*

2. $$\mathbf{F} = \mathbf{P}\mathbf{M}^*\mathbf{V}^{-1}, \qquad 0 = \mathbf{P}\mathbf{A}^* + \mathbf{A}\mathbf{P} - \mathbf{P}\mathbf{M}^*\mathbf{V}^{-1}\mathbf{M}\mathbf{P} + \mathbf{D}\mathbf{W}\mathbf{D}^* \tag{9.10}$$

for any $\mathbf{V} > \mathbf{0}$, $\mathbf{W} > \mathbf{0}$.

This procedure and the pole assignment procedure are both ad hoc methods to construct state esimators. In the following section we choose a more formal basis for the design of the estimator.

9.1 Optimal State Estimation

Subtracting (9.7) from (9.2) yields the estimation error $\tilde{\mathbf{x}}$,

$$\dot{\tilde{\mathbf{x}}} = [\mathbf{A} - \mathbf{F}\mathbf{M}]\tilde{\mathbf{x}} + \mathbf{D}\mathbf{w} - \mathbf{F}\mathbf{v}, \qquad \tilde{\mathbf{x}}(0) = \mathbf{x}(0) - \hat{\mathbf{x}}(0),$$
$$\tilde{\mathbf{y}} = \mathbf{C}\tilde{\mathbf{x}}. \tag{9.11}$$

This is a linear system driven by impulsive inputs, and $\tilde{\mathbf{y}} = \mathbf{y} - \hat{\mathbf{y}} = \mathbf{Cx} - \mathbf{C}\hat{\mathbf{x}}$ denotes the estimation errors of particular interest. Such systems were analyzed in Chapter 4, where it was found convenient to evaluate cost functions of the form

$$\mathscr{V} = \sum_{i=1}^{n_x+n_w+n_v} \int_0^{\infty} \tilde{\mathbf{y}}^{i*}\mathbf{Q}\tilde{\mathbf{y}}^{i}\, dt, \tag{9.12}$$

where $\tilde{\mathbf{y}}^i(t)$ is the response of (9.11) with only the ith element of the vector $(\mathbf{w}^*, \mathbf{v}^*, \tilde{\mathbf{x}}^*(0))$ acting. In terms of the state covariance matrix

$$\tilde{\mathbf{X}} \triangleq \sum_{i=1}^{n_w+n_v} \int_0^{\infty} \tilde{\mathbf{x}}^i(t)\tilde{\mathbf{x}}^{i*}(t)\, dt,$$

from (4.167) and (4.156) it follows that

$$\mathscr{V} = \operatorname{tr} \mathbf{QC}\tilde{\mathbf{X}}\mathbf{C}^*, \tag{9.13}$$

where, for system (9.11) $\tilde{\mathbf{X}}$ satisfies

$$\mathbf{0} = \tilde{\mathbf{X}}(\mathbf{A} - \mathbf{FM})^* + (\mathbf{A} - \mathbf{FM})\tilde{\mathbf{X}} + \mathbf{DWD}^* + \mathbf{FVF}^* + \tilde{\mathbf{X}}_0, \tag{9.14}$$

where $\mathbf{W} \triangleq \operatorname{diag}[\ldots w_i^2 \ldots]$, $\mathbf{V} \triangleq \operatorname{diag}[\ldots v_i^2 \ldots]$, $\tilde{\mathbf{X}}_0 \triangleq \operatorname{diag}[\ldots \tilde{x}_i^2(0) \ldots]$. Consider now the task of finding the $\mathbf{F}$ that minimizes (9.12). This is equivalent to the problem of minimizing (9.13) subject to the constraints (9.14), or the equivalent unconstrained problem

$$\min_{\Lambda, \tilde{\mathbf{X}}, \mathbf{F}} \hat{\mathscr{V}},$$

where

$$\hat{\mathscr{V}} \triangleq \operatorname{tr} \mathbf{QC}\tilde{\mathbf{X}}\mathbf{C}^* + \operatorname{tr} \Lambda\left[\tilde{\mathbf{X}}(\mathbf{A} - \mathbf{FM})^* + (\mathbf{A} - \mathbf{FM})\tilde{\mathbf{X}} + \mathbf{DWD}^* + \mathbf{FVF}^* + \tilde{\mathbf{X}}_0\right]. \tag{9.15}$$

The necessary conditions for a minimum are

$$\frac{\partial \hat{\mathscr{V}}}{\partial \Lambda} = \mathbf{0} = \tilde{\mathbf{X}}(\mathbf{A} - \mathbf{FM})^* + (\mathbf{A} - \mathbf{FM})\tilde{\mathbf{X}} + \mathbf{FVF}^* + \mathbf{DWD}^* + \tilde{\mathbf{X}}_0, \tag{9.16a}$$

$$\frac{\partial \hat{\mathscr{V}}}{\partial \tilde{\mathbf{X}}} = \mathbf{0} = \mathbf{C}^*\mathbf{QC} + (\mathbf{A} - \mathbf{FM})^*\Lambda + \Lambda(\mathbf{A} - \mathbf{FM}), \tag{9.16b}$$

$$\frac{\partial \hat{\mathscr{V}}}{\partial \mathbf{F}} = \mathbf{0} = -2\Lambda\tilde{\mathbf{X}}\mathbf{M}^* + 2\Lambda\mathbf{FV}. \tag{9.16c}$$

From (9.16c) it follows that

$$\mathbf{F} = \tilde{\mathbf{X}}\mathbf{M}^*\mathbf{V}^{-1}. \tag{9.16d}$$

Substitution of (9.16d) into (9.16a) yields

$$\mathbf{0} = \tilde{\mathbf{X}}\mathbf{A}^* + \mathbf{A}\tilde{\mathbf{X}} - \tilde{\mathbf{X}}\mathbf{M}^*\mathbf{V}^{-1}\mathbf{M}\tilde{\mathbf{X}} + \mathbf{D}\mathbf{W}\mathbf{D}^* + \tilde{\mathbf{X}}_0. \tag{9.17}$$

Equation (9.16d) and the Riccati equation (9.17) have the identical form of (8.49b). Also note that the optimal estimation problem is a "dual" of the optimal state feedback control problem (8.70b) in the sense that the same computer code may be used to solve the Riccati equation associated with both problems, by a simple change of variables. The reader should note that (9.16d) is a necessary condition if $\Lambda > \mathbf{0}$ in (9.16c). Otherwise, other solutions may satisfy (9.16c) besides (9.16d). From (9.16b), the condition for $\Lambda > \mathbf{0}$ is that $(\mathbf{A} - \mathbf{FM}, \mathbf{C})$ is observable. The choice $\mathbf{C} = \mathbf{I}$ (which means we have an interest in *all* states, $\tilde{\mathbf{y}} = \tilde{\mathbf{x}}$) guarantees this. The results of this section are summarized as follows.

Theorem 9.3

For any $\mathcal{Q} \geq \mathbf{0}$, the state estimator for (9.2) that minimizes

$$\sum_{i=1}^{n_x+n_v+n_w} \int_0^\infty (\mathbf{x} - \hat{\mathbf{x}})^{i*} \mathcal{Q} (\mathbf{x} - \hat{\mathbf{x}})^i \, dt, \qquad x_i(0) - \hat{x}_i(0) = \tilde{x}_{i0}, \tag{9.18}$$

where $(\mathbf{x} - \hat{\mathbf{x}})^i$ denotes the response of $\dot{\tilde{\mathbf{x}}} = (\mathbf{A} - \mathbf{FM})\tilde{\mathbf{x}} + \mathbf{D}\mathbf{w} - \mathbf{F}\mathbf{v}$ due to excitations $w_i = \omega_i \delta(t)$, $v_i = \upsilon_i \delta(t)$, $\tilde{x}_i(0) = \tilde{x}_{i0}$ applied one at a time, is given by

$$\dot{\hat{\mathbf{x}}} = \mathbf{A}\hat{\mathbf{x}} = \mathbf{B}\mathbf{u} + \mathbf{F}(\mathbf{z} - \mathbf{M}\hat{\mathbf{x}}), \qquad \hat{\mathbf{x}}(0) = \hat{\mathbf{x}}_0, \tag{9.19}$$

$$\mathbf{F} = \tilde{\mathbf{X}}\mathbf{M}^*\mathbf{V}^{-1}, \tag{9.20}$$

$$\mathbf{0} = \tilde{\mathbf{X}}\mathbf{A}^* + \mathbf{A}\tilde{\mathbf{X}} - \tilde{\mathbf{X}}\mathbf{M}^*\mathbf{V}^{-1}\tilde{\mathbf{X}} + \mathbf{D}\mathbf{W}\mathbf{D}^* + \tilde{\mathbf{X}}_0, \tag{9.21}$$

where

$$\mathbf{V} = \operatorname{diag}[\ldots \upsilon_i^2 \ldots], \quad \mathbf{W} = \operatorname{diag}[\ldots w_i^2 \ldots], \quad \tilde{\mathbf{X}}_0 = \operatorname{diag}[\ldots \tilde{x}_i^2(0) \ldots],$$
$$w_i(t) = \omega_i \delta(t), \qquad v_i(t) = \upsilon_i \delta(t), \tag{9.22}$$

and the minimum value of the estimation error is

$$\mathscr{V} = \sum_{i=1}^{n_x+n_v+n_w} \int_0^\infty (\mathbf{x} - \hat{\mathbf{x}})^{i*} \mathcal{Q} (x - \hat{x})^i \, dt = \operatorname{tr} \mathcal{Q}\tilde{\mathbf{X}}. \tag{9.23}$$

EXAMPLE 9.1

A certain theory of learning suggests that

$$G = \mu(L - E),$$

$$W = \beta(G_e - G) - D,$$

$$\dot{L} = \alpha W,$$

where $L(t)$ is the amount learned, $G_e(t)$ is the grade the student expects of himself, $D(t)$ is the distraction (effort not applied to the subject), $G(t)$ is the grade, and E is the error in the grading process.

For this learning model, find the optimal estimate of the amount learned L, assuming that the measurement errors are represented by an impulse of weight e, $E = e\delta(t)$, and the effect of distractions is such that $D = d\delta(t)$. The model is given by

$$\dot{L} = -\alpha\beta\mu L + \alpha\beta G_e + \alpha\beta\mu E - \alpha D,$$

$$L(0) = 0 \text{ (starting without knowledge of the subject)},$$

$$= -\alpha\beta\mu L + \alpha\beta G_e + (\alpha\beta\mu e - \alpha d)\delta(t)$$

$$= AL + BG_e + w(t), \quad A \triangleq -\alpha\beta\mu, \quad B \triangleq \alpha\beta, \tag{9.24}$$

$$w \triangleq \alpha\beta\mu e - \alpha d,$$

$$z \triangleq G = \mu L - \mu E = ML + v(t), \qquad M \triangleq \mu, \tag{9.25}$$

$$v \triangleq -\mu e,$$

where $V = \mu^2 e^2$ and $W = \alpha^2(\beta\mu e - d)^2$.

Solution: The optimal estimator is given by Theorem 9.2, and the calculations are

$$0 = 2\tilde{X}A - \tilde{X}^2 M^2 V^{-1} + D^2 W$$

or

$$\tilde{X} = \frac{-1}{2M^2V^{-1}}\left[-2A \pm \sqrt{4A^2 + 4\frac{D^2WM^2}{V}}\right]$$

$$= \frac{VA}{M^2}\left[1 \pm \sqrt{1 + \frac{D^2M^2W}{A^2V}}\right].$$

Or, using the data from (9.24) and (9.25),

$$\tilde{X} = -\alpha\beta\mu e^2 + \alpha\sqrt{2\mu^2\beta^2 + \frac{d}{e}\left(\frac{d}{e} - 2\beta\mu\right)}. \tag{9.26}$$

From (9.20),

$$F = -\alpha\beta + \frac{\alpha}{\mu e^2}\sqrt{2\mu^2\beta^2 + \frac{d}{e}\left(\frac{d}{e} - 2\beta\mu\right)}. \tag{9.27}$$

The speed of the estimator is governed by the eigenvalues of $A - FM$, which in this case yields

$$A - FM = -\frac{\alpha}{e^2}\sqrt{2\mu^2\beta^2 + \frac{d}{e}\left(\frac{d}{e} - 2\beta\mu\right)}. \tag{9.28}$$

■

These conclusions are obvious from the location of the $A - FM$ eigenvalue:

1. All other constants being fixed, higher student aptitude ($\alpha \gg 0$) and lower grading errors e allow more rapid estimates of knowledge (amount learned L).
2. Increasing motivation (β) or instructor difficulty (μ) will increase the speed with which knowledge can be determined from grades if d/e is small.
3. Increasing d/e (smaller grading errors) will increase the speed with which knowledge can be determined from grades, since for large d/e, $A - FM = -\alpha d/e^3$.

Note that for any given set of initial conditions, $x_i(0)$, $\hat{x}_i(0)$ applied simultaneously, (9.18) is minimized by the state estimator (9.19) with gains (9.20) and (9.21), where $[\tilde{X}_0]_{ii} = [x_i(0) - \hat{x}_i(0)]^2$. This is quite a different optimization problem than that posed in (8.48) where initial conditions $x_i(0)$ and $\hat{x}_i(0)$ where applied one at a time. However, see that $\tilde{\mathbf{X}}$ in (9.21) is the same matrix as $\mathbf{X}_{11}$ in (8.49b) if $\hat{x}_i(0) = 0$ for all $i = 1, 2, \ldots, n_x$. In this event the **F** in (9.20) is the same as the **F** in (8.49b). This proves the following.

Corollary 1 to Theorem 9.3
Let $\hat{\mathbf{x}}(0) = \mathbf{0}$ *for the state estimator*

$$\dot{\hat{\mathbf{x}}} = \mathbf{A}\hat{\mathbf{x}} + \mathbf{B}\mathbf{u} + \mathbf{F}(\mathbf{z} - \mathbf{M}\hat{\mathbf{x}}),$$

where **F** *is given by* (9.20) *and* (9.21) *with* $\tilde{\mathbf{X}}_0 = \mathbf{X}_0 = \mathrm{diag}[\ldots x_i^2(0)\ldots]$. *Then, the optimal LQI controller* (8.48), (8.49) *contains a state estimator which minimizes* (9.18).

Corollary 2 to Theorem 9.3

The optimal state estimator (9.19) *and* (9.21) *yields the smallest value of* (9.18) *when the initial conditions* $\hat{x}_i(0) = x_i(0)$ *are chosen.*

Corollary 2 follows from the fact that $\tilde{\mathbf{X}}$ in (9.21) and (9.23) takes on its smallest value when $\tilde{X}_0 = 0$ in (9.21), in the sense that $[\tilde{\mathbf{X}}(\tilde{\mathbf{X}}_0 = \mathbf{0})] < [\tilde{\mathbf{X}}(\tilde{\mathbf{X}}_0 > 0)]$. Of course, Corollary 2 is not useful in practice since we usually do not know $x(0)$.

9.2 Minimal-Order State Estimators

Now, we treat two questions which are logical extensions of earlier work. Note that the state estimator (9.19) is of order n_x. That is, the estimator contains one integrator for each state variable x_i, $i = 1 \to n_x$. Since the measurement $\mathbf{z} = \mathbf{M}\mathbf{x}$ is a linear combination of state variables, the reader may ask why we should have to estimate the states that we already measure.

Suppose now that the measurement vector $\mathbf{z}$ contains some disturbance-free measurements $\mathbf{z}_R$ and some other measurements $\mathbf{z}_T$:

$$\mathbf{z} = \begin{pmatrix} \mathbf{z}_R \\ \mathbf{z}_T \end{pmatrix} = \begin{bmatrix} \mathbf{M}_R \\ \mathbf{M}_T \end{bmatrix} \mathbf{x} + \begin{pmatrix} \mathbf{0} \\ v_T \end{pmatrix} \delta(t). \tag{9.29}$$

In vector form, (9.29) is

$$\mathbf{z} = \mathbf{M}\mathbf{x} + \mathbf{v}, \quad \mathbf{v}(t) = v\delta(t), \quad v^* = (0, 0, \ldots, 0, v_T^*),$$

$$\mathbf{M} = \begin{bmatrix} \mathbf{M}_R \\ \mathbf{M}_T \end{bmatrix}. \tag{9.30}$$

We cannot use Theorem 9.3 in a rigorous sense since $\mathbf{V}$ is not now invertible. Actually, this is good news. This is a warning signal that we are making the problem harder than it should be. Some of the states are being measured exactly in (9.29). We must reduce $\mathbf{V}$ to an invertible matrix. Our efforts will be rewarded by a lower-order estimator. The idea that accomplishes these reduced-order estimators is to use the states which are measured and only estimate those $(n_x - n_z)$ states not measured exactly.

The first step is to convert the linear combination of state variables $\mathbf{M}_R\mathbf{x}$ (9.29) to state variables. This can be accomplished by a coordinate transformation. Define a transformation $\mathbf{x} = \mathbf{T}\boldsymbol{x}$ such that $\boldsymbol{x}$ is separated into states which are measured $\mathbf{z}_R$ and $\boldsymbol{\gamma}$ state variables which are not:

$$\mathbf{x} = \mathbf{T}\boldsymbol{x} = [\mathbf{T}_R \quad \mathbf{T}_T] \begin{pmatrix} \mathbf{z}_R \\ \boldsymbol{\gamma} \end{pmatrix} \tag{9.31}$$

and the inverse relationship

$$\begin{pmatrix} \mathbf{z}_R \\ \boldsymbol{\gamma} \end{pmatrix} = \begin{bmatrix} \mathbf{L}_R \\ \mathbf{L}_T \end{bmatrix} \mathbf{x}, \tag{9.32}$$

where $\mathbf{L}_R = \mathbf{M}_R$ because of (9.29). For invertible T,

$$\begin{bmatrix} \mathbf{L}_R \\ \mathbf{L}_T \end{bmatrix} [\mathbf{T}_R \quad \mathbf{T}_T] = \begin{bmatrix} \mathbf{I} & \mathbf{0} \\ \mathbf{0} & \mathbf{I} \end{bmatrix}. \tag{9.33}$$

The state transformation of

$$\begin{aligned} \dot{\mathbf{x}} &= \mathbf{A}\mathbf{x} + \mathbf{B}\mathbf{u} + \mathbf{D}\mathbf{w}, \\ \mathbf{x} &= \mathbf{M}\mathbf{z} + \mathbf{v} \end{aligned} \tag{9.34}$$

yields

$$\begin{aligned} \dot{\mathscr{x}} &= \mathbf{T}^{-1}\mathbf{A}\mathbf{T}\mathscr{x} + \mathbf{T}^{-1}\mathbf{B}\mathbf{u} + \mathbf{T}^{-1}\mathbf{D}\mathbf{w}, \\ \mathbf{z} &= \mathbf{M}\mathbf{T}\mathscr{x} + \mathbf{v} \end{aligned}$$

or

$$\begin{aligned} \begin{pmatrix} \dot{\mathbf{z}}_R \\ \dot{\boldsymbol{\gamma}} \end{pmatrix} &= \begin{bmatrix} \mathbf{L}_R \\ \mathbf{L}_T \end{bmatrix} \mathbf{A} [\mathbf{T}_R \quad \mathbf{T}_T] \begin{pmatrix} \mathbf{z}_R \\ \boldsymbol{\gamma} \end{pmatrix} + \begin{bmatrix} \mathbf{L}_R \\ \mathbf{L}_T \end{bmatrix} \mathbf{B}\mathbf{u} + \begin{bmatrix} \mathbf{L}_R \\ \mathbf{L}_T \end{bmatrix} \mathbf{D}\mathbf{w}, \\ \begin{pmatrix} \mathbf{z}_R \\ \mathbf{z}_T \end{pmatrix} &= \begin{bmatrix} \mathbf{M}_R \\ \mathbf{M}_T \end{bmatrix} [\mathbf{T}_R \quad \mathbf{T}_T] \begin{pmatrix} \mathbf{z}_R \\ \boldsymbol{\gamma} \end{pmatrix} + \begin{pmatrix} \mathbf{0} \\ v_T \end{pmatrix} \delta(t). \end{aligned} \tag{9.35}$$

But since $\mathbf{z}_R$ is *available as a measurement*, we need not integrate (9.35) to obtain $\mathbf{z}_R$. Hence, in the new state variables only $\boldsymbol{\gamma}$ needs to be estimated. Note from (9.33) that $\mathbf{M}_R[\mathbf{T}_R \quad \mathbf{T}_T] = [\mathbf{I} \quad \mathbf{0}]$. Then, from (9.35),

$$\begin{aligned} \dot{\boldsymbol{\gamma}} &= \mathbf{L}_T\mathbf{A}\mathbf{L}_R\mathbf{z}_R + \mathbf{L}_T\mathbf{A}\mathbf{T}_T\boldsymbol{\gamma} + \mathbf{L}_T\mathbf{B}\mathbf{u} + \mathbf{L}_T\mathbf{D}\mathbf{w}, \\ \mathbf{z}_R &= \mathbf{z}_R, \\ \mathbf{z}_T &= \mathbf{M}_T\mathbf{T}_R\mathbf{z}_R + \mathbf{M}_T\mathbf{T}_T\boldsymbol{\gamma} + \mathbf{v}_T, \end{aligned} \tag{9.36}$$

which is rearranged in the form

$$\begin{aligned} \dot{\boldsymbol{\gamma}} &= \mathscr{A}\boldsymbol{\gamma} + \mathbf{f}_1 + \mathscr{D}\mathbf{w}, \\ \tilde{\mathbf{z}} &= \mathscr{M}\boldsymbol{\gamma} + \mathbf{v}_T, \end{aligned} \qquad \begin{aligned} \mathscr{A} &= \mathbf{L}_T\mathbf{A}\mathbf{T}_T, \\ \mathscr{D} &= \mathbf{L}_T\mathbf{D}, \\ \mathbf{f}_1 &= \mathbf{L}_T\mathbf{B}\mathbf{u} + \mathbf{L}_T\mathbf{A}\mathbf{T}_R\mathbf{z}_R, \\ \mathscr{M} &= \mathbf{M}_T\mathbf{T}_T, \\ \tilde{\mathbf{z}} &= \mathbf{z}_T - \mathbf{M}_T\mathbf{T}_R\mathbf{z}_R, \end{aligned} \tag{9.37}$$

where $\mathbf{f}_1$ and $\tilde{\mathbf{z}}$ are known quantities and $\boldsymbol{\gamma}$ is to be estimated. Using the state estimation theory, the estimate of $\boldsymbol{\gamma}$ is obtained from

$$\dot{\hat{\boldsymbol{\gamma}}} = \mathscr{A}\hat{\boldsymbol{\gamma}} + \mathbf{f}_1 + \mathbf{F}(\tilde{\mathbf{z}} - \mathscr{M}\hat{\boldsymbol{\gamma}}), \tag{9.38a}$$

where $\mathbf{F}$ may be chosen by Theorem 9.1 or 9.2, with the substitution of $(\mathscr{A}, \mathscr{M}, \mathbf{f}_1)$ for $(\mathbf{A}, \mathbf{M}, \mathbf{Bu})$. The estimate of $\mathbf{x}$ follows from (9.31):

$$\hat{\mathbf{x}} = \mathbf{T}_R\mathbf{z}_R + \mathbf{T}_T\hat{\boldsymbol{\gamma}}. \tag{9.38b}$$

Hence, (9.38) provides the state estimate for $\mathbf{x}$, but the order of the estimator is $n_\gamma = n_x - n_{z_R}$, where $\gamma \in \mathscr{R}^{n_\gamma}$. The results are summarized as follows.

Theorem 9.4

Given

$$\begin{aligned} \dot{\mathbf{x}} &= \mathbf{Ax} + \mathbf{Bu} + \mathbf{Dw}, \qquad && \mathbf{x} \in \mathscr{R}^{n_x}, \\ \mathbf{z} &= \mathbf{Mx} + \mathbf{v}, && \mathbf{z} \in \mathscr{R}^{n_x}, \end{aligned} \tag{9.39}$$

with n_{z_R} error-free measurements

$$\mathbf{v} = \begin{pmatrix} 0 \\ v_T \end{pmatrix} \delta(t), \quad \mathbf{v} \in \mathscr{R}^{n_z}, \quad v_T \in \mathscr{R}^{n_z - n_{z_R}}, \tag{9.40}$$

an $(n_x - n_{z_R})$th order estimator of $\mathbf{x}$ is given by

$$\hat{\mathbf{x}} = \mathbf{T}_R\mathbf{z}_R + \mathbf{T}_T\hat{\boldsymbol{\gamma}}, \tag{9.41}$$

where $\mathbf{z}_R$ represents the error-free measurements

$$\mathbf{z} = \begin{pmatrix} \mathbf{z}_R \\ \mathbf{z}_T \end{pmatrix} = \begin{bmatrix} \mathbf{M}_R \\ \mathbf{M}_T \end{bmatrix} \mathbf{x} + \begin{pmatrix} 0 \\ v_T \end{pmatrix} \delta(t) \tag{9.42}$$

and $\hat{\boldsymbol{\gamma}}$ is the solution of

$$\dot{\hat{\boldsymbol{\gamma}}} = \mathbf{L}_T\mathbf{A}\mathbf{T}_T\hat{\boldsymbol{\gamma}} + \mathbf{L}_T(\mathbf{Bu} + \mathbf{A}\mathbf{T}_R\mathbf{z}_R) + \mathbf{F}[\mathbf{z}_T - \mathbf{M}_T(\mathbf{T}_R\mathbf{z}_R + \mathbf{T}_T\hat{\boldsymbol{\gamma}})], \tag{9.43}$$

where $\mathbf{L}_T$, $\mathbf{T}_T$, and $\mathbf{T}_R$ are any matrices satisfying

$$\begin{bmatrix} \mathbf{M}_R \\ \mathbf{L}_T \end{bmatrix} [\mathbf{T}_R \quad \mathbf{T}_T] = \begin{bmatrix} \mathbf{I} & \mathbf{0} \\ \mathbf{0} & \mathbf{I} \end{bmatrix} \tag{9.44}$$

and the estimation error is governed by

$$\begin{aligned} \dot{\tilde{\boldsymbol{\gamma}}} &= [\mathbf{L}_T\mathbf{A} - \mathbf{F}\mathbf{M}_T]\mathbf{T}_T\tilde{\boldsymbol{\gamma}} + \mathbf{L}_T\mathbf{Dw} - \mathbf{F}\mathbf{v}_T, \qquad \tilde{\boldsymbol{\gamma}} \triangleq \boldsymbol{\gamma} - \hat{\boldsymbol{\gamma}}, \\ \tilde{\mathbf{x}} &= \mathbf{T}_T\tilde{\boldsymbol{\gamma}}, \end{aligned} \tag{9.45}$$

provided $\mathbf{F}$ is chosen to stabilize $[\mathbf{L}_T\mathbf{A} - \mathbf{F}\mathbf{M}_T]\mathbf{T}_T$.

Note that (9.45) follows from the difference between (9.37) and (9.38a). The pioneering work of Luenberger [9.5] developed the idea of state estimators of the type discussed here. He chose to call them "observers," when some noise-free measurements are available.

Exercise 9.2

Show that minimal-order estimators can be placed in the form

$$\hat{\mathbf{x}} = [\mathbf{T}_R \quad \mathbf{0}]\mathbf{z} + \mathbf{T}_T\hat{\boldsymbol{\gamma}},$$

$$\dot{\hat{\boldsymbol{\gamma}}} = [\mathbf{L}_T\mathbf{A} - \mathbf{F}\mathbf{M}_T]\mathbf{T}_T\hat{\boldsymbol{\gamma}} + \mathbf{L}_T\mathbf{B}\mathbf{u} + \left[(\mathbf{L}_T\mathbf{A} - \mathbf{F}\mathbf{M}_T)\mathbf{T}_R, \mathbf{F}\right]\mathbf{z} \tag{9.46}$$

and that the block diagram has a feed-forward structure. Find $\mathbf{C}_1$, $\mathbf{C}_2$, $\mathbf{C}_3$, $\mathbf{C}_4$, and $\mathbf{C}_5$ in Fig. 9.1.

Exercise 9.3

For the spinning spacecraft described by (3.84), suppose only one rate gyro is available and measures ω_1 perfectly ($v = 0$) and $J_{ij} = 0$, $i \neq j$, $J_{22} = J_{11} > J_{33}$. Only one torquer is active, u_2. Find a first-order state estimator for the velocity vector $\omega^T = (\omega_1 \quad \omega_2)$. Explain in terms of the dynamics of the system why the rate gyro measuring ω_1 provides information about ω_2 (allows ω_2 to be estimated), even though the rate gyro is mounted about an axis which is perpendicular to the ω_2 axis.

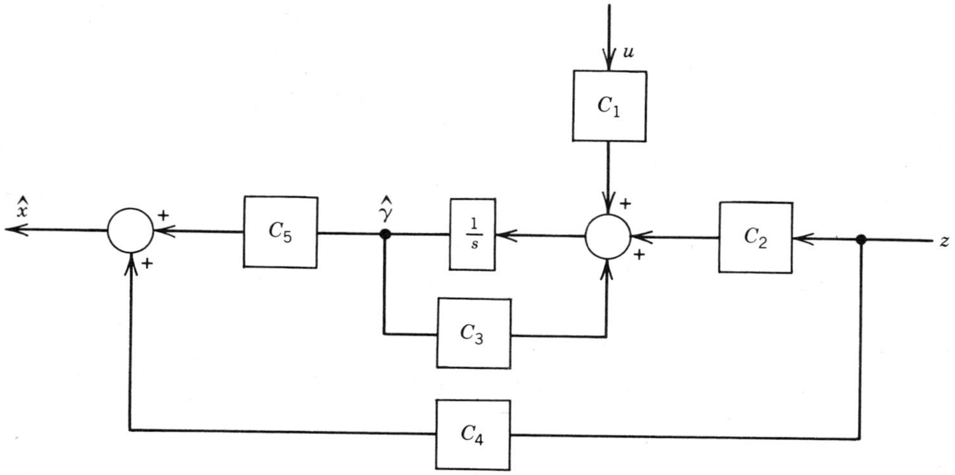

Figure 9.1 *Feedforward Structure of the Minimal-Order State Estimator*

9.3 Closed-Loop Behavior with Estimator-Based Controllers

Consider now any state estimator of the form (9.7) in combination with the state feedback control law ($\mathbf{u} = \mathbf{Gx}$). The estimate $\hat{\mathbf{x}}$ *replaces* $\mathbf{x}$ in the control law to give

$$\begin{aligned} &\dot{\mathbf{x}} = \mathbf{Ax} + \mathbf{Bu} + \mathbf{Dw}, \qquad \dot{\hat{\mathbf{x}}} = \mathbf{A}\hat{\mathbf{x}} + \mathbf{Bu} + \mathbf{F}(\mathbf{z} - \mathbf{M}\hat{\mathbf{x}}), \\ &\mathbf{y} = \mathbf{Cx}, \qquad \mathbf{u} = \mathbf{G}\hat{\mathbf{x}}, \end{aligned} \tag{9.47}$$

or

$$\begin{aligned} \begin{pmatrix} \dot{\mathbf{x}} \\ \dot{\hat{\mathbf{x}}} \end{pmatrix} &= \begin{bmatrix} \mathbf{A} & \mathbf{BG} \\ \mathbf{FM} & \mathbf{A} + \mathbf{BG} - \mathbf{FM} \end{bmatrix} \begin{pmatrix} \mathbf{x} \\ \hat{\mathbf{x}} \end{pmatrix} + \begin{bmatrix} \mathbf{D} & \mathbf{0} \\ \mathbf{0} & \mathbf{F} \end{bmatrix} \begin{pmatrix} \mathbf{w} \\ \mathbf{v} \end{pmatrix}, \\ \begin{pmatrix} \mathbf{y} \\ \mathbf{u} \end{pmatrix} &= \begin{bmatrix} \mathbf{C} & \mathbf{0} \\ \mathbf{0} & \mathbf{G} \end{bmatrix} \begin{pmatrix} \mathbf{x} \\ \hat{\mathbf{x}} \end{pmatrix}. \end{aligned} \tag{9.48}$$

The eigenvalues of the closed-loop system are not evident from (9.48). A coordinate transformation makes them more obvious. Consider the transformation

$$\begin{pmatrix} \mathbf{x} \\ \mathbf{x} - \hat{\mathbf{x}} \end{pmatrix} = \begin{bmatrix} \mathbf{I} & \mathbf{0} \\ \mathbf{I} & -\mathbf{I} \end{bmatrix} \begin{pmatrix} \mathbf{x} \\ \hat{\mathbf{x}} \end{pmatrix} = \mathbf{T} \begin{pmatrix} \mathbf{x} \\ \hat{\mathbf{x}} \end{pmatrix} \tag{9.49}$$

applied to (9.48). This yields

$$\begin{aligned} \begin{pmatrix} \dot{\mathbf{x}} \\ \dot{\mathbf{x}} - \dot{\hat{\mathbf{x}}} \end{pmatrix} &= \begin{bmatrix} \mathbf{A} + \mathbf{BG} & \mathbf{BG} \\ \mathbf{0} & \mathbf{A} - \mathbf{FM} \end{bmatrix} \begin{pmatrix} \mathbf{x} \\ \mathbf{x} - \hat{\mathbf{x}} \end{pmatrix} + \begin{bmatrix} \mathbf{D} & \mathbf{0} \\ \mathbf{D} & -\mathbf{F} \end{bmatrix} \begin{pmatrix} \mathbf{w} \\ \mathbf{v} \end{pmatrix}, \\ \begin{pmatrix} \mathbf{y} \\ \mathbf{u} \end{pmatrix} &= \begin{bmatrix} \mathbf{C} & \mathbf{0} \\ \mathbf{G} & -\mathbf{G} \end{bmatrix} \begin{pmatrix} \mathbf{x} \\ \mathbf{x} - \hat{\mathbf{x}} \end{pmatrix}. \end{aligned} \tag{9.50}$$

Hence, the closed-loop system has the n_x eigenvalues of $[\mathbf{A} + \mathbf{BG}]$ and the n_x eigenvalues of $[\mathbf{A} - \mathbf{FM}]$. It is striking to note that the eigenvalues $[\mathbf{A} + \mathbf{BG}]$ are precisely those obtained by perfect state feedback ($\mathbf{u} = \mathbf{Gx}$), and the eigenvalues of $[\mathbf{A} - \mathbf{FM}]$ are precisely those obtained from state estimation without regard for the particular use of the estimate. This "separation principle" holds also for reduced-order estimators.

The closed-loop system using the control law $\mathbf{u} = \mathbf{G}\hat{\mathbf{x}}$ and the minimal-order estimator (9.43) yields

$$\begin{aligned} \begin{pmatrix} \dot{\mathbf{x}} \\ \dot{\hat{\boldsymbol{\gamma}}} \end{pmatrix} = &\begin{bmatrix} \mathbf{A} + \mathbf{BGT}_R\mathbf{M}_R & \mathbf{BGT}_T \\ \mathbf{L}_T(\mathbf{A} + \mathbf{BG})\mathbf{T}_R\mathbf{M}_R + \mathbf{FM}_T(\mathbf{I} - \mathbf{T}_R\mathbf{M}_R) & [\mathbf{L}_T(\mathbf{A} + \mathbf{BG}) - \mathbf{FM}_T]\mathbf{T}_T \end{bmatrix} \\ &\times \begin{pmatrix} \mathbf{x} \\ \hat{\boldsymbol{\gamma}} \end{pmatrix} + \begin{bmatrix} \mathbf{D} & \mathbf{0} \\ \mathbf{0} & \mathbf{F} \end{bmatrix} \begin{pmatrix} \mathbf{w} \\ \dot{\mathbf{v}}_T \end{pmatrix} \end{aligned} \tag{9.51}$$

or the transformation

$$\begin{pmatrix} \mathbf{x} \\ \boldsymbol{\gamma} - \hat{\boldsymbol{\gamma}} \end{pmatrix} = \begin{bmatrix} \mathbf{I} & \mathbf{0} \\ \mathbf{L}_T & -\mathbf{I} \end{bmatrix} \begin{pmatrix} \mathbf{x} \\ \hat{\boldsymbol{\gamma}} \end{pmatrix} \tag{9.52}$$

yields

$$\begin{aligned} \begin{pmatrix} \dot{\mathbf{x}} \\ \dot{\boldsymbol{\gamma}} - \dot{\hat{\boldsymbol{\gamma}}} \end{pmatrix} &= \begin{bmatrix} \mathbf{A} + \mathbf{B}\mathbf{G} & -\mathbf{B}\mathbf{G}\mathbf{T}_T \\ \mathbf{0} & [\mathbf{L}_T\mathbf{A} - \mathbf{F}\mathbf{M}_T]\mathbf{T}_T \end{bmatrix} \begin{pmatrix} \mathbf{x} \\ \boldsymbol{\gamma} - \hat{\boldsymbol{\gamma}} \end{pmatrix} + \begin{bmatrix} \mathbf{D} & \mathbf{0} \\ \mathbf{L}_T\mathbf{D} & -\mathbf{F} \end{bmatrix} \begin{pmatrix} \mathbf{w} \\ \mathbf{v} \end{pmatrix}, \\ \begin{pmatrix} \mathbf{y} \\ \mathbf{u} \end{pmatrix} &= \begin{bmatrix} \mathbf{C} & \mathbf{0} \\ \mathbf{G}\mathbf{T}_R\mathbf{M}_R & \mathbf{G}\mathbf{T}_T \end{bmatrix} \begin{pmatrix} \mathbf{x} \\ \boldsymbol{\gamma} - \hat{\boldsymbol{\gamma}} \end{pmatrix}. \end{aligned} \tag{9.53}$$

This form reveals that the eigenvalues of the closed-loop system (9.53) are those of $[\mathbf{A} + \mathbf{BG}]$ and $[\mathbf{L}_T\mathbf{A} - \mathbf{FM}_T]\mathbf{T}_T$. Again, these are the eigenvalues available by perfect state feedback and just state estimation, respectively. This establishes the following result.

Theorem 9.5 (The Separation Principle of Estimation and Control)
When the control law $\mathbf{u} = \mathbf{G}\hat{\mathbf{x}}$ is used in conjunction with either a full-order or reduced-order state estimator for

$$\dot{\mathbf{x}} = \mathbf{Ax} + \mathbf{Bu} + \mathbf{Dw},$$

$$\mathbf{z} = \mathbf{Mx} + \mathbf{v},$$

the control gain $\mathbf{G}$ does not influence the eigenvalues of the state estimator, and the choice of the estimator gain $\mathbf{F}$ does not influence the remaining eigenvalues, which are $\lambda_i[\mathbf{A} + \mathbf{BG}]$, $i = 1 \to n_x$.

This theorem assures the separation principle for any control and estimator gains ($\mathbf{G}$ and $\mathbf{F}$) one might wish to use. Chapter 8 establishes the optimal $\mathbf{G}$ and $\mathbf{F}$ according to a quadratic cost function. References [9.1] through [9.3] prove the separation principle for stochastic problems.

Exercise 9.5

For the spinning symmetric spacecraft (3.32) with $1 = J_{11} = J_{22} > J_{33}$ and $z = \omega_1 + v$, $v(t) = \nu\delta(t)$, $\nu = 1$, *find a closed-loop controller which minimizes*

$$\mathscr{V} = \sum_{i=1}^{n_z + n_u} \int_0^{\infty} \left(\omega_1^{i^2} + \omega_2^{i^2} + \rho u_2^{i^2} \right) dt$$

with disturbance intensity $\mathscr{W} = 1$.

(a) For your *closed*-loop design, plot $\mathscr{V}_y \triangleq \int_0^\infty (\omega_1^2 + \omega_2^2)\, dt$ versus $\mathscr{V}_u = \int_0^\infty u_2^2\, dt$ as $\rho = 10^{-2}, 1, 10, 10^2$. Discuss the engineering significance of this plot. Can your controller (for any ρ) satisfy the performance constraints $\mathscr{V}_y \leqq 1$, $\mathscr{V}_u \leqq 1$?

(b) Repeat (a) for the optimal state feedback controller $\mathbf{u} = \mathbf{G}\mathbf{x}$. The comparison of performance plots (a) and (b) reflects the effect of the state estimator in closed-loop performance.

(c) Compute the component cost of the controller states. Compare the component cost of the controller state with the difference between the two curves plotted in (b). Discuss.

EXAMPLE 9.2

For the learning model of Example 9.1 find the optimal LQI controller for G_e to minimize

$$\mathscr{V} = \int_0^\infty \left(L - \frac{1}{\mu} G_e \right)^2 dt, \qquad 0 < \mu \leqq 1.$$

For the estimator design, use Theorem 9.3 with model parameters $E(t) = 1\delta(t)$, $D(t) = 1\delta(t)$, $\alpha = 1 = \beta$.

Solution: The estimator was designed in Example 9.1 from (9.27) and (9.28),

$$F = -1 + \frac{1}{\mu}\sqrt{2\mu^2 - 2\mu + 1},$$

$$A - FM = -\sqrt{2\mu^2 - 2\mu + 1}.$$

From Exercise 8.4,

$$u = -R^{-1}(B^*K + N^*C)\hat{x}, \qquad \dot{\hat{x}} = (A - FM + BG)\hat{x} + Fz,$$

$$0 = K(A - BR^{-1}N^*C) + (A - BR^{-1}N^*C)^*K - KBR^{-1}B^*K + C^*(Q - NR^{-1}N^*)C.$$

Now, $C = 1$, $Q = 1$, $R = \mu^{-2}$, $N = -\mu^{-1}$. It follows that $A - BR^{-1}N^*C = 0$, $K = 0$, and, therefore,

$$u = \mu\hat{x}, \qquad \dot{\hat{x}} = \left(-\sqrt{2\mu^2 - 2\mu + 1} + \mu\right)\hat{x} + \left(-1 + \frac{1}{\mu}\sqrt{2\mu^2 - 2\mu + 1}\right)z.$$

Now, from

$$\begin{pmatrix} \dot{x} \\ \dot{\hat{x}} \end{pmatrix} = \begin{bmatrix} -\mu & \mu \\ -p & p \end{bmatrix} \begin{pmatrix} x \\ \hat{x} \end{pmatrix} + \begin{bmatrix} 1 & 0 \\ 0 & -1 + \dfrac{1}{\mu}\sqrt{2\mu^2 - 2\mu + 1} \end{bmatrix} \begin{pmatrix} w \\ v \end{pmatrix},$$

$$p \triangleq \mu - \sqrt{2\mu^2 - 2\mu + 1},$$

see that the state estimation error $\tilde{x} \triangleq x - \hat{x}$ goes to zero according to

$$\dot{\tilde{x}} = -\sqrt{2\mu^2 - 2\mu + 1}\,\tilde{x} + w - \mu p v, \quad w(t) = W\delta(t), \quad v(t) = V\delta(t). \quad \blacksquare$$

Closure

This chapter derives optimal state estimators from a completely deterministic point of view. These estimators are observed to match those contained in the LQI controllers of Chapter 8, hence the separation principle of optimal estimation and control. When some of the measurements are exact (no disturbances), then the state estimator can be reduced in order by the number of disturbance-free measurements.

References

9.1 P. D. Joseph and J. T. Tou, "On Linear Control Theory," *Trans. AIEE*, Pt. II, 80, 193, 1961.

9.2 H. Kwakernaak and R. Sivan, *Linear Optimal Control Systems*. Wiley-Interscience, New York, 1972.

9.3 R. F. Stengel, *Stochastic Optimal Control*. Wiley-Interscience, New York, 1986.

9.4 J. Meditch, *Stochastic Optimal Linear Estimation and Control*. McGraw-Hill, New York, 1969.

9.5 D. G. Luenberger, "Observing the State of a Linear System," *IEEE Trans. Military Electronics*, MIL8, April 1964.

CHAPTER 10

Model Error Concepts and Compensation

There is much literature on the *modeling* of dynamic systems, and there is much literature on the *control* of dynamic systems. A student's traditional education in engineering certainly presents both topics. There is the unstated *implication* in this literature that real-world control problems can and ought to be treated by judicious juxtaposition of the available *theories of modeling* with the available *theories of control*. The purpose of this chapter is to point out that there are theoretical issues which are not addressed by such attitudes. The modeling and control problems are not issues that can truly be separated. Theories of modeling dynamic systems according to the first principles of physics has developed to a sophisticated level of maturity, but the premises upon which these theorems are based often collapse when merged with the control problem.

Analysts developing mathematical models for physical phenomena know very well that the models are always incomplete. That is, the models never *exactly* describe the physical phenomena. There are always assumptions which the analyst impose to make the task tractable. For example, wind turbulence might be neglected in the rocket problem of Section 3.3. A flat plate might be idealized by a number of different assumptions. The plate might be assumed to be rigid or flexible. The mass density of the plate might be assumed uniform and the disturbances (aerodynamic disturbances from air currents, thermal effects) might be assumed zero. The *set of assumptions* form the *idealization* of the system.

The steps in the derivation of a mathematical model from the first principles of physics are these. The analyst must first decide upon some *idealization* of the system

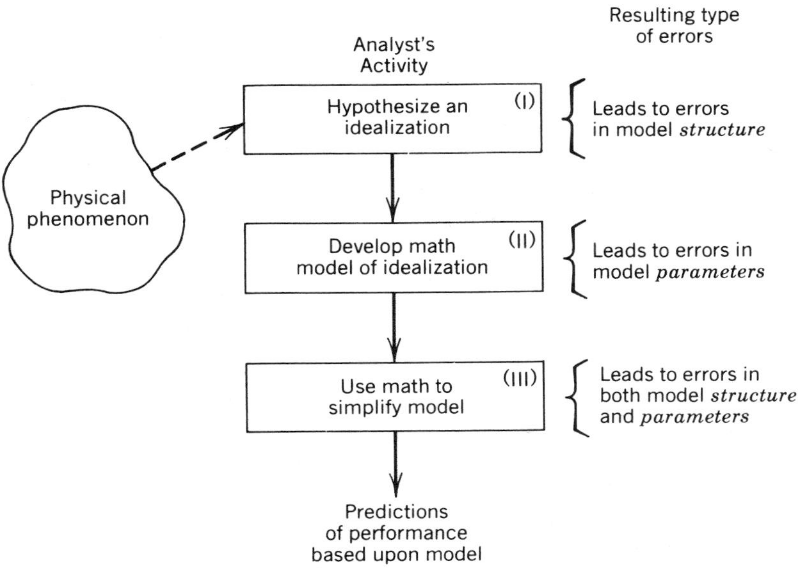

Figure 10.1 *Steps in the Modeling Process*

(rigid plate, elastic plate with uniform mass density, etc.). Then, the analyst must apply known physical laws to develop a *mathematical model* for this *idealization* of the system. Now, the given *mathematical model* may also differ from his *idealization* due to uncertainties in the parameters associated with his idealization. (For the idealization of the plate as a rigid body, Newton's laws are perfectly correct (ignoring relativistic effects) but the assumed mass properties of the rigid body might be in error.) Fig. 10.1 illustrates the logical steps the analyst takes to produce a final model. Note that the resulting model order and the *parameters* of the idealization may be in error. Further errors are imposed by step III, but usually *all* these steps are necessary for tractability, or for synthesis of resulting control laws or filtering devices. The controller may also be reduced after the initial design of the controller. Recall from Chapter 8 that optimal controllers of a high dimension (= order of plant) are easy to construct and it is more difficult to construct lower-order controllers.

To illustrate the points in this chapter in the simplest way, consider the following examples. Each example illustrates a different point highlighted by italics.

EXAMPLE 10.1

The most appropriate model of the plant depends upon the controller.

Consider that some optimal identification procedure has been used to determine $\hat{\mathscr{G}}(s)$ an approximation of the system $\mathscr{G}(s)$. Now, $\hat{\mathscr{G}}(s) \neq \mathscr{G}(s)$ since identification is never perfect. Suppose that $\mathscr{G}(s)$ is a closed-loop system with a known controller $H(s)$. Then,

$$\mathscr{G}(s) = \frac{G(s)}{1 + G(s)H(s)},$$

where $G(s)$ represents the actual plant which we do not know, but, of course, the approximation $\hat{G}(s)$ of the plant satisfies

$$\hat{\mathscr{G}}(s) = \frac{\hat{G}(s)}{1 + \hat{G}(s)H(s)}.$$

Hence, from the given data $\hat{\mathscr{G}}(s)$, $H(s)$, the plant approximation is given by

$$\hat{G}(s) = \frac{\hat{\mathscr{G}}(s)}{1 - \hat{\mathscr{G}}(s)H(s)}.$$

Observe that $\hat{G}(s)$ is a function of the controller $H(s)$. Therefore, it is clear that *an appropriate model of the plant depends upon the controller to be used.*

EXAMPLE 10.2

Small open-loop modeling errors can lead to large closed-loop errors.

For the system described by the transfer function $G(s)$,

$$G(s) = \frac{1}{(1 + s)(s + \varepsilon s)^{\alpha}}, \qquad Y(s) = G(s)u(s), \tag{10.1}$$

$\varepsilon > 0$ is small, possibly representing fact actuator or sensor dynamics. If these fast dynamics are ignored (a common approach in practice), then the control design model becomes

$$G_R(s) = \frac{1}{1 + s}, \qquad y_R(s) = G_R(s)u(s), \tag{10.2}$$

where, for $\alpha = 1$, it may be shown that the step input erorrs between $G(s)$ and $G_R(s)$ are bounded by

$$|y(t) - y_R(t)| = \mathscr{L}^{-1}\left\{\frac{1}{s}[G(s) - G_R(s)]\right\} < \frac{2\varepsilon e^{-t}}{1 - \varepsilon}, \qquad t > 0. \tag{10.3}$$

Hence, the modeling errors (as measured by the open-loop response) are arbitrarily small if ε is arbitrarily small. Assume an output feedback controller with $u = -Ky$.

If

$$K \ll \frac{1}{4\varepsilon}, \tag{10.4}$$

then the model $G_R(s)$ is also useful for predicting closed-loop behavior,

$$y_R(s) = \left[\frac{G_R(s)}{1 + KG_R(s)}\right]\frac{1}{s} \simeq y(s).$$

The interesting observation about (10.4) is that it tells that *the usefulness of the control design model $G_R(s)$ for predicting closed-loop behavior depends both upon the modeling error (characterized by ε) and the controller gain K.* This illustrates the theme of this entire chapter; *the modeling and control problems are not independent.*

Now, suppose the requirements on modeling error and control design (10.4) are violated to the extent that

$$K \gg \frac{1}{4\varepsilon}. \tag{10.5}$$

(The reader may verify that $K \simeq 1/4\varepsilon$ corresponds to the breakaway point on the root locus.) Now, the model $G_R(s)$ and controller (10.5) predict a closed-loop system with no overshoot, no oscillations, and a small steady-state error. However, the actual response will yield severely underdamped oscillations. The point of this example is that arbitrarily small modeling errors do not necessarily lead to small errors in the closed-loop predictions.

EXAMPLE 10.3

Large open-loop modeling errors do not preclude small closed-loop prediction errors.

Most of the available theories on model reduction try to achieve as accurate a model as possible according to some open-loop criterion for minimization. (Example 10.2 may even provide some motivation for this goal.) The purpose of Example 10.3 is to illustrate that it is not necessary to have small modeling errors if the control scheme has the right characteristics (remember the theme of the chapter). Consider a plant described by

$$G(s) = \frac{1}{s+1}, \qquad y(s) = G(s)u(s), \tag{10.6}$$

and an approximate model

$$G_R(s) = \frac{1}{s}, \qquad y_R(s) = G_R(s)u(s). \tag{10.7}$$

The controller again is output feedback with negative gain of magnitude K. Note that the actual system is asymptotically stable, whereas the approximation is not. $G_R(s)$ would not be deemed a good approximation of $G(s)$ by any of the model reduction theories available in the literature (see Chapter 6). It is interesting, however, to ask, "In what sense is $G_R(s)$ a good model for predicting *closed-loop* performance of the plant?"

Stability of $G_R(s)$ under the feedback gain K will also yield stability of $G(s)$ under this K. Errors in predicting performance of the step response are given by

$$|y(t) - y_R(t)| < \frac{1}{1+K}, \qquad t > 0. \tag{10.8}$$

Hence, the errors in the closed-loop response predictions can be made as small as desired by choosing an appropriate K, despite the fact that open-loop errors are large. Thus, closed-loop errors may be smaller than open-loop errors. The conclusion from this example is that the modeling errors must be *appropriate* for the controller design and not necessarily *small*. Arranging errors that are *appropriate* is not a straightforward task.

EXAMPLE 10.4

Modeling errors can lead to large errors in optimality predictions.

Consider Example 10.2 with $\alpha = 2$ and the same $G_R(s) = 1/(s+1)$. Let K be the optimal control gain for the model $G_R(s)$ so as to minimize

$$\mathscr{V} = \int_0^\infty [y_R^2(t) + \rho u^2(t)]\, dt, \qquad y_R(s) = G_R(s)u(s). \tag{10.9}$$

Then,

$$K = -1 + \sqrt{1+\rho^{-1}}, \tag{10.10}$$

and the performance of the closed-loop system $\{y(s) = G(s)u(s),\ u(s) = Ky(s)\}$ is described in Fig. 10.2 in terms of $\mathscr{V}_y \triangleq \int_0^\infty y^2(t)\, dt$ versus $\mathscr{V}_u \triangleq \int_0^\infty u^2(t)\, dt$ as ρ varies from ∞ to 0.

For $\rho < \varepsilon^2/4$, the actual closed-loop system is unstable, whereas the predicted behavior based on $G_R(s)$ approaches its maximal accuracy, $\mathscr{V}_y(\text{predicted}) = 0$, as ρ goes to 0. Hence, in the neighborhood of *maximal* accuracy predictions for $G_R(s)$ in Fig. 10.2 (this is where $\mathscr{V}_u$ is large), the actual system delivers its worst accuracy.

Now, consider the "absurd" design model

$$G_R'(s) = \frac{1}{1+\varepsilon s} \tag{10.11}$$

in lieu of $G_R(s) = 1/(s+1)$. The model (10.11) is "absurd" to the extent that the

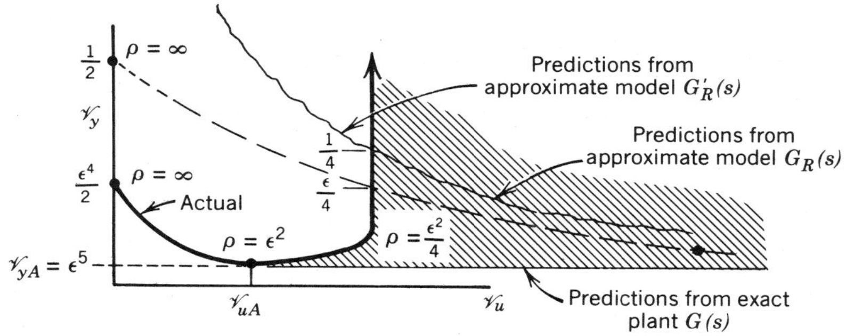

Figure 10.2 *Errors in Optimal Controller (Not to Scale)*

open-loop step response error is not necessarily small. The optimal control for this realization of the plant ($A = -1/\varepsilon$, $B = 1/\varepsilon$, $C = 1$) yields the optimal control $u = -Ky$, $K = -1 + \sqrt{1 + \rho^{-1}}$, (the same as (10.10)) which is precisely the same K as for model $G_R(s) = 1/(s+1)$! Hence, the *actual* performance is the same as that described in Fig. 10.2 for design modal $G_R(s)$. The performance *predicted* by model $G_R'(s)$ is shown in Fig. 10.2 to be arbitrarily far from the actual performance for large values of ρ *and* for small values of ρ, and the predictions are close to those of model $G_R(s)$ for small ρ.

In Fig. 10.2, the comparison between the *actual* and *predictions* from $G(s)$ indicate that for small gains (small $\mathscr{V}_u$, large ρ), the actual performance agrees with the performance which would be optimal using the exact model $G(s)$. That is, model errors do no damage for small control efforts. The actual performance has a "best" performance at a particular value of control effort ($\rho = \varepsilon^2$), and increasing control efforts beyond this point degrades performance leading eventually to instability ($\rho < \varepsilon^2/4$). Practically all controllers will drive the system unstable as the control effort is increased without bound. The shaded area in Fig. 10.2 describes the difference in performance between that which would be optimal for the exact plant $G(s)$ and that which results from controllers which are optimal with respect to an erroneous model [$G_R(s)$ or $G_R'(s)$ in our example]. This property is generic, according to the following conjecture.

Imagine the physical process $\mathscr{S}_0$, an evaluation model $\mathscr{S}_1$, which is necessarily in error but represents the best information available (from analytical or experimental data), and finally another model $\mathscr{S}_2$ which might be used for design of the controller $\mathscr{S}_c$. See Fig. 10.3. Control theory has a sophisticated level of maturity *assuming* that the model $\mathscr{S}_2$ is specified *a priori*; or, if not given, then the assumption is that the model is something that *exists* in an absolute sense ($\mathscr{S}_1 = \mathscr{S}_2 = \mathscr{S}_0$) irrespective of the control policy (which is yet to be developed). The thesis of this chapter is that the model and the control policy must be developed *together* and that no meaning (in an absolute sense) can be attached to *either* one in isolation. These concepts require careful definitions of *robustness*, as well as *identification*.

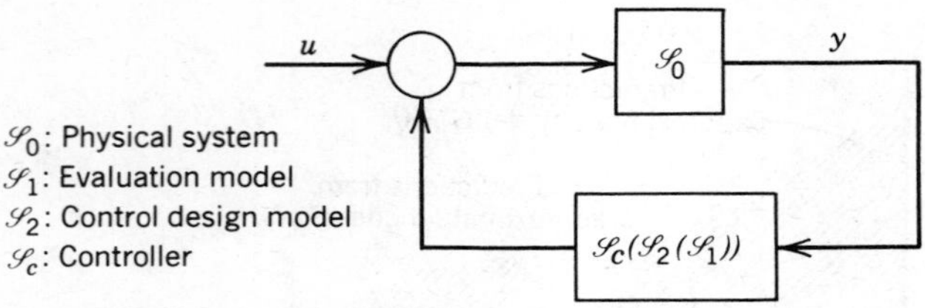

Figure 10.3 *Controller Data Base*

Let $\mathscr{S}_c$ be a linear dynamic controller that is optimal for the linear model $\mathscr{S}_2$. Let the actual linear plant be $\mathscr{S}_0$ and its optimal controller be $\mathscr{S}_{c0}$ for $\mathscr{V} = \mathscr{V}_y + \rho\mathscr{V}_u$. Since mathematical models are always approximations of the physical plant, $\mathscr{S}_0 \neq \mathscr{S}_2$. Let $\mathscr{V}_y = \int_0^\infty \mathbf{y}^*\mathbf{Q}\mathbf{y}\, dt$ denote the output norm when $\mathscr{S}_c$ drives $\mathscr{S}_0$. Let $\mathscr{V}_{y0}$ denote the output norm when $\mathscr{S}_{c0}$ drives $\mathscr{S}_0$. Denote by $\underline{\mathscr{V}}_y$ the smallest value that $\mathscr{V}_y$ takes on, and under this circumstance the control norm has value $\underline{\mathscr{V}}_u$. $\mathscr{V}_{y0}$ takes on its smallest value $\underline{\mathscr{V}}_{y0}$ when $\mathscr{V}_u$ is arbitrarily large $\underline{\mathscr{V}}_{u0}$. See the example in Fig. 10.2.

Conjecture

Controller S_c (optimal for any model S_2) driving the stable plant S_0 always yields:

$$\lim_{\rho \to 0} \mathscr{V}_y(\rho) = \infty, \tag{10.12a}$$

$$\lim_{\rho \to \infty} \left[\mathscr{V}_y(\rho) - \mathscr{V}_{y0}(\rho)\right] = 0, \tag{10.12b}$$

$$\underline{\mathscr{V}}_y > \underline{\mathscr{V}}_{y0}, \tag{10.12c}$$

$$\underline{\mathscr{V}}_u < \underline{\mathscr{V}}_{u0} \tag{10.12d}$$

The conjecture (10.12d) indicates that the actual maximal accuracy $\underline{\mathscr{V}}_y$ always occurs at a lower value of control effort $\underline{\mathscr{V}}_u$ than the control effort $\mathscr{V}_{u0}$ at the maximal accuracy of the truly optimal system $\underline{\mathscr{V}}_{y0}$ which occurs at a value of $\underline{\mathscr{V}}_{u0} = \infty$. Conjecture (10.12a) asserts that controllers (based upon erroneous models) will always be bad (unstable) for large control effort $\mathscr{V}_u$, and conjecture (10.12b) asserts that controllers will always be as good as the optimal controller $\mathscr{S}_{c0}$ for arbitrarily small control effort (10.12b). This suggests that a meaningful comparison of candidate controllers must be conducted in the neighborhood of the maximal accuracy $\underline{\mathscr{V}}_y$ for each controller. Each controller design, based perhaps upon different reduced-order models, will yield a different $\underline{\mathscr{V}}_y$. Furthermore, one cannot say *a priori* which is the best reduced-order model (as the above examples illustrate). This dependence between the modeling and control problems is the theme of this chapter.

This chapter is designed to show certain generic properties of modeling errors and their subsequent effect in closed-loop feedback control. Section 10.1 characterizes the general model error vectors, and Sections 10.2 and 10.3 show the behavior of the system under first-order perturbations and sensitivity calculations. Sections 10.4 and 10.5 show control schemes which accommodate a certain class of modeling errors.

10.1 The Structure of Modeling Errors

For the sake of discussion only, imagine that the mathematical model II of Fig. 10.1 is so accurate as to mimic the physical phenomena for all practical purposes. Then, in the following arguments we refer (by a slight abuse of language in this section only) to the model

$$\dot{\mathscr{x}} = \mathscr{A}\mathscr{x} + \mathscr{B}\mathbf{u} + \mathscr{w} + \mathbf{f}(\mathscr{x}, \mathbf{u}, t), \tag{10.13a}$$

$$\mathbf{y} = \mathscr{C}\mathscr{x} \tag{10.13b}$$

$$\mathbf{z} = \mathscr{M}\mathscr{x} + \mathscr{v} + \mathbf{g}(\mathscr{x}, \mathbf{u}, t) \tag{10.13c}$$

as the "physical system" with control inputs $\mathbf{u} \in \mathscr{R}^{n_u}$, disturbance inputs $\mathbf{w} \in \mathscr{R}^{n_w}$ and measurements $\mathscr{z} \in \mathscr{R}^{n_z}$. The output of interest for performance evaluation is $\mathbf{y} \in \mathscr{R}^{n_y}$. The terms $\mathbf{f}(\mathscr{x}, \mathbf{u}, t)$, $\mathbf{h}(\mathscr{x}, \mathbf{u}, t)$, and $\mathbf{g}(\mathscr{x}, \mathbf{u}, t)$ represent nonlinearities, and $\mathscr{w}(t)$ represents any time-varying disturbances ($\mathscr{w}(t)$ is not a function of $\mathscr{x}$ or $\mathbf{u}$). Of course, to attribute "physical system" status to (10.13) the dimension of the state $\mathscr{x}$ is arbitrarily large. Now, with respect to *any* mathematical model of the form

$$\dot{\mathbf{x}} = \mathbf{A}\mathbf{x} + \mathbf{B}\mathbf{u} + \mathbf{w}, \qquad \mathbf{x} \in \mathscr{R}^{n_x} \tag{10.14a}$$

$$\mathbf{z} = \mathbf{M}\mathbf{x} + \mathbf{v}, \qquad \mathbf{z} \in \mathscr{R}^{n_z}, \tag{10.14b}$$

that might be used to model the actual system (10.13), we wish to completely characterize the model errors of (10.14) with respect to the "truth" (10.13). For this purpose, rewrite (10.13) in the partitioned form

$$\begin{pmatrix} \dot{\mathscr{x}}_R \\ \dot{\mathscr{x}}_T \end{pmatrix} = \begin{bmatrix} \mathscr{A}_R & \mathscr{A}_{RT} \\ \mathscr{A}_{TR} & \mathscr{A}_T \end{bmatrix} \begin{pmatrix} \mathscr{x}_R \\ \mathscr{x}_T \end{pmatrix} + \begin{bmatrix} \mathscr{B}_R \\ \mathscr{B}_T \end{bmatrix} \mathbf{u} + \begin{pmatrix} \mathscr{w}_R \\ \mathscr{w}_T \end{pmatrix} + \begin{pmatrix} \mathbf{f}_R(\mathscr{x}, \mathbf{u}, t) \\ \mathbf{f}_T(\mathscr{x}, \mathbf{u}, t) \end{pmatrix}, \tag{10.15}$$

$$\mathbf{z} = [\mathscr{M}_R \quad \mathscr{M}_T] \begin{pmatrix} \mathscr{x}_R \\ \mathscr{x}_T \end{pmatrix} + \mathscr{v} + \mathbf{g}(\mathscr{x}, \mathbf{u}, t)$$

where the dimension of $\mathscr{x}_R \in \mathscr{R}^{n_x}$ is chosen to be the dimension of $\mathbf{x}$. Using the following definitions,

$$\Delta\mathbf{A} \triangleq \mathscr{A}_R - \mathbf{A}, \quad \Delta\mathbf{B} \triangleq \mathscr{B}_R - \mathbf{B}, \quad \Delta\mathbf{M} \triangleq \mathscr{M}_R - \mathbf{M}, \tag{10.16}$$

the equation for $\mathscr{x}_R$ and $\mathbf{z}$ can be written, from (10.15) and (10.16), as

$$\begin{aligned}\dot{\mathscr{x}}_R &= (\mathbf{A} + \Delta\mathbf{A})\mathscr{x}_R + \mathscr{A}_{RT}\mathscr{x}_T + (\mathbf{B} + \Delta\mathbf{B})\mathbf{u} + \mathscr{w}_R + \mathbf{f}_R(\mathscr{x}, \mathbf{u}, t),\\ \mathbf{z} &= (\mathbf{M} + \Delta\mathbf{M})\mathscr{x}_R + \mathscr{M}_T\mathscr{x}_T + \mathscr{v} + \mathbf{g}(\mathscr{x}, \mathbf{u}, t)\end{aligned} \tag{10.17}$$

or, simply,

$$\begin{aligned}\dot{\mathscr{x}}_R &= \mathbf{A}\mathscr{x}_R + \mathbf{B}\mathbf{u} + \mathbf{w} + \mathbf{e}_x,\\ \mathbf{z} &= \mathbf{M}\mathscr{x}_R + \mathbf{v} + \mathbf{e}_z,\end{aligned} \tag{10.18}$$

where the "model error vectors" $\mathbf{e}_x$ and $\mathbf{e}_x$ are given by

$$\mathbf{e} \triangleq \begin{pmatrix}\mathbf{e}_x\\ \mathbf{e}_z\end{pmatrix} = \underbrace{\begin{bmatrix}\Delta\mathbf{A}\\ \Delta\mathbf{M}\end{bmatrix}\mathscr{x}_R + \begin{bmatrix}\Delta\mathbf{B}\\ \mathbf{0}\end{bmatrix}\mathbf{u}}_{\mathbf{e}_p} + \underbrace{\begin{bmatrix}\mathscr{A}_{RT}\\ \mathscr{M}_T\end{bmatrix}\mathscr{x}_T}_{\mathbf{e}_0}$$

$$+ \underbrace{\begin{pmatrix}\mathscr{w}_R - \mathbf{w}\\ \mathscr{v} - \mathbf{v}\end{pmatrix}}_{\mathbf{e}_d} + \underbrace{\begin{pmatrix}\mathbf{f}_R(\mathscr{x}_R, \mathscr{x}_T, \mathbf{u}, t)\\ \mathbf{g}(\mathscr{x}_R, \mathscr{x}_T, \mathbf{u}, t)\end{pmatrix}}_{\mathbf{e}_n}. \tag{10.19}$$

We now have the following conclusion from (10.19).

Proposition 10.1

For every model of the form (10.14), *there exists "model error vectors"* $\mathbf{e}_x$ *and* $\mathbf{e}_z$ *which represent "corrections" to the state equation* (10.14a) *and measurement equation* (10.14b), *respectively, such that* $\mathscr{z}$ *evolving from* (10.14) *matches the measurements* $\mathbf{z}$ *from the physical system* (10.13). *Furthermore, the sum of four kinds of errors are always present: parameter errors* $\mathbf{e}_p$, *errors in model order* $\mathbf{e}_0$, *neglected disturbances* $\mathbf{e}_d$, *and nonlinearities* $\mathbf{e}_n$.

Note that all four types of modeling errors, $\mathbf{e}_p$, $\mathbf{e}_0$, $\mathbf{e}_d$, and $\mathbf{e}_n$, are always present in any mathematical characterization of a physical plant. We wish also to declare early in this chapter that no control theories exist that can promise satisfactory control in the simultaneous presence of all four categories of modeling error. We can only describe techniques which have made some progress in the accommodation of a *subset* of these four types of modeling errors.

Certain other conclusions are also obvious from (10.16). The partitioning of the state vector (10.15) was necessary to define "parameter errors" with respect to a specified model of lower order. The matrices $\mathscr{A}_R$, $\mathscr{B}_R$, and $\mathscr{M}_R$ result from a partitioning of (10.13) after (10.13) is written in some coordinate frame. It should be clear from (10.16) that since the parameters $\mathscr{A}_R$, $\mathscr{B}_R$, and $\mathscr{M}_R$ depend upon the coordinates in (10.13), the phrase "parameter errors" in dynamic systems has no precise meaning. The definitions (10.16) are as precise as one can be, yet these definitions are arbitrary to within a coordinate transformation on (10.15). Note that if one chooses a canonical structure for $(\mathbf{A}, \mathbf{B}, \mathbf{M})$ in (10.16), such as phase variable or Hessenberg form, and so forth, the parameters of $(\mathscr{A}_R, \mathscr{B}_R, \mathscr{M}_R)$ may not have the same structure. Hence, $(\Delta \mathbf{A}, \Delta \mathbf{B}, \Delta \mathbf{M})$ does not generally have a canonical structure. This suggests that parameter adaptive control or identification schemes might not converge if a canonical structure for $(\Delta \mathbf{A}, \Delta \mathbf{B}, \Delta \mathbf{M})$ has been presumed. Also, note from (10.19) that a change in coordinates which changes the $\mathbf{e}_p$ term will also change the $\mathbf{e}_0$ and $\mathbf{e}_n$ terms. Hence, the individual terms $\mathbf{e}_p$, $\mathbf{e}_0$, $\mathbf{e}_d$, and $\mathbf{e}_n$ in the model error decomposition (10.19) are not unique. Furthermore schemes that reduce the impact of one of these four types of error may in fact increase the impact of another type of error.

It will not always serve our purpose to characterize the model error vector explicitly in terms of higher-order model states $\boldsymbol{x}_T$. For the reduced model (10.18), the vector functions $\mathbf{e}_x$ and $\mathbf{e}_z$ can be considered functions only of $\boldsymbol{x}_R$ and $\mathbf{u}$ and t. Equation (10.18) shows that $\mathbf{e}_0$ is a function of $\boldsymbol{x}_T$, but $\boldsymbol{x}_T$ is in turn a function of $\boldsymbol{x}_R$, $\mathbf{u}$, and t, as the solution of the second equation in (10.15) reveals:

$$\begin{aligned}\boldsymbol{x}_T(\boldsymbol{x}_R, \mathbf{u}, t) &= \boldsymbol{\Phi}_T(t,0)\boldsymbol{x}_T(0)\\ &\quad + \int_0^t \boldsymbol{\Phi}_T(t-\sigma)\big(\mathscr{A}_{TR}\boldsymbol{x}_R + \mathscr{B}_T\mathbf{u} + \mathbf{w}_T + \mathbf{f}_T(\boldsymbol{x}_R, \boldsymbol{x}_T, \mathbf{u}, \sigma)\big)\, d\sigma,\end{aligned} \tag{10.20}$$

where $\boldsymbol{\Phi}_T$ is the state transition matrix for $\mathscr{A}_T$. Hence, for (10.18) we may consider $\mathbf{e}_x$ to depend only on $(\boldsymbol{x}_R, \mathbf{u}, t)$; and when we need to do so, we shall write $\mathbf{e}_x(\boldsymbol{x}_R, \mathbf{u}, t)$. In fact, using (10.20), the model order error $\mathbf{e}_0$ of (10.19) may be characterized by

$$\begin{aligned}\mathbf{e}_0 = {} & \underbrace{\begin{bmatrix}\mathscr{A}_{RT}\\ \mathscr{M}_T\end{bmatrix}\big[\boldsymbol{\Phi}_T(t,0)\boldsymbol{x}_T(0)\big]}_{\mathbf{e}_{0t}} + \underbrace{\begin{bmatrix}\mathscr{A}_{RT}\\ \mathscr{M}_T\end{bmatrix}\left[\int_0^t \boldsymbol{\Phi}_T(t-\sigma)\boldsymbol{w}_T(\sigma)\, d\sigma\right]}_{\mathbf{e}_{0d}}\\ & + \underbrace{\begin{bmatrix}\mathscr{A}_{RT}\\ \mathscr{M}_T\end{bmatrix}\left[\int_0^t \boldsymbol{\Phi}_T(t-\sigma)\mathscr{A}_{TR}\boldsymbol{x}_R\, d\sigma\right]}_{\mathbf{e}_{0x}} + \underbrace{\begin{bmatrix}\mathscr{A}_{RT}\\ \mathscr{M}_T\end{bmatrix}\left[\int_0^t \boldsymbol{\Phi}_T(t-\sigma)\mathscr{B}_T\mathbf{u}\, d\sigma\right]}_{\mathbf{e}_{0u}}\\ & + \underbrace{\begin{bmatrix}\mathscr{A}_{RT}\\ \mathscr{M}_T\end{bmatrix}\left[\int_0^t \boldsymbol{\Phi}_T(t-\sigma)\mathbf{f}_T(\boldsymbol{x}_R, \boldsymbol{x}_T, \mathbf{u}, \sigma)\, d\sigma\right]}_{\mathbf{e}_{0n}}.\end{aligned} \tag{10.21}$$

From (10.19) and (10.4), conclude that $\mathbf{e}$ may be written

$$\mathbf{e} = \{\mathbf{e}_{px} + \mathbf{e}_{0x}\} + \{\mathbf{e}_{pu} + \mathbf{e}_{0u}\} + \{\mathbf{e}_{0t} + \mathbf{e}_{0d} + \mathbf{e}_d\} + \{\mathbf{e}_n + \mathbf{e}_{0n}\}. \tag{10.22}$$

With respect to the reduced model (10.14), the first bracketed term in (10.22) denotes errors which depend on the state $\boldsymbol{x}_R$ ($\mathbf{e}_{px}$ is a linear function of $\boldsymbol{x}_R$, whereas $\mathbf{e}_{0x}$ is an integral operator on $\boldsymbol{x}_R$), the second bracket denotes errors which depend on the control $\mathbf{u}$ ($\mathbf{e}_{pu}$ is a linear function of $\mathscr{x}_R$, whereas $\mathbf{e}_{0u}$ is an integral operator on $\mathbf{u}$), the third bracketed term in (10.22) denotes errors which depend only on time, and finally the last bracketet term in (10.22) denotes the errors due to nonlinearities.

It is important to note that the model error vector $\mathbf{e}_x$ in (10.18) depends upon the *integral* of the input $\mathbf{u}$. *Hence, one cannot assess the impact of modeling errors* $\mathbf{e}_x$ *without knowledge of the nature of the controls* $\mathbf{u}(t)$. Even small inputs $\mathbf{u}$ can have an arbitrarily large effect in $\mathbf{e}_x(t)$. To see that this is possible, suppose A_T is a positive scalar and $u(t) = \varepsilon =$ constant. Then $e_{0u}(t) = (A_{RT}B_T\varepsilon/A_T)(e^{A_Tt} - 1)$ gets arbitrarily large. Hence, *the homogeneous part of the system may be modeled arbitrarily closely* ($\mathbf{e}_p$ arbitrarily small or zero), *and yet the model may not be acceptable for control design.*

10.1.1 THE MODEL ERROR SYSTEM

The model error structure (10.21) can be further detailed. See that $\mathbf{e}_{0x}$ satisfies the differential equations

$$\dot{\mathbf{e}}_{0x} = \begin{bmatrix} \mathscr{A}_{RT} \\ \mathscr{M}_T \end{bmatrix} \mathscr{A}_{TR}\mathscr{x}_R + \begin{bmatrix} \mathscr{A}_{RT} \\ \mathscr{M}_T \end{bmatrix} \int_0^t \mathscr{A}_T \Phi_T(t-\sigma) \mathscr{A}_{TR}\mathscr{x}_R \, d\sigma \tag{10.23}$$

and the definition

$$\mathbf{e}_{ix} \triangleq \begin{bmatrix} \mathscr{A}_{RT} \\ \mathscr{M}_T \end{bmatrix} \int_0^t \mathscr{A}_T^i \Phi_T(t-\sigma) \mathscr{A}_{TR}\mathscr{x}_R \, d\sigma \tag{10.24}$$

allows (10.23) to be written

$$\dot{\mathbf{e}}_{0x} = \begin{bmatrix} \mathscr{A}_{RT} \\ \mathscr{M}_T \end{bmatrix} \mathscr{A}_{TR}\mathscr{x}_R + \mathbf{e}_{1x} \tag{10.25}$$

and, likewise for $\mathbf{e}_{1x}$ and $\mathbf{e}_{ix}$,

$$\begin{aligned} \dot{\mathbf{e}}_{1x} &= \begin{bmatrix} \mathscr{A}_{RT} \\ \mathscr{M}_T \end{bmatrix} \mathscr{A}_T \mathscr{A}_{TR}\mathscr{x}_R + \mathbf{e}_{2x}, \\ \dot{\mathbf{e}}_{ix} &= \begin{bmatrix} \mathscr{A}_{RT} \\ \mathscr{M}_T \end{bmatrix} \mathscr{A}_T^i \mathscr{A}_{TR}\mathscr{x}_R + \mathbf{e}_{(i+1)x}, \qquad i = 0, 1, 2, \ldots . \end{aligned} \tag{10.26}$$

Similarly, for $\mathbf{e}_{0u}$ in (10.21),

$$\begin{aligned} \dot{\mathbf{e}}_{0u} &= \begin{bmatrix} \mathscr{A}_{RT} \\ \mathscr{M}_T \end{bmatrix} \mathscr{B}_T \mathbf{u} + \mathbf{e}_{1u}, \\ \dot{\mathbf{e}}_{iu} &= \begin{bmatrix} \mathscr{A}_{RT} \\ \mathscr{M}_T \end{bmatrix} \mathscr{A}_T^i \mathscr{B}_T \mathbf{u} + \mathbf{e}_{(i+1)u}, \qquad i = 0, 1, 2, \ldots . \end{aligned} \tag{10.27}$$

Equations (10.26), (10.27), and (10.22) are now combined with (10.17) to give the exact structure of the model error system:

$$\dot{\boldsymbol{x}}_R = \mathbf{A}\boldsymbol{x}_R + \mathbf{B}\mathbf{u} + \mathbf{w} + \mathbf{E}_{10}\big[(\mathbf{e}_{0x} + \mathbf{e}_{0u}) + (\mathbf{e}_{px} + \mathbf{e}_{pu}) + \mathbf{e}_t + \mathbf{e}_N\big], \quad (10.28a)$$

$$\dot{\mathbf{e}}_{ix} = \mathbf{P}_i\boldsymbol{x}_R + \mathbf{e}_{(i+1)x}, \qquad i = 0, 1, 2, \ldots . \quad (10.28b)$$

$$\dot{\mathbf{e}}_{iu} = \mathbf{Q}_i\mathbf{u} + \mathbf{e}_{(i+1)u}, \qquad i = 0, 1, 2, \ldots . \quad (10.28c)$$

$$\mathbf{z} = \mathbf{M}\boldsymbol{x}_R + \mathbf{v} + \mathbf{E}_{01}\big[(\mathbf{e}_{0x} + \mathbf{e}_{0u}) + (\mathbf{e}_{px} + \mathbf{e}_{pu}) + \mathbf{e}_t + \mathbf{e}_N\big], \quad (10.28d)$$

where $\mathbf{E}_{10} \triangleq [\mathbf{I} \quad \mathbf{0}]$, $\mathbf{E}_{01} \triangleq [\mathbf{0} \quad \mathbf{I}]$,

$$\mathbf{P}_i \triangleq \begin{bmatrix} \mathscr{A}_{RT} \\ \mathscr{M}_T \end{bmatrix} \mathscr{A}^i_T \mathscr{A}_{TR}, \qquad \mathbf{Q}_i \triangleq \begin{bmatrix} \mathscr{A}_{RT} \\ \mathscr{M}_T \end{bmatrix} \mathscr{A}^i_T \mathscr{B}_T, \quad (10.29)$$

and

$$\mathbf{e}_t \triangleq \mathbf{e}_{0t} + \mathbf{e}_{0d} + \mathbf{e}_d, \qquad \mathbf{e}_N \triangleq \mathbf{e}_n + \mathbf{e}_{0n}, \quad (10.30)$$

using the definitions of $\mathbf{e}_{0t}$, $\mathbf{e}_{0d}$, $\mathbf{e}_d$, $\mathbf{e}_n$, and $\mathbf{e}_{0n}$ in (10.19) and (10.21).

In state form, (10.28) becomes

$$\begin{pmatrix} \dot{\boldsymbol{x}}_R \\ \dot{\mathbf{e}}_{0x} \\ \dot{\mathbf{e}}_{0u} \\ \dot{\mathbf{e}}_{1x} \\ \dot{\mathbf{e}}_{1u} \\ \dot{\mathbf{e}}_{2x} \\ \dot{\mathbf{e}}_{2u} \\ \vdots \\ \mathbf{z} \end{pmatrix} = \begin{bmatrix} \mathbf{A} + \Delta\mathbf{A} & \mathbf{E}_{10} & \mathbf{E}_{10} & \mathbf{0} & \mathbf{0} & \mathbf{0} & \mathbf{0} & \mathbf{0} & \mathbf{0} & \cdots & \\ \mathbf{P}_0 & \mathbf{0} & \mathbf{0} & \mathbf{I} & \mathbf{0} & \mathbf{0} & \mathbf{0} & \mathbf{0} & \mathbf{0} & \cdots & \\ \mathbf{0} & \mathbf{0} & \mathbf{0} & \mathbf{0} & \mathbf{I} & \mathbf{0} & \mathbf{0} & \mathbf{0} & \mathbf{0} & \cdots & \\ \mathbf{P}_1 & \mathbf{0} & \mathbf{0} & \mathbf{0} & \mathbf{0} & \mathbf{I} & \mathbf{0} & \mathbf{0} & \mathbf{0} & \cdots & \\ \mathbf{0} & \mathbf{0} & \mathbf{0} & \mathbf{0} & \mathbf{0} & \mathbf{0} & \mathbf{I} & \mathbf{0} & \mathbf{0} & \cdots & \\ \mathbf{P}_2 & \mathbf{0} & \mathbf{0} & \mathbf{0} & \mathbf{0} & \mathbf{0} & \mathbf{0} & \mathbf{I} & \mathbf{0} & \cdots & \\ \mathbf{0} & \mathbf{0} & \mathbf{0} & \mathbf{0} & \mathbf{0} & \mathbf{0} & \mathbf{0} & \mathbf{0} & \mathbf{I} & \cdots & \\ \vdots & \vdots & \vdots & \vdots & \vdots & \vdots & \vdots & \vdots & \vdots & & \\ \mathbf{M} + \Delta\mathbf{M} & \mathbf{E}_{01} & \mathbf{E}_{01} & \mathbf{0} & \mathbf{0} & \mathbf{0} & \mathbf{0} & \mathbf{0} & \mathbf{0} & \cdots & \mathbf{0} \end{bmatrix} \begin{pmatrix} \boldsymbol{x}_R \\ \mathbf{e}_{0x} \\ \mathbf{e}_{0u} \\ \mathbf{e}_{1x} \\ \mathbf{e}_{1u} \\ \mathbf{e}_{2x} \\ \mathbf{e}_{2u} \\ \vdots \\ \vdots \end{pmatrix}$$

$$+ \begin{bmatrix} \mathbf{B} + \Delta\mathbf{B} \\ \mathbf{0} \\ \mathbf{Q}_0 \\ \mathbf{0} \\ \mathbf{Q}_1 \\ \mathbf{0} \\ \mathbf{Q}_2 \\ \vdots \\ \mathbf{0} \end{bmatrix} \mathbf{u} + \begin{bmatrix} \mathbf{w} + \mathbf{E}_{10}(\mathbf{e}_t + \mathbf{e}_N) \\ \mathbf{0} \\ \mathbf{0} \\ \mathbf{0} \\ \mathbf{0} \\ \mathbf{0} \\ \mathbf{0} \\ \vdots \\ \mathbf{v} + \mathbf{E}_{01}(\mathbf{e}_t + \mathbf{e}_N) \end{bmatrix}. \quad (10.31)$$

Note that the matrices that are *unknown* are $\mathbf{P}_i$, $\mathbf{Q}_i$, $\Delta\mathbf{A}$, and $\Delta\mathbf{B}$. Knowledge of the model error structure (10.31) can be very useful in analysis (predictions of performance) and control design. Note that the assumption of parameter errors only, ignores the $\mathbf{e}_{ix}(t)$ and $\mathbf{e}_{iu}(t)$.

The transfer functions of (10.28) are developed as follows. In the context of linear systems, we shall ignore $\mathbf{e}_N$. Take the Laplace transform of (10.28) and see that

$$\mathbf{z}(s) = \left(\mathbf{M} + \Delta\mathbf{M} + \mathbf{E}_{01}\left(\sum_{i=0}^{\infty} \frac{\mathbf{P}_i}{s^{i+1}}\right)\right)\left[(s\mathbf{I} - \mathbf{A}) - \Delta\mathbf{A} - \mathbf{E}_{10}\left(\sum_{i=0}^{\infty} \frac{\mathbf{P}_i}{s^{i+1}}\right)\right]^{-1}$$

$$\times\left[\left(\mathbf{B} + \Delta\mathbf{B} + \mathbf{E}_{10}\left(\sum_{i=0}^{\infty} \frac{\mathbf{Q}_i}{s^i}\right)\right)\mathbf{u}(s) + \mathbf{E}_{10}\mathbf{e}_t + \mathbf{w}\right]$$

$$+\mathbf{E}_{01}\left(\sum_{i=0}^{\infty} \frac{\mathbf{Q}_i}{s^{i+1}}\right)\mathbf{u}(s) + \mathbf{v}(s) + \mathbf{E}_{10}\mathbf{e}_t(s).$$

To simplify this expression, define

$$\mathscr{P}_{ix} \triangleq \begin{cases} \Delta\mathbf{A} & \text{when } i = 0, \\ \mathbf{E}_{10}\mathbf{P}_{i-1} & \text{when } i > 0, \end{cases} \qquad \mathscr{P}_{iz} = \begin{cases} \Delta\mathbf{M} & \text{when } i = 0, \\ \mathbf{E}_{01}\mathbf{P}_{i-1} & \text{when } i > 0, \end{cases} \tag{10.32a}$$

$$\mathscr{Q}_{ix} \triangleq \begin{cases} \Delta\mathbf{B} & \text{when } i = 0, \\ \mathbf{E}_{10}\mathbf{Q}_{i-1} & \text{when } i > 0 \end{cases} \qquad \mathscr{Q}_{iz} \triangleq \mathbf{E}_{01}\mathbf{Q}_{i-1}$$

$$\boldsymbol{\Phi}(s) \triangleq (s\mathbf{I} - \mathbf{A})^{-1}. \tag{10.32b}$$

Then,

$$\mathbf{z}(s) = \left\{\left[\mathbf{M} + \sum_{i=0}^{\infty} \frac{\mathscr{P}_{iz}}{s^i}\right]\left[\boldsymbol{\Phi}^{-1}(s) - \sum_{i=0}^{\infty} \frac{\mathscr{P}_{ix}}{s^i}\right]^{-1}\left[\mathbf{B} + \sum_{i=0}^{\infty} \frac{\mathscr{Q}_{ix}}{s^i}\right]\right.$$

$$\left. + \sum_{i=1}^{\infty} \frac{\mathscr{Q}_{iz}}{s^i}\right\}\mathbf{u}(s) + \boldsymbol{\varepsilon}_t(s), \tag{10.33}$$

where $\boldsymbol{\varepsilon}_t(t)$ represents the effects of only *time*-dependent terms,

$$\boldsymbol{\varepsilon}_t(s) \triangleq \left[\mathbf{M} + \sum_{i=0}^{\infty} \frac{\mathscr{P}_{iz}}{s^i}\right]\left[\boldsymbol{\Phi}^{-1}(s) - \sum_{i=0}^{\infty} \frac{\mathscr{P}_{ix}}{s^i}\right]^{-1}[\mathbf{w}(s) + \mathbf{E}_{10}\mathbf{e}_t(s)]$$

$$+\mathbf{E}_{01}\mathbf{e}_t(s) + \mathbf{v}(s).$$

Let the inverse of the sum of two matrices be written

$$[\boldsymbol{\Phi}^{-1} + \boldsymbol{\Psi}]^{-1} = \boldsymbol{\Phi} \sum_{j=0}^{\infty} \left[(-1)^j (\boldsymbol{\Psi\Phi})^j\right].$$

This can be verified by writing $[\boldsymbol{\Phi}^{-1} + \boldsymbol{\Psi}]\mathbf{x} = \mathbf{y}$ as $\boldsymbol{\Phi}^{-1}\mathbf{x} = (\mathbf{y} - \boldsymbol{\Psi}\mathbf{x})$ or

$$\begin{aligned} \mathbf{x} &= \boldsymbol{\Phi}(\mathbf{y} - \boldsymbol{\Psi}\mathbf{x}) \\ &= \boldsymbol{\Phi}\mathbf{y} - \boldsymbol{\Phi\Psi\Phi}(\mathbf{y} - \boldsymbol{\Psi}\mathbf{x}) \\ &= \text{etc.} \end{aligned}$$

Now, let

$$\boldsymbol{\Psi}_1 \triangleq -\sum_{i=0}^{\infty} \mathscr{P}_{ix}/s^i, \quad \boldsymbol{\Psi}_2 \triangleq \sum_{i=0}^{\infty} \mathscr{P}_{iz}/s^i, \quad \boldsymbol{\Psi}_3 \triangleq \sum_{i=0}^{\infty} \mathscr{Q}_{ix}/s^i, \quad \boldsymbol{\Psi}_4 \triangleq \sum_{i=0}^{\infty} \mathscr{Q}_{iz}/s^i,$$

$$\mathbf{J}(s) \triangleq \sum_{j=1}^{\infty} (-1)^j \boldsymbol{\Psi}_1^j(s) \boldsymbol{\Phi}^j(s).$$

Then,

$$\begin{aligned} \mathbf{z}(s) &= \{[\mathbf{M} + \boldsymbol{\Psi}_2(s)]\boldsymbol{\Phi}(s)[\mathbf{I} + \mathbf{J}(s)][\mathbf{B} + \boldsymbol{\Psi}_3(s)] + \boldsymbol{\Psi}_4(s)\}\mathbf{u}(s) + \boldsymbol{e}_t(s) \\ &= [\mathbf{G}(s) + \Delta\mathbf{G}(s)]\mathbf{u}(s) + \boldsymbol{e}_t(s), \end{aligned} \tag{10.34}$$

expressed in terms of the (known) transfer function $\mathbf{G}(s) = \mathbf{M}\boldsymbol{\Phi}(s)\mathbf{B}$, where the unknown terms are:

$$\begin{aligned} \Delta\mathbf{G}(s) &\triangleq \boldsymbol{\Psi}_4(s) + \mathbf{M}\boldsymbol{\Phi}(s)\boldsymbol{\Psi}_3(s) + [\mathbf{M}\boldsymbol{\Phi}(s)\mathbf{J}(s) + \boldsymbol{\Psi}_2(s)\boldsymbol{\Phi}(s) \\ &\qquad + \boldsymbol{\Psi}_2(s)\boldsymbol{\Phi}(s)\mathbf{J}(s)][\mathbf{B} + \boldsymbol{\Psi}_3(s)], \\ \boldsymbol{e}_t(s) &= [\mathbf{M} + \boldsymbol{\Psi}_2(s)]\boldsymbol{\Phi}(s)[\mathbf{I} + \mathbf{J}(s)][\mathbf{w}(s) + \mathbf{E}_{10}\mathbf{e}_t(s)] \\ &\qquad + \mathbf{E}_{01}\mathbf{e}_t(s) + \mathbf{v}(s). \end{aligned} \tag{10.35}$$

Both expressions (10.31) and (10.35) simplify greatly by special choices of coordinates of (10.15). Without loss of generality, one can take $\Delta\mathbf{M} = 0$ and $\mathscr{M}_T = 0$ if $\dim \mathbf{z} \leq \dim \boldsymbol{x}_R$. To see this, note that a similarity transformation on the state $\boldsymbol{x} = \mathbf{T}\boldsymbol{\eta}$ can always take the measurement $\mathbf{z} = \mathscr{M}\boldsymbol{x}$ to $\mathbf{z} = \mathscr{M}\mathbf{T}\boldsymbol{\eta} = [\mathbf{I} \quad \mathbf{0}]\boldsymbol{\eta}$ if rank $\mathscr{M} = \dim \mathbf{z} \leq \dim \boldsymbol{x}$. Of course, we cannot *construct* this $\mathbf{T}$ (since we don't know $\mathscr{M}$), but we know that it exists. Therefore, by assuming that the states of our model are not fewer than the number of measurements, we can, from (10.29) and (10.32), without loss of generality, set

$$\mathbf{E}_{01}\mathbf{P}_i = \mathbf{0}, \qquad \mathbf{E}_{01}\mathbf{Q}_i = \mathbf{0}.$$

Hence, from (10.32), $\mathscr{Q}_{iz} = \mathbf{0}$, $\forall i$, and $\mathscr{P}_{iz} = \mathbf{0}$, $\forall i$. This simplifies (10.35), the modeling errors in any linear system to

$$\begin{aligned}\Delta\mathbf{G}(s) &= \mathbf{M}\mathbf{\Phi}(s)\mathbf{\Psi}_3(s) + [\mathbf{M} + \mathbf{\Psi}_2(s)]\mathbf{\Phi}(s)\mathbf{J}(s)[\mathbf{B} + \mathbf{\Psi}_3(s)],\\ \boldsymbol{e}_t(s) &= \mathbf{M}\mathbf{\Phi}(s)[\mathbf{I} + \mathbf{J}(s)][\mathbf{w}(s) + \mathbf{E}_{10}\mathbf{e}_t(s)] + \mathbf{v}(s).\end{aligned} \tag{10.36}$$

The state equations (10.31) also simplify in an obvious way when $\Delta\mathbf{M} = \mathbf{0}$. The term $\boldsymbol{e}_t$ in (10.33) represents unknown excitations, and in the literature, [10.1] and [10.2], upper bounds on the magnitude of $\boldsymbol{e}_t$ (before losing stability) have been established. However, in Chapters 1 through 9 of this text, the $\Delta\mathbf{G}(s)$ and $\boldsymbol{e}_t(s)$ terms in (10.36) have been *ignored* in the control design. See that $\mathbf{u}(s)$ multiplies $\Delta\mathbf{G}(s)$. Hence, it should be emphasized that *the effects of model error cannot be assessed independently of the control law* $u(\cdot)$, and this is the fundamental pitfall that prevents the modeling problems and the control problems from being separable! Most texts on modeling dynamic systems (circuits, mechanical systems, thermodynamics, etc.) and most texts on control (including the first nine chapters of this one) are written as though the modeling and control problems were separable. The student of control must break down barriers that have been established along traditional lines.

The following is the established *idea* of the MODELING PROBLEM:

> Find a set of differential equations $\mathscr{S}_1$ describing the dynamic relationships between the response $\mathbf{y}(t)$ and the (unspecified) inputs {controls $\mathbf{u}(t)$, disturbances $\mathbf{w}(t)$, and initial conditions}.

The flaw in this task statement is the presumption that *there exists* a set of differential equations which relate $\mathbf{y}(t)$ and $\mathbf{u}(t)$, irrespective of $\mathbf{u}(t)$. We argue that *any* set of differential equations $\mathscr{S}_1$ is only an approximation of the physical phenomenon $\mathscr{S}_0$, and it was shown in (10.28) that the errors associated with this approximation cannot be assessed, qualitatively or quantitatively, independently of $\mathbf{u}(t)$. In other words, knowledge of the *control* inputs $\mathbf{u}(\cdot)$ are required in any assessment of *model* fidelity! In modeling and identification literature we read and write about model errors with respect to a *truth* model, $\mathscr{S}$. There is no truth model ($\mathscr{S}_1 \neq \mathscr{S}_0$). The model and its controller should be discussed as a *pair*. They have no significance separately.

The following is the established *idea* of the CONTROL PROBLEM:

> Given the set of models Σ which describe the dynamic process, find an appropriate control $u(t)$ or controller $u(z(t), t)$ to meet a specified set of control objectives.

The flaw in this task statement is the presumption that the class of models Σ which appropriately describe the process can be defined *a priori*, independently of knowledge of $\mathbf{u}(t)$ or, consequently, of knowledge of the controller generating $\mathbf{u}(t)$. Now, the control law cannot logically be specified prior to model development. Thus, if one wishes to squeeze the best possible performance from the controller design, then one cannot ignore the following:

Modeling and Control Inseparability Principle

The modeling and control problems are not separable and are necessarily iterative.

This means that for any given $\mathscr{S}_0$, in Fig. 10.3 the development of $\mathscr{S}_1$ (and possibly $\mathscr{S}_2$ if a reduced-order model from $\mathscr{S}_1$ is required) and $\mathscr{S}_c$ is an iterative process.

Several implications of the *modeling and control inseparability principle* are:

(a) The phrase "model of the plant" is a misnomer. We must refer to a model as being appropriate *under the influence of a particular controller*. Hence, we must refer to a (model, controller) *pair* as appropriate or inappropriate for each other.

(b) Only "local" properties can be stated concerning the model and the controller. This means that the interpretation of both classical and modern control theory must be tempered with this knowledge, since parameters of *neither* plant model nor controller can be taken to infinity (or wide ranges). Three examples follow.

The root locus theory presumes a *fixed* plant while the controller gain goes to infinity [10.3]. But the fidelity of the plant model *depends* upon the control gain. Hence, the same model of the plant is not appropriate at *both* the vicinity of the open-loop poles and the open-loop zeros.

The Nyquist plots are reliable only over a limited frequency range and are certainly not reliable in the vicinity of the origin, where $\omega \to \infty$. If this region of uncertainty extends to a *unit* radius around the origin, then even the stability results of the Nyquist plot are suspect. See Fig. 10.4.

In LQI theory it is presumed that the model is *fixed* and that the weights in the performance index may be varied over wide ranges. See Section 8.5. This generates the theoretical predictions of maximal accuracy in Fig. 10.5 (solid curve). However, the actual performance follows the dotted curve, due to modeling errors (note the decreasing effect of modeling errors as the control effort decreases). Thus, the deviation from the theoretical is *greatest* where maximal accuracy predictions are made. Hence, there is often a large discrepancy between achieved and predicted maximal accuracy, and large errors in the value of the control effort at which maximal accuracy occurs. These inequalities hold from conjectures (10.12c), (10.12d)

$$u_a^2 < u_t^2, \qquad y_a^2 > y_t^2.$$

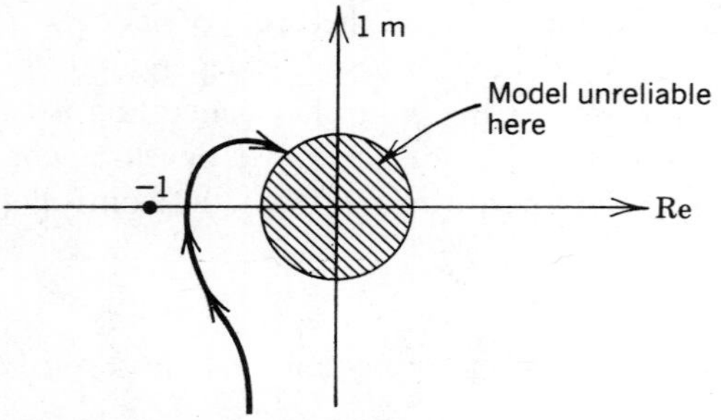

Figure 10.4 Nyquist Diagram

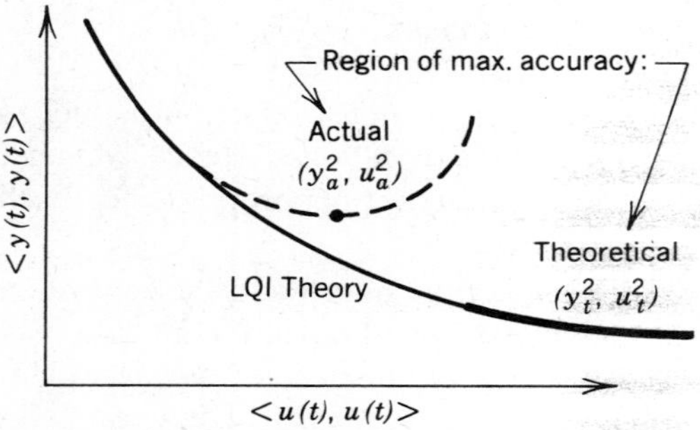

Figure 10.5 Performance Evaluation with "Truth" Model

10.1.2 THE STRUCTURE OF ERRORS IN THE CLOSED-LOOP SYSTEM

Suppose our hypothetical "physical system" (10.13) (with nonlinear terms f, g neglected) is now driven by the linear dynamic controller with transfer matrix $\mathbf{H}(s)$:

$$\mathbf{u}(s) = \mathbf{H}(s)\mathbf{z}(s). \tag{10.37}$$

Without loss of generality, we may associate a state space realization with (10.37) and write

$$\mathbf{H}(s) = \mathbf{G}(s\mathbf{I} - \mathbf{A}_c)^{-1}\mathbf{F} \tag{10.38}$$

or, in state form,

$$\begin{aligned} \dot{\mathbf{x}}_c &= \mathbf{A}_c\mathbf{x}_c + \mathbf{F}\mathbf{z}, \qquad \mathbf{x}_c \in \mathscr{R}^{n_x}, \\ \mathbf{u} &= \mathbf{G}\mathbf{x}_c, \qquad \mathbf{u} \in \mathscr{R}^{n_u}. \end{aligned} \tag{10.39}$$

In this section we ask how the poles of the closed-loop system behave as either **G** or **F** approaches zero. Arbitrary gains **G** and **F** are next considered but with a restrictive assumption about parameter errors ($\mathbf{e}_p = \mathbf{0}$). It will prove convenient to write results in terms of a set of matrices $(\mathbf{A}, \mathbf{B}, \mathbf{M})$ which can *always* be found satisfying

$$\mathbf{A} + \mathbf{BG} - \mathbf{FM} \triangleq \mathbf{A}_c \tag{10.40}$$

for any given $\mathbf{A}_c$, **G**, and **F**.

The system (10.13) is now driven by controller (10.39), (10.40). Subtracting (10.39) from (10.17) gives a differential equation for $\tilde{\mathbf{x}} \triangleq \boldsymbol{x}_R - \mathbf{x}_C$. Now, writing (10.13) and (10.39) in terms of the states $\boldsymbol{x}$ and $\tilde{\mathbf{x}}$ yields the homogeneous part of the closed-loop system:

$$\begin{pmatrix} \dot{\boldsymbol{x}} \\ \dot{\tilde{\mathbf{x}}} \end{pmatrix} = \left[\begin{array}{c|c} \mathscr{A} + \mathscr{B}\mathbf{GE}_{10} & -\mathscr{B}\mathbf{G} \\ \hline (\mathscr{A}_{RT} - \mathbf{F}\mathscr{M}_T)\mathbf{E}_{01} + (\Delta\mathbf{A} - \mathbf{F}\Delta\mathbf{M} + \Delta\mathbf{BG})\mathbf{E}_{10} & \mathbf{AM} - \mathbf{FM} - \Delta\mathbf{BG} \end{array}\right] \begin{pmatrix} \boldsymbol{x} \\ \tilde{\mathbf{x}} \end{pmatrix},$$

$$\begin{pmatrix} \mathbf{y} \\ \mathbf{u} \end{pmatrix} = \begin{bmatrix} \mathscr{C} & \mathbf{0} \\ \mathbf{GE}_{10} & -\mathbf{G} \end{bmatrix} \begin{pmatrix} \boldsymbol{x} \\ \tilde{\mathbf{x}} \end{pmatrix}, \tag{10.41}$$

where $\mathbf{E}_{10} \triangleq [\mathbf{I}_{n_x} \quad \mathbf{0}]$ and $\mathbf{E}_{01} \triangleq [\mathbf{0}_{n_x} \quad \mathbf{I}]$.

Exercise 10.1

Verify (10.41).

In the limit as $\mathbf{G} \to \mathbf{0}$, the eigenvalues of (10.41) become those of the block-diagonal matrices $\mathscr{A}$ and $\mathbf{A} - \mathbf{FM}$. This conclusion is summarized as follows:

Theorem 10.1

In the limit as $\mathbf{G} \to \mathbf{0}$, *any linear system* (10.13) *driven by any controller of the form* (10.39) *has the eigenvalues of* $\mathscr{A}$ *and* $(\mathbf{A} - \mathbf{FM})$.

Theorem 10.1 suggests that the low-bandwidth controller (characterized by small **G**) is stable if the open-loop system is stable and if the matrix $(\mathbf{A} - \mathbf{FM})$ (which is entirely under the design of the analyst) is stable.

Now, multiply (10.39) by $\mathbf{E}_{10}^T$, and subtract this equation from (10.13). This defines the vector $\tilde{\boldsymbol{x}} \triangleq \boldsymbol{x} - \mathbf{E}_{10}^T\mathbf{x}_c$. The homogeneous part of the closed-loop system (10.13), (10.39) can now be described in terms of states $\tilde{\boldsymbol{x}}$ and $\mathbf{x}_c$, yielding

$$\begin{pmatrix} \dot{\tilde{\boldsymbol{x}}} \\ \dot{\mathbf{x}}_c \end{pmatrix} = \begin{bmatrix} \mathscr{A} - \mathbf{E}_{10}^T\mathbf{F}\mathscr{M} & \mathbf{E}_{10}^T(\Delta\mathbf{A} + \Delta\mathbf{BG} - \mathbf{F}\Delta\mathbf{M}) + \mathbf{E}_{01}^T(\mathscr{A}_{TR} + \mathscr{B}_T\mathbf{G}) \\ \mathbf{F}\mathscr{M} & \mathbf{A} + \mathbf{BG} + \mathbf{F}\Delta\mathbf{M} \end{bmatrix} \begin{pmatrix} \tilde{\boldsymbol{x}} \\ \mathbf{x}_c \end{pmatrix},$$

$$\begin{pmatrix} \mathbf{y} \\ \mathbf{u} \end{pmatrix} = \begin{bmatrix} \mathscr{C} & \mathscr{C}\mathbf{E}_{10}^T \\ \mathbf{0} & \mathbf{G} \end{bmatrix} \begin{pmatrix} \tilde{\boldsymbol{x}} \\ \mathbf{x}_c \end{pmatrix}. \tag{10.42}$$

Exercise 10.2

Verify (10.42)

In the limit $\mathbf{F} \to \mathbf{0}$, the eigenvalues of (10.42) become those of the block-diagonal matrices $\mathscr{A}$ and $(\mathbf{A} + \mathbf{BG})$. This conclusion is summarized as follows.

Theorem 10.2

In the limit as $\mathbf{F} \to \mathbf{0}$, *any linear system* (10.13) *driven by any linear controller of the form* (10.39) *has the eigenvalues of* $\mathscr{A}$ *and* $(\mathbf{A} + \mathbf{BG})$.

Again, the reader is reminded that $(\mathbf{A} + \mathbf{BG})$ is under the design of the analyst.

For further insight into the effects of modeling errors, suppose that the gains $\mathbf{G}$ and $\mathbf{F}$ are not small but that the parameter errors are zero ($\Delta\mathbf{A} = \mathbf{0}$, $\Delta\mathbf{B} = \mathbf{0}$, $\Delta\mathbf{M} = \mathbf{0}$). Then, the following is true.

Theorem 10.3

In the absence of parameter errors ($\Delta\mathbf{A} = \mathbf{0}$, $\Delta\mathbf{B} = \mathbf{0}$, $\Delta\mathbf{M} = \mathbf{0}$), *the closed-loop system eigenvalues are those of* $(\mathbf{A} - \mathbf{FM})$, $\mathscr{A}_T$, *and* $(\mathbf{A} + \mathbf{BG})$, *if* $\tilde{x}_T$ *is either uncontrollable or unobservable in the measurements.*

Proof: Letting ($\Delta\mathbf{A} = \mathbf{0}$, $\Delta\mathbf{B} = \mathbf{0}$, $\Delta\mathbf{M} = \mathbf{0}$) in (10.42) yields

$$\begin{pmatrix} \dot{\tilde{x}} \\ \dot{\mathbf{x}}_c \end{pmatrix} = \begin{bmatrix} \mathscr{A} - \mathbf{E}_{10}^T\mathbf{F}\mathscr{M} & \mathbf{E}_{01}^T(\mathscr{A}_{TR} + \mathscr{B}_T\mathbf{G}) \\ \mathbf{F}\mathscr{M} & \mathbf{A} + \mathbf{BG} \end{bmatrix} \begin{pmatrix} \tilde{x} \\ \mathbf{x}_c \end{pmatrix},$$

$$\begin{pmatrix} \mathbf{y} \\ \mathbf{u} \end{pmatrix} = \begin{bmatrix} \mathscr{C} & \mathscr{C}\mathbf{E}_{10}^T \\ \mathbf{0} & \mathbf{G} \end{bmatrix} \begin{pmatrix} \tilde{x} \\ \mathbf{x}_c \end{pmatrix}. \tag{10.43}$$

The block diagram of the homogenous part of the closed-loop system using the notation of (10.15) and (10.39) is given in Fig. 10.6. Using the definitions of $\mathbf{E}_{10}$ and $\mathbf{E}_{01}$, (10.43) is further expanded as follows:

$$\begin{pmatrix} \dot{\tilde{x}}_R \\ \dot{\tilde{x}}_T \\ \dot{\mathbf{x}}_c \end{pmatrix} = \left[\begin{array}{c|c:c} \mathscr{A}_R - \mathbf{F}\mathscr{M}_R & \mathscr{A}_{RT} - \mathbf{F}\mathscr{M}_T & 0 \\ \hline \mathscr{A}_{TR} & \mathscr{A}_T & \mathscr{A}_{TR} + \mathscr{B}_T G \\ \hdashline \mathbf{F}\mathscr{M}_R & \mathbf{F}\mathscr{M}_T & \mathbf{A} + \mathbf{BG} \end{array}\right] \begin{pmatrix} \tilde{x}_R \\ \tilde{x}_T \\ \mathbf{x}_c \end{pmatrix}, \tag{10.44}$$

$$\begin{pmatrix} \mathbf{y} \\ \mathbf{u} \end{pmatrix} = \begin{bmatrix} \mathscr{C}_R & \mathscr{C}_T & \mathscr{C}_R \\ \mathbf{0} & \mathbf{0} & \mathbf{G} \end{bmatrix} \begin{pmatrix} \tilde{x}_R \\ \tilde{x}_T \\ \mathbf{x}_c \end{pmatrix}.$$

We shall first let $\tilde{x}_T$ in (10.15) be uncontrollable. This is equivalent to the statement that ($\mathscr{A}_{TR} = \mathbf{0}$, $\mathscr{B}_T = \mathbf{0}$) in (10.15) and in Fig. 10.6. (Recall the

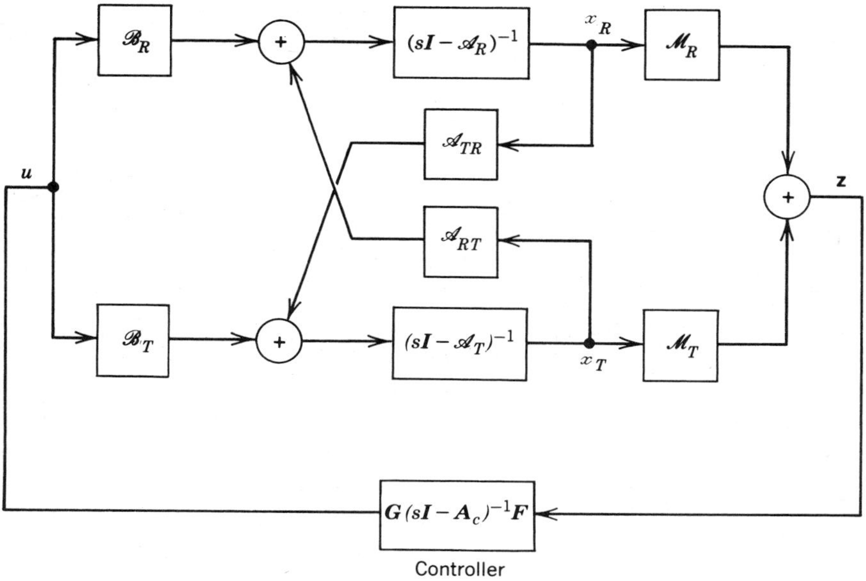

Figure 10.6 *Structure of Linear System Control with No Parameter Errors*

controllable canonical form of (5.53).) This makes the upper right matrix of (10.44) zero (using the *dotted* line partitions). In this event the eigenvalues of (10.44) become those of ($\mathbf{A} - \mathbf{BG}$) and those of the upper left block matrix (using again the *dotted* line partitions). But since $\mathscr{A}_{TR} = \mathbf{0}$, this upper left matrix is now also block diagonal and therefore has the eigenvalues of $(\mathscr{A}_R - \mathbf{F}\mathscr{M}_R)$ and $\mathscr{A}_T$. In the absence of parameter error $\mathscr{A}_R = \mathbf{A}$, $\mathscr{M}_R = \mathbf{M}$, and the first part of the theorem is proved.

For the second part of the proof, let x_T in (10.15) be unmeasurable (unobservable from the measurement $\mathbf{z}$). This is equivalent to the statement that $\mathscr{A}_{RT} = \mathbf{0}$, $\mathscr{M}_T = \mathbf{0}$ in (10.15) and in Fig. 10.6. Recall the observable canonical form of (5.106).) This makes the upper right matrix of (10.44) zero using the *solid* line partitions. In this event, the eigenvalues of (10.44) become those of $(\mathscr{A}_R - \mathbf{F}\mathscr{M}_R)$ and those of the lower right partition, again using the *solid* line partitions. But since $\mathscr{M}_T = \mathbf{0}$, this lower right matrix is now also block diagonal and therefore has the eigenvalues of $\mathscr{A}_T$ and ($\mathbf{A} + \mathbf{BG}$). Now, in the absence of parameter errors $\mathscr{A}_R = \mathbf{A}$, $\mathscr{M}_R = \mathbf{M}$. This gives the same set of eigenvalues [of ($\mathbf{A} - \mathbf{FM}$), $\mathscr{A}_T$, and ($\mathbf{A} + \mathbf{BG}$)] as in the first part of the proof. Hence, the theorem is proved. ■

Theorems 10.1 through 10.3 suggest that it is important that the matrices $\{(\mathbf{A} - \mathbf{FM}), (\mathbf{A} + \mathbf{BG}), \mathscr{A}_T\}$ be asymptotically stable and, more significantly, that this conclusion remain *independent* of the methods by which the controller parameters $(\mathbf{A}_c, \mathbf{F}, \mathbf{G})$ were derived! [Note that the optimal LQI controller stabilized $(\mathbf{A} - \mathbf{FM}), (\mathbf{A} + \mathbf{BG})$, and this section shows that this is a desirable feature of LQI controllers. They have very good stability robustness at low gain.] Suppose the controller (10.39) was designed based upon the assumed model $(\mathscr{A}_R, \mathscr{M}_R, \mathscr{B}_R)$ so that this model had certain desired behavior in the closed loop. Figure 10.6 makes it clear that, in the absence of parameter errors, the fundamental cause of deviation in system behavior from predicted behavior is the *relative degree of controllability and measurability of the states* $\mathscr{x}_T$. While controllability and measurability prove to be of great benefit for optimal control in the *absence* of model errors (Chapter 8), complete controllability and measurability would be a serious handicap in the real world. Control designers are indeed *fortunate* that most of the "neglected states" associated with "real-world $\mathscr{x}_T$" are uncontrollable and unmeasurable as indicated in the following proposition. Otherwise, there would be even fewer successful control designs to celebrate in practice.

Proposition 10.2 Uncontrollability and Unobservability of Dynamic Systems
Using any number of sensors and actuators, the physical plant will not be completely observable or controllable.

This proposition requires some explanation, since observability and controllability are mathematical properties associated with a mathematical model, whereas the "physical plant" defies exact description by any mathematical model. Suppose one improves a given mathematical representation of the physical plant by adding additional dynamics which were originally ignored in the model. As one continues this process, adding more and more details so that the new model more accurately models the physical plant, the mathematical model eventually becomes both uncontrollable and unobservable. In other words, we claim that an uncontrollable, unobservable model can always be constructed to provide a closer representation of the physical plant than any controllable, observable model. It takes little convincing to see that this argument is correct. Taken to an extreme, it is obvious that the molecular motions in an aircraft wing will not be completely controlled by aileron actions or observed by rate gyros. For control design the molecular motions may not be important, but note that several examples in Chapter 3 were proved uncontrollable or unobservable without resorting to such extreme discussions as molecular motions. The important impact of the lack of observability is that the stability proofs which rely on observability cannot be used to assure that the *physical plant* will be stable. Indeed, stability is a mathematical concept relating to a mathematical model and the physical plant can never be *proved* stable by mathematics. We can only say that the *model* is stable subject to a given range of parameter values or a given magnitude of the model error vector $\mathbf{e}_x$.

The model error concepts of this chapter were developed in refs. [10.4] and [10.5]. The special interpretations of these results in modal coordinates without parameter errors now follows.

Now, we examine (10.44) when $\mathscr{A}$ of (10.13) is diagonal and treat only model errors of the type $(\mathbf{e}_{0x}, \mathbf{e}_{0u})$ in (10.19) through (10.21). In this case, (10.44) becomes

$$\begin{pmatrix} \dot{\tilde{x}}_R \\ \dot{\tilde{x}}_T \\ \dot{\mathbf{x}}_c \end{pmatrix} = \begin{bmatrix} \mathscr{A}_R - \mathbf{F}\mathscr{M}_R & -\mathbf{F}\mathscr{M}_T & \mathbf{0} \\ \mathbf{0} & \mathscr{A}_T & \mathscr{B}_T\mathbf{G} \\ \mathbf{F}\mathscr{M}_R & \mathbf{F}\mathscr{M}_T & \mathbf{A}+\mathbf{B}\mathbf{G} \end{bmatrix} \begin{pmatrix} \tilde{x}_R \\ \tilde{x}_T \\ \mathbf{x}_c \end{pmatrix},$$

$$\begin{pmatrix} \mathbf{y} \\ \mathbf{u} \end{pmatrix} = \begin{bmatrix} \mathscr{C}_R & \mathscr{C}_T & \mathscr{C}_R \\ \mathbf{0} & \mathbf{0} & \mathbf{G} \end{bmatrix} \begin{pmatrix} \tilde{x}_R \\ \tilde{x}_T \\ \mathbf{x}_c \end{pmatrix}, \tag{10.45}$$

where $\mathscr{A}_R = \mathbf{A}$, $\mathscr{M}_R = \mathbf{M}$, and $\mathscr{B}_R = B$. Note that Fig. 10.6 is simplified in this case, since $\mathscr{A}_{RT} = \mathbf{0}$ and $\mathscr{A}_{TR} = \mathbf{0}$. Since $\mathscr{A}_{RT} = \mathbf{0}$, $\mathscr{A}_{TR} = \mathbf{0}$ in (10.15), it is clear that the entire substate x_T is uncontrollable if $\mathscr{B}_T = \mathbf{0}$ and that the entire substate x_T is unmeasurable if $\mathscr{M}_T = \mathbf{0}$ (see Chapter 5). Hence, the term $\mathscr{B}_T\mathbf{u}$ is due to the controllability of x_T and is often called "control spillover" [10.2]. The term $\mathscr{M}_T x_T$ is due to the observability of x_T in $\mathbf{z}$ and is often called "observation spillover" [10.2].

If x_T is either unmeasurable ($\mathscr{M}_T = \mathbf{0}$) or uncontrollable ($\mathscr{B}_T = \mathbf{0}$), then the eigenvalues of (10.45) are those of $(\mathscr{A}_R - \mathbf{F}\mathscr{M}_R)$, $\mathscr{A}_T$, $(\mathscr{A}_R + \mathscr{B}_R\mathbf{G})$ due to the subsequent block-diagonal structure of (10.45). This, of course, is in agreement with the more general Theorem 10.3. Now, consider the truncated modes x_T to be both controllable and measurable ($\mathscr{B}_T \neq \mathbf{0}$, $\mathscr{M}_T \neq \mathbf{0}$). Suppose that $(\mathscr{A}_R, \mathscr{B}_R, \mathscr{M}_R)$ represents any subset of modes of the system $(\mathscr{A}, \mathscr{B}, \mathscr{M})$ and recall that $\mathscr{A}_R = \mathbf{A} = \text{diag}$.

The following exercise shows the upper bound on controllability and observability of a truncated model to assure stability of the closed-loop system.

Exercise 10.3

Using any first-order controller with the properties

$$u(s) = \left[G(sI - A_c)^{-1}F\right]z(s),$$

$$A_c = \mathscr{A}_R + \mathscr{B}_R G - F\mathscr{M}_R,$$

where G is chosen so that

$$\mathscr{A}_R + \mathscr{B}_R G < 0$$

and F is chosen so that

$$\mathscr{A}_R - F\mathscr{M}_R < 0,$$

show that stability of the second-order system

$$\begin{pmatrix} \dot{x}_R \\ \dot{x}_T \end{pmatrix} = \begin{bmatrix} \mathscr{A}_R & 0 \\ 0 & \mathscr{A}_T \end{bmatrix} \begin{pmatrix} x_R \\ x_T \end{pmatrix} + \begin{bmatrix} \mathscr{B}_R \\ \mathscr{B}_T \end{bmatrix} u, \qquad z = (\mathscr{M}_R, \mathscr{M}_T)x,$$

requires the product of controllability and observability of x_T to be limited by

$$|\mathscr{M}_T \mathscr{B}_T| < |\mathscr{A}_R - F\mathscr{M}_R| \, |\mathscr{A}_T| \, |\mathscr{A}_R + \mathscr{B}_R G| \frac{1}{|G\mathscr{A}_R F|}. \tag{10.46}$$

Recall that the balancing and modal cost methods of model reduction in Chapter 6 delete coordinates on the basis of small products of observability, controllability measures. Equation (10.46) supports this philosophy. This simple second-order example provides important insight into the necessary and sufficient condition for stability of modal controllers. The closed loop will be stable if the neglected mode is unobservable ($\mathscr{M}_T = 0$) or uncontrollable ($\mathscr{B}_T = 0$), as promised by Theorem 10.3. But (10.46) also shows the upper bound on $|\mathscr{M}_T \mathscr{B}_T|$ which will allow stability even if neither $\mathscr{M}_T$ nor $\mathscr{B}_T$ is zero. We wish to generalize this result in the following section.

10.1.3 STABILITY MARGINS IN CONTROL

In classical control theory, gain and phase margins are defined on the Nyquist plot. We cannot associate gain and phase margins with tolerance of *any one* of the four categories of model error in (10.19), although errors in model order ($x_T \neq 0$) will certainly modify the phase of the system. It can be expected that larger phase margins may allow the design to be less sensitive to errors in model order. Gain margins, on the other hand, do not necessarily provide tolerance to either parameter error or model order error. Refer to Fig. 10.7a and $\Delta G(s)$ in (10.36) to see that the Nyquist test for stability is satisfied if the length of the vector $1 + HG(j\omega)$ is larger than the length of the vector $H\Delta G(j\omega)$, over all frequencies ω, since the -1 point cannot be encircled in this case. See Fig. 10.7b. Hence, a sufficient condition for stability is

$$1 + HG(j\omega) > H\Delta G(j\omega), \qquad \forall \omega,$$

or

$$1 > [1 + HG(j\omega)]^{-1}[H\Delta G(j\omega)], \qquad \forall \omega. \tag{10.47a}$$

This condition may be extended to the matrix case as follows, [10.6] through [10.9],

$$1 > \max_{\omega} \lambda\{[I + \mathbf{HG}(j\omega)]^{-1}\mathbf{H}\,\Delta\mathbf{G}(j\omega)\}, \tag{10.47b}$$

where $\lambda\{\cdot\}$ denotes eigenvalue of matrix $\{\cdot\}$. Note, however, that this can be an extremely conservative condition, since it is possible for the vector $H\Delta G(j\omega)$ to be much longer than $1 + HG(j\omega)$ without causing an encirclement. See Fig. 10.7c for a stable situation which violates (10.47). However, a more fundamental limitation of

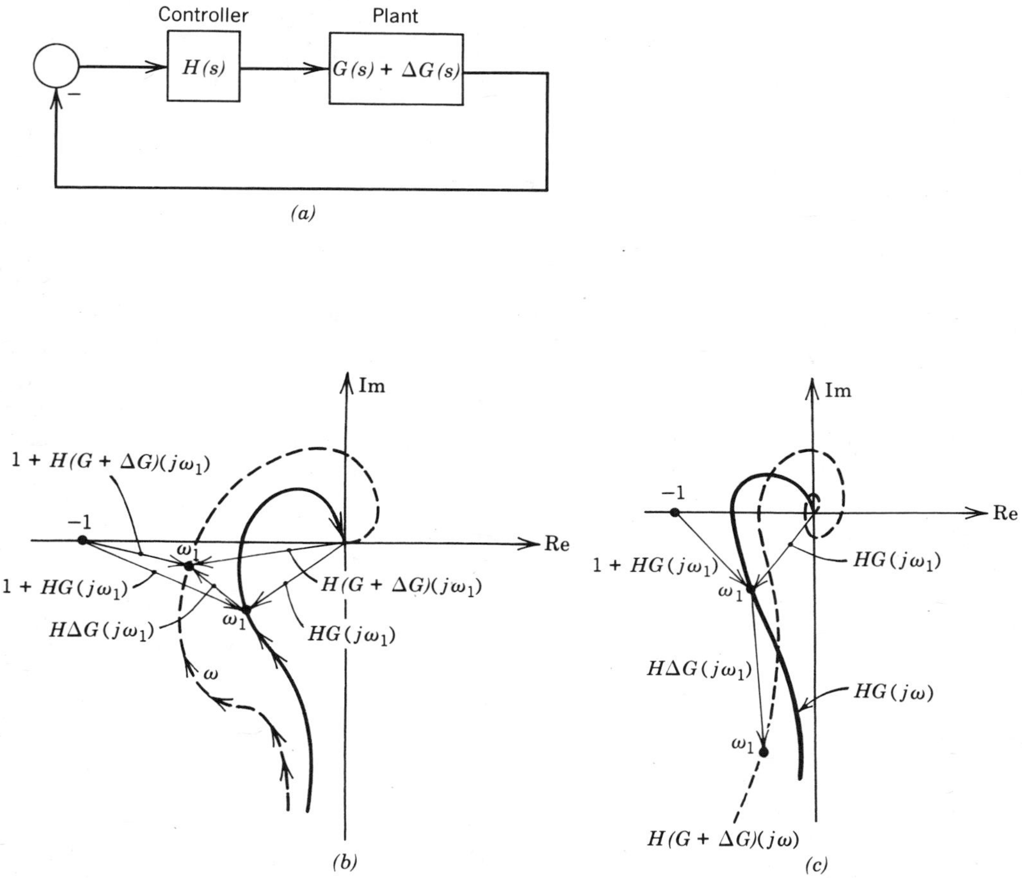

Figure 10.7 *Nyquist Plot of Perturbed System*

these results is due to the fact that Fig. 10.7a does *not* describe most physical situations, since the $e_t(s)$ term in (10.36) has been ignored. This is significant since e_t is composed of the *same source* of errors which make up $\Delta G(s)$. [See that $J(s)$ is a model-order error term which appears in both equations of (10.36).]

It is possible to develop stability margins due to *specific* parameter errors and model-order errors. Toward this end we need the following result.

Theorem 10.4

$\mathcal{Re}\,\lambda_i[\mathbf{A}] < -\eta$ for all i iff there exists a nonsingular similarity transformation matrix **T** *such that*

$$\mathbf{T}^{-1}\mathbf{A}\mathbf{T} + (\mathbf{T}^{-1}\mathbf{A}\mathbf{T})^* + 2\eta\mathbf{I} < \mathbf{0}. \tag{10.48}$$

Proof: In the proof we replace $\mathbf{A}$ by $\hat{\mathbf{A}} = \mathbf{A} + \eta\mathbf{I}$ and ask whether $\mathscr{Re}\,\lambda[\hat{\mathbf{A}}] < 0$ iff $\mathbf{T}^{-1}\hat{\mathbf{A}}\mathbf{T} + (\mathbf{T}^{-1}\hat{\mathbf{A}}\mathbf{T})^* < \mathbf{0}$. We will show that (10.48) is equivalent to the Liapunov Theorem 7.3, which assures the existence of a Liapunov function

$$\mathscr{V}(\mathbf{x}(t)) = \mathbf{x}^*(t)\mathbf{K}\mathbf{x}(t) > 0 \qquad (\text{i.e., } \mathbf{K} > \mathbf{0}), \tag{10.49a}$$

such that its derivative

$$\dot{\mathscr{V}}(\mathbf{x}(t)) = \mathbf{x}^*(t)[\mathbf{K}\hat{\mathbf{A}} + \hat{\mathbf{A}}^*\mathbf{K}]\mathbf{x}(t) < 0 \tag{10.49b}$$

is negative definite, where $\hat{\mathbf{A}} \triangleq \mathbf{A} + \eta\mathbf{I}$. Note from (7.18) that $\dot{\mathscr{V}}(\mathbf{x}(t)) = -\mathbf{x}^*(t)\mathbf{C}^*\mathbf{Q}\mathbf{C}\mathbf{x}(t)$ can be negative semidefinite if we impose the further restriction that $(\hat{\mathbf{A}}, \mathbf{C})$ is observable and $\mathbf{Q} > \mathbf{0}$. However, even in the absence of knowledge of $\hat{\mathbf{A}}$, one can be assured that $(\hat{\mathbf{A}}, \mathbf{C})$ is observable if $\mathbf{C}$ is chosen nonsingular. So, let $\mathbf{C}^*\mathbf{Q}\mathbf{C}$ be any positive definite matrix. This ensures the observability condition, and Theorem 7.3 guarantees the existence of a $\mathbf{K} > \mathbf{0}$ if $\hat{\mathbf{A}}$ is asymptotically stable. Consider the coordinate transformation $\mathbf{x} = \mathbf{T}\boldsymbol{\gamma}$, where $\mathbf{K}^{-1} = \mathbf{T}\mathbf{T}^*$. Then, $\dot{\boldsymbol{\gamma}} = \mathbf{T}^{-1}\hat{\mathbf{A}}\mathbf{T}\boldsymbol{\gamma}$ and (10.48) becomes

$$\begin{aligned}\mathscr{V}(\boldsymbol{\gamma}(t)) &= \boldsymbol{\gamma}^*\mathbf{T}^*\mathbf{K}\mathbf{T}\boldsymbol{\gamma} = \boldsymbol{\gamma}^*\mathbf{T}^*\mathbf{T}^{-*}\mathbf{T}^{-1}\mathbf{T}\boldsymbol{\gamma} = \boldsymbol{\gamma}^*\boldsymbol{\gamma} > 0,\\ \dot{\mathscr{V}}(\boldsymbol{\gamma}(t)) &= \boldsymbol{\gamma}^*\mathbf{T}^*[\mathbf{T}^{-*}\mathbf{T}^{-1}\hat{\mathbf{A}} + \hat{\mathbf{A}}^*\mathbf{T}^{-*}\mathbf{T}^{-1}]\mathbf{T}\boldsymbol{\gamma}\\ &= \boldsymbol{\gamma}^*\left[\mathbf{T}^{-1}\hat{\mathbf{A}}\mathbf{T} + (\mathbf{T}^{-1}\hat{\mathbf{A}}\mathbf{T})^*\right]\boldsymbol{\gamma} < 0.\end{aligned} \tag{10.50}$$

Hence, Theorem 10.4 relates to Liapunov theory as follows. If there exists a $\mathbf{T}$ such that (10.50) holds true, then

$$-\mathbf{C}^*\mathbf{Q}\mathbf{C} \triangleq \mathbf{T}^{-1}\hat{\mathbf{A}}\mathbf{T} + (\mathbf{T}^{-1}\hat{\mathbf{A}}\mathbf{T})^* < \mathbf{0} \tag{10.51}$$

and Theorem 7.3 assures stability of $\hat{\mathbf{A}}$. Of course, stability of $\hat{A}$ implies that $\mathbf{A} = \hat{\mathbf{A}} - \eta\mathbf{I}$ has all eigenvalues to the left of $-\eta$. If no such $\mathbf{T}$ exists satisfying (10.51), then no positive definite solution of (7.18) exists. ■

Our next result provides a stability margin.

Theorem 10.5

$\mathscr{Re}\,\lambda_i[\mathbf{A} + \Delta\mathbf{A}] < 0\ \forall i$ *if Theorem* 10.4 *holds true and*

$$\|\Delta\mathbf{A}\| \le \frac{\|\eta\mathbf{I}\| n_x}{\|\mathbf{T}^{-1}\|\,\|\mathbf{T}\|}, \qquad \|\mathbf{T}\|^2 = \operatorname{tr}\mathbf{T}^*\mathbf{T}, \tag{10.52a}$$

or

$$\|\Delta\mathbf{A}\|_2 \le \frac{\eta}{\|\mathbf{T}^{-1}\|_2\|\mathbf{T}\|_2}, \qquad \|T\|_2 \triangleq \textit{max singular value of } T. \tag{10.52b}$$

Proof: To prove (10.52b), we must show that

$$\|\Delta \mathbf{A}\|_2 \|\mathbf{T}^{-1}\|_2 \|\mathbf{T}\|_2 \le \|\eta \mathbf{I}\|_2 \tag{10.53}$$

implies that

$$\mathbf{T}^{-1}(\mathbf{A} + \Delta \mathbf{A})\mathbf{T} + \left[\mathbf{T}^{-1}(\mathbf{A} + \Delta \mathbf{A})\mathbf{T}^*\right] < \mathbf{0}, \tag{10.54}$$

which guarantees stability of $\mathbf{A} + \Delta\mathbf{A}$ via Theorem 10.4. Now, from (10.53), note that

$$\|\mathbf{T}^{-1} \Delta \mathbf{A}\,\mathbf{T}\|_2 \le \|\mathbf{T}^{-1}\|_2 \|\mathbf{T}\|_2 \|\Delta \mathbf{A}\|_2 \le \|\eta \mathbf{I}\|_2 = \eta \|\mathbf{I}\|_2 \tag{10.55}$$

and that

$$\left\|\mathbf{T}^{-1}\Delta\mathbf{A}\,\mathbf{T} + (\mathbf{T}^{-1}\Delta\mathbf{A}\,\mathbf{T})^*\right\|_2 = 2\|\mathbf{T}^{-1}\Delta\mathbf{A}\,\mathbf{T}\|_2 \le 2\|\eta\mathbf{I}\|_2 = 2\eta\|\mathbf{I}\|_2 \tag{10.56}$$

Hence, $\mathbf{x}^*[\mathbf{T}^{-1}\Delta\mathbf{A}\,\mathbf{T} + (\mathbf{T}^{-1}\Delta\mathbf{A}\,\mathbf{T})^*]\mathbf{x} \le \mathbf{x}^*[2\eta\mathbf{I}]\mathbf{x}$ means that

$$-\left[\mathbf{T}^{-1}\Delta\mathbf{A}\,\mathbf{T} + (\mathbf{T}^{-1}\Delta\mathbf{A}\,\mathbf{T})^*\right] \ge -2\eta\mathbf{I}.$$

Now, we are given that $\mathbf{A}$ satisfies

$$\mathbf{T}^{-1}\mathbf{A}\mathbf{T} + (\mathbf{T}^{-1}\mathbf{A}\mathbf{T})^* < -2\eta\mathbf{I}.$$

The last two statements imply that

$$-\left[\mathbf{T}^{-1}\Delta\mathbf{A}\,\mathbf{T} + (\mathbf{T}^{-1}\Delta\mathbf{A}\,\mathbf{T})^*\right] > \mathbf{T}^{-1}\mathbf{A}\mathbf{T} + (\mathbf{T}^{-1}\mathbf{A}\mathbf{T})^* \tag{10.57}$$

or

$$\mathbf{0} > \mathbf{T}^{-1}\Delta\mathbf{A}\,\mathbf{T} + (\mathbf{T}^{-1}\Delta\mathbf{A}\,\mathbf{T})^* + \mathbf{T}^{-1}\mathbf{A}\mathbf{T} + (\mathbf{T}^{-1}\mathbf{A}\mathbf{T})^*$$

or

$$\mathbf{0} > \mathbf{T}^{-1}(\mathbf{A} + \Delta\mathbf{A})\mathbf{T} + \left[\mathbf{T}^{-1}(\mathbf{A} + \Delta\mathbf{A})\mathbf{T}\right]^*,$$

which completes the proof of (10.52b). Now, to prove (10.52a), consider that the Frobenius norm $\|\mathbf{T}\|^2 = \operatorname{tr}\mathbf{T}^*\mathbf{T}$ has a relationship with the 2-norm $\|\mathbf{T}\|_2^2 = \sigma_1^2$ as follows [10.10]. Consider the singular value decomposition of $\mathbf{T} = \mathbf{U\Sigma V}^*$, where $\Sigma = \operatorname{diag}[\sigma_1, \sigma_2, \ldots, \sigma_{n_x}]$ and $\sigma_i \ge \sigma_{i+1}$. Then,

$$\|\mathbf{T}\|^2 = \operatorname{tr}\mathbf{V\Sigma U^* U\Sigma V^*} = \operatorname{tr}\Sigma^2 = \sum_{i=1}^{n_x} \sigma_i^2$$

and

$$\|\mathbf{T}\|_2^2 = \sigma_1^2.$$

Now, since $\sigma_1^2 \geq \sigma_2^2 \geq \cdots \geq \sigma_{n_x}^2$, we have

$$\|\mathbf{T}\|^2 = \|\mathbf{T}\|_2^2 + \sigma_2^2 + \sigma_3^2 + \cdots + \sigma_{n_x}^2 \leq n_x \|\mathbf{T}\|_2^2.$$

Hence, $\|\mathbf{T}\| \leq \sqrt{n_x}\|\mathbf{T}\|_2$ and (10.52b) yields

$$\left(\frac{1}{\sqrt{n_x}}\right)^3 \|\Delta\mathbf{A}\|\,\|\mathbf{T}^{-1}\|\,\|\mathbf{T}\| \leq \|\Delta\mathbf{A}\|_2\|\mathbf{T}^{-1}\|_2\|\mathbf{T}\|_2 \leq \eta.$$

Hence,

$$\|\Delta\mathbf{A}\| \leq \frac{\eta\left(\sqrt{n_x}\right)^3}{\|\mathbf{T}\|\,\|\mathbf{T}^{-1}\|},$$

which verifies (10.52a), since $\|\mathbf{I}\| = \sqrt{n_x}$. ■

The condition number of a matrix $\mathbf{T}$ is $\text{cond}\,(\mathbf{T}) \triangleq \|\mathbf{T}\|\,\|\mathbf{T}^{-1}\|$. Let the SVD of $\mathbf{T}$ be $\mathbf{T} = \mathbf{U\Sigma V}^*$. Then, the Frobenius norm $\|\mathbf{T}\|^2 \triangleq \text{tr}\,\mathbf{T}^*\mathbf{T}$ yields

$$\begin{aligned}[\text{cond}\,(\mathbf{T})]^2 &= \text{tr}\,\mathbf{T}^*\mathbf{T}\,\text{tr}\,\mathbf{T}^{-1*}\mathbf{T}^{-1} = \text{tr}\,\boldsymbol{\Sigma}^2\,\text{tr}\,\boldsymbol{\Sigma}^{-2} \\ &= \left(\sum_{i=1}^{n_x}\sigma_i^2\right)\left(\sum_{i=1}^{n_x}\sigma_i^{-2}\right), \qquad \Sigma_{ij} = \delta_{ij}\sigma_i.\end{aligned}$$

Hence, the largest possible value of the condition number is ∞ (when some singular value of $\mathbf{T}$ approaches zero) and the smallest is $\text{cond}\,(\mathbf{T}) = n_x$ (which happens if $\sigma_i = 1$, $\forall i = 1, 2, \ldots, n_x$). Hence, a unitary matrix has the smallest $\text{cond}\,(\mathbf{T}) = n_x$, and in this case $\mathbf{T} = \mathbf{T}^*$, $\mathbf{V} = \mathbf{U}$, yielding orthogonal eigenvectors of $\mathbf{T}$. From (10.52), the largest possible upper bound on $\|\Delta\mathbf{A}\|$ is therefore

$$\|\Delta\mathbf{A}\| \leq \frac{\|\eta\mathbf{I}\|}{\text{cond}\,(\mathbf{T})} = \frac{\eta\sqrt{n_x}}{n_x} = \frac{\eta}{\sqrt{n_x}}.$$

Using the norm $\|\mathbf{T}\|_2 \triangleq \sigma_1 \triangleq$ max singular value of $\mathbf{T}$, (10.52) yields the largest upper bound (since for unitary $\mathbf{T}$, $\text{cond}\,(\mathbf{T})_2 = 1$):

$$\|\Delta\mathbf{A}\|_2 \leq \frac{\|\eta\mathbf{I}\|_2}{\text{cond}\,(\mathbf{T})_2} = \frac{\eta}{1} = \eta.$$

Hence, the 2-norm gives a less conservative (better) bound. It may be shown that the Frobenius norm of a matrix cannot be smaller than the 2-norm $\|\mathbf{T}\|_2 \leq \|\mathbf{T}\| \leq \sqrt{n}\,\|\mathbf{T}\|_2$, [10.10].

EXAMPLE 10.4

Return to Exercise 10.3 and find an upper bound on $|\mathscr{M}_T\mathscr{B}_T|$ for stability using Theorem 10.5.

Solution: The closed-loop system plant matrix is described by (10.45), where we choose to use Theorem 10.5 with these definitions

$$\mathbf{A} \triangleq \begin{bmatrix} \mathscr{A}_R - F\mathscr{M}_R & 0 & 0 \\ 0 & \mathscr{A}_T & 0 \\ F\mathscr{M}_R & 0 & \mathscr{A}_R + \mathscr{B}_R G \end{bmatrix}, \qquad \Delta\mathbf{A} \triangleq \begin{bmatrix} 0 & -F\mathscr{M}_T & 0 \\ 0 & 0 & \mathscr{B}_T G \\ 0 & F\mathscr{M}_T & 0 \end{bmatrix},$$

where

$$\begin{Bmatrix} \mathscr{A}_R - F\mathscr{M}_R = -\alpha_1^2 \\ \mathscr{A}_T = -\alpha_2^2 \\ \mathscr{A}_R + \mathscr{B}_R G = -\alpha_3^2 \end{Bmatrix}$$

make $\mathbf{A}$ stable and $\eta \triangleq \min\{\alpha_1^2, \alpha_2^2, \alpha_3^2\}$. A $\mathbf{T}$ that satisfies (10.48) is the modal matrix for $\mathbf{A}$:

$$T = \begin{bmatrix} 1 & 0 & 0 \\ 0 & 1 & 0 \\ F\mathscr{M}_R(\alpha_3^2 - \alpha_1^2)^{-1} & 0 & 1 \end{bmatrix},$$

since then

$$\mathbf{T}^{-1}\mathbf{A}\mathbf{T} + (\mathbf{T}^{-1}\mathbf{A}\mathbf{T})^* = \begin{bmatrix} -2\alpha_1^2 & 0 & 0 \\ 0 & -2\alpha_2^2 & 0 \\ 0 & 0 & -2\alpha_3^2 \end{bmatrix} \leq -2\eta\mathbf{I}$$

and

$$\|\mathbf{T}\| = \sqrt{\operatorname{tr}\mathbf{T}^*\mathbf{T}} = \left[3 + \left(\frac{F\mathscr{M}_R}{\alpha_3^2 - \alpha_1^2}\right)^2\right]^{1/2},$$

$$\mathbf{T}^{-1} = \begin{bmatrix} 1 & 0 & 0 \\ 0 & 1 & 0 \\ -F\mathscr{M}_R(\alpha_3^2 - \alpha_1^2)^{-1} & 0 & 1 \end{bmatrix}, \qquad \|\mathbf{T}^{-1}\| = \left[3 + \left(\frac{F\mathscr{M}_R}{\alpha_3^2 - \alpha_1^2}\right)^2\right]^{1/2}.$$

Now,

$$\|\Delta\mathbf{A}\|^2 = \operatorname{tr}\Delta\mathbf{A}^*\Delta\mathbf{A} = \operatorname{tr}[G^*\mathscr{B}_T^*\mathscr{B}_T G + 2\mathscr{M}_T^* F^* F\mathscr{M}_T]$$

$$= \|\mathscr{B}_T\|^2_{GG^*} + 2\|\mathscr{M}_T\|^2_{F^*F}. \qquad \blacksquare$$

Thus, the conclusion from Theorem 10.5 is that the second-order system driven by the first-order controller is stable if

$$\|\mathscr{B}_T\|^2_{GG^*} + 2\|\mathscr{M}_T\|^2_{F^*F}$$
$$\leq \frac{\min\{-(\mathscr{A}_R + F\mathscr{M}_R), -\mathscr{A}_T, -(\mathscr{A}_R + \mathscr{B}_R G)\}(3)^{3/2}[-\mathscr{B}_R G - F\mathscr{M}_R]^4}{\{[F\mathscr{M}_R]^2_+[-(\mathscr{A}_R + \mathscr{B}_R G) + \mathscr{A}_R - F\mathscr{M}_R]^2 3\}^2}$$

or

$$\|\mathscr{B}_T\|^2_{GG^*} + 2\|\mathscr{M}_T\|^2_{F^*F}$$
$$\leq \frac{\min\{|\mathscr{A}_R - F\mathscr{M}_R|, |\mathscr{A}_T|, |\mathscr{A}_R + \mathscr{B}_R G|\}(3)^{3/2}(\mathscr{B}_R G + F\mathscr{M}_R)^4}{\{F^2\mathscr{M}_R^2 + (\mathscr{B}_R G + F\mathscr{M}_R)^2 3\}^2}. \qquad (10.58)$$

Now, see how conservative is the result (10.58) compared to the *necessary and sufficient* result (10.46). The system will be stable if *either* $\mathscr{B}_T$ or $\mathscr{M}_T$ is zero [see (10.46)]. For example, the system is stable if $\mathscr{B}_T = 0$ and $\mathscr{M}_T$ is arbitrarily large, but these conditions do not satisfy the sufficient condition (10.58).

It is worth repeating that the acceptable modeling errors (e.g., $\mathscr{M}_T$, $\mathscr{B}_T$) *depend* upon the control gains F and G, and both (10.46) and (10.58) illustrate this dependence.

Exercise 10.4

Use Theorem 10.5 to develop an upper bound involving matrices ($\mathscr{A}_{RT}$, $\mathscr{M}_T$, $\mathscr{A}_{RT}$, $\mathscr{B}_T$) so that (10.44) is stable. Verify that the requirement involves the inequality

$$\|\mathscr{A}_{TR}\|^2 + \|\mathscr{A}_{TR} + \mathscr{B}_T G\|^2 + \|F\mathscr{M}_T\|^2 + \|\mathscr{A}_{RT} - F\mathscr{M}_T\|^2 \leq \frac{\|\eta\mathbf{I}\|^2}{\|\mathbf{T}\|^2\|\mathbf{T}^{-1}\|^2}.$$

Yedavalli [10.20] has extended Theorem 10.5 to provide bounds on *each* element of ΔA.

10.2 First-Order Perturbations of Modal Data

In this section only errors of the class $\mathbf{e}_{px}$ in (10.19) are considered. We consider how the eigenvalues and eigenvectors of a matrix $\mathbf{A}$ change with respect to small changes in the elements of the matrix **A**. The system

$$\dot{\mathbf{x}} = \mathbf{A}\mathbf{x} + \mathbf{e}_x, \qquad \mathbf{x} \in \mathscr{R}^{n_x}, \tag{10.59}$$

may represent an open- or closed-loop system. The model error vector is assumed to have the structure

$$\mathbf{e}_x = \Delta\mathbf{A}\,\mathbf{x}. \tag{10.60}$$

Note that disturbances $\mathbf{e}_d$ in (10.19) can be written in state form and augmented to the plant to give the structure of errors in (10.59), (10.60). Hence, the actual perturbed system is

$$\dot{\mathbf{x}} = (\mathbf{A} + \Delta\mathbf{A})\mathbf{x}, \tag{10.61}$$

and the question of interest is, "If $\Delta\mathbf{A}$ is small, how are the eigenvalues and eigenvectors of **A** perturbed?"

10.2.1 ROOT PERTURBATIONS IN STATE MODELS

To answer this question, let us consider the definition of left eigenvectors $\mathbf{l}_i^*$, right eigenvectors $\mathbf{e}_i$, and eigenvalues λ_i given by Chapter 2. It will be assumed that both **A** and $(\mathbf{A} + \Delta\mathbf{A})$ are nondefective. Hence, the Jordan forms of **A** and $\mathbf{A} + \Delta\mathbf{A}$ are diagonal, and

$$\mathbf{A}\mathbf{E} = \mathbf{E}\Lambda, \quad \mathbf{E}^{-1} = \begin{bmatrix} \mathbf{l}_1^* \\ \vdots \\ \mathbf{l}_n^* \end{bmatrix}, \quad \mathbf{E} = [\mathbf{e}_1, \ldots, \mathbf{e}_n], \quad \Lambda = \operatorname{diag}\{\lambda_1, \ldots, \lambda_n\}, \tag{10.62}$$

$$\begin{aligned} [\mathbf{A} + \Delta\mathbf{A}][\mathbf{E} + \Delta\mathbf{E}] &= [\mathbf{E} + \Delta\mathbf{E}][\Lambda + \Delta\Lambda], \\ \mathbf{E} + \Delta\mathbf{E} &= [\mathbf{e}_1 + \Delta\mathbf{e}_1, \ldots, \mathbf{e}_n + \Delta\mathbf{e}_n], \\ \Lambda + \Delta\Lambda &= \operatorname{diag}\{\lambda_1 + \Delta\lambda_1, \ldots, \lambda_n + \Delta\lambda_n\}. \end{aligned} \tag{10.63}$$

To consider only first-order perturbations, the products of small terms $(\Delta\mathbf{A}\Delta\mathbf{E})$ and $(\Delta\mathbf{E}\Delta\Lambda)$ are ignored in (10.63), leaving

$$\mathbf{A}\mathbf{E} + \mathbf{A}\Delta\mathbf{E} + \Delta\mathbf{A}\mathbf{E} = \mathbf{E}\Lambda + \mathbf{E}\Delta\Lambda + \Delta\mathbf{E}\Lambda. \tag{10.64}$$

Solving for $\Delta\Lambda$ yields

$$\Delta\Lambda = \mathbf{E}^{-1}\Delta\mathbf{A}\mathbf{E} + \mathbf{E}^{-1}\mathbf{A}\Delta\mathbf{E} - \mathbf{E}^{-1}\Delta\mathbf{E}\Lambda. \tag{10.65}$$

Now, consider the manner in which the eigenvectors are perturbed. Figure 10.8

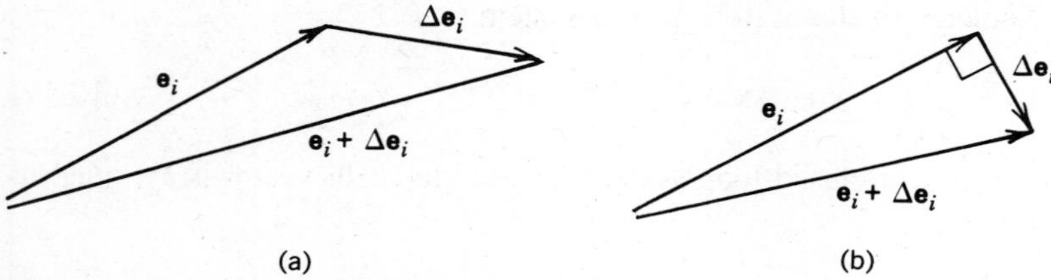

Figure 10.8 *Perturbations of Eigenvectors*

illustrates the eigenvector $\mathbf{e}_i$ and a possible eigenvector $(\mathbf{e}_i + \Delta\mathbf{e}_i)$. The eigenvector can be multiplied by any nonzero constant to change the length. If we shorten the eigenvector $(\mathbf{e}_i + \Delta\mathbf{e}_i)$ in Fig. 10.8a just so that $\Delta\mathbf{e}_i$ will be *perpendicular* to $\mathbf{e}_i$ as in Fig. 10.8b, then there is one less *direction* to worry about in the perturbation $\Delta\mathbf{e}_i$. That is, since the original eigenvectors form a basis for the entire n_x-dimensional space, $\Delta\mathbf{e}_i$ can be written as a linear combination of the original eigenvectors $\mathbf{e}_j$, $j = 1, \ldots, n_x$, except that $j \neq i$ since $\Delta\mathbf{e}_i$ is *not* (by proper normalization of Fig. 10.8) stretched in direction $\mathbf{e}_i$. Thus, for some constants H_{ij},

$$\Delta\mathbf{e}_i = \sum_{j=1}^{n} H_{ji}\mathbf{e}_j, \qquad H_{ii} = 0,$$

or, in matrix form,

$$\Delta\mathbf{E} = \mathbf{EH}, \quad H_{ii} = 0, \qquad \forall i. \tag{10.66}$$

Substitution of (10.66) into (10.65) yields

$$\Delta\mathbf{\Lambda} = \mathbf{E}^{-1}\Delta\mathbf{AE} + \mathbf{E}^{-1}\mathbf{AEH} - \mathbf{E}^{-1}\mathbf{EH\Lambda}.$$

Using the identity $\mathbf{E}^{-1}\mathbf{AE} = \mathbf{\Lambda}$,

$$\Delta\mathbf{\Lambda} = \mathbf{E}^{-1}\Delta\mathbf{AE} + \mathbf{\Lambda H} - \mathbf{H\Lambda}. \tag{10.67}$$

The matrix $\Delta\mathbf{\Lambda}$ is presumed diagonal [by nondefective assumption on $(\mathbf{A} + \Delta\mathbf{A})$]. Since the diagonal elements of the last two matrix products in (10.67) are zero, the *ii* element of the matrix equation (10.67) leads to

$$\Delta\lambda_i = [\Delta\mathbf{\Lambda}]_{ii} = [\mathbf{E}^{-1}\Delta\mathbf{AE}]_{ii} = \mathbf{l}_i^*\Delta\mathbf{A}\mathbf{e}_i. \tag{10.68}$$

Now, investigate the *ij* element of the matrix equation (10.67):

$$[\Delta\mathbf{\Lambda}]_{ij} = 0 = \mathbf{l}_i^*\Delta\mathbf{A}\mathbf{e}_j + \Lambda_{ii}H_{ij} - H_{ij}\Lambda_{jj}.$$

Solving for the unknown H_{ij} leads to

$$H_{ij} = (\lambda_j - \lambda_i)^{-1}(\mathbf{l}_i^* \Delta \mathbf{A} \mathbf{e}_j), \qquad i \neq j, \tag{10.69}$$

which requires the additional assumption of distinct λ_i, $i = 1, \ldots, n$. These results are summarized as follows.

Theorem 10.6
If the matrix $\mathbf{A}$ *has distinct eigenvalues* λ_i, $i = 1, \ldots, n$, *and left and right eigenvectors* $\mathbf{l}_i, \mathbf{e}_i$, *respectively, then the first-order approximation for the eigenvalues and eigenvectors of the matrix* $(\mathbf{A} + \Delta\mathbf{A})$ *is*

$$\Delta\lambda_i = \mathbf{l}_i^* \, \Delta\mathbf{A}\mathbf{e}_i, \tag{10.70a}$$

$$\Delta\mathbf{E} \doteq \mathbf{EH}, \quad H_{ij} = (\lambda_j - \lambda_i)^{-1} \mathbf{l}_i^* \, \Delta\mathbf{A}\mathbf{e}_j, \quad H_{ii} = 0. \tag{10.70b}$$

Similar results were originally obtained in the early work of Jacobi [10.11].

Exercise 10.5
For Exercise 10.3 and Example 10.4, show that the use of Theorem 10.6 yields these first-order approximations of the modal data $\Delta\lambda_i = 0$, $i = 1, 2, 3$,

$$\Delta\mathbf{E} = \mathbf{EH} = \begin{bmatrix} 1 & 0 & 0 \\ 0 & 1 & 0 \\ \dfrac{F\mathcal{M}_R}{\alpha_3^2 - \alpha_1^2} & 0 & 1 \end{bmatrix}$$

$$\times \begin{bmatrix} 0 & \dfrac{F\mathcal{M}_T}{\alpha_2^2 - \alpha_1^2} & 0 \\ \dfrac{\mathcal{B}_T\mathcal{M}_R GF}{(\alpha_2^2 - \alpha_1^2)(\alpha_3^2 - \alpha_1^2)} & 0 & \dfrac{\mathcal{B}_T G}{\alpha_2^2 - \alpha_3^2} \\ 0 & \dfrac{F\mathcal{M}_T}{\alpha_3^2 - \alpha_2^2}\left[1 + \dfrac{F\mathcal{M}_R}{\alpha_3^2 - \alpha_1^2}\right] & 0 \end{bmatrix}$$

$$= \begin{bmatrix} 0 & \dfrac{F\mathcal{M}_T}{\alpha_2^2 - \alpha_1^2} & 0 \\ \dfrac{\mathcal{B}_T\mathcal{M}_R GF}{(\alpha_2^2 - \alpha_1^2)(\alpha_3^2 - \alpha_1^2)} & 0 & \dfrac{\mathcal{B}_T G}{\alpha_2^2 - \alpha_3^2} \\ 0 & \dfrac{F\mathcal{M}_T}{\alpha_3^2 - \alpha_2^2}\left[\dfrac{F\mathcal{M}_R(2\alpha_3^2 - \alpha_2^2 - \alpha_1^2)}{(\alpha_3^2 - \alpha_1)^2} + 1\right] & 0 \end{bmatrix}.$$

Exercise 10.6

Suppose the control policy $\mathbf{u} = \mathbf{G}\mathbf{x}$ is used for the system

$$\begin{pmatrix} \dot{\mathbf{x}} \\ \dot{\mathbf{x}}_T \end{pmatrix} = \begin{bmatrix} \mathscr{A}_R & \mathbf{0} \\ \mathbf{0} & \mathscr{A}_T \end{bmatrix} \begin{pmatrix} \mathbf{x} \\ \mathbf{x}_T \end{pmatrix} + \begin{bmatrix} \mathscr{B}_R \\ \mathscr{B}_T \end{bmatrix} \mathbf{u},$$

$$\mathscr{A}_R = \text{diag}\left\{ \lambda_1, \ldots, \lambda_{n_x} \right\}, \tag{10.71}$$

$$\mathscr{A}_T = \text{diag}\left\{ \lambda_{n_x+1} \ldots \lambda_{n_x+t} \right\}.$$

Within a first-order approximation, show that the shifts of the closed-loop eigenvalues from their open-loop positions are

For $i = 1, \ldots, n$: $\Delta\lambda_i = \mathbf{b}_i^* \mathbf{g}_i$, $\mathscr{B}_R = \begin{bmatrix} \mathbf{b}_1^* \\ \vdots \\ \mathbf{b}_n^* \end{bmatrix}$, $\mathbf{G} = [\mathbf{g}_1, \ldots \mathbf{g}_n]$,

For $i = (n_x + 1, \ldots, n_x + t)$: $\Delta\lambda_i = 0.$ (10.72)

Hint: Using Theorem 10.6, let

$$\mathbf{E} = \mathbf{I}, \quad \mathbf{A} = \begin{bmatrix} \mathscr{A}_R & \mathbf{0} \\ \mathbf{0} & \mathscr{A}_T \end{bmatrix}, \quad \Delta\mathbf{A} = \begin{bmatrix} \mathscr{B}_R \mathbf{G} & \mathbf{0} \\ \mathscr{B}_T \mathbf{G} & \mathbf{0} \end{bmatrix}.$$

EXAMPLE 10.5

Repeat Exercise 10.6 for the system (10.71) using the measurement feedback control $\mathbf{u} = \mathscr{G}\mathbf{z}$, $\mathbf{z} = [\mathscr{M}_R, \mathscr{M}_T] \begin{pmatrix} \mathbf{x} \\ \mathbf{x}_T \end{pmatrix}$.

Solution: In this case,

$$\mathbf{E} = \mathbf{I}, \quad \mathbf{A} = \begin{bmatrix} \mathscr{A}_R & \mathbf{0} \\ \mathbf{0} & \mathscr{A}_T \end{bmatrix}, \qquad \Delta\mathbf{A} = \begin{bmatrix} \mathscr{B}_R \mathscr{G} \mathscr{M}_R & \mathscr{B}_R \mathscr{G} \mathscr{M}_T \\ \mathscr{B}_T \mathscr{G} \mathscr{M}_R & \mathscr{B}_T \mathscr{G} \mathscr{M}_T \end{bmatrix}$$

and, from Theorem 10.6,

$$\begin{aligned} \Delta\lambda_i &= \mathbf{b}_i^* \mathscr{G} \mathbf{m}_i, \qquad i = 1, \ldots, n, \\ \Delta\lambda_i &= \mathbf{b}_{T\alpha}^* \mathscr{G} \mathbf{m}_{T\alpha}, \qquad i = n + \alpha, \quad \alpha = 1, \ldots, t, \end{aligned} \tag{10.73}$$

where

$$\begin{aligned} \mathscr{B}_R^* &= [\mathbf{b}_1, \ldots, \mathbf{b}_n], \qquad \mathscr{B}_T^* = [\mathbf{b}_{T1}, \ldots, \mathbf{b}_{Tt}], \\ \mathscr{M}_R &= [\mathbf{m}_1, \ldots, \mathbf{m}_n], \qquad \mathscr{M}_T = [\mathbf{m}_{T1}, \ldots, \mathbf{m}_{Tt}]. \end{aligned} \tag{10.74}$$

Note from Chapter 5 that $\mathbf{b}_i^*, \mathbf{b}_{Ti}^*$ are the controllability vectors associated with modal coordinates x_i and x_{Ti}, respectively, and $\mathbf{m}_i, \mathbf{m}_{Ti}$ are the observability vectors associated with modal coordinates x_i and x_{Ti}, respectively. Thus, (10.73) reveals that the root perturbation caused by measurement feedback is the inner product of the controllability and observability vectors with the gain $\mathscr{G}$ as a weighting matrix. Hence, an eigenvalue will not be perturbed if it is either uncontrollable or unobservable. ■

EXAMPLE 10.6

Let $\mathscr{A}_R = -1$ and $\mathscr{A}_T = -10$ in the previous example, and $\mathscr{M}_R = 1$, $\mathscr{M}_T = -0.2$, $\mathscr{B}_R = 1$, $\mathscr{B}_T = 70$. Then, from (10.73),

$$\Delta\lambda_1 = \mathscr{G}, \qquad \lambda_1 + \Delta\lambda_1 = -1 + \mathscr{G}, \tag{10.75a}$$

$$\Delta\lambda_2 = -14\mathscr{G}, \qquad \lambda_2 + \Delta\lambda_2 = -10 - 14\mathscr{G}. \tag{10.75b}$$

Note that since $(\mathscr{A}_R, \mathscr{B}_R, \mathscr{M}_R)$ is completely controllable and observable, the poles for the design model $(\mathscr{A}_R, \mathscr{B}_R, \mathscr{M}_R)$ can be arbitrarily assigned $\lambda[\mathscr{A}_R + \mathscr{B}_R\mathscr{G}\mathscr{M}_R] = -1 + \mathscr{G}$. But in the presence of errors of model order, this example shows that the pole of the design model cannot be moved further to the left than $\lambda_1 + \Delta\lambda_1 = -1\frac{5}{7}$ before the "unmodeled mode" is driven unstable ($\lambda_2 + \Delta\lambda_2 \geq 0$ for $\mathscr{G} < -\frac{5}{7}$). (Mode 1 is driven unstable for $\mathscr{G} > 1$.) These are *first-order* predictions. Show that the *exact* stability constraint on $\mathscr{G}$ is

$$-\infty < \mathscr{G} < 1 \qquad \text{(exact for } \textit{design model}\text{)},$$

$$-\tfrac{5}{7} < \mathscr{G} < 1 \qquad \text{(first order } \textit{approximation} \text{ for complete system)},$$

$$-\tfrac{11}{13} < \mathscr{G} < \infty \qquad \text{(exact for } \textit{complete system}\text{)}$$

and show that the corresponding root locus for negative $\mathscr{G}$ is given by Fig. 10.9.

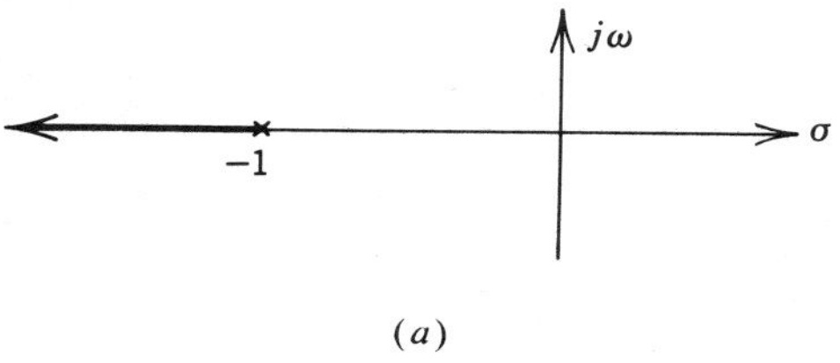

Figure 10.9a *Model Root Locus for Design Model*

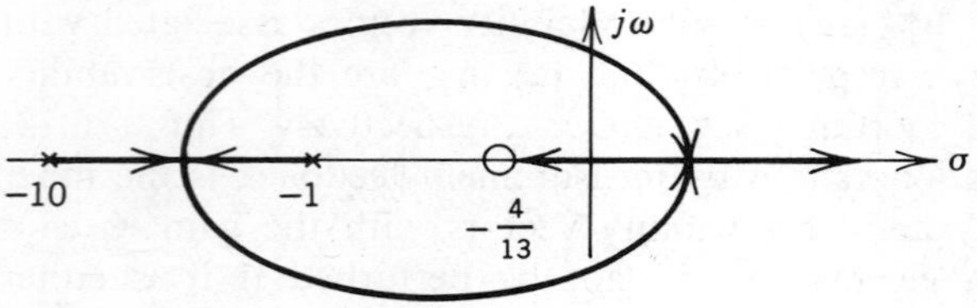

(b)

Figure 10.9b Actual Root Locus (Complete System)

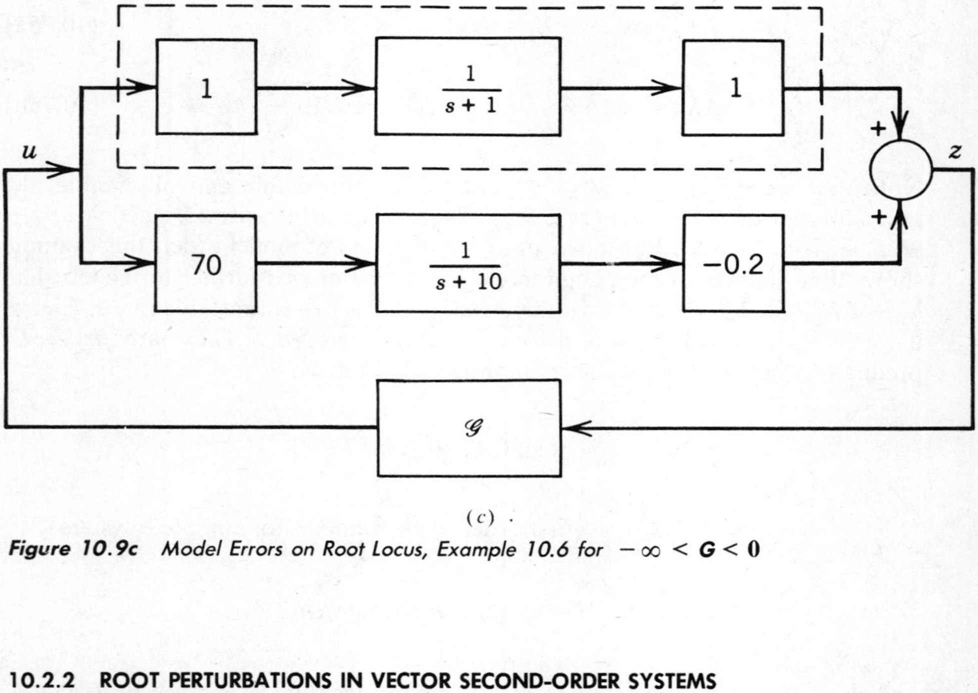

(c)

Figure 10.9c Model Errors on Root Locus, Example 10.6 for $-\infty < G < 0$

10.2.2 ROOT PERTURBATIONS IN VECTOR SECOND-ORDER SYSTEMS

The results of the previous section are now applied to systems of the form

$$\ddot{\boldsymbol{\eta}} + 2\zeta\boldsymbol{\Omega}\dot{\boldsymbol{\eta}} + \boldsymbol{\Omega}^2\boldsymbol{\eta} = \mathscr{B}\mathbf{u}, \quad \boldsymbol{\Omega}^2 = \begin{bmatrix} \omega_1^2 & & \\ & \ddots & \\ & & \omega_N^2 \end{bmatrix}, \qquad \begin{matrix} \boldsymbol{\eta} \in \mathscr{R}^N, \\ \mathbf{u} \in \mathscr{R}^{n_u}, \end{matrix} \tag{10.76}$$

$$\mathbf{z} = \mathscr{M}_p\boldsymbol{\eta} + \mathscr{M}_r\dot{\boldsymbol{\eta}}, \qquad \mathbf{z} \in \mathscr{R}^{n_z}.$$

With diagonal *damping* matrix,

$$\boldsymbol{\zeta} = \begin{bmatrix} \zeta_1 & & \\ & \ddots & \\ & & \zeta_N \end{bmatrix}. \tag{10.77}$$

This eliminates the possibility of *gyroscopic* systems. For dynamic systems with gyroscopic terms

$$\ddot{\boldsymbol{\eta}} + (2\boldsymbol{\zeta\Omega} + \mathscr{G})\dot{\boldsymbol{\eta}} + \boldsymbol{\Omega}^2\boldsymbol{\eta} = \mathscr{B}\mathbf{u},$$

$$\mathbf{z} = \mathscr{M}_p\boldsymbol{\eta} + \mathscr{M}_r\dot{\boldsymbol{\eta}}, \tag{10.78}$$

the theory of the previous section may be used after putting (10.53) into state-variable form. Gyroscopic systems are dynamic systems with part or all of the structure spinning with respect to an inertial reference. In such cases, $\mathscr{G}$ is skew-symmetric ($\mathscr{G} = -\mathscr{G}^T$). Clearly, $2\boldsymbol{\zeta\Omega} + \mathscr{G}$ cannot be diagonal if skew-symmetric terms are present. Hence, such $\mathscr{G}$ terms are excluded in this section.

For present purposes the system (10.76) and (10.77) is first put into the state form

$$\dot{\mathbf{x}} = \mathbf{A}\mathbf{x} + \mathbf{B}\mathbf{u}, \qquad \mathbf{x} \in \mathscr{R}^{2N}, \tag{10.79}$$

where the ith equation of (10.76) is

$$\ddot{\eta}_i + 2\zeta_i\omega_i\dot{\eta}_i + \omega_i^2\eta_i = \mathbf{b}_i^T\mathbf{u} \tag{10.80}$$

and the state

$$\mathbf{x}^T = (\eta_1, \ldots, \eta_c | \eta_{c+1}, \ldots, \eta_N | \dot{\eta}_1, \ldots, \dot{\eta}_c | \dot{\eta}_{c+1}, \ldots, \dot{\eta}_N) \tag{10.81}$$

is composed of c modal variables $(\eta_1, \ldots, \eta_c)$ associated with complex roots (with $\zeta_i < 1,\ i = 1, \ldots, c$) and $(N - c)$ modal variables $(\eta_{c+1}, \ldots, \eta_N)$ associated with real roots $\zeta_i > 1,\ i = c + 1, \ldots, N$. This leads to an $\mathbf{A}$ matrix of the form

$$\mathbf{A} = \begin{bmatrix} \mathbf{0} & \mathbf{0} & \mathbf{I} & \mathbf{0} \\ \mathbf{0} & \mathbf{0} & \mathbf{0} & \mathbf{I} \\ -\boldsymbol{\Omega}_c^2 & \mathbf{0} & -2\boldsymbol{\zeta}_c\boldsymbol{\Omega}_c & \mathbf{0} \\ \mathbf{0} & -\boldsymbol{\Omega}_r^2 & \mathbf{0} & -2\boldsymbol{\zeta}_r\boldsymbol{\Omega}_r \end{bmatrix}, \tag{10.82}$$

where $\boldsymbol{\Omega}_c \triangleq \text{diag}\{\omega_1, \ldots, \omega_c\}$, $\Omega_r \triangleq \text{diag}\{\omega_{c+1}, \ldots, \omega_N\}$, $\boldsymbol{\zeta}_c = \text{diag}\{\zeta_1, \ldots, \zeta_c\} < \mathbf{I}$,

$\zeta_r \triangleq \operatorname{diag}\{\zeta_{c+1}, \ldots, \zeta_N\} > \mathbf{I}$, and the $\mathbf{B}$ matrix for (10.79) is

$$\mathbf{B} = \begin{bmatrix} \mathbf{0} \\ \mathbf{0} \\ \mathscr{B}_c \\ \mathscr{B}_r \end{bmatrix} \qquad \begin{aligned} \mathscr{B}_c^T &= [\mathbf{b}_1, \ldots, \mathbf{b}_c], \\ \mathscr{B}_r^T &= [\mathbf{b}_{c+1}, \ldots, \mathbf{b}_N]. \end{aligned} \tag{10.83}$$

For this $\mathbf{A}$ matrix the modal matrix of right eigenvectors is

$$\mathbf{E} = \begin{bmatrix} \mathbf{I} & \mathbf{I} & 0 & 0 \\ 0 & 0 & \mathbf{I} & \mathbf{I} \\ \Lambda_c & \overline{\Lambda}_c & 0 & 0 \\ 0 & 0 & \Lambda_{r1} & \Lambda_{r2} \end{bmatrix}, \qquad \begin{aligned} \Lambda_c &= -\zeta_c\Omega_c + j\Omega_c\sqrt{\mathbf{I} - \zeta_c^2}, \\ \overline{\Lambda}_c &= -\zeta_c\Omega_c - j\Omega_c\sqrt{\mathbf{I} - \zeta_c^2}, \\ \Lambda_{r1} &= -\zeta_r\Omega_r - \Omega_r\sqrt{\zeta_r^2 - \mathbf{I}}, \\ \Lambda_{r2} &= -\zeta_r\Omega_r + \Omega_r\sqrt{\zeta_r^2 - \mathbf{I}}, \end{aligned} \tag{10.84}$$

where the square root of a diagonal matrix is composed of the square root of the diagonal elements. The inverse of $\mathbf{E}$ gives the matrix of left eigenvectors

$$\mathbf{E}^{-1} =$$

$$\begin{bmatrix} \frac{1}{2}\left[\mathbf{I} - j(\mathbf{I} - \zeta_c^2)^{-1/2}\zeta_c\right] & 0 & -\frac{1}{2}j(\mathbf{I} - \zeta_c^2)^{-1/2}\Omega_c^{-1} & 0 \\ \frac{1}{2}\left[\mathbf{I} + j(\mathbf{I} - \zeta_c^2)^{-1/2}\zeta_c\right] & 0 & \frac{1}{2}j(\mathbf{I} - \zeta_c^2)^{-1/2}\Omega_c^{-1} & 0 \\ 0 & \frac{1}{2}\left[\mathbf{I} - (\zeta_r^2 - \mathbf{I})^{-1/2}\zeta_r\right] & 0 & -\frac{1}{2}(\zeta_r^2 - \mathbf{I})^{-1/2}\Omega_r^{-1} \\ 0 & \frac{1}{2}\left[\mathbf{I} + (\zeta_r^2 - \mathbf{I})^{-1/2}\zeta_r\right] & 0 & \frac{1}{2}(\zeta_r^2 - \mathbf{I})^{-1/2}\Omega_r^{-1} \end{bmatrix}. \tag{10.85}$$

Exercise 10.7

Verify that (10.84) and (10.85) satisfy

$$\mathbf{E}^{-1}\mathbf{E} = \mathbf{I}, \qquad \mathbf{E}^{-1}\mathbf{A}\mathbf{E} = \operatorname{diag}\{\Lambda_c, \overline{\Lambda}_c, \Lambda_{r1}, \Lambda_{r2}\}.$$

Now, consider that the difference between the eigenvalues of the perturbed system

$$\ddot{\eta} + [\mathscr{D} + \Delta\mathscr{D}]\dot{\eta} + [\Omega^2 + \Delta\Omega^2]\eta = \mathbf{0} \tag{10.86}$$

and the nominal system

$$\ddot{\eta} + \mathscr{D}\dot{\eta} + \Omega^2\eta = \mathbf{0} \tag{10.87}$$

can be obtained from Theorem 10.6. To simplify the results, we shall assume that the *number* of real roots does not change from (10.86) to (10.87). This assumes

$\Delta\mathscr{D}$ = block diag $\{\Delta\mathscr{D}_c, \Delta\mathscr{D}_r\}$, $\Delta\Omega^2$ = block diag $\{\Delta\Omega_c^2, \Delta\Omega_r^2\}$. Then, the structure of $\mathbf{A}$ in (10.82) may be used to obtain the following corollary of Theorem 10.6. From (10.82), (10.86), and (10.87), see that

$$\Delta\mathbf{A} \triangleq \begin{bmatrix} \mathbf{0} & \mathbf{0} & \mathbf{0} & \mathbf{0} \\ \mathbf{0} & \mathbf{0} & \mathbf{0} & \mathbf{0} \\ -\Delta\Omega_c^2 & \mathbf{0} & -\Delta\mathscr{D}_c & \mathbf{0} \\ \mathbf{0} & -\Delta\Omega_r^2 & \mathbf{0} & -\Delta\mathscr{D}_r \end{bmatrix}. \tag{10.88}$$

Corollary 1 of Theorem 10.6

For the system (10.86) *with* $2c$ *complex eigenvalues*

$$\lambda_i = -\zeta_i\omega_i \pm j\omega_i\sqrt{1-\zeta_i^2}, \qquad i = 1, 2, \ldots, c, \tag{10.89a}$$

and $N - 2c$ *real eigenvalues*

$$\lambda_i = -\zeta_i\omega_i \mp \omega_i\sqrt{\zeta_i^2 - 1}, \qquad i = c+1, \ldots, N, \tag{10.89b}$$

the first-order effects of $\Delta\mathscr{D}$ *and* $\Delta\Omega^2$ *in* (10.86) *are*

$$\Delta\lambda_i = -\frac{1}{2}[\Delta\mathscr{D}]_{ii} \mp j\frac{1}{2\sqrt{1-\zeta_i^2}}\left[\zeta_i[\Delta\mathscr{D}]_{ii} + \frac{1}{\omega_i}[\Delta\Omega^2]_{ii}\right], \qquad i = 1, 2, \ldots, c, \tag{10.90a}$$

and

$$\Delta\lambda_i = -\frac{1}{2}[\Delta\mathscr{D}]_{ii} + \frac{1}{2\sqrt{\zeta_i^2 - 1}}\left[\pm\zeta_i[\Delta\mathscr{D}]_{ii} \mp \frac{1}{\omega_i}[\Delta\Omega^2]_{ii}\right], \qquad i = c+1, \ldots, N, \tag{10.90b}$$

where the first-order approximation of the eigenvalues of (10.61) *are* $\lambda_i + \Delta\lambda_i$.

Proof: The proof follows by direct substitution of $\mathbf{e}_i$ from (10.84), $\mathbf{l}_i^*$ from (10.85), and $\Delta\mathbf{A}$ from (10.88) into (10.70a). ■

Now, let the feedback control terms be considered a perturbation on the parameters in the equations of motion. The results of the corollary to Theorem 10.6 may then be applied to investigate the effects of control on the eigenvalues. Consider measurement feedback control using colocated sensors and actuators. In this event the measurement and control equations in (10.76) become

$$\mathbf{z} = \begin{pmatrix} \mathbf{z}_p \\ \mathbf{z}_r \end{pmatrix} = \begin{bmatrix} \mathscr{B}^* \\ \mathbf{0} \end{bmatrix}\boldsymbol{\eta} + \begin{bmatrix} \mathbf{0} \\ \mathscr{B}^* \end{bmatrix}\dot{\boldsymbol{\eta}}, \qquad \mathbf{u} = \mathscr{G}_p\mathbf{z}_p + \mathscr{G}_r\mathbf{z}_r. \tag{10.91}$$

Corollary 2 to Theorem 10.6

Given the system (10.76) *with all complex* ($c = N/2$) *eigenvalues* (10.89), *consider the closed-loop system using control* (10.91). *The first-order approximation of the closed-loop eigenvalues of*

$$\ddot{\eta} + (2\zeta\Omega + \mathscr{B}\mathscr{G}_r\mathscr{B}^*)\dot{\eta} + (\Omega^2 + \mathscr{B}\mathscr{G}_p\mathscr{B}^*)\eta = 0 \tag{10.92}$$

is $\lambda_i + \Delta\lambda_i$, *where*

$$\Delta\lambda_i = -\frac{1}{2}[\mathscr{B}\mathscr{G}_r\mathscr{B}^*]_{ii} \mp j\frac{1}{2\omega_i}[\mathscr{B}\mathscr{G}_p\mathscr{B}^*]_{ii}, \qquad i = 1, 2, \ldots, N,$$

$$= -\frac{1}{2}\|\mathbf{b}_i\|^2_{\mathscr{G}_r} \mp j\frac{1}{2\omega_i}\|\mathbf{b}_i\|^2_{\mathscr{G}_p}, \qquad \mathscr{B}^* = [\mathbf{b}_1, \ldots, \mathbf{b}_N]. \tag{10.93}$$

From (10.99) and Chapter 2, observe that the real part of the eigenvalue perturbation is a norm of the *controllability* vector $\|\mathbf{b}_i\|^2_{\mathscr{G}_r}$, with a weighting $\mathscr{G}_r$. Likewise, the imaginary part of the eigenvalue perturbation is a norm of the modified controllability vector $(1/\omega_i)\|\mathbf{b}_i\|^2_{\mathscr{G}_p}$ with weighting $\mathscr{G}_p$. Note that $\mathbf{b}_i$ is also the *observability* vector of mode i since the sensors and actuators are collocated.

A very interesting connection exists between the modal costs (5.161) and the first-order approximations $\Delta\lambda_i$ of the eigenvalues when the system is driven by controller (10.91). The proof of the following involves a straightforward comparison of (10.93) and (5.161) when $\mathbf{y} = \mathbf{z}_p$ or $\mathbf{y} = \mathbf{z}_r$.

Theorem 10.7

Let $\mathscr{V}_{\eta_i}$ *denote the modal cost of the lightly damped system*

$$\ddot{\eta}_i + 2\zeta_i\omega_i\dot{\eta}_i + \omega_i^2\eta_i = \mathbf{b}_i^*\mathbf{u}, \qquad i = 1, 2, \ldots, N, \tag{10.94}$$

$$\mathbf{z}_p = \sum_{i=1}^{N}\mathbf{b}_i\eta_i, \qquad \mathbf{z}_r = \sum_{i=1}^{N}\mathbf{b}_i\dot{\eta}_i, \tag{10.95}$$

when the cost function is either $\mathscr{V}_p$ *or* $\mathscr{V}_r$,

$$\mathscr{V}_p = \sum_{i=1}^{n_u}\int_0^\infty \mathbf{z}_p^{i*}\mathbf{Q}\mathbf{z}_p^i\,dt, \tag{10.96}$$

$$\mathscr{V}_r = \sum_{i=1}^{n_u}\int_0^\infty \mathbf{z}_r^{i*}\mathbf{Q}\mathbf{z}_r^i\,dt, \tag{10.97}$$

where the strength of each impulse $u_i(t) = u_i\delta(t)$ *in this open-loop calculation is* u_i, *and* $\mathscr{U} \triangleq \operatorname{diag}[\ldots u_i^2 \ldots]$. *Let* $\Delta\lambda_i$ *represent the first order approximation of the shift in the open-loop eigenvalue* λ_i *when either of the collocated controls*

$$\mathbf{u}_p = \mathbf{G}_p \mathbf{z}_p, \tag{10.96}$$

$$\mathbf{u}_r = \mathbf{G}_r \mathbf{z}_r \tag{10.97}$$

are applied to (10.94). *Then under either* $(\mathbf{u}_p, \mathscr{V}_p)$ *or* $(\mathbf{u}_r, \mathscr{V}_r)$ *the open loop modal cost* $\mathscr{V}_{\eta_i}$ *is related to the closed-loop performance by*

$$\mathscr{V}_{\eta_i} = \frac{|\Delta\lambda_i|^2}{|\mathscr{Re}\,\lambda_i|}. \tag{10.98}$$

provided $\mathbf{Q} = \mathscr{U} = \mathbf{G}_p$ *or* $\mathbf{G}_r$, *depending on case* (10.96) *or* (10.97), *respectively.*

Theorem 10.7 provides a rare circumstance where open loop calculations can predict closed loop performance.

10.3 Sensitivity Analysis and Control

In this section we consider the analysis and control of first order sensitivity of linear systems. In Section 10.3.1, we define and calculate root sensitivity. In Section 10.3.2, a different sensitivity is calculated; the sensitivity of the state trajectory with respect to parameters in the model. Section 10.3.3 reduces the order of the trajectory sensitivity model, and Section 10.3.4 describes a controller design for the trajectory sensitivity model.

10.3.1 MINIMAL ROOT SENSITIVITY IN LINEAR SYSTEMS

The previous sections of this chapter described first order *perturbations* of modal data. Now, we consider first order *sensitivity* of the modal data.

We are interested in the smallest possible sensitivity of eigenvalues λ_i with respect to the independent plant parameters in linear systems of the form

$$\dot{\mathbf{x}} = \mathbf{A}\mathbf{x}, \qquad \mathbf{x} \in \mathscr{R}^{n_x}. \tag{10.99}$$

That is, the norm of the root sensitivity matrix

$$\frac{\partial\lambda_i}{\partial\mathbf{A}} = \begin{bmatrix} \dfrac{\partial\lambda_i}{\partial A_{11}} & \cdots & \dfrac{\partial\lambda_i}{\partial A_{1n}} \\ \vdots & \cdots & \vdots \\ \dfrac{\partial\lambda_i}{\partial A_{n1}} & \cdots & \dfrac{\partial\lambda_i}{\partial A_{nn}} \end{bmatrix} \tag{10.100}$$

and the lower bound of its norm are of interest. The norm of a matrix shall be the Frobenius norm,

$$\|[\cdot]\|^2 = \operatorname{tr}[\cdot]^*[\cdot], \tag{10.101}$$

and the norm of a vector shall be denoted by

$$\|(\cdot)\|^2 = (\cdot)^*(\cdot), \tag{10.102}$$

where * denotes complex conjugate transpose. Results herein are limited to the case of distinct eigenvalues for A.

Construction of a Root Sensitivity Metric

The sensitivity of the ith eigenvalue $\partial\lambda_i/\partial\mathbf{A}$ is an $n_x \times n_x$ matrix denoted by $\mathbf{S}_i \triangleq \partial\lambda_i/\partial\mathbf{A}$. The norm of $\mathbf{S}_i$ from (10.101) is

$$\|\mathbf{S}_i\|^2 \triangleq \operatorname{tr}\mathbf{S}_i^*\mathbf{S}_i = \sum_{\alpha=1}^{n}\sum_{\beta=1}^{n}\left(\frac{\partial\lambda_i}{\partial A_{\alpha\beta}}\right)^2. \tag{10.103}$$

The complete root sensitivity metric of interest is

$$\hat{s} \triangleq \sum_{i=1}^{n}\|\mathbf{S}_i\|^2 \triangleq \sum_{i=1}^{n}\left\|\frac{\partial\lambda_i}{\partial\mathbf{A}}\right\|^2. \tag{10.104}$$

Thus, from the point of view of root sensitivity, a system design with a large value of $\hat{s}$ might be considered less desirable than a system design with a small value of $\hat{s}$. This would be true if the analyst is *specifically* concerned that root locations remain fixed in the presence of parameter uncertainties.

To compute the root sensitivity metric, we shall assume that $\mathbf{A}$ has a linearly independent set of eigenvectors $\mathbf{e}_i$:

$$\mathbf{A}\mathbf{e}_i = \mathbf{e}_i\lambda_i, \qquad i = 1,\ldots,n. \tag{10.105}$$

The reciprocal basis vectors ℓ_i are defined by

$$\mathbf{E} \triangleq [\mathbf{e}_1,\ldots,\mathbf{e}_n], \qquad \begin{bmatrix}\ell_1^* \\ \vdots \\ \ell_n^*\end{bmatrix} \triangleq \mathbf{E}^{-1}. \tag{10.106}$$

Hence, $\ell_i^*\mathbf{e}_j = \delta_{ij}$. Multiplying (10.105) from the left by ℓ_i^*, using (10.106), yields the eigenvalues in terms of $\mathbf{A}$, its eigenvectors, and its reciprocal basis vectors:

$$\lambda_i = \ell_i^*\mathbf{A}\mathbf{e}_i. \tag{10.107}$$

Differentiation of the scalar (10.107) with respect to **A** provides the required sensitivity $\partial\lambda_i/\partial\mathbf{A}$. To derive this result, two identities from Section 2.7.1.3 are required:

$$\operatorname{tr}\mathbf{AB} = \operatorname{tr}\mathbf{BA}, \tag{10.108a}$$

$$\frac{\partial}{\partial\mathbf{A}}(\operatorname{tr}\mathbf{AB}) = \frac{\partial}{\partial\mathbf{A}}(\operatorname{tr}\mathbf{BA}) = \mathbf{B}^T, \tag{10.108b}$$

where equations (10.108) hold for real or complex matrices **B** and **A**, and (10.108) holds if the elements of **A** are independent. Hence, from (10.107), using (10.108),

$$\mathbf{S}_i = \frac{\partial\lambda_i}{\partial\mathbf{A}} = \frac{\partial}{\partial\mathbf{A}}\left[\operatorname{tr}\mathbf{A}(\mathbf{e}_i\ell_i^*)\right] = (\mathbf{e}_i\ell_i^*)^T = \bar{\ell}_i\mathbf{e}_i^T, \tag{10.109}$$

where the overbar denotes complex conjugate.

Result (10.109) can also be obtained from (10.70a) and (10.107),

$$\Delta\lambda_i = \ell_i^*\,\Delta\mathbf{A}\mathbf{e}_i, \tag{10.110}$$

where $\Delta\lambda_i$ is the change (to first-order approximation) of λ_i in the presence of a perturbation of **A** to $\mathbf{A} + \Delta\mathbf{A}$. The norm (10.103) may now be written from (10.109),

$$\begin{aligned}\|\mathbf{S}_i\|^2 &\triangleq \operatorname{tr}\mathbf{S}_i^*\mathbf{S}_i = \operatorname{tr}\left(\bar{\ell}_i e_i^T\right)^*\left(\bar{\ell}_i\mathbf{e}_i^T\right)\\ &= \|\ell_i\|^2\|\mathbf{e}_i\|^2,\end{aligned} \tag{10.111}$$

where the last equality requires use of identity (10.108) again, and where

$$\|\ell_i\|^2 = \ell_i^*\ell_i, \qquad \|\mathbf{e}_i\|^2 = \mathbf{e}_i^*\mathbf{e}_i. \tag{10.112}$$

We seek necessary and sufficient conditions for minimality of (10.104). The Cauchy–Schwarz inequality (2.31) holds for *any* two vectors

$$|\ell_i^*\mathbf{e}_i| \le \|\ell_i\|\,\|\mathbf{e}_i\|. \tag{10.113}$$

Since the *particular* vectors $\ell_i, \mathbf{e}_i$ are related by (10.106),

$$\ell_i^*\mathbf{e}_i = 1. \tag{10.114}$$

Equations (10.113) and (10.114) lead immediately to

$$\|\ell_i\|\,\|\mathbf{e}_i\| \ge 1. \tag{10.115}$$

Squaring both sides of (10.115) and using (10.111) leads to

$$\|\mathbf{S}_i\|^2 \ge 1. \tag{10.116}$$

The equality in (10.115), and in (10.116) holds if and only if ℓ_i and $\mathbf{e}_i$ are colinear ($\ell_i = e_i$). From linear algebra [10.12], $\ell_i = \mathbf{e}_i$ if and only if $\mathbf{A}$ is normal ($\mathbf{AA}^* = \mathbf{A}^*\mathbf{A}$). Thus, the main theoretical results of this section are summarized as follows.

Theorem 10.8

Let $(\lambda_i, \mathbf{e}_i, \ell_i)$ be the ith eigenvalue, the ith eigenvector, and its reciprocal basis vector associated with the real $n_x \times n_x$ matrix $\mathbf{A}$. If $\mathbf{A}$ has a linearly independent set of eigenvectors $\mathbf{e}_i$, $i = 1, \ldots, n$, then

$$\left\| \frac{\partial \lambda_i}{\partial \mathbf{A}} \right\|^2 \geq 1, \qquad i = 1, 2, \ldots, n, \tag{10.117}$$

where the lower bound

$$\left\| \frac{\partial \lambda_i}{\partial \mathbf{A}} \right\|^2 = 1, \qquad i = 1, 2, \ldots, n,$$

is achieved if and only if $\mathbf{AA}^* = \mathbf{A}^*\mathbf{A}$. *The sensitivity metric* (10.104) *is bounded from below by*

$$\hat{s} \geq n_x, \tag{10.118}$$

and the minimum sensitivity $\hat{s} = n_x$ is achieved if and only if $\mathbf{A}$ is normal ($\mathbf{AA}^* = \mathbf{A}^*\mathbf{A}$).

The theorem provides necessary and sufficient conditions for minimum root sensitivity. If one wishes to keep roots relatively fixed in the presence of parameter variations, Theorem 10.8 indicates that normality of $\mathbf{A}$ is a necessary and sufficient condition for a globally minimal value of $\|\partial \lambda_i / \partial \mathbf{A}\|$ for all i. Gilbert [10.13] has shown that symmetry of $\mathbf{A}$ is a sufficient condition for a globally minimal value of $\|\partial \mathcal{Re}\, \lambda_i / \partial \mathbf{A}\|$. This is in agreement with Theorem 10.8, since a symmetric $\mathbf{A}$ is also normal and only has real eigenvalues.

One might be motivated by Theorem 10.8 to declare that normality of the closed-loop system matrix should be a design goal. The following example will dampen such enthusiasm. The conclusion here is that the global minimum of $\hat{s}$ occurs at the same place where a global minimum of abnormality $\|\mathbf{AA}^* - \mathbf{A}^*\mathbf{A}\|$ occurs. However, "nearly" normally does not imply "nearly" insensitive, since $\|\mathbf{AA}^* - \mathbf{A}^*\mathbf{A}\|$ can be a nice convex function of parameters even when $\hat{s}$ is not convex. Hence, $\hat{s}$ can be arbitrarily large when $\|\mathbf{AA}^* - \mathbf{A}^*\mathbf{A}\|$ is small, but not zero.

Let α = angle of attack, q = pitch rate, u = elevator angle, τ = lifting time constant, ω_0 = the undamped pitch natural frequency, $\mathcal{Q}$ = elevator effectiveness = 1. Then, the pitch motion of a rigid aircraft is governed by, [10.18].

$$\begin{pmatrix} \dot{\alpha} \\ \dot{q} \end{pmatrix} = \begin{bmatrix} -1/\tau & 1 \\ -\omega_0^2 & 0 \end{bmatrix} \begin{pmatrix} \alpha \\ q \end{pmatrix} + \begin{bmatrix} 0 \\ \omega_0^2 \mathcal{Q} \end{bmatrix} u, \qquad \alpha = [1 \quad 0] \begin{pmatrix} \alpha \\ q \end{pmatrix} = \mathbf{Mx}.$$

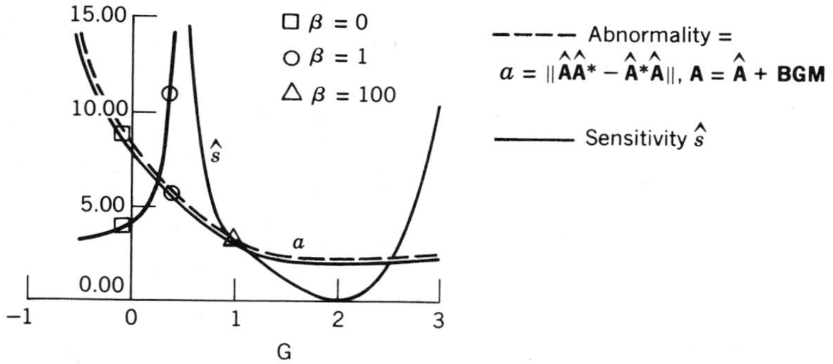

Figure 10.10 *Root Sensitivity and Abnormality*

The angle of attack feedback control $u = G\alpha$ leads to the closed-loop matrix

$$\mathscr{A} + \mathbf{B}G\mathbf{M} = \begin{bmatrix} -1/\tau & 1 \\ \omega_0^2(\mathscr{Q}G - 1) & 0 \end{bmatrix},$$

which is normal iff $G = (\omega_0^2 + 1)/\mathscr{Q}\omega_0^2$. A plot of the "abnormality" of $[\mathbf{A} + \mathbf{B}G\mathbf{M}]$ is shown versus G in Fig. 10.10, superimposed upon a plot of $\hat{s}$. The reader will note that the global minimum of both functions occurs at the *same* value $G = 2 = (\omega_0^2 + 1)/\omega_0^2\mathscr{Q}$. However, *away* from the global minimum, the measure of abnormality and $\hat{s}$ lead to contradictory suggestions. To make $\mathbf{A} + \mathbf{B}G\mathbf{M}$ more normal from an initial value of $G = 0$, one must increase G. But, at the same time as G increases from zero to 0.5 (and abnormality *decreases* from 8.0 to 5.0), the actual sensitivity $\hat{s}$ *increases* from 4.0 to ∞. Of course, the difficulty here is that the abnormality is a convex function of G while $\hat{s}$ is not. Hence, caution is required to not read too much into Theorem 10.8. "Nearly" normal does not imply "nearly" insensitive, and "*normality*" can be a bad design goal.

10.3.2 TRAJECTORY SENSITIVITY ANALYSIS AND CONTROL

In this section we look at output sensitivity instead of root sensitivity. This could be more important when the output performance is critical (and one might argue that this is the usual case). Suppose there are h parameters $p_1, p_2, \ldots, p_h$ that are considered uncertain. Define

$$\mathbf{p} \triangleq \begin{pmatrix} p_1 \\ p_2 \\ \vdots \\ p_h \end{pmatrix}, \quad \mathbf{y}_p \triangleq \begin{pmatrix} \partial \mathbf{y}/\partial p_1 \\ \partial \mathbf{y}/\partial p_2 \\ \vdots \\ \partial \mathbf{y}/\partial p_h \end{pmatrix}, \quad \tilde{\mathbf{A}} \triangleq \begin{bmatrix} \mathbf{A} & & & \\ & \mathbf{A} & & \\ & & \ddots & \\ & & & \mathbf{A} \end{bmatrix}, \tag{10.119}$$

and then note that the differentiation of the product of two matrices with respect to vector $\mathbf{p}$ is

$$[\mathbf{MN}]_p = \mathbf{M}_p\mathbf{N} + \tilde{\mathbf{M}}\mathbf{N}_p, \qquad \mathbf{N}_p \triangleq \begin{bmatrix} \partial\mathbf{N}/\partial p_1 \\ \vdots \\ \partial\mathbf{N}/\partial p_h \end{bmatrix}, \tag{10.120}$$

and also $\widetilde{\mathbf{MN}} = \tilde{\mathbf{M}}\tilde{\mathbf{N}}$. The reader should pause to verify this identity before proceeding. Now, the differentiation of the system equations

$$\begin{aligned} \dot{\mathbf{x}}(\mathbf{p}, t) &= \mathbf{A}(\mathbf{p})\mathbf{x}(\mathbf{p}, t) + \mathbf{B}(\mathbf{p})\mathbf{u}(\mathbf{p}, t) + \mathbf{D}(\mathbf{p})\mathbf{w}(t), \\ \mathbf{y}(\mathbf{p}, t) &= \mathbf{C}(\mathbf{p})\mathbf{x}(\mathbf{p}, t) \end{aligned} \tag{10.121}$$

with respect to the vector $\mathbf{p}$ yields

$$\begin{aligned} \dot{\mathbf{x}}_p(\mathbf{p}, t) &= \mathbf{A}_p(\bar{\mathbf{p}})\mathbf{x}(\mathbf{p}, t) + \tilde{\mathbf{A}}(\bar{\mathbf{p}})\mathbf{x}_p(\mathbf{p}, t) + \mathbf{B}_p(\bar{\mathbf{p}})\mathbf{u}(\mathbf{p}, t) + \tilde{\mathbf{B}}(\bar{\mathbf{p}})\mathbf{u}_p(\mathbf{p}, t) \\ &\quad + \mathbf{D}_p(\bar{\mathbf{p}})\mathbf{w}(t), \\ \mathbf{y}_p(\mathbf{p}, t) &= \mathbf{C}_p(\bar{\mathbf{p}})\mathbf{x}(\mathbf{p}, t) + \tilde{\mathbf{C}}(\bar{\mathbf{p}})\mathbf{x}_p(\mathbf{p}, t), \end{aligned} \tag{10.122}$$

where $\mathbf{p} = \bar{\mathbf{p}}$ is the nominal vector value of $\mathbf{p}$, and we have assumed that $\mathbf{w}(t)$ is not a function of $\mathbf{p}$. We abbreviate the notation to read simply

$$\begin{aligned} \dot{\mathbf{x}}_p &= \mathbf{A}_p\mathbf{x} + \tilde{\mathbf{A}}\mathbf{x}_p + \mathbf{B}_p\mathbf{u} + \tilde{\mathbf{B}}\mathbf{u}_p + \mathbf{D}_p\mathbf{w}, \\ \mathbf{y}_p &= \mathbf{C}_p x + \tilde{\mathbf{C}}\mathbf{x}_p. \end{aligned} \tag{10.123}$$

Now, augment the trajectory sensitivity model (10.123) to the plant model (10.121) to obtain the sensitivity system. Note that $\mathbf{u}_p$ cannot be determined exactly, prior to knowledge of the control function $\mathbf{u}(\mathbf{y}(\mathbf{p}), t)$. We *approximate* $\mathbf{u}_p$ as $\mathbf{u}_p = \mathbf{G}_p\mathbf{x} + \tilde{\mathbf{G}}\mathbf{x}_p$ to reflect our interest in sensitivity information about the nominal optimal control, where $\mathbf{A}(\bar{p}), \mathbf{B}(\bar{p}), \mathbf{C}(\bar{p})$ are evaluated at $p = \bar{p}$,

$$\mathbf{u}_0 = \mathbf{G}\mathbf{x}, \quad \mathbf{G} = -\mathbf{R}^{-1}\mathbf{B}^*\mathbf{K}, \quad \mathbf{0} = \mathbf{KA} + \mathbf{A}^*\mathbf{K} - \mathbf{KBR}^{-1}\mathbf{B}^*\mathbf{K} + \mathbf{C}^*\mathbf{QC}. \tag{10.124}$$

Hence, *for the purposes of approximating* $\mathbf{u}_p$ the control sensitivity $\mathbf{u}_p$ has the structure

$$\mathbf{u}_p \approx \mathbf{u}_{0_p} = \mathbf{G}_p\mathbf{x} + \tilde{\mathbf{G}}\mathbf{x}_p \tag{10.125}$$

The matrix $\mathbf{G}_p$ appears in (10.125) and may be computed as follows. Differentiate (10.124) with respect to p to obtain

$$\mathbf{G}_p = -\tilde{\mathbf{R}}^{-1}\mathbf{B}_p^*\mathbf{K} - \tilde{\mathbf{R}}^{-1}\tilde{\mathbf{B}}^*\mathbf{K}_p,$$

$$\mathbf{0} = \mathbf{K}_p\mathbf{A} + \tilde{\mathbf{K}}\mathbf{A}_p + \mathbf{A}_p^*\mathbf{K} + \tilde{\mathbf{A}}^*\mathbf{K}_p - \mathbf{K}_p\mathbf{B}\mathbf{R}^{-1}\mathbf{B}^*\mathbf{K} - \tilde{\mathbf{K}}\mathbf{B}_p\mathbf{R}^{-1}\mathbf{B}^*\mathbf{K} \quad (10.126)$$

$$-\tilde{\mathbf{K}}\tilde{\mathbf{B}}\tilde{\mathbf{R}}^{-1}\mathbf{B}_p^*\mathbf{K} - \tilde{\mathbf{K}}\tilde{\mathbf{B}}\tilde{\mathbf{R}}^{-1}\mathbf{B}^*\mathbf{K}_p + \mathbf{C}_p^*\mathbf{Q}\mathbf{C} + \tilde{\mathbf{C}}^*\tilde{\mathbf{Q}}\mathbf{C}_p.$$

The latter equation is *linear* in $\mathbf{K}_p$ and can be written

$$0 = \mathbf{K}_p[\mathbf{A} + \mathbf{B}\mathbf{G}] - [\tilde{\mathbf{A}} + \tilde{\mathbf{B}}\tilde{\mathbf{G}}]^*\mathbf{K}_p + \mathbf{Q}_0, \quad (10.127)$$

where

$$\mathbf{Q}_0 \triangleq \mathbf{C}_p^*\mathbf{Q}\mathbf{C} + \tilde{\mathbf{C}}^*\tilde{\mathbf{Q}}\mathbf{C}_p - \tilde{\mathbf{K}}\left[\mathbf{B}_p\mathbf{R}^{-1}\mathbf{B}^* + \tilde{\mathbf{B}}\tilde{\mathbf{R}}^{-1}\mathbf{B}_p^*\right]\mathbf{K} + \tilde{\mathbf{K}}\mathbf{A}_p + \mathbf{A}_p^*\mathbf{K}.$$

Now, (10.127) is equivalent to

$$\mathbf{0} = \mathbf{K}_{p_i}[\mathbf{A} + \mathbf{B}\mathbf{G}] + [\mathbf{A} + \mathbf{B}\mathbf{G}]^*\mathbf{K}_{p_i} + \mathbf{Q}_i,$$

$$\mathbf{K}_p = \begin{bmatrix} \mathbf{K}_{p1} \\ \mathbf{K}_{p2} \\ \vdots \\ \mathbf{K}_{ph} \end{bmatrix}, \quad (10.128)$$

$$\mathbf{Q}_i \triangleq \mathbf{C}_{p_i}^*\mathbf{Q}\mathbf{C} + \mathbf{C}^*\mathbf{Q}\mathbf{C}_{p_i} - \mathbf{K}\left(\mathbf{B}_{p_i}\mathbf{R}^{-1}\mathbf{B}^* + \mathbf{B}\mathbf{R}^{-1}\mathbf{B}_{p_i}^*\right)\mathbf{K} + \mathbf{K}\mathbf{A}_{p_i} + \mathbf{A}_{p_i}^*\mathbf{K},$$

where the advantage in the computation (10.128) is obvious since h solutions of $\mathbf{K}_{p_i}$, $i = 1, 2, \ldots, h$, can be computed one at a time, in lieu of one solution of the $n_x(1 + h) \times n_x$ matrix $\mathbf{K}_p$ in (10.127).

The system and its sensitivity model is now composed of (10.121), (10.123), and (10.125), which may be combined to give

$$\begin{bmatrix} \dot{\mathbf{x}} \\ \dot{\mathbf{x}}_p \end{bmatrix} = \begin{bmatrix} \mathbf{A} & \mathbf{0} \\ [\mathbf{A} + \mathbf{B}\mathbf{G}]_p & [\tilde{\mathbf{A}} + \tilde{\mathbf{B}}\tilde{\mathbf{G}}] \end{bmatrix} \begin{bmatrix} \mathbf{x} \\ \mathbf{x}_p \end{bmatrix} + \begin{bmatrix} \mathbf{B} \\ \mathbf{B}_p \end{bmatrix} \mathbf{u} + \begin{bmatrix} \mathbf{D} \\ \mathbf{D}_p \end{bmatrix} \mathbf{w}, \quad (10.129a)$$

$$\begin{bmatrix} \mathbf{y} \\ \mathbf{y}_p \end{bmatrix} = \begin{bmatrix} \mathbf{C} & \mathbf{0} \\ \mathbf{C}_p & \tilde{\mathbf{C}} \end{bmatrix} \begin{pmatrix} \mathbf{x} \\ \mathbf{x}_p \end{pmatrix}, \qquad \mathbf{u}_{0p} = \begin{bmatrix} \mathbf{G}_p & \tilde{\mathbf{G}} \end{bmatrix} \begin{pmatrix} \mathbf{x} \\ \mathbf{x}_p \end{pmatrix}$$

which is written compactly

$$\dot{\mathbf{x}}_s = \mathbf{A}_s\mathbf{x}_s + \mathbf{B}_s\mathbf{u} + \mathbf{D}_s\mathbf{w},$$
$$\mathbf{y}_s = \mathbf{C}_s\mathbf{x}_s, \quad \mathbf{y}_s^* \triangleq (\mathbf{y}^*, \mathbf{y}_p^*), \quad \mathbf{u}_s^* \triangleq (\mathbf{u}^*, \mathbf{u}_{0p}^*), \quad \mathbf{u}_{0p} = \mathbf{G}_{2s}\mathbf{x}_s \tag{10.129b}$$
$$\mathbf{z} = [\mathbf{M} \quad \mathbf{0}]\mathbf{x}_s + \mathbf{v} = \mathbf{M}_s\mathbf{x}_s + \mathbf{v},$$

where we have added the measurement $\mathbf{z}$ in preparation for control design. Now, using the cost function

$$\mathscr{V} = \sum_{i=1}^{n_w+n_v} \int_0^\infty [\mathbf{y}_s^*\mathbf{Q}_s\mathbf{y}_s + \mathbf{u}_s^*\mathbf{R}_s\mathbf{u}_s]\, dt, \quad = \mathscr{V}_0 + \beta\mathscr{V}_s \tag{10.130}$$

where $\mathscr{V}_s$ collects all terms multiplied by β and $\mathscr{V}_0$ contains the remaining terms, and

$$\mathbf{Q}_s = \begin{bmatrix} \mathbf{Q} & \mathbf{0} \\ \mathbf{0} & \beta\tilde{\mathbf{Q}} \end{bmatrix}, \quad \mathbf{R}_s = \begin{bmatrix} \mathbf{R} & \mathbf{0} \\ \mathbf{0} & \beta\tilde{\mathbf{R}} \end{bmatrix}. \tag{10.131}$$

The LQI solution to this problem is

$$\mathbf{u}_s = \mathbf{G}_s\mathbf{x}_c, \quad \dot{\mathbf{x}}_c = \mathbf{A}_c\mathbf{x}_c + \mathbf{F}_s\mathbf{z}, \quad \mathbf{A}_c = \mathbf{A}_s + \mathbf{B}_s\mathbf{G}_s - \mathbf{F}_s\mathbf{M}_s$$
$$\mathbf{G}_s = -\mathbf{R}_s^{-1}\mathbf{B}_s^*\mathbf{K}_s, \quad \mathbf{0} = \mathbf{K}_s\mathbf{A}_s + \mathbf{A}_s\mathbf{K}_s - \mathbf{K}_s\mathbf{B}_s\mathbf{R}_s^{-1}\mathbf{B}_s^*\mathbf{K}_s + \mathbf{C}_s^*\mathbf{Q}_s\mathbf{C}_s, \tag{10.132}$$
$$\mathbf{F}_s = \mathbf{P}_s\mathbf{M}_s^*\mathbf{V}_s^{-1}, \quad \mathbf{0} = \mathbf{P}_s\mathbf{A}_s^* + \mathbf{A}_s\mathbf{P}_s - \mathbf{P}_s\mathbf{M}_s^*\mathbf{V}^{-1}\mathbf{M}_s\mathbf{P}_s + \mathbf{D}_s\mathbf{W}\mathbf{D}_s^*.$$

Controller (10.132) will be called the linear quadratic impulse sensitivity (LQIS) controller.

Now, the weight β in (10.130) trades nominal performance $\mathscr{V}_0$ versus sensitivity reduction $\mathscr{V}_s$. For example, β might be chosen such that the nominal optimal performance $\mathscr{V}_0$ is compromised no more than 10% (when compared to a nominal design with $\beta = 0$, $\mathbf{x}_p = \mathbf{0}$). That is, β could be indexed $\beta = 0, 1, 10, 100$, and so on, to iteratively find a satisfactory value of β to trade performance versus sensitivity reduction.

Exercise 10.8

Given

$$\dot{x} = x + p(u + w),$$
$$y = x,$$
$$z = y + v.$$

Find the LQIS controller associated with the cost function

$$\mathscr{V} = \int_0^\infty \left[y^2 + u^2 + \beta\left(y_p^2 + u_{0_p}^2 \right) \right] dt = \mathscr{V}_0 + \beta \mathscr{V}_s,$$

where $u_0 = Gx$ is the LQI control law when $p \equiv \bar{p} = 1$. Plot $\mathscr{V}_0$ versus $\mathscr{V}_s$ as $\beta = 0, 1, 10, 100$. What value of β would you choose?

Solution Hints: The solution is developed from (10.126) through (10.132), where

$$A_s = \begin{bmatrix} -\sqrt{2} & 0 \\ -\dfrac{1}{\sqrt{2}} & -\sqrt{2} \end{bmatrix}, \qquad B_s = D_s = \begin{pmatrix} 1 \\ 1 \end{pmatrix},$$

$$\begin{matrix} G = -1 - \sqrt{2}, \\ M_s = [1 \quad 0], \end{matrix} \qquad G_p = \frac{1 + \sqrt{2}}{\sqrt{2}}, \quad C_p = 0, \quad C = 1.$$ ■

It is noted that the matrix pair $(\mathbf{A}_s, \mathbf{M}_s)$ in (10.129) is detectable but not observable (hence, $\mathbf{P}_s$ will exist). The pair $(\mathbf{A}_s, \mathbf{D}_s)$ is stabilizable but might not be controllable (hence, $\mathbf{P}_s \geq \mathbf{0}$ or $\mathbf{P}_s > \mathbf{0}$). $(\mathbf{A}_s, \mathbf{C}_s)$ is observable if $(\mathbf{A} + \mathbf{BG}, \mathbf{C})$ is; $(\mathbf{A}_s, \mathbf{B}_s)$ is stabilizable but might not be controllable (hence, $\mathbf{K}_s$ exists). It may also be noted that the LQIS controller is of order $n_x(1 + h)$. This high order is a serious disadvantage which can be overcome by applying any of the model reduction techniques of Chapter 6. For example, a q-Markov COVER of (10.129b) will preserve the RMS values of all sensitivity outputs $(y_p)_i$ and all the correlations with y at $t = 0$, since the matrix $\int_0^\infty y_s y_s^* \, dt = R_0$ is preserved, [10.19].

10.4 Model Error Estimation

The point of view in the previous section is to minimize sensitivity to errors of the type $\mathbf{e}_p$ in (10.19). Errors in model order were not directly treated, except to start with a high-order model and reduce the model and its sensitivity together. This has a certain appeal but the computational burdens are great. Reducing sensitivity is a methodology that also prejudges the model error to be "bad." If the system "changes for the better" so as to yield better than nominal performance, the sensitivity controller LQIS will not recognize this. The control approach of the last section (LQIS) will insist on minimizing the effect of these system changes even if the changes are helpful.

In this section we attempt to estimate a more general model error vector than just the $\mathbf{e}_p$ term considered before. In this way we may get more information about the errors for use in control. We will find that this information also comes with a price, and the resulting controllers are not necessarily better than those of the previous section.

To begin this study, we assume $\mathbf{e}_z = 0$ in (10.18), (10.19). This is without loss of generality since all model errors can be lumped into the $\mathbf{e}_x$ term in (10.18), with a coordinate choice such that $\Delta\mathbf{M} = \mathbf{0}$, $\mathbf{M}_T = \mathbf{0}$. (This is always possible if $\dim \mathbf{x}_R \geq \dim \mathbf{z}$.) Hence, we have the model error system (10.31) with $\mathbf{E}_{10} = \mathbf{I}$, $\mathbf{E}_{01} = \mathbf{0}$. One additional step we take here is to approximate the time-varying term $\mathbf{w} + \mathbf{E}_{10}(\mathbf{e}_t + \mathbf{e}_N) = \mathbf{w} + \mathbf{e}_t$ by the dynamic model

$$\mathbf{w} + \mathbf{e}_t = \mathbf{C}_w \mathbf{x}_w, \qquad \dot{\mathbf{x}}_w = \mathbf{A}_w \mathbf{x}_w, \tag{10.133}$$

where the system excitation is from intial conditions on the state

$$\mathbf{x}_e^* = (\mathbf{x}_R^*, \mathbf{e}_{0x}^*, \mathbf{e}_{0u}^*, \mathbf{e}_{1x}^*, \mathbf{e}_{1u}^*, \ldots, \mathbf{x}_w^*).$$

Hence, our complete model error system (10.31), (10.133) has the form

$$\dot{\mathbf{x}}_e = \mathbf{A}_e \mathbf{x}_e + \mathbf{B}_e \mathbf{u},$$

$$\mathbf{z} = \mathbf{M}_e \mathbf{x}_e.$$

Of course, $\mathbf{x}_e$ must be truncated to a finite dimension. To illustrate the concepts of the section, we truncate $\mathbf{x}_e$ to include only the first four model error vectors $\mathbf{e}_{0x}, \mathbf{e}_{0u}, \mathbf{e}_{1x}, \mathbf{e}_{1u}$ so that

$$\mathbf{x}_e^* = (\mathbf{x}_R^*, \mathbf{e}_{0x}^*, \mathbf{e}_{0u}^*, \mathbf{e}_{1x}, \mathbf{e}_{1u}, \mathbf{x}_w^*),$$

$$\mathbf{A}_e = \begin{bmatrix} \mathbf{A} + \Delta\mathbf{A} & \mathbf{I} & \mathbf{I} & \mathbf{0} & \mathbf{0} & \mathbf{C}_w \\ \mathbf{P}_0 & \mathbf{0} & \mathbf{0} & \mathbf{I} & \mathbf{0} & \mathbf{0} \\ \mathbf{0} & \mathbf{0} & \mathbf{0} & \mathbf{0} & \mathbf{I} & \mathbf{0} \\ \mathbf{P}_1 & \mathbf{0} & \mathbf{0} & \mathbf{0} & \mathbf{0} & \mathbf{0} \\ \mathbf{0} & \mathbf{0} & \mathbf{0} & \mathbf{0} & \mathbf{0} & \mathbf{0} \\ \mathbf{0} & \mathbf{0} & \mathbf{0} & \mathbf{0} & \mathbf{0} & \mathbf{A}_w \end{bmatrix}, \qquad \mathbf{B}_c = \begin{bmatrix} \mathbf{B} + \Delta\mathbf{B} \\ \mathbf{0} \\ \mathbf{Q}_0 \\ \mathbf{0} \\ \mathbf{Q}_1 \\ \mathbf{0} \end{bmatrix}, \tag{10.134}$$

$$\mathbf{M}_e = [\mathbf{M} \;\; \mathbf{0} \;\; \mathbf{0} \;\; \mathbf{0} \;\; \mathbf{0} \;\; \mathbf{0}].$$

We presume a controller of the form

$$\dot{\hat{\mathbf{x}}}_e = \hat{\mathbf{A}}_e \hat{\mathbf{x}}_e + \hat{\mathbf{B}}_e \mathbf{u} + \mathbf{F}_e(\mathbf{z} - \mathbf{M}_e \hat{\mathbf{x}}_e), \qquad \mathbf{u} = \mathbf{G}_e \hat{\mathbf{x}}_e, \tag{10.135a}$$

where the parameters of $\mathbf{A}_e$ are unknown and are estimated by the rule

$$\dot{\hat{\mathbf{A}}}_e = \mathbf{Q}_A^{-1} \mathbf{M}_e^* \mathbf{Q}_z \tilde{\mathbf{z}} \hat{\mathbf{x}}_e^*, \qquad \tilde{\mathbf{z}} \triangleq \mathbf{z} - \mathbf{M}_e \hat{\mathbf{x}}_e,$$

$$\dot{\hat{\mathbf{B}}}_e = \mathbf{Q}_B^{-1} \mathbf{M}_e^* \mathbf{Q}_z \tilde{\mathbf{z}} \mathbf{u}^*. \tag{10.135b}$$

To motivate this choice, consider the dynamics of the closed-loop system, defining

$$\tilde{\mathbf{x}}_e(t) \triangleq \mathbf{x}_e(t) - \hat{\mathbf{x}}_e(t), \quad \tilde{\mathbf{A}}_e(t) \triangleq \mathbf{A}_e - \hat{\mathbf{A}}_e(t), \quad \tilde{\mathbf{B}}_e(t) \triangleq \mathbf{B}_e - \hat{\mathbf{B}}_e(t).$$

Then,

$$\dot{\tilde{\mathbf{A}}}_e(t) = -\dot{\hat{\mathbf{A}}}_e(t) = -\mathbf{Q}_A^{-1}\mathbf{M}_e^*\mathbf{Q}_z\tilde{\mathbf{z}}(t)\hat{\mathbf{x}}_e^*,$$

$$\dot{\tilde{\mathbf{B}}}_e(t) = -\dot{\hat{\mathbf{B}}}_e(t) = -\mathbf{Q}_B^{-1}\mathbf{M}_e^*\mathbf{Q}_z\tilde{\mathbf{z}}(t)\mathbf{u}^*,$$

and

$$\dot{\tilde{\mathbf{x}}}_e = (\mathbf{A}_e - \mathbf{F}_e\mathbf{M}_e)\tilde{\mathbf{x}}_e + (\tilde{\mathbf{A}}_e + \tilde{\mathbf{B}}_e\mathbf{G}_e)\hat{\mathbf{x}}_e, \qquad \tilde{\mathbf{z}} = \mathbf{M}_e\tilde{\mathbf{x}}_e.$$

Using these relationships, the Liapunov function

$$\mathscr{V}(\mathbf{x}_e, t) = \tilde{\mathbf{z}}^*\mathbf{Q}_z\tilde{\mathbf{z}} + \operatorname{tr}\tilde{\mathbf{A}}_e^*\mathbf{Q}_A\tilde{\mathbf{A}}_e + \operatorname{tr}\tilde{\mathbf{B}}_e^*\mathbf{Q}_B\tilde{\mathbf{B}}_e$$

has the derivative

$$\begin{aligned}
\dot{\mathscr{V}}(\mathbf{x}_e, t) &= 2\tilde{\mathbf{z}}^*\mathbf{Q}_z\dot{\tilde{\mathbf{z}}} + 2\operatorname{tr}\dot{\tilde{\mathbf{A}}}_e^*\tilde{\mathbf{A}}_e + 2\operatorname{tr}\dot{\tilde{\mathbf{B}}}_e^*\mathbf{Q}_B\tilde{\mathbf{B}}_e \\
&= 2\tilde{\mathbf{x}}_e^*\mathbf{M}_e^*\mathbf{Q}_z\mathbf{M}_e\left[(\mathbf{A}_e - \mathbf{F}_e\mathbf{M}_e)\tilde{\mathbf{x}}_e + (\tilde{\mathbf{A}}_e + \tilde{\mathbf{B}}_e\mathbf{G}_e)\hat{\mathbf{x}}_e\right] \\
&\quad -2\operatorname{tr}\hat{\mathbf{x}}_e\tilde{\mathbf{x}}_e^*\mathbf{M}_e^*\mathbf{Q}_z\mathbf{M}_e\mathbf{Q}_A^{-1}\mathbf{Q}_A\tilde{\mathbf{A}}_e - 2\operatorname{tr}\mathbf{G}_e\hat{\mathbf{x}}_e\tilde{\mathbf{x}}_e^*\mathbf{M}_e^*\mathbf{Q}_z\mathbf{M}_e\mathbf{Q}_B^{-1}\mathbf{Q}_B\tilde{\mathbf{B}}_e \\
&= \tilde{\mathbf{x}}_e^*\left[\mathbf{M}_e^*\mathbf{Q}_z\mathbf{M}_e(\mathbf{A}_e - \mathbf{F}_e\mathbf{M}_e) + (\mathbf{A}_e - \mathbf{F}_e\mathbf{M}_e)^*\mathbf{M}_e^*\mathbf{Q}_z\mathbf{M}_e\right]\tilde{\mathbf{x}}_e,
\end{aligned}$$

which is negative semidefinite if there exists a matrix $\mathbf{L}_e$ such that

$$\mathbf{M}_e^*\mathbf{Q}_z\mathbf{M}_e(\mathbf{A}_e - \mathbf{F}_e\mathbf{M}_e) + (\mathbf{A}_e - \mathbf{F}_e\mathbf{M}_e)^*\mathbf{M}_e^*\mathbf{Q}_z\mathbf{M}_e + \mathbf{L}_e^*\mathbf{L}_e = \mathbf{0}, \quad (10.136)$$

since in this event

$$\dot{\mathscr{V}}(\mathbf{x}_e, t) = -\tilde{\mathbf{x}}_e^*\mathbf{L}_e^*\mathbf{L}_e\tilde{\mathbf{x}}_e \le 0.$$

Note that the motivation for the choice for the parameter estimation rule (10.135) is to cancel the terms containing the unknowns $\tilde{\mathbf{A}}_e$, $\tilde{\mathbf{B}}_e$ in the expression for $\dot{\mathscr{V}}(\mathbf{x}_e, t)$. Condition (10.136) cannot always be verified since $\mathbf{A}_e$ is unknown.

Theorem 10.9

Let the "state and parameter estimator" defined by

$$\dot{\hat{\mathbf{x}}}_e = \hat{\mathbf{A}}_e\hat{\mathbf{x}}_e + \mathbf{F}_e(\mathbf{z} - \mathbf{M}_e\hat{\mathbf{x}}_e) + \hat{\mathbf{B}}_e\mathbf{u}, \tag{10.137a}$$

$$\dot{\hat{\mathbf{A}}}_e = \mathbf{Q}_A^{-1}\mathbf{M}_e^*\mathbf{Q}_z(\mathbf{z} - \mathbf{M}_e\hat{\mathbf{x}}_e)\hat{\mathbf{x}}_e^*, \tag{10.137b}$$

$$\dot{\hat{\mathbf{B}}}_e = \mathbf{Q}_B^{-1}\mathbf{M}_e^*\mathbf{Q}_z(\mathbf{z} - \mathbf{M}_e\hat{\mathbf{x}}_e)\mathbf{u}^* \tag{10.137c}$$

be used to identify the system

$$\dot{\mathbf{x}}_e = \mathbf{A}_e\mathbf{x}_e + \mathbf{B}_e\mathbf{u}, \qquad \mathbf{z} = \mathbf{M}_e\mathbf{x}_e,$$

where $\mathbf{M}_e$ is known and $\mathbf{A}_e, \mathbf{B}_e$ are completely unknown but constant. Now, if there exists an $\mathbf{F}_e$ such that (10.136) *holds for some $\mathbf{L}_e$, then the function*

$$\mathscr{V}(\mathbf{x}_e, t) = (\mathbf{z} - \mathbf{M}_e\hat{\mathbf{x}}_e)^*\mathbf{Q}_z(\mathbf{z} - \mathbf{M}_e\hat{\mathbf{x}}_e) + \operatorname{tr}(\mathbf{A}_e - \hat{\mathbf{A}}_e)^*\mathbf{Q}_A(\mathbf{A}_e - \hat{\mathbf{A}}_e) + \operatorname{tr}(\mathbf{B}_e - \hat{\mathbf{B}}_e)^*\mathbf{Q}_B(\mathbf{B}_e - \hat{\mathbf{B}}_e)$$

is strictly nonincreasing in the sense that $\dot{\mathscr{V}}(\mathbf{x}_e, t) \le 0$.

Now, to find practical use of Theorem 10.9, we must satisfy (10.136). The matrix $\mathbf{A}_e$ is unknown, hence the existence of an $\mathbf{L}_e$ satisfying (10.136) can only be guaranteed if $\mathbf{M}_e^*\mathbf{Q}_z\mathbf{M}_e = \mathbf{M}^*\mathbf{Q}_z\mathbf{M} > \mathbf{0}$. To see this, note from Theorem 7.3, equation (7.18), that $\mathbf{K}_0 > \mathbf{0}$ from

$$\mathbf{K}_0\mathbf{A}_0 + \mathbf{A}_0^*\mathbf{K}_0 + \mathbf{C}_0^*\mathbf{C}_0 = \mathbf{0}$$

implies that $\mathbf{A}_0$ is stable if $(\mathbf{A}_0, \mathbf{C}_0)$ is observable. Letting $\mathbf{K}_0 \triangleq \mathbf{M}^*\mathbf{Q}_z\mathbf{M}$, $\mathbf{A}_0 \triangleq \mathbf{A}_e - \mathbf{F}_e\mathbf{M}_e$, $\mathbf{C}_0 = \mathbf{L}_e = \mathbf{M}_e$, we see that $(\mathbf{A}_0, \mathbf{C}_0)$ can only be observable (for arbitrary $\mathbf{A}_e$) if $\mathbf{C}_0 = \mathbf{L}_e = \mathbf{M}_e$ is a nonsingular matrix. The choice $\mathbf{L}_e \triangleq \mathbf{M}_e$ yields an observable pair, $(\mathbf{A}_e - \mathbf{F}_e\mathbf{M}_e, \mathbf{M}_e)$, for arbitrary $\mathbf{A}_e$ if and only if $\mathbf{M}_e$ is nonsingular. Note from the construction of $\mathbf{M}_e$ in (10.134) that $\mathbf{M}_e$ can be nonsingular only if $\mathbf{M}_e$ is $(\dim x_R) \times (\dim x_R)$. Hence, the "state and parameter estimator" of the form (10.137) can only accommodate parameter errors (accommodation of model order errors $\mathbf{e}_0$ are not guaranteed). This yields the following result.

Corollary to Theorem 10.19

Let the "state and parameter estimator"

$$\dot{\hat{\mathbf{x}}} = \hat{\mathbf{A}}\hat{\mathbf{x}} + \mathbf{F}(\mathbf{z} - \mathbf{M}\hat{\mathbf{x}}) + \hat{\mathbf{B}}\mathbf{u}, \tag{10.138a}$$

$$\dot{\hat{\mathbf{A}}} = \mathbf{Q}_A^{-1}\mathbf{M}^*\mathbf{Q}_z(\mathbf{z} - \mathbf{M}\hat{\mathbf{x}})\hat{\mathbf{x}}^*, \qquad \mathbf{Q}_A > \mathbf{0}, \quad \mathbf{Q}_z > \mathbf{0}, \tag{10.138b}$$

$$\dot{\hat{\mathbf{B}}} = \mathbf{Q}_B^{-1}\mathbf{M}^*\mathbf{Q}_z(\mathbf{z} - \mathbf{M}\hat{\mathbf{x}})\mathbf{u}^*, \qquad \mathbf{Q}_B > \mathbf{0}, \tag{10.138c}$$

$$\mathbf{F} = \hat{\mathbf{A}}\mathbf{M}^{-1} + \tfrac{1}{2}\mathbf{M}^{-1}\mathbf{Q}_z^{-1}(\mathbf{I} + \mathbf{S}), \qquad (\mathbf{S} = -\mathbf{S}^*, \text{arbitrary}), \tag{10.138d}$$

be used to identify the system

$$\dot{\mathbf{x}} = \mathbf{A}\mathbf{x} + \mathbf{B}\mathbf{u}, \qquad \mathbf{z} = \mathbf{M}\mathbf{x},$$

where **A** *and* **B** *are unknown but* **M** *is known and nonsingular. Then, the scalar function*

$$\mathscr{V}(\mathbf{x}, t) \triangleq (\mathbf{z} - \mathbf{M}\hat{\mathbf{x}})^*\mathbf{Q}_z(\mathbf{z} - \mathbf{M}\hat{\mathbf{x}}) + \operatorname{tr}(\mathbf{A} - \hat{\mathbf{A}})^*\mathbf{Q}_A(\mathbf{A} - \hat{\mathbf{A}}) + \operatorname{tr}(\mathbf{B} - \hat{\mathbf{B}})^*\mathbf{Q}_B(\mathbf{B} - \hat{\mathbf{B}})$$

has the derivative

$$\dot{\mathscr{V}}(\mathbf{x}, t) = \tilde{\mathbf{x}}^*\big[\mathbf{M}^*\mathbf{Q}_z\mathbf{M}\tilde{\mathbf{A}} + \tilde{\mathbf{A}}^*\mathbf{M}^*\mathbf{Q}_z\mathbf{M}\big]\tilde{\mathbf{x}} - \tilde{\mathbf{x}}^*\mathbf{M}^*\mathbf{M}\tilde{\mathbf{x}}. \tag{10.139}$$

Proof: The choice of $\mathscr{V}(\mathbf{x}, t)$ follows from Theorem 10.9. The choice of **F** follows from Theorem 7.14, whose proof is repeated here for clarity. We seek an **F** such that

$$\mathbf{M}^*\mathbf{Q}_z\mathbf{M}(\mathbf{A} - \mathbf{F}\mathbf{M}) + (\mathbf{A} - \mathbf{F}\mathbf{M})^*\mathbf{M}^*\mathbf{Q}_z\mathbf{M} + \mathbf{M}^*\mathbf{M} = \mathbf{0}.$$

All solutions are characterized by

$$\mathbf{M}^*\mathbf{Q}_z\mathbf{M}(\mathbf{A} - \mathbf{F}\mathbf{M}) = -\tfrac{1}{2}(\mathbf{M}^*\mathbf{M} + \mathscr{S}), \tag{10.140}$$

where $\mathscr{S} = -\mathscr{S}^*$ is an arbitrary skew-symmetric matrix. Now, since **M** is nonsingular, (10.140) can be solved immediately to obtain

$$\mathbf{F} = \mathbf{A}\mathbf{M}^{-1} + \tfrac{1}{2}\mathbf{M}^{-1}\mathbf{Q}_z^{-1}(\mathbf{I} + \mathbf{S}), \qquad \mathbf{S} \triangleq \mathbf{M}^{-*}\mathscr{S}\mathbf{M}^{-1}. \tag{10.141}$$

Now, since **A** is not known, the estimate of **A** from (10.138b) is used in the formula (10.141). Now,

$$\dot{\mathscr{V}}(\mathbf{x}, \mathbf{t}) = \tilde{\mathbf{x}}\big[\mathbf{M}^*\mathbf{Q}_z\mathbf{M}(\mathbf{A} - \mathbf{F}\mathbf{M}) + (\mathbf{A} - \mathbf{F}\mathbf{M})^*\mathbf{M}^*\mathbf{Q}_z\mathbf{M}\big]\tilde{\mathbf{x}},$$

but **F** satisfies

$$\mathbf{M}^*\mathbf{Q}_z\mathbf{M}(\hat{\mathbf{A}} - \mathbf{F}\mathbf{M}) + (\hat{\mathbf{A}} - \mathbf{F}\mathbf{M})^*\mathbf{M}^*\mathbf{Q}_z\mathbf{M} + \mathbf{M}^*\mathbf{M} = \mathbf{0}.$$

Hence, use $\mathbf{A} = \tilde{\mathbf{A}} + \hat{\mathbf{A}}$ to see that

$$\begin{aligned}\mathbf{M}^*\mathbf{Q}_z\mathbf{M}(\mathbf{A} - \mathbf{F}\mathbf{M}) &+ (\mathbf{A} - \mathbf{F}\mathbf{M})^*\mathbf{M}^*\mathbf{Q}_z\mathbf{M} \\ &= \mathbf{M}^*\mathbf{Q}_z\mathbf{M}\tilde{\mathbf{A}} + \tilde{\mathbf{A}}^*\mathbf{M}^*\mathbf{Q}_z\mathbf{M} - \mathbf{M}^*\mathbf{M}.\end{aligned}$$

Equation (10.139) then follows. ■

Note that the above results remain independent of the choice of control **u**. Hence, there exists a certain separation between the *identification* and *control* problems when there are no errors of model order. Otherwise, the two problems are not separable.

The parameter estimation in this section can be used to construct indirect adaptive controls, [10.14] and [10.15], but a rigorous study of adaptive control is beyond the scope of this introductory book. Suffice it to say that such adaptive controls can accommodate parameter errors but generally cannot accommodate errors in model order. In the following section we accommodate a class of model errors which lie in the column space of the **B** matrix $\mathbf{e}_x = \mathbf{B}\mathbf{e}(\mathbf{x}, t)$ for some $\mathbf{e}(\mathbf{x}, t)$. Such errors do not include the class of all parameter errors in **A** (since $\Delta\mathbf{A}$ might not be equal to $\Delta\mathbf{A} = \mathbf{BN}$ for any **N**) but can include errors in disturbances and errors in model order if they satisfy $\mathbf{e}_x = \mathbf{B}\mathbf{e}(\mathbf{x}, t)$.

10.5 Compensation for a Class of Model Errors

Now, suppose we consider the special class of errors $\mathbf{e}_x(\mathbf{x}_R, \mathbf{u}, t)$ such that

$$\dot{\mathbf{x}}_R = \mathbf{A}_R\mathbf{x}_R + \mathbf{B}_R\mathbf{u} + \mathbf{e}_x(\mathbf{x}_R, \mathbf{u}, t),$$

$$\mathbf{z} = \mathbf{M}_R\mathbf{x}_R$$

$$\mathbf{e}_x(\mathbf{x}_R, \mathbf{u}, t) = \mathbf{B}_R\mathbf{e}(\mathbf{x}_R, \mathbf{u}, t),$$

$$\|\mathbf{e}(\mathbf{x}_R, \mathbf{u}, t)\| \le \beta_1\|\mathbf{x}_R\| + \beta_2\|\mathbf{u}\|, \qquad \beta_2 < 1, \tag{10.142}$$

for some known constants β_1, β_2, and where $\mathbf{A}_R$ is assumed stable (otherwise, we break **u** into two parts $\mathbf{u} = \mathbf{u}_1 + \mathbf{u}_2$ and begin our design with $\mathbf{u}_1 = \mathbf{G}_1\mathbf{x}_R$ such that $\mathbf{A}_R + \mathbf{B}_R\mathbf{G}_1$ is stable, and we seek to find $\mathbf{u}_2$). Since $\mathbf{A}_R$ is stable, we know from Theorem 7.3 that $\mathbf{P} > \mathbf{0}$ is the solution of

$$\mathbf{P}\mathbf{A}_R + \mathbf{A}_R^*\mathbf{P} + \mathbf{Q} = \mathbf{0}, \qquad \mathbf{Q} > \mathbf{0},$$

for any $\mathbf{Q} > \mathbf{0}$. Hence, the derivative of the Liapunov function

$$\mathscr{V}(\mathbf{x}_R) = \mathbf{x}_R^*\mathbf{P}\mathbf{x}_R$$

yields

$$\begin{aligned}\dot{\mathscr{V}}(\mathbf{x}_R) &= 2\mathbf{x}_R^*\mathbf{P}[\mathbf{A}_R\mathbf{x}_R + \mathbf{B}_R(\mathbf{u} + \mathbf{e})] \\ &= \mathbf{x}_R^*[\mathbf{P}\mathbf{A}_R + \mathbf{A}_R^*\mathbf{P}]\mathbf{x}_R + 2\mathbf{x}_R^*\mathbf{P}\mathbf{B}_R(\mathbf{u} + \mathbf{e}),\end{aligned}$$

where $\mathbf{P}\mathbf{A}_R + \mathbf{A}_R^*\mathbf{P} = -\mathbf{Q}$, and a control of the form $\mathbf{u} = -\gamma\mathbf{B}_R^*\mathbf{P}\mathbf{x}_R$ yields

$$\dot{\mathscr{V}}(\mathbf{x}_R) = -\mathbf{x}_R^*\mathbf{Q}\mathbf{x}_R - 2\gamma(\mathbf{x}_R^*\mathbf{P}\mathbf{B}_R\mathbf{B}_R^*\mathbf{P}\mathbf{x}_R) + 2\mathbf{x}_R^*\mathbf{P}\mathbf{B}_R\mathbf{e}.$$

From the Cauchy–Schwarz inequality

$$|2\mathbf{x}_R^*\mathbf{P}\mathbf{B}_R\mathbf{e}| \le 2\|\mathbf{x}_R^*\mathbf{P}\mathbf{B}_R\|\,\|\mathbf{e}\|,$$

which leads to these steps, using the special assumption (10.142),

$$\begin{aligned}\dot{\mathscr{V}}(\mathbf{x}_R) &= -\|\mathbf{x}_R\|_{\mathbf{Q}}^2 - 2\gamma\|\mathbf{x}_R^*\mathbf{P}\mathbf{B}_R\|^2 + 2\mathbf{x}_R^*\mathbf{P}\mathbf{B}_R\mathbf{e}\\ &\le -\|\mathbf{x}_R\|_{\mathbf{Q}}^2 - 2\gamma\|\mathbf{x}_R^*\mathbf{P}\mathbf{B}_R\|^2 + 2\|\mathbf{x}_R^*\mathbf{P}\mathbf{B}_R\|\,\|\mathbf{e}\|\\ &\le -\|\mathbf{x}_R\|_{\mathbf{Q}}^2 - 2\|\mathbf{x}_R^*\mathbf{P}\mathbf{B}_R\|(\gamma\|\mathbf{x}_R^*\mathbf{P}\mathbf{B}_R\| - \beta_1\|\mathbf{x}_R\| - \beta_2\|\mathbf{u}\|)\\ &= -\|\mathbf{x}_R\|_{\mathbf{Q}}^2 - 2\|\mathbf{x}_R^*\mathbf{P}\mathbf{B}_R\|(\gamma\|\mathbf{x}_R^*\mathbf{P}\mathbf{B}_R\| - \beta_1\|\mathbf{x}_R\| - \beta_2\gamma\|\mathbf{x}_R^*\mathbf{P}\mathbf{B}_R\|).\end{aligned}$$

Let ($\lambda_1 \triangleq$ min eigenvalue of $\mathbf{Q}$). Then, note that

$$\mathbf{x}_R^*\mathbf{Q}\mathbf{x}_R = \sum_{i=1}^{n_x} q_i^2\lambda_i \ge \sum_{i=1}^{n_x} q_i^2\lambda_1 = \|\mathbf{q}\|^2\lambda_1,$$

where the spectral decomposition of $\mathbf{Q}$ is $\mathbf{Q} = \mathbf{E}\Lambda\mathbf{E}^*$, where $\mathbf{E}^*\mathbf{E} = \mathbf{I}$, and $\mathbf{q} \triangleq \mathbf{E}^*\mathbf{x}_R$. Hence, we have $\|\mathbf{q}\| = \|\mathbf{x}_R\|$, and $\|\mathbf{x}_R\|_{\mathbf{Q}}^2 \ge \|\mathbf{q}\|^2\lambda_1$ allows us to write

$$\begin{aligned}\dot{\mathscr{V}}(\mathbf{x}_R) &\le -\|\mathbf{x}_R\|^2\lambda_1 + 2\beta_1\|\mathbf{x}_R^*\mathbf{P}\mathbf{B}_R\|\cdot\|\mathbf{x}_R\| - \|\mathbf{x}_R^*\mathbf{P}\mathbf{B}_R\|^2\\ &= -\begin{bmatrix}\|\mathbf{x}_R\|\\ \|\mathbf{x}_R^*\mathbf{P}\mathbf{B}_R\|\end{bmatrix}^T\begin{bmatrix}\lambda_1 & -\beta_1\\ -\beta_1 & 2\gamma(1-\beta_2)\end{bmatrix}\begin{bmatrix}\|\mathbf{x}_R\|\\ \|\mathbf{x}_R^*\mathbf{P}\mathbf{B}_R\|\end{bmatrix}. \qquad (10.143)\end{aligned}$$

Hence, $\dot{\mathscr{V}}(\mathbf{x}_R) < 0$ if the square 2×2 matrix in (10.143) is positive definite:

(i) $\lambda_1 > 0$,
(ii) $2\gamma(1 - \beta_2) > 0$,
(iii) $2\gamma\lambda_1(1 - \beta_2) - \beta_1^2 > 0$.

Now, (i) holds since $\mathbf{Q} > \mathbf{0}$, (ii) holds by virtue of assumptions $\gamma > 0$, $\beta_2 < 1$, and (iii) holds by restricting γ as follows:

$$\gamma > \frac{\beta_1^2}{2\lambda_1(1 - \beta_2)}.$$

Hence, we have the following result:

Theorem 10.10
Suppose $\mathbf{A}_R$ *is stable and* $\mathbf{P} > \mathbf{0}$ *is the solution of*

$$\mathbf{0} = \mathbf{P}\mathbf{A}_R + \mathbf{A}_R^*\mathbf{P} + \mathbf{Q}, \qquad (10.144)$$

where $\mathbf{Q} > \mathbf{0}$ *and* λ_1 = min *eigenvalue of* $\mathbf{Q}$. *Suppose also that* β_1 *and* β_2 *are known such that*

$$\dot{\mathbf{x}}_R = \mathbf{A}_R\mathbf{x}_R + \mathbf{B}_R(\mathbf{u} + \mathbf{e}), \tag{10.145}$$

$$\|\mathbf{e}\| < \beta_1\|\mathbf{x}_R\| + \beta_2\|\mathbf{u}\|, \qquad \beta_2 < 1. \tag{10.146}$$

Then, the control

$$\mathbf{u} = -\gamma\mathbf{B}_R^*\mathbf{P}\mathbf{x}_R, \qquad \gamma > \frac{\beta_1^2}{2\lambda_1(1-\beta_2)} \tag{10.147}$$

stabilizes (10.145) *for arbitrary* $\mathbf{e}(\mathbf{x}_R, \mathbf{u}, t)$ *satisfying* (10.146).

EXAMPLE 10.7

Find a linear state feedback control that will stabilize

$$\dot{x} = -x + bu + e(x, u, t),$$

where

$$e(x, u, t) = \Delta_1(t)x + \Delta_2(t)u,$$

$$\max_t \Delta_1(t) = \bar{\Delta}_1, \qquad \max_t \Delta_2(t) = \bar{\Delta}_2 < 1.$$

Solution: Using Theorem 10.10, we may find β_1, β_2 from

$$\beta_1 = \bar{\Delta}_1, \qquad \beta_2 = \bar{\Delta}_2.$$

Then, the stabilizing controller is

$$u = -\gamma b\frac{Q}{2}x, \qquad \gamma > \frac{\bar{\Delta}_1^2}{2Q(1-\bar{\Delta}_2)}$$

for any choice of $Q > 0$. By choosing

$$\gamma = \frac{\bar{\Delta}_1^2}{2Q(1-\bar{\Delta}_2)} + \frac{1}{Q},$$

the reader should show that

$$u = -\frac{b}{2}\left[1 + \frac{\bar{\Delta}_1^2}{2(1-\bar{\Delta}_2)}\right]x$$

stabilizes this system. ■

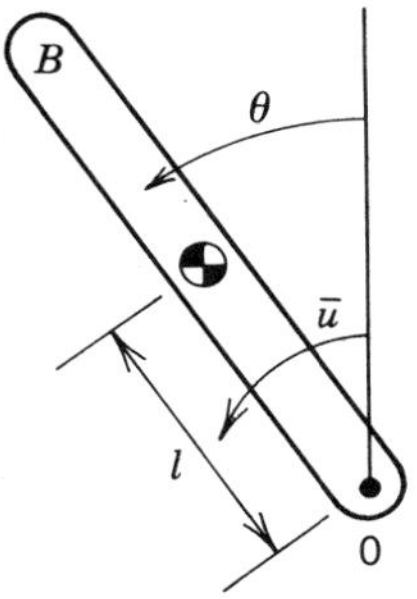

Figure 10.11 Inverted Pendulum

EXAMPLE 10.8 (Corless, [10.16])

In Fig. 10.11, B is constrained to rotate about a fixed horizontal axis through 0; $\bar{u}$ is a control torque acting on B; $m \triangleq$ mass of B; $I \triangleq$ moment of inertia of B about its axis of rotation.

Letting $x_1 \triangleq \theta$ and $x_2 \triangleq \dot{\theta}$, the motion of B is described by

$$\dot{x}_1 = x_2,$$

$$\dot{x}_2 = a \sin x_1 + b\bar{u},$$

where

$$a \triangleq I^{-1}mgl, \qquad b \triangleq I^{-1},$$

and g is the Earth's gravitational acceleration constant.

Letting

$$\bar{u} = -l_1 x_1 - l_2 x_2 + u,$$

with $l_1, l_2 > 0$, the system is described by

$$\dot{x}_1 = x_2,$$

$$\dot{x}_2 = -a_1 x_1 - a_2 x_2 + b[u + d \sin x_1],$$

where

$$a_1 \triangleq bl_1, \qquad a_2 \triangleq bl_2,$$

and

$$d \triangleq b^{-1}a = mgl.$$

The above description is in the form of (10.145) with

$$A_R = \begin{pmatrix} 0 & 1 \\ -a_1 & -a_2 \end{pmatrix}, \quad B_R = \begin{pmatrix} 0 \\ b \end{pmatrix}, \quad e = d \sin x_1.$$

Thus, (10.146) is satisfied with $\beta_2 = 0$ and with β_1 being any upper bound on d. Choosing

$$Q = \begin{pmatrix} 2 & 0 \\ 0 & 2 \end{pmatrix},$$

solving (10.144), and utilizing (10.147) yields

$$P = \begin{bmatrix} \dfrac{a_1(a_1+1)+a_2^2}{a_1 a_2} & a_1^{-1} \\ a_1^{-1} & (1+a_1^{-1})a_2^{-1} \end{bmatrix}$$

$$u = -\gamma(k_1 x_1 + k_2 x_2),$$

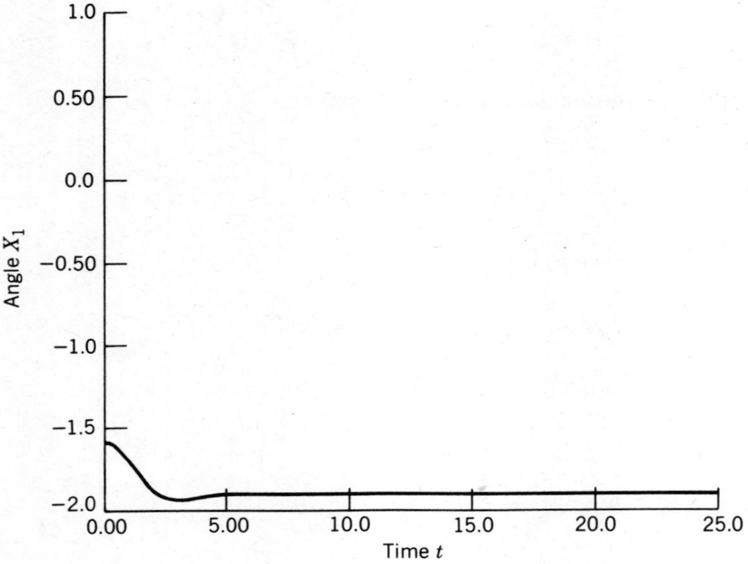

(*a*)

Figure 10.12a *Case I: Pendulum Angle Without Model Error Control*

where

$$k_1 \triangleq l_1^{-1}, \qquad k_2 \triangleq l_2^{-1}(1 + b^{-1}l_1^{-1})$$

and

$$\gamma > \tfrac{1}{4}\beta_1^2 \qquad \beta_1 \triangleq \max\left(\frac{a}{b}\right) = \max(mgl)$$

Simulations

$$a = 2, \quad b = 1, \quad l_1 = 1, \quad l_2 = \sqrt{2}.$$

Initial conditions:

$$x_1(0) = -\frac{\pi}{2}, \qquad x_2(0) = 0.$$

Case I: $\gamma = 0$ (see Fig. 10.12).
Case II: $\gamma = 2$ (see Fig. 10.13).

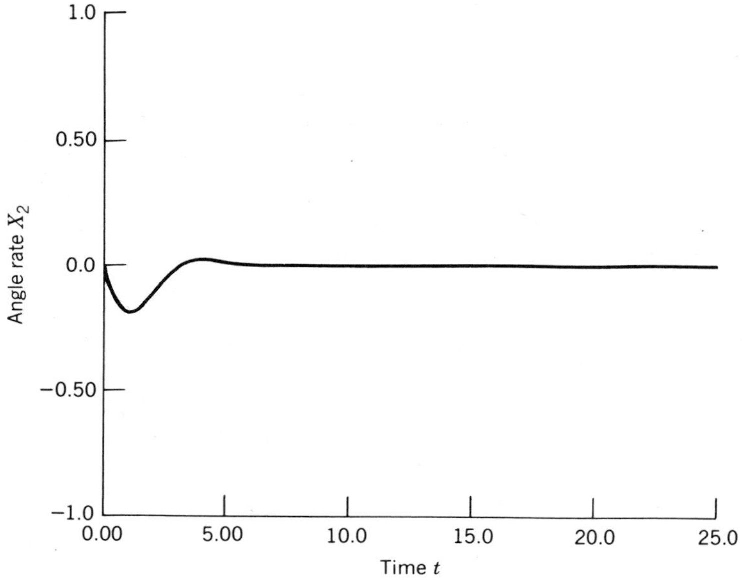

(*b*)

Figure 10.12b *Case I: Pendulum Rate Without Model Error Control*

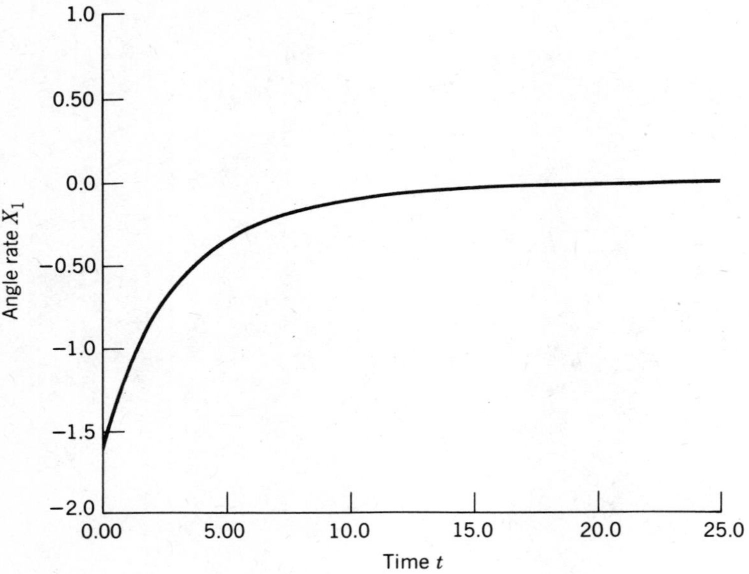

(*a*)

Figure 10.13a *Case II: Pendulum Angle With Model Error Control*

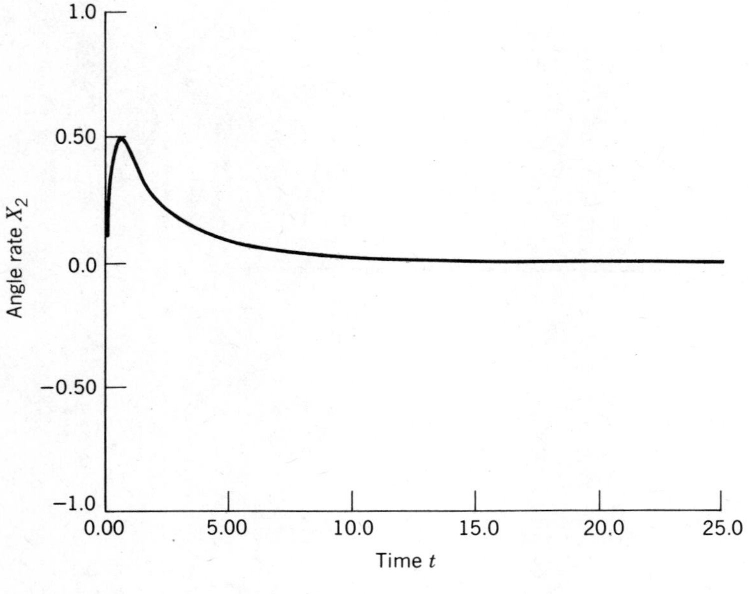

(*b*)

Figure 10.13b *Case II: Pendulum Rate With Model Error Control*

Note from case II that the controller (10.147) stabilizes the pendulum in the inverted position in the presence of modeling errors $e = mgl \sin x_1$, whereas the standard state feedback approach (case I) does not.

Closure

This chapter categorizes model errors into four classes: parameter errors, errors in model error, disturbance errors, and neglected nonlinearities. In general, because there is no control theory that can cope with the simultaneous presence of all four types of errors, certain restrictions on the assumed errors allow the use of special theories.

The root perturbations of Section 10.2 are taken from ref. [10.5], the root sensitivity of Section 10.3.1 from [10.17], and the sensitivity model reduction from ref. [10.19]. The stability margins of Section 10.1.3 have very conservative restrictions on the size of the perturbations. Proofs in this section were motivated by refs. [10.6], [10.7] and [10.8]. The controller of Section 10.5 permits large errors if they lie in the column space of the input matrix. The proof in this section is taken from ref. [10.16]. Recommended additional reading for control in the presence of model uncertainty includes [10.19] through [10.26].

References

10.1 D. D. Siljak, *Large-Scale Dynamic Systems, Stability and Structure*. North-Holland, 1978.

10.2 M. J. Balas, "Toward a More Practical Control Theory for Distributed Parameter Systems," *Control and Dynamic Systems*, 361–421. Academic Press, New York, 1900.

10.3 F. H. Raven, *Automatic Control Engineering*, McGraw Hill, New York, 1978.

10.4 R. E. Skelton and D. H. Owens, eds., *Model Error Concepts and Compensation*. Pergamon Press, New York, 1986.

10.5 R. E. Skelton, "Control Design of Flexible Spacecraft," *Theory and Applications of Optimal Control in Aerospace Systems*, J. P. Kant, ed., AGARD-AG-251, NATO publication, 7 Rue Ancelle, 92200 Neurilly-Sur-Seine, France.

10.6 K. Glover, "All Optimal Hankel Norm Approximations of Linear Multivariable Systems and Their L^∞ Error Bound," *Int. J. Control*, 39, 1115–1193 (1984).

10.7 D. Enns, "Model Reduction for Control Design," Ph.D. Thesis, 1984, EE Dept. Stanford University, Stanford, California.

10.8 Yi Liu and B. D. O. Anderson, "Controller Reduction via Stable Factorization and Balancing," *Int. J. Control*, 1986, 44, pp. 507–531.

10.9 J. C. Doyle, "Quantitative Feedback Theory and Robust Control," *Proceedings ACC*, Seattle, 664–665 (1986).

10.10 G. H. Goloub, C. F. Van Loan, *Matrix Computations*. Johns Hopkins Press, Baltimore, 1983.

10.11 C. G. J. Jacobi, *Crelle's Journal für die Reine und Angewandte Mathematik*, Vol. 30. DeGruyter, Berlin, 1846, pp. 51–95.

10.12 C. G. Cullen, *Matrices and Linear Transformations*. Addison-Wesley, Reading, Mass., 1967.

10.13 E. G. Gilbert, "Conditions for Minimizing the Norm Sensitivity of Characteristic Roots," submitted for publication in *IEEE Trans. Autom. Control*, 1983.

10.14 K. J. Aström and B. Wittenmark, "On Self-Tuning Regulators," *Automatica*, Vol. 9, pp. 195–199, 1973.

10.15 G. C. Goodwin, P. J. Ramadge, and P. E. Caines, "Discrete-Time Multivariable Adaptive Control," *IEEE Trans. Auto. Control*, Vol. AC-25, pp. 449–456, 1980.

10.16 B. R. Barmish, M. Corless, G. Leitmann, "A New Class of Stabilizing Controllers for Uncertain Dynamical Systems," *SIAM J. Control*, 21(2), 246–255 (1983).

10.17 R. E. Skelton and D. A. Wagie, "Minimal Root Sensitivity in Linear Systems," *J. Guidance, Control, Dynamics*, 7(5), 570–574 (Oct. 1984).

10.18 A. Bryson and Y. C. Ho, *Applied Optimal Control*. Hemisphere Publishing, Washington, D.C., 1975.

10.19 D. A. Wagie and R. E. Skelton, "Model Reduction and Controller Synthesis in the Prescence of Parameter Uncertainty," *Automatica*, 22(3), 295–308 (1986).

10.20 R. K. Yedavalli, "Perturbation Bounds for Robust Stability in Linear State Space Models," *Int. J. Control* Vol. 42, pp. 1507–1517, Dec. 1985.

10.21 P. C. Byrne and M. Burke, "Optimization with Trajectory Sensitivity Considerations," *IEEE Trans Autom. Control*, AC-21, 282–283 (1976).

10.22 A. G. J. MacFarlane and Y. S. Hung, "A Quasi-Classical Approach to Multivariable Feedback Systems Design," *Proceedings of the Second IFAC Symposium on Computer Aided Design of Multivariable Technological Systems*, West Lafayette, Indiana, Sept. 1982, pp. 39–48.

10.23 R. V. Patel and M. Toda, "Quantitative Measures of Robustness for Multivariable Systems," *Proceedings of the 1980 Joint Autom. Control Conf.*, San Francisco, California, August 1980.

10.24 I. Postlethwaite, "Sensitivity of the Characteristic Gain Loci," *Proceedings of the Second IFAC Symposium on Computer-Aided Design of Multivariable Technological Systems*, West Lafayette, Indiana, September 1982, pp. 153–158.

10.25 J. E. Van Ness, J. M. Boyle, and F. P. Imad, "Sensitivities of Large, Multiple-Loop Control Systems," *IEEE Trans. Autom. Control*, AC-10, 308–315 (1965).

10.26 L. Ljung and T. Söderström, *Theory and Practice of Recursive Identification*. MIT Press, Cambridge, MA, 1987.

APPENDIX A

Axiomatic Definition of a Linear Space

We wish to describe the algebraic rules for manipulation of vectors such as $\mathbf{x}$ in equation (2.14), $\mathbf{a}_i$ in (2.5), or h_i in (2.6). It can be verified that such vectors obey the following rules of algebra, called the Axiomatic Definition of a Linear Vector Space:

1. For any two vectors $\mathbf{x} \in \mathscr{C}^n$, $\mathbf{y} \in \mathscr{C}^n$, $\mathbf{x} + \mathbf{y} = \mathbf{v}$, where $\mathbf{v} \in \mathscr{C}^n$.
2. Addition is commutative $\mathbf{x} + \mathbf{y} = \mathbf{y} + \mathbf{x}$.
3. Addition is associative, $(\mathbf{x} + \mathbf{y}) + \mathbf{v} = \mathbf{x} + (\mathbf{y} + \mathbf{v})$.
4. There is a zero vector $\mathbf{0} \in \mathscr{C}^n$ satisfying $\mathbf{x} + \mathbf{0} = \mathbf{0} + \mathbf{x} = \mathbf{x}$.
5. For every $\mathbf{x} \in \mathscr{C}^n$, there exists a unique vector $\mathbf{y} \in \mathscr{C}^n$ such that $\mathbf{x} + \mathbf{y} = \mathbf{0}$.
6. For every $\mathbf{x} \in \mathscr{C}^n$ and for every scalar $\alpha \in \mathscr{C}^1$, the product is $\alpha\mathbf{x} \in \mathscr{C}^n$.
7. For every $\alpha \in \mathscr{C}^1$ and $\beta \in \mathscr{C}^1$ and $\mathbf{x} \in \mathscr{C}^n$, $\alpha(\beta\mathbf{x}) = (\alpha\beta)\mathbf{x}$.
8. Multiplication by scalars is distributive, $(\alpha + \beta)\mathbf{x} = \alpha\mathbf{x} + \beta\mathbf{x}$ and $\alpha(\mathbf{x} + \mathbf{y}) = \alpha\mathbf{x} + \alpha\mathbf{y}$.

The vectors $\mathbf{x}(t)$ in (2.14), $\mathbf{a}_i$ in (2.5), and $\mathbf{b}_i$ in (2.6) can have real or complex values, and the above rules 1 through 8 have been written to include the complex case. It is convenient to think of the vector $\mathbf{x} \in \mathscr{C}^n$ as a point in an n-dimensional space. (Of course, one cannot *visualize* such a space if $n > 3$.) The notation $\mathscr{C}^n$ is used to denote a complex n-dimensional linear vector space, and $\mathscr{R}^n$ is used to denote a real n-dimensional linear vector space. The above axioms 1 through 8 provide a definition of a linear space [2.3]. We consider the space of all rational functions in this book. The n-dimensional linear vector spaces $\mathscr{C}^n$ and $\mathscr{R}^n$ are special cases of the space of all rational functions.

APPENDIX B

The Four Fundamental Subspaces of Matrix Theory

The geometrical ideas of linear vector spaces have led to the concepts of "spanning a space" and a "basis for a space." The idea of this section is to define four important *subspaces* which are useful. The entire linear vector space of a specific problem can be decomposed into the sum of these subspaces.

Definition B.1

The "column space" of **A** *is the space spanned by the columns of* **A**. *The "column space" of* **A** *is also called the "range space" of* **A** *and is denoted by* $\mathscr{R}[\mathbf{A}]$. *The "row space" of* **A** *is the space spanned by the rows of* **A**.

EXAMPLE B.1

Determine whether $\mathbf{y} = \begin{pmatrix} 1 \\ -1 \end{pmatrix}$ lies in the column space of

$$\mathbf{A} = \begin{bmatrix} 1 & -1 & -3 \\ 0 & 10 & 0 \end{bmatrix}.$$

Solution: The vector **y** lies in the column space of **A** if **y** can be expressed as a linear combination of the columns of **A**. In other words,

$$\mathbf{y} = \begin{pmatrix} 1 \\ -1 \end{pmatrix} \stackrel{?}{=} \begin{pmatrix} 1 \\ 0 \end{pmatrix} x_1 + \begin{pmatrix} -1 \\ 10 \end{pmatrix} x_2 + \begin{pmatrix} -3 \\ 0 \end{pmatrix} x_3 \qquad \text{for some } x_1, x_2, x_3.$$

In matrix form,

$$\mathbf{Ax} = \mathbf{y} \qquad \text{form some } \mathbf{x}.$$

Since the columns of **A** span the entire two dimensional space, the answer is *yes* for *any y*. In the present example, one can choose, for instance, $x_1 = 9/10$, $x_2 = -1/10$, $x_3 = 0$. ■

EXAMPLE B.2

Does the vector $\mathbf{x} = \begin{pmatrix} 1 \\ 10 \\ 0 \end{pmatrix}$ lie within the row space of the **A** in Example B.1?

Solution: Note from the definition of row space that the row space of **A** is the column space of $\mathbf{A}^T$, denoted by $\mathscr{R}[A^T]$. For complex matrices, we write $[A^*]$. Thus, the question reduces to "Does **x** lie in the space spanned by the rows of **A**?" or "$\mathbf{A}^*\boldsymbol{\beta} = \mathbf{x}$ for some $\boldsymbol{\beta}$?"

$$\begin{bmatrix} 1 & 0 \\ -1 & 10 \\ -3 & 0 \end{bmatrix} \begin{pmatrix} \beta_1 \\ \beta_2 \end{pmatrix} \overset{?}{=} \begin{pmatrix} 1 \\ 10 \\ 0 \end{pmatrix}.$$

In three-dimensional space the row space of **A** is the plane spanned by the two vectors $\begin{pmatrix} 1 \\ -1 \\ -3 \end{pmatrix}$ and $\begin{pmatrix} 0 \\ 10 \\ 0 \end{pmatrix}$. The vector $\begin{pmatrix} 1 \\ 10 \\ 0 \end{pmatrix}$ is not in this plane. Hence, the answer is *no*. See from the scalar equations,

$$\beta_1 = 1,$$

$$-\beta_1 + 10\beta_2 = 10,$$

$$-3\beta_1 = 0,$$

that the first and third equations are contradictory. ■

Since the column rank of a matrix is the dimension of the space spanned by the columns and the row rank is the dimension of the space spanned by the rows, it is clear from Theorem 2.1 that the spaces $\mathscr{R}[\mathbf{A}]$ and $\mathscr{R}[\mathbf{A}^*]$ have the same dimension $r = \operatorname{rank} \mathbf{A}$.

Definition B.2

The "right null space" of **A** *is the space spanned by all vectors* **x** *that satisfy* $\mathbf{Ax} = \mathbf{0}$, *and is denoted* $\mathscr{N}[\mathbf{A}]$. *The right null space of* **A** *is also called the "kernel" of* **A**. *The "left null space" of* **A** *is the space spanned by all vectors* **y** *satisfying* $\mathbf{y}^*\mathbf{A} = \mathbf{0}$. *This space is denoted* $\mathscr{N}[\mathbf{A}^*]$, *since it is also characterized by* $\mathbf{A}^*\mathbf{y} = \mathbf{0}$.

EXAMPLE B.3

Does $\mathbf{x} = \begin{pmatrix} 1 \\ 10 \\ 0 \end{pmatrix}$ lie in the right null space of the **A** in Example B.1?

Solution: The question reduces to

$$\mathbf{Ax} \stackrel{?}{=} \mathbf{0}.$$

Specifically,

$$\begin{bmatrix} 1 & -1 & -3 \\ 0 & 10 & 0 \end{bmatrix}\begin{pmatrix} 1 \\ 10 \\ 0 \end{pmatrix} = \begin{pmatrix} -9 \\ 100 \end{pmatrix} \neq \begin{pmatrix} 0 \\ 0 \end{pmatrix},$$

so the answer is *no*. ■

The dimensions of the four subspaces $\mathscr{R}[\mathbf{A}]$, $\mathscr{R}[\mathbf{A}^*]$, $\mathscr{N}[\mathbf{A}]$, and $\mathscr{N}[\mathbf{A}^*]$ are to be determined now. Since **A** is $k \times n$, we have the following:

$$r \triangleq \operatorname{rank} \mathbf{A} = \text{dimension of column space } \mathscr{R}[\mathbf{A}],$$

$$\dim \mathscr{N}[\mathbf{A}] \triangleq \text{dimension of null space } \mathscr{N}[\mathbf{A}],$$

$$n \triangleq \text{total number of columns of } \mathbf{A}.$$

Hence,

$$r + \dim \mathscr{N}[\mathbf{A}] = n$$

yields the dimension of the null space $\mathscr{N}[A]$,

$$\boxed{\dim \mathscr{N}[\mathbf{A}] = n - r.}$$

Now, do the same for $\mathbf{A}^*$, using the fact (Theorem 2.1) that $\operatorname{rank} \mathbf{A} = \operatorname{rank} \mathbf{A}^*$,

$$r = \operatorname{rank}[\mathbf{A}^*] = \text{dimension of row space } \mathscr{R}[\mathbf{A}^*],$$

$$\dim \mathscr{N}[\mathbf{A}^*] = \text{dimension of left null space, } \mathscr{N}[\mathbf{A}^*],$$

$$k = \text{total number of rows of } \mathbf{A}.$$

Hence,

$$r + \dim \mathscr{N}[\mathbf{A}^*] = k$$

yields

$$\boxed{\dim \mathcal{N}[\mathbf{A}^*] = k - r.}$$

These facts are summarized below:

$$\begin{aligned} \mathcal{R}[\mathbf{A}^*] &= \text{row space of } \mathbf{A}\text{: dimension } r, \\ \mathcal{N}[\mathbf{A}] &= \text{right null space of } \mathbf{A}\text{: dimension } n - r. \end{aligned} \tag{B.1a}$$

$$\begin{aligned} \mathcal{R}[\mathbf{A}] &= \text{column space of } \mathbf{A}\text{: dimension } r, \\ \mathcal{N}[\mathbf{A}^*] &= \text{left null space of } \mathbf{A}\text{: dimension } k - r. \end{aligned} \tag{B.1b}$$

Note from (B.1) that the entire n-dimensional space can be decomposed into the sum of the two subspaces $\mathcal{R}[\mathbf{A}^*]$ and $\mathcal{N}[\mathbf{A}]$. Note from (B.1b) that the entire k-dimensional space can be decomposed into the sum of the two subspaces $\mathcal{R}[\mathbf{A}]$ and $\mathcal{N}[\mathbf{A}^*]$.

EXAMPLE B.4

For the **A** in Example B.1, express the vector $\mathbf{x} = \begin{pmatrix} 1 \\ 10 \\ 0 \end{pmatrix}$ as the sum of vectors in $\mathcal{R}[\mathbf{A}^*]$ and $\mathcal{N}[\mathbf{A}]$.

Solution:

$$\mathbf{x} = \alpha_1 \mathbf{x}^1 + \alpha_2 \mathbf{x}^2 + \alpha_3 \mathbf{x}^3,$$

where $\mathbf{x}^i$, $i = 1, 2, 3$, are vectors in either $\mathcal{R}[\mathbf{A}^*]$ or $\mathcal{N}[\mathbf{A}]$. To find out how many are in each subspace, compute $\dim \mathcal{R}[\mathbf{A}^*] = \text{rank } A = 2$, and $\dim \mathcal{N}[\mathbf{A}] = 3 - 2 = 1$. Hence, $\mathbf{x}^1$ and $\mathbf{x}^2$ will be in $\mathcal{R}[\mathbf{A}^*]$ and $\mathbf{x}^3$ will be in $\mathcal{N}[\mathbf{A}]$. By definition, $\mathbf{x}^3$ satisfies

$$\mathbf{A}\mathbf{x}^3 = \mathbf{0} \Rightarrow \begin{Bmatrix} x_1^3 + x_2^3 - 3x_3^3 = 0 \\ 10x_2^3 = 0 \end{Bmatrix}$$

$$\Rightarrow \mathbf{x}^3 = \begin{pmatrix} 3 \\ 0 \\ 1 \end{pmatrix}.$$

By definition, $\mathbf{x}^1, \mathbf{x}^2$ satisfy

$$\mathbf{A}^*\boldsymbol{\beta}^1 = \mathbf{x}^1,$$

$$\mathbf{A}^*\boldsymbol{\beta}^2 = \mathbf{x}^2$$

for some $\boldsymbol{\beta}^1$, $\boldsymbol{\beta}^2$. Independent $\mathbf{x}^1$ and $\mathbf{x}^2$ are obtained by the choice $\boldsymbol{\beta}^1 = (1,0)^*$, $\boldsymbol{\beta}^2 = (0,1)^*$:

$$\mathbf{A}^*\boldsymbol{\beta}^1 = \begin{pmatrix} 1 \\ -1 \\ -3 \end{pmatrix} = \mathbf{x}^1,$$

$$\mathbf{A}^*\boldsymbol{\beta}^2 = \begin{pmatrix} 0 \\ 10 \\ 0 \end{pmatrix} = \mathbf{x}^2.$$

Now, $\mathscr{N}[\mathbf{A}]$ is one-dimensional and is spanned by $\mathbf{x}^3$, and $\mathscr{R}[\mathbf{A}^*]$ is two-dimensional and is spanned by $\mathbf{x}^1, \mathbf{x}^2$. Solve now for α_1, α_2, α_3:

$$\mathbf{x} = \begin{pmatrix} 1 \\ 10 \\ 0 \end{pmatrix} = [\mathbf{x}^1 \mathbf{x}^2 \mathbf{x}^3] \begin{pmatrix} \alpha_1 \\ \alpha_2 \\ \alpha_3 \end{pmatrix} = \begin{bmatrix} 1 & 0 & 3 \\ 1 & 10 & 0 \\ -3 & 0 & 1 \end{bmatrix} \begin{pmatrix} \alpha_1 \\ \alpha_2 \\ \alpha_3 \end{pmatrix},$$

$$\begin{pmatrix} \alpha_1 \\ \alpha_2 \\ \alpha_3 \end{pmatrix} = \begin{bmatrix} 1 & 0 & 3 \\ 1 & 10 & 0 \\ -3 & 0 & 1 \end{bmatrix}^{-1} \begin{pmatrix} 1 \\ 10 \\ 0 \end{pmatrix} = \begin{pmatrix} \frac{1}{10} \\ \frac{99}{100} \\ \frac{3}{10} \end{pmatrix}.$$

Note From Example B.4 that $\mathbf{x}^3$ is orthogonal to both $\mathbf{x}^1$ and $\mathbf{x}^2$:

$$\mathbf{x}^{3*}\mathbf{x}^2 = 0, \qquad \mathbf{x}^{3*}\mathbf{x}^1 = 0.$$

The following shows that this is no accident. ■

Theorem B.1

$\mathscr{N}[\mathbf{A}]$ and $\mathscr{R}[\mathbf{A}^*]$ are orthogonal subspaces. This fact is denoted by $\mathscr{R}[\mathbf{A}^*]^\perp = \mathscr{N}[\mathbf{A}]$.

Proof: The meaning of Theorem B.1 is that every vector in $\mathscr{N}[\mathbf{A}]$ is orthogonal to every vector in $\mathscr{R}[\mathbf{A}^*]$. To prove this, show that $\mathbf{x} \in \mathscr{N}[\mathbf{A}]$ satisfying

$$\mathbf{A}\mathbf{x} = \mathbf{0} \tag{B.2}$$

is orthogonal to $\mathbf{z} \in \mathscr{R}[\mathbf{A}^*]$ satisfying

$$\mathbf{A}^*\boldsymbol{\beta} = \mathbf{z}. \tag{B.3}$$

This is accomplished as follows:

$$\mathbf{x}^*\mathbf{z} = \mathbf{x}^*\mathbf{A}^*\boldsymbol{\beta} = (\mathbf{A}\mathbf{x})^*\boldsymbol{\beta} = 0,$$

where (B.3) is used to obtain the second equality and (B.2) is used to obtain the last equality. ■

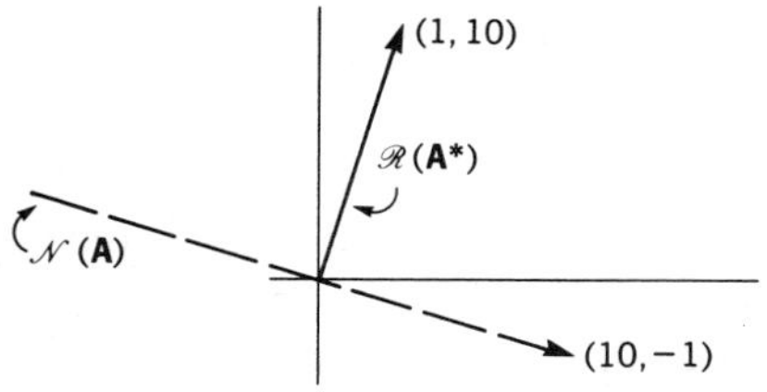

Figure B.1 *The Null Space and Row Space of* ***A***

It should now be clear how to construct a basis for $\mathscr{R}[\mathbf{A}^*]$ and $\mathscr{N}[\mathbf{A}]$. Simply take the independent rows of **A** as a basis for $\mathscr{R}[\mathbf{A}^*]$ and take the vectors perpendicular to these independent rows of **A** as a basis for $\mathscr{N}[\mathbf{A}]$.

EXAMPLE B.5

Graphically show the spaces $\mathscr{N}[\mathbf{A}]$ and $\mathscr{R}[\mathbf{A}^*]$ if $\mathbf{A} = \begin{bmatrix} 1 & 10 \\ 0 & 0 \end{bmatrix}$.

Solution: Since $r = \operatorname{rank} \mathbf{A} = 1$, both $\mathscr{N}[\mathbf{A}]$ and $\mathscr{R}[\mathbf{A}^*]$ are one-dimensional: $\dim \mathscr{N}[\mathbf{A}] = n - r = 2 - 1 = 1$, $\dim \mathscr{R}[\mathbf{A}^*] = r = 1$. The independent rows of **A** form the basis for $\mathscr{R}[\mathbf{A}^*]$ as shown in Fig. B.1. Any vector perpendicular to this is a basis for $\mathscr{N}[\mathbf{A}]$. ■

Theorem B.2

$\mathscr{R}[\mathbf{A}]$ and $\mathscr{N}[\mathbf{A}^*]$ are orthogonal subspaces. This fact is denoted by $\mathscr{R}[\mathbf{A}]^{\perp} = \mathscr{N}[\mathbf{A}^*]$.

Proof: $\mathbf{x} \in \mathscr{R}[\mathbf{A}]$ satisfies

$$\mathbf{A}\boldsymbol{\beta} = \mathbf{x},$$

and $\mathbf{z} \in \mathscr{N}[\mathbf{A}^*]$ satisfies

$$\mathbf{A}^*\mathbf{z} = \mathbf{0}.$$

Then,

$$\mathbf{x}^*\mathbf{z} = \boldsymbol{\beta}^*(\mathbf{A}^*\mathbf{z}) = \boldsymbol{\beta}^*(\mathbf{0}) = \mathbf{0}.$$ ■

Exercise B.1

Graphically show the spaces $\mathscr{R}[\mathbf{A}]$ and $\mathscr{N}[\mathbf{A}^*]$ if

$$\mathbf{A} = \begin{bmatrix} 1 & -10 \\ 1 & -10 \end{bmatrix}.$$

Express $\mathbf{x} = \begin{pmatrix} 1 \\ 7 \end{pmatrix}$ as the sum of vectors in $\mathscr{R}[\mathbf{A}]$ and $\mathscr{N}[\mathbf{A}^*]$. Does $\mathbf{x} = \begin{pmatrix} 1 \\ -1 \end{pmatrix}$ lie in the left null space of $\mathbf{A}$?

This section gives powerful methods for accomplishing in a neat and compact manner many of the tasks which lie ahead. Some books [2.3, 2.6, 2.7] are written in this language. The advantage of such abstract methods is the economy of thought. Unfortunately, this economy comes with a price. The *shortest* way to accomplish a proof is not always the most informative, nor is it always constructive. There are many tasks in this book which could be completed with the abstract methods of this section, but we shall often choose to use more traditional tools when warranted by the added insight afforded.

APPENDIX C

Calculus of Complex Vectors and Matrices

Suppose $f(\mathbf{x})$ is a real or complex scalar function of a complex vector $\mathbf{x} \in \mathscr{C}^n$. In this situation, f has twice as many variables: $\mathbf{x}$ and its conjugate $\bar{\mathbf{x}}$. Hence, the expansion (2.73) becomes

$$\delta f(\mathbf{x}) = \left[\left(\frac{\partial f}{\partial \mathbf{x}}\right)^T, \left(\frac{\partial f}{\partial \bar{\mathbf{x}}}\right)^T\right]\begin{pmatrix}\delta \mathbf{x}\\ \delta \bar{\mathbf{x}}\end{pmatrix} + \frac{1}{2}(\delta \mathbf{x}^T, \delta \bar{\mathbf{x}}^T)\left[\begin{array}{c|c}\partial^2 f/\partial \mathbf{x}^2 & \partial^2 f/\partial \mathbf{x}\,\partial \bar{\mathbf{x}}\\ \hline \partial^2 f/\partial \bar{\mathbf{x}}\,\partial \mathbf{x} & \partial^2 f/\partial \bar{\mathbf{x}}^2\end{array}\right]\begin{pmatrix}\delta \mathbf{x}\\ \delta \bar{\mathbf{x}}\end{pmatrix}, \tag{C.1}$$

where the variation $\delta \mathbf{x}$ is the complex conjugate of $\delta \bar{\mathbf{x}}$.

Only the first-order terms of the Taylor series expansion appear in the differential we need for calculus. Let us study, therefore, the first term of (C.1) rewritten as the differential

$$df(\mathbf{x}) = \left(\frac{\partial f}{\partial \mathbf{x}}\right)^T d\mathbf{x} + \left(\frac{\partial f}{\partial \bar{\mathbf{x}}}\right)^T d\bar{\mathbf{x}}. \tag{C.2}$$

Substituting

$$d\mathbf{x} = d\mathbf{x}_R + j\,d\mathbf{x}_I, \qquad d\bar{\mathbf{x}} = d\mathbf{x}_R - j\,d\mathbf{x}_I \tag{C.3}$$

yields

$$df(\mathbf{x}) = \left(\frac{\partial f}{\partial \mathbf{x}}\right)^T (d\mathbf{x}_R + j\,d\mathbf{x}_I) + \left(\frac{\partial f}{\partial \bar{\mathbf{x}}}\right)^T (d\mathbf{x}_R - j\,d\mathbf{x}_I)$$

$$= \left[\frac{\partial f}{\partial \mathbf{x}} + \frac{\partial f}{\partial \bar{\mathbf{x}}}\right]^T d\mathbf{x}_R + j\left[\frac{\partial f}{\partial \mathbf{x}} - \frac{\partial f}{\partial \bar{\mathbf{x}}}\right]^T d\mathbf{x}_I. \tag{C.4}$$

By defining the symbols

$$\frac{\partial f}{\partial \mathbf{x}_R} \triangleq \frac{\partial f}{\partial \mathbf{x}} + \frac{\partial f}{\partial \bar{\mathbf{x}}}, \tag{C.5a}$$

$$\frac{\partial f}{\partial \mathbf{x}_I} \triangleq j\left(\frac{\partial f}{\partial \mathbf{x}} - \frac{\partial f}{\partial \bar{\mathbf{x}}}\right), \tag{C.5b}$$

(C.4) can be written

$$df(\mathbf{x}) = \left(\frac{\partial f}{\partial \mathbf{x}_R}\right)^T d\mathbf{x}_R + \left(\frac{\partial f}{\partial \mathbf{x}_I}\right)^T d\mathbf{x}_I. \tag{C.6}$$

Hence, (C.2) and (C.6) are equivalent expressions. The inverse of relationships (C.5) are

$$\frac{\partial f}{\partial \mathbf{x}} = \frac{1}{2}\left(\frac{\partial f}{\partial \mathbf{x}_R} - j\frac{\partial f}{\partial \mathbf{x}_I}\right), \tag{C.7a}$$

$$\frac{\partial f}{\partial \bar{\mathbf{x}}} = \frac{1}{2}\left(\frac{\partial f}{\partial \mathbf{x}_R} + j\frac{\partial f}{\partial \mathbf{x}_I}\right). \tag{C.7b}$$

Note also that (C.3) is equivalent to

$$d\mathbf{x}_R \triangleq \frac{1}{2}(d\mathbf{x} + d\bar{\mathbf{x}}), \qquad d\mathbf{x}_I \triangleq \frac{d\mathbf{x} - d\bar{\mathbf{x}}}{2j}, \tag{C.8}$$

allowing one to go back and forth from (C.2) and (C.6) with ease.

If $f(\mathbf{x})$ is a *complex* scalar

$$f(\mathbf{x}) = f_R(\mathbf{x}) + jf_I(\mathbf{x}), \tag{C.9}$$

then (C.7b) becomes

$$\frac{\partial f}{\partial \bar{\mathbf{x}}} = \frac{1}{2}\left(\frac{\partial f_R(\mathbf{x})}{\partial \mathbf{x}_R} + j\frac{\partial f_I(\mathbf{x})}{\partial \mathbf{x}_R}\right) + \frac{1}{2}j\left(\frac{\partial f_R(\mathbf{x})}{\partial \mathbf{x}_I} + j\frac{\partial f_I(\mathbf{x})}{\partial \mathbf{x}_I}\right)$$

$$= \frac{1}{2}\left(\frac{\partial f_R(\mathbf{x})}{\partial \mathbf{x}_R} - \frac{\partial f_I(\mathbf{x})}{\partial \mathbf{x}_I}\right) + \frac{1}{2}j\left(\frac{\partial f_I(\mathbf{x})}{\partial \mathbf{x}_R} + \frac{\partial f_R(\mathbf{x})}{\partial \mathbf{x}_I}\right). \tag{C.10}$$

This result is now compared to the standard Cauchy–Riemann conditions of differential calculus [2.10].

Definition 2.18

The complex scalar function $f(\mathbf{x})$ is said to be "analytic" if the Cauchy–Riemann conditions (C.11) *hold true.*

$$\frac{\partial f_R(\mathbf{x})}{\partial \mathbf{x}_R} = \frac{\partial f_I(\mathbf{x})}{\partial \mathbf{x}_I}, \qquad f(\mathbf{x}) = f_R(\mathbf{x}) + jf_I(\mathbf{x}), \tag{C.11a}$$

$$\frac{\partial f_I(\mathbf{x})}{\partial \mathbf{x}_R} = -\frac{\partial f_R(\mathbf{x})}{\partial \mathbf{x}_I}, \qquad \mathbf{x} = \mathbf{x}_R + j\mathbf{x}_I. \tag{C.11b}$$

From (C.10) and (C.11) it is clear that the condition $\partial f/\partial \bar{\mathbf{x}} = \mathbf{0}$ is equivalent to the Cauchy–Riemann conditions (C.11). *Hence, a function $f(\mathbf{x})$ is analytic iff*

$$\frac{\partial f}{\partial \bar{\mathbf{x}}} = \mathbf{0}. \tag{C.12}$$

The Cauchy–Riemann conditions (C.11) have a simpler explanation. We state without proof that the derivative of the function $f(\mathbf{x})$,

$$\frac{df(\mathbf{x})}{dx_i} \triangleq \lim_{\Delta x_i \to 0} \frac{f(\mathbf{x} + \Delta\mathbf{x}) - f(\mathbf{x})}{\Delta x_i},$$

is independent of the particular path $\Delta\mathbf{x}$,

$$\Delta\mathbf{x} = \Delta\mathbf{x}_R + j\Delta\mathbf{x}_I, \qquad \Delta x_i \triangleq (\Delta x)_i,$$

if the derivative taken along $\Delta\mathbf{x} = \Delta\mathbf{x}_R$ is equal to the derivative taken along $\Delta\mathbf{x} = j\Delta\mathbf{x}_I$. We will show that this condition leads to (C.11). Note that

$$\frac{df(\mathbf{x})}{dx_i} = \lim_{\Delta x_{R_i} \to 0} \frac{f_R(\mathbf{x} + \Delta\mathbf{x}) + jf_I(\mathbf{x} + \Delta\mathbf{x}) - f_R(\mathbf{x}) - jf_I(\mathbf{x})}{\Delta x_{R_i}}$$

$$= \frac{\partial f_R(\mathbf{x})}{\partial x_{R_i}} + j\frac{\partial f_I(\mathbf{x})}{\partial x_{R_i}}$$

and also

$$\frac{df(\mathbf{x})}{dx_i} = \lim_{j\Delta x_{I_i}\to 0} \frac{f_R(\mathbf{x}+\Delta\mathbf{x}) - f_R(\mathbf{x})}{j\Delta x_{I_i}} + j\frac{f_I(\mathbf{x}+\Delta\mathbf{x}) - f_I(\mathbf{x})}{j\Delta x_{I_i}}$$

$$= -j\frac{\partial f_R(\mathbf{x})}{\partial x_{I_i}} + \frac{\partial f_I(\mathbf{x})}{\partial x_{I_i}}.$$

Hence, in vector notation,

$$\frac{df(\mathbf{x})}{d\mathbf{x}} = \frac{\partial f_R(\mathbf{x})}{\partial \mathbf{x}_R} + j\frac{\partial f_I(\mathbf{x})}{\partial \mathbf{x}_R} = -j\frac{\partial f_R(\mathbf{x})}{\partial \mathbf{x}_I} + \frac{\partial f_I(\mathbf{x})}{\partial \mathbf{x}_I},$$

which yields (C.11). A function $f(\mathbf{x})$ is analytic in a region of the complex plane if all its derivatives exist in that region, and the derivatives are independent of path if the Cauchy–Riemann conditions (C.11) hold.

Suppose one wishes to choose $\mathbf{x}$ so as to minimize a scalar function $f(\mathbf{x})$. A necessary condition is that small perturbations in $f(\mathbf{x})$ are not negative, and $df(\mathbf{x}) \geq 0$ in the vicinity of the optimal value of $\mathbf{x}$. See Fig. 2.8.

The necessary condition is

$$df(\mathbf{x}) = \left(\frac{\partial f}{\partial \mathbf{x}}\right)^T d\mathbf{x} + \left(\frac{\partial f}{\partial \bar{\mathbf{x}}}\right)^T d\bar{\mathbf{x}} \geq 0 \tag{C.13a}$$

or, employing (C.6),

$$df(\mathbf{x}) = \left(\frac{\partial f}{\partial \mathbf{x}_R}\right)^T d\mathbf{x}_R + \left(\frac{\partial f}{\partial \mathbf{x}_I}\right)^T d\mathbf{x}_I \geq 0. \tag{C.13b}$$

Equation (C.13b) is easier to work with since $d\mathbf{x}_R$ and $d\mathbf{x}_I$ are independent variations. Hence, the condition (C.13b) can only be assured if

$$\frac{\partial f}{\partial \mathbf{x}_R} = \mathbf{0}, \qquad \frac{\partial f}{\partial \mathbf{x}_I} = \mathbf{0}. \tag{C.14}$$

Note that deriving such conditions from (C.13b) is not so straightforward since $d\mathbf{x}$ and $d\bar{\mathbf{x}}$ are *not* independent. (They are related by $d\mathbf{x} = d\mathbf{x}_R + j\,d\mathbf{x}_I$, $d\bar{\mathbf{x}} = d\mathbf{x}_R - j\,d\mathbf{x}_I$).

Combine (C.14) and (C.5) to obtain

$$\frac{\partial f}{\partial \mathbf{x}} + \frac{\partial f}{\partial \bar{\mathbf{x}}} = \mathbf{0}, \tag{C.15a}$$

$$\frac{\partial f}{\partial \mathbf{x}} - \frac{\partial f}{\partial \bar{\mathbf{x}}} = \mathbf{0} \tag{C.15b}$$

or, equivalently,

$$2\frac{\partial f}{\partial \mathbf{x}} = \mathbf{0}, \tag{C.16a}$$

$$\frac{\partial f}{\partial \mathbf{x}} = \overline{\frac{\partial f}{\partial \overline{\mathbf{x}}}}. \tag{C.16b}$$

This proves the following:

Theorem C.1

If the partial derivatives of the complex scalar function $f(\mathbf{x})$ exist, then a necessary condition for $f(\mathbf{x})$ to have a local minimum at $\mathbf{x}$ is

$$\frac{\partial f(\mathbf{x})}{\partial \mathbf{x}} = \mathbf{0}, \qquad \frac{\partial f(\mathbf{x})}{\partial \overline{\mathbf{x}}} = \mathbf{0}. \tag{C.17}$$

In other words, the correct result is obtained by setting the coefficients of $d\mathbf{x}$ and $d\overline{\mathbf{x}}$ to zero in (C.13a) even though $d\mathbf{x}$ and $d\overline{\mathbf{x}}$ are not independent.

EXAMPLE C.1

Write the differential of the scalar function

$$f(\mathbf{x}) \triangleq \|\mathbf{x}\|_{\mathbf{Q}}^2 = \mathbf{x}^*\mathbf{Q}\mathbf{x}, \qquad \mathbf{x} \in \mathscr{C}^n, \quad \mathbf{Q} \in \mathscr{C}^{n \times n}. \tag{C.18}$$

Solution: From (C.2),

$$d(\mathbf{x}^*\mathbf{Q}\mathbf{x}) = \left[\frac{\partial}{\partial \mathbf{x}}(\mathbf{x}^*\mathbf{Q}\mathbf{x})\right]^T d\mathbf{x} + \left[\frac{\partial}{\partial \overline{\mathbf{x}}}(\mathbf{x}^*\mathbf{Q}\mathbf{x})\right]^T d\overline{\mathbf{x}}, \tag{C.19}$$

where,

$$\begin{aligned}
\left[\frac{\partial}{\partial \mathbf{x}}(\mathbf{x}^*\mathbf{Q}\mathbf{x})\right]^T &= \left[\frac{\partial}{\partial \mathbf{x}}(\overline{\mathbf{x}}^T\mathbf{Q}\mathbf{x})\right]^T = \left[\frac{\partial}{\partial \mathbf{x}}\left(\sum_{i,j=1}^{n} \bar{x}_i Q_{ij} x_j\right)\right]^T \\
&= \left[\frac{\partial}{\partial x_1}\left(\sum_{i,j=1}^{n} \bar{x}_i Q_{ij} x_j\right), \ldots, \frac{\partial}{\partial x_n}\left(\sum_{i,j=1}^{n} \bar{x}_i Q_{ij} x_j\right)\right] \\
&= \left[\sum_{i=1}^{n} \bar{x}_i Q_{i1}, \ldots, \sum_{i=1}^{n} \bar{x}_i Q_{in}\right] \\
&= \overline{\mathbf{x}}^T\mathbf{Q}.
\end{aligned} \tag{C.20}$$

This proves a general result worth remembering:

$$\left[\frac{\partial}{\partial \mathbf{x}}(\mathbf{y}^T\mathbf{Q}\mathbf{x})\right]^T = \mathbf{y}^T\mathbf{Q}$$

or

$$\boxed{\frac{\partial}{\partial \mathbf{x}}(\mathbf{y}^T\mathbf{Q}\mathbf{x}) = \mathbf{Q}^T\mathbf{y}.} \tag{C.21}$$

Now, compute

$$\left[\frac{\partial}{\partial \bar{\mathbf{x}}}(\mathbf{x}^*\mathbf{Q}\mathbf{x})\right]^T = \left[\frac{\partial}{\partial \bar{\mathbf{x}}}(\bar{\mathbf{x}}^T\mathbf{Q}\mathbf{x})\right]^T = \left[\frac{\partial}{\partial \bar{\mathbf{x}}}(\mathbf{x}^T\mathbf{Q}^T\bar{\mathbf{x}})\right]^T. \tag{C.22}$$

Using rule (C.21), (C.22) becomes

$$\left[\frac{\partial}{\partial \bar{\mathbf{x}}}(\mathbf{x}^*\mathbf{Q}\mathbf{x})\right]^T = \mathbf{x}^T\mathbf{Q}^T. \tag{C.23}$$

Note from (C.23) and (C.13a) that $\|\mathbf{x}\|_{\mathbf{Q}}^2$ is *not* an analytic function, unless $\mathbf{Q}\mathbf{x} = \mathbf{0}$. Equation (C.2) now becomes

$$d\left(\|\mathbf{x}\|_{\mathbf{Q}}^2\right) = \bar{\mathbf{x}}^T\mathbf{Q}\,d\mathbf{x} + \mathbf{x}^T\mathbf{Q}^T\,d\bar{\mathbf{x}} = \mathbf{x}^*\mathbf{Q}\,d\mathbf{x} + d\mathbf{x}^*\mathbf{Q}\mathbf{x}. \tag{C.24}$$

Equation (C.24) can also be written in the more general language of (2.20):

$$d(\langle \mathbf{x}, \mathbf{Q}\mathbf{x}\rangle) = \langle \mathbf{x}, \mathbf{Q}\,d\mathbf{x}\rangle + \langle d\mathbf{x}, \mathbf{Q}\mathbf{x}\rangle. \tag{C.25}$$

■

EXAMPLE C.2

Repeat Example C.1 using (C.6) instead of (C.2):

$$d(\mathbf{x}^*\mathbf{Q}\mathbf{x}) = \left[\frac{\partial}{\partial \mathbf{x}_R}(\mathbf{x}^*\mathbf{Q}\mathbf{x})\right]^T d\mathbf{x}_R + \left[\frac{\partial}{\partial \mathbf{x}_I}(\mathbf{x}^*\mathbf{Q}\mathbf{x})\right]^T d\mathbf{x}_I. \tag{C.26}$$

Solution: Using the rule (C.21),

$$\begin{aligned}\frac{\partial}{\partial \mathbf{x}_R}(\mathbf{x}_R^T - j\mathbf{x}_I^T)\mathbf{Q}(\mathbf{x}_R + j\mathbf{x}_I) &= \mathbf{Q}\mathbf{x}_R + \mathbf{Q}^T\mathbf{x}_R - j\mathbf{Q}^T\mathbf{x}_I + j\mathbf{Q}\mathbf{x}_I \\ &= (\mathbf{Q} + \mathbf{Q}^T)\mathbf{x}_R + j(\mathbf{Q} - \mathbf{Q}^T)\mathbf{x}_I, \\ \frac{\partial}{\partial \mathbf{x}_I}(\mathbf{x}_R^T - j\mathbf{x}_I^T)\mathbf{Q}(\mathbf{x}_R + j\mathbf{x}_I) &= \mathbf{Q}\mathbf{x}_I + \mathbf{Q}^T\mathbf{x}_I - j\mathbf{Q}\mathbf{x}_R + j\mathbf{Q}^T\mathbf{x}_R \\ &= (\mathbf{Q} + \mathbf{Q}^T)\mathbf{x}_I + j(\mathbf{Q}^T - \mathbf{Q})\mathbf{x}_R.\end{aligned}$$

Thus, (C.26) becomes

$$d\left(\|\mathbf{x}\|_{\mathbf{Q}}^2\right) = \left[\mathbf{x}_R^T(\mathbf{Q} + \mathbf{Q}^T) + j\mathbf{x}_I^T(\mathbf{Q}^T - \mathbf{Q})\right] d\mathbf{x}_R$$

$$+\left[\mathbf{x}_I^T(\mathbf{Q} + \mathbf{Q}^T) + j\mathbf{x}_R^T(\mathbf{Q} - \mathbf{Q}^T)\right] d\mathbf{x}_I. \qquad \text{(C.27)}$$

To show that (C.27) is equivalent to (C.24), substitute $\mathbf{x} = \mathbf{x}_R + j\mathbf{x}_I$, $d\mathbf{x} = d\mathbf{x}_R + j\,d\mathbf{x}_I$ into (C.24):

$$d\left(\|\mathbf{x}\|_Q^2\right) = (\mathbf{x}_R^T - j\mathbf{x}_I^T)\mathbf{Q}(d\mathbf{x}_R + j\mathbf{x}_I) + (\mathbf{x}_R^T + j\mathbf{x}_I^T)\mathbf{Q}^T(d\mathbf{x}_R - j\,d\mathbf{x}_I)$$

$$= \mathbf{x}_R^T\mathbf{Q}\,d\mathbf{x}_R + \mathbf{x}_I^T\mathbf{Q}\,d\mathbf{x}_I + j\big(\mathbf{x}_R^T\mathbf{Q}\,d\mathbf{x}_I - \mathbf{x}_I^T\mathbf{Q}\,d\mathbf{x}_R$$

$$+\mathbf{x}_I^T\mathbf{Q}^T\,d\mathbf{x}_R - \mathbf{x}_R^T\mathbf{Q}^T\,d\mathbf{x}_I\big)$$

$$+\mathbf{x}_R^T\mathbf{Q}^T\,d\mathbf{x}_R + \mathbf{x}_I^T\mathbf{Q}^T\,d\mathbf{x}_I$$

$$= \mathbf{x}_R^T(\mathbf{Q} + \mathbf{Q}^T)\,d\mathbf{x}_R + \mathbf{x}_I^T(\mathbf{Q} + \mathbf{Q}^T)\,d\mathbf{x}_I$$

$$+j\left\{\mathbf{x}_R^T(\mathbf{Q} - \mathbf{Q}^T)\,d\mathbf{x}_I + \mathbf{x}_I^T(\mathbf{Q}^T - \mathbf{Q})\,d\mathbf{x}_R\right\},$$

which agrees with (C.27). ∎

Note that if $\mathbf{Q} = \mathbf{Q}^T$, then $d(\|\mathbf{x}\|_Q^2) = 2(\mathbf{x}_R^T\mathbf{Q}\mathbf{x}_R + \mathbf{x}_I^T\mathbf{Q}\mathbf{x}_I)$.
Other identities will prove useful. Note that

$$\frac{\partial}{\partial \mathbf{x}_R}(\mathbf{x}^*\mathbf{Q}\mathbf{y}) = \frac{\partial}{\partial \mathbf{x}_R}(\mathbf{x}_R^T - j\mathbf{x}_I^T)\mathbf{Q}\mathbf{y} = \frac{\partial}{\partial \mathbf{x}_R}(\mathbf{x}_R^T\mathbf{Q}\mathbf{y} - j\mathbf{x}_I^T\mathbf{Q}\mathbf{y}) = \mathbf{Q}\mathbf{y} \qquad \text{(C.28a)}$$

and

$$\frac{\partial}{\partial \mathbf{x}_I}(\mathbf{x}^*\mathbf{Q}\mathbf{y}) = -j\mathbf{Q}\mathbf{y}. \qquad \text{(C.28b)}$$

Similarly,

$$\frac{\partial}{\partial \mathbf{x}_R}(\mathbf{y}^*\mathbf{Q}\mathbf{x}) = (\mathbf{y}^*\mathbf{Q})^T = \mathbf{Q}^T\bar{\mathbf{y}}, \qquad \text{(C.28c)}$$

$$\frac{\partial}{\partial \mathbf{x}_I}(\mathbf{y}^*\mathbf{Q}\mathbf{x}) = -j(\mathbf{y}^*\mathbf{Q})^T = -j\mathbf{Q}^T\bar{\mathbf{y}}. \qquad \text{(C.28d)}$$

APPENDIX D

Solution of the Linear Matrix Equations $0 = AX + XB + Q$

Consider the linear algebraic equations of the form

$$\mathbf{0} = \mathbf{AX} + \mathbf{XB} + \mathbf{Q}, \tag{D.1}$$

where all matrices $\mathbf{A}, \mathbf{B}, \mathbf{X}, \mathbf{Q}$ are $n \times n$. Certainly, this is much different than the linear equation $\mathbf{AXB} = \mathbf{Y}$ considered in Chapter 2. To illustrate the conditions under which a unique solution to (D.1) exists, substitute the spectral decomposition for $\mathbf{A}$ and $\mathbf{B}$ into (D.1):

$$\mathbf{0} = \mathbf{E}_A \boldsymbol{\Lambda}_A \mathbf{E}_A^{-1} \mathbf{X} + \mathbf{X} \mathbf{E}_B \boldsymbol{\Lambda}_B \mathbf{E}_B^{-1} + \mathbf{Q}. \tag{D.2}$$

Multiply by $\mathbf{E}_A^{-1}$ from the left and $\mathbf{E}_B$ from the right to obtain

$$\mathbf{0} = \boldsymbol{\Lambda}_A \left(\mathbf{E}_A^{-1} \mathbf{X} \mathbf{E}_B\right) + \left(\mathbf{E}_A^{-1} \mathbf{X} \mathbf{E}_B\right) \boldsymbol{\Lambda}_B + \mathbf{E}_A^{-1} \mathbf{Q} \mathbf{E}_B. \tag{D.3}$$

Define $\mathbf{X}' \triangleq \mathbf{E}_A^{-1} \mathbf{X} \mathbf{E}_B$ and write the ij element of the matrix equation (D.3):

$$0_{ij} = \lambda_{Ai} X'_{ij} + X'_{ij} \lambda_{Bj} + \left(\mathbf{E}_A^{-1} \mathbf{Q} \mathbf{E}_B\right)_{ij}, \tag{D.4}$$

where λ_{Ai}, $i = 1, 2, \ldots, n$, and λ_{Bj}, $j = 1, 2, \ldots, n$, are the diagonal entries of $\boldsymbol{\Lambda}_A$ and $\boldsymbol{\Lambda}_B$, respectively. (We have assumed nondefective $\mathbf{A}$ and $\mathbf{B}$ so that $\boldsymbol{\Lambda}_A$ and $\boldsymbol{\Lambda}_B$

are diagonal.) Solving (D.4) for X'_{ij} yields

$$X'_{ij} = (\lambda_{Ai} + \lambda_{Bj})^{-1}(E_A^{-1}QE_B)_{ij}, \qquad X = E_A X' E_B^{-1}, \tag{D.5}$$

provided $\lambda_{Ai} + \lambda_{Bj} \neq 0$ for any i, j. The following summarizes the situation:

Theorem D.1
The linear equation

$$\mathbf{0} = \mathbf{AX} + \mathbf{XB} + \mathbf{Q}, \quad \mathbf{A} \in \mathscr{R}^{n \times n}, \quad \mathbf{B} \in \mathscr{R}^{n \times n} \tag{D.6}$$

has a unique solution $\mathbf{X}$ *iff* $(\mathbf{A})$ *and* $(-\mathbf{B})$ *have no common eigenvalues. Furthermore, if* $\mathscr{Re}\,\lambda_{Ai} + \mathscr{Re}\,\lambda_{Bj} < 0$ *for all* i, j, *then the unique solution can be written as*

$$\mathbf{X} = \int_0^\infty e^{\mathbf{A}t}\mathbf{Q}e^{\mathbf{B}t}\,dt \tag{D.7}$$

where

$$e^{\mathbf{A}t} \triangleq \sum_{i=0}^{\infty} \frac{\mathbf{A}^i t^i}{i!}, \qquad e^{\mathbf{B}t} \triangleq \sum_{i=0}^{\infty} \frac{\mathbf{B}^i t^i}{i!} \tag{D.8}$$

The proof of this result may be found in Chapter 4.

Exercise D.1
Solve $\mathbf{0} = \mathbf{AX} + \mathbf{XB} + \mathbf{Q}$ for $\mathbf{X}$ if

$$\mathbf{A} = \begin{bmatrix} 1 & 1 \\ 0 & 2 \end{bmatrix}, \quad \mathbf{B} = \begin{bmatrix} -3 & 0 \\ 1 & 3 \end{bmatrix}, \quad \mathbf{Q} = \begin{bmatrix} 1 & 0 \\ 0 & 1 \end{bmatrix}.$$

Can $\int_0^\infty e^{\mathbf{A}t}\mathbf{Q}e^{\mathbf{B}t}$ be computed in this case? If so, compute it.

Exercise D.2
Repeat Exercise D.1 if

$$\mathbf{A} = \mathbf{B}^T = \begin{bmatrix} -1 & 1 \\ 0 & -2 \end{bmatrix}, \quad \mathbf{Q} = \begin{bmatrix} 1 & 0 \\ 0 & 1 \end{bmatrix}.$$

Exercise D.3
Let $\mathbf{B} = \mathbf{A}^*$ in (D.6), and let $n = 2$. Show that if $\mathscr{Re}\,\lambda_i[A] < 0$ for all $i = 1, 2$, then the solution of

$$\mathbf{0} = \mathbf{XA}^* + \mathbf{AX} + \mathbf{Q} \tag{D.9}$$

is as follows (define $D_i \triangleq \sqrt{Q_{ii}}$):

If $A_{ii} \neq 0$ for either ($i = 2,\ j = 1$) or ($i = 1,\ j = 2$).

$$X_{jj} = \frac{-1}{2|\mathbf{A}|\,\mathrm{tr}\,\mathbf{A}}[D_j^2(|\mathbf{A}| + A_{ii}^2) + A_{ji}D_i(A_{ji}D_i - 2A_{ii}D_j)], \tag{D.10a}$$

$$X_{ij} = \frac{-A_{ii}\left(A_{ij}X_{jj} + D_iD_j\right) + \frac{1}{2}A_{ji}D_i^2}{|\mathbf{A}| + A_{ii}^2}, \tag{D.10b}$$

$$X_{ii} = \frac{D_i^2 + 2A_{ij}X_{ij}}{-2A_{ii}}. \tag{D.10c}$$

If $A_{ij} \neq 0$ for either ($i = 2,\ j = 1$) or ($i = 1,\ j = 2$)

$$X_{ii} = \frac{-1}{2|\mathbf{A}|\,\mathrm{tr}\,\mathbf{A}}[D_i^2|\mathbf{A}| + (A_{ij}D_j - A_{jj}D_i)^2, \tag{D.11a}$$

$$X_{ij} = \frac{D_i^2 + 2A_{ii}X_{ii}}{-2A_{ij}}, \tag{D.11b}$$

$$X_{jj} = \frac{1}{2A_{ij}^2}[D_i(D_i\,\mathrm{tr}\,\mathbf{A} - 2A_{ij}D_j) + 2[|\mathbf{A}| + A_{ii}^2]X_{ii}]. \tag{D.11c}$$

These equations provide the analytic solution to all second-order equations of the form (D.9), providing a solution exists (see further discussion in Chapter 4). To use these equations, find the first nonzero element of the given **A** matrix and use the appropriate equations (D.10) or (D.11).

APPENDIX **E**

Laplace Transforms

*I*n the notation

$$x(s) \triangleq \mathscr{L}[x(t)] \triangleq \int_0^\infty e^{-st}x(t)\,dt \tag{E.1}$$

will denote the "Laplace transform of $x(t)$," where $s = \sigma + j\omega$ is a complex number. The smallest value of σ for which the integral exists is called the *abscissa of convergence*. If the integral (E.1) is infinite for every value of σ, the Laplace transform of $x(t)$ does not exist. In such cases the abscissa of convergence is $+\infty$.

Exercise E.1

Show that the abscissa of convergence of the function e^{t^2} is $+\infty$ (hence, e^{t^2} is not Laplace transformable).

In the study of linear differential equations, the functions of interest are usually Laplace transformable. Some difficulties arise in the definition (E.1) when $x(t)$ is discontinuous at $t = 0$, the two notable examples are the step function $1(t)$, which is 0 for $t = 0^-$ and 1 for $t = 0^+$, and the Dirac delta impulse function $\delta(t)$, which is defined by $\int_{-\infty}^{\infty}\delta(t)\,dt = 1$, and $\delta(t) = 0$ for both $t < 0^-$ and $t > 0^+$. There is a variety of such functions described in the mathematical literature. However, for all of our purposes, we may choose $\delta(t)$ to be the function plotted in Fig. E.1 in the limit as $\varepsilon \to 0$. Hence, in both the cases of step inputs $1(t)$ and impulsive inputs $\delta(t)$, one may take the lower limit on the integral in (E.1) to be 0^+. Using this convention, the Laplace transforms of some simple functions are given in Table E.1, where listed functions $x(t)$ are assumed to be zero for all $t < 0$.

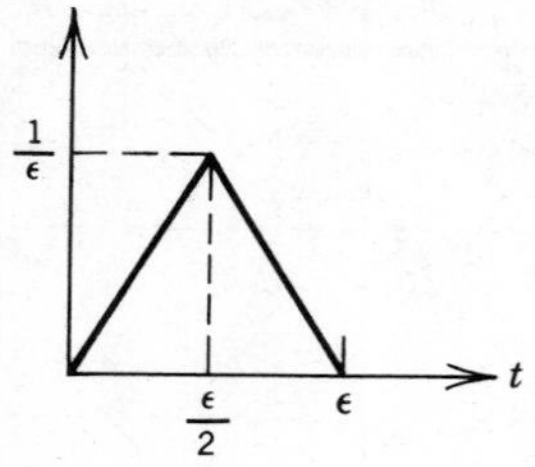

Figure E.1 *A Function Leading to $\delta(t)$*

TABLE E.1 Laplace Transforms

	$x(t)$	$x(s)$
1.	$1(t)$	$1/s$
2.	$\delta(t)$	1
3.	$\dot{x}(t)$	$sx(s) - x(0)$
4.	$t^k/k!$	$1/s^{k+1}$
5.	e^{-at}	$1/(s+a)$
6.	$x(t-T)^{\dagger}$	$e^{-Ts}x(s)$
7.	$e^{-\sigma t}\sin\omega t$	$\omega/[(s+\sigma)^2+\omega^2]$
8.	$e^{-\sigma t}\cos\omega t$	$(s+\sigma)/[(s+\sigma)^2+\omega^2]$
9.	$(1/\omega_n\sqrt{1-\zeta^2})e^{-\zeta\omega_n t}\sin\omega_n\sqrt{1-\zeta^2}\,t,\ (0<\zeta<1)$	$1/(s^2+2\zeta\omega_n s+\omega_n^2)$
10.	$\lim_{t\to\infty} x(t)$	$\lim_{s\to 0} sx(s)$
11.	$\lim_{t\to 0} x(t)$	$\lim_{s\to\infty} sx(s)$
12.	$x(t)e^{-at}$	$x(s+a)$
13.	$\int_0^T x_1(\sigma)x_2(t-\sigma)\,d\sigma,\ T\ge t$	$x_1(s)x_2(s)$
14.	$x_1(t)x_2(t)$	$\int_0^{\infty} x_1(r)x_2(s-r)\,dr$

† The notation $x(t-T)$ will mean a function which is zero until $t \ge T$. Hence, e^{-Ts} is sometimes called a "delay" operator.

EXAMPLE E.1

Solve

$$\dot{x} + x = \delta(t) \tag{E.2}$$

using Laplace transforms.

Solution: The Laplace transform of (E.2) yields

$$sx(s) - x(0) + x(s) = 1, \tag{E.3}$$

which leads to

$$x(s) = \frac{1 + x(0)}{s + 1}. \tag{E.4}$$

From Table E.1, we have the result

$$x(t) = [1 + x(0)]e^{-t}. \tag{E.5}$$

This example illustrates the role of Laplace transforms to change the *calculus* problem of solving (E.2) to the *algebra* problem of solving (E.3), which yields (E.4). However, interpretations of (E.4) can only be fully appreciated by a *thorough* understanding of the time domain properties (E.5) that are associated with the Laplace domain properties of (E.4). The world we live in and the motions of systems we observe in experiments take place in real *time*. Thus, it is helpful for the student to develop an understanding of *time* domain methods of solving differential equations in addition to the more advanced notions of *complex* domain methods. For more treatment of Laplace and frequency domain methods, see refs. [4.6] through [4.8]. ■

Index

Abnormality of a matrix, 456
Abscissa of convergence, 493
Actuator costs, 365
Actuators, 3, 184
Adjoint matrix, 10, 128
Adjoint systems and duality, 238
All-pass factor, 284
All-pass networks, 119, 284
Anagnostou, K. E., 120
Analytic functions, 485
Anderson, B. D. O., 263, 300, 327, 397, 473
Angles between vectors, 22
Angular momentum, 91, 103
Angular velocities, 72, 96, 103
Arbib, M. A., 83
Arnold, W. F., 263
Aström, K. J., 474
Astronaut backpack, 95
Asymptotically stable, 259, 302
Athens, M., 397
Auslander, D. M., 199
Axisymmetric spacecraft, 103

Backpack, dynamics of, 94
Balanced coordinates, 250, 255, 287, 289
Balanced model reduction, 286
Balanced realization algorithm, 252
Balas, M. J., 473
Bandwidth, 192
Barmish, B. R., 474
Base vector, 66
Basis:
  functions, 57, 108
  orthogonal, 12
  for a space, 12, 17, 476
Bernstein, D. S., 300, 397
Bessel's inequality, 55
Bilateral Laplace transform, 189
Bilinear systems, 88
Bliss, G. A., 120
Block diagonal, 433
Block diagrams, 123
Blum, E. K., 83
Bode, H. W., 4
Boyle, J. M., 474
Brani, C., 120
Brockett, Roger W., 327
Brogan, W. L., 263
Bryson, A., 474
Burke, M., 474
Butterworth pattern of poles, 364
Byrne, P. C., 474

Caines, P. E., 474
Calculus of matrices, 39
Calculus of variations, 105
Canonical equations, Hamilton's, 5
Canonical structures of matrices, 58, 422
Cauchy–Riemann conditions, 485
Cauchy–Schwarz inequality, 16, 21, 79, 455, 467
Causal system, 89, 128
Cayley–Hamilton theorem, 62, 160, 208
Center of mass, 3
Center of pressure, 3
Characteristic equation, 59
Characteristic polynomial, 45, 52, 128, 132, 144
Chebyshev polynomials, 154
Chen, C. T., 83
Cholesky decomposition, 82
Circular orbits, 102
Classical control, 4
Closed-loop system, 150
Clough, R. W., 263
Cofactor of a matrix, 10, 128
Collins, E., 327
Collocated sensors and actuators, 367, 451

Column rank, 17
Column space, 50, 476
Compensation for a class of model errors, 466
Complete basis, 56
Complex conjugate, 10
Complex elements of a matrix, 10
Complex scalar function, 487
Component cost, 187, 195, 412
Component cost analysis, 186, 285, 289, 292
Condition number of a matrix, 440
Configuration variables, 12, 85, 94
Constrained optimization problems, 330
Continuum model, 105
Control:
  covariance, *see* Covariance, control
  design, 7
  policy, 419
  problem, 428
  spillover, 435
  vector, 89
Controllability, 343
  degree of, and observability, 240
  grammian, 205, 219
  of modal coordinates, 213
  of transformed coordinates, 212
  of vector second-order systems, 222
  vectors, 447
Controllable canonical coordinates, 219, 220
Controllable Hessenberg coordinates, 255
Controllable modes, 251
Controller:
  dynamic input constrained (DIC), 392
  dynamic output constrained (DOC), 392
  estimator-based, 410
Controller parameters, 336
Controller poles and zeros, 340
Controller state, 338
Convolution integral, 189
Coordinate transformations, 97, 143, 144
Corless, M., 474
Correlation, time, 174, 194
Cost:
  decomposition, 184
  decoupled coordinates, 255, 280, 282
  equivalent realizations, 279, 304
  function, 182, 195, 331
Covariance:
  assignment, 321, 328, 399
  control, 322
  equivalent realization, 271
  matrix, 177
  matrix, state, 177, 178, 194, 195
  of the output, 179, 183
  parameters, 180, 187, 194, 297
Cover, covariance equivalent realization, 271
Cross product, 91
Cullen, C. G., 474
Current law, Kirchoff's, 118

Damping matrix, 167
Daniel, J., 83
Decomposition of matrices, 58
Decoupling, 97
Defective matrices, 61
Degree of controllability and observability, 240
Delorenzo, M., 199
Design, control, 7
Desoer, C. A., 83
Detectability, 314, 317, 349, 374
Determinant of a matrix, 10
Deterministic systems, 338
Diagonal matrix, 450
Differentiation:
  of a scalar with respect to a matrix, 40
  of a scalar with respect to a vector, 39
Dimensions of a matrix, 9
Di Pillo, G., 120
Dirac delta impulse function, 493
Distinct eigenvalues, 61, 69
Disturbance:
  accommodation, 377
  basis functions, 152
  cancellation control, 381, 382
  modeling principle, 150, 152, 380
  models, 148, 377
  rejection, 209
  utilization control, 378, 381
  vector, 89
Dorney, C. Nelson, 83
Dot product, 97
Doyle, J. C., 473
Drag force, 91
Dynamic controller, 395, 420
Dynamic system, 85, 100, 128

Eckhaus, W., 300
Eigenvalue, 11
Eigenvalues of the optimal LQI system, 337
Eigenvectors, right, left, 11
Elastic structures, 104
Electrical circuits, 116
Element of a matrix, 9
Enns, D., 473
Equality constraints, 388
Equilibrium solution, 102
Equivalent realizations, 265
Error-free measurements, 408

Errors:
  in model order, 422
  in the closed-loop system, 430
Estimation error, 403, 408
Estimator-based controllers, 410
Euler, L., 4
Euler–Bernoulli beam, 107, 110, 181, 194, 246, 369
Evans, W. R., 4
Exponential stability, 314

Feedback linearization, 90
Feedforward path, 128, 138, 409
Final value theorem, 150
First-order approximation, 445
  of the closed-loop eigenvalues, 452
First-order perturbations, 442
First variation, 105
Fix, G. J., 120
Flexible beam, 194
Fortmann, T. E., 199, 397
Fourier series, 57
Fourier transform, 189, 190, 193
Frazho, A. E., 120
Frequency domain, 189
Friedland, G. B., 300
Frobenius norm, 441
Function minimization, 40
Functions, 56
  orthogonal, 113
Fundamental matrix, 156
Fundamental subspaces of matrix theory, 476

Gain margin, 356
Garg, S. C., 199
Generalized (non-conservative) forces, 96
Generalized coordinates, 96
Generalized eigenvectors, 63, 64, 65, 351
Generalized LQI problem, 342
Gibbs, J. W., 9, 12
Gilbert, E. G., 263, 266, 300, 474
Global minimum, 456
Glover, K., 300, 473
Goodwin, G. C., 474
Gouloub, G. H., 473
Gradient, 39
Gram–Schmidt procedure, 31
Gravity constant, 102
Gravity vector, 91
Gregory, C., 263
Gyroscopic systems, 223

Hablani, H. B., 263
Hamilton, W. R., 4, 120
Hamilton's principle, 105
Harvey, A. H., 397
Hermitian matrices, 10, 24, 68, 69, 81
Hessenberg coordinates, 255
Hessian matrix, 39, 87
Hitz, K. L., 199, 397
Ho, Y. C., 474
Homeopathic plants, 341
Hotz, A., 327
Hu, A., 263
Hughes, P. C., 120, 199, 263
Hung, Y. S., 474
Hutton, M. F., 300
Hyland, D. C., 300, 397

Idealization of the system, 414
Identification, 293, 415
  frequency domain, 299
  1-Cover, 296
Imad, F. P., 474
Impulse function, 150
Impulse response, 127, 145
Independent constraints, 329
Inductance, 116
Inertia dyadic, 103
Inertial space, 91
Inner product of two vectors, 21
Input, 174, 191
  constrained (IC) controller, 389
  correlation, 174
  cost, 185
  cost analysis, 185
  covariance, 326
  output cost analysis, 364
Integral controller, 380
Internal model of the disturbances, 150
Invariance of the characteristic polynomial, 144
Inverse of a matrix, 11
Inverse square gravitational field, 101
Inversion property, 157
Inverted pendulum, 469
Iterative modeling, 2
Iterative schemes, 346

Jacobi, C. G. J., 445, 473
Johnson, C. D., 199, 397
Johnson, T. L., 397
Jordan block, 63
Jordan form, 63, 64, 246, 550
Jordan modal coordinates, 246
Joseph, P. D., 413

Kailath, T., 83, 120, 300
Kalman's controllable canonical form, 222

Kalman, R. E., 263, 266, 300
Kernel, 477
Kinematic variables, 91
Kinetic energy, 97, 105, 385
Kirchoff's laws, 116
Koch, G., 120
Kokotovic, P. V., 300
Kung, S. Y., 300
Kwakernaak, H., 413

Lagrange, J. L., 4
Lagrange multiplier, 47, 329, 387
Lagrange's equation, 95
Laplace transform, 93, 493
Laplace transform, bilateral, 189
Laub, A. J., 263, 397
Law of cosines, 22
Learning model, 412
Least squares theory, 38, 50, 153
Left eigenvectors, 11, 58, 450
Left inverse, 36
Left null space, 477
Leitmann, G., 474
Levine, W. S., 397
Liapunov equations, 346
Liapunov function, 302, 328, 338, 448, 463, 466
Liapunov stability, 301, 328
Liapunov transformations, 156, 314
Lift forces, 91
Lightly damped structures, 169, 452
Likins, P. W., 83, 120, 199, 327
Lin, D. W., 300
Linear algebra, 8, 35
Linear dynamic systems, 89
Linear independence, 13
  of constant vectors, 27
  geometrical interpretations of, 16
  tests for determining, 18
  of time-varying vectors, 25
  of vectors, 16
Linearization, 86, 89
Linearized model, 87
Linearly independent eigenvectors, 60
Linearly independent on an interval, 14, 29, 54
Linear quadratic impulse (LQI) optimal control, 331
Linear quadratic impulse sensitivity (LQIS) controller, 460
Linear quadratic (LQ) optimal measurement feedback control, 345
Linear vector space, 13
  axiomatic definition of a, 475
Lipschitz condition, 89, 157
Liu, Yi, 473
Ljung, L., 474
Lowest bound on output performance, 363
LQ controller, 396
LQI optimal controller, 338
Luenberger, D. G., 263, 327, 409, 413

Macfarlane, A. G. J., 199, 474
Markov parameters, 131, 144, 145, 180, 297
Mass matrix, 109, 112
Mathematical models, 420
Matrix:
  blode diagonal, 433
  calculus, 39
  cofactor of, 9, 10, 128
  column space, 50, 476
  condition number of, 440
  covariance, 177
  determinant, 9
  diagonal, 450
  eigenvalues, 11
  eigenvectors, 11
  elements, 9
  equations, 35, 490
  exponential, 160, 162
  factor, 80
  Hermitian, 10, 24, 68, 69, 81
  Hessian, 39, 87
  inverse, 11
  minor, 9
  modal, 58, 63, 67, 76, 163, 166
  negative definite, 23
  negative semidefinite, 23
  norm of, 79, 441
  normal, 9, 11
  null space of, 478
  orthonormal, 9
  partitioned, 334
  positive definite, 23
  positive semidefinite, 23, 82, 205
  range space of, 476
  rank, 17
  representations, 8
  row space, 476
  singular, 11
    value decomposition of, 58, 72, 75, 77, 254
  skew-Hermitian, 68, 70
  skew-symmetric, 10, 68, 70, 104, 465
  spectral decomposition of, 116, 163, 251, 313, 350
  state transition, 156, 157, 172
  symmetric, 68
  trace of, 11, 44
  unitary, 9
Maximal accuracy, 418, 429

Mean squared value, 194
Measurement:
  feedback, 368
    LQ problem, 345
    control, 393, 395, 446
  impulsive disturbances, 331
Medanic, J. V., 300
Meditch, J., 413
Meirovitch, L., 120
Miller, R. A., 199
Minimal energy control, 353
Minimal-order realizations, 265
Minimal transfer equivalent realization, 266
Minimization problem, 45
Minimum norm, 49, 51
Minimum phase margin of state feedback controllers, 359
Minor of a matrix, 10
Mishra, R. N., 300
Modal coordinates, 146, 255
Modal cost, 240, 241, 242, 244, 247, 370, 452
Modal cost analysis, 240
Modal data, 58
Modal form, 141
Modal frequencies, 112
Modal matrix, 57, 58, 63, 76, 163, 166
Modal methods for solving Riccati equations, 349
Mode shapes, 112
Mode slope, 114
Model error concepts, 414
Model error estimation, 461
Model error system, 424
Model error vectors, 422
Model reduction, 249, 287
Modeling:
  and control inseparability principle, 429
  and control problems, 85
  of dynamic systems, 92, 414
  errors, 370, 416
  problem, 428
  process, 415
Modified balanced realization algorithm, 253
Modulus of elasticity, 107
Momenta, generalized, 4
Momentum vector, 91
Moore, B. C., 83, 263, 300
Moore, J. B., 397
Moore–Penrose inverse, 37
Mullis, C. T., 263
Multiobjective control, 386
Multiple inputs, outputs, 3

Nachbin, L., 83
n-dimensional linear vector spaces, 475
Necessary conditions for a minimum, 48, 330, 402
Negative definite, 23
Negative semidefinite, 23
Neglected disturbances, 422
Neighboring trajectory, 301
Newton's laws, 91, 101
Noble, B., 83
Nodes, 114
Noncausal, 128
Nonconservative forces, 105
Nondefective matrices, 59, 61, 69, 490
Nongyroscopic systems, 235, 366
Nonlinear equations, 104
Nonlinear system, 88,100
Nonlinearities, 422
Nonsingular matrix, 11, 70
Normal matrices, 9, 11
Normalized observable Hessenberg algorithm, 260
Normalized observable Hessenberg form, 259
Norm of a matrix, 79
Norm of a vector, 20
Null solution of differential equations, 302
Null space, 51, 59, 78, 478
Nutation control, 389
Nyquist, H., 4
Nyquist plot, 357, 429

Objective function, 331
Observability, 226, 343
  grammians, 253, 269, 348
  of modal coordinates, 230
  of observable canonical coordinates, 232
  of transformed coordinates, 230
  vectors, 447
  of vector second-order systems, 235
Observable canonical coordinates, 234
Observable Hessenberg coordinates, 259, 275
Observable modes, 251
Observation spillover, 435
Observers, 409
O'Malley, R. E., Jr., 300
Open loop control, 2
Optimal:
  control, 328
  control of time-invariant systems, 364
  control of vector second-order systems, 366, 371
  covariance control, 329, 331
  LQI controller, 339
  LQ state feedback control, 346
  reduced order controllers, 393

Optimal (*Continued*)
  state estimation, 401
  state feedback control, 366
  tracking and servomechanisms, 383
Orbiting spacecraft, 101
Orthogonal, 9, 11, 19, 69, 480
  axes, 72
  bases, 28
  basis, 32
  filter, 382
  functions, 29, 30, 55, 153
  matrices, 72
  polynomials, 30
  subspaces, 51, 480
Orthogonalization procedure, 33
Orthonormal, 30
  functions, 31
  matrices, 9
Outer product, 53, 54
Output, 100, 174, 178, 179, 183, 194
  constrained (OC) controller, 386
  controllability, 201, 203
  of the controller, 335
  correlation, 174, 178, 194, 271
  correlation equivalent realization, 271
  cost analysis, 188
  covariance, 194, 206
  disturbability, 208, 209
  feedback controller, 416
  function, 195
  of the plant, 335
  tracking, 207
Output cost, 188
Owens, D. H., 473

Padulo, L., 83
Parameter errors, 422, 432
Parameter estimator, 464
Parnebo, L., 300
Parseval's equality, 56, 113, 154
Parseval's theorem, 189
Partial differential equations, 104, 106
Partial fraction expansion, 132
Partitioned matrices, 334
Patel, R. V., 474
Penzien, J., 263
Performance, 4
  evaluation, 430
Periodic linear systems, 171, 172
Perkins, W. R., 300
Perturbations of eigenvectors, 444
Phase margin, 356
Phase-variable form, 139
Physical plant, 420, 434
Pimenides, T. G., 120
Pinned elastic beam, 106
Plant state, 338
Plant trajectory, 338
Pole assignment, 318, 328, 386, 399
Pole-zero cancellations, 128
Poles (of a transfer function), 132
Popov, V. M., 263
Positive definite, 23
  function, 308
Positive semidefinite, 23, 82, 205
Postlelhwaite, I., 474
Potential energy, 97, 105
Power series, 160
Power spectral density, 191, 194
Principal axes, 104
Principal minors, 11
Projection of the vector, 32
Proper transfer functions, 128, 129
Pseudo inverses, 35, 37, 320

Q-COVER, 272
Q-Markov COVER, 272, 273, 275
Quadratic function of the outputs, 331

Rabins, M. J., 199
Ramadge, P. J., 474
Range space, 476
Rank of a matrix, 17
Rate gyro, 409
Rate measurements, 367
Rational functions, 13, 475
Raven, F. H., 473
Rayleigh, J. W. S., 263
Rayleigh damping, 167, 243, 247
Real elements of a matrix, 10
Realization, state space, 121, 123, 198
Realizing models from output data, 293
Real Jordan form, 171
Reciprocal basis vectors, 66, 454
Rectangular form, 137
Rectangular method, 137
Redheffer, R. M., 83
Reduced-order estimators, 406, 410
Reduced-order model, 288, 375
Relative controllability, observability, 239
Residues, 132, 144, 146, 241
Resistance, 116
Resolvent algorithm, 161, 319
Resolvent matrix, 160, 162, 164
Riccati, Count, 347
Riccati equation, 347, 356, 377, 403
Right eigenvectors, 70
Right inverse, 36

Right null space, 477
Rigid body, 96
  in space, 103
Rissanen, J., 300
Ritz method, 109
Roberts, R. A., 263
Robustness, 419
Rocket dynamics, 91
Roll attitude regulator for a missle, 325
Root locus from LQI controllers, 341
Root locus of the LQ optimal state feedback controller, 351
Root perturbations, 443
  in vector second-order systems, 448
Root sensitivity matrix, 453
Root sensitivity metric, 454
Row rank, 17
Row space, 476
Rozenoer, L. I., 209, 263
Rules of differentiation, 106

Sannuti, P., 300
Satellite, 101
Scalar input/scalar output (SISO) systems, 135, 352
Second-order systems, 169
Sensitivity analysis and control, 453
Sensors, 3
  cost, 363
  dynamics, 416
  effectiveness, 363
Separation:
  of the estimation and control problems, 338
  principle of estimation and control, 410, 411
  principle of modeling and control, 3
  of variables, 108, 110
Signal-to-noise ratio, 364
Sign definiteness of matrices, 23
Siljak, D. D., 327, 473
Silverman, L. M., 300
Similarity transformation, 143, 266, 351, 437
Simply supported beam, 107
Single-input/single-output systems, 135, 352
Singular matrix, 11
Singular perturbation, 289, 290
Singular value decomposition, 58, 72, 75, 77, 254, 273, 439
Singular values, 73, 253
Sinusoidal basis functions, 152
Sinusoidal disturbance, 148
Sivan, R., 413
Skelton, R. E., 199, 263, 300, 327, 397, 473
Skew-Hermitian, 68, 70
Skew-Hermitian matrices, 10, 71, 322
Skew-symmetric matrices, 10, 68, 70, 92, 104, 449, 465
Söderström, T., 474
Sokolnikoff, I. S., 83
Sounding rocket, 378
Space backpack, 94
Spacecraft, 71
  attitude control problem, 375
  simplified orbital dynamics, 101
  spinning, 188, 409, 411
Spanning a space, 16, 476
Spectral decomposition, 58, 66, 67, 74, 116, 163, 251, 313, 350, 490
Spectrum arbitrary, 318
Spinning govenor system, 372
Square-integrable function, 108
Square root decomposition, 58, 80
Stability, 4, 301
  bounded-input/bounded-output, 311
  margin of the LQ controller, 356
  margins in control, 436
Stabilizability, 316, 374
Stable in the sense of Liapunov, 301
State:
  controllability, 210
  correlation, 174, 194
  Covariance, 177, 181, 194, 195, 205, 271, 402
  estimation, 398
  estimator, 338, 403
    minimal-order, 409
  feedback control, 337, 346, 388
  and parameter estimator, 464
  space realization, 121
  transition matrix, 156, 157, 172
    for vector second-order systems, 168
  variables, 12, 13, 85
  vector, 85, 100, 174, 175, 194
Stein, G., 397
Stengel, R. F., 413
Stiffness matrix, 109, 112
Strang, G., 83, 120
Strength of the impulse, 158
Strictly proper transfer function, 128, 129
Structure of modeling errors, 415, 421
Subspaces, 476
Subsystems, 130
Superposition property of linear systems, 89
Symmetric matrices, 10

Takahasi, Y., 199
Taylor's series, 39, 86
Thrusting, 101
Time:
  correlation, 144, 173, 174, 190

Time (*Continued*)
  moments, 145
  -invariant systems, 90
  -varying linear systems, 90
Titchmarsh, E. C., 199
Toda, M., 474
Torque, 109
Tou, J. T., 413
Trace identities, 44
Tracking error, 150
Trajectory sensitivity, 457
Transfer decomposition, 195
Transfer equivalent realizations, 266, 395
Transfer function error, 287
Transfer functions, 93, 118, 126, 127, 144, 145, 195
Transfer matrices, 127, 195
Transformation, coordinate, 143
Transformation, similarity, 143
Transform methods, 7
Transition property, 157
Transmission zeros, 133
Triangle inequality, 23
Truth model, 428
Tse, E. C. Y., 300
Tzafestas, S. G., 120

Udink Ten Cate, A. J., 263
Unconstrained optimization, 48
Uncontrollability and unobservability of dynamic systems, 432, 434
Uncontrollable modes, 347
Undamped natural frequency, 375
Uniform mass density, 107
Uniform stability, 314
Unitary matrices, 9, 11, 69, 72, 440
Unit intensity impulse inputs, 219
Unmodeled mode, 447
Unobservable, 432
Unstabilizability, 378
Unstable controllers, 341
Unstable mode, 347
Unstable plants, 341

Van Amerongen, J., 263
Van Loan, C. F., 473
Van Ness, J. E., 474
Vector(s), 9, 11, 19, 39
  perpendicular, 33
Vector first-order form (state form), 94
Vector inner products, 19, 22
Vector norms, 20
Vector outer products, 19
Vector second-order (VSO) systems, 85, 93, 115, 166
Verriest, E. I., 300
Vigneron, F. R., 199
Virtual work, 96
Voltage drop, 117
Voltage source, 116

Wagie, D. A., 300, 474
Walsh functions, 31
Weighted norm, 20
Weight selection algorithm:
  for the dynamic input constrained (DIC) controller, 392
  for the dynamic output constrained (DOC) controller, 392
  for the IC controller, 389
  for the OC controller, 388
Weight selections in the LQ and LQI problems, 385
Weinberg, I., 397
White noise, 192
Wilson, D. A., 300
Wittenmark, B., 474
Wonham, W. M., 263, 327

Yedavalli, R. K., 474
Yousuff, A., 199, 263, 300

Zadeh, L. A., 83
Zeros of the transfer function, 132